Evolution

LEAVING CERTIFICATE BIOLOGY

for New Senior Cycle

HUMPHREY JONES AND
YVONNE HIGGINS-VALLELY

GILL EDUCATION

Gill Education
Hume Avenue
Park West
Dublin 12

www.gilleducation.ie

Gill Education is an imprint of M.H. Gill & Co.

ISBN: 978-1-8045-82749

Design: Sarah McCoy

Print origination: emc

Illustrations: Derry Dillon, Andrii Yankovskyi and emc design Ltd

For permission to reproduce images, the authors and publisher gratefully acknowledge the following:

© Adobe Stock: vi, 6T, 7TR, 7BL, 7BR, 9C, 11L, 11C, 11R, 21T, 21B, 22, 24T, 24B, 25, 41C, 46, 47, 48TL, 48TC, 48TR, 48UL, 48UC, 48UR, 48LL, 48LC, 48LR, 48BL, 48BR, 50TL, 53, 55, 59, 62, 64, 65TCL, 65TCR, 65CL, 65CCL, 65CCR, 65CR, 65BL, 65BCL, 65BCR, 65BR, 66, 68, 71, 72T, 72B, 74, 75L, 75C, 75R, 76TL, 76B, 81, 84T, 87, 88, 90, 91T, 91CT, 91CB, 91B, 106T, 111T, 111L, 111B, 111R, 114, 115T, 115a, 115b, 115d, 115e, 115f, 115g, 118T, 121, 122a, 122b, 122c, 122d, 122e 122f, 122g, 122h, 122i, 122j, 122k, 122l, 122m, 122n, 122o, 122p, 122q, 122r, 125, 126, 129T, 129B, 133a, 133b, 133c, 133d, 135, 136, 138T, 138CT, 138CB, 138B, 139L, 139C, 139R, 140L, 140C, 145, 147, 151, 154T, 155c, 155d, 156T, 156CB, 157B, 164T, 164CL, 164CCL, 164C, 164CCR, 164CR, 164BL, 164BCL, 164BC, 164BCR, 164BR, 165T1, 165T2, 165CT1, 165CL, 165CCL, 165C, 165CCR, 165CR, 165BL, 165BCL, 165BC, 165BCR, 165BR, 166, 167TL, 167TCL, 167TC, 167TCR, 167TR, 167BL, 167BCL, 167BC, 167BCR, 167BR, 168, 170, 176, 180, 181T, 181BL, 181BC, 181BR, 185, 192, 193, 195TL/R, 195TC, 195CL, 157T, 195CR, 195BL, 195BR, 197R, 199T, 199BL, 199BR, 201TL, 201BL, 202TL, 202TR, 202BL, 219, 231, 232, 234, 235, 245, 247, 251, 252, 253, 254, 264, 266, 268, 271L, 271R, 280, 281, 282L, 282CL, 282CR, 282R, 286, 288, 290TL, 290TR, 290B, 294, 296, 297, 303, 304, 305L, 305C, 305R, 308TL, 308TR, 308CL, 308CR, 308BL, 308BR, 309TL, 308TR, 309BL, 308BR, 316, 317T, 317B, 318L, 318R, 319, 320TL, 320TR, 320B, 325, 326, 329, 336, 343, 344B, 351, 361, 367, 382, 383T, 383BL, 382BC, 400, 414, 421, 427, 432, 439, 442, 453, 465, 469, 472B, 477, 485a, 491, 496, 516, 521T, 521B, 523, 524, 527, 528, 534, 536, 539, 540, 569, 570, 575T, 577T, 577B, 578B, 579T, 581B, 582; © Alamy: 3, 4C, 15, 40, 48BC, 65TR, 113L, 155e, 201TR, 344T, 383BR; © Biodiversity Ireland: 28; © Biophoto Associates/ Science Photo Library: 201BR; © Carolina Biological Supply Company/Science Photo Library: 256B; © CNRI/ Science Photo Library: 256T; © Courtesy of MIT: 377; © Courtesy of the National Library of Medicine (https:// www.ncbi.nlm.nih.gov): 419, 420; Crown Copyright/ Imperial War Museum: 42T; © David M. Phillips/Science Photo Library: 240; © Dennis Kunkel Microscopy/Science Photo Library: 236, 237, 238, 239; © Don Fawcett/Science Photo Library: 237; © Dr Gopal Murty/Science Photo Library: 236; © Dr Jeremy Burgess/Science Photo Library: 240, 241; © Eye of Science/Science Photo Library: 9T; © Hervé Conge/Science Photo Library: 255; © HSE: 107, 498; © Humphrey Jones: 480, 481; © IDA Ireland: 42B; © International Rice Research Institute (IRRI): 410; © Irish Cancer Society: 261; © iStock: 1, 4T, 7TL, 37, 140R, 154B, 155a, 155f, 165CT2, 182, 244, 356, 501, 502, 529; © Keith Chambers/Science Photo Library: 489; © Lennart Nilsson/Boehringer Ingelheim/TT/Science Photo Library: 468, 488; © Marilyn Schaller/Science Photo Library: 240; © Martyn F. Chillmaid/Science Photo Library: 156C, 210; © Microscape/Science Photo Library: 238; © Milwaukee Instruments: 155g; © Nina Sesina, via Wikimedia Commons: 563L; © Pascal Goetgheluck/ Science Photo Library: 563C; © Research Ireland: 14; © SafeFood: 194; © Science Photo Library: 230; © Sheila Terry/Science Photo Library: 50TR; © Shutterstock: 16, 50B, 70, 76TR, 86, 106C, 106B, 115c, 122T, 150, 155b, 156B, 184, 197L, 218, 220, 247, 472T, 575B; © Tree Council of Ireland: 157; © Wellcome Trust: 41T, 51T, 112; © World Health Organization: 41B.

5 4

For Niamh and Caragh (HJ)

For Andrew, Eoin, Zara, Nuala and Edmond (YHV)

CONTENTS

List of Specified Investigations

Introduction

Welcome to *Evolution*, your new textbook for Leaving Certificate Biology!

Evolution is divided into 12 Units of Learning. These 12 units have been carefully created to meet the Learning Outcomes from across the four strands of the new Leaving Certificate Biology specification:

Unifying Strand: Nature of Science

Strand 1: Organisation of Life

Strand 2: Structures and Processes of Life

Strand 3: Interactions of Life

The chapters are broken down into a series of **bite-sized lessons**, all clearly mapped to the new **Learning Outcomes**, and include the following useful features:

- An **Eye-opener** grabs your attention and gets you thinking about the topic.
- Indicators for the **Senior Cycle Key Competencies** are shown by the following icons:

Thinking and Solving Problems	Working with Others	Participating in Society
Being Creative	Literacies	Managing Learning and Self
Communicating	Cultivating Wellbeing	Numeracy

- **FOCUS ON** panels highlight where further depth of knowledge is needed to meet the Learning Outcome.
- **Definitions** are pulled out from the text and are accompanied by an audio soundbite to help you with pronunciation of tricky subject vocabulary.
- **Higher Level only** content is clearly indicated throughout. HL
- **RESEARCH TASKS** and **Think, Pair, Share activities** encourage pair work, investigative inquiry and discussion.
- **SCIENCE IN SOCIETY** case studies link to the cross-cutting themes of **Health**, **Sustainability** and **Technology** and showcase real-life applications of Biology.

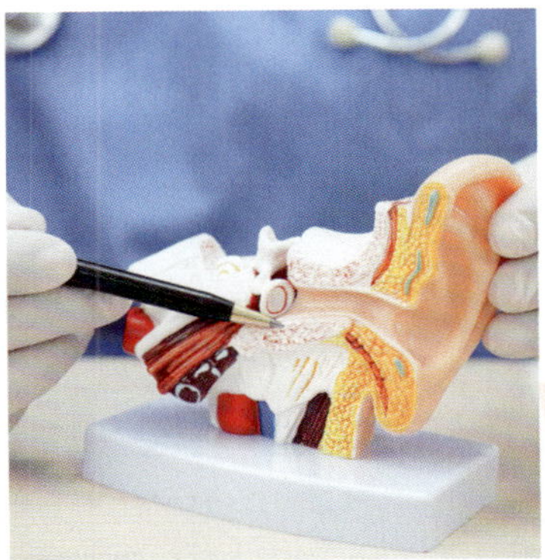

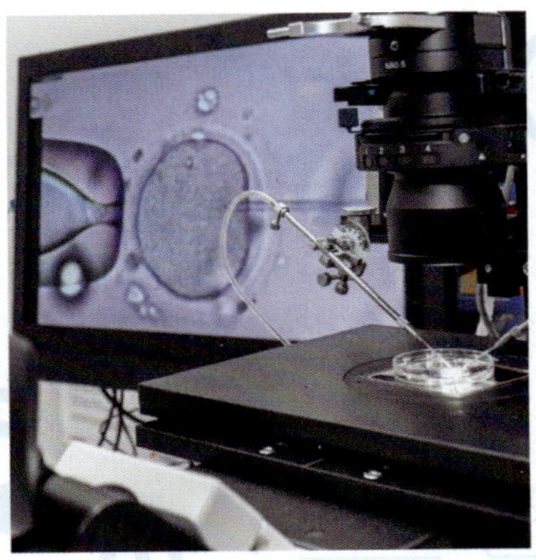

- **Specified Investigations** are presented step-by-step with visuals and accompanying videos that are easy to access via QR codes.

Each chapter ends with a **Summary** to help you with your revision and an **End-of-Chapter Assessment** which challenges you to Remember, Understand, Apply, Analyse, Evaluate and Create.

At the end of this textbook, you will find detailed guidelines on completing your **Biology in Practice Investigation**, which will account for 40% of your final grade.

2-in-1 Workbook: Learning Portfolio and Investigative Log

Throughout the textbook you will find links to the 2-in-1 Learning Portfolio and Investigative Log where you can complete homework, classwork and lab work.

PORTFOLIO LINK **Go to your Learning Portfolio and complete Learning Check 1 to assess your knowledge and understanding of the topic so far.**

INVESTIGATIVE LOG LINK **Go to your Investigative Log and complete your report on the ecosystem you studied.**

- **Learning Checks** to provide interesting homework tasks and help your teacher to assess how well you have understood the key learning from the lesson.
- **Great variety of activity types** including definition practice, diagram labelling, data analysis, assessment of scientific articles and much more.
- **Exam questions** from the new SEC sample exam papers to help you to prepare for the written exam.
- All your **lab notes** in one place and **questions on the experiments** to reinforce learning.
- Space to log your ideas, research and experimentation for the **Biology in Practice Investigation**.

Digital Resources

Evolution is accompanied by a suite of digital resources, all easily available through your eBook and on www.gillclassroom.ie. These include experiment videos, YouTube videos, PowerPoints, interactive quizzes and more!

UNIT 1

BIOLOGY – THE LIVING SCIENCE

CHAPTERS

STRANDS

Unifying Strand: Nature of Science

Strand 1: Organisation of Life

1 What Is Biology?

LEARNING OUTCOMES: U1.1, U4.2, 1.1.1

At the end of this chapter, you should be able to:

- Discuss how life can be defined.
- Outline the characteristics of all living things.
- Evaluate biology as a science.
- Describe the wide nature of biology as a science.

LINKS TO:

Chapter 2	Scientific Knowledge and Research
Chapter 3	Experiments and Data Analysis
Chapter 4	Classification and Taxonomy
Chapter 9	Evolution and Natural Selection
Chapter 10	Biodiversity and Ecosystems
Chapter 14	Biomolecules
Chapter 15	Metabolism and Enzymes
Chapter 16	Cell Structure and Diversity
Chapter 23	Respiration

SUBJECT VOCABULARY

life, cell, metabolism, biomolecules, heredity, autotrophic, heterotrophic, respiration, excretion, homeostasis, response, continuity of life, biology, hypothesis

EYE-OPENER

If you're reading this then you're probably a living thing (also called an **organism**) and, while it's rude to assume, you're also most likely human. As a living breathing human you move around your world interacting with other humans, with animals, with plants and a wide variety of other organisms. You are made of cells – trillions of them – and there are around 200 different types of them in your body. You eat, sleep, excrete, breathe in and breathe out. Your heart beats over 70 times a minute while your brain ensures your body is in perfect balance. However, as a living thing, you're not alone. There are over 8 billion more humans on the planet and we are just one of between 10-14 million different kinds of living things (species).

Billions of chemical reactions are taking place inside your body right now – some build new materials while others break materials down. You use energy to power your body – energy harnessed from the food you eat. You were made by other living things (your parents) and you may, in time, produce new living things. If you are curious about living things, yourself or others, then you are a biologist. You are life's way of looking at itself.

Section 1: What Is Life?

At the end of this section, you should be able to:

- Discuss different ways life can be defined.
- Name the characteristics common to all living things.
- Outline the characteristics of living things.

Unpacked from LO 1.1.1

There is a certain level of debate around what it means to be **living**. Many simply point to a set of characteristics that all living things share:

- They all comprise cells which organise themselves into various tissues and organs.
- They all contain DNA.
- They all have a metabolism (i.e. they are all chemically active).
- They can all reproduce.
- They all utilise energy.
- They all grow and are sensitive to their surroundings.

This can be a useful way of looking at things, but some think it is oversimplistic.

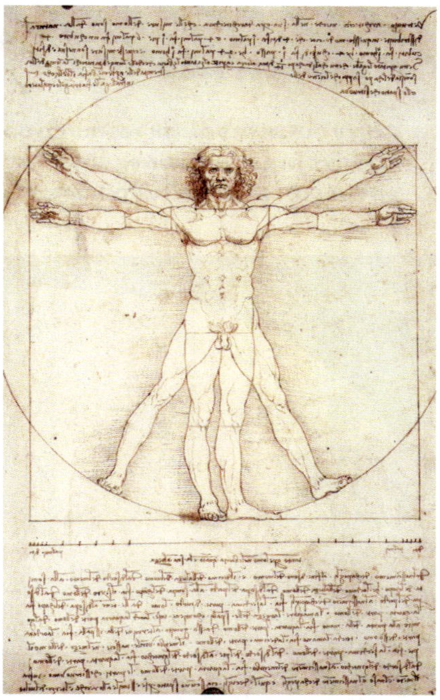

Figure 1.1: Leonardo da Vinci's 'Vitruvian Man'. Da Vinci was a keen biologist.

Some scientists would look at the chemicals that are common to all living things. All living things on Earth are predominantly made of the elements carbon, hydrogen, oxygen and nitrogen, and share identical molecules in their cells and tissues (e.g. DNA, RNA, phospholipids). These molecules interact in similar ways in all living things. We call these chemical building blocks biomolecules.

Physicists would argue that living things are a type of open thermodynamic system, capable of bringing order to matter while harnessing and converting various forms of energy within their cells. Living things, made of complex chemicals, are bound to the same laws of physics as non-living things. However, they are also capable of self-replicating and evolving. Living things can convert energy to different forms and utilise it for their own development, while non-living things can't.

Life is a condition whereby organic matter has the capacity to use energy for its development.

The laws of thermodynamics suggest the universe is always increasing in disorder, with all types of energy eventually turning into heat. The same laws also explain why all living things eventually die, as the contents of cells become increasingly disordered over time. However, many living things develop high levels of organisation over their lifetime, which actually goes against the laws of thermodynamics.

However one defines it, life thrives in virtually every corner of our planet – in extreme heat, extreme cold, extreme pressure and extreme dryness.

Figure 1.2: Turritopsis dohrnii, aka 'the immortal jellyfish', is able to revert to an immature stage of life if damaged or threatened.

RESEARCH ACTIVITY

Carry out some quick research on extremophiles – organisms that can live in extreme environments. How do they adapt to live in those environments? Share your results with your class.

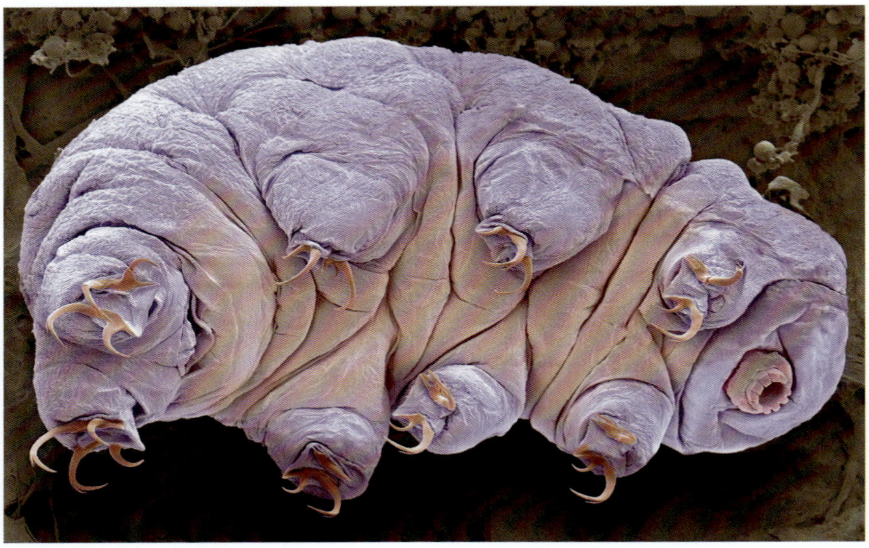

Figure 1.3: The tardigrade is a tiny living thing which can survive in the most extreme conditions, including the vacuum of space.

Go to YouTube.com and search for 'The 5 core principles of life | Nobel Prize-winner Paul Nurse' (7:36) by Big Think.

Such ambiguity around the meaning of the term life means there are still some grey areas. For example, viruses exhibit many of the features of living things common to animals, plants and micro-organisms but not all of these features. Many see them on the edge of life, despite the fact that they contain genetic material (DNA or RNA) and evolve via natural selection.

Figure 1.4: Erwin Schrödinger.

RESEARCH ACTIVITY

Carry out some research on the physicist Erwin Schrödinger. In 1943, while living in Ireland, Schrodinger gave a speech in Trinity College entitled 'What Is Life?'. His speech inspired a new generation of biologists to explore the chemical and physical nature of living things.

Create a poster summarising your research.

CHARACTERISTICS OF LIVING THINGS

'Life' is difficult to define, but the following list of characteristics is a useful way to distinguish between living and non-living things. Some non-living things might satisfy one or more of these, but living things satisfy *all* of the characteristics. We will explore all of these elements in more detail later in the book.

Cells

All living things contain **cells**. Cells are the functional units of living things. Some living things are composed of a single cell and are known as **unicellular** (e.g. bacteria), while others are **multicellular**, and contain possibly trillions of cells (e.g. humans).

In the most basic sense, a cell contains two basic parts: a cell membrane and a cytoplasm. The cytoplasm contains genetic material (e.g. DNA), for the most part found in a nucleus, and may contain other structures called **organelles**, each with a particular function. Chapter 16 explores cells in more detail.

- The **cell** is the basic unit of life.
- **Metabolism** is the collective term for all the chemical reactions that occur within an organism.
- **Biomolecules** are the chemical building blocks of cells.

Growth

Cells are special because they can divide and replicate. This replication allows organisms to grow. Growth requires more of the chemical building blocks of cells, which are often referred to as **food**. We will explore cell division, the main process involved in growth, in Chapter 17.

Metabolism

All living things are chemically active with billions of chemical reactions taking place within the organism every second. The term **metabolism** refers to the complete set of reactions that occurs in an organism. We will learn more about metabolism in Chapter 15.

Organisation

Living things display a high level of organisation, especially multicellular organisms. Atoms are arranged into **biomolecules**: chemicals found in all living things, which arrange themselves into cells. Cells are specialised for a particular function and come together to form **tissues**, which come together to form **organs**, which come together to form **organ systems**. The organ systems work collaboratively to form complex organisms like humans or oak trees. An organism's ability to organise biomolecules, cells, tissues and organs is only temporary; all organisms eventually die. We will further explore the organisation of living things in Chapter 16.

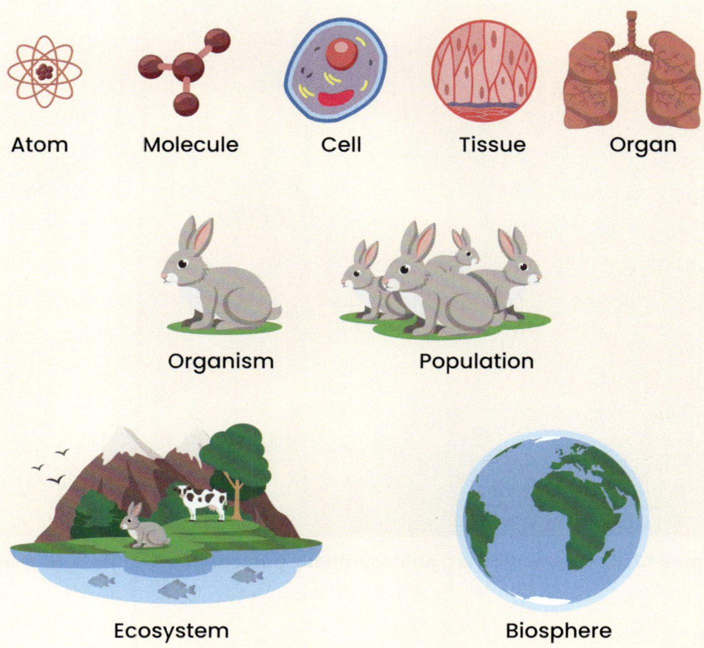

Figure 1.5: Organisation of living things.

DNA and heredity

There are many chemicals common to living things but DNA is the most important. It is the main genetic material in all living things, storing information in the cell and allowing that information to pass from one generation to the next. A cell can copy DNA, making it perfect for living things to pass on information from one cell to another. We will learn more about DNA and how it works in Chapter 24.

The process of passing information, using DNA, from one generation to the next is called **heredity**. This is explored more in Chapter 25.

- **Heredity** is the process of passing information, using DNA, from one generation to the next.
- **Autotrophic** organisms have the ability to produce their own food, e.g. photosynthesis.
- **Heterotrophic** organisms must consume food to obtain energy and the raw materials for growth.
- **Respiration** is the release of energy from food.

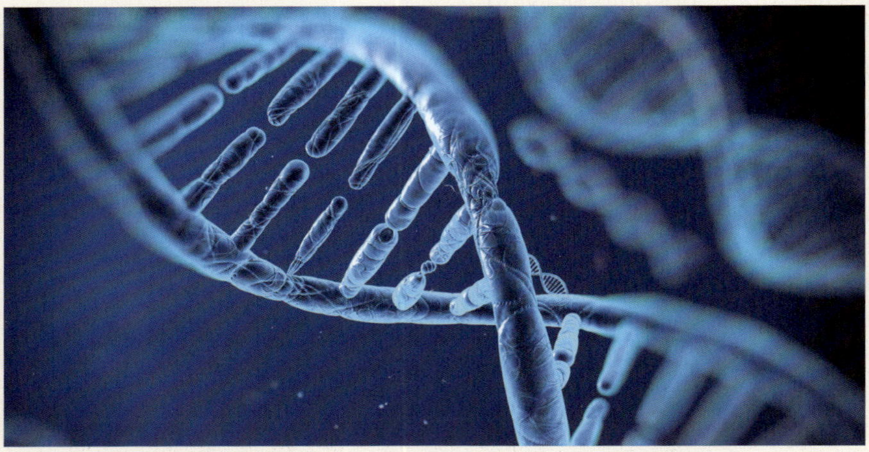

Figure 1.6: DNA is common to all living things and carries instructions for the organism.

Nutrition and respiration

Living things require food. Food contains the building blocks of new cells and a source of energy to carry out the organism's chemical reactions. Plants and some bacteria can make their own food, harnessing light (or chemical) energy to make glucose in a process called **photosynthesis** (or chemosynthesis). Such organisms are known as **autotrophic**. Other organisms acquire food by consuming the cells and tissues of other living things. These are known as **heterotrophic**.

The organism uses food to build new cells (growth) or to power their metabolism. The release of the energy from food, for use by the organism, is known as **respiration**.

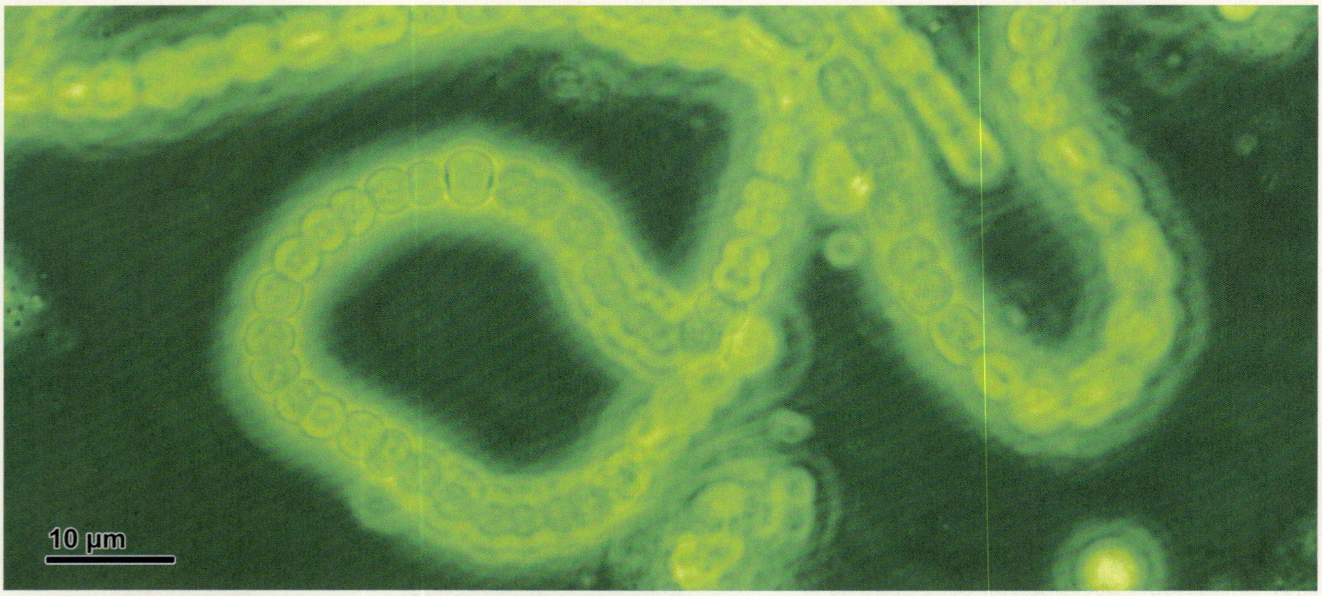

Figure 1.7: Cyanobacteria is a photosynthetic bacterium that uses light energy to make food.

Homeostasis and excretion

As living things metabolise (carry out their chemical reactions), they produce some wastes (e.g. we produce CO_2, H_2O, urea and more) that must be removed through **excretion**.

Organisms need to be able to maintain a set of conditions that allows these reactions to occur so these wastes need to be removed. An organism's ability to maintain the conditions that allow metabolism to take place is known as **homeostasis**. Excretion plays a big role in homeostasis.

- **Excretion** is the removal of the waste products of metabolism.
- **Homeostasis** is the maintenance of a constant internal environment within an organism.
- **Response** is the ability of an organism to detect changes in its surroundings and react to increase their chance of survival.

Response

All living things can **detect** changes in their environment (where they live) and **respond** to these changes to improve their chance of survival. Humans have sophisticated sense organs (eyes, ears, skin, tongue and nose) to detect light, sound, touch, temperature, taste and smell. Some animals are capable of detecting an even wider range of stimuli. Humans and other animals also use a complex hormonal system to detect changes to their external and internal environments. Examples of animals' responses include moving – either away from a threat or towards potential prey – or creating vocalisations, such as talking, roaring or chirping.

THINK, PAIR, SHARE

Working in pairs, brainstorm various types of responses plants can make to changes in their surroundings. Think of a wide variety of plants.

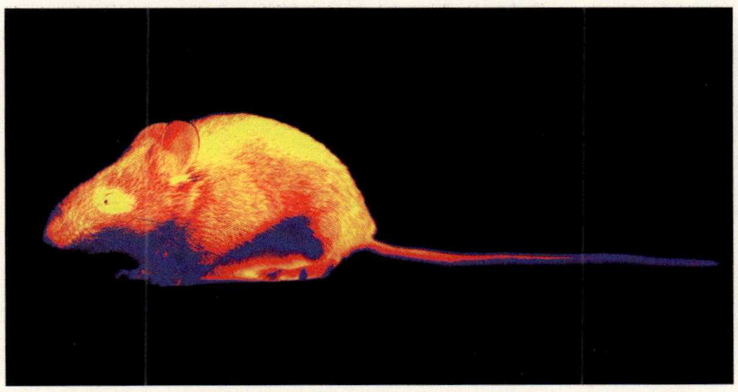

Figure 1.8: Snakes can see in infrared, so they can detect the body heat of their prey in the dark.

Figure 1.9: Birds can detect the Earth's magnetic field, which helps them to navigate their yearly migrations.

Plants can detect their surroundings too. They can sense the direction of light, the duration of daylight hours, the temperature, the presence of a wall or obstacle and the direction of gravity. Plants' responses are much slower than animal responses and generally involve growth towards or away from the stimulus. Again, these responses give the plant a greater chance of survival.

Figure 1.10: The Venus flytrap responds to touch.

Figure 1.11: Sunflowers respond to the direction of light.

Reproduction

All life on Earth shares a common ancestor – the very first cell that formed spontaneously in the soupy waters of the early Earth around 4 billion years ago. All living things derive from other living things (all cells derive from other cells). This is known as the **continuity of life**.

Reproduction is the ability of an organism to create new living things. Creating new living things – or new cells – is a key feature of all living things. There are two general types of reproduction: sexual and asexual.

- **Sexual reproduction** occurs in more advanced organisms and introduces **variation** into the offspring (the offspring's DNA is different from their parents'). Animals and plants generally reproduce sexually.
- **Asexual reproduction** occurs in more primitive organisms and involves organisms creating identical copies of themselves (the offspring's DNA is the same as their parents').

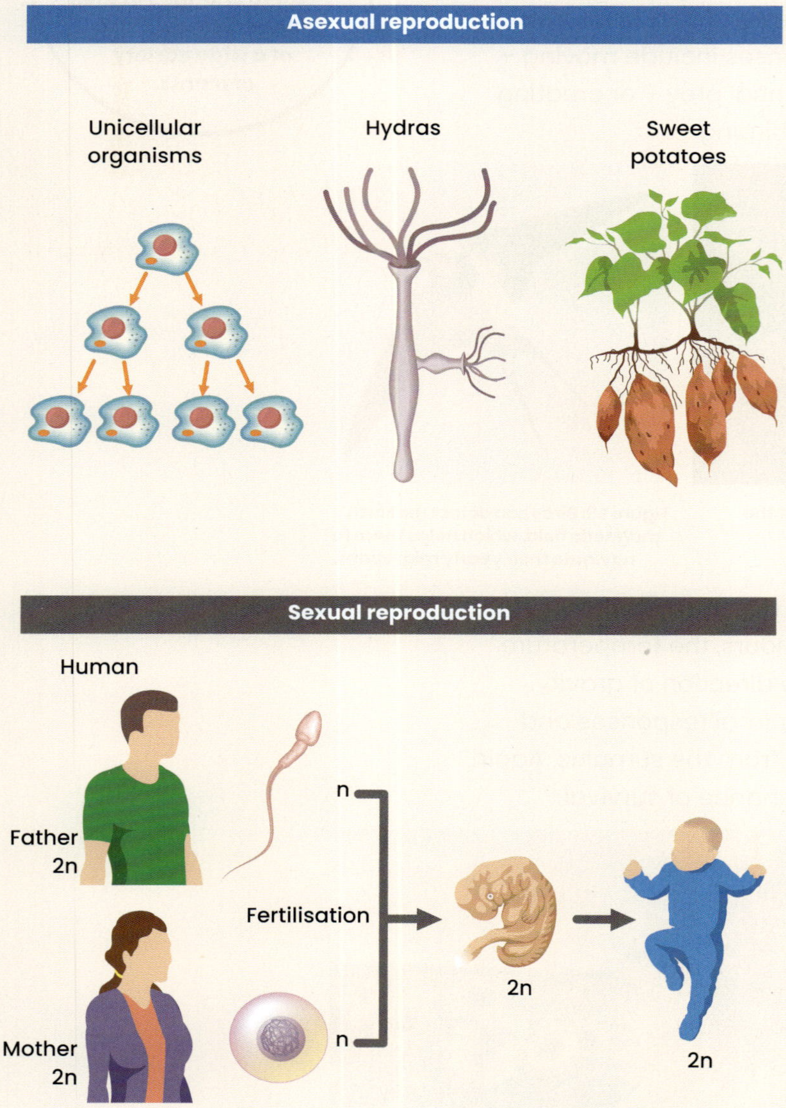

Figure 1.12: Asexual vs sexual reproduction.

Go to YouTube.com and search for 'Introduction to Biology: Crash Course Biology #1' (13:26) by CrashCourse. Watch the section on life characteristics beginning at 2:14.

Continuity of life explains that all life can only derive from other living things.

THE FIRST LIVING THING

The first living thing appeared on Earth over 4 billion years ago, roughly half a billion years after Earth was formed. Before that first primitive cell appeared, the building blocks of cells were already forming in the early Earth's seas. The first living thing was likely just a simple cell known as *archaea*, containing some genetic material – DNA – which carried instructions for how the cell might function.

The first few hundred million years of Earth's existence were not suitable for the formation of life – it would have been too hot and too dry, with a lot of volcanic and asteroid activity. However, all the ingredients that were needed were present. Water, ammonia and methane, along with elements like hydrogen, oxygen and carbon were in abundance in the early Earth, and these somehow came together to form amino acids, the building blocks of proteins. This would have required an energy source, which we can assume was from hydrothermal vents in the deep ocean or perhaps even lightning. The process of creating life from inorganic materials is referred to as **abiogenesis**.

This theory was famously tested in the Miller-Urey experiment.

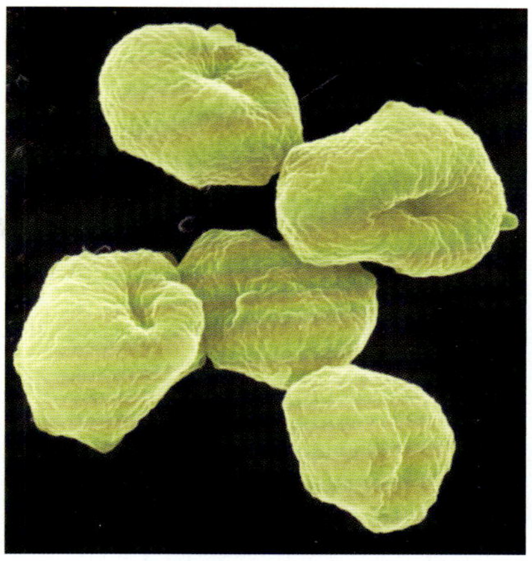

Figure 1.13: Archaea are very ancient lifeforms.

Figure 1.14: Early Earth was too hot and violent to host life.

The Miller-Urey experiment

In 1952, Stanley Miller and Harold Urey carried out some experiments to see if they could make the initial building blocks of life – substances that are usually made by living things – using only the conditions thought to have existed on the early Earth, before life as we know it existed. The scientists sealed a mixture of water, ammonia, methane and hydrogen in a sterile flask. The mixture was heated to evaporate the water, producing water vapour. Sparks of electricity were passed through the mixture of water vapour and gases, simulating lightning. After one week, the contents of the flask were analysed. Amino acids, the building blocks for proteins, were found.

But let's not jump to conclusions; while the scientists did demonstrate how amino acids – building blocks of life – could have been formed on early Earth, this experiment did not explain how these amino acids then formed the first living thing.

Since that first cell was formed 4 billion years ago, life has evolved and diversified to create the huge variety we have now on Earth.

DISCUSSION

How might a changing world (pollution, climate change, etc.) affect the future of life on Earth?

PORTFOLIO LINK

Go to your Learning Portfolio and complete Learning Check 1 to assess your knowledge and understanding of the topic so far.

Section 2: The Study of Life

At the end of this section, you should be able to:

- Define biology as the study of life.
- Evaluate biology as a science.
- Give examples of different branches within biology.
- Discuss the importance of studying the living world.

Unpacked from LOs U1.1, U4.2, 1.1.1

Now that we know what living things are, we can say that **biology is the study of life and living things**. It is often referred to simply as **life science**.

Biology is the study of life and living things.

The huge variety of life on Earth means that biology is a very broad field of study. One can spend a lifetime studying one organism, or indeed one process within one organism, and still not fully understand it after 20 years. However, the inherent variety that makes biology so challenging is also what makes it a hugely enjoyable subject.

Over the course of your studies, we will look at the core elements of biology but, most importantly of all, we will encourage you to consider both the impact of biology on your life and your own impact on the living world.

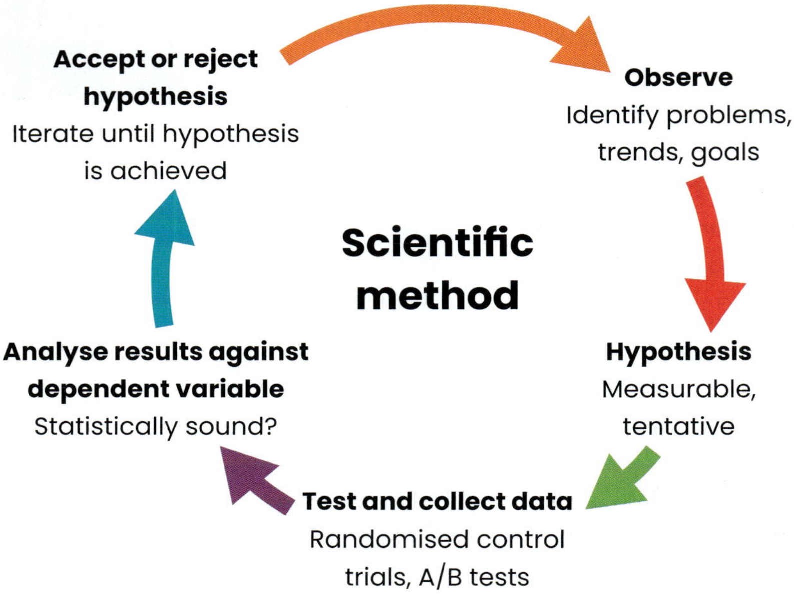

Figure 1.15: Biologists, like all scientists, follow the scientific method.

Being a scientist is not about collecting and memorising facts – it's a process of continuously finding out more about the world around us. Biology as a subject is constantly changing, as our understanding of the living world changes. Biologists are carrying out a wide range of experiments, leaning on the findings of other biologists before them, and widening our understanding of the living world. There are millions of people around the world studying biology and hundreds of thousands of scholarly papers are published by biologists every year.

Biology is widening in scope, too, and the lines are being blurred with other subjects like chemistry, physics and geography. Over the course of the next two years, we will focus on the processes that underpin science and apply them to the biological world.

BRANCHES OF BIOLOGY

Below is a brief list of some of the main branches of biology and a short description of each. You'll notice that biology is beginning to incorporate other subjects like chemistry, geography, engineering, computer science, agriculture, medicine and more. If you choose to study biology at Third Level, you may choose to specialise in one of these fields.

- **Anatomy** – the study of the internal structures of animals.
- **Bacteriology** – the study of bacteria and their applications.
- **Biochemistry** – the study of the chemical reactions that occur within living things.
- **Bioengineering** – the creation of useful products by utilising biological processes.
- **Biogeography** – the study of the distribution of living things around the planet.
- **Bioinformatics** – the use of technology to store biological data (like genomes).
- **Biomathematics** – the application of maths to the living world.
- **Biomedicine** – the study of human health and disease.
- **Botany** – the study of plants and their structures.
- **Cytology** – the study of cells and the processes within them.
- **Ecology** – the study of how living things interact with each other and their environment.
- **Embryology** – the study of the unborn.
- **Evolutionary biology** – the study of how life has evolved on planet Earth.
- **Genetics** – the study of inheritance.
- **Genomics** – the study of genomes (the genes) in humans and other organisms.
- **Marine biology** – the study of living things within the oceans.
- **Nanobiology** – the study of biological functions at the nanoscale.
- **Physiology** – the study of how organisms work.
- **Taxonomy** – the study of classification of living things and their evolution.
- **Virology** – the study of viruses.
- **Zoology** – the study of animals.

Figure 1.16: Biomathematics looks at the application of maths to the living world.

Figure 1.17: Nanobiology looks at living structures on the nanoscale like these spindle fibres – tiny filaments found in cells.

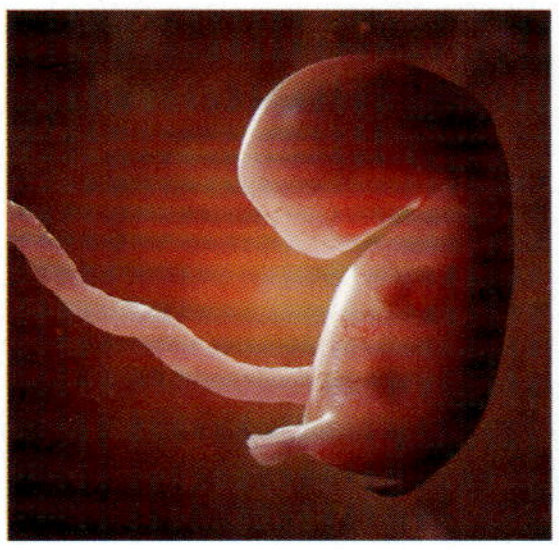

Figure 1.18: Embryologists study the unborn to help understand our development from a single fertilised egg.

RESEARCH ACTIVITY

In pairs, choose one of the branches of biology from the list above (or a branch not mentioned here) and create a short presentation for the rest of your class.

WHY STUDY BIOLOGY?

If you're reading this, you've already decided to study biology for your Leaving Certificate. Well, you've chosen wisely. Yes, you will be challenged, but you will learn a whole new subject vocabulary, explore some complex concepts, carry out research, design your own investigations and learn new practical laboratory skills.

Biology is an extremely practical subject. Furthering your knowledge and understanding of your body – and how it works – will help you understand what is happening if you get injured or sick. A knowledge of biology will help you understand the principles of a healthy lifestyle and will help you understand and appreciate nature and your role in maintaining it. It may answer some of the questions you already have about your body and the living world, and will most likely generate further questions, too. It may provide you with a foundation for further Third-Level study in medicine, veterinary science, engineering, general science or other subjects. Biology also opens up a wide range of fulfilling careers.

Figure 1.19: There are many varied biology-related careers in biology and many more that don't exist now but will in 10 years' time.

By studying biology, we hope you will cast a more critical eye on the world around you, ask questions, develop your own **hypotheses** (educated guesses), design investigations to test those hypotheses, learn to solve problems when they arise, evaluate your results and draw conclusions. We hope you learn to analyse and evaluate the science and biology presented to you in the news, on social media, from politicians and organisations and that you learn to recognise bias.

A **hypothesis** is an educated guess based on observation.

Ultimately, you will learn to think critically and, with a greater knowledge and appreciation for life in all its forms, you will be a more useful, understanding, empathic and environmentally conscious member of society.

PORTFOLIO LINK **Go to your Learning Portfolio and complete the Learning Check 2 to assess your knowledge and understanding of the topic so far.**

Go to YouTube.com and search for 'What Biologists Do: Crash Course Biology #3' (12:37) by CrashCourse.

Chapter Summary

- **Life** is difficult to define, but can be described as a condition whereby organic matter has the capacity to use energy for its development.
- All living things have the following characteristics:
  - They all have a **metabolism**.
  - They have a high level of organisation.
  - The all contain DNA, which is passed from one generation to the next (**heredity**).
  - They all require food for energy and to grow.
  - They all produce waste, which is removed or **excreted**.
  - They can all **respond** to their surroundings and react to increase their chance of survival.
  - They can all **reproduce** and create new living things.
- Life has existed on Earth for around 4 billion years.
- Biology is the study of life and living things.

- Biology is a broad and interesting subject with many branches of study.
- Biology is a useful subject which helps people understand how the human body works and how humans impact on other living things.
- There are a wide range of biology-related careers.

END OF CHAPTER ASSESSMENT

REMEMBER

1. List **five** characteristics of all living things.
2. What is meant by the term *metabolism*?
3. Name the chemical used as genetic material in all living things.
4. Name the process by which plants make their own food.
5. What is excretion?
6. What is meant by the *continuity of life*?
7. How can you define *biology*?
8. Define the following branches of biology: bioinformatics, cytology, anatomy, botany.
9. What is a hypothesis?

UNDERSTAND

10. Briefly describe **three** common characteristics of all living things.
11. Distinguish between autotrophic and heterotrophic organisms.
12. Why do all living things need food?

APPLY

13. Why are viruses often considered non-living despite having many characteristics of life?
14. Why do you think all living things on Earth use DNA as their genetic material?
15. Discuss the need for living things to maintain homeostasis – a steady internal environment.
16. Compare sexual and asexual reproduction.

ANALYSE

17. Using a table, compare and contrast responses in plants and animals.
18. Research and compare **two** branches of study within biology.

EVALUATE

19. Which of the characteristics of life do you think is most important to consider when evaluating if something is living or non-living?
20. Design a survey to evaluate the understanding of life amongst the students in your school.
21. Artificially intelligent (AI) computers and devices can learn and adapt more quickly than humans. Could AI be considered living? Discuss.
22. Debate the following motion: *Studying and understanding the living world is crucial to a sustainable world.*
23. Do you think life on other planets, if it exists, might have the same characteristics of living things found on Earth?

CREATE

24. Construct a mind map to summarise the characteristics of life.
25. Design your own fictional 'living thing'. Show how it aligns with the characteristics of life.

2 Scientific Knowledge and Research

LEARNING OUTCOMES: U1.1, U1.2, U3.1, U3.2, U4.1, U4.2

EYE-OPENER

Taighde Éireann – Research Ireland is the national competitive research and innovation funding agency in Ireland. It funds research carried out in Ireland in areas such as sustainable living, health, the bioeconomy, climate action and artificial intelligence, among others. It aims to ensure Ireland will be a global innovation leader in research of all disciplines, for the benefit of Ireland's economy and society. (Source: Research Ireland)

At the end of this chapter, you should be able to:

- Understand that communication within the science community is on a global basis.
- Understand that research carried out by scientists must meet international standards and is subject to peer review and reproducibility.
- Appreciate that scientific ideas are modified over time.
- Research a scientific issue using different sources of information including data produced by others.
- Evaluate different sources of scientific information including science and technology related articles in the media for relevance, accuracy, bias and lack of detail.
- Relate science and scientists to society by considering economic, social, sustainability, environmental and ethical factors.
- Research and present on the impact that scientific discovery and invention by scientists makes to society.
- Explain biological phenomena using appropriate means, e.g. systems, interdependence, form fits function, transfer of information.

SUBJECT VOCABULARY

scientific community, SI units, peer review, reproducibility, relevance, bias, primary data, secondary data, representative, random, sustainability, ethics, system, interdependence, form fits function

LINKS TO:

The scientific practices outlined in this unit apply to all Learning Outcomes throughout the rest of the Leaving Certificate Biology course.

Section 1: Scientific Knowledge

At the end of this section, you should be able to:

- Describe how scientists work.
- Understand how scientific ideas are altered and revised over time as knowledge builds up.
- Understand that the scientific community is a global one.
- Give examples of how communication is carried out within the scientific community.
- Explain why international standard rules or conventions are used when communicating data within the scientific community.
- Explain what is meant by 'peer review' and 'reproducibility' and why they are important within the scientific community.

Unpacked from LO U1.1

MODIFICATION OF SCIENTIFIC IDEAS OVER TIME

What's exciting about science is that it is never 'finished' – scientific knowledge is constantly being revised and expanded. Changes in our knowledge can result from scientific observations that weren't previously possible and ideas that were once accepted by society can be modified or rejected based on new evidence.

For example, in the 1600s, it was believed that a parent's sperm contained a miniature infant called a homunculus, which simply had to grow larger. This idea was called 'preformation'. In 1677, Antonie van Leeuwenhoek developed a microscope that allowed him to discover and observe the human sperm cell. However, it wasn't until the 1800s when biologist Karl Ernst von Baer discovered the female egg that preformation was disproved.

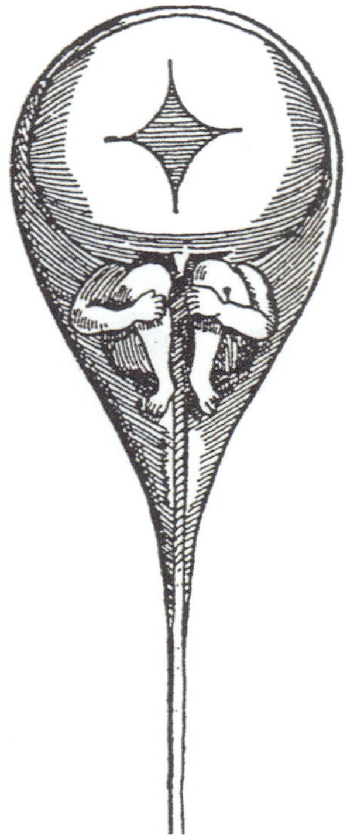

Figure 2.1: Human homunculus in the theory of preformation.

Antonie van Leeuwenhoek and Karl Ernst von Baer worked in a particular way. They made observations and asked questions about the world around them. These questions were testable and could be answered by carrying out investigations and collecting evidence. They communicated their research with other scientists. This ultimately led to ongoing research into the subject.

COMMUNICATION IN THE SCIENCE COMMUNITY

Scientists work together on research. This interacting network of scientists, known as the **scientific community**, stretches across the entire world. One of the greatest examples of the collaborative work of scientists is the Human Genome Project (HGP). The HGP was founded in 1990 to map all the genes of a human being. This project, involving over 2,800 scientists from six countries, ran until 2003 when the entire human gene sequence was first released. Working collaboratively allows scientists with different areas of expertise to research a

The **scientific community** is the global interacting network of scientists working on research.

problem. Scientists need to have good writing and oral communication skills as they publish their research in scientific journals and present it at meetings called conferences. These conferences allow scientists from around the world to learn about new scientific developments or technologies, meet with other scientists and review each other's work.

International conventions

Different measurement systems exist around the world, e.g. length can be measured in metres or feet. To ensure that global scientific measurements are standardised, an international system was set up. This is known as the **International System of Units (SI)**. This decimal measurement system defines the units which should be used by scientists when recording measurements and is based on the metric system. There are seven base SI units from which other units are derived (see Figure 2.2 below).

QUANTITY	SI UNIT
Time (t)	second (s)
Length (l)	metre (m)
Mass (m)	kilogram (kg)
Temperature (T)	Kelvin (K)
Luminous intensity (Iv)	candela (cd)
Amount of a substance (n)	mole (mol)
Electric current (I)	ampere (A)

Figure 2.2: The seven base SI units of measurement.

SI units are standard scientific measurement units based on the metric system.

Figure 2.3: A scientist presenting their research at a scientific conference.

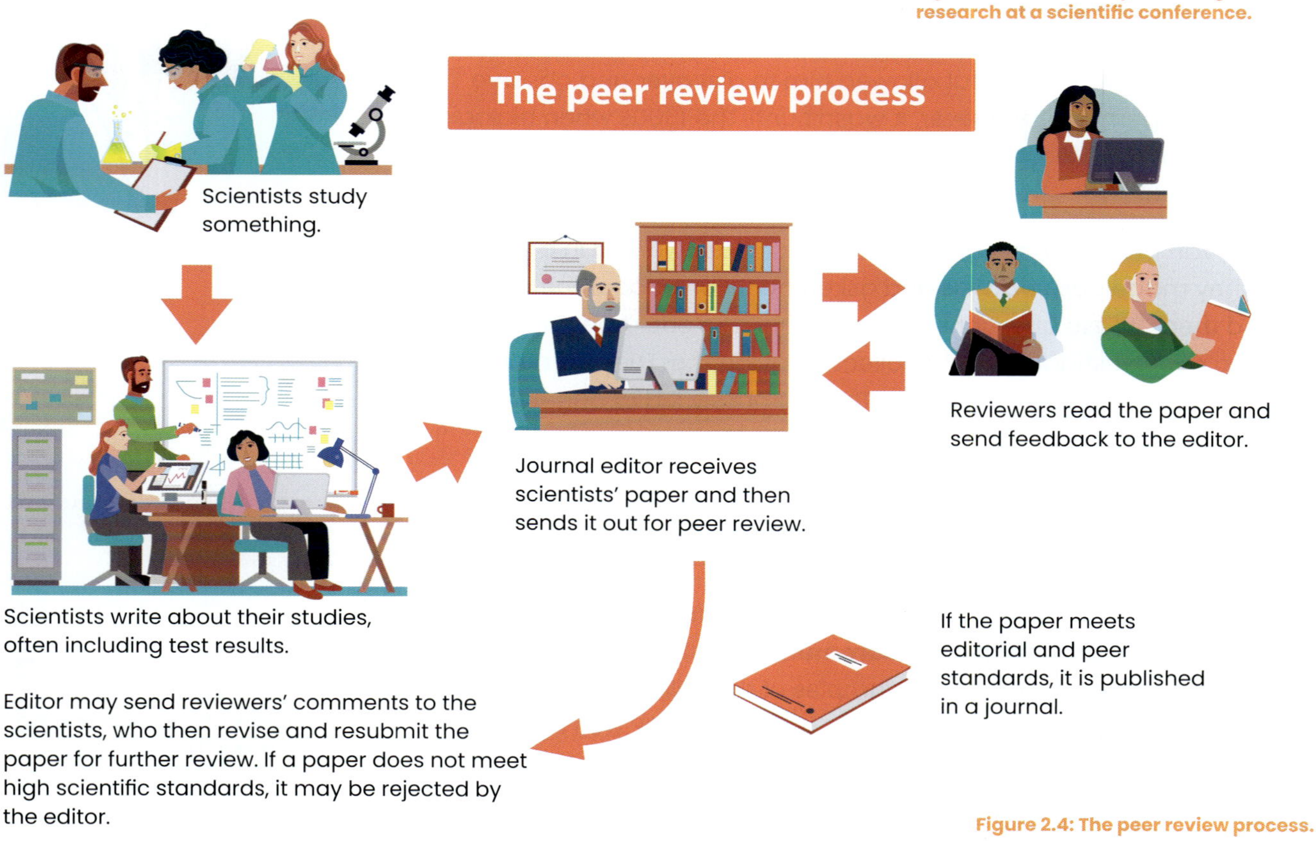

Figure 2.4: The peer review process.

Peer review

When a group of scientists completes a study, they write their research in the form of a 'paper' and submit it to a journal for publication. This paper is sent to other scientists who are experts on the subject. These peer reviewers provide feedback on whether the study is well designed, has accurate conclusions, is free from bias, not plagiarised and meets high scientific standards. The editor may return the paper to the authors with this feedback so that they can revise it. The **peer review** process prevents inaccurate studies from reaching the scientific community and helps ensure that papers published in scientific journals can be trusted.

- **Peer review** refers to the evaluation of a scientist's work by others who are experts in the same field.
- **Reproducibility** describes whether an investigation can be carried out by a different research team using the same method outlined in a scientific paper and obtain the same results.

Reproducibility

Another reason why clear communication is needed in scientific research is to ensure **reproducibility**. When a paper is published in a journal, other scientists can carry out the same investigation in their own laboratory following the method outlined in the paper and obtain the same results. This is to ensure the results obtained are not related to a particular setup in one laboratory. Reproducing studies helps eliminate mistakes due to human error and guards against rare cases of fraud. Clear communication, the use of standard measurement units, peer review and reproducibility help ensure that the build-up of scientific knowledge is accurate.

Go to YouTube.com and search for 'How does peer review make research more trustworthy?' (2:31) by Cambridge University Press.

PORTFOLIO LINK **Go to your Learning Portfolio and complete Learning Check 1 to assess your knowledge and understanding of the topic so far.**

Section 2: Scientific Research

At the end of this section, you should be able to:

- Research a scientific issue.
- Evaluate different sources of information including secondary data produced by others.
- Recognise when an information source lacks detail.
- Identify bias in a source of information related to science and technology.
- Evaluate the accuracy and relevance of a source of information related to science and technology.

Unpacked from LOs U1.1, U1.2, U3.1

EVALUATING RESEARCH

As part of your study of Leaving Certificate Biology, you will complete a Biology in Practice Investigation for examination purposes. This investigation will provide you with an opportunity to work as a scientist. In the first phase of the investigation, you will gather, process and evaluate information from research produced by others. When you are reading this or any other scientific research, it is important to ask yourself some of the following questions:

- Does the research lack detail? Would it be possible to reproduce the study using the information provided?
- Is the research **relevant** – does it contribute something valuable to scientific knowledge?
- Is the research **biased**? Any deviation from the truth in scientific research is known as 'bias'. All researchers must avoid bias when carrying out their own research and data analysis. Data collected by a researcher themselves is called **primary data**. It is also important to be aware of bias when analysing research carried out by others. Data collected by others is referred to as **secondary data**.

There are many different forms of bias which can affect scientific research.

- **Relevant** research contributes something valuable to scientific knowledge.
- **Bias** refers to any deviation from the truth in data collection, analysis, interpretation and publication that leads to false conclusions.
- **Primary data** is data collected by the researcher themselves.
- **Secondary data** is data collected by another researcher.
- A **representative** sample is a subset of the population that reflects the characteristics of the entire population.
- **Random** selection means that all individuals have an equal chance of being selected to take part in a study.

1. Bias in data collection

When carrying out studies, it is often not possible to study an entire population. Instead, a sample that is large enough to reflect the characteristics of the population is studied. Such a sample is said to be **representative** of the population. The sample should be selected at **random** to avoid bias; this means that all individuals have an equal chance of being selected to take part in the study. A sample that does not reflect the total population would be considered unrepresentative or biased (Figure 2.5).

Bias can be introduced to studies in the following ways:

- Survey questions that encourage a particular answer, e.g. 'You exercise regularly, don't you?'
- Non-random selection of a sample, e.g. selecting a sample with a particular background
- Sample size that is too small
- Faulty equipment, e.g. a balance that isn't measuring mass accurately
- Errors in the method, e.g. not including a control group.

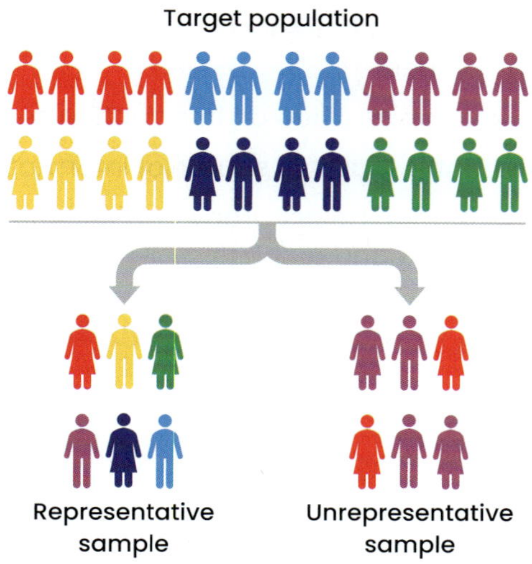

Figure 2.5: Sample selection.

Consider scientists researching a gluten intolerance test who invited colleagues to take part in the study. The people who volunteered to take part in the study may suspect that they are gluten intolerant. In this way, gluten intolerant people could be overrepresented in the study. This research could not be used to draw conclusions about the general population.

2. Bias in data analysis

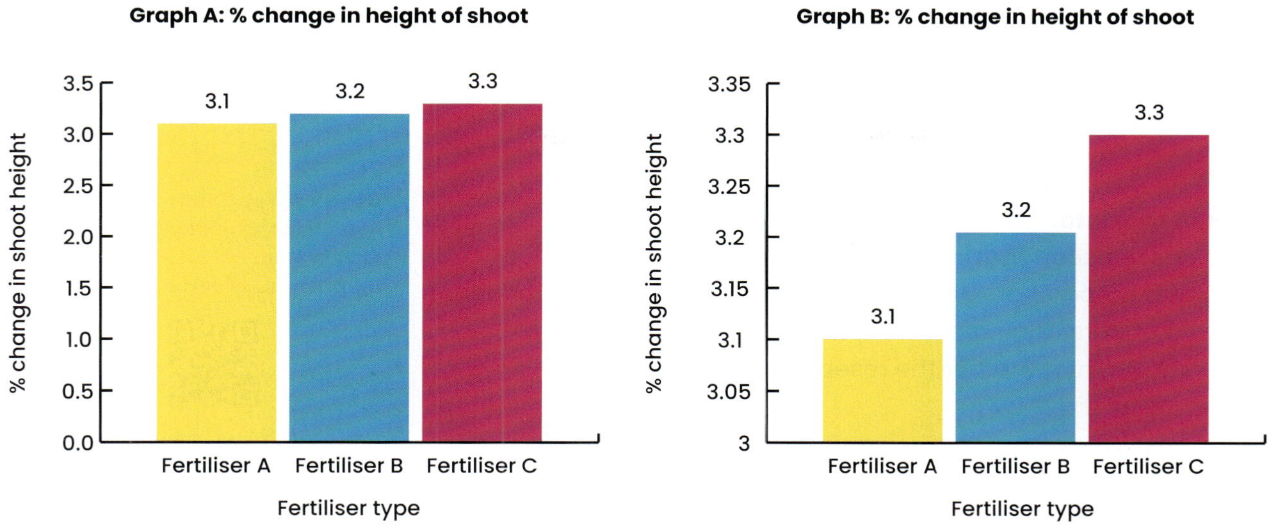

Figure 2.6: Misleading data: truncated y-axis.

There are several ways that a researcher can introduce bias to a study when analysing data:

- Reporting fake data
- Excluding data which does not support their original hypothesis, e.g. extreme values
- Using inappropriate methods to analyse the data.

Study the two charts in Figure 2.6 above. Both graphs represent the same data values, but the y-axis in Graph B does not start at zero, resulting in a skewed view of the data.

3. Bias in data interpretation

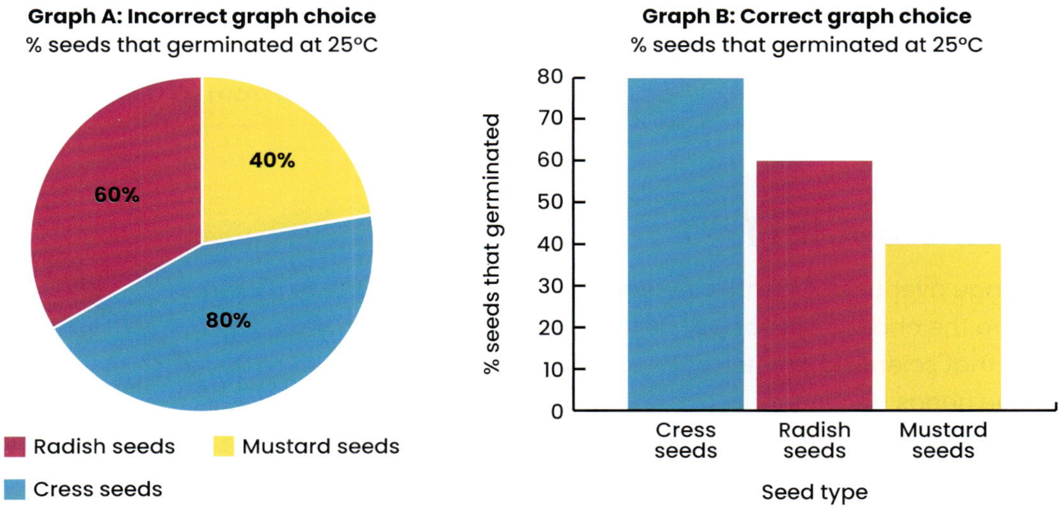

Figure 2.7: Misleading graph choice.

When interpreting results from a study, it is important to check for bias.

- **Method:** Would the equipment used allow accurate measurements to be taken? You will learn more about accuracy in the next chapter.
- **Results:** Are the results presented correctly? In Figure 2.7, a pie-chart is not an appropriate means of presenting the results. In Graph A, the total of the percentages is over 100%. Graph B, the bar chart, is a more appropriate choice here.
- **Range of data:** Drawing conclusions about values outside the range of data observed in the study can introduce bias, e.g. estimating plant growth at temperatures outside those the study measured.

- **Overgeneralisation of the study:** Results may not be generalised to the entire population, e.g. the fact that certain bacteria in probiotics impart health benefits does not mean that all do.
- **Biased language/Lack of detail:** Is the person presenting all the evidence or just the information that agrees with their point of view?
- **Source of the data:** Is the article from a peer reviewed journal or a popular science magazine? What is the source of the data: a university, a corporation, a government body or a special interest group? An understanding of who funded the research may provide insight into whether the research is biased or not.

RESEARCH ACTIVITY

Research the history of the tobacco industry in cancer research.

Go to YouTube.com and search for 'Objectivity and Bias in Science' (3:38) by Purposeful Universe.

PORTFOLIO LINK

Go to your Learning Portfolio and complete Learning Check 2 to assess your knowledge and understanding of the topic so far.

Section 3: Scientific Discovery and Invention

At the end of this part, you should be able to:

- Research and present information on the role of scientists in scientific discovery and invention.
- Give examples of the impact that scientific discovery and invention has made to society.
- Give examples of how science impacts on economic development in society.
- Give examples of how science is used to create a more sustainable society.
- Give examples of the role of ethics in scientific decision making.

Unpacked from LO U3.2

SCIENCE AND SOCIETY

Societies change over time. Scientific research should be relevant and respond to the changes in society as they arise. It is important to remember that scientists are normal people who are often sensitive to the needs and interests of the society in which they live. This influences the research they carry out, whether it is driven by curiosity, economics, or both. Societal needs also influence what research gets funding. For instance, the devastating impact of World War II on human life resulted in improvements in medical techniques such as blood transfusions and skin grafts. Today, in response to climate change, a lot of investment is going into research on topics like renewable energy.

THINK, PAIR, SHARE

How have scientific discoveries and inventions impacted on your daily life? Think about scientific developments in the areas of agriculture, medicine, technology and energy.

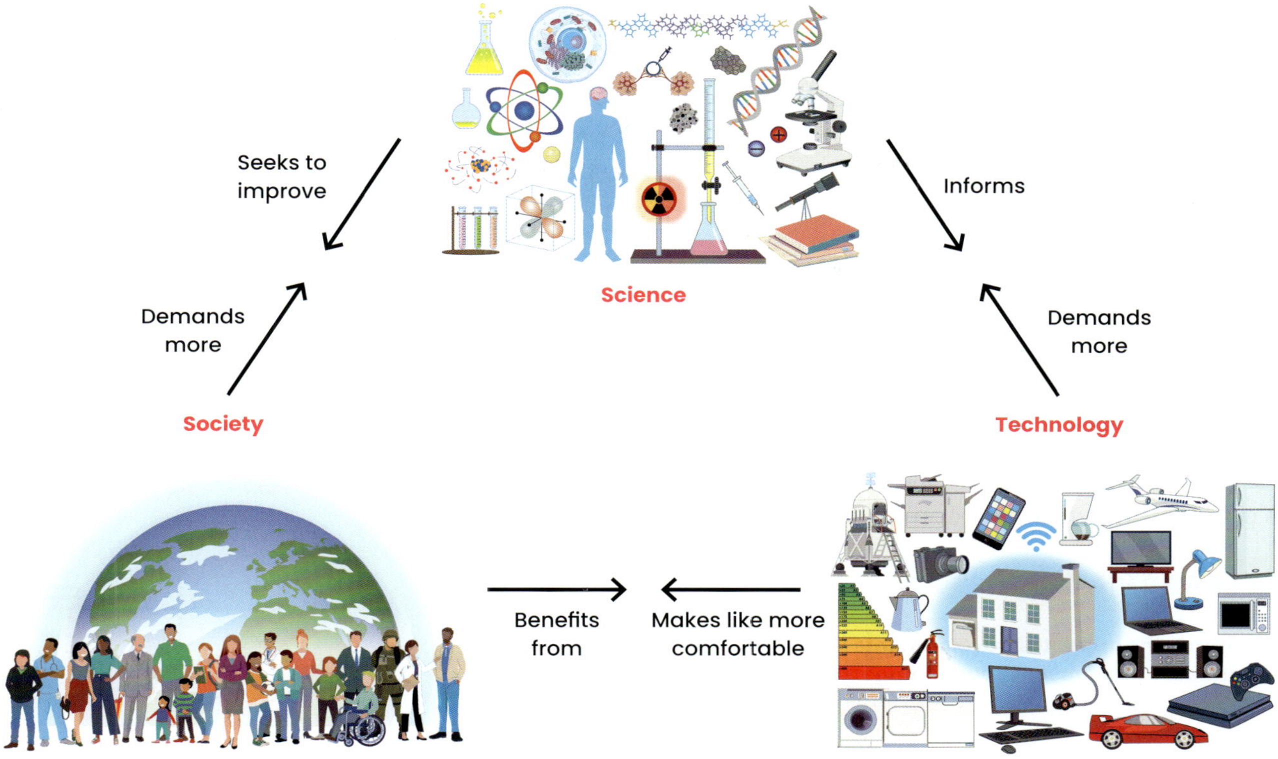

Figure 2.8: Interaction between science, technology and society.

Impact of scientific discovery and invention on society

Developments in science affect our everyday lives. Public policy on issues such as energy provision, agriculture, healthcare, transport, sustainability and economics – as well as our own personal decisions – are all informed by science. From when we get up in the morning until we go to bed at night, scientific advancements impact all aspects of our lives.

Agriculture

The world's population now exceeds 8 billion people. From the agricultural revolution in the 1700s when crop rotation was introduced to maintain soil fertility, to the development of artificial fertilisers in the early 1900s and pesticides in the 1930s, to the introduction of genetically modified, disease resistant, higher yield crops in the 1990s, science has allowed agriculture to keep up with the demands of a growing population on food supply.

Figure 2.9: Science has impacted greatly on food production.

Medicine

In 1928, Alexander Fleming discovered penicillin, the world's first antibiotic. Antibiotics have since saved millions of lives and are a vital tool in everyday medical care.

Vaccines are another key scientific innovation in medicine. Although Edward Jenner developed the first vaccine in 1796 against smallpox, it was not until the 1940s that scientific knowledge allowed large-scale vaccine production. It is estimated that vaccines now prevent 2–3 million deaths each year.

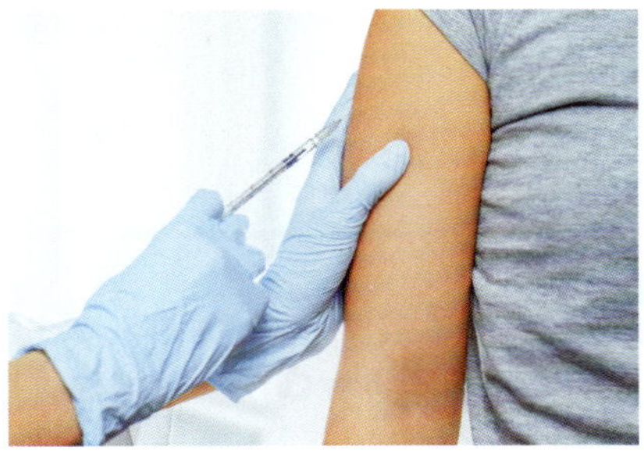

Figure 2.10: Science has enhanced medical care.

Science and economic development

Our increasing knowledge of science and technology results in the development of new products. The sale of these new products is of benefit to the economy. Scientific knowledge can solve problems in areas such as transport and manufacturing, allowing them to become more efficient and productive, generating wealth.

The development of Covid-19 vaccines is an important example of how science impacts on the economy. Not only did the vaccines save millions of lives, but they also enabled economies around the world to reopen, safeguarding countless jobs.

On the other hand, the more that nations invest in science and research in socially relevant issues like health and climate change, the greater the demand becomes on economic wealth.

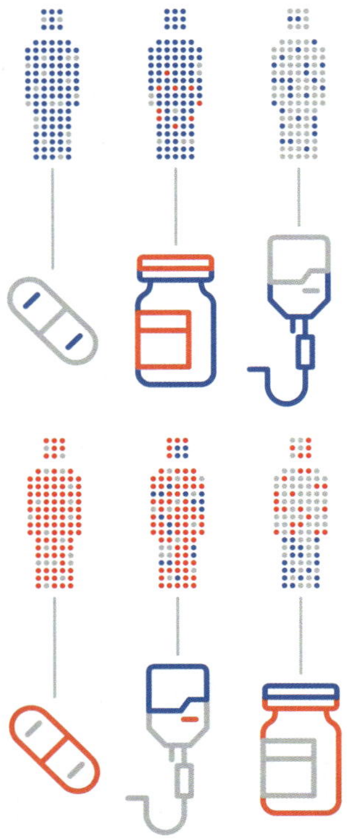

RESEARCH ACTIVITY

Research the impact of developments in science on personalised healthcare and present your findings to your class.

SCIENCE IN SOCIETY: SUSTAINABILITY

Sustainability refers to the ability of biological systems to remain healthy, diverse and productive over time. Many environmental challenges face humanity today: climate change, energy security, pollution and loss of biodiversity. Scientific research has a critical role to play in addressing these issues. Sustainable development should not cause harm to the environment or deplete natural resources. It must also be economically viable, and this raises social and economic concerns. For example, how can scientific research assist society in reducing our carbon emissions and our reliance on fossil fuels without affecting standards of living? How can modern agriculture continue to feed a growing world population without impacting ecosystems? Science is central to finding new solutions to these problems.

Sustainability is the ability of biological systems to remain healthy, diverse and productive over time.

ETHICAL ISSUES IN SCIENTIFIC DECISION MAKING

Ethics refers to whether something is right or wrong. Scientists must make ethical decisions at all stages of scientific research.

Some of the most common ethical issues in science are:

- The use of children, patients and other vulnerable people in experiments
- The use of human embryonic stem cells
- Assisted reproduction in humans
- The use of animals in experiments
- Genetically modified organisms
- Privacy and data protection issue.

Ethical standards also require scientists not to fabricate (make up) results or copy (plagiarise) another person's work. Scientists must strive to avoid bias when designing experiments and analysing data. They must also disclose any personal or financial interests that may affect their research.

Ethics is the study of standards of right and wrong – dealing with moral conduct and judgement.

Go to YouTube.com and search for 'Ethics | Biology | FuseSchool' (5:30) by FuseSchool – Global Education.

PORTFOLIO LINK **Go to your Learning Portfolio and complete Learning Check 3 to assess your knowledge and understanding of the topic so far.**

Section 4: Biological Phenomena

At the end of this section, you should be able to:

- Understand that simplified models can be used to explain complex systems and that these models can be modified as more data becomes available.
- Give examples of how systems are used to explain biological events.
- Explain the term 'form fits function'.
- Give examples of how the design of biological structures relates to their function.
- Give examples of how living things are interdependent.
- Give examples of cycles that transfer matter between organisms and their environment.
- Explain the role of genetic codes in transferring information in living organisms.

Unpacked from LOs U4.1, U4.2

USING SYSTEMS TO EXPLAIN BIOLOGICAL EVENTS

Living things are very complex with many integrated parts that interact in many ways. The human body is composed of approximately 37 trillion (37,000,000,000,000) cells. To make sense of how these cells work together, scientists study them at different structural levels of organisation: cells, tissues, organs and biological **systems**. A biological system is a group of organs working together to carry out a particular function, e.g. the digestive system and respiratory system. Taking a systems approach allows us to understand the bigger picture of how the organism works more clearly.

A **system** is the organisation of interdependent or interrelated objects working together as a whole, e.g. a biological system or the classification system.

Scientists have estimated that there are about 8.7 million different species of plants and animals on the planet. To make sense of the relationships and shared characteristics between these species, biologists use another organisational system: a classification system. You will learn more about classification systems in Chapter 4.

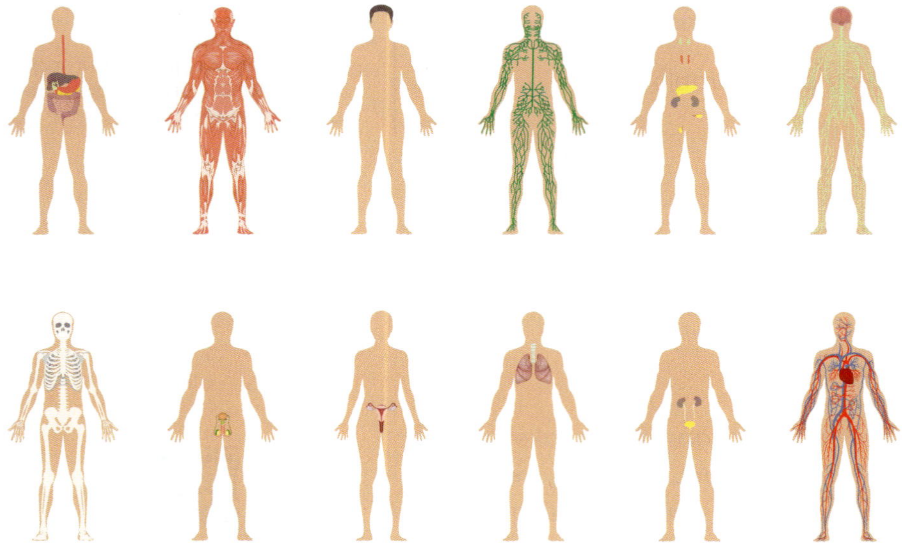

Figure 2.11: Organ systems of the human body – an example of a classification system in biology.

USING MODELS TO DESCRIBE SYSTEMS

As biological systems are very complex, we use simplified models or representations to describe them. These models can be used to predict the behaviour of a biological system. As scientists learn more about the biological system, they can modify the model to include new information or data. For example, our understanding of how the human brain works is constantly being updated as scientists gain more knowledge. In the 1800s, scientists proposed that the shape of a person's brain influenced the shape of their skull and their personality, this was known as phrenology. Although phrenology is now discredited; it is an early example of how scientists use models to simplify complex biological systems. Data collected from various medical imaging techniques has deepened our understanding of the brain and allowed the model of the brain to be revised and developed.

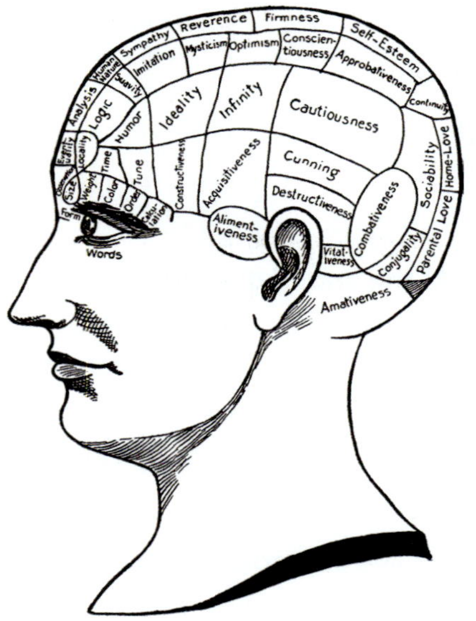

Figure 2.12: Phrenology in the 1800s: an early model of the human brain.

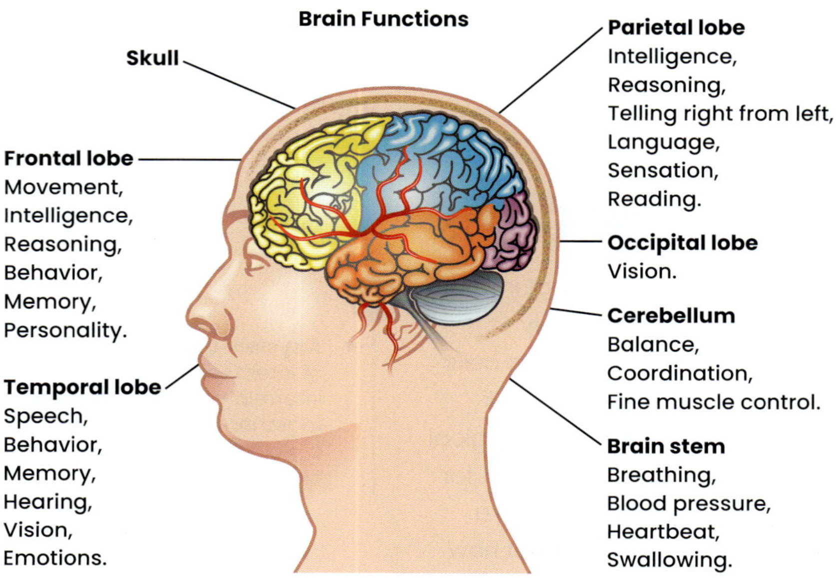

Figure 2.13: Modern model of the human brain.

'FORM FITS FUNCTION'

In biology, the term **'form fits function'** is used to describe when the structure of a body part is well designed for its function or the job it carries out. This helps the living thing to survive. Examples include your incisor teeth that are sharp and pointed for tearing food or the opposable thumb in your hand that allows you to grasp items. These adaptations have evolved due to natural selection which you will study in more detail in Chapter 9.

'Form fits function' exists when a structure of a body part is well designed for the job it carries out.

Figure 2.14: The aerodynamic design of an eagle's wing is an example of 'form fitting function'.

DIVERSITY, INTERDEPENDENCE AND UNITY OF LIFE

One of the most incredible things about life is the many different types of species on this planet, including plants, animals, bacteria and fungi. These species are dependent on each other for their survival – they are **interdependent**. For example, plants provide animals with oxygen to breathe, while animals exhale carbon dioxide and water vapour which plants use to make food.

Despite the diverse number of species, they are united by the fact that they all share a genetic code DNA based on the same four building blocks or bases: adenine, thymine, guanine and cytosine.

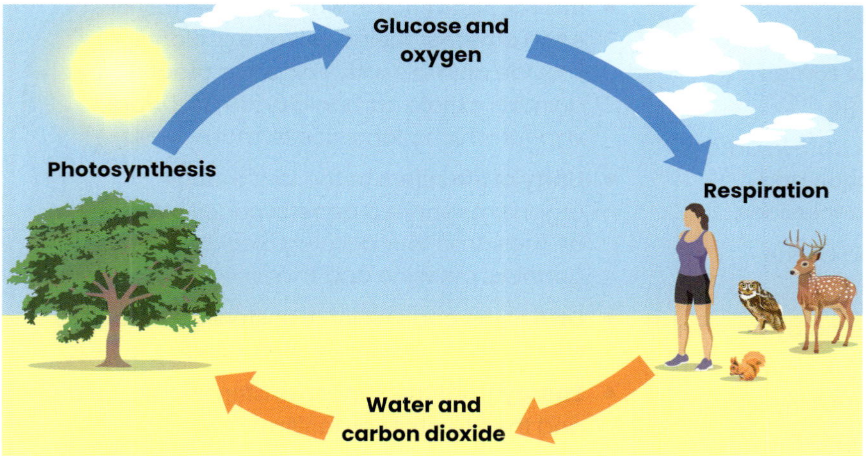

Figure 2.15: All living things are interdependent.

Interdependence describes how all species rely on each other for their survival.

Transfer of information

DNA is the code which transfers genetic information from one generation to the next. You will learn more about how DNA and its copy RNA transfer information in Chapter 24.

Transfer of matter and energy

Energy from the Sun is transferred to plants and then to other organisms through feeding. Matter such as carbon and nitrogen moves from non-living locations, e.g. soil, to living organisms and back again in continuous cycles. In this way, all the matter on Earth is recycled. Energy is also transferred in these cycles. You will learn more about these cycles in Chapter 12.

PORTFOLIO LINK **Go to your Learning Portfolio and complete Learning Check 4 to assess your knowledge and understanding of the topic so far.**

Chapter Summary

- **Scientific knowledge** is constantly revised and expanded over time.
- **The scientific community** is the global interacting network of scientists working on research.
- Scientists publish their research in scientific journals and present it at meetings called conferences.
- **The International System of Units (SI)** is a decimal measurement system that defines the units which should be used by scientists when recording measurements e.g. mass in kg, time in s, length in m.
- **Peer review** is the evaluation of a scientist's work by others who are experts in the same field. Scientists submit a paper on their research to a journal. This is sent to other expert scientists who evaluate it and provide feedback. The authors of the paper may revise it based on this feedback.
- **Reproducibility** is a measure of whether an investigation can be carried out by a different research team using the same methods as outlined in a scientific paper and obtain the same results.
- **Relevant** research contributes something valuable to scientific knowledge.
- **Bias** is any deviation from the truth in data collection, analysis, interpretation and publication leading to false conclusions.
- **Primary data** is data collected by the researcher themselves.
- **Secondary data** is data collected by another researcher.
- **Random** selection means that all individuals have an equal chance of being selected to take part in a study.
- When evaluating a study, it is important to check for bias in the method, results, range of data, detail given, language used and source of the data.
- **Overgeneralisation** is when results of a study may not be generalised to the entire general population.
- Developments in science affect our everyday lives, e.g. genetically modified crops in agriculture, vaccines and antibiotics in medicine.
- Scientific knowledge can solve problems in areas such as transport and manufacturing allowing them to be more efficient and productive, generating wealth.
- **Sustainability** is the ability of biological systems to remain healthy, diverse and productive over time.
- **Ethics** is the study of standards of right and wrong, dealing with moral conduct and judgement, e.g. use of human embryonic stem cells, genetically modified organisms, assisted reproduction in humans.
- A **system** is an organisation of interdependent or interrelated objects working together as a whole, e.g. a biological system or the classification system.
- **'Form fits function'** describes when the structure of a body part is well designed for its role.
- **Interdependence** is when species rely on each other for their survival, e.g. plants provide animals with oxygen to breathe and animals exhale carbon dioxide and water vapour which plants use to make food.
- **Unity of life** refers to the fact that all organisms share a genetic code DNA based on the same four building blocks (adenine, guanine, cytosine and thymine).
- **DNA** is the code which transfers genetic information from one generation to the next.
- **Energy** from the Sun is transferred to plants and then to other organisms.
- Matter moves from non-living locations, e.g. soil, to living organisms and back again in continuous cycles. Examples include the carbon and nitrogen cycles.

END OF CHAPTER ASSESSMENT

REMEMBER

1. What is the scientific community?

2. State the SI unit of (a) length (b) mass (c) time.

3. Define the term *peer review*.

4. Define the term *reproducibility* in relation to scientific research.

5. Define the term *bias* in relation to scientific research.

6. What is meant by *ethics* in scientific research?

7. Define the term *sustainability* in relation to science.

UNDERSTAND

8. Give examples of how bias can be introduced to scientific research during (a) data collection (b) data analysis (c) data interpretation.

9. Describe **two** ways in which members of the scientific community communicate with each other.

10. Summarise the peer review process.

11. What measures are taken by the global scientific community to ensure that the build-up of scientific knowledge is accurate?

12. Give examples of how science has impacted on (a) economic development (b) society.

13. Explain the importance of ethics in scientific research.

APPLY

14. In 1926, the average life expectancy in Ireland was 57.6 years, but in 2019, this had jumped to 82.3 years. Give **two** possible reasons for this increase.

15. A group of scientists are investigating the effect of a new multivitamin on mood. They test the effect of the medication on ten females, all aged between 25 and 35 years, from their own department. The participants in the study were given a single multivitamin a day for one week. They completed a questionnaire on their mood before and after the week.

(a) Comment on the size of the sample. Is the sample representative of the general population? Why?

(b) What advice would you give to these scientists to improve the design of their investigation?

16. John is investigating the average height of students in his class. There are two members of the basketball team in his class who are over six feet tall. John decides not to include the height of these students when calculating the average height of the students in his class. What do you think of John's method?

ANALYSE

17. Distinguish between primary and secondary data.

18. Which of the following are examples of 'form fits function'?

(a) Birds' bones have a honey-comb structure that is strong but light.

(b) Leaves are flat and thin to maximise the absorption of light.

(c) Elite athletes have stronger heart muscle.

EVALUATE

19. What are the main challenges to sustainability on our planet?

20. Research and evaluate the ethical issues in each of the following.

(a) The use of animals when testing cancer drugs.

(b) The use of embryonic stem cells to treat diseases such as Alzheimer's or Parkinson's disease.

CREATE

21. Research a scientific discovery or invention and present your findings to your class.

22. Carry out some research on interdependence in the living world. Create a short presentation on how your chosen species depend on each other for survival.

3 Experiments and Data Analysis

LEARNING OUTCOMES: U2.1, U2.2, U2.3, U2.4, U2.5, U2.6

EYE-OPENER

Biodiversity Maps
Mapping Ireland's Wildlife

The National Biodiversity Data Centre collects data to monitor and record changes in Ireland's wildlife species in terrestrial, freshwater, marine, river and wetland environments. Citizens can submit data they have collected to the data centre. This data is recorded on biodiversity maps, which are verified and validated, to ensure that it is a trustworthy source of high-quality data. Any errors identified are corrected. The data collected is used to help identify conservation priorities.

At the end of this chapter, you should be able to:

- Recognise appropriate questions for scientific investigation.
- Pose testable hypotheses.
- Evaluate and compare strategies for investigating hypotheses.
- Design, plan and conduct investigations.
- Explain how reliability, accuracy, precision, error, fairness, safety, integrity and the selection of suitable equipment have been considered in the investigation design.
- Identify sources of random and systematic error in an investigation.
- Produce and select qualitative data and quantitative data.
- Identify patterns, relationships and anomalous observations in data.
- Evaluate data in terms of repeatability and reproducibility.
- Draw and justify conclusions.
- Use investigative skills and learning to solve problems in unfamiliar situations.
- Organise and communicate research in a variety of ways.

SUBJECT VOCABULARY

scientific practice | observation | hypothesis | sample group | sample size | controlled experiment | variable | dependent variable | independent variable | control group | experimental group | placebo | double blind testing | error | precision | accuracy | reliability | validity | data | integrity | quantitative data | qualitative data | repeatability | conclusion | anomaly

LINKS TO:

The scientific practices outlined in this unit apply to all Learning Outcomes throughout the rest of the Leaving Certificate Biology course.

Section 1: Questioning and Predicting

At the end of this section, you should be able to:

- Write a question that can be investigated in a scientific manner.
- Distinguish between questions that can and cannot be investigated in a scientific manner.
- Explain the term 'hypothesis'.
- Write a hypothesis for an investigation that can be tested in a scientific manner.
- Consider different methods to test a hypothesis and decide on the best one.

Unpacked from LOs U2.1, U2.2

SCIENTIFIC PRACTICE

When studying the world around them, scientists are susceptible to bias from their own personal beliefs and experience. This may result in them failing to accurately account for errors or ignoring data that does not support their hypothesis. To minimise the influence of personal bias, biologists employ a standardised approach to conducting investigations: the **scientific method** or **scientific practice**. Biologists use scientific practice to construct an accurate, reliable and valid picture of the world around them.

Scientific practice is the set of steps followed by scientists when carrying out investigative research:

1. Propose a possible explanation for the observation – hypothesis.
2. Design an experiment to test the hypothesis.
3. Evaluate the experimental evidence and form a conclusion.
4. Communicate the experimental result.

We will look at each stage of the scientific practice in turn.

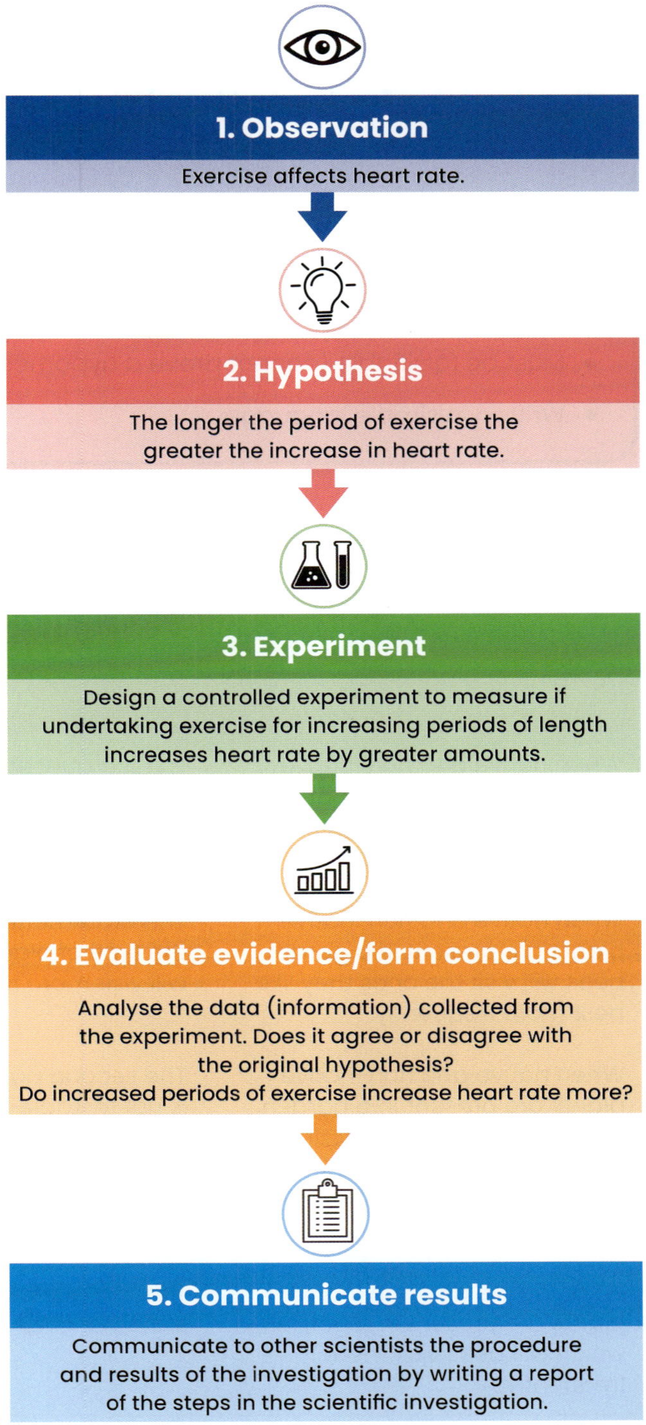

Figure 3.1: Stages of scientific practice.

HYPOTHESIS

When a biologist makes an **observation**, e.g. that a certain drug appears to reduce the chance of a person developing bowel cancer, the researcher may propose an explanation for this observation. This is called a **hypothesis**.

Observation is the collection of information or data from a primary source, e.g. an experiment.

A **hypothesis** is an educated guess based on observation.

To form a hypothesis, a biologist researches all background information available on the topic of interest. They then decide what is the main question they wish to answer. The biologist's hypothesis may or may not be the correct explanation for the observation.

To find out if the hypothesis is correct, it must be possible to test it using a controlled experiment. The evidence collected from the experiment will either agree with the original hypothesis or will prove it wrong. If the original hypothesis is proved wrong, the biologist can put forward and test a different hypothesis, following the steps of the scientific process again.

Features of a hypothesis

- Statement that describes a relationship between two or more variables
- Outlines the expected result
- Not a question
- Can test using a controlled experiment
- Must be falsifiable (cannot prove a hypothesis true in every situation)
- Written in simple language

What makes a good hypothesis?

Imagine that you are an ecologist. You want to find out if bees in domestic gardens prefer native plants to non-native plants. There are many different hypotheses you could form for the study:

SUGGESTED HYPOTHESIS	CRITERIA	SUITABLE HYPOTHESIS?
When native and non-native plants are present in a garden, bees will visit native plants more frequently.	Is not a question ✓ Testable ✓ Expected result predicted ✓ Simple language ✓	A suitable hypothesis
When native and non-native plants are present in a garden, bees will visit the ones they like best more frequently.	This statement does not predict the type of plant which the bees will visit more frequently. ✗	Not a suitable hypothesis
When native and non-native plants are present in a garden, bees will preferentially visit native plants because they prefer their appearance.	The second part of this statement is not testable – you can measure whether the bees visit native plants more than non-native plants, but your results will not tell you why. ✗	Not a suitable hypothesis
When native and non-native plants are present in a garden, which will bees visit more frequently?	This is a question, not a predictive statement. ✗	Not a hypothesis

Figure 3.2: Examples of suitable and unsuitable hypotheses.

Go to YouTube.com and search for 'Scientific Method Basic Summary-Leaving Cert Biology (updated)' (2:11) by Biology Bugbears.

PORTFOLIO LINK Go to your Learning Portfolio and complete Learning Check 1 to assess your knowledge and understanding of the topic so far.

Section 2: Conducting Investigations

At the end of this section, you should be able to:

- Design, plan and carry out an investigation
- Identify variables in an investigation – independent, dependent and control variables
- Explain each of the following:
  - how your selected method of investigation is fair
  - the reasons behind your choice of equipment
  - how you ensured the (i) accuracy (ii) precision (iii) reliability and integrity of the collected data
- Identify any safety procedures you followed
- Explain the difference between 'random error' and 'systematic error' and identify possible sources of each in your investigation
- Explain the difference between 'accuracy' and 'precision'
- Apply learning and skills from investigations to unfamiliar contexts

Unpacked from LOs U2.3, U2.5

MAKING PREDICTIONS AND GENERALISATIONS BASED ON AVAILABLE EVIDENCE

In an ideal world, a scientist would test every member of a population – every plant, animal, human or whatever it is they are studying. However, in most cases, it is not possible to do this. The scientist must choose a **sample group** to study. This group must have characteristics that reflect or are representative of the whole population. This allows the researcher to make **predictions** and **generalisations** about the entire population from their observations with the sample.

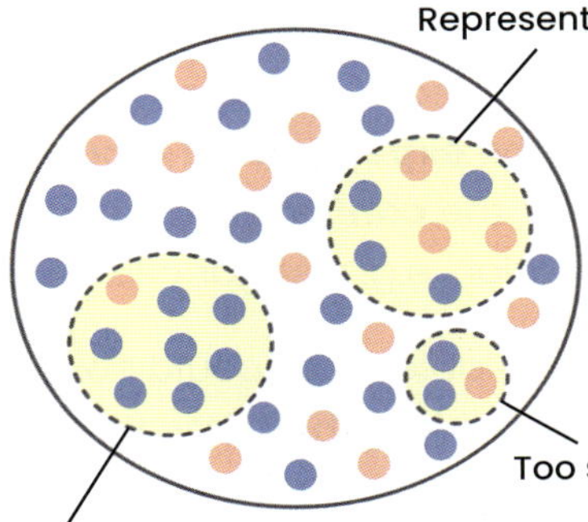

Figure 3.3: Types of sample.

- A **sample group** is a subset with characteristics reflective of its population.
- A **prediction** is the expected result of a test that is derived from a hypothesis.
- **Generalisation** refers to an overall conclusion about a population based on a sample.
- **Sample size** refers to the number of individuals or items tested in an experiment.

It is important that the **sample size** chosen is large enough to be representative of the population being studied.

Go to YouTube.com and search for 'The Scientific Method Part 3: Setting up an Experiment' (4:05) by By: Rachel Taylor.

EXPERIMENTING

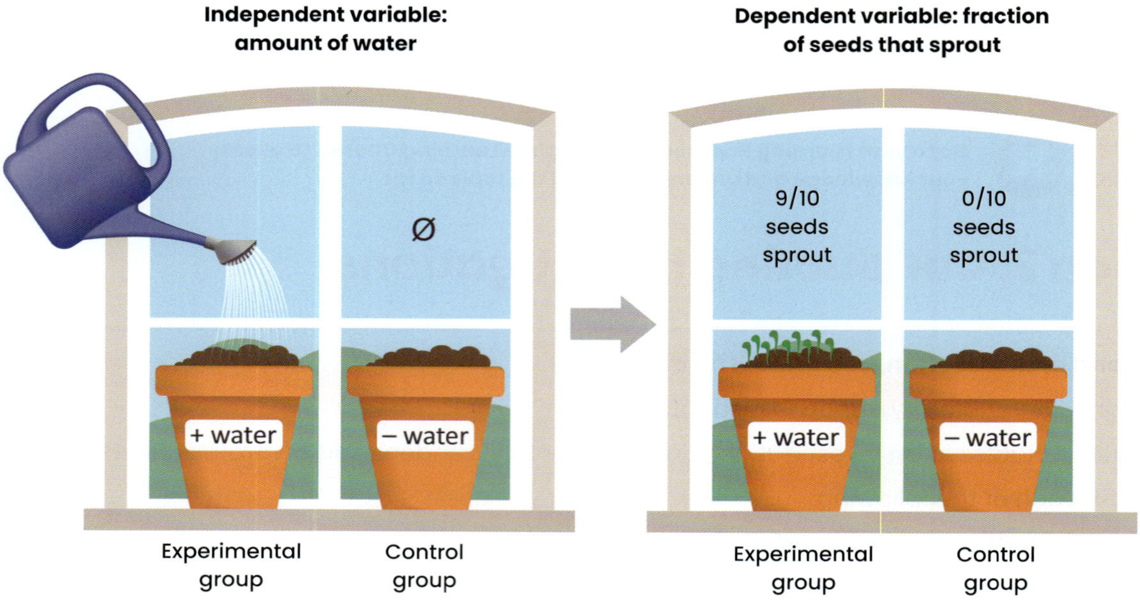

Figure 3.4: Experimental design.

A **controlled experiment** must be carried out to test the hypothesis. Different types of **variables** are used in scientific experiments. The variable that is changed by the scientist is called the **independent variable**. The variable that alters as the independent variable is changed is called the **dependent variable**. The dependent variable is measured or observed by the scientist during the investigation. In a **controlled experiment** or **fair test**, the **experimental group** and the **control group** should only differ in one variable, the one we want to test. All other variables should be kept constant, e.g. if a scientist was investigating the effect of a blood pressure medication, they would test two groups of people – one group, the **experimental group**, would be given the medication and another, the **control group**, would be given a placebo or fake medication, e.g. a sugar pill.

Other variables they could consider include patient age and gender, the length of time they take the treatment for and the number of doses to be taken each day. To prevent bias when testing medications, it is important that neither the researcher nor the patient should know which patient is receiving the drug or the **placebo** (a fake treatment, e.g. a sugar pill). This is known as **double blind testing**.

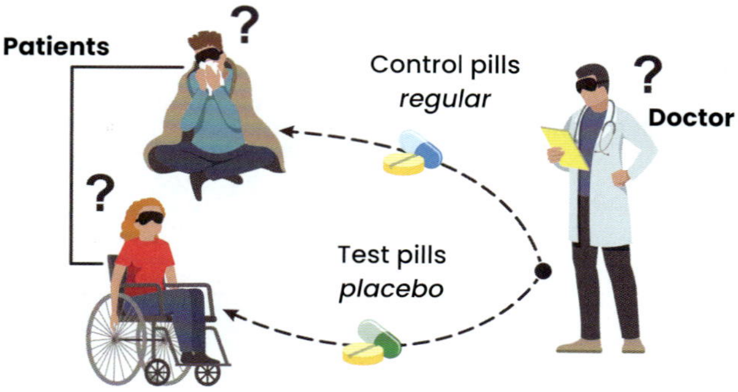

Figure 3.5: Double blind testing.

- A **controlled experiment** (or fair test) is one where only one variable is changed at a time and all other conditions are kept constant.
- A **variable** is a factor or condition that can be changed or controlled in an investigation.
- The **independent variable** is the variable that is changed by the scientist.
- The **dependent variable** is a variable that is observed or measured during the experiment.
- The **control group** is a group not exposed to the variable under study and used as a comparison to the experimental group.
- The **experimental group** is a group exposed to the variable under study.
- A **placebo** is a fake treatment that is used with a control group in a medical trial.
- In **double blind testing**, neither the researcher nor the test subject knows who is receiving the test drug or the placebo in a clinical trial.

Uncertainty

Biologists often cannot measure the exact or true value of a quantity. In this case, when reporting measurements, they provide an estimate of the **uncertainty** related to the measured value. Uncertainty and **error** are not the same.

- **Uncertainty** refers to the range of possible values within which the true value of the measurement lies.
- **Error** refers to the difference between a measurement and its true value.

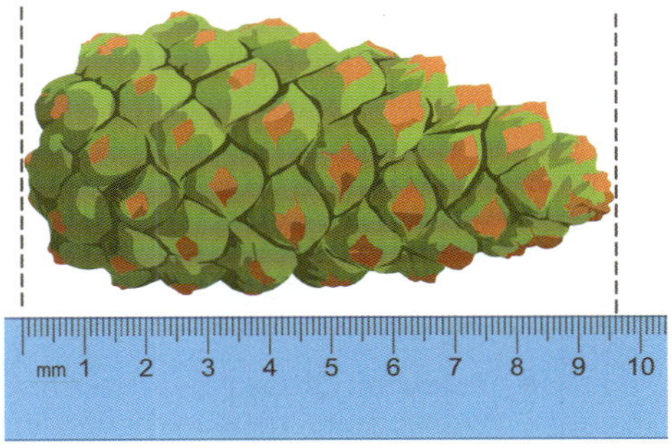

As the ruler only measures accurately to the nearest mm, we take the uncertainty range to be +/– half of this value. Length of pinecone = 96mm +/– 0.5mm.

Figure 3.6: An example of uncertainty: measurement +/– (best estimate uncertainty) units.

The error in a measurement can only be calculated when the exact or true value of a measurement is known. Including the uncertainty linked with a measurement allows other scientists to judge the quality of the experiment and to make comparisons with data from other experiments.

STATISTICAL AND SYSTEMATIC UNCERTAINTY

There are two types of uncertainty: statistical and systematic.

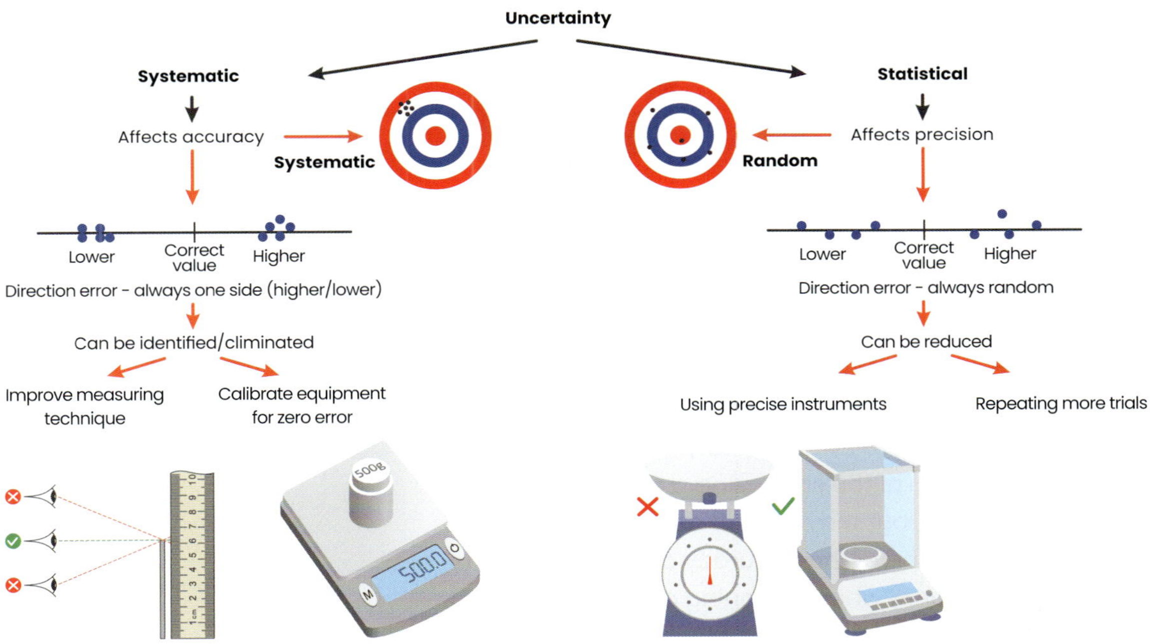

Figure 3.7: Systematic and statistical uncertainty.

Statistical uncertainty arises from random variations in a measurement, e.g. you may measure the mass of a soil sample three times and get slightly different values each time. Variation in these measurements may be due to random vibrations, drafts, changes in temperature or effects from nearby apparatus. It is not due to a fault in the measuring instrument. Random variations may be removed by averaging many measurements. Using a more precise instrument, e.g. a digital balance that reads to a greater number of decimal places, helps to reduce statistical uncertainty.

- **Statistical uncertainty** is a measure of the range of random variation in the results obtained when repeated measurements of the same event are recorded.

Systematic uncertainty can be difficult to detect because all the recorded data is either too high or too low. Systematic uncertainty often arises from defects in the calibration of measuring instruments, e.g. if an electronic balance is not calibrated correctly, it may read all measurements at 95% of their true values. This means that it has a systematic uncertainty of 5%. Systematic errors can be identified and fixed.

Systematic uncertainty arises from defects in the measurement apparatus, equipment or method used by the experimenter.

Accuracy and precision

Two important considerations that a scientist considers when recording data are **accuracy** and **precision**.

Accuracy describes how close a measured value is to the actual value.

Precision describes how close two or more measurements are to each other.

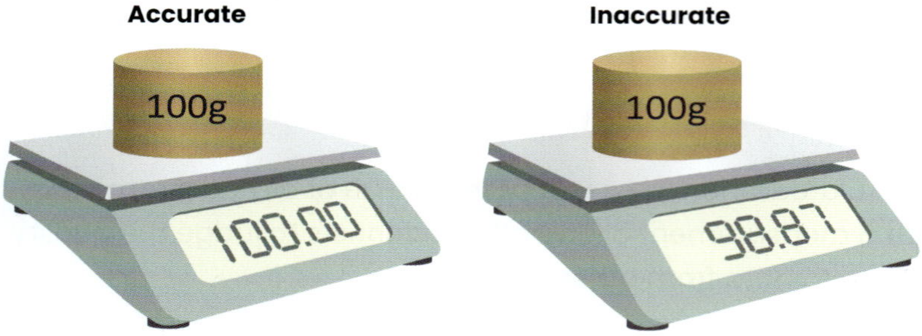

Figure 3.8: Accuracy.

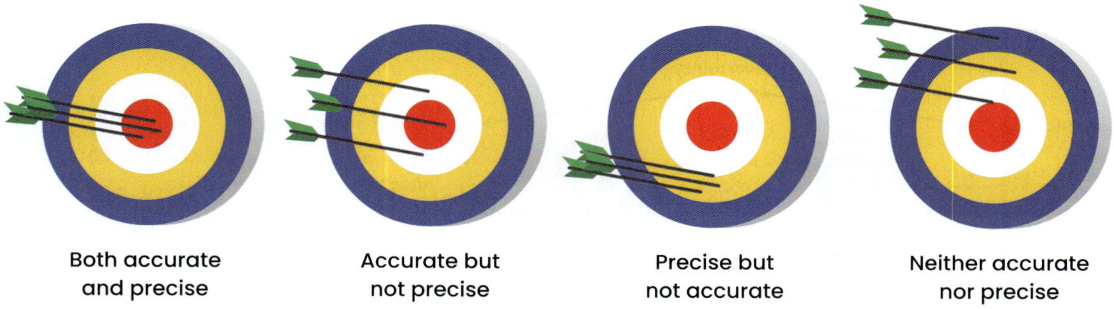

Figure 3.9: The difference between accuracy and precision.

For example, imagine that you measure the length of a snail's shell using a ruler to be 3.2cm, but you then measure it as 3.28cm using Vernier callipers. The measurement using the Vernier callipers is more accurate as it measures to two decimal places. If you repeatedly measure the length of the shell using a ruler and get 3.2cm, your results are precise, but they are not as accurate as those obtained from the Vernier callipers. The accuracy of the instruments to be used should be considered when designing an investigation.

Figure 3.10: Measuring biological samples with a ruler is less accurate than measuring them with Vernier callipers.

COMPARING PRECISION OF DATA SETS

Imagine you recorded the mass of two soil samples and obtained the sets of data below. To decide which set of values is more precise, calculate the **range** of the data by subtracting the lowest value from the highest value. The sample with the lowest range is the most precise. In Figure 3.10 below, the set of data for sample 2 is more precise than that for sample 1.

							RANGE
SAMPLE 1	4.56g	4.55g	4.53g	4.57g	4.58g	4.58g	4.58g – 4.53g = 0.05g
SAMPLE 2	2.12g	2.11g	2.10g	2.13g	2.10g	2.14g	2.14g – 2.10g = 0.04g

Figure 3.11: Precision in data sets.

Reliability

When a biologist carries out an investigation, it is important that other researchers can repeat the experiment under the same conditions to get the same results. In this way, the original hypothesis becomes accepted by the scientific community. **Reliability** is measured on a scale where 100% reliable indicates that identical results are obtained every time the experiment is repeated. To maintain reliability, scientists take repeats of measurements. In general, a test-retest correlation of +0.80 or greater is considered to indicate good reliability. For example, if physicians measured the VO_2 max (maximal oxygen consumption) of students two weeks apart and the results of their test-retest indicate a correlation of 0.98 between the measurements each week, this would indicate that the test is reliable (see Figure 3.12).

- **Range** = highest data value – lowest data value.
- **Reliability** is the ability of an instrument or method to measure stable and consistent results.

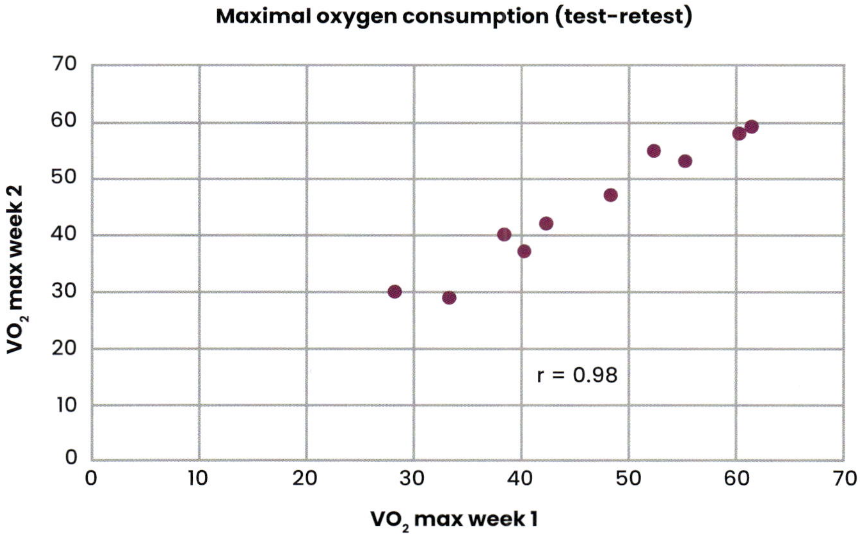

Figure 3.12: Correlation between two sets of results indicates the level of reliability of an experimental test.

Validity

Validity refers to whether a test measures what it is supposed to measure. A test may be reliable and still not be valid. For example, if a balance repeatedly measures the mass of a soil sample as 3g, it is reliable, but if the balance was not calibrated correctly and the sample really weighs 4g, then the test is not valid.

Figure 3.13: Reliability and validity.

Go to YouTube.com and search for 'Systematic and Random Error' (2:25) by Prof. Essa.

To ensure validity in tests, sample groups must be random, and controls must be included.

Safety

When carrying out an investigation, the **safety** of any participants in the study must always be ensured. Participants should not be subjected to any measures which would cause them harm.

Figure 3.14: Safety in scientific practices.

- **Validity** indicates how well a test measures what it intends to measure.
- **Safety** is keeping researchers and participants free from harm in an investigation.
- **Integrity** is maintaining honesty at all stages of scientific practice.

Integrity

Scientists must maintain honesty in all stages of scientific practice. This means that they are not biased, do not fabricate data or plagiarise another scientist's report and are honest about potential conflicts of interest. We call this scientific **integrity**.

PORTFOLIO LINK **Go to your Learning Portfolio and complete Learning Check 2 to assess your knowledge and understanding of the topic so far.**

Section 3: Collecting and Evaluating Data

At the end of this section, you should be able to:

- Judge whether the data collected from an investigation may be collected repeatedly using the same equipment and conditions.
- Judge whether an investigation can be carried out by others under the same conditions to give the same data (i.e. is it reproducible?).
- Collect numerical (quantitative) and non-numerical (qualitative) data from an investigation.
- Identify patterns and relationships that exist between data collected.
- Identify outliers (anomalous) data.
- Use the collected data to form a conclusion related to your original hypothesis.

Unpacked from LOs U2.3, U2.4

DATA

The measurements or observations collected in an investigation are known as **data**. Data can be classified depending on (i) who collects the data and (ii) the type of data that is collected. We learned in the last chapter that data collected by the researcher is referred to as primary data. Secondary data refers to data collected by someone else.

Quantitative data can be expressed in numerical form e.g. measuring the amount (numbers) of organisms or the length of a plant root. **Qualitative data** is non-numerical, e.g. the presence or absence of a particular organism.

Quantitative

Seven snails

Qualitative

Pink, yellow and orange butterflies

Figure 3.15: Quantitative and qualitative data.

- **Data** refers to observations or measurements recorded in an experimental test.
- **Quantitative data** is data that can be expressed in numerical form.
- **Qualitative data** is non-numerical data.

REPEATABILITY

In Chapter 2, we learned the importance of reproducibility in a study: whether an investigation can be carried out by a different research team using the same methods and obtain the same results. **Repeatability** refers to a scientist getting the same result each time they make a particular measurement under the same conditions. A repeat of an experiment is called a **replicate**.

- **Repeatability** describes getting the same result when a scientist makes a particular measurement under the same conditions.
- A **replicate** is a repeat of an experiment.

Figure 3.16: Replicates in an investigation on plants.

CONCLUSION

Scientists use the data collected from an experiment to form a conclusion related to their original hypothesis. From analysing the data collected in an experiment, there are two possible outcomes: the results agree or disagree with the original hypothesis.

A **conclusion** summarises how the results of an investigation support or contradict your original hypothesis.

Consider the hypothesis '*Patients with high blood pressure who take drug Brand-A do not differ in levels of blood pressure from patients who take a placebo*'. Imagine the results indicate that the Brand-A drug significantly reduces blood pressure compared to the placebo. The original hypothesis is restated in a form that agrees with the results: '*The use of drug Brand-A reduces levels of high blood pressure compared to the use of a placebo*'. The hypothesis can be accepted as true based on the evidence collected but it cannot be proven to be true.

There are always other possible explanations that could explain the results observed – maybe the patients who took drug Brand-A ate a yoghurt every day and this lowered their blood pressure. Researchers would carry out more tests to find out if the hypothesis is wrong. If these tests also fail to disprove the hypothesis, it can be accepted as true because it withstood several attempts to show it is false.

CORRELATION AND CAUSATION

Scientists look for relationships between sets of data when analysing information collected in their research. This is called **correlation**. For example, they might investigate if there is a relationship between:

Correlation is a numerical measure of the degree of a relationship between two variables.

- dissolved oxygen content of a river and the species diversity present
- eating more vegetables and having lower blood pressure
- the mass of a cat and the number of kittens they have
- hormone replacement therapy and coronary heart disease.

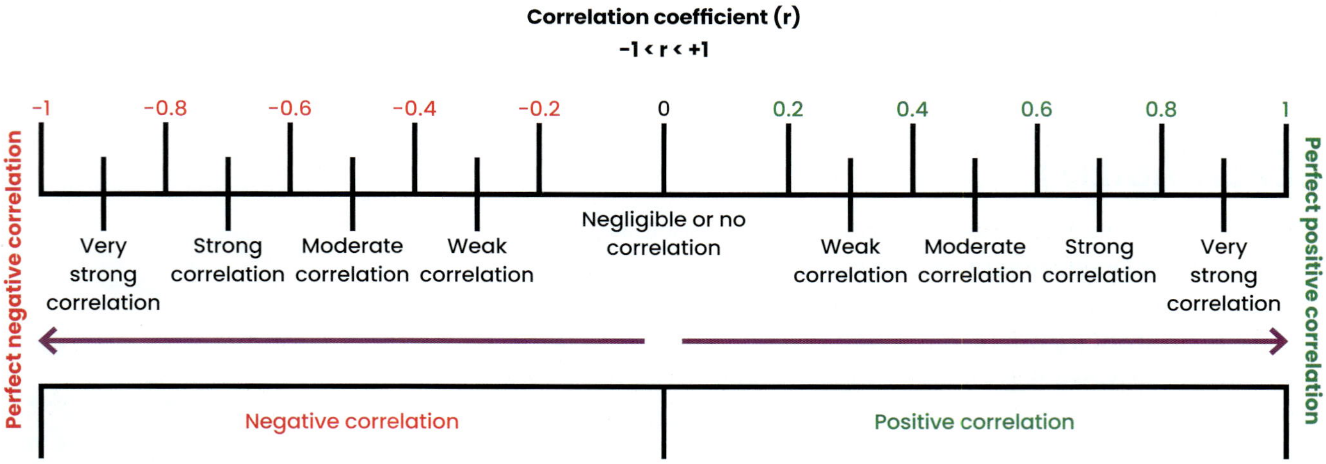

Figure 3.17: Correlation coefficients.

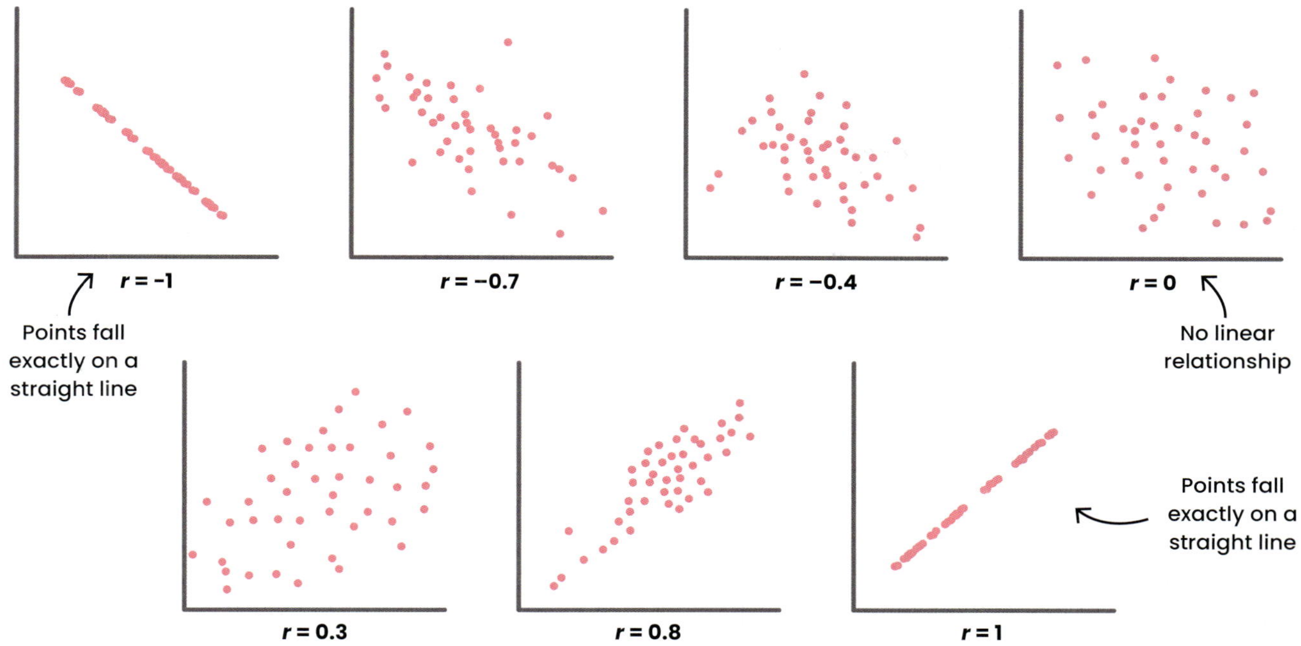

Figure 3.18: Correlation coefficients.

Relationships between variables can be positive or negative. A positive relationship exists when both variables increase together. A negative relationship exists when one variable increases, the other decreases. The degree of the relationship is measured by a correlation coefficient (r) which has values between +1.0 and −1.0.

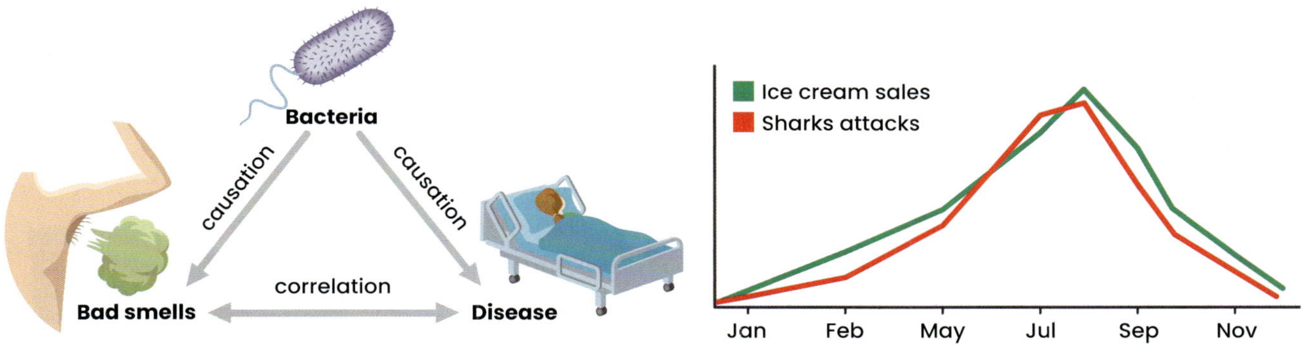

Figure 3.19: Correlation vs causation.

If a correlation exists between variables, it does not mean that the change in one variable is the cause of the change in the values of the other variable. A correlation may exist between smoking and alcoholism, but this does not mean that smoking causes alcoholism. Likewise, studies indicate that a correlation exists between fruit and vegetable intake and lower blood pressure. However, it has not been proven that fruit and vegetable intake causes lower blood pressure. There are many other factors which may contribute to the correlation between these variables: people who eat more fruit and vegetables may exercise more, lowering their blood pressure; or people who can afford to buy more fruit and vegetables may have less stressful lives than people who cannot, resulting in lower blood pressure. A biologist studying this would have to carry out more investigations to control for other potential variables and prove **causation**.

Causation indicates that one event is the result of the occurrence of another event.

Go to YouTube.com and search for 'Lifestyle and Disease: Cause and Correlation' (2:07) by SnapRevise.

ANOMALOUS DATA

Anomalous data or an **anomaly** is a result that doesn't fit the pattern of the other data. Anomalous data can be easily spotted and should be left out when calculating an average.

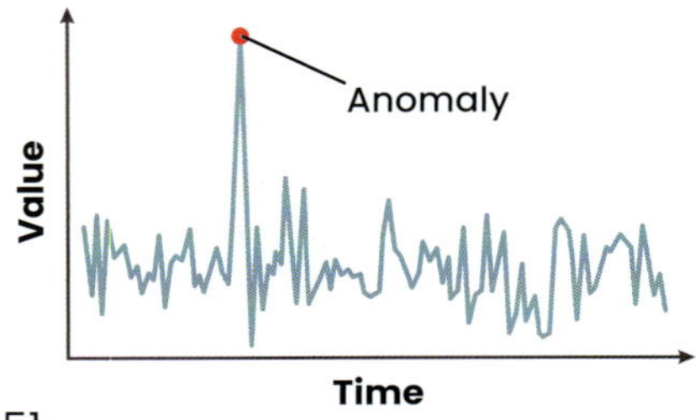

Figure 3.20: Anomalous data.

Example:
A student took the following six measurements: 0.57, 0.54, 0.48, 0.62, 1.5, 0.51.

The 1.5 value is too high compared to all the other data and is an anomaly.

PORTFOLIO LINK **Go to your Learning Portfolio and complete Learning Check 3 to assess your knowledge and understanding of the topic so far.**

An **anomaly** is an irregular piece of data that does not fit the pattern of a particular data group.

Section 4: Communicating Results and Limitations of the Scientific Method

At the end of this section, you should be able to:

- Communicate a report of a scientific investigation using a variety of methods: written lab report, scientific poster, verbal presentation, written article.
- Present findings from a scientific investigation using a variety of methods: tables, graphs, charts, diagrams.
- List different methods used within the science community to communicate scientific research.
- Give examples of the limitations of the scientific method.

Unpacked from LO U2.6

We saw in Chapter 2 that scientists publish their findings in scientific journals and books, e.g. the *Journal of Molecular Biology*, *Nature*, *The Journal of Biochemistry*. As you study biology, you will carry out several investigations which you will write up in the form of laboratory reports. You will also present your findings to your class using a variety of methods, e.g. poster, verbal presentation, written articles. To illustrate the results that you collected you may use tables, graphs, charts or diagrams to summarise the data.

Figure 3.21: Scientific journals.

LIMITATIONS OF SCIENTIFIC EVIDENCE

While the scientific method is an important part of studying biology, it does have some limitations.

1. Extent of existing knowledge

Limitations in knowledge and instrumentation affect the observations scientists make. Until the 19th century, diseases such as cholera were thought to be caused by 'bad air' from rotting organic material. It was not until the late 19th century that discoveries by scientists such as Louis Pasteur led to the development of germ theory: micro-organisms – and not 'bad air' – cause disease.

2. Basis of the investigation

Sometimes equipment may not be available to test a particular hypothesis. Prior to the invention of the microscope, scientists had no way of viewing cells. Today, continued advancements in technology provide scientists with novel tools to make new discoveries. The invention by Antonie van Leeuwenhoek (1632–1723) of the first practical microscope led to some of the most important discoveries in biology – bacteria, sperm cells and blood cells.

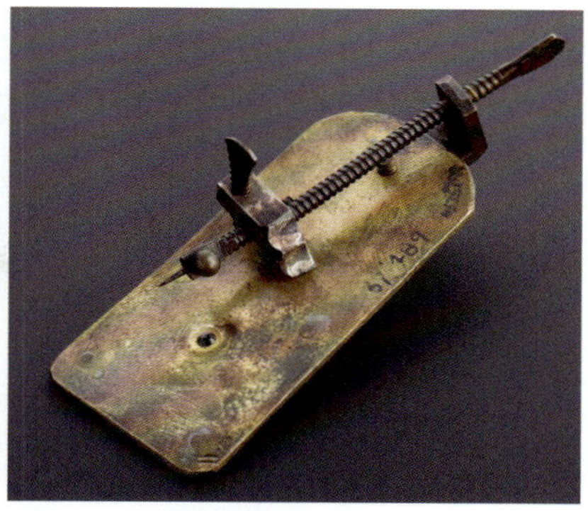

Figure 3.22: Van Leeuwenhoek microscope, circa late 1600s.

3. Ability to interpret the results

The ability of a scientist to interpret the results of an experiment depends on the knowledge they already have. The less knowledge a scientist has, the more difficult it is to interpret the results of an experiment correctly. If the results of an experiment are misinterpreted, a wrong conclusion may be drawn. The use of artificial intelligence may increase our capacity to interpret large amounts of data in the future.

Figure 3.23: Scientists' ability to interpret data continues to evolve.

4. Changes in the natural world

The scientific method must be applied repeatedly to keep up with changes in the world. For example, in today's world, antibiotic resistance is a global concern. Micro-organisms such as bacteria are developing resistance to antibiotics which were once effective against them. This process happens over time due to genetic mutations. Antibiotic resistance threatens our ability to treat infections and increases the risk with medical procedures such as surgery and chemotherapy.

Figure 3.24: Antibiotic resistance – changes in the natural world.

ACCIDENTAL DISCOVERIES

Figure 3.25: Alexander Fleming.

Many great breakthroughs in science occurred accidentally. Alexander Fleming discovered the first antibiotic penicillin in 1928 when he noticed that a petri dish containing Staphylococcus bacteria had become contaminated by a fungus, *Penicillium notatum*. Fleming observed that the bacteria were killed in the areas around the fungus. The fungus had produced the antibiotic we now know as penicillin.

Go to YouTube.com and search for 'The Scientific Method Not So Perfect After All' (4:00) by ScienceTelly.

SCIENCE IN SOCIETY: HEALTH AND TECHNOLOGY

Ireland's medical technology (MedTech) sector employs over 40,000 people in more than 300 companies, making Ireland the largest employer of MedTech professionals per capita in Europe. MedTech accounts for 8% of exports from Ireland and a value of approximately €13 billion to the Irish economy. Ireland is home to 14 of the top 15 MedTech companies, many of which have multiple sites around Ireland. The MedTech community in Ireland spans cutting-edge industry, academic and clinical research groups and government agencies.

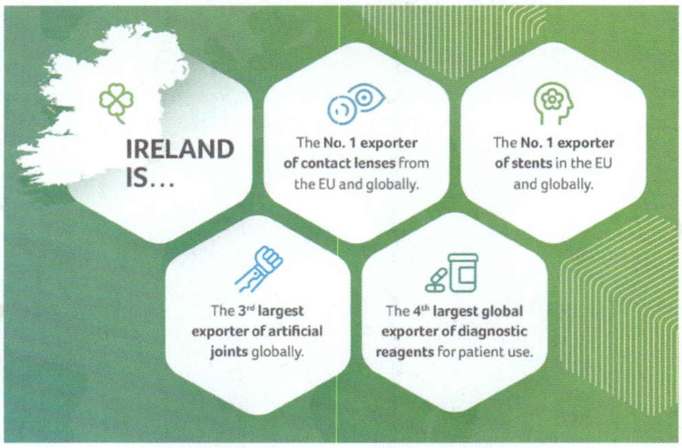

Figure 3.26: The MedTech industry in Ireland.

(*Adapted from IDA Ireland*)

PORTFOLIO LINK Go to your Learning Portfolio and complete Learning Check 4 to assess your knowledge and understanding of the topic so far.

Chapter Summary

- The **scientific process** is a set of steps followed by scientists when carrying out investigative research: (a) propose a possible explanation for the observation 'hypothesis' (b) design an experiment to test the hypothesis (c) evaluate the experimental evidence (d) communicate the experimental result.
- **Observation** is the collection of information or data from a primary source, e.g. an experiment.
- A **hypothesis** is an educated guess based on observation.
- A **prediction** is the expected result of a test that is derived from a hypothesis.
- **Generalisation** means drawing a general conclusion from a specific observation, e.g. generalising about a large population from a much smaller sample.
- A **sample group** is a subset with characteristics reflective of its population.
- A **controlled experiment** is a scientific test where only one variable is changed at a time and all other conditions are kept constant.
- A **variable** is a factor or condition that can be changed or controlled in a scientific investigation.
- A **control variable** is a variable that is kept the same by the experimenter.
- An **independent variable** is a variable that is changed by the scientist, while a **dependent variable** is a variable that is observed or measured during the experiment.
- A **control group** is a group that is not exposed to the variable under study.
- In **double blind testing**, neither the researcher nor the test subject knows who is receiving the test drug or the **placebo** (a sugar/starch pill) in a clinical trial.
- A repeat of an experiment is known as a **replicate**.
- **Data** refers to observations or measurements recorded in an experimental test.
- **Primary data** means results collected by the person who conducted the experiment.
- **Secondary data** is data collected by someone other than the user.
- **Qualitative data** is non-numerical data, e.g. colour, smell, species name, texture.
- **Quantitative data** is data that can be expressed in numerical form, e.g. length, mass, time, area.
- **Range** = Highest data value − Lowest data value.
- **A conclusion** is a summary of how the results of an investigation support or contradict your original hypothesis.
- **Causation** indicates that one event is the result of the occurrence of another event.
- **Correlation** is a numerical measure of the degree to which a relationship exists between two variables.
- **Uncertainty** is the range of possible values within which the true value of the measurement lies. Measurement = (best estimate ± uncertainty) units.
- **Systematic uncertainty** arises from defects in the measurement apparatus, equipment or method used by the experimenter.
- **Statistical uncertainty** is a measure of the range of the random variation in the set of results obtained when repeated measurements of the same event are recorded.
- **Error** is the difference between a measurement and its true value.
- **Accuracy** is how close a measured value is to the true actual value.
- **Precision** is how close two or more measurements are to each other.
- **Reliability** is the ability of an instrument or method to measure stable and consistent results.
- Getting the same result when a scientist makes a particular measurement under the same conditions is known as **repeatability**.
- **Validity** is how well a test measures what it intends to measure.
- **Integrity** means maintaining honesty at all stages of scientific practice.
- An **anomaly** is an irregular piece of data that does not fit the pattern of a particular data group.
- **Limitations of the scientific process** are the extent of existing knowledge, basis of the investigation, ability to interpret the results, changes in the natural world, accidental discoveries.

END OF CHAPTER ASSESSMENT

REMEMBER

1. List the stages in the scientific process.

2. Define each of the following terms.
(a) hypothesis (b) placebo (c) double blind testing (d) replicate (e) control group.

3. Identify the qualitative data in the following list: length, species, time, mass, colour.

4. State a method you could use to reduce (a) statistical uncertainty (b) systematic uncertainty.

5. Where do scientists communicate their scientific research and its results?

6. State **three** limitations of the scientific method.

UNDERSTAND

7. Distinguish between each of the following sets.

(a) Causation and correlation

(b) Statistical and systematic uncertainty

(c) Accuracy and precision

(d) Reliability and validity.

8. Distinguish between each of the following sets.

(a) Primary and secondary data

(b) Independent and dependent variables

(c) Error and uncertainty

(d) Qualitative and quantitative data.

9. A scientist examined the relationship between two variables and drew the following graph from their data. How would you describe the relationship between the two variables?

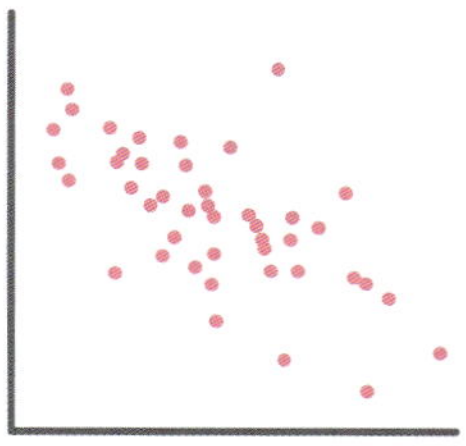

10. Select the piece of equipment you would use from the following list in each of the situations below. Give a reason for your answer in each case.

ruler   Vernier callipers
balance to one decimal place
balance to two decimal places
stopwatch   clock   digital thermometer
alcohol thermometer

(a) Measuring the diameter of a stem

(b) Measuring the length of time taken for a solution to change colour in a chemical test

(c) Measuring the mass of a food sample.

APPLY

11. A student measured the mass of a mouse five times on two different electronic balances. Her results were as follows:

BALANCE 1	17.8g	18.1g	18.0g	17.9g	18.0g
BALANCE 2	19.0g	19.0g	18.5g	18.7g	19.0g

Which balance is more precise? Explain your answer.

12. A student measured the length of leaves collected on an ecology trip using a ruler. One leaf had a length of 10cm. What is the uncertainty range associated with this measurement?

13. A student was checking the validity of the measurements taken by an electronic balance. He placed a known mass of 100g on the balance. The balance read the mass as 99.7g. Calculate the error associated with this reading.

14. Calculate the uncertainty for the following pieces of equipment:

(a) A thermometer with 1°C divisions

(b) A measuring cylinder with 5cm divisions.

15. You wish to investigate the effect of the amount of a fertiliser on the growth of radish seeds. Apply your knowledge of the scientific process to answer the following.

(a) Identify the independent variable.

(b) Identify the dependent variable.

(c) What control would you use?

(d) How would you employ replicates?

(e) What factors would you keep constant?

ANALYSE

16. A student investigates the effect of length of exercise on a person's heart rate. Identify the different variables that the student would need to consider in this study and classify them under the following headings:

(a) Independent variable

(b) Dependent variable

(c) Control variables.

17. A scientist is carrying out background research when developing a hypothesis. Categorise the following sources of information as either a source or an unreliable source. Give a reason for your choice in each case.

(a) Wikipedia

(b) *Nature* journal

(c) Central Statistics Office (CSO)

(d) A website ending in .com

(e) A website ending in .gov

(f) A website ending in .edu

(g) A scientific textbook.

EVALUATE

18. Which of the following is the best hypothesis for an investigation? Give a reason for your choice.

(a) If you refrigerate apples, they will last longer before going bad.

(b) It's better to store apples in a fridge than in a fruit bowl.

(c) Where do apples last longer, in a fridge or in a fruit bowl?

CREATE

19. A student wishes to investigate the effect of different types of cooking on Vitamin C levels in potatoes.

(a) Write a suitable hypothesis for this research.

(b) Design a controlled experiment to test this hypothesis.

(c) How would the student analyse their results?

UNIT 2

THE LIVING WORLD

CHAPTERS

STRANDS

Unifying Strand: Nature of Science

Strand 1: Organisation of Life

Strand 2: Structures and Processes of Life

Strand 3: Interactions of Life

4 Classification and Taxonomy

LEARNING OUTCOMES: U1.1, U3.2, U4.1, U4.2, 1.1.3, 1.1.4, 1.3.1

EYE-OPENER

Earth is the only planet we know of where life exists – and that life is extremely varied. A study in 2011 estimated that there are between 10 and 14 million different species of living things on our planet, although only 1.2 million have been documented and named. That means that over 85% of the variety of life on Earth hasn't been discovered or documented yet! And that doesn't include any species that have gone extinct – scientists estimate that 99.9% of all species that once lived on Earth no longer exist.

At the end of this chapter, you should be able to:

- Outline the origins and evolution of the science of classification – taxonomy.
- Describe the contributions of key scientists (Aristotle, Linnaeus, Darwin and Mendel) in the development and evolution of classification.
- Describe the three domains of life and the kingdoms within them.
- Compare phenetic (trait-based) and phylogenetic systems of classification.
- Use classification systems to identify and classify organisms.

LINKS TO:

Chapter 2	**Scientific Knowledge and Research**
Chapter 5	**The Animal Kingdom**
Chapter 6	**The Plant Kingdom**
Chapter 7	**Bacteria and Fungi**
Chapter 8	**Viruses**
Chapter 9	**Evolution and Natural Selection**
Chapter 10	**Biodiversity and Ecosystems**
Chapter 16	**Cell Structure and Diversity**

SUBJECT VOCABULARY

species, biodiversity, taxonomy, binomial nomenclature, variation, genome, prokaryote, eukaryote, archaea, phenetic (trait-based) classification, phylogeny, clade, most recent common ancestor, nucleoid, plasmid, murein, binary fission, ribosome, extremophile, organelle, last universal common ancestor

Section 1: Classification and Taxonomy

At the end of this section, you should be able to:

- Define the terms 'species' and 'biodiversity'.
- Explain what is meant by 'taxonomy'.
- Describe the contributions of Aristotle to early methods of classification.
- Describe binomial nomenclature and give examples.
- Describe the work of Carl Linnaeus.

Unpacked from LOs U1.1, U3.2, U4.1, U4.2, 1.1.4

BIODIVERSITY

Biodiversity describes the variety of life on Earth. Each *type of living thing* on Earth is called a **species** and it is more accurately defined as a group of living things that can reproduce and produce offspring, which can go on to reproduce themselves. The variety of life on Earth captivated philosophers and scientists for hundreds if not thousands of years and various efforts have been made to make sense of this diversity. Thus, the science of **taxonomy** formed – the science of placing living things into ordered groups, better known as classification.

- A **species** is defined as a population of organisms that can interbreed successfully and produce fertile offspring.
- **Biodiversity** describes the variety of life on Earth or in a particular habitat.
- **Taxonomy** is the science of classification.

Go to YouTube.com and search for 'Why is biodiversity important – with Sir David Attenborough' (5:39) by The Royal Society.

Figure 4.1: Living and extinct species. How many do you recognise?

EARLY METHODS OF CLASSIFICATION

The science of classification began with the work of the philosopher Aristotle in the 4th century BCE. He developed the *scala naturae* (or 'ladder of being') where he tried to organise all things in the natural world into various groups. He deemed some matter to be of a 'higher order' (like humans) and other matter (like rocks and other inanimate objects) to be of a 'lower order'. He also distinguished between plants and animals, based on their ability to move, whether or not they had blood and whether they lived in water or on land.

The *scala naturae* was the first real attempt to make order of the natural world and it had many flaws. Aristotle didn't include micro-organisms, because he had no way of seeing them or knowing they existed, and he also believed that living things were fixed and unchanging. However, the method of classification and ideas of the *scala naturae* persisted for nearly 2,000 years.

Figure 4.2: Aristotle's *scala naturae*.

RESEARCH ACTIVITY

Aristotle was an ancient Greek philosopher and scientist and is considered one of the greatest intellectual figures in the history of humanity. He contributed to the arts, science, politics, psychology, philosophy and even poetry. Conduct some research on Aristotle's contributions to the sciences and create a mind map summarising the main points. Be sure to note your references.

Figure 4.3: Aristotle.

CARL LINNAEUS AND BINOMIAL NOMENCLATURE

There was little development in the science of classification until the 1600s, when the voyages of exploration unearthed a vast variety of plants and animals previously unknown to Western scientists. Scientists showed great interest in placing these new species within existing classifications and, when problems arose, created new methods of classification.

Gaspard Bauhin, a Swiss doctor, classified thousands of plants and was the first to use a two-name system (**binomial nomenclature**) for naming plants and animals (think *Homo sapiens* or *Tyrannosaurus rex*). Up until that point species names were extremely long and complicated, but Bauhin decided to just give each organism a Latin name composed of a genus (e.g. *Homo*) and a species (e.g. *sapiens*).

Binomial nomenclature is a system of using two names (a genus and a species) to name and classify an organism.

Figure 4.4: *Tyrannosaurus rex*: genus *Tyrannosaurus*, species *rex*.

Figure 4.5: Carl Linnaeus.

The work of Bauhin influenced **Carl Linnaeus**, who, in 1735, wrote a book called *Systema Naturae,* or *The System of Nature*. In his book, Linnaeus proposed placing all living things within three 'kingdoms' (Animalia or animals, Plantae or plants and Minerae or minerals). Each kingdom would then contain smaller groups called 'classes', then smaller groups called 'orders' and finally 'genus' and 'species'. This was the first time living things were placed into groups according to their similarities.

Linnaeus continued to develop his *Systema Naturae* and classified over 4,000 animals and 7,000 plants using his system. He tweaked it as he found errors (a good example of the scientific method) and was the first to use the term *Homo sapiens* (thinking man) to describe modern humans. His system is still in use today, although it now has six kingdoms and groups called 'phyla' (singular 'phylum').

!

When writing genus and species names, we always type them in italics. The genus is always given a capital letter while the species name gets a small letter, e.g. *Homo sapiens*. If you're handwriting a species name, you should underline it, e.g. Homo sapiens.

Sometimes the genus name can be replaced with just a letter for convenience, e.g. *T. rex* or *E. coli*.

Scientists are known to have a sense of humour, and assigning names to newly discovered species is a real chance for scientists to show their creative side. There are rules for assigning names, but they're not very strict, so many unusual scientific names exist. For example, one now-extinct species of trilobite is named *Han solo* after the Star Wars character. The world of Harry Potter influenced the name of the dinosaur species *Dracorex hogwartsia*, while J.R.R. Tolkien's *Lord of the Rings* trilogy influenced the naming of the insects *Gollumjapyx smeagol* and *Macrostyphlus frodo*.

What would you name a new species?

Figure 4.6: A recently discovered Australian fly was named *Scaptia beyonceae* as its distinctive rear end reminded researchers of the song 'Bootylicious'.

DID YOU KNOW?

PORTFOLIO LINK

Go to your Learning Portfolio and complete Learning Check 1 to assess your knowledge and understanding of the topic so far.

Section 2: Building a Model of Classification

At the end of this section, you should be able to:

- Discuss the influence of Darwin and Mendel on our understanding of classification.
- Describe the current three domain system of classification and discuss how it developed from Linnaeus's kingdoms.

Unpacked from LOs U1.1, U3.2, 1.1.3

INFLUENCE OF DARWIN AND MENDEL

Aristotle and Linnaeus both worked on the assumption that living things were unchanged and were created (by God) in their current form. They assumed that all the living things that existed then had always existed and will always exist. They placed organisms in the same order or class simply because they resembled each other, and didn't explore if species were related in some way.

In the 1800s, the work of **Charles Darwin**, who pioneered the principle of evolution, and later **Gregor Mendel**, who is known as the Father of Genetics, greatly changed our understanding of classification and how animal and plant species are related.

Figure 4.7: Charles Darwin.

Darwin proposed that populations of living things contain **variation** (differences between each individual animal or plant) and that these variations can give some members of the species advantages within their environment. He suggested that, because of these minor differences, new species can be formed from older ones over a long period of time. For the first time, scientists started looking at how organisms might be related to each other and whether they shared common ancestors.

Figure 4.8: Gregor Mendel.

Darwin's theory of natural selection, which we will explore in more detail in Chapter 9, explains the principles of how new species can form from old ones. The classification of organisms helped him to find connections between species but also altered our perception of classification as a science.

Gregor Mendel, an Austrian monk, laid the foundation for modern genetics. He carried out experiments on pea plants and his results led him to theorise that organisms passed on information from one generation to the next. He didn't know it at the time, but he was referring to the genetic makeup (**genome**) of the organism. We now know that species that are similar share genetic sequences and genes. The more closely related organisms are, the more similar their genetic makeup. Scientists can now use sophisticated DNA analysis of any newly discovered organisms to establish their exact classification.

- **Variation** refers to differences between members of the same species.
- An organism's **genome** is the complete set of genes present in their cells. Humans have over 20,000 genes.

CURRENT MODEL OF CLASSIFICATION

The system of classification used by scientists today is built upon the foundations laid by Linnaeus, but draws from the work of Darwin, Mendel and other evolutionary biologists and geneticists. With the invention of the microscope, and the subsequent discovery of bacteria, fungi and single cell organisms (some of which behaved like plants and others like animals), scientists needed to alter Linnaeus's three kingdom system.

- A **prokaryote** is an organism (or cell) without a membrane-bound nucleus or organelles.
- A **eukaryote** is an organism (or cell) with a membrane-bound nucleus and organelles.
- **Archaea** are an 'ancient' type of bacteria that are more closely related to eukaryotes than other 'true' bacteria.

The Animalia and Plantae kingdoms remained in use. A third kingdom, Protista, was introduced in the late 1800s to classify all single-celled organisms. Then it was discovered that some single-celled organisms didn't have a nucleus like most animal and plant cells; those without a nucleus would now be called **prokaryotes** while those with a nucleus were referred to as **eukaryotes**. This would later lead to the creation of kingdom Prokaryotae (sometimes called Monera). In the 1960s, fungi were given a kingdom of their own – kingdom Fungi – which also includes moulds and mildews. The five kingdom model – Animalia, Plantae, Fungi, Protista and Prokaryotae – was now complete.

However, in the 1990s it was discovered that bacteria within kingdom Prokaryotae contained two distinct types – true bacteria and ancient bacteria. These ancient bacteria, now known as **archaea**, had different structures, genes and biochemistry to most bacteria and were therefore reclassified. Archaea are actually more closely related to eukaryotes (like plants and animals) than bacteria. Their discovery led to a major shift in Linnaeus's system of classification and the creation of 'domains' above 'kingdoms' in the classification hierarchy.

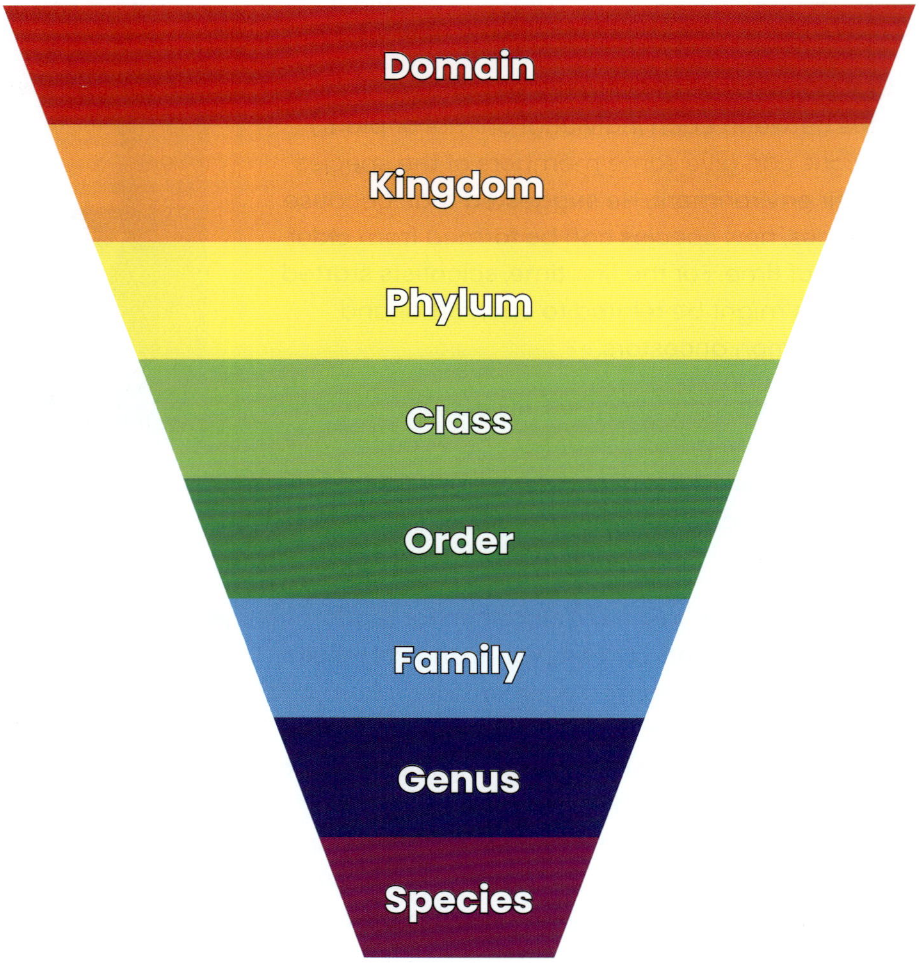

Figure 4.9: The current model of classification.

The current system of classification is referred to as a **three domain system**. It works like this:

- Organisms are first placed into one of three **domains** – Bacteria, Archaea and Eukaryota.
- Organisms are then placed within one of six **kingdoms** – Bacteria (only found in the Bacteria domain), Archaebacteria (the new kingdom, only found in the Archaea domain), Protista, Fungi, Plantae and Animalia (all found in the Eukaryota domain).
- Organisms are then assigned to a **phylum**, then a **class**, then an **order** and **family**.
- Finally, the organism is assigned a **genus** and **species** name, using the binomial system.

Go to YouTube.com and search for 'Classification' (7:41) by Amoeba Sisters.

Use this handy mnemonic to remember the various stages of classification.

Do Kings Play Chess On Fine Green Silk?

Domain Kingdom Phylum Class Order Family Genus Species

Human Classification

Using the system above, humans would be classified as follows:

Domain	**Eukaryota**	We are eukaryotic and our cells contain a membrane-bound nucleus.
Kingdom	**Animalia**	We are animals and we feed on other living things. We have sense organs to detect the world around us and muscle cells to move.
Phylum	**Chordata**	We are animals with a backbone, or spine.
Class	**Mammalia**	We are mammals. We can produce milk from specialised cells and generally have hair.
Order	**Primate**	We have hands, hand-like feet, large brains and eyes that generally face forward. All monkeys and apes are primates.
Family	**Hominidae**	We are hominids, meaning we are apes. Apes have large brains, no tails, the same dental formula and are omnivorous. Hominids include humans, chimpanzees, gorillas and orangutans.
Genus	**Homo**	We are human, and the only remaining human species. Extinct members of the *Homo* genus included *Homo erectus* ('upright walking man'), *Homo neanderthalensis* (Neanderthal) *and Homo habilis* ('handy man').
Species	**Sapiens**	We are modern humans. *Homo sapiens* means 'thinking man'. We are characterised by our large brains and social constructs.

Figure 4.10: *Homo sapiens.*

PORTFOLIO LINK **Go to your Learning Portfolio and complete Learning Check 2 to assess your knowledge and understanding of the topic so far.**

Section 3: Phenetic and Phylogenetic Classification

At the end of this section, you should be able to:

- Define the terms 'trait-based' and 'phylogenetic' classification and distinguish between them.
- Define and draw a simple phylogenetic tree.
- Discuss why classification is important.

Unpacked from LOs U4.1, U4.2, 1.1.4

PHENETIC VS PHYLOGENETIC

Classification and taxonomy played a large part in our understanding of evolution, including human evolution, and taxonomists and evolutionary biologists now share common goals. Both want to accurately classify organisms, but also want to understand how they evolved, from what, and who they are related to genetically.

Darwin studied the similarities between organisms to find species that were related or might have shared a common ancestor. This **trait-based or phenetic system** of classification classifies organisms according to similarities in their structures, e.g. bone structure in the arm. The analysis of these structures can help form an understanding of the evolutionary history of the organism – what organisms have the same structures, either extinct or currently living (**extant**). This system allows scientists create a good understanding of the evolutionary history of the organism but it doesn't work for all species.

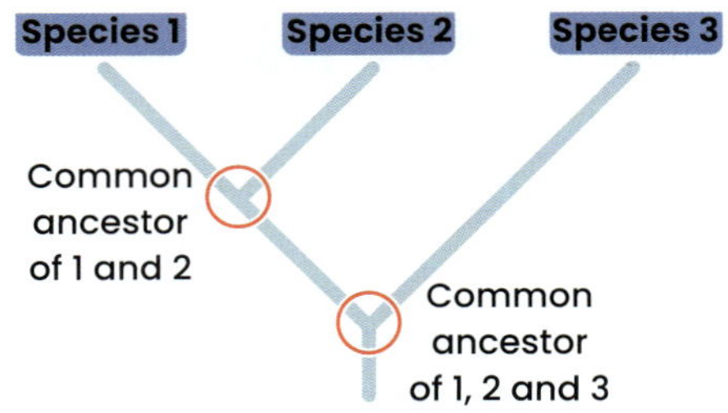

Figure 4.11: A phylogenetic tree.

However, a new method known as **phylogeny** or **phylogenetic classification** uses similarities in the structures of organisms, but also accounts for similarities in their DNA or genetic makeup, behaviours and biochemistry to create a more accurate picture of the organism's evolutionary development. In phylogenetic classification, the evolutionary relationship between different species is often shown graphically in diagrams called **phylogenetic trees**. These diagrams, which are often hypothetical, can be useful tools to analyse the evolution of species. They are similar in a lot of ways to family trees, which show who descended from whom.

- **Phenetics (or trait-based) classification** organises organisms into groups according to similarities in their structure, i.e. number of legs or petals.
- **Phylogeny (phylogenetic classification)** classifies organisms based on their evolutionary development (genetics).
- The **most recent common ancestor (MRCA)** of a set of organisms is the most recent individual from which all organisms in that set are descended from.
- A **clade** is a grouping on a phylogenetic tree that share a common ancestor.

A simple phylogenetic tree is shown in Figure 4.11. The tips of the branches represent genetically related species, and common ancestors are represented where the branches meet. Species 1 and 2 share a more recent common ancestor with each other than with Species 3. Therefore, Species 1 and 2 are more closely related to one another than to Species 3. The **most recent common ancestor (MRCA)** of any set of individuals is the **most** recent individual from which all the individuals in the group are directly descended. A **clade** is a grouping on a phylogenetic tree that share a common ancestor.

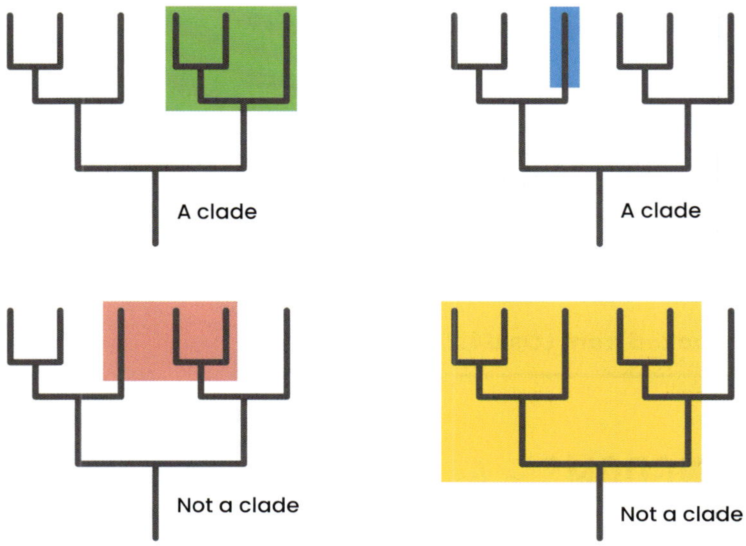

Go to YouTube.com and search for 'Phylogeny: How We're All Related' (13:50) by CrashCourse.

Figure 4.12: A clade is a group of organisms that share a common ancestor.

WHY IS CLASSIFICATION IMPORTANT?

Taxonomy has been a central feature in the study of biology for a long time and, as we've seen, it has changed as our understanding of the living world has evolved. It is important for many reasons.

- It makes it easier to study large numbers of organisms at the same time as they may share the same DNA or genes, biochemistry and common ancestors.
- It helps us understand how life evolved on Earth and what the future of life on Earth might be like.
- It allows scientists understand the human genome (DNA) and can help find cures to common diseases and genetic conditions.
- Classification ensures we can record and appreciate the variety of life on Earth, especially as we face a crisis of biodiversity.
- Classification helps us make better decisions about conservation of species and sustainable development.
- Newly discovered species can be named and classified more easily.
- The binomial system ensures scientists can share information on a particular species to other scientists around the world, despite having different names in their own language.

Figure 4.13: Classification helps us to understand how life evolved on Earth.

PORTFOLIO LINK

Go to your Learning Portfolio and complete Learning Check 3 to assess your knowledge and understanding of the topic so far.

Section 4: Three Domains of Life

At the end of this section, you should be able to:

- Outline the main properties of each of the domains and kingdoms.
- Classify organisms based on your knowledge of the domains and kingdoms.
- Describe how the three domains are linked to LUCA – the last universal common ancestor.

Unpacked from LOs U4.2, 1.1.3, 1.1.4, 1.3.1

THE THREE DOMAINS AND SIX KINGDOMS – AT A GLANCE

Over the course of the next two years, you will study the biology of a wide variety of living organisms, across the three domains. For now, here is a quick breakdown of the features of each of the three domains and the six kingdoms that occupy them.

Domain Bacteria

Bacteria are all single-celled prokaryotic organisms. They are the most abundant organisms on the planet; there are trillions of bacteria living in and on you right now. The domain Bacteria contains organisms within the kingdom Bacteria, sometimes called Eubacteria meaning 'true bacteria'. The properties of Eubacteria are as follows:

- They are all incredibly small, single-celled (unicellular) and prokaryotic (no nucleus).
- They can be autotrophic (make their own food) or heterotrophic (acquire food from other organisms).
- Instead of a nucleus, they have a large circular strand of DNA called a **nucleoid** and some smaller pieces of DNA called **plasmids**.
- They do not have any membrane-bound organelles (small structures within the cells) but do have protein-building **ribosomes**.
- They have cell walls made of **murein** – a complex chemical made from sugars and protein – and phospholipid membranes.
- They may have a flagellum or tail.
- They reproduce asexually, by **binary fission**.

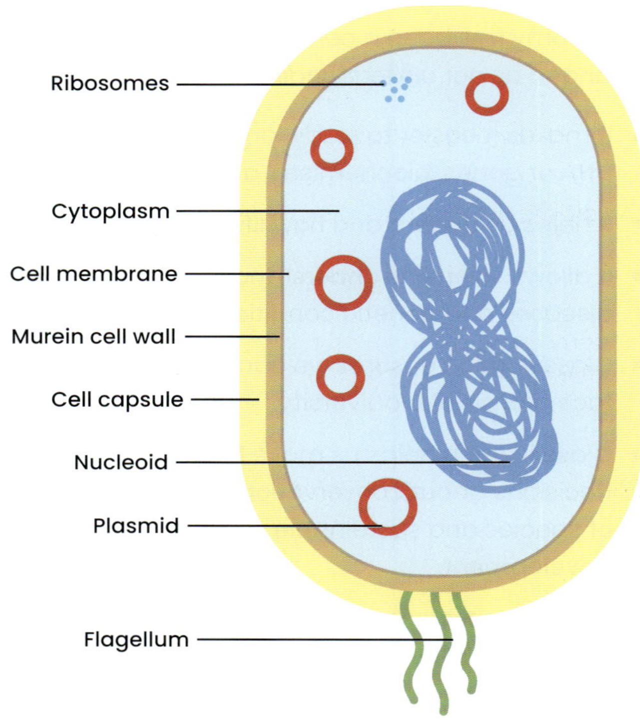

Figure 4.14: Bacteria and archaea cells look very similar but have different biochemistry.

- A **nucleoid** is a structure within prokaryotic cells where the DNA is found.
- A **plasmid** is a small ring of DNA found in prokaryotic cells.
- **Binary fission** is the way prokaryotic cells divide and reproduce.
- **Ribosomes** are structures within cells that make protein.
- **Murein** is a complex of protein and sugars which forms the cell walls of eubacteria.

Domain Archaea

Archaea, containing members of kingdom Archaebacteria, share many of the same properties as bacteria above but with some subtle differences in the chemicals they are made from and the reactions that occur within their cells. Some of the differences between the members of kingdoms Archaea and Bacteria include:

- The cell walls of archaebacteria do not contain murein. Instead, their cells walls are made from complex sugars.
- The cell membranes of archaea are slightly different from bacterial membranes.
- Archaea contain genes that are more similar to Eukaryota – it is believed the Eukaryota share a common ancestor with archaea.
- The protein-building ribosomes within archaea are different from bacteria, and again resemble those of Eukaryota.

There are three main groups: the **halophiles**, the **thermophiles** and the **methanogens**. Halophiles thrive in really salty environments, like the Dead Sea, while thermophiles live in areas of extreme heat like geysers or hydrothermal vents. Methanogens are anaerobic (don't need oxygen) and live in sewage, at the bottom of ponds and in the intestines of animals, especially ruminants like cattle or horses. They release methane as they respire.

Many members of the domain Archaea are **extremophiles** – organisms that love to live in areas where most living things cannot survive. The archaea are some of the oldest organisms on the planet.

RESEARCH ACTIVITY

Carry out some research on extremophiles. What unusual adaptations do they have to allow them live in their environment?

Extremophiles are micro-organisms that thrive in harsh environments, e.g. extreme heat or pH.

Domain Eukaryota

The domain Eukaryota contains four kingdoms – Protista, Fungi, Plantae and Animalia. There are many similarities between each of the four kingdoms:

- All contain cells with a membrane-bound nucleus that contains their DNA ('eukarya' means 'true nucleus').
- Their cells also contain membrane-bound **organelles** – small cellular machines with specific functions – like mitochondria, the cell's internal engine.
- The cells have a complex method of division, known as mitosis, where the DNA is copied in the nucleus before splitting to form two identical cells.
- Unlike bacteria and archaea, eukarya can reproduce sexually by forming special cells called **gametes**.

An **organelle** is a structure within eukaryotic cells with a particular function, e.g. mitochondria.

We will explore the characteristics of each of the four Eukaryota kingdoms in the upcoming chapters, but this chart gives a rough breakdown of the major differences.

KINGDOM	ORGANISATION	TYPE OF NUTRITION	REPRESENTATIVE ORGANISMS				
Protista	Complex single cell, some multicellular	Absorb, photosynthesise or ingest food	Paramecium 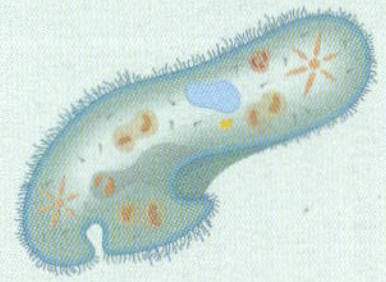	Euglenoid 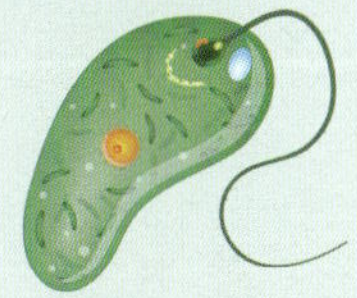	Slime mould 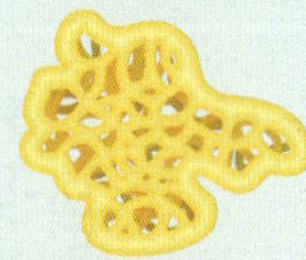	Dinoflagellate 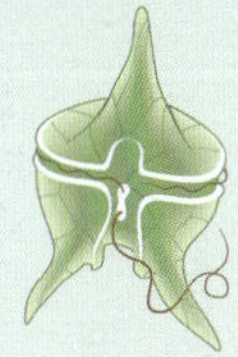	Protozoans, algae, water moulds, slime moulds
Fungi	Some unicellular, most multicellular filamentous forms with specialised complex cells	Absorb food	Black bread mould 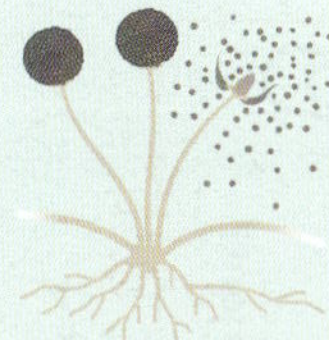	Yeast 	Mushroom 	Bracket fungus 	Moulds, yeast, mushrooms
Plantae	Multicellular forms with specialised complex cells	Photosynthesise food	Moss 	Fern 	Pine tree 	Non-woody flowering plant 	Mosses, ferns, non-woody and woody flowering plants
Animalia	Multicellular forms with specialised complex cells	Ingest food	Coral 	Earthworm 	Crow 	Squirrel 	Invertebrates, fishes, reptiles, amphibians, birds, mammals

Figure 4.15: The properties of the four Eukarya kingdoms.

LUCA – THE LAST UNIVERSAL COMMON ANCESTOR

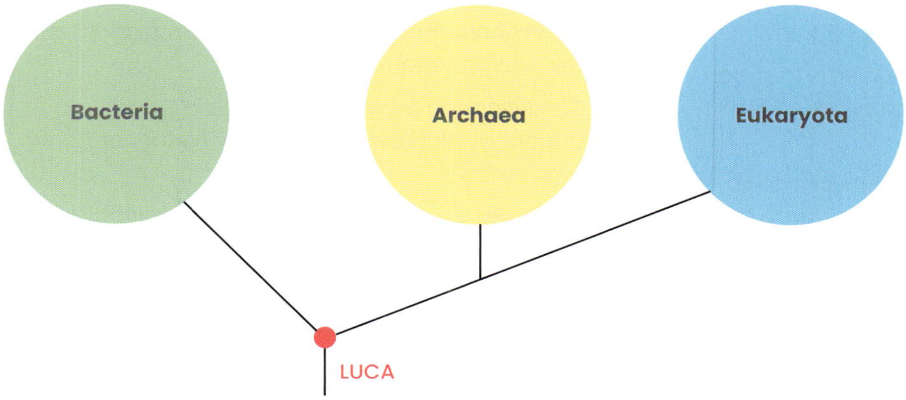

Figure 4.16: LUCA represented on a phylogenetic tree.

When you look at the phylogenetic tree above, you can see that the three domains all belong to one clade. This means they all share a common ancestor. Since all life on Earth exists within one of the three domains, we can say that all life on Earth has a common ancestor. We call this the **last universal common ancestor** or LUCA for short.

LUCA – the **last universal common ancestor** – is the organism/cell that all life derived from.

LUCA is thought to have existed around 4 billion years ago. It was a single-celled autotrophic organism and likely lived around hydrothermal vents at the bottom of the ocean. It's a remarkable thought that every single living thing on the planet is related to that single ancient cell.

When studying the clade, you might have noticed that archaea and Eukaryota share a branch on the phylogenetic tree. As mentioned previously, these two domains share some biochemical traits and are thought to be more closely related.

HEY, WHAT ABOUT VIRUSES?

You may have noticed that viruses have not been mentioned so far and don't seem to fall into any of the three domains of life. As mentioned earlier in Chapter 1, viruses are not actually considered to be 'living things'. They don't satisfy all of the characteristics of life and crucially, from a classification perspective, they don't have cells. Perhaps someday biologists will re-evaluate the classification of viruses but, for now, they belong on the edge of life. They are important biological entities, though, so we will study them in more detail in Chapter 8.

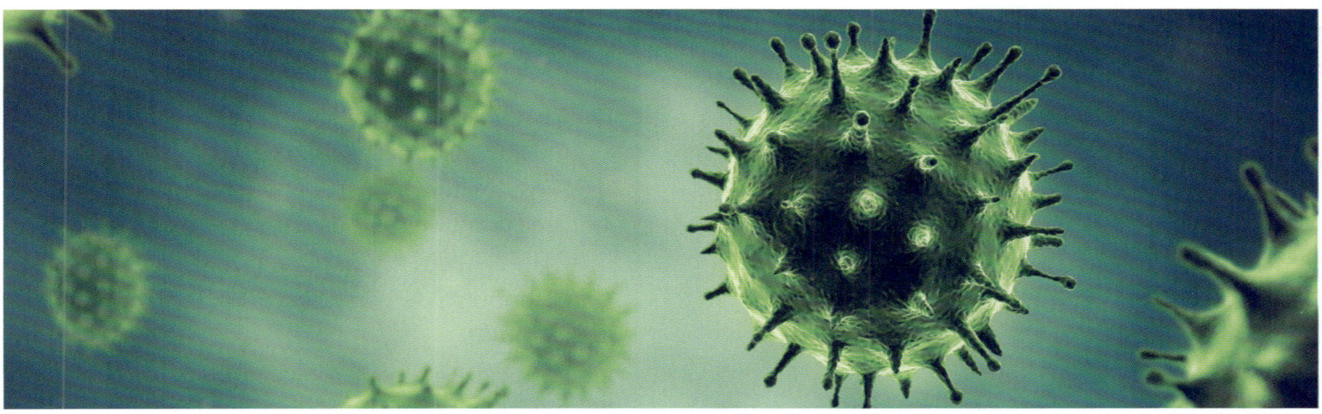

Figure 4.17: Viruses are not considered to be living organisms.

PORTFOLIO LINK **Go to your Learning Portfolio and complete Learning Check 4 to assess your knowledge and understanding of the topic so far.**

Chapter Summary

- Biodiversity describes the variety of life on Earth or a particular habitat.
- A species is a group of organisms that can interbreed successfully and produce fertile offspring.
- Taxonomy is the science of classification – the placing of organisms into ordered groups.
- The Greek philosopher Aristotle was the first person to classify living things.
- **Carl Linnaeus** developed a system of classification where organisms were placed into kingdoms then smaller groups within.
- Both Aristotle and Linnaeus believed that species did not change, were fixed and were unrelated.
- Charles Darwin, in his book *On the Origin of Species*, proposed that organisms were not fixed and evolved from older species.
- The current model of classification is called the **three domain system**.
- Organisms are first placed into one of three **domains** – Archaea, Bacteria and Eukaryota.
- Organisms are then placed within one of six **kingdoms** – Archaebacteria, Bacteria, Protista, Fungi, Plantae and Animalia.
- Organisms are then assigned to a **phylum** then a **class**, then an **order** and **family**.
- Finally, organisms are given a genus and species name, using the binomial system, e.g. *Homo sapiens*.
- **D**o **K**ings **P**lay **C**hess **O**n **F**ine **G**reen **S**ilk is a mnemonic used to remember the steps on the classification ladder.
- Classification plays a major role in understanding the evolutionary history of organisms.
- **Trait-based** classification is the classification of organisms into groups according to similarities in their structure, i.e. number of legs or petals.
- **Phylogeny** or **phylogenetic classification** is the classification of organisms based on their evolutionary development. This includes analysis of similarities in their structure, genetic makeup, biochemistry, behaviour and other factors.
- Phylogenetic trees are diagrams which illustrate how different organisms, extinct and existing, are related.
- There are subtle differences between the biochemistry and structure of bacteria and archaea.
- Eukaryota, which contain the kingdoms Protista, Fungi, Plantae and Animalia, are a diverse group, containing 8 million species.
- All members of the domain Eukaryota have cells with a true nucleus and membrane-bound organelles like mitochondria.
- The three domains of life share a phylogenetic clade so have a common ancestor, known as LUCA – the **last universal common ancestor** – from which all life on Earth derived.

END OF CHAPTER ASSESSMENT

REMEMBER

1. Define the terms *biodiversity*, *species* and *taxonomy*.
2. Name the scientist credited with the 'kingdom' system of classification.
3. Give the genus and species name for humans.
4. List the steps on the current model of classification.
5. Name the three domains and the six kingdoms within them.
6. Define the term *archaea*.
7. Select the correctly written name from options a–c below. Explain your answer.
   (a) *Homo Sapiens* (b) *Homo sapiens*
   (c) Homo sapiens
8. What does *LUCA* stand for?

UNDERSTAND

9. Compare prokaryotes and eukaryotes.
10. Describe the influence of Aristotle on our current system of classification.
11. Explain why 'domains' were added above 'kingdoms' in the current system of classification.
12. Distinguish between archaea and 'true' bacteria.
13. Summarise the current system of classification, in your own words.
14. Explain how phylogenetic classification is different from trait-based classification.
15. What is LUCA?

APPLY

16. Why are viruses not considered to be living things?

17. Discuss why classification is important to scientists.

18. Darwin suggested that members of the same species have variation. Using a named animal or plant, describe the variation that can exist within a species.

19. List 10 living things around your school or home. Classify them into their domain and kingdom, based on what you have learned so far.

ANALYSE

20. Study the diagram below and answer the following questions.

(a) At which point can we find the common ancestor between mammals and reptiles?

(b) At which point can we find the common ancestor from which feathered animals formed?

(c) What physical features do these animals all have in common?

(d) Are birds more closely related to reptiles or mammals? Explain your answer.

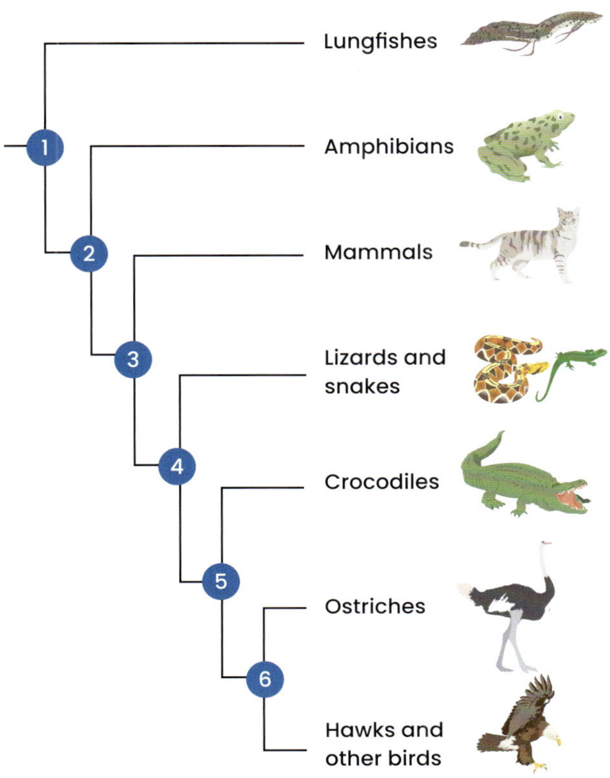

21. Study the phylogenetic tree shown below and answer the following questions.

(a) To which species is A most closely related to?

(b) C is more closely related to D than to B. True or false? Explain your answer.

(c) E shares a common ancestor with A. True or false? Explain your answer.

(d) Sketch the tree into your copy and indicate where the common ancestor of all five species can be found.

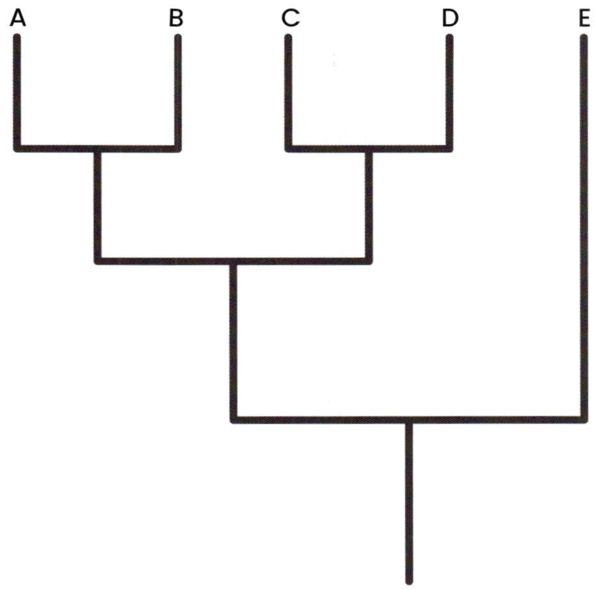

EVALUATE

22. Compare and contrast the features of the kingdoms bacteria and archaebacteria.

23. Fungi were once considered plants. Suggest why fungi were once in the plant kingdom and why they were later given a kingdom of their own.

CREATE

24. Classify the items in your pencil case or bedroom according to their physical features (e.g. what they're made from, colour, texture, etc.) and build a theoretical phylogenetic tree based on their differences and similarities.

25. Using a mind map, summarise the features of each of the six kingdoms of the living world.

5 The Animal Kingdom

LEARNING OUTCOMES: U4.1, U4.2, 1.1.4, 1.3.1

EYE-OPENER

Found in the warm southern waters of Australia, the sea dragon surely must be one of Earth's weirdest and most wonderful animals. By far the most ornately camouflaged members of the animal kingdom, they blend seamlessly into the seaweed and corals that dominate their habitat. As is the case with seahorses, it is the male sea dragon that carries the young, holding them in a spongy brood patch on the underside of the tail where females deposit their bright-pink eggs during mating. The eggs are fertilised during the transfer from the female to the male. The males incubate the eggs and carry them to term, releasing miniature sea dragons into the water four to six weeks later.

There are two species of sea dragon: the leafy sea dragon (*Phycodurus eques*) and weedy sea dragon (*Phyllopteryx taeniolatus*). 'Leafies' are generally brown to yellow in body colour with olive-tinted appendages. 'Weedies' have less flamboyant projections but are more colourful, typically red with yellow spots.

At the end of this chapter, you should be able to:

- Define the common features of all animals.
- Briefly outline the evolution of animals.
- Classify animals into the various animal phyla based on their body structures and characteristics.
- Describe the evolution of animal life on Earth and relate it to phylogenetic trees.

LINKS TO:

Chapter 2	**Scientific Knowledge and Research**
Chapter 4	**Classification and Taxonomy**
Chapter 6	**The Plant Kingdom**
Chapter 9	**Evolution and Natural Selection**
Chapter 10	**Biodiversity and Ecosystems**
Chapter 16	**Cell Structure and Diversity**

SUBJECT VOCABULARY

animal | metazoan | zoology | multicellular | sessile | motile | tetrapod | exoskeleton | vertebrate | mammal

Section 1: Animal Design

At the end of this section, you should be able to:

- Describe the characteristics of all animals.
- Discuss the multicellular nature of animals.
- Distinguish between sessile and motile animals, giving examples of both.

Unpacked from LOs U4.2, 1.1.4, 1.3.1

WHAT IS AN ANIMAL?

The word **animal** comes from the Latin word meaning 'to have breath' or to 'have life' (think of the word *animate*) but that tells us little about what it means to be an animal. Within the three domains of the living world, the animal kingdom sits on one of the branches within the Eukaryota domain. It's a hugely diverse kingdom, with up to 8 million species (we've identified around 1 million so far, and most of them are insects), yet every member of the animal kingdom has the following features in common.

- **Animals** are multicellular eukaryotic and heterotrophic organisms. Animals are multicellular, with cells without walls.
- **Zoology** is the study of animals.
- **Multicellular** refers to organisms which contain many cells, usually of different types.

Animals are also called **metazoans**, and the study of animals is called **zoology**.

- All animals are **multicellular**, their bodies containing many specialised cells with a defined **nucleus** and **no cell wall**.
- Animals are **heterotrophic**, meaning they consume other organisms for food.
- The vast majority of animals **respire aerobically**, meaning they need oxygen to survive.
- All animals can move (they are said to be **motile**), at least at some stage of their life cycle. Some animals, like sponges and coral, lose the ability to move later in their life cycle.
- All animals exhibit high levels of **cell diversity**, meaning they have many different types of cells designed for specific purposes, e.g. bone, muscle or nerves. These cells are organised into tissues – groups of similar cells.
- Most animals contain **muscle cells** for movement and **nerve cells** to transmit messages around their bodies.
- All animals have skeletons of some sort, but not all have bones.
- Almost all animals reproduce sexually. They produce gametes (sperm and eggs) by a special type of cell division called meiosis and the gametes fuse together to form a zygote.
- Animals form bodies, in one of a number of various forms, but generally they have a digestive tract.

FOCUS ON ... REPRODUCTION IN ANIMALS

While almost all animals reproduce sexually by fusing sperm and egg cells, some species of animal can reproduce asexually. Asexual reproduction creates clones – genetically identical offspring – which can be advantageous when the animals live in a stable environment.

Some species of flatworm can split their bodies into head and tail regions, with each part regenerating the missing part. Corals can grow buds on the side of their body which can, when they fall off, grow into a new coral. Some insects can reproduce in a process called parthenogenesis, which allows for eggs to grow into adults without being fertilised. Bees produce male drones in this way while worker and queen bees are produced from fertilised eggs.

RESEARCH ACTIVITY

Carry out some research on the three main body forms in animals. Also consider whether all animals have a body cavity or not.

Go to YouTube and search for 'What is an Animal? Crash Course Zoology #1' (15:25) by CrashCourse.

PORTFOLIO LINK

Go to your Learning Portfolio and complete Learning Check 1 to assess your knowledge and understanding of the topic so far.

Section 2: Animal Classification

At the end of this section, you should be able to:

- Outline how animals are classified.
- Briefly outline the timeline of animal evolution.
- Apply our knowledge of phylogenetic trees to animal evolution.
- Classify animals into the various phyla, based on physical features.

Unpacked from LOs U4.1, U4.2, 1.1.4, 1.3.1

CLASSIFICATION OF ANIMALS

Classification of animals is primarily based on their body forms although, more recently, DNA analysis is allowing us to refine their classification. Later, we will look at some phylogenetic trees within the animal kingdom but first let's look at the first subdivisions of the animal kingdom – the phyla.

Like all kingdoms, species are first placed into one of a number of phyla (singular: phylum). The most important animal phyla, their characteristics and signature species are outlined below.

FOCUS ON ... ANIMAL PHYLA

PHYLUM PORIFERA

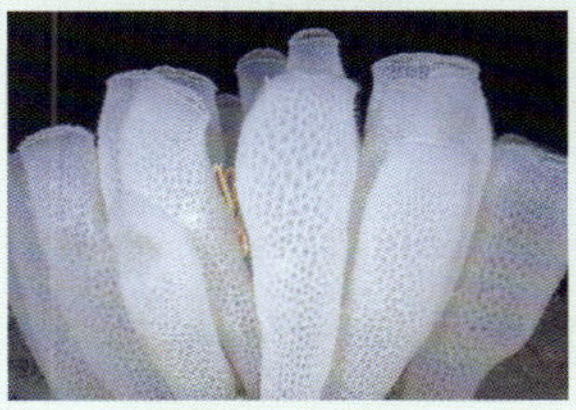

The Porifera contain the sponges and are named after the pores on their bodies, by which they channel water through in the hope of capturing food. They are sessile animals, meaning they don't move, although do produce motile sperm.

PHYLUM CNIDARIA

The Cnidaria (the 'C' is silent) are named after the specialised stinging cells found in these animals, which include the corals, jellyfish and sea anemones. They can be either immobile (like corals) or motile (like jellyfish).

PHYLUM PLATYHELMINTHES

Commonly called flatworms, these animals have an internal body cavity. They include species that live independently (free-living) but also parasitic species like tapeworms or liver fluke.

PHYLUM NEMATODA

Nematodes, or roundworms, have a body cavity. Again, there are free-living and parasitic varieties and most have a complete digestive tract running through their bodies. Examples include pinworms (common in children) and stomach worms (found in cattle).

PHYLUM ANNELIDA

These are the segmented worms, whose bodies are characterised by having various similar sections on their bodies. Examples include earthworms and leeches.

PHYLUM ARTHROPODA

The vast majority of animals belong to phylum Arthropoda. They all have a hard shell called an exoskeleton, which they moult as they grow, segmented bodies and jointed limbs, which greatly increases their ability to move. They include the insects, spiders and crustaceans.

PHYLUM MOLLUSCA

This is the second largest phylum in the animal kingdom and contains the snails, octopi and mussels.

PHYLUM ECHINODERMATA

The echinoderms include starfish and sea urchins. Their bodies are usually divided into five sections.

PHYLUM CHORDATA

Animals in phylum Chordata all have a string of nerve tissue along their backs.The most diverse group of chordates are the **vertebrates**. These animals have a bony spinal column that protects the nerves within, as well as a ventral heart and pairs of limbs. In humans, the spinal cord (nervous tissue) is protected by bones called vertebrae (which make up the spine). Not all chordates have vertebrae. Chordata include fish, amphibians (e.g. frogs and newts), reptiles, birds and mammals.

Vertebrates are animals with a bony spinal column, or vertebrae.

Go to YouTube.com and search for 'Animals: Tour of 9 Phyla' (12:20) by Amoeba Sisters.

By classifying animals and building phylogenetic (evolutionary) trees, we can more fully understand their origin and evolution.

ORIGIN AND EVOLUTION OF ANIMALS

Scientists believe animals first evolved around 600 million years ago, although many of these don't feature in the fossil record probably due to having soft bodies.

Animals first evolved in the seas. Oxygen levels in the atmosphere at that time were much lower than they are now and ultraviolet radiation levels were much higher. The evolution of animals was quite slow, until around 500 million years ago, in a period known as the **Cambrian explosion**, when the diversity of animal life on Earth began to increase exponentially. The cause of this explosion in diversity is still heavily debated, but some suggest it was due to increasing oxygen levels in the atmosphere, while others suggest it was down to the emergence of specific genes.

Animals may have first stepped onto land around 500 million years ago, but only temporarily – it wasn't until much later that animals began to colonise and live permanently on the land. Arthropods were the first to live permanently on land, probably around 410 million years ago. They were closely followed by **tetrapods**: four-legged animals from which amphibians, reptiles, dinosaurs, birds and mammals all evolved.

Tetrapods are four-limbed animals.

Figure 5.1: The fossilised footprints of a tetrapod on Valentia Island, Co. Kerry.

Insects evolved around 400 million years ago, reptiles and dinosaurs around 300 million years ago, mammals around 200 million years ago, birds 150 million years ago, bees 100 million year ago, primates (monkeys and apes) 60 million years ago and the early human species emerged around 1 million years ago. Modern humans appeared just 50,000 years ago.

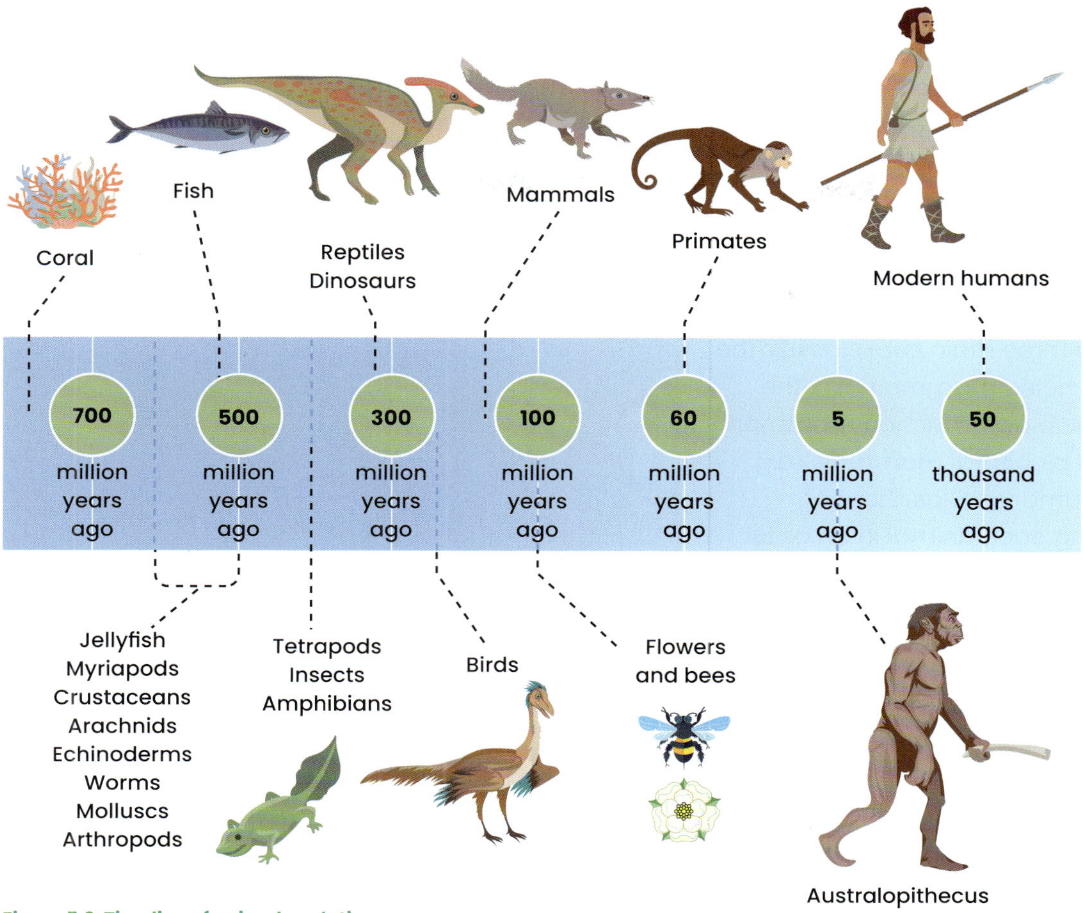

Figure 5.2: Timeline of animal evolution.

In the context of the history of life on Earth, animals are a relative newcomer and humans especially so, but their importance to ecosystems and the evolution of other species has been critical, e.g. flowering plants wouldn't have evolved without the evolution of pollinators, and vice versa.

EVOLUTION OF ANIMALS AND PHYLOGENY

A simple phylogenetic tree of the animal kingdom is shown in Figure 5.3. We can see that all animal life has a common ancestor. There are three major clades, branches of the tree, based on body forms, and the nine phyla we studied can be found on these.

The Porifera and Cnidaria were some of the first animals to evolve and are more distantly related to the animals in the other phyla. All the other phyla have a common ancestor at point 1 on the tree. Clades 2 and 3 are separated based on differences in how their embryos form, while those in Clade 4 (Arthopods and Nematodes) all have have an outer protective layer called an **exoskeleton**.

An **exoskeleton** is a strong supportive layer outside of the body of an animal.

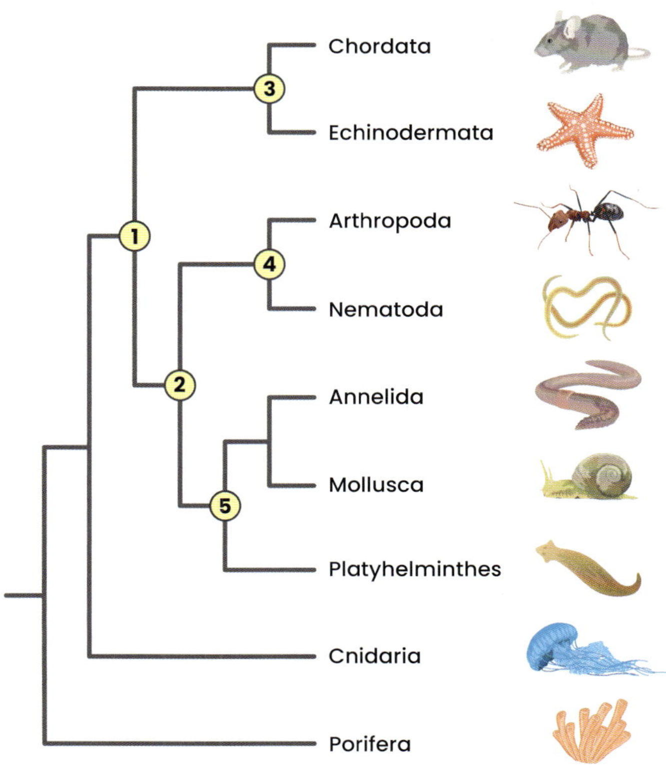

Figure 5.3: The animal kingdom phylogenetic tree.

FURTHER CLASSIFICATION

As outlined in Chapter 4, each phylum can be divided into classes and classes into orders. For example, within the phylum Chordata we find class Mammalia – the **mammals**. Mammals are backboned animals that can all produce milk, have hair and can regulate their body temperature. Mammals are very diverse so they can be divided into three main sub-classes: the monotremes (one-holed mammals, e.g. the platypus), the marsupials (pouched mammals, e.g. kangaroo) and placental mammals (e.g. whales, dolphins, humans). Classification continues by placing each animal in an order, then family and finally assigning it a genus and species name.

Mammals are animals that can produce milk, have hair and regulate their body temperatures.

Figure 5.4: The platypus is a mammal that can be further classified as a monotreme.

Example: Classifying the red fox

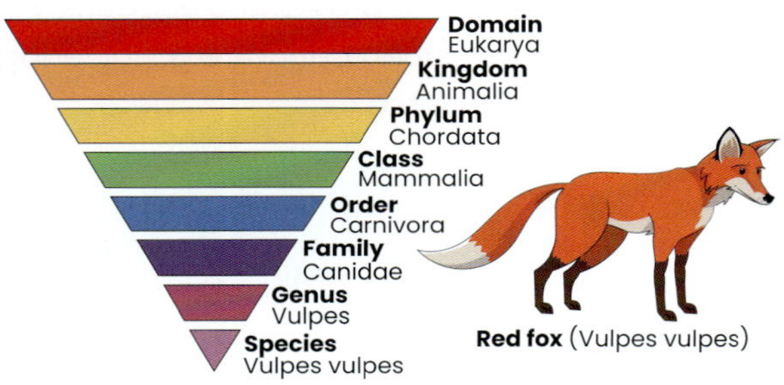

Figure 5.5: Classification of the red fox, *Vulpes vulpes*.

We can use the red fox as an example of animal classification. It's an animal, so it sits in the domain Eukaryota and kingdom Animalia. It has a backbone, so it is placed in phylum Chordata. Based on its prominent canine teeth, the fox is placed within order Carnivora, and within the dog family (Canidae). Finally, it is given the genus name *Vulpes* and species name *vulpes*, making the name of the red fox *Vulpes vulpes*.

There are other species of fox with the same genus name, e.g. the Arctic fox, *Vulpes lagopus* and the fennec fox, *Vulpes zerda*.

Modern humans, *Homo sapiens*, are of course animals too, and most of what we will study in Leaving Certificate Biology will relate to humans. In later chapters, we will study animal cells and tissues in greater detail and learn how the advanced organs and organ systems of humans work to ensure we stay alive. We will also consider the role of other animals in ecological systems and explore how their relationships (with each other, different species of animals, plants and humans) can affect ecosystems. We will also explore how animals can cause disease in humans and how our immune system protects us from invaders.

PORTFOLIO LINK

Go to your Learning Portfolio and complete Learning Check 2 to assess your knowledge and understanding of the topic so far.

Chapter Summary

- There are likely 8 million species of animal or **metazoans**, most insects, all with the following common features:
  - Animals are **multicellular**.
  - Animals are **heterotrophic** – they consume other organisms for food.
  - They respire **aerobically**.
  - All animals are **motile**, at least at some stage in their life cycle.
  - Animals have high levels of **cell diversity**, including muscle and nerve cells.
  - They mostly reproduce **sexually**.
- Animals are classified based on their body plans but also using DNA analysis.
- The animal kingdom is broken down into phyla, the most important of which include:
  - Porifera – sponges
  - Cnidaria – jellyfish and corals
  - Platyhelminthes – flatworms
  - Nematoda – round worms
  - Annelida – segmented worms
  - Arthropoda – insects, spiders and crustaceans
  - Mollusca – snails, octopi and mussels
  - Echinodermata – starfish and sea urchins
  - Chordata – **vertebrates**, e.g. fish, amphibians, reptiles, birds and mammals.
- Animals evolved around 600 million years ago, in the sea, but most phyla emerged during the Cambrian explosion around 500 million years ago.
- Animal phyla are broken into classes, orders and families before every animal receives a genus and species name.

END OF CHAPTER ASSESSMENT

REMEMBER

1. Define the term *animal*.
2. What is zoology?
3. Define the term *multicellular*.
4. What term is used to describe an organism that obtains its food from consuming other organisms?
5. What is a tetrapod?
6. Name the phylum to which sponges belong.
7. What is a vertebrate? Name **four** examples of vertebrates.

UNDERSTAND

8. What is meant by the term *sexual reproduction*?
9. What is meant by the term *aerobic*?
10. What is meant by the term *motile* and why is it an important term for animal classification?
11. Name **four** different species of tetrapod.

APPLY

12. *More advanced animals have more diverse tissues and organs.*
    Discuss the validity of this statement.
13. Outline the functions of (a) muscle cells and (b) nerve cells in an animal.
14. Describe the features of phylum Chordata and suggest why they're the most advanced animals on the planet.
15. Using your knowledge of the main animal phyla, classify the following animals in as much detail as you can: tarantula, polar bear, slug.
16. Which of the following animals do you think are the closest relatives? Justify your answer.

    beetle mouse crab snail jellyfish

ANALYSE

17. Animals have high levels of cell diversity. What do you think are the advantages of having diverse types of cells?
18. The arthropods are the largest group of animals. Suggest a reason why they have been so successful.
19. Only 4% of the total biomass (combined mass of bodies) of mammals on Earth are wild. The remaining 96% contains humans (34%) and domesticated animals (cattle, sheep, etc. – 62%). Comment on these statistics. (Source: www.ourworldindata.org)
20. Suggest a reason why cnidarians have stinging cells in their bodies.

EVALUATE

21. An amoeba is a single-celled organism that is heterotrophic. It was once considered a primitive animal but is now classified as a protist. Suggest reasons why the amoeba is no longer classified as an animal.

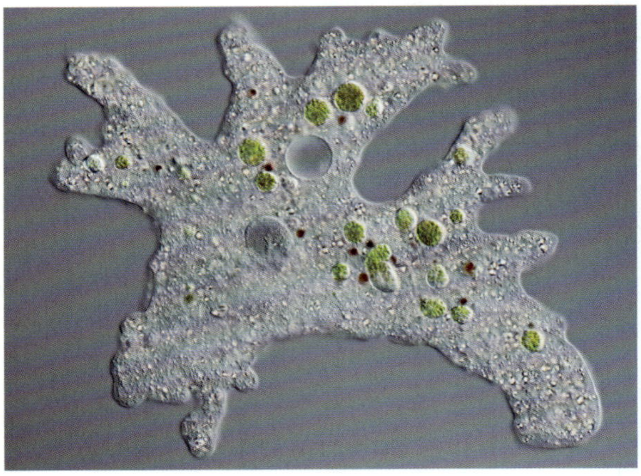

22. Below is a phylogenetic tree showing the evolutionary relationship between different groups of vertebrates.

(a) What do all these groups have in common, apart from being vertebrates?

(b) The coelacanth, thought once extinct, are a type of fish with strong, leg-like appendages. What evidence from the tree suggests the coelacanth may be a distant relative to all tetrapods?

(c) *Lizards are more closely related to mammals than to frogs.* Comment on this statement, making reference to the tree.

(d) Study the tree and outline **three** items that you find notable or interesting.

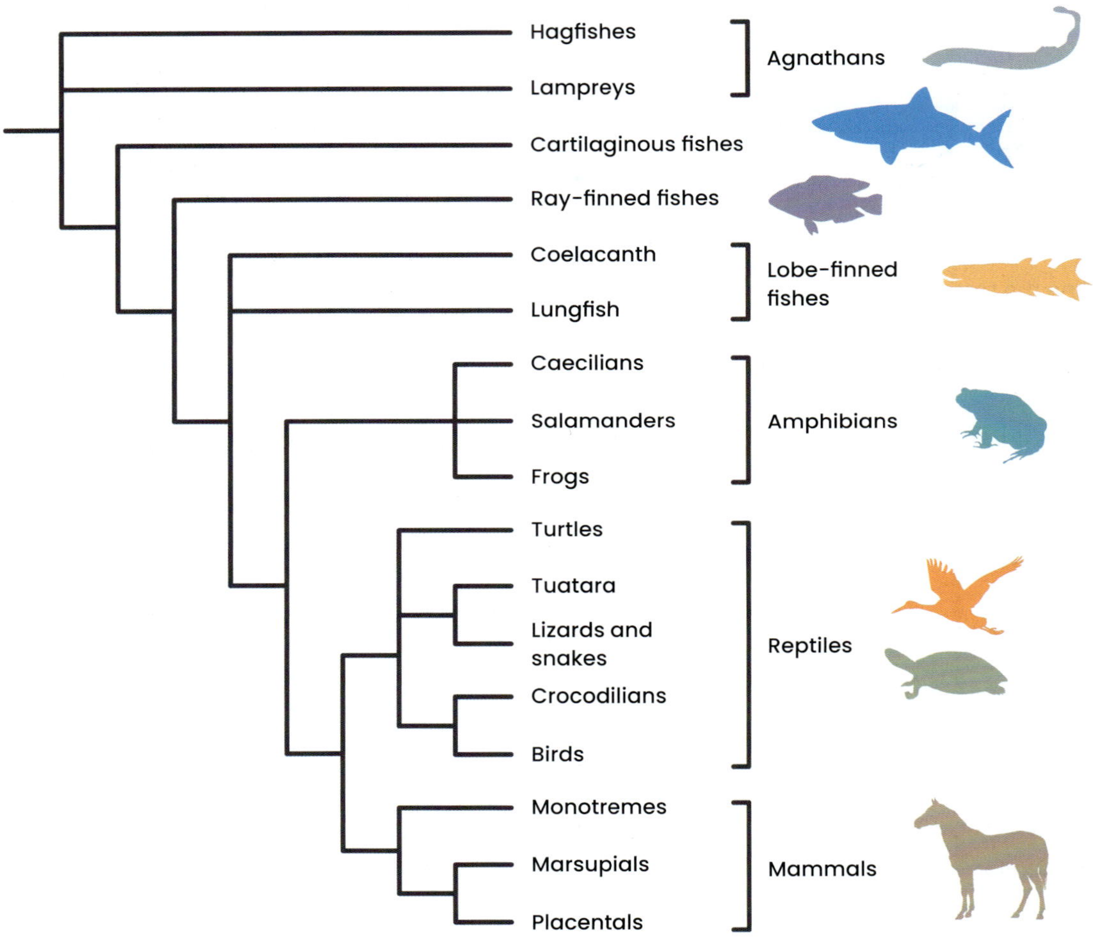

CREATE

23. Create a timeline diagram of the evolution of animal life, from 600 million years ago to the arrival of modern humans.

6 The Plant Kingdom

LEARNING OUTCOMES: U4.1, U4.2, 1.1.4, 1.3.1

EYE-OPENER

The Methuselah Tree is a bristlecone pine tree in the White Mountains in California. What makes it special is it's the oldest living thing on the planet – it is nearly 5,000 years old. Named after the biblical figure of the same name, who was said to have lived to 969 years of age, the Methuselah Tree's exact location is a carefully guarded secret. The tree was well-established by the time the Egyptians built the pyramids.

At the end of this chapter, you should be able to:

- Define the common features of all plants.
- Describe how land plants are broadly classified.
- Describe the features of the four main types of plants.
- Discuss the classification of algae.
- Outline the origin and evolution of plants.
- Classify flowering plants into their plant family, based on their physical features.

LINKS TO:

Chapter 2	**Scientific Knowledge and Research**
Chapter 4	**Classification and Taxonomy**
Chapter 5	**The Animal Kingdom**
Chapter 9	**Evolution and Natural Selection**
Chapter 10	**Biodiversity and Ecosystems**
Chapter 16	**Cell Structure and Diversity**

SUBJECT VOCABULARY

plants | botany | autotrophic | chloroplast | differentiation | vascular tissue | bryophytes | pteridophytes | gymnosperms | angiosperms | co-evolution

Section 1: The Plant Kingdom

At the end of this section, you should be able to:

- Describe the features all plants have in common.
- Distinguish between vascular and non-vascular plants.
- Briefly describe the features used to classify plants.

Unpacked from LOs U4.1, U4.2, 1.1.4, 1.3.1

WHAT IS A PLANT?

Plants are a diverse group of organisms found all over the world, in practically every habitat. There are close to half a million plant species and each of them have the following characteristics in common:

- All plants are **multicellular** and **eukaryotic**, their structures containing cells with a prominent **nucleus** and **cell walls** made of **cellulose**, a complex carbohydrate (sugar).
- Plants are **autotrophic**, meaning they can make their own food through the process of **photosynthesis**.
- Plant cells contain **chloroplasts**, specialised organelles for photosynthesis, which contain the pigment **chlorophyll**.
- Plants can reproduce both sexually and asexually and generally respire aerobically.

The study of plants is called **botany**.

- **Plants** are multicellular, eukaryotic and autotrophic organisms. Their cells contain cell walls and chloroplasts.
- **Botany** is the study of plants.
- **Autotrophic** organisms can make their own food, e.g. photosynthesis.
- **Chloroplasts** are cell organelles that allow plants to make food – they contain chlorophyll.

Figure 6.1: Plant diversity.

Go to YouTube.com and search for 'What is Botany?' (13:40) by CrashCourse.

FOCUS ON ... HOW PLANTS FEED

Figure 6.2: Sundew.

The vast majority of plant species are autotrophic, making their food by photosynthesis. In some instances, though, plants may need additional nutrients and can become partial heterotrophs. One of the best known examples is the Venus flytrap, which lives in nutrient-poor soil and catches flies to supplement its diet. It can be described as insectivorous as it 'eats' insects. Photosynthesis is still the main source of food for insectivorous plants, but essential nutrients like nitrogen and phosphorous are provided by digesting the insects. There are insectivorous plants in Ireland too, including the beautiful sundew plant, which catches small insects on the droplets at the end of its flowers.

PLANT DIVERSITY

There are an estimated 500,000 species of plants. Despite this diversity, plants can be placed into groups according to their level of differentiation, whether they have **vascular** or transport tissues, if they produce **seeds** and whether they produce **fruits**.

Differentiation

Differentiation refers to the variety in the kinds of cells within a plant's structures. Plants with little or no differentiation can lack roots, stems and leaves, while more advanced plants, with lots of different cell types, have these parts.

Vascular tissue

In plants, **vascular tissues** include **xylem** (for transporting water) and **phloem** (for transporting food). Whether or not a plant contains vascular tissue can be used to classify it. Mosses have no vascular tissue, while ferns do.

- **Differentiation** refers to the variety of cell types in an organism.
- **Vascular tissue** refers to cells that transport materials, e.g. xylem transports water in plants.

Seeds

One might assume that all plants make seeds but that is not the case. More primitive plants produce **spores**, like fungi, or may even reproduce asexually. Mosses and ferns both reproduce using spores while grasses produce seeds.

Flowers and fruits

Some plants produce flowers as reproductive organs and, later, fruits which surround their seeds protecting them and aiding their dispersal. Again, we can use this to help classify plants.

In the simple phylogenetic tree shown in Figure 6.3, we can see how we can create divisions in the plant kingdom using the above features. By analysing plant DNA, we can get a greater indication of how species are related.

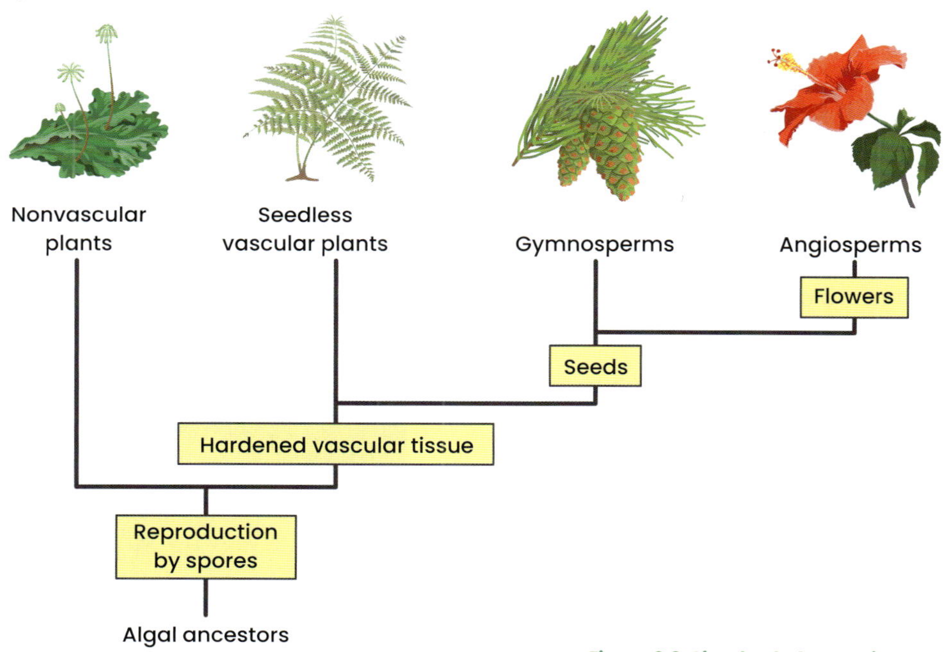

Figure 6.3: Simple phylogenetic tree of the plant kingdom.

PORTFOLIO LINK **Go to your Learning Portfolio and complete Learning Check 1 to assess your knowledge and understanding of the topic so far.**

Section 2: Classification of Plants

At the end of this section, you should be able to:

- Name the main divisions within kingdom Plantae and briefly describe their features.
- Briefly outline the origin and evolution of plants.
- Describe the diversity amongst flowering plants.

Unpacked from LOs U4.1, U4.2, 1.1.4, 1.3.1

Plant classification is rather complex and different biologists classify plants in different ways; there is still debate on whether algae should be included in the plant kingdom (more on this later). Most agree that there are four main divisions: bryophyta (mosses), pteridophyta (ferns), gymnosperms (conifers) and angiosperms (flowering plants).

Figure 6.4: The four main divisions of land plants.

Some scientists would include 10 other divisions (including Chlorophyta – green algae), but the four groups above are the most significant and certainly the most common. In fact, it is estimated that around 94% of all plant species are angiosperms (flowering plants).

FEATURES OF THE MAIN PLANT DIVISIONS

Bryophytes

- **Bryophytes** (mosses) are **non-vascular**, meaning they don't have the **xylem** or **phloem** that more advanced plants use to transport water and food around their structures.
- For this reason, mosses need to live in areas with high levels of water.
- They also lack roots, instead anchoring themselves using rhizoids – these root-like threads can't absorb water or nutrients like roots can.
- Like fungi, mosses reproduce using **spores**.
- Mosses are important for the formation of soils.

Bryophytes are mosses: non-vascular green plants that reproduce using spores.

Figure 6.5: An example of a bryophyte.

Figure 6.6: An example of a pteridophyte..

Figure 6.7: An example of a gymnosperm.

Figure 6.8: An example of an angiosperm.

Pteridophytes

- **Pteridophytes** (ferns) do have a vascular system, transporting water from the roots via xylem and food from the leaves via phloem.
- They have stems, leaves and roots. Their roots can absorb water and minerals.
- Like mosses and fungi, ferns still reproduce using spores.

Gymnosperms

- **Gymnosperms**, along with angiosperms, are seed-producing plants.
- Gymnosperm seeds are found in cone-like structures – hence they are often called conifers.
- Their seeds are **not** surrounded by a fruit.
- Typically, conifers have needle-like leaves that don't fall off in winter, but there are some exceptions.

Angiosperms

- **Angiosperms** are plants that produce flowers and there are nearly 370,000 species identified.
- From their flowers, they produce seeds that are surrounded by a fruit, which aids seed dispersal and protects the young plant within.
- Angiosperms include all broadleaf trees, grasses, roses, daisies, orchids and many more.
- Flowers are the reproductive part of the plant, designed to distribute pollen from one plant to another.
- Some flowers have colourful petals, nectar and fragrant scents to attract insects and other pollinators to help spread their pollen.
- Plants that use the wind to spread their pollen lack colourful petals, nectar and fragrant scents.

Pteridophytes (pronounced *ter-id-o-fights*) are ferns: vascular green plants that reproduce using spores.

Gymnosperms are conifers: vascular green plants that produce seeds in cone-like structures.

Angiosperms are green plants that produce flowers to aid their reproduction.

Go to YouTube.com and search for 'Plant Classification | Evolution | Biology | FuseSchool' (3:13) by FuseSchool – Global Education.

FOCUS ON ... SPORES VS SEEDS

SPORES	SEEDS
 Figure 6.9: Spores on the underside of a fern.	 Figure 6.10: Barley seeds.
• Contain a single cell	• Contain lots of cells
• Haploid (only one set of chromosomes)	• Diploid (the embryo contains two sets of chromosomes in its cells)
• No food reserve	• A food reserve for the embryo until it can photosynthesise

What about algae?

Figure 6.11: Ulva or sea lettuce is a green alga.

Whether algae – seaweed – are plants is still a hotly debated topic in taxonomy. Some scientists place algae in the plant kingdom, while most consider them protists, despite having cells (with cell walls and chloroplasts) like all other plants. For others, it's more complicated because algae are a very diverse group; there are microscopic single-celled algae and large multicellular species, too. Some are green, while others are red or brown. For simplicity, we will place them in kingdom Protista, but it is important to know that plants and green algae share a common ancestor, and some green algae can be classified within the plant kingdom too.

By classifying plants and building phylogenetic (evolutionary) trees, we can more fully understand their origin and evolution.

ORIGIN AND EVOLUTION OF PLANTS

Land plants (mosses, ferns, gymnosperms and angiosperms) evolved from green algae around 500 million years ago. If you recall, this was around the same time animal life was leaving the sea for the first time and making their way onto land.

The first land plants were mosses, which dominated the landscape at the time. Vascular plants evolved from these around 400 million years ago before the first seeded plants, gymnosperms, appeared around 320 million years ago. Flowering plants first appeared around 150 million years ago, around the same time as the bees and other insects. This is an example of **co-evolution**, where two groups of organisms evolve at around the same time and become **interdependent**, relying on each other to feed and reproduce.

Co-evolution is evolution of two or more species, creating an interdependent relationship.

Interdependence occurs when two species rely on each other for survival.

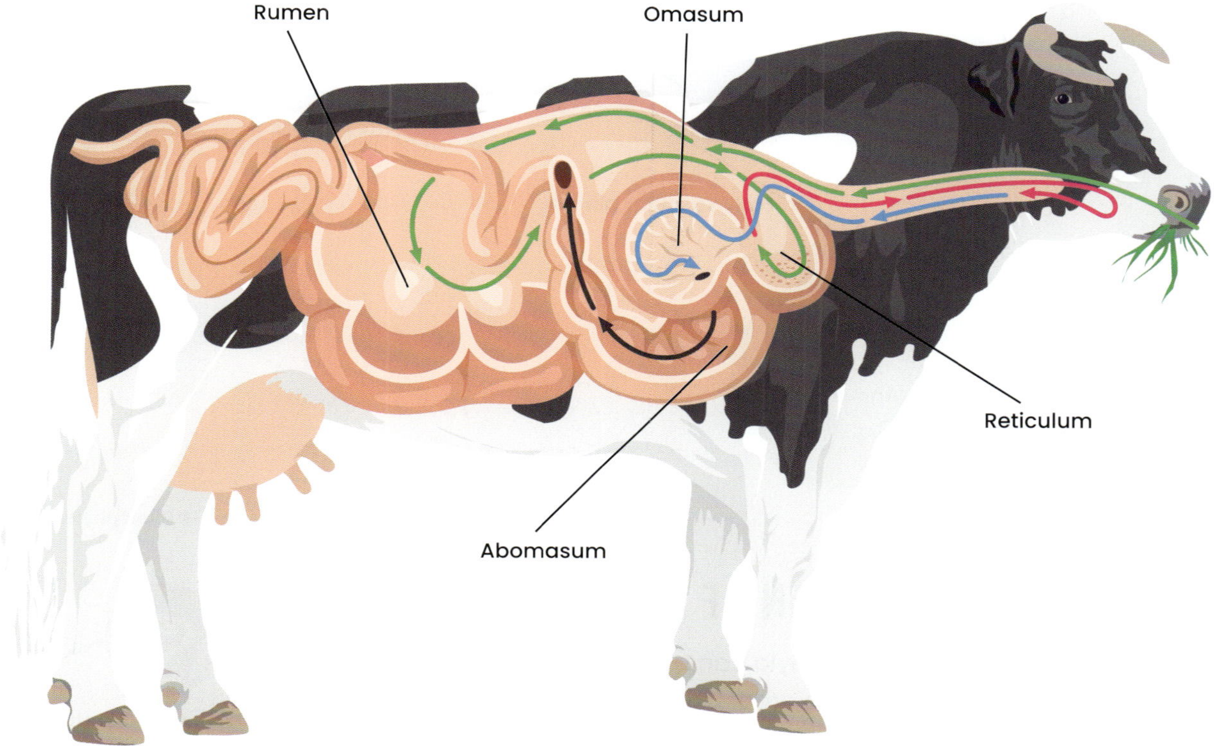

Figure 6.12: Ruminants contain a four-chambered stomach.

Grass species are thought to have evolved around 60 million years ago – after the age of the dinosaurs. This coincides with the evolution of a group of animals known as ruminants. Ruminants include modern horses, cattle and sheep and have a four-chambered stomach to help digest grass. Again, this is often used as an example of co-evolution. We will explore evolution in more detail in Chapter 9.

RESEARCH ACTIVITY

Carry out some research on ruminants. What common domestic and wild species of animals are ruminants? How do their diets differ from humans? How are they able to digest the cellulose in the plant cells?

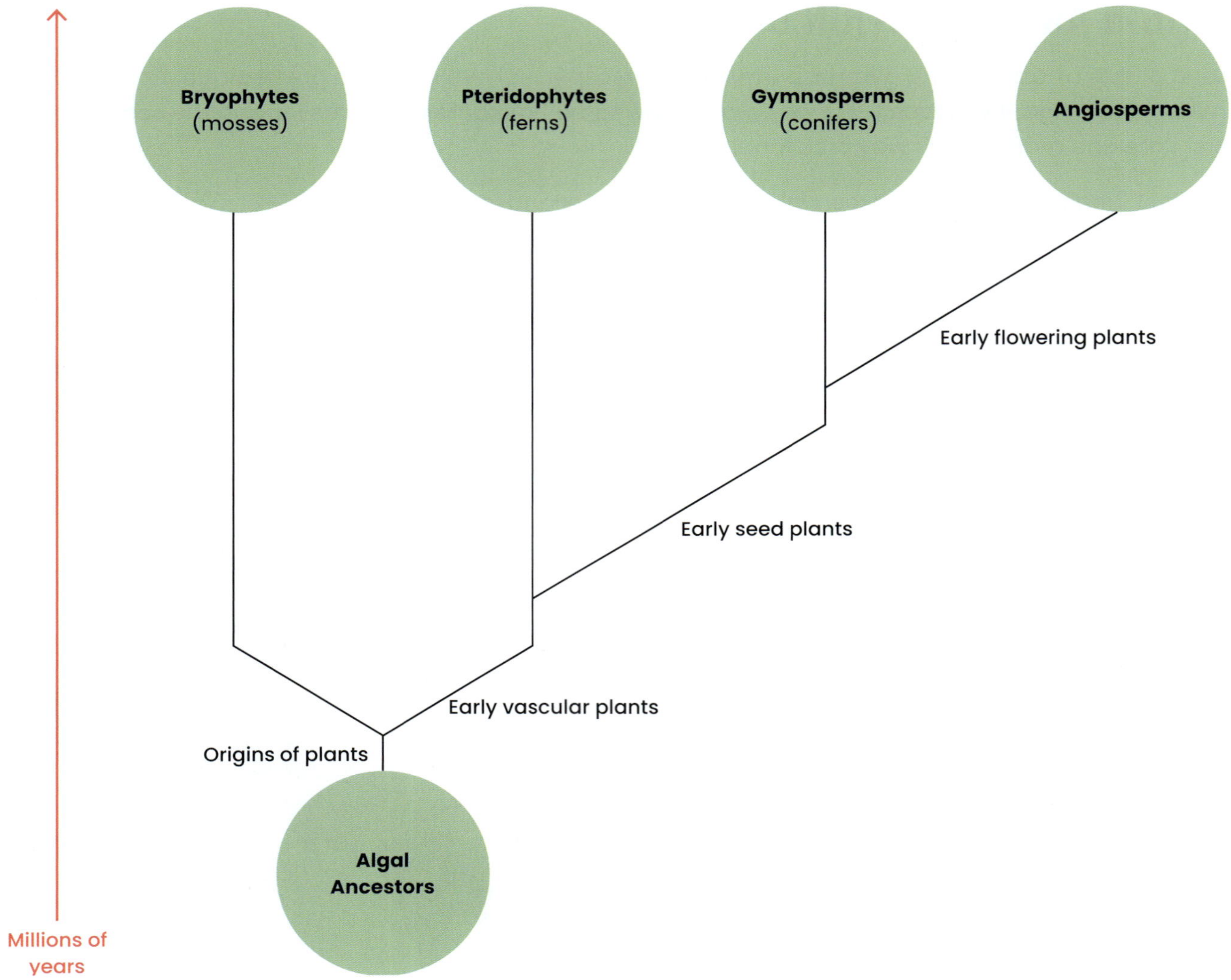

Figure 6.13: Plant phylogenetic tree.

FURTHER CLASSIFICATION OF ANGIOSPERMS

As we saw in the animal kingdom, phyla can be divided into small groups called classes, and then into orders, which in turn can be divided into families.

There are four orders within the angiosperms, but we can simplify them into two main types – the **monocots** and **dicots**. (We will learn more about these in Chapter 19.) The orders divide into over 400 different plant families, some of which you'll be familiar with. The most important plant families include roses, orchids, legumes (peas), brassicas (cabbages) and grasses.

In later chapters, we will explore the structure of flowering plants in greater detail. We will also look at how they transport materials and produce new plants.

THINK, PAIR, SHARE

Walk around your school grounds, local park or woodland area and take photos of the plant species you find (depending on the time of year, many plants may not be in flower). Create a collage (print or digital) of the plants you found. Classify them as bryophytes, pteridophytes, gymnosperms and angiosperms.

PORTFOLIO LINK

Go to your Learning Portfolio and complete Learning Check 2 to assess your knowledge and understanding of the topic so far.

Chapter Summary

- There are nearly 500,000 species of plants but all have the following in common:
  - Multicellular
  - **Autotrophic** – they can photosynthesise
  - Contain **chloroplasts** with the pigment **chlorophyll** for photosynthesis.
- Plants are classified based on:
  - Similarities in their structures
  - The level of cell differentiation
  - Whether they have vascular tissue
  - Whether they produce seeds
  - Whether they have flowers and fruits.
- All plants share a common ancestor, a type of green alga.
- Plants are classified into four main groups: bryophytes (mosses), pteridophytes (ferns), gymnosperms (conifers) and angiosperms (flowering plants).
  - Mosses are non-vascular plants that reproduce using spores.
  - Ferns are vascular plants that reproduce using spores.
  - Gymnosperms produce seeds that form cone-like structures.
  - Angiosperms produce flowers and seeds surrounded by a fruit.
- Plants evolved from their algal ancestors around 500 million years ago.
- Many plant and animal species evolved at the same time, forming interdependent relationships – this is called co-evolution.
- Angiosperms are divided further into classes, orders and then families.

END OF CHAPTER ASSESSMENT

REMEMBER

1. Name the process by which plants make their food.
2. Name the part of the plant cell responsible for this process.
3. What kind of plant tissue are xylem and phloem?
4. Name **two** types of plants that don't make seeds.
5. Name the two main groups that all flowering plants are divided into.

UNDERSTAND

6. Describe the common features of all plants.
7. From which type of organism did plants evolve from?
8. Describe the features of flowering plants (angiosperms).
9. Suggest a reason why gymnosperms got their name.
10. Give **two** examples of where plants and animals have an interdependent relationship.

APPLY

11. How is the level of cell differentiation used in the classification of plants?

12. Discuss why plant classification is important to scientists.

13. Why are algae not considered to be in the plant kingdom, despite sharing many of the characteristics with plants?

14. What features in Bryophytes suggest early plants shared a common ancestor with fungi?

ANALYSE

15. A plant species has the following characteristics: it reproduces using spores, it does not have xylem, it does not have flowers. What type of plant is it?

16. Suggest a reason why flowering plants and bees evolved at around the same time.

17. What features of flowers do you think help them attract insects?

18. A group of students conducted an experiment to investigate the effect of light intensity on the rate of photosynthesis in a common houseplant. They measured the rate of oxygen production (ml/hour) at different light intensities (lux). The data they collected is presented in the table below:

LIGHT INTENSITY (lux)	OXYGEN PRODUCTION (ml/hour)
0	0.2
1000	1.5
2000	2.8
3000	3.9
4000	4.7
5000	5.2
6000	5.4

Using this data, answer the following questions.

(a) Create a line graph to visualise the relationship between light intensity and oxygen production.

(a) Describe the general trend you observe in the graph. How does oxygen production change as light intensity increases?

(c) Calculate the change in oxygen production between 0 and 3000 lux. What does this tell you about the plant's response to light in this range?

(d) At what light intensity does the rate of oxygen production appear to slow down or level off? What might this suggest about the plant's photosynthetic capacity?

(e) If you were to predict the oxygen production at 7000 lux, would you expect it to be much higher than at 6000 lux? Explain your reasoning based on the data trend.

EVALUATE

19. Compare and contrast mosses and ferns.

20. Outline the reasons why fungi are not considered plants.

21. Study the labelled diagram of a single-celled organism called euglena, and answer the questions that follow.

(a) Suggest a reason why euglena could be considered a plant.

(b) Give **three** reasons why euglena should not be considered a plant.

(c) Is there anything in the diagram to suggest euglena has animal-like properties?

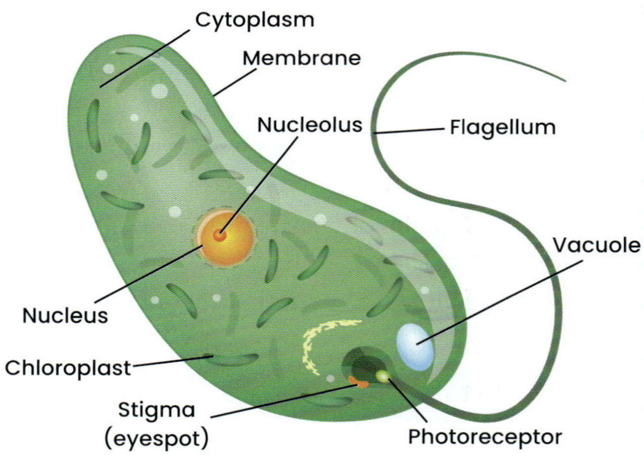

22. *Seeds have many advantages over spores.* Discuss the validity of this statement.

CREATE

23. Create a timeline diagram of the evolution of plants.

24. Design a mind map to summarise the characteristics and classification of plants.

7 Bacteria and Fungi

LEARNING OUTCOMES: U2.3, U2.5, 3.2.1, 3.2.2, 3.2.3, 3.2.4

At the end of this chapter, you should be able to:

- Distinguish between bacteria and fungi in terms of structure, nutrition and cellular nature.
- Investigate factors affecting microbial growth, use primary and secondary data to support conclusions.
- Outline factors influencing growth of micro-organisms.
- Draw and explain micro-organism growth curves. HL
- Discuss the importance of micro-organisms in industries.
- Outline the concept of the microbiome and its role in human health and nutrient cycling in soils.

EYE-OPENER

It is estimated that 1 gram of soil can contain up to 10 billion micro-organisms!

LINKS TO:

Chapter 4	**Classification and Taxonomy**
Chapter 9	**Evolution and Natural Selection**
Chapter 12	**Nutrient Recycling**
Chapter 16	**Cell Structure and Diversity**
Chapter 18	**Movement through Cell Membranes**
Chapter 26	**Genetic Technology and Genomics**
Chapter 30	**The Digestive System**
Chapter 32	**Infectious Disease and the Human Immune Response**

SUBJECT VOCABULARY

autotrophic, heterotrophic, photosynthetic, chemosynthetic, saprophytic, parasitic, pathogen, antibiotic, aseptic, sterile

Section 1: Bacteria

At the end of this section, you should be able to:

- Draw a labelled diagram of a bacterial cell.
- State the function of each of the parts of a bacterial cell.
- Describe the autotrophic, heterotrophic, saprophytic and parasitic forms of nutrition in bacteria.
- State the prokaryote cellular structure found in bacteria.

Unpacked from LO 3.2.1

Micro-organisms such as bacteria and fungi are found everywhere in the Earth's biosphere – air, soil and water. Bacteria and fungi play an important role in nature as decomposers, breaking down dead organic matter and recycling nutrients such as carbon and nitrogen (see Chapter 12).

MICROBIAL STRUCTURE

Structure and cellular nature of bacterial cells

Bacteria are unicellular micro-organisms that may be viewed under a microscope. Bacteria only measure 1–5μm in length. Bacterial cells are classified as **prokaryotic**, as they do not have a nucleus, or any other membrane-bound organelles such as mitochondria or chloroplasts (see Chapter 16).

A **prokaryotic** cell does not have a nucleus or other membrane-bound organelles.

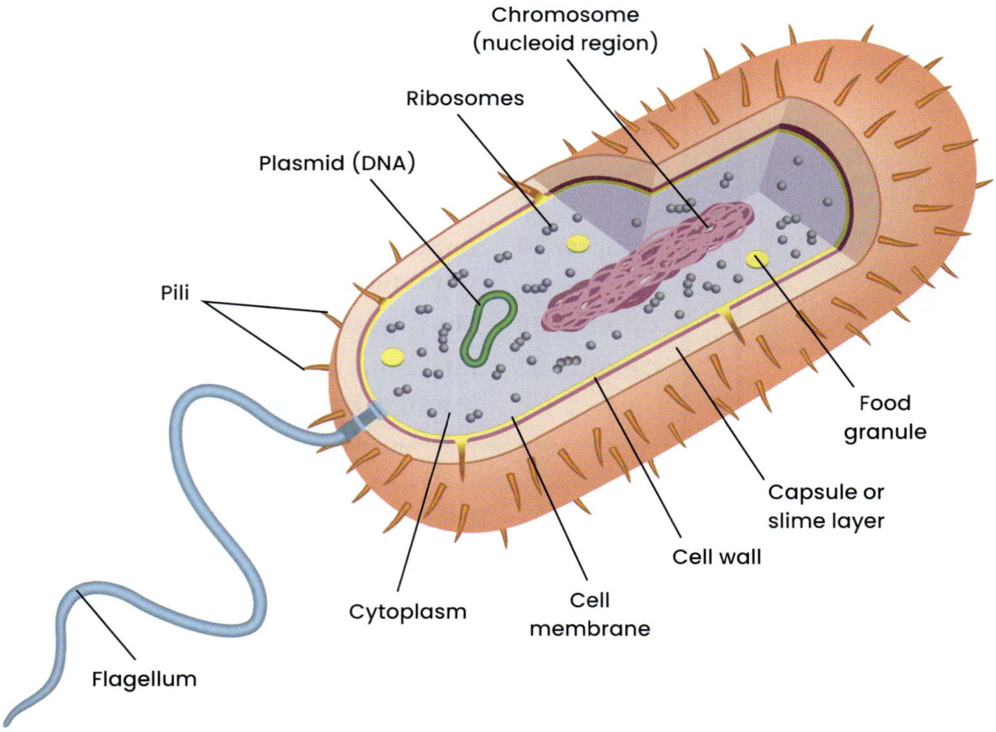

Figure 7.1: Structure of a bacterial cell.

A micrometre (μm) is 1×10^{-6}m, or one-millionth of a metre.

DID YOU KNOW?

STRUCTURE	FUNCTION	ALL BACTERIA
Chromosomal DNA	Bacterial DNA is loose in the cytoplasm and does not have any associated proteins.	✓
Plasmid DNA	A small, circular piece of DNA which can move from one bacterial cell to another. Often has genes for antibiotic resistance.	✗
Cell (plasma) membrane	Selectively permeable cell membrane that controls the passage of substances in and out of the cell. Composed of phospholipids and proteins.	✓
Cell wall	Gives shape and strength to the bacterial cell. Made of peptidoglycan.	✓
Capsule	Slimy layer outside the cell wall that helps prevent the cell from drying out and protects it from phagocytosis (being engulfed) by white blood cells.	✗
Flagellum	Hair-like flagella help the bacterial cell to move.	✗
Pili	Small hair-like pili (singular: pilus) help the bacteria to attach to other cells and surfaces.	✗
Cytoplasm	Liquid filling the cell in which ribosomes, food granules and chromosomes are suspended.	✓

Figure 7.2: The parts of a bacterial cell.

THINK, PAIR, SHARE

How are bacterial cells adapted to their function?

Bacterial shapes

Bacteria are commonly found in three different shapes: spherical (coccus), rod-shaped (bacillus) and spiral (spirillum).

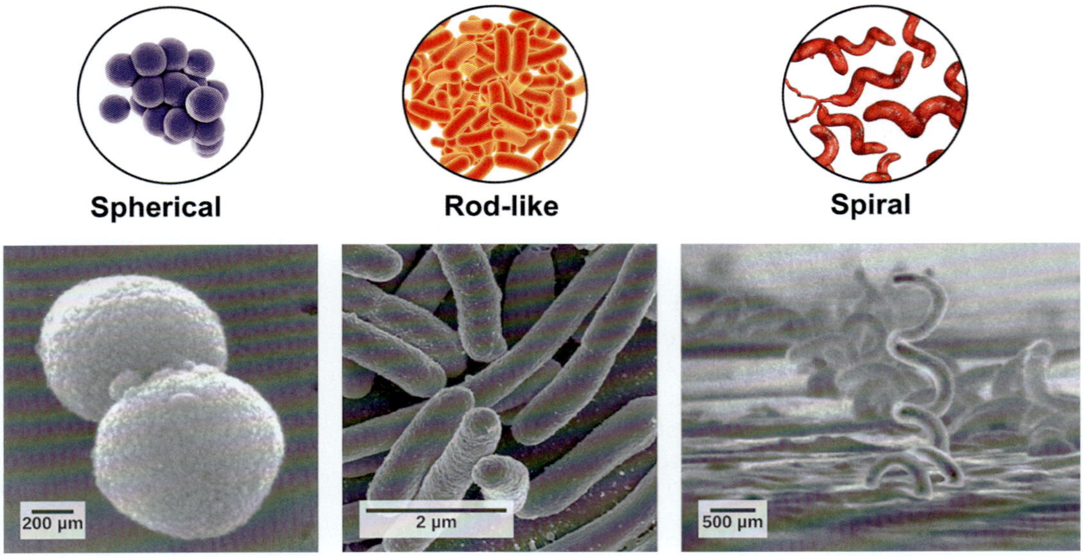

Figure 7.3: Bacterial shapes.

Microbial nutrition

Microbes can be classified as autotrophs or heterotrophs depending on how they obtain their nutrition. Autotrophs can make their own food, whereas heterotrophs must take in food from other organisms.

Bacterial nutrition

Most bacteria are heterotrophic, although some bacteria are autotrophic. **Autotrophic bacteria** make their own food using energy from the Sun (photosynthetic bacteria) or from chemical reactions (chemosynthetic bacteria). **Heterotrophic bacteria** cannot make their own food and must take in food from other organisms. Heterotrophic bacteria include both saprophytic and parasitic bacteria.

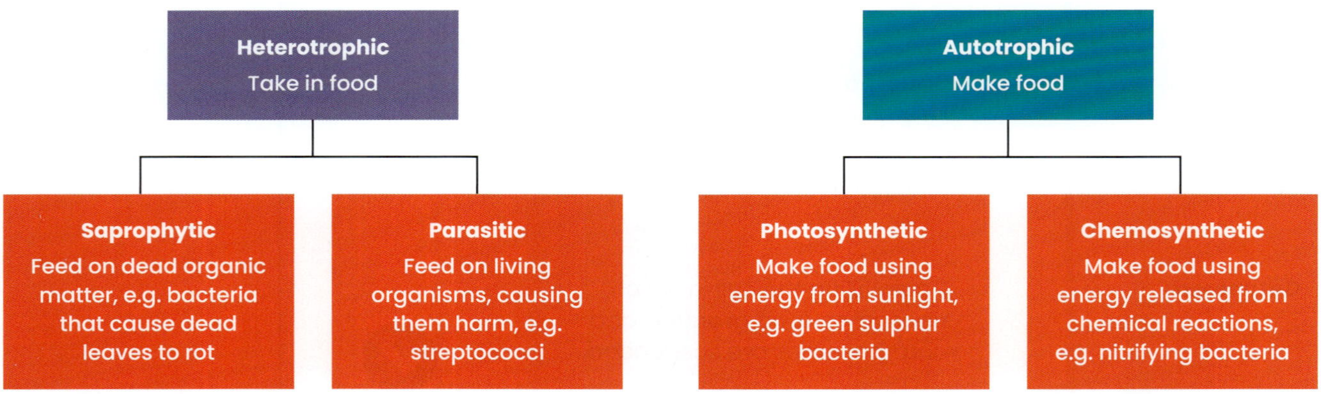

Figure 7.4: Bacterial nutrition.

PORTFOLIO LINK

Go to your Learning Portfolio and complete Learning Check 1 to assess your knowledge and understanding of the topic so far.

Section 2: Fungi

At the end of this section, you should be able to:

- Draw a labelled diagram of a multicellular fungus – Rhizopus.
- State the function of each of the parts of the Rhizopus structure.
- Describe how fungi undergo heterotrophic nutrition.
- State the eukaryote cellular structure found in fungi.

Unpacked from LO 3.2.1

STRUCTURE AND CELLULAR NATURE OF FUNGI

Most fungi are multicellular in nature but some, e.g. yeast, are unicellular. Fungal cells are larger than bacterial cells, measuring 10–100μm in length. Fungal cell walls are composed of the polysaccharide chitin. Rhizopus (bread mould) is an example of a multicellular fungus. Unlike bacteria, fungi are classified as **eukaryotic** as their cells contain a membrane-bound nucleus and organelles.

A **eukaryotic** cell has a nucleus and other membrane-bound organelles.

Multicellular fungi are composed of filamentous structures called hyphae. The functions of each of the parts of a multicellular fungus are outlined in Figure 7.5 opposite.

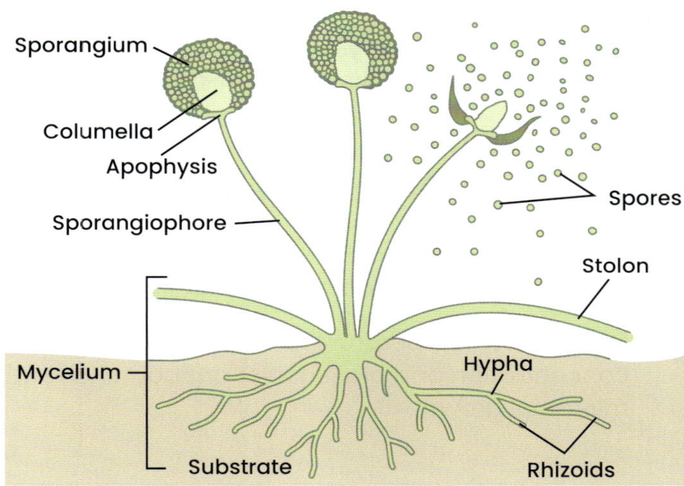

Figure 7.5: Rhizopus – a multicellular fungus.

STRUCTURE	FUNCTION
Mycelium	Network of hyphae
Hyphae	Long filamentous structures
Rhizoid	Hyphae that anchor fungus in its substrate and absorb nutrients
Sporangiophore	Aerial hyphae that support sporangium
Sporangium	Structure in which spores for reproduction form
Spores	Asexual reproduction
Stolon	Aerial hyphae that allow the fungus to colonise more substrate
Columella	Aid spore release
Apophysis	Support the sporangium

Figure 7.6: The structure of a multicellular fungus.

FUNGAL NUTRITION

Unlike bacteria, there are no autotrophic fungi as they do not contain any photosynthetic pigments. Fungi are heterotrophic with most being saprophytic, feeding on dead organic matter. Digestion in fungi occurs outside the cells. The fungus secretes enzymes directly into their food to digest it and the digested food products are absorbed. This is how Rhizopus obtains its nutrition from foods such as bread. Carbohydrates may be stored by fungal cells in the form of glycogen, just as plant cells store carbohydrate in the form of starch.

Figure 7.7: ***Rhizopus stolonifer*** **feeding on bread.**

Some fungi are parasitic, feeding off organisms causing them harm, e.g. athlete's foot, ringworm and *Phytophthora infestans* (potato blight).

Go to YouTube.com and search for 'Leaving Cert Fungi 2023' (8:10) by Biology Bugbears.

PORTFOLIO LINK **Go to your Learning Portfolio and complete Learning Check 2 to assess your knowledge and understanding of the topic so far.**

Section 3: Microbial Growth

At the end of this section, you should be able to:

- Explain how temperature, pH, nutrient supply, water and external solute concentration and presence of antibacterial/antifungal chemicals affect microbial growth.

Unpacked from LOs 3.2.2, 3.2.4

FACTORS THAT AFFECT MICRO-ORGANISM GROWTH

1. Temperature

Most bacteria and fungi have optimum growth between 20°C and 45°C. Cold-loving micro-organisms, such as those found in the Arctic, grow between -5°C and 15°C. Heat-loving micro-organisms such as those found in hot springs grow between 45°C and 70°C. Increasing the temperature increases microbial growth rate because it causes the rate of metabolism to increase. Maximum growth rate occurs at the optimal temperature. At temperatures above the optimum, enzymes become denatured, and the bacteria no longer divide. At temperatures below the optimum, cell activity slows down and microbial growth decreases.

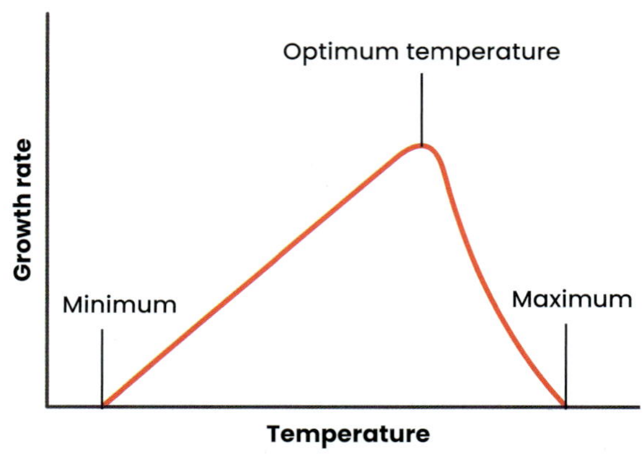

Figure 7.8: Effect of temperature on microbial growth.

2. pH

Most microbes grow best at a neutral pH range of 5–8. However, certain acid-loving microbes grow best at pH values below 5. Alkali-loving microbes grow best at pH values above 8.5. If a microbial cell is exposed to a pH outside its preferred range, its enzymes will permanently lose their shape and will no longer function. They are said to be 'denatured'.

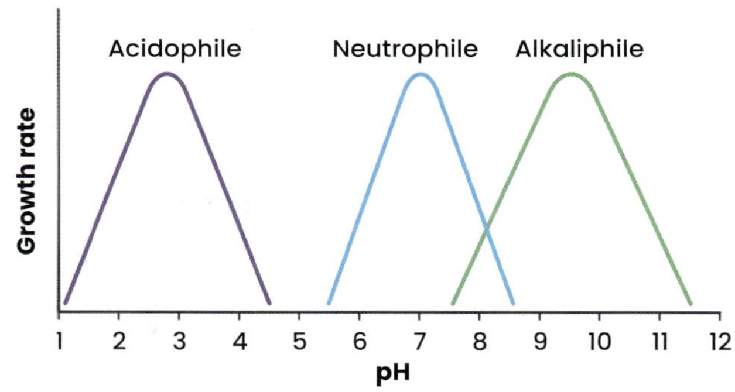

Figure 7.9: Effect of pH on microbial growth.

3. Nutrient supply

Bacteria and fungi vary in the type and amount of nutrients (carbohydrates, fats, proteins, vitamin, and minerals) needed for growth and metabolism. Microbial growth is affected by nutrient supply. If nutrient supply is limited, microbial growth will slow down.

4. Water

Microbial cell cytoplasm needs to have the correct water concentration to ensure that the cell membrane is pressed firmly against the cell wall.

5. External solute concentration

If the concentration of the external solution is higher than that of the microbial cell cytoplasm, water will move out of the cell by osmosis, causing it to dehydrate and die. Osmosis is the passive movement of water from an area of high water concentration to an area of low water concentration across a semi-permeable membrane. Some food preservation techniques, e.g. cured bacon, are based on this method. However, if the concentration of the external solution is lower than that of the microbial cell cytoplasm, water will enter the cell by osmosis. In most cases, the cell wall prevents the microbial cell from bursting upon uptake of large amounts of water. You will study this in more detail in Chapter 18.

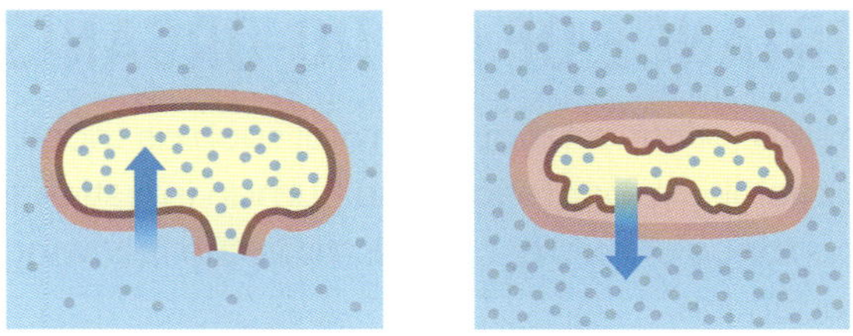

External solution is less concentrated

Water moves into the cell and may cause the cell to burst if the wall is weak or damaged (osmotic lysis).

External solution is more concentrated

Water moves out of the cell, causing its cytoplasm to shrink (plasmolysis).

Figure 7.10: Effect of external solution concentration on microbes.

6. Presence of antibacterial/ antifungal chemicals

Antimicrobial agents are natural or synthetic chemicals with the ability to inhibit or kill microbial **pathogens**, e.g. antibiotics are used against pathogenic bacteria and antifungal chemicals are used to kill fungi.

A **pathogen** is a disease-causing micro-organism.

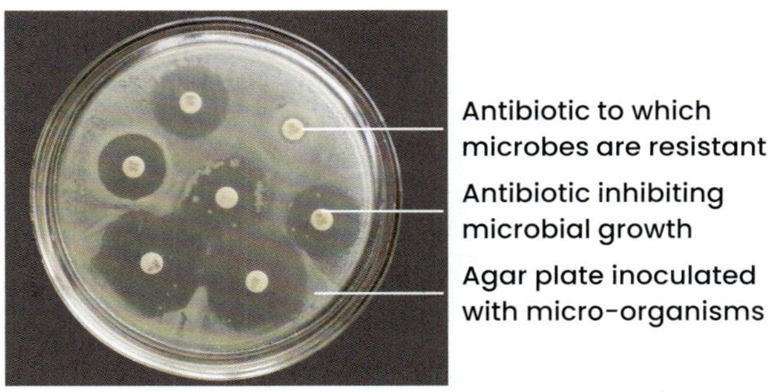

Figure 7.11: Testing the effect of antibiotics on microbial growth.

PORTFOLIO LINK

Go to your Learning Portfolio and complete Learning Check 3 to assess your knowledge and understanding of the topic so far.

Section 4: Importance of micro-organisms in industries

At the end of this section, you should be able to:

- Discuss the importance of micro-organisms in industries (pharmaceutical, agricultural, medical and food production).
- Outline the concept of the microbiome and its role in human health and nutrient cycling in soils.
- Describe the stages in microbial growth with the aid of a labelled graph. HL
- Describe endospore formation with the aid of a labelled diagram.
- List precautions that should be taken when handling micro-organisms.
- Distinguish between sterile and aseptic conditions.
- Describe with the aid of labelled diagrams how you could investigate the effect of different factors on microbial growth.

Unpacked from LOs 3.2.2, 3.2.3, 3.2.4

IMPORTANCE OF MICRO-ORGANISMS

Some micro-organisms are harmful such as those that cause food spoilage or diseases. However, others are beneficial, e.g. bacteria in the human intestine that produce Vitamin K. We will now take a closer look at the importance of micro-organisms in the following industries: pharmaceutical, medical, agricultural and food production.

MICRO-ORGANISMS IN THE PHARMACEUTICAL INDUSTRY

Antibiotics

Bacteria and fungi produce chemical '**antibiotics**' to kill other bacteria, giving them a competitive advantage for food, water and other resources. In medicine, bacterial infections are treated using antibiotics as they kill bacterial cells but not human cells. **Antibiotics have no effect on viral infections**, e.g. the common cold or flu. You will learn more about antibiotic resistance in Chapter 8.

Antibiotics are chemicals produced by micro-organisms such as bacteria and fungi that are toxic to bacteria.

Vaccines

Bacterial vaccines contain killed or weakened bacteria that stimulate the immune system to produce antibodies, e.g. whooping cough (pertussis) vaccine. You will learn more about vaccines in Chapter 32.

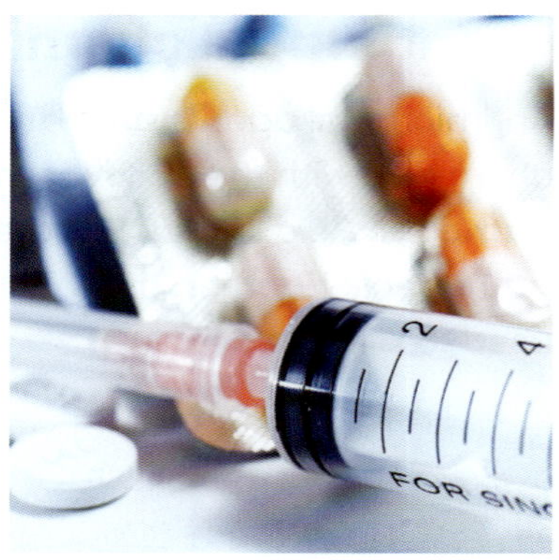

Figure 7.12: Micro-organisms are used to produce antibiotics and vaccines.

Manufacture of insulin

Bacteria cells have been genetically modified to carry the human insulin gene causing them to produce insulin. Bacteria reproduce rapidly, allowing the commercial production of insulin for treatment of diabetics. You will learn more about genetic engineering in Chapter 26.

MICRO-ORGANISMS IN MEDICINE

Many bacteria and fungi cause disease; tuberculosis, botulism and chlamydia are all caused by bacterial infections. Parasitic fungi can cause infections such as athlete's foot and ringworm.

FOCUS ON ... THE MICROBIOME AND HUMAN HEALTH

The **microbiome** is the community of micro-organisms (bacteria, fungi and viruses) that exists in a particular environment. In humans, it often refers to the micro-organisms that live in or on a particular part of the body, e.g. mouth, skin or gastrointestinal tract. Each person's microbiome is unique to them. The microbiome dynamic changes in response to environmental factors, such as exercise, diet and medication. The microbiome plays a significant role in human health within the gastrointestinal (GI) tract:

- **Digestion:** Bacteria that live in your gastrointestinal tract help break down certain carbohydrates that the human body cannot break down on its own. Short-chain fatty acids are produced in the process.
- **Immune signalling:** Beneficial microbes in the GI tract help the immune system to distinguish between beneficial microbes and harmful, pathogenic microbes. These bacteria also help prevent pathogenic bacteria colonising the gut. Short-chain fatty acids produced during digestion by GI bacteria help prevent inflammatory reactions by the immune system.
- **Metabolic health:** Bacteria in the gastrointestinal tract provide enzymes needed to produce certain vitamins, including Vitamin B12 and Vitamin K. Gut bacteria also metabolise bile so that bile acids can be reabsorbed and recycled by the liver. If bile was not recycled in this way, the digestive system would be unable to digest fats and cholesterol from the breakdown of bile would build up in the blood.

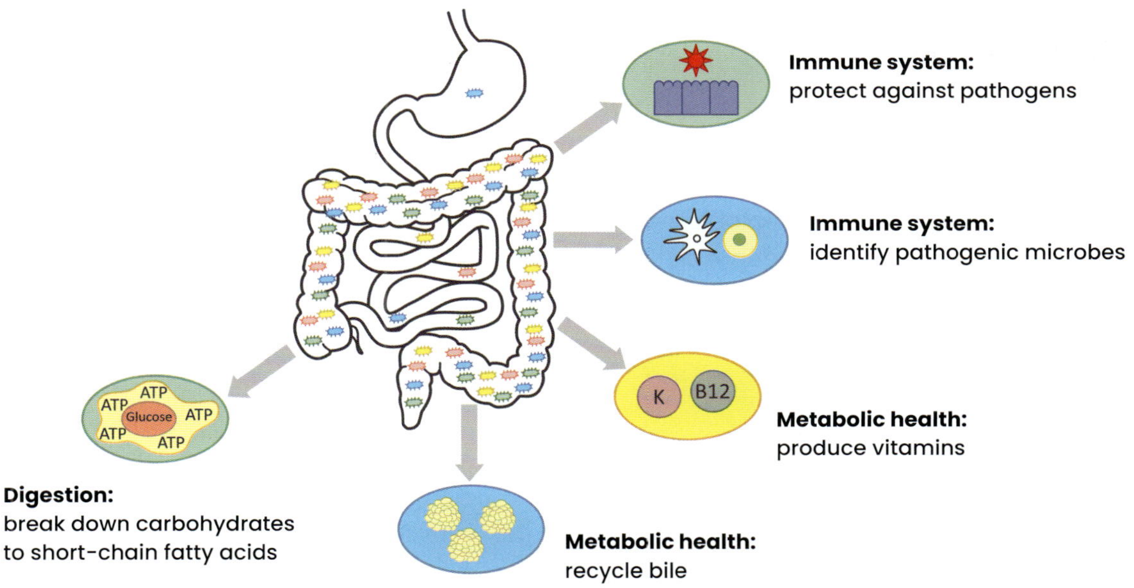

Figure 7.13: Role of microbiome in gastrointestinal tract in human health.

The **microbiome** is the community of micro-organisms that live in a particular environment.

MICRO-ORGANISMS IN AGRICULTURE

Bacteria play an important role in agriculture. Bacteria are used as biopesticides to protect crops against insects and mite pests, while fungi and nitrogen-fixing bacteria help maintain soil fertility. We will learn more about the nitrogen cycle in Chapter 12.

Go to YouTube.com and search for 'The Role of Microbiology in Sustainable Development – Science Summit at UNGA76' (58:27) by American Society for Microbiology to watch the video conference.

SCIENCE IN SOCIETY: SUSTAINABILITY

Microbes such as bacteria and fungi play integral roles in food production; some soil microbes help fertilise the soil, aiding plant growth. Microbes can potentially increase plant yields to aid feeding a growing world population. In 2021, microbiologists held an international virtual panel to discuss the role and contribution of science to achieving the United Nations Sustainable Development Goals.

Figure 7.14: Fertile soil.

FOCUS ON ... THE MICROBIOME AND NUTRIENT CYCLING IN SOILS

The soil microbiome consists of millions of different species of micro-organisms, such as bacteria, fungi, archaea and protists living in soil. Microbes are responsible for cycling multiple soil nutrients and help the processes to occur at a faster rate. Climate warming influences the functioning of soil microbiomes. Increased temperatures cause previously dormant microbes in the soil to become active and respire more. This, in turn, increases the release of carbon into the atmosphere.

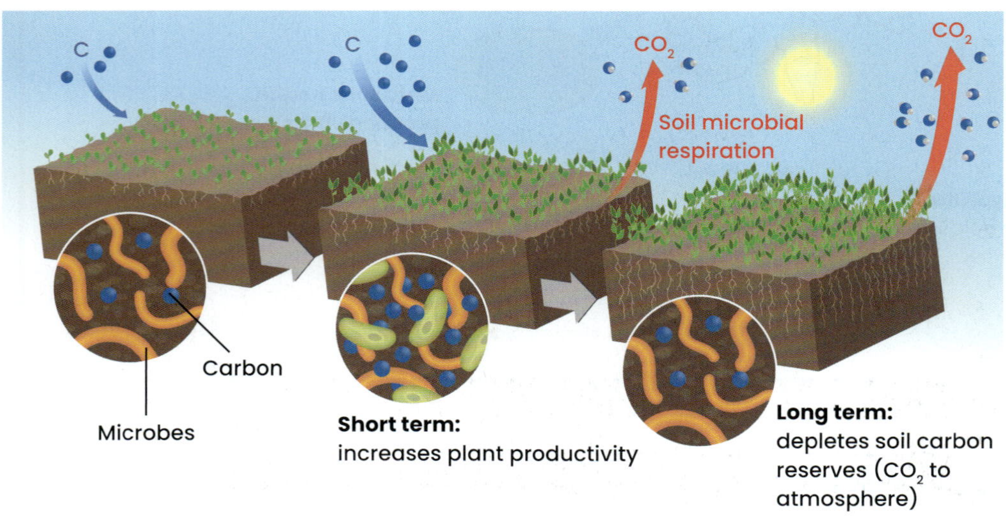

Figure 7.15: Effect of climate warming on soil microbiome.

MICRO-ORGANISMS IN FOOD PRODUCTION

Foods

Mushrooms are an edible form of fungi. Care needs to be taken when eating mushrooms as some forms are poisonous. Bacteria such as *Lactobacillus* are used to make yoghurt from milk. Different bacteria and fungi are also used in cheese production.

Brewing and baking

Yeast, a fungus, is used to ferment glucose, forming ethanol and carbon dioxide, in the manufacture of beer and wine. Yeast is also used in baking. The carbon dioxide produced during fermentation causes bread to rise; the ethanol evaporates during baking.

Food spoilage

Bacteria and fungi can cause food to spoil and become unsuitable for human consumption. The food industry tries to reduce food spoilage by storing food under conditions that minimise microbial growth.

Figure 7.16: Micro-organisms such as bacteria and fungi are used in producing cheese, yoghurt and in brewing and baking.

Figure 7.17: Forms of edible mushrooms.

Figure 7.18: Food spoilage caused by Rhizopus.

HL

MICRO-ORGANISM (BACTERIAL/FUNGAL) GROWTH

Micro-organism growth refers to an increase in the numbers of micro-organisms present. If initial conditions are favourable in a closed container such as a petri dish, micro-organism growth follows a set number of stages:

1. Lag phase
Micro-organism growth is slow as they take time to adjust to their new environment.

2. Log (exponential) phase
Micro-organisms divide continuously, and their numbers increase exponentially as nutrients, oxygen and space are abundant.

3. Stationary phase
Micro-organism numbers are constant as cell division and cell death occur at the same rate.

4. Decline/Death phase
Number of micro-organisms decreases exponentially as the growth rate is slower than death rate. Death is caused by a shortage of nutrients and oxygen and a build-up of toxic wastes.

5. Survival
Some bacteria survive until favourable conditions return by forming endospores.

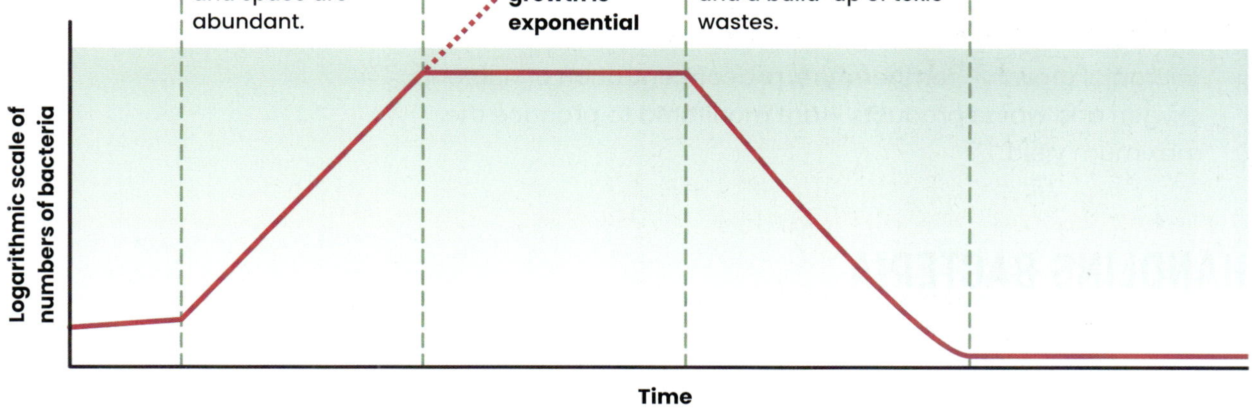

Figure 7.19: Stages of micro-organism growth.

Endospores

During unfavourable conditions, some bacteria can form spores during the late log phase or early stationary phase. Spores remain dormant until favourable conditions return. The spore then germinates, and the bacterium will reproduce once again.

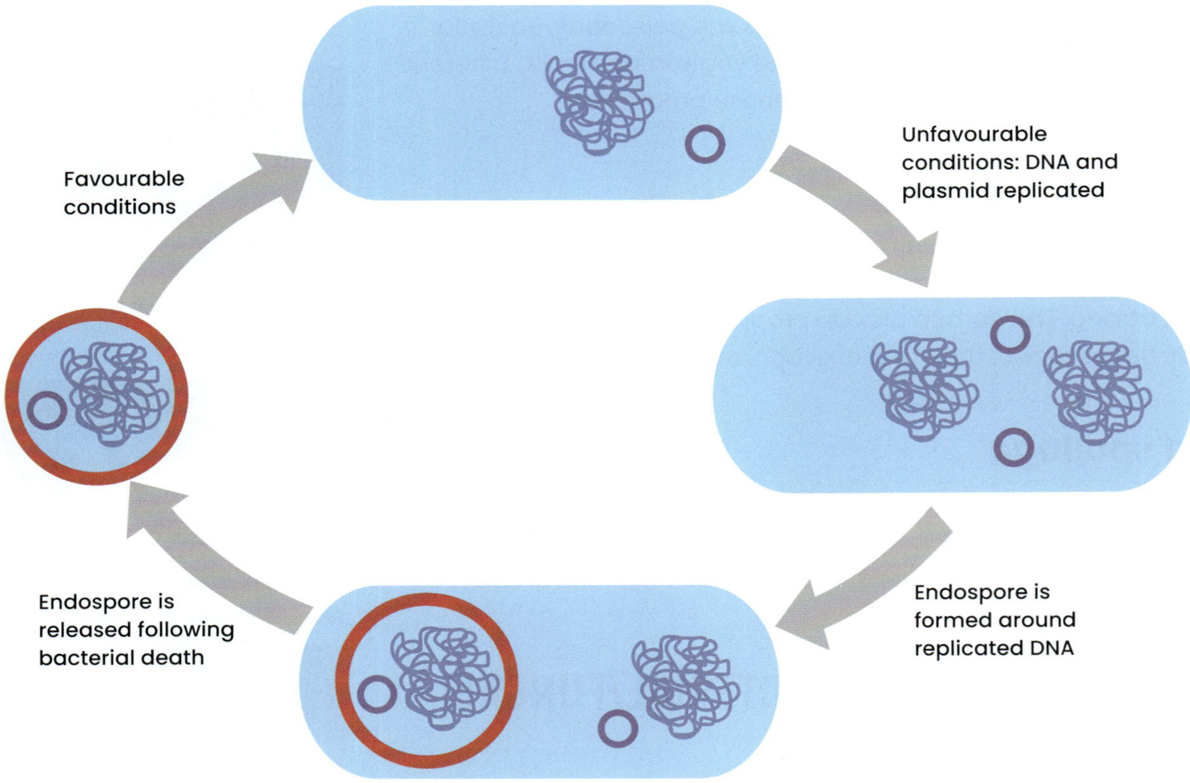

Figure 7.20: Endospore formation.

HL

Microbial growth curves and product yields

In industries that use living organisms to produce products such as yoghurt, cheese, alcohol, etc., knowledge of microbial growth curves allows an increase in product yield to be achieved. If a micro-organism is provided with nutrients, the micro-organism will follow the stages in the microbial growth curve (lag, log, stationary, death, decline). The greatest amount of product will be formed when microbial numbers are at their highest, i.e. in the upper log/stationary phase. Thus, the process will either be stopped or maintained at this phase. Other factors that affect microbial growth – temperature, pH, concentration of nutrients, oxygen and waste products – are monitored to produce the maximum yield.

Go to YouTube and search for 'Bacterial Growth Curve in 2 minutes | 4 Phases of Microbial Growth Curve' (1:48) by biologyexams4u.

HANDLING BACTERIA

When handling micro-organisms in the laboratory, it is important to take precautions to protect oneself. It is not possible to remove all micro-organisms to create **sterile** conditions, but efforts should be made to remove as many micro-organisms as possible, especially those causing disease. This is called the **aseptic** technique.

- If something is **sterile**, it is free from all micro-organisms.
- If something is **aseptic**, it is free from pathogenic disease-causing micro-organisms.

Some of the precautions that should be taken in the laboratory are outlined below.

- All micro-organisms should be treated as potential disease-causing pathogens.
- Long hair should be tied back.
- Wear goggles, gloves and a lab coat when working with micro-organisms.
- Wash your hands thoroughly with disinfectant soap before and after the experiment.
- Disinfect all work areas before and after use.
- Do not eat or drink in the laboratory.
- Incubate all agar plates upside-down so that condensation collects on the lid of the plate.
- Autoclave or disinfect all agar plates after use.
- Sterilise all equipment after use in an autoclave.

PORTFOLIO LINK **Go to your Learning Portfolio and complete Learning Check 4 to assess your knowledge and understanding of the topic so far.**

Specified Investigation 1

Investigate factors affecting microbial growth. (LO 3.2.2)

Scan the QR code to watch a video demonstration of this investigation.

A. TEMPERATURE

Hypothesis: If bacteria cultures are incubated at different temperatures (4°C and 20°C), they will grow in greatest numbers at 20°C.

Equipment: Sterile agar plates, sterile swab, Parafilm, disinfectant, incubator, autoclave, milk samples.

VARIABLES

Independent variable: The temperature at which the agar plate is incubated.

Dependent variable: The % of the agar plate that is covered with bacteria after 2–3 days.

Control: Unopened agar plate.

METHOD

1. Using a sterile swab, take a sample of milk. Slightly open the lid of agar plate. Streak the swab across the surface of the agar. Keep the plate open for the minimum amount of time. Repeat this step for another plate.

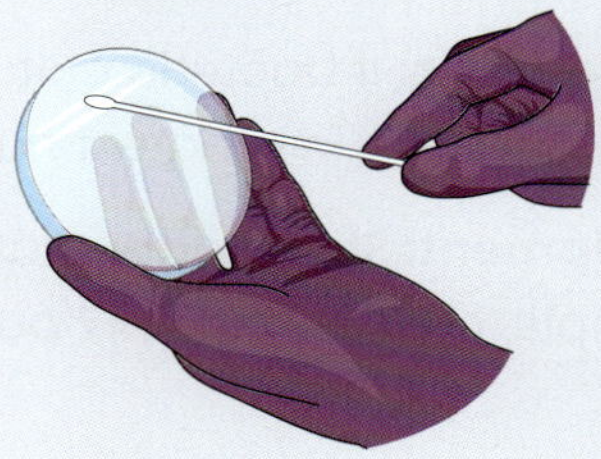

2. Label and seal the two petri dishes.

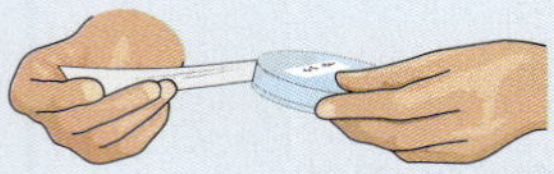

3. Place one plate upside-down in a fridge at 4°C, incubate the second at 20°C. Leave the plates for 2–3 days. The plates are incubated upside-down to prevent any condensation falling onto the agar. As a control, place an unopened plate at each of the temperatures.

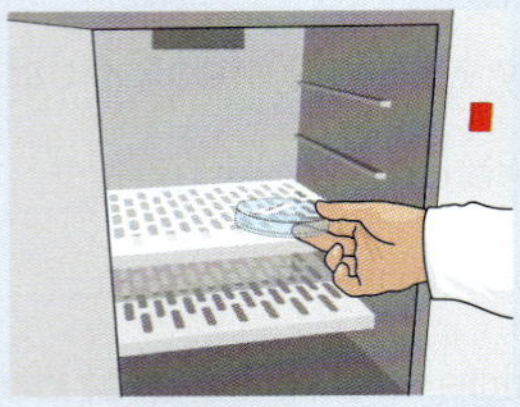

4. Compare the area of each plate covered by bacteria. Record your results in a table.

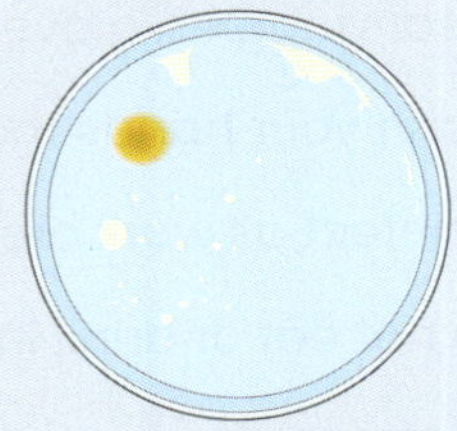

5. Sterilise the agar plates to kill all bacteria using an autoclave (or pressure cooker) at 120°C for 15 minutes or soak in disinfectant for 24 hours.

RESULTS

	TEMPERATURE	4°C	20°C
PLATE WITH SWAB	% of plate covered with bacteria		
UNOPENED PLATE	% of plate covered with bacteria		

Bacteria did not grow on the unopened plates. Very few bacteria grew on the plate at 4°C but a large number of bacteria grew at 20°C.

Safety precautions: Do not open agar plates once they have been inoculated to avoid releasing any potentially harmful bacteria to the environment. Plates were not incubated at body temperature or any temperature above 30°C, to avoid growing bacteria that may be pathogenic to humans.

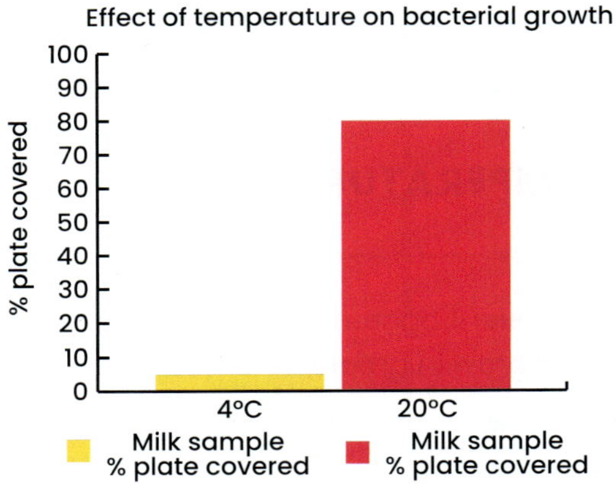

Conclusion: Increasing the temperature from 4°C to 20°C increased the growth of bacteria. To produce individual colonies and avoid the colonies overgrowing, the milk sample may be diluted, e.g. a 1 ml in 1,000 ml distilled water. A countable range of 30–300 colonies per plate is recommended. When counting the number of colonies, remember to multiply by the dilution factor to determine the original value.

B. pH

Hypothesis: If bacteria cultures in milk are incubated at different pH levels (pH 4, pH 7, pH 9), they will grow in greatest numbers at pH 7.

Equipment: Sterile agar plates, sterile swab, Parafilm, disinfectant, incubator, autoclave, milk sample, pH 4 buffer, pH 7 buffer, pH 9 buffer, universal indicator paper or a pH probe.

VARIABLES

Independent variable: The pH at which the milk sample is inoculated.

Dependent variable: The % of the agar plate that is covered with bacteria after 2–3 days.

Control: Unopened agar plate.

METHOD

1. Adjust the pH of the first milk sample to pH 4, by adding pH 4 buffer 1 ml at a time to a 50 ml sample of milk. Check the pH of the sample after each addition, using universal indicator paper or a pH probe.
2. Using a sterile swab, take a sample from the milk. Slightly open the lid of agar plate away from you. Streak the swab across the surface of the agar. Keep the plate open for the minimum amount of time.
3. Repeat Steps 1 and 2 for two more plates – one at pH 7 and a second at pH 9.
4. Label and seal the three petri dishes.
5. Place the plates upside-down in an incubator at 20°C. As a control, place an unopened plate in the incubator. Leave the plates for 2–3 days.
6. Compare the area of each plate covered by bacteria. Record your results in a table.
7. Sterilise the agar plates to kill all bacteria using an autoclave (or pressure cooker) at 120°C for 15 minutes or soak in disinfectant for 24 hours.

RESULTS

PLATE WITH MILK SWAB	% of plate covered with bacteria	pH 4 pH 7 pH 9	
UNOPENED PLATE	% of plate covered with bacteria		

Bacteria did not grow on the unopened plates. Very few bacteria grew on the plates at pH 4 or pH 9 but a large amount of bacteria grew at pH 7.

Safety precautions: Do not open agar plates once they have been inoculated to avoid releasing any potentially harmful bacteria to the environment. Plates were not incubated at body temperature or any temperature above 30°C, to avoid growing bacteria that may be pathogenic to humans.

Conclusion: pH 7 is the optimal pH for the growth of the bacteria found in milk as it gave the greatest growth on the plate. pH values lower or higher than the optimal, e.g. pH 4 or pH 9, resulted in reduced bacterial growth.

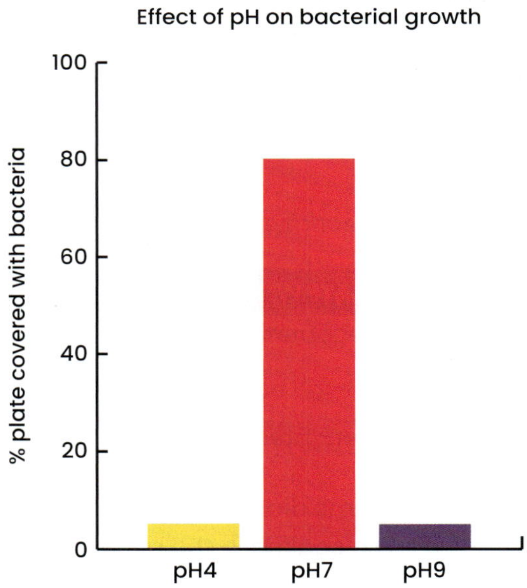

C. WATER AND EXTERNAL SOLUTE CONCENTRATIONS

Scan the QR code to watch a video demonstration of this investigation.

Hypothesis: If bacterial cultures in milk are exposed to concentrated external solutions, they will grow in lower numbers.

Equipment: Sterile agar plates, sterile swab, Parafilm, disinfectant, incubator, autoclave, milk sample, salt, cooled boiled water, Bunsen burner, cork-borer, forceps, filter paper, ruler.

VARIABLES

Independent variable: Concentration of the external solution to which the bacteria are exposed to.

Dependent variable: Percentage of the agar plate that is covered with bacteria after 2–3 days.

Control: Unopened agar plate.

METHOD

1. Cut three discs from a piece of filter paper using a cork-borer previously sterilised in a Bunsen burner flame.
2. Prepare the three filter paper discs as follows:
   A: Left dry
   B: Soak in cooled boiled water
   C: Soak in a saturated solution of salt in cooled boiled water.

   Divide the bottom of a sterile agar plate into three equal parts and label as A, B and C.
3. Using a sterile swab, streak a milk sample across the agar plate.
4. Using a forceps previously flamed in a Bunsen burner, place the three filter-paper discs to the centre of the areas labelled A, B and C on the agar plate. Re-flame the forceps between each use.
5. Place the plate upside-down in an incubator at 20°C. As a control, place an unopened plate in the incubator.
6. Leave the plates for 2–3 days.
7. Using a ruler, measure the maximum diameter of any zones around each disc that are free from bacterial growth. Record your results in a table.
8. Sterilise and safely dispose of the agar plates.

RESULTS

PLATE WITH MILK SWAB	DIAMETER OF ZONE OF INHIBITION
A: Dry disc	
B: Disc with cooled boiled water	
C: Disc with saturated salt solution	

Micro-organisms did not grow on the unopened plate. The disc soaked in the saturated salt solution produced the largest clear zone of inhibition, free from microbial growth.

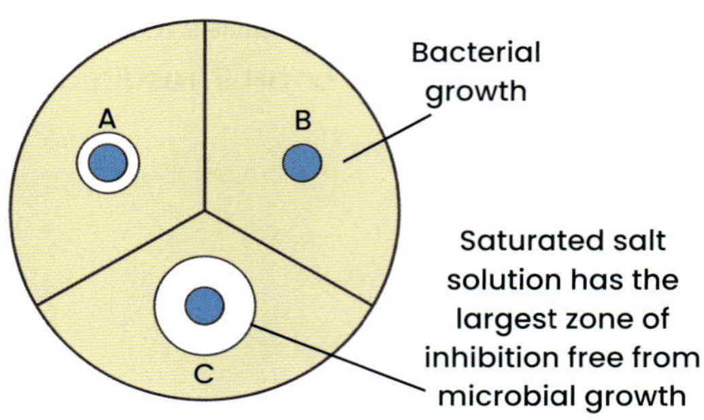

Safety precautions: Do not open agar plates once they have been inoculated to avoid releasing any potentially harmful bacteria to the environment.

Conclusion: The disc soaked in the saturated salt solution produced the largest clear zone of inhibition as concentrated salt solutions kill micro-organisms. The greatest microbial growth was observed near the disc soaked in water as micro-organisms require water to reproduce.

Note: You could also test salt solutions of different concentrations, e.g. 10g salt/100ml, 20g salt/100ml, 30g salt/100ml.

D. ANTIBACTERIAL/ANTIFUNGAL CHEMICALS

METHOD

Repeat the above investigation using the following paper discs:

A: Dry paper disc.

B: Disc soaked in antibacterial/antifungal solution.

Note: Different brands of antibacterial chemicals may be tested. The larger the zone of inhibition, the more effective the antibacterial/antifungal solution is.

E. NUTRIENTS

METHOD

Repeat the above investigation using the following paper discs on separate agar plates:

A: Dry paper disc.

B: Disc soaked in sugar solution.

C: Disc soaked in starch solution.

RESULTS

In this case, the sugar and starch solutions will not inhibit microbial growth but will act as a food source promoting growth. The percentage of the plate covered by microbial growth can be compared in each case.

Note: Different concentrations of sugar or starch solutions may also be tested. Microbial growth on different agar plates could also be compared, e.g. a malt agar plate vs a starch agar plate. Different micro-organisms prefer different food sources, so more fungi may grow on a malt agar plate than on a starch agar plate. The greater the concentration of the nutrient solution, the greater the microbial growth.

INVESTIGATIVE LOG LINK

Go to your Investigative Log to record your lab work and answer questions on the experiments in this chapter.

- Bacteria and fungi:

	BACTERIA	FUNGI
STRUCTURE	• Unicellular • 1–5μm • **Chromosomal DNA:** bacterial DNA found loose in the cytoplasm • **Plasmid DNA:** small circular piece of DNA often containing antibiotic resistance genes • **Cell membrane:** selectively permeable membrane controls passage of substances in and out of bacterial cell • **Cell wall:** peptidoglycan cell wall gives cell shape and strength • **Capsule:** slimy layer that prevents the cell drying out • **Flagellum:** help bacteria to move • **Pili:** help bacteria to attach to other cells/surfaces • **Endospore:** dormant spores formed by some bacteria in unfavourable conditions. When favourable conditions return, the spore germinates, and the bacterium reproduces	• Yeast is unicellular but most fungi are multicellular • 10–100μm • **Cell wall:** containing chitin • **Hyphae:** tubular filaments • **Rhizoids:** hyphae that anchor fungus in substrate • **Mycelium:** network of hyphae • **Sporangiophore:** aerial hyphae that support sporangium • **Sporangium:** produce spores for reproduction • **Stolon:** aerial hyphae that colonise new substrate areas
CELLULAR NATURE	**Prokaryote:** no nucleus or membrane-bound organelles	**Eukaryote:** nucleus and membrane-bound organelles
NUTRITION	**Heterotrophic:** • Saprophytic: feed on dead organic matter • Parasitic: feed on living organisms, causing them harm **Autotrophic:** • Photosynthetic: make food using energy from sunlight • Chemosynthetic: make food using energy from chemical reactions	**Heterotrophic:** mainly saprophytic

- Importance of micro-organisms:

FOOD PRODUCTION	AGRICULTURE	MEDICAL	PHARMACEUTICAL
• Yoghurt and cheese production, baking, brewing • Food and food spoilage	• Agriculture – biopesticides • Soil microbiome	• Disease and illness • Gastrointestinal tract microbiome	• Manufacture of insulin • Antibiotics • Bacterial vaccines

Chapter Summary

- A **microbiome** is a community of micro-organisms that live in a particular environment.
- Factors affecting micro-organism growth:

TEMPERATURE	Most microbes have optimum growth at 25–45°C. Enzymes denature at temperatures above optimum. Temperature below optimum – microbial growth decreases.
PH	Most microbes grow best at pH 5–8. Enzymes denature outside preferred pH range.
NUTRIENT SUPPLY	If nutrient supply is limited, microbial growth slows down.
WATER	Bacterial cell cytoplasm needs to have correct amount of water to ensure cell membrane is pressed firmly against the cell wall. Bacteria gain/lose water by osmosis.
EXTERNAL SOLUTE CONCENTRATION	Higher external solution concentrations cause water to move out of the microbial cell by osmosis, causing it to die, e.g. in food preservation – cured bacon. Lower external solution concentration causes water to enter the microbial cell by osmosis – cell wall prevents the microbial cell from bursting.
ANTIBACTERIAL/ ANTIFUNGAL CHEMICALS	Antibiotics kill pathogenic bacteria. Antifungal chemicals kill fungi.

HL

- Microbe growth:

1. LAG PHASE	2. LOG (EXPONENTIAL) PHASE	3. STATIONARY PHASE	4. DECLINE/DEATH PHASE	5. SURVIVAL
Micro-organism growth is slow as they take time to adjust to their new environment.	Micro-organisms divide continuously, and their numbers increase exponentially as nutrients, oxygen and space are abundant.	Micro-organism numbers are constant as cell division and cell death occur at the same rate.	Number of micro-organisms decreases exponentially as the growth rate is slower than death rate. Death is caused by a shortage of nutrients and oxygen and a build-up of toxic wastes.	Some bacteria survive unfavourable conditions by forming endospores.

- Industries that use microbes to produce products maintain the process in the log phase to maximise product yield.

END OF CHAPTER ASSESSMENT

REMEMBER

1. (a) Label the following bacterial cell parts A–J and give a function for each part.
   (b) Identify the parts of the bacterial cell that are (i) always present (ii) sometimes present.

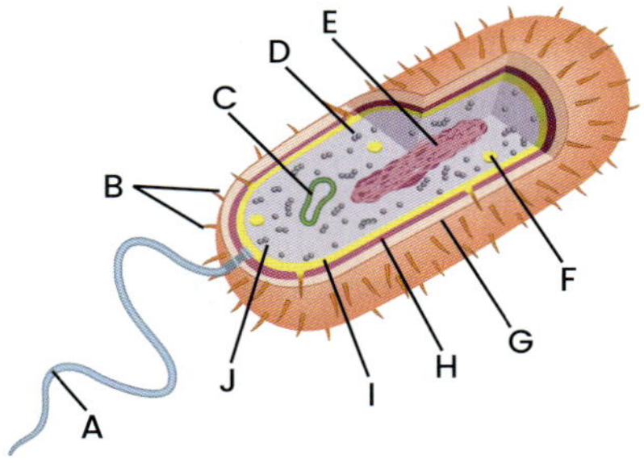

2. List **five** factors that affect microbial growth.
3. Name the following parts of the Rhizopus fungus (A–H) and give a function of each.

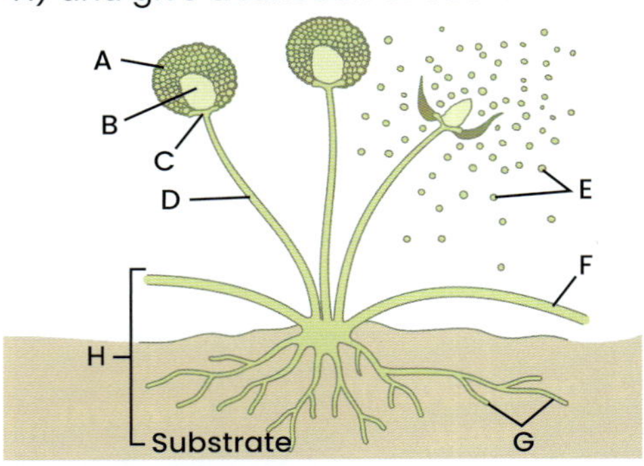

4. Distinguish between the terms prokaryotic and eukaryotic.
5. Define each of the following.
   (a) autotroph
   (b) heterotroph
   (c) saprophyte
   (d) parasite

UNDERSTAND

6. Discuss the importance of micro-organisms in the following industries.
   (a) Pharmaceutical
   (b) Medical
   (c) Agriculture
   (d) Food production
7. (a) Explain the term *microbiome*.
   (b) Describe the role of the gastrointestinal tract microbiome in human health.
   (c) Why is the soil microbiome important in agriculture?

APPLY

8. How may knowledge of bacterial growth curves be used in industry to improve product yields? HL
9. Sketch a bacterial growth curve for a situation with an unlimited supply of food and resources. HL

ANALYSE

10. A student investigated the effect of three different antibacterial/antifungal chemicals (alcohol, hydrogen peroxide and iodine) on the growth of *Lactobacillus* bacteria. Agar plates were inoculated with the bacterium. A different antibacterial/antifungal chemical was spread onto each plate. A plate with no antibacterial or antifungal chemical was used as a control. The plates were incubated at 20°C for 48 hours.

	LACTOBACILLUS BACTERIA
ALCOHOL	Very little growth
HYDROGEN PEROXIDE	Very little growth
IODINE	Some growth
CONTROL PLATE	A lot of growth

   (a) What conclusion can you draw from these results about the effectiveness of each of the disinfectant chemicals?
   (b) Why was a control used in the investigation?
   (c) How and why was the temperature kept constant at 20°C?
   (d) Why would chemicals such as these be used in the home?

EVALUATE

11. Evaluate the following statement:
    Bacteria are necessary for human life on Earth.

CREATE

12. Design an experiment to determine the most effective concentration of salt solution to inhibit microbial growth.

8 Viruses

LEARNING OUTCOMES: 1.1.2, 2.4.13

EYE-OPENER

Number of lives saved by vaccinations from 1974 to 2024, World

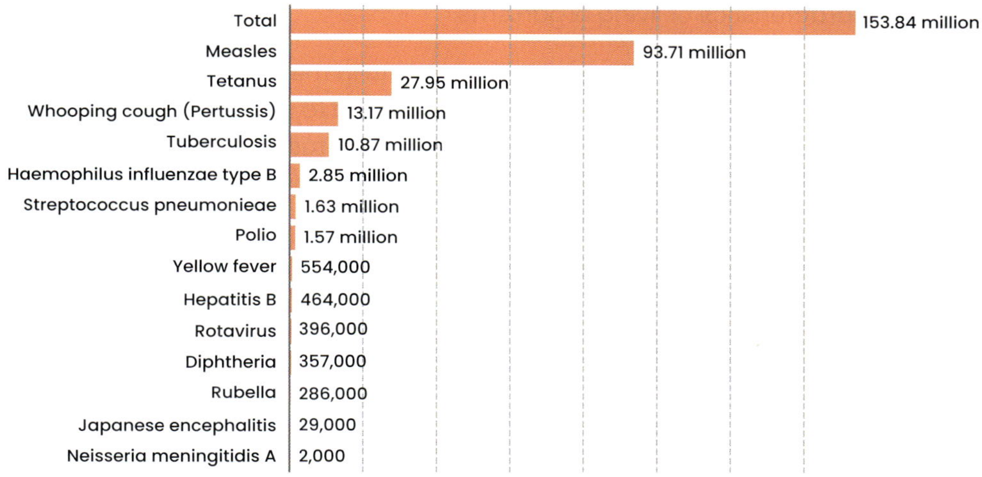

Data source: Shattock et al. (2024).

A 2024 scientific study estimated that vaccines have saved 154 million lives in the past 50 years.

(Source: *The Lancet*)

At the end of this chapter, you should be able to:

- Describe the structure of a virus and give the function of each part.
- Describe the difficulty of defining viruses.
- Explain how viruses replicate within cells.
- Outline the modes of action of viruses.
- Distinguish between RNA and DNA viruses.
- Discuss the economic and medical importance of viruses.

LINKS TO:

Chapter 1 What Is Biology?

Chapter 24 Information Flow in the Cell

Chapter 32 Infectious Disease and the Human Immune Response

SUBJECT VOCABULARY

virus | bacteriophage | capsid

Section 1: Structure and Living/Non-Living Characteristics

At the end of this section, you should be able to:

- Draw a labelled diagram of a virus (DNA, RNA, protein coat).
- State a function of each part of a virus.
- List the 'living' and 'non-living' characteristics of viruses.

Unpacked from LO 1.1.2

The onset of the Covid-19 pandemic in 2020 was a sharp reminder to humanity of how our immune systems are constantly under attack from **viruses**. You may have noticed in Chapter 4: Classification and Taxonomy that viruses are not included in one of the three domains of life. Viruses are not referred to as organisms. This is because they do not display all the characteristics of living organisms.

A **virus** is a microscopic obligate parasite that can only replicate in a living host cell.

Viruses are microscopic parasitic agents that can only replicate in the living cells of plants, animals, fungi or bacteria. When not in a host cell, a virus is dormant and referred to as a **virion**. Once a virion infects a host cell, it is activated and is now known as a virus. The largest viruses, measuring about 500nm in diameter, can only be seen at the highest resolution with a light microscope. At no more than 20nm in diameter, the smallest viruses are smaller than a ribosome.

A nanometre (nm) is 1×10^{-9}m, or one billionth of a metre.

DID YOU KNOW?

VIRUS STRUCTURE

Viruses are not made of cells and do not contain cell organelles such as ribosomes. Viruses contain a nucleic acid, either single- or double-stranded DNA or RNA, surrounded by a protein coat called a **capsid**. Some viruses have an envelope outside the capsid which is formed from proteins and phospholipids from the host cell.

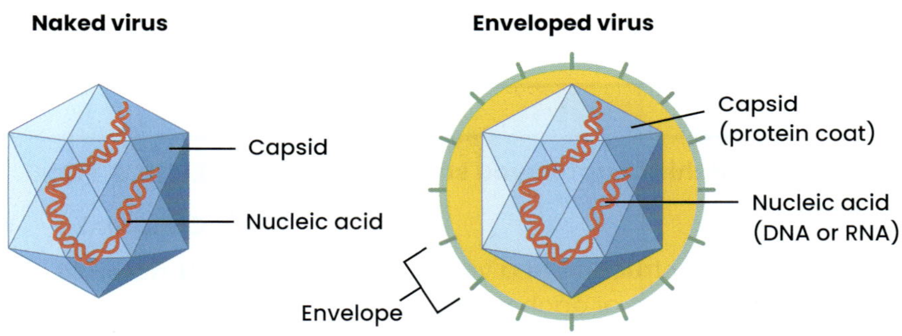

Figure 8.1: Virus structure.

The **capsid** is a protein coat that surrounds the nucleic acid in a virus.

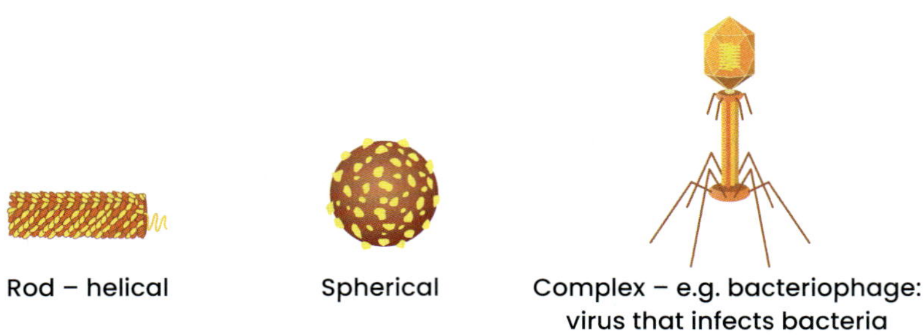

Figure 8.2: Virus shapes.

Viruses can have different shapes depending on their capsid: rod, spherical or complex. The functions of each part of a virus are outlined in Figure 8.3 below.

	FUNCTION
NUCLEIC ACID (DNA/RNA)	Viral genome consisting of single- or double-stranded DNA or RNA containing the virus's genetic information.
CAPSID	• Protein coat that protects the genetic material (DNA/RNA) from digestion by enzymes. • Allows the virus to attach to host cells. • Has proteins that enable virus to penetrate host cell membrane.
ENVELOPE	Enables the virus to bind to host cells. Enveloped viruses are less stable than non-enveloped ones.

Figure 8.3: Function of viral structures.

ARE VIRUSES ALIVE?

When viruses were first discovered in the late 1800s, they were thought to be alive. However, when the structure of the tobacco mosaic virus was determined in 1935, it was found not to be composed of cells, but of protein and RNA.

The difficulty in classifying viruses results from the fact that they possess both living and non-living characteristics. Figure 8.4 below outlines the living and non-living characteristics displayed by viruses.

Go to YouTube.com and search for 'Viral Structure and Functions' (6:46) by Osmosis from Elsevier.

LIVING CHARACTERISTICS	NON-LIVING CHARACTERISTICS
Genetic material, either DNA or RNA, is present	Only contain one type of nucleic acid (living cells have both DNA and RNA)
More complex level of organisation compared to inorganic substances, e.g. rocks	Organelles, e.g. mitochondria, nucleus, are not present
Can replicate inside a living host	Cannot replicate outside of a living host cell
Protein in structure	Non-cellular
Genetic information is passed on to next generation	Cannot carry out metabolism independently

Figure 8.4: Living and non-living characteristics of viruses.

PORTFOLIO LINK

Go to your Learning Portfolio and complete Learning Check 1 to assess your knowledge and understanding of the topic.

Section 2: Viral Replication

At the end of this section, you should be able to:

- Outline the modes of action of viruses.
- Describe with the aid of a labelled diagram how a virus replicates.
- State the differences between DNA and RNA viruses.
- Give examples of the economic importance of viruses.
- Give examples of the medical importance of viruses.

Unpacked from LOs 1.1.2, 2.4.13

HOW A VIRUS REPLICATES

Viruses do not have enzymes needed for metabolism or ribosomes to make their own proteins; they are obligate parasites. Each virus only infects a specific range of cells, for example, the common cold only infects cells lining the upper respiratory tract in humans. A virus identifies its host by matching proteins on its surface to specific receptors on the host cell surface. Viruses can only replicate their genetic material and proteins inside a host cell. These are then assembled to form new virions. The new virions are released from the host cell and may infect other cells. The length of this replication process depends on the virus and can vary from days to weeks. The stages in viral replication are shown in Figure 8.5 below.

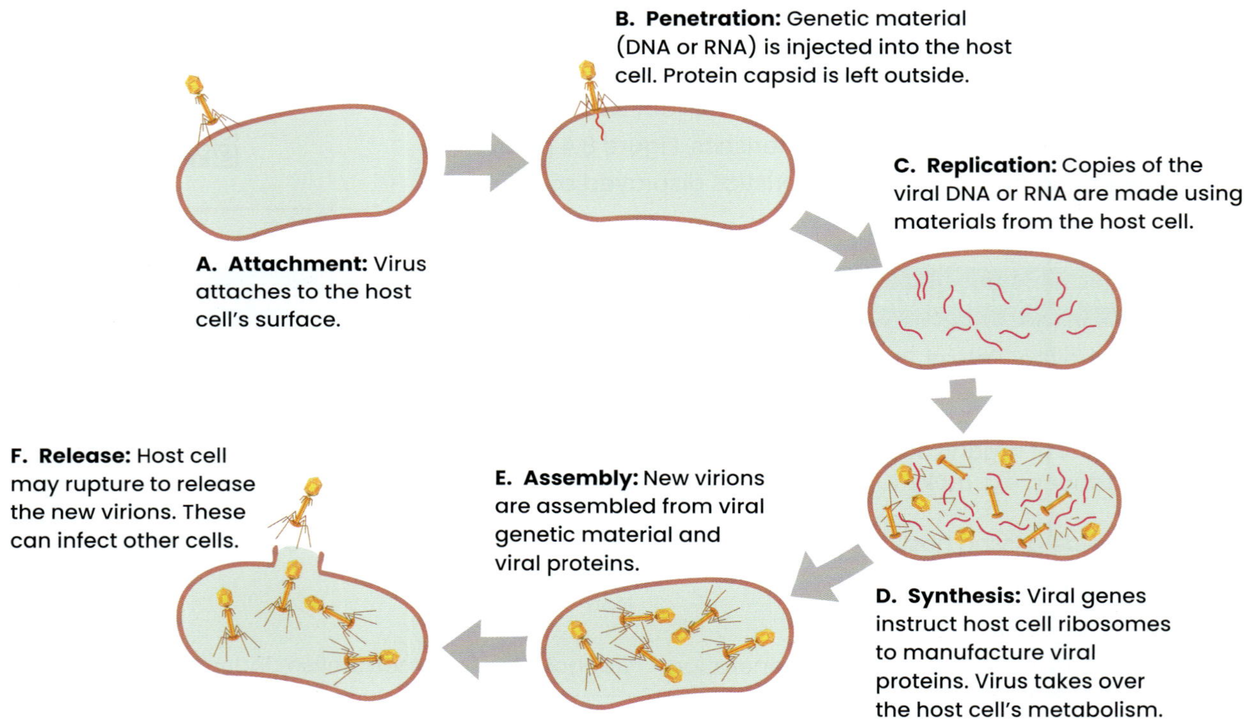

Figure 8.5: The stages of viral replication.

Go to YouTube.com and search for 'The Viral Life Cycle' (2:28) by RicochetScience.

DNA AND RNA VIRUSES

Viruses can be divided into two categories based on their genetic material: DNA viruses or RNA viruses. DNA and RNA viruses infect both eukaryotic and prokaryotic cells.

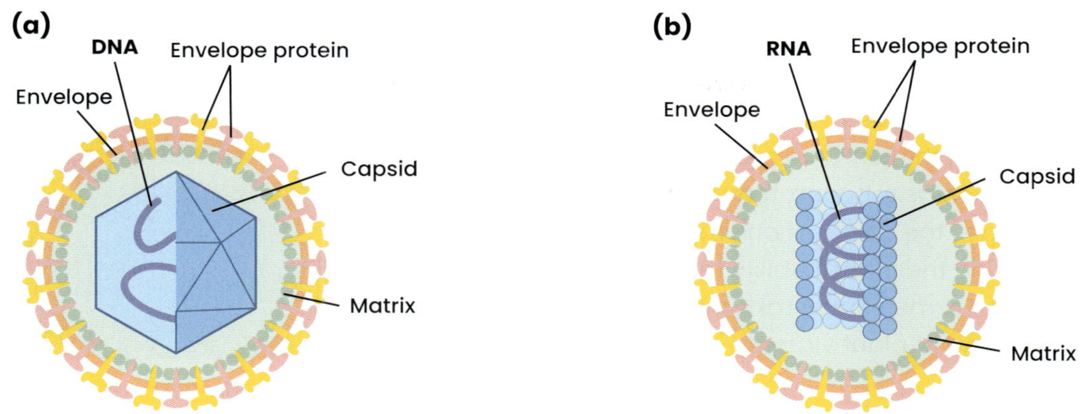

Figure 8.6: (a) DNA virus and (b) RNA virus.

DNA viruses
DNA viruses contain mainly double-stranded DNA but some can contain single-stranded DNA. DNA viruses are more stable than and don't mutate as often as RNA viruses, making it easier to develop vaccines against them. Examples of DNA viruses include herpesviruses and the human papillomavirus (HPV).

RNA viruses
RNA viruses contain either double-stranded or single-stranded RNA. RNA viruses have a higher mutation rate than DNA viruses, making it more difficult to vaccinate against them. Examples of RNA viruses include influenza, coronavirus, polio and measles.

THINK, PAIR, SHARE
How can mutations affect a virus?

	DNA VIRUSES	RNA VIRUSES
NUCLEIC ACID	Deoxyribonucleic acid (mainly double-stranded DNA, but single-stranded DNA also exists)	Ribonucleic acid (mainly single-stranded RNA, but double-stranded RNA also exists)
GENOME SIZE	Large genome	Small genome
ASSEMBLY SITE	In the nucleus of the host cell	In the cytoplasm of the host cell
MUTATION RATE	Very stable – low mutation rate	Less stable – high mutation rate
VACCINATION	Easier to vaccinate against	More difficult to vaccinate against

Figure 8.7: Differences between DNA and RNA viruses.

MODES OF ACTION OF VIRUSES

Viruses can impact on living organisms in different ways:

- **Cancer causing viruses:** Certain viruses, such as the human papillomavirus (HPV), can alter their host cell's function and cause cancer.
- **Infectious disease:** Viruses can cause infectious diseases such as influenza and Covid-19 in humans.
- **Latency:** Some viruses, such as the herpesvirus, can remain dormant or latent within host cells for long periods of time before reactivating, causing recurrent infections.

ECONOMIC IMPORTANCE OF VIRUSES

The Covid-19 pandemic has had huge economic consequences globally. Efforts by governments to control the spread of the disease included shutting workplaces for periods of time, which directly affected people's income. Also, as people contracted the virus, workplace absences increased, and healthcare facilities were placed under greater strain. Governments around the world placed economic measures in place to assist those who lost their jobs due to the pandemic and provided extra funding to healthcare systems. Manufacturing productivity was reduced, disrupting supply-chains and pushing up prices of goods.

PLANT DISEASES

Plant viruses are of economic importance to farmers as they can cause disease in their crops, reducing yield. For example, the tobacco mosaic virus, which infects plants such as tobacco, tomatoes and peppers, changes the colour of chloroplasts to yellow or white, reducing the plant's ability to undergo photosynthesis. Farmers must take measures to reduce the spread of plant viruses.

Figure 8.8: A tomato plant infected with the tobacco mosaic virus.

ANIMAL DISEASES

Viral diseases also affect animals. Foot-and-mouth disease is a contagious viral disease that affects animals such as cattle, pigs and sheep. The disease is of huge economic importance due to animal deaths, loss in milk production and veterinary care costs. An outbreak of the disease in a country can affect its trade with other countries, as its animals and animal products will not be accepted by other countries.

Figure 8.9: An animal infected with foot-and-mouth disease.

MEDICAL IMPORTANCE OF VIRUSES

Human disease

Diseases such as Covid-19, chickenpox, measles, mumps, rubella, influenza and AIDS are caused by viruses. Viral infections are not affected by antibiotics, but vaccines may be used to prevent the spread of some serious viral infections, including measles, mumps and rubella. It is not possible to vaccinate against some viral infections including the common cold, as they can be caused by over 160 different strains of virus.

Viral infections can spread in different ways: some can travel through the air when a person coughs or sneezes, some are spread through sexual contact or through blood and other bodily fluids. If a virus lands on a solid surface, it can generally survive for up to 24 hours, though some viruses can survive for longer periods of time.

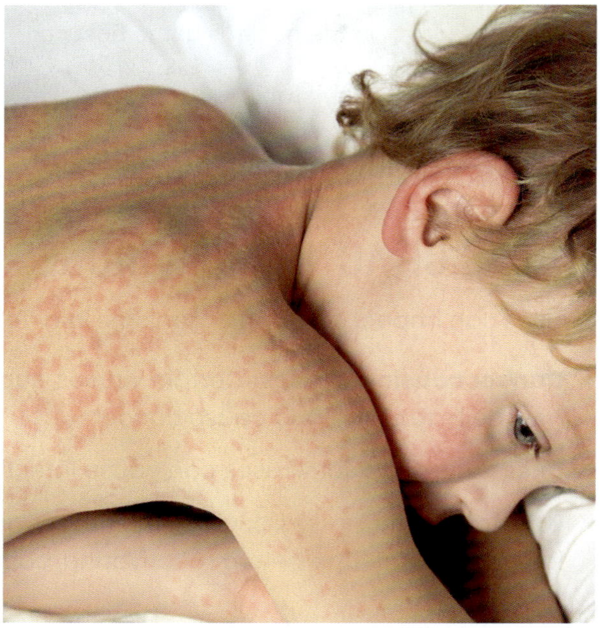

Figure 8.10: Rash caused by the rubella virus.

Vaccines

Viruses, or parts of their structure such as proteins or their genetic material (DNA or RNA), are used to manufacture vaccines. Vaccines stimulate the body's immune system to form antibodies against the virus, without infecting the person with the disease. You will learn more about the different types of vaccine and the human immune system in Chapter 32.

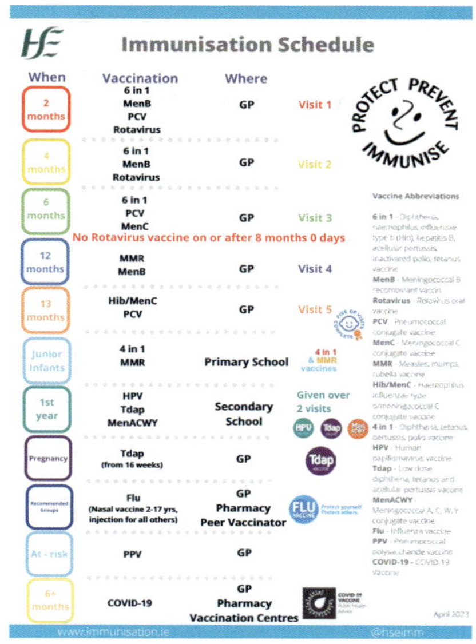

Immunisation Schedule

When	Vaccination	Where	
2 months	6 in 1 MenB PCV Rotavirus	GP	Visit 1
4 months	6 in 1 MenB Rotavirus	GP	Visit 2
6 months	6 in 1 PCV MenC	GP	Visit 3
	No Rotavirus vaccine on or after 8 months 0 days		
12 months	MMR MenB	GP	Visit 4
13 months	Hib/MenC PCV	GP	Visit 5
Junior Infants	4 in 1 MMR	Primary School	4 in 1 & MMR vaccines
1st year	HPV Tdap MenACWY	Secondary School	Given over 2 visits
Pregnancy	Tdap (from 16 weeks)	GP	
Recommended Groups	Flu (Nasal vaccine 2-17 yrs, injection for all others)	GP Pharmacy Peer Vaccinator	
At - risk	PPV	GP	
6+ months	COVID-19	GP Pharmacy Vaccination Centres	

PROTECT PREVENT IMMUNISE

Vaccine Abbreviations

6 in 1 - Diphtheria, haemophilus influenzae type b (Hib), hepatitis B, acellular pertussis, inactivated polio, tetanus vaccine
MenB - Meningococcal B recombinant vaccine
Rotavirus - Rotavirus oral vaccine
PCV - Pneumococcal conjugate vaccine
MenC - Meningococcal C conjugate vaccine
MMR - Measles, mumps, rubella vaccine
Hib/MenC - Haemophilus influenzae type b/meningococcal C conjugate vaccine
4 in 1 - Diphtheria, tetanus, pertussis, polio vaccine
HPV - Human papillomavirus vaccine
Tdap - Low dose diphtheria, tetanus and acellular pertussis vaccine
MenACWY - Meningococcal A, C, W, Y conjugate vaccine
Flu - Influenza vaccine
PPV - Pneumococcal polysaccharide vaccine
COVID-19 - COVID-19 Vaccine

April 2023

www.immunisation.ie @hseimm

Figure 8.11: Immunisation programme offered to babies in Ireland.

Viral vectors in gene therapy

Scientists are currently researching the use of viruses to carry necessary genes into a human cell in the treatment of diseases such as cancer. This is known as gene therapy. The viruses are modified so that they don't cause disease.

Use of bacteriophages to treat infections caused by bacteria

As antibiotic resistance amongst bacteria is rising, scientists are researching the use of **bacteriophages** to treat infections caused by bacteria. Bacteriophages target and kill specific bacteria, promising an alternative to antibiotics. In this way, bacteriophages are considered a beneficial virus.

A **bacteriophage** is a virus that replicates inside a bacterial cell.

SCIENCE IN SOCIETY: SUSTAINABILITY

Plankton, bacteria and bacteriophages

Bacteriophages (viruses that infect bacteria) are extremely important to our environment, especially in ocean ecosystems. These viruses help keep bacterial populations in check.

Plankton is the word used to describe plants and animals that are carried by the sea's tides and currents rather than by their own movements. Some plankton are microscopic algae in the oceans that are estimated to produce about 50% of the world's oxygen. When these plankton die, bacteria feed on them, using up oxygen as they respire. By killing bacteria, viruses help to ensure that the levels of oxygen in the ocean are maintained.

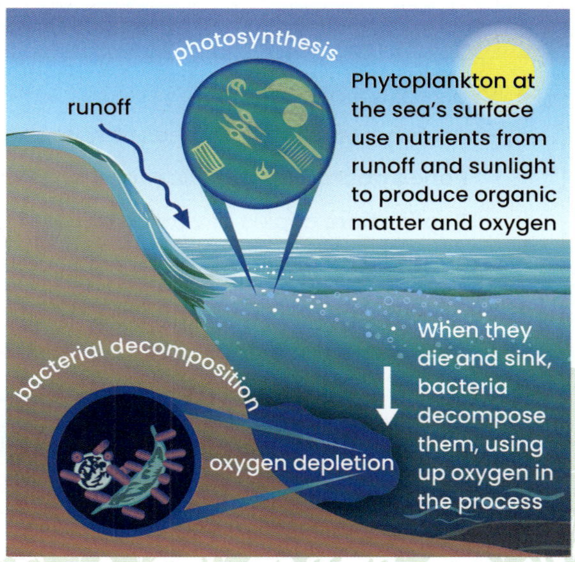

PORTFOLIO LINK

Go to your Learning Portfolio and complete Learning Check 2 to assess your knowledge and understanding of the topic.

- Viruses are obligate parasites and can only replicate inside a host cell.

LIVING CHARACTERISTICS	NON-LIVING CHARACTERISTICS
Genetic material, either DNA or RNA, is present	Only contain one type of nucleic acid (living cells have both DNA and RNA)
More complex level of organisation compared to inorganic substances, e.g. rocks	Organelles, e.g. mitochondria, nucleus, are not present
Can replicate inside a living host cell	Cannot replicate outside of a living host cell
Protein in structure	Non-cellular
Genetic information is passed on to next generation	Cannot carry out metabolism independently

- Virus structure: nucleic acid (single- or double-stranded DNA or RNA), surrounded by a protein coat 'capsid'. Some viruses have an outer envelope formed from proteins and host cell phospholipids.
- Virus shapes: rod, spherical, complex.

	FUNCTION
Nucleic acid (DNA/RNA)	Single- or double-stranded DNA or RNA containing the virus's genetic information.
Capsid	Protein coat that protects the genetic material (DNA/RNA) and delivers it to its host.
Envelope	Enables the virus to bind to host cells. Enveloped viruses are less stable than non-enveloped ones.

- Viral replication:
  - **Attachment:** virus attaches to host cell's surface.
  - **Penetration:** viral genetic material (DNA or RNA) is injected into the host cell.
  - **Replication:** copies of viral DNA or RNA are made by the host cell.
  - **Synthesis:** viral genes instruct host cell ribosomes to manufacture viral proteins.
  - **Assembly:** new virions are assembled from viral genetic material and viral proteins.
  - **Release:** host cell may rupture to release the new virions which can infect other cells.

	DNA VIRUSES	RNA VIRUSES
Nucleic acid	Deoxyribonucleic acid (mainly double-stranded DNA)	Ribonucleic acid (mainly single-stranded RNA)
Genome size	Large genome	Small genome
Assembly site	In the nucleus of the host cell	In the cytoplasm of the host cell
Mutation rate	Low mutation rate – easier to vaccinate against	High mutation rate – more difficult to vaccinate against

- Economic and medical importance of viruses.

ECONOMIC	MEDICAL
Covid-19 pandemic: Huge global economic consequences, job losses affecting people's income, healthcare facilities under greater strain, reduced manufacturing productivity, disrupting supply-chains and pushing up prices of goods.	**Vaccines:** Viruses, or parts of their structure such as proteins or their genetic material (DNA/RNA), are used to manufacture vaccines. Vaccines stimulate the body's immune system to form antibodies against viral infections, without infecting the person with the disease.
Animal diseases: Foot-and-mouth disease is a contagious disease that affects animals such as cattle, pigs and sheep, causing animal deaths, loss in milk production and veterinary care costs.	**Human disease:** Diseases such as chickenpox, measles, mumps and influenza caused by viruses. Viral infections are not affected by antibiotics. Vaccines used to prevent viral infections.
Plant diseases: Plant viruses are of economic importance to farmers as they can cause disease in their crops, reducing yield, e.g. tobacco mosaic virus.	**Viral vectors in gene therapy:** Scientists are researching the use of viruses to carry necessary genes into a human cell in the treatment of diseases such as cancer. The viruses are modified so that they don't cause disease.
	Use of bacteriophages to treat infections caused by bacteria.

END OF CHAPTER ASSESSMENT

REMEMBER

1. Label parts A, B and C in the diagram of a virus below and give a function of each part.

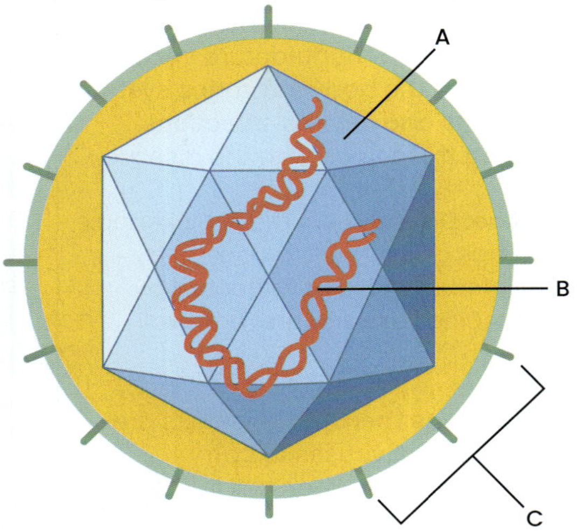

2. Name the **three** different shapes of viruses.
3. List the stages in viral replication.
4. Select the diseases that are caused by viruses from the following list:
   measles, tuberculosis, mumps, AIDS, influenza, cholera.
5. State (i) **three** living characteristics (ii) **three** non-living characteristics of viruses.

UNDERSTAND

6. Explain why viruses are often described as obligate parasites.
7. Classify each of the following viral traits as either living or non-living.
   (a) Protein present
   (b) Non-cellular
   (c) Can replicate inside a living host cell
   (d) Nucleic acid present
   (e) Only DNA or RNA present, but not both
   (f) No organelles present
   (g) Genetic information passed on to the next generation
8. Describe, with the aid of a labelled diagram, the stages in viral replication.
9. Discuss the economic and medical importance of viruses.
10. Explain how a vaccine works.

APPLY

11. Could you grow a virus in nutrient agar like can be done with bacteria? Explain your answer.
12. The Spanish Flu pandemic, which broke out in 1918, infected 500 million people around the world and resulted in the deaths of between 50 and 100 million people (3–5% of the world's population at the time). What measures would you recommend today to prevent the spread of a similar pandemic?

ANALYSE

13. Why do you think antibiotics can be used to treat bacterial infections but have no effect on viruses?
14. Compare and contrast a virus with an animal cell.
15. Research the pros and cons of using bacteriophages to combat bacterial infections.

EVALUATE

16. In some countries, vaccinations against certain diseases are required by law. Do you think that vaccinations should be made mandatory in Ireland?
17. Why do countries not vaccinate against all viral diseases?

CREATE

18. Research and design an information sheet for parents of young children to explain how vaccines work.

9 Evolution and Natural Selection

LEARNING OUTCOMES: 1.5.1, 1.5.2, 1.5.3

EYE-OPENER

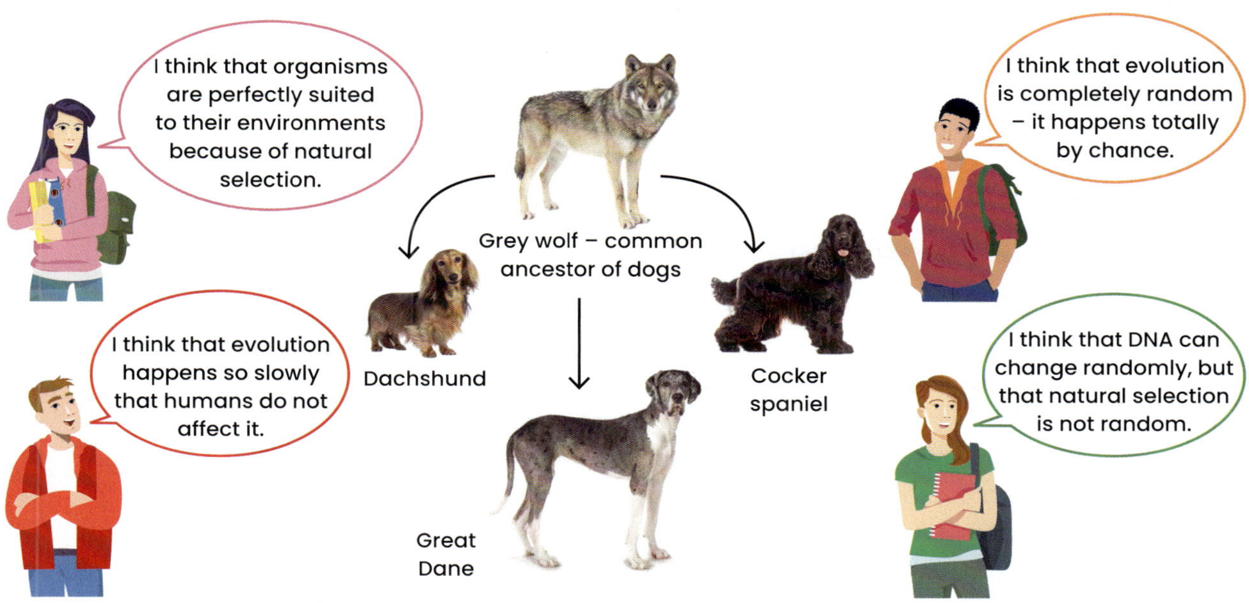

What do you think of each of the above statements? Discuss them in pairs.

At the end of this chapter, you should be able to:

- Understand the concept of a theory.
- Discuss the theory of evolution by natural selection, evidence for the theory and its value in the modern world.
- Use evolutionary theory to explain common origins and predict future directions for organisms.
- Explain how sexual reproduction and mutations (gene and chromosome) give rise to variation.
- Discuss natural and artificial selection.

LINKS TO:

Chapter 4	**Classification and Taxonomy**
Chapter 11	**Modelling an Ecosytem**
Chapter 17	**Cell Continuity**
Chapter 24	**Information Flow in the Cell**

SUBJECT VOCABULARY

variation, species, evolution, natural selection, mutation, theory, antibiotic resistance, artificial selection

Section 1: Theory of Evolution by Natural Selection

At the end of this section, you should be able to:

- Define the terms natural selection, species and evolution.
- Give examples of natural selection in action.
- Give an example of a characteristic that makes a species more adapted or suited to its environment.
- Discuss the main points of Darwin's theory of evolution.

Unpacked from LO 1.5.2

In Chapter 4, we learned about the efforts of biologists such as Carl Linnaeus in the 1700s to classify organisms into groups. Linnaeus believed that God had created all organisms in fixed, unchangeable species. However, the work of Charles Darwin resulted in an understanding that the classification of organisms could be better explained by the basis of a common ancestor. Darwin's research overthrew Linnaeus's idea of unchangeable species.

DARWIN AND WALLACE

Charles Darwin (1809–1882)

Charles Darwin was an English naturalist. In 1835, he visited the Galapagos Islands, off the coast of Ecuador. Here, Darwin observed similar but non-identical species of plants, birds and reptiles on the different nearby islands. For example, he observed that finches on one island that fed on insects had narrow, sharp beaks, while finches on a different island that fed on large seeds had larger, tough beaks. Darwin returned to England and spent the next two years putting together a theory to explain the variation in organisms that he had observed. This became the basis of his ground-breaking book, *On the Origin of Species*, which he did not publish for another 20 years. In this book, Darwin outlined his theory of evolution by natural selection.

Figure 9.1: Charles Darwin.

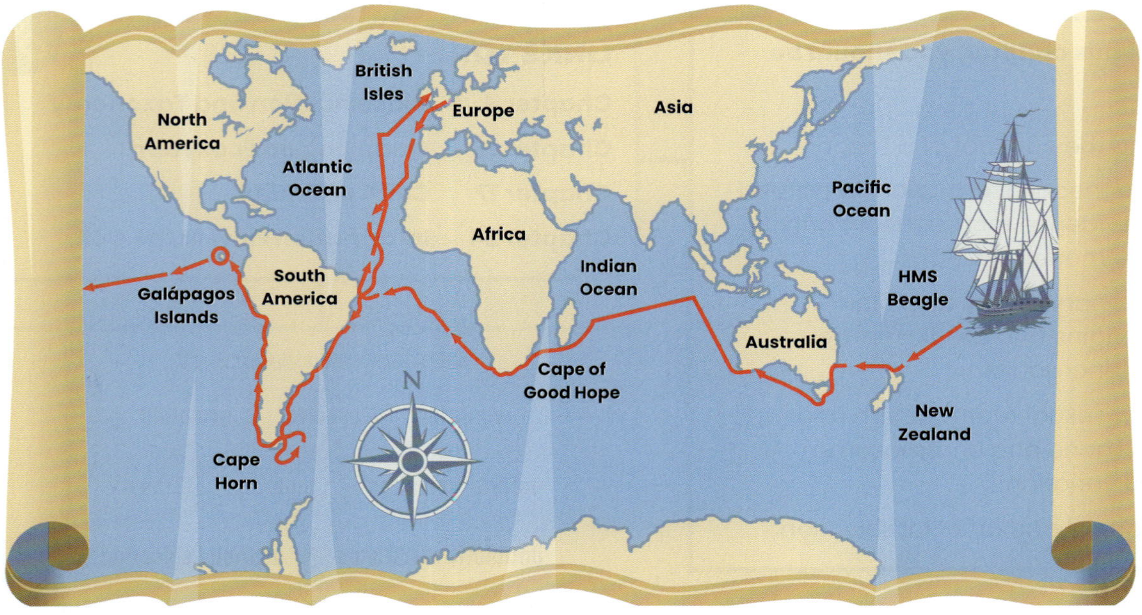

Figure 9.2: Voyage of Darwin's expedition ship, the HMS *Beagle*.

Alfred Russel Wallace (1823–1913)

Alfred Russel Wallace was an English naturalist. In 1858, while in Indonesia, Wallace came up with a theory to explain the variations he observed in the animal life of the islands. Wallace had independently come up with the theory of evolution by natural selection. He wrote to Darwin to tell him about his theory. Darwin and Wallace presented their work to the scientific community in 1858. In 1859, Darwin published *On the Origin of Species* while Wallace continued his travels. Since then, the theory of evolution by natural selection has been associated primarily with Darwin's name.

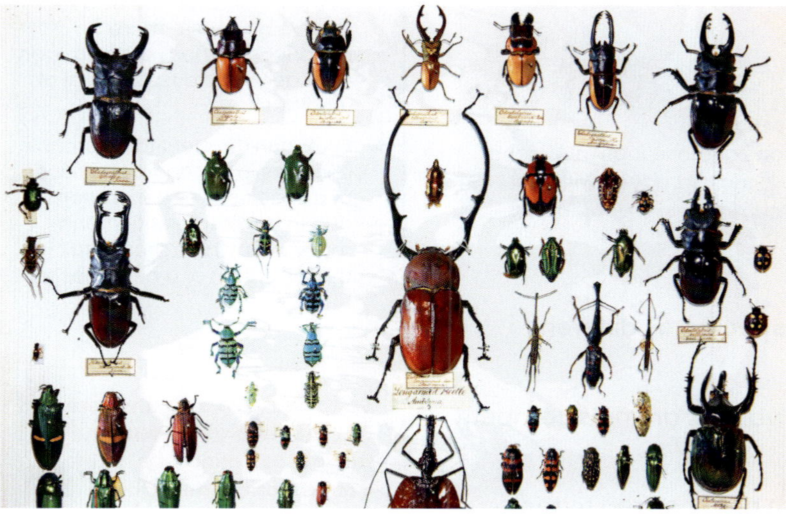

Figure 9.3: Beetles collected by Alfred Wallace.

Figure 9.4: English naturalist and explorer Alfred Russel Wallace.

EVOLUTION AND NATURAL SELECTION

We have mentioned that Darwin observed variation in the animals on the Galapagos Islands. For example, he observed that finches on the different islands had different shaped beaks depending on the island environment and what they fed on. Darwin believed that these finches had a common ancestor from which the different species had evolved. He proposed the process of natural selection to explain how these changes had evolved.

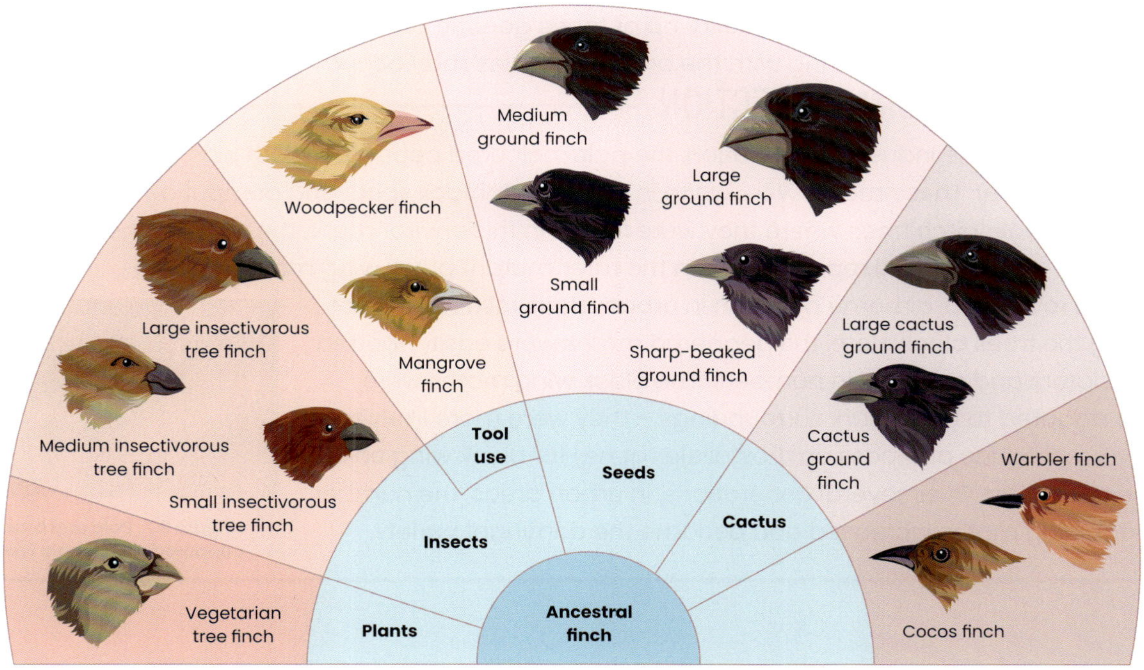

Figure 9.5: Darwin's finches – showing a common ancestor and adaptive traits of new species.

Theory of evolution by natural selection

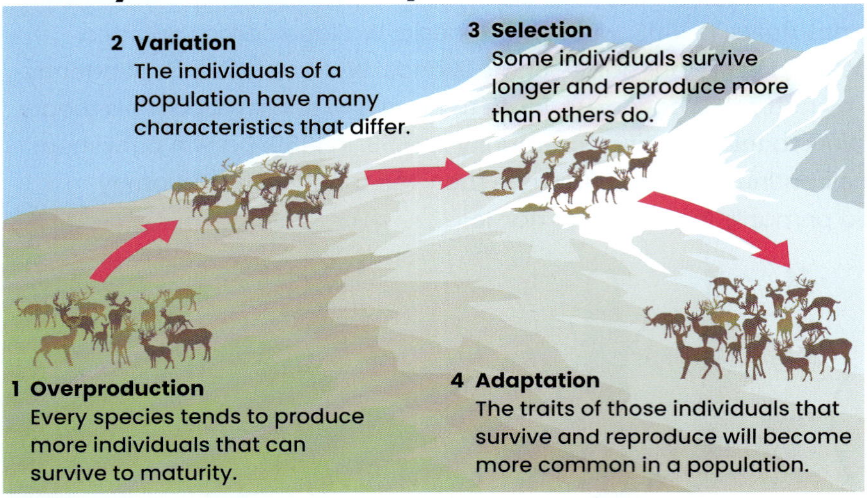

Figure 9.6: Theory of evolution by natural selection.

- A **species** is defined as a population of organisms that can interbreed successfully and produce fertile offspring.
- **Natural selection** is the process by which species members with genetic characteristics better adapted to their environment tend to survive and have more offspring to whom they pass on their genetic traits: 'survival of the fittest'.
- **Evolution** refers to inheritable changes in the characteristics of a species over generations due to natural selection and mutations, which may result in the formation of a new species.

- Variation exists between members of a **species**, as different individuals have different genes.
- Members of a species compete for survival, e.g. animals compete for food, plants for light.
- Individuals with features that make them better suited to their environment are more likely to survive and reproduce. Their genetic characteristics are passed to their offspring. For example, predators that can run faster and catch more prey are more likely to survive and reproduce. This is known as 'survival of the fittest' or '**natural selection**'.
- Individuals who are more poorly adapted to their environment are less likely to survive or reproduce. Their genes are less likely to be passed on to the next generation.
- Over time and successive generations, the genes (alleles) of the more successful individuals in a species become more common.
- Given enough time, and due to the build-up of these genetic changes, **evolution** of a new species which is unable to form offspring with the original species may occur.

Go to YouTube.com and search for 'The Theory of Evolution (by Natural Selection)' (3:28) by Cornerstones Education.

FOCUS ON ... NATURAL SELECTION

In Britain, before the Industrial Revolution, the pale-coloured peppered moth was the dominant variety. These moths were white in colour with black spots and camouflaged easily against the pale birch trees where they were found. In this environment, the pale moths were more likely to survive and reproduce than the rarer mutant black wing moth. With the industrial revolution, airborne pollution in urban areas resulted in soot staining the trees black. Here, the peppered moths were easily spotted by predators and declined in numbers. The black wing moths were better adapted to these dark surroundings – they were more likely to survive, reproduce and pass on their allele (gene) for black wing colour to their offspring. Over several generations, in urban areas, the number of black wing moths increased and became the dominant variety.

Figure 9.7: Pale and black peppered moths on a tree.

PORTFOLIO LINK **Go to your Learning Portfolio and complete Learning Check 1 to assess your knowledge and understanding of the topic so far.**

Section 2: Variation

At the end of this section, you should be able to:

- Define the term 'variation'.
- Explain that sexual reproduction gives rise to variation within a species.
- Explain that mutations (gene or chromosome) give rise to variation within a species.

Unpacked from LOs 1.5.1, 1.5.2

VARIATION FROM SEXUAL REPRODUCTION

In humans, somatic (non-sex) cells contain 46 chromosomes. Gametes or sex cells (egg and sperm) in humans contain 23 chromosomes. These haploid gametes form by a type of cell division called **meiosis**. When a sperm fertilises an egg during sexual reproduction, a diploid zygote containing 46 chromosomes is formed. This zygote grows to an adult by **mitosis**, another form of cell division.

Offspring inherit half of their chromosomes from their mother and half from their father. Each parent can form 2^{23} different sex cells; the diploid zygote formed in sexual reproduction is one of over 64,000,000,000,000 possible combinations that could form from the genes of the two parents! Thus, meiosis and sexual reproduction is a source of **variation** in a species.

Variation refers to differences between members of the same species.

Examples of genetic variation in humans include eye colour, blood group, skin colour and earlobe shape. These inherited variations are controlled by genes. Other differences between humans, known as acquired variation, are learned during life, e.g. ability to play the piano or speak a language. In Chapter 17, we will learn more about the two forms of cell division: meiosis and mitosis.

Figure 9.8: Children inherit genetic characteristics from both their parents.

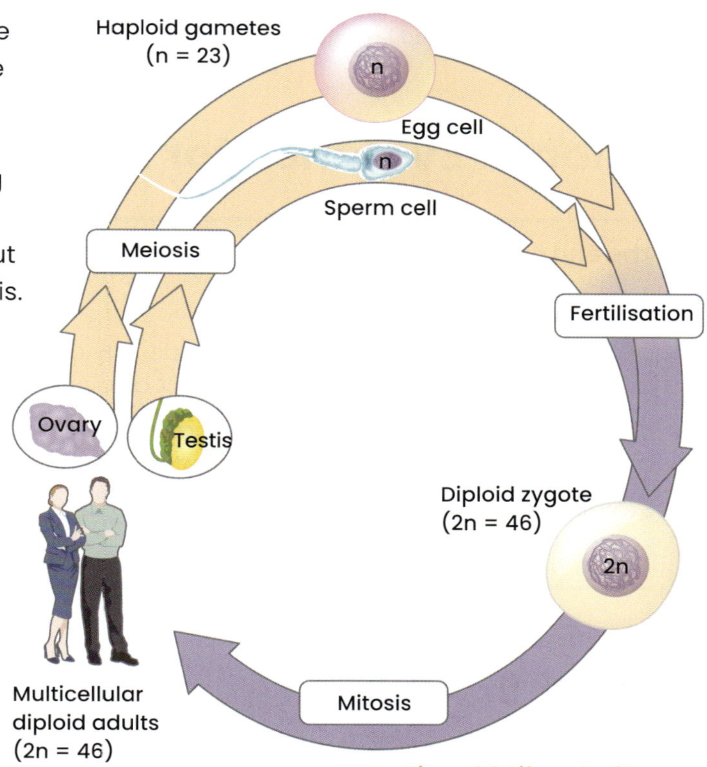

Figure 9.9: Life cycle of humans.

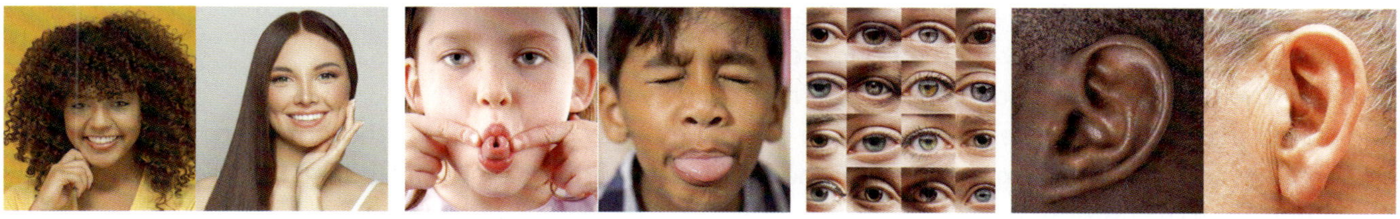

Figure 9.10: Variation in genetic traits in humans (L–R): hair texture, tongue mobility, eye colour, earlobe shape.

VARIATION FROM MUTATION

An organism's DNA contains the instructions for it to grow, develop and reproduce. The information in DNA is used to build proteins. A section of DNA that codes for a protein is called a gene. In Chapter 24, we will learn about the structure of DNA. Random changes in DNA are called **mutations**. Mutations that affect the DNA in egg and sperm cells may be inherited and passed on to offspring. In this way, mutations bring about genetic variation within a population. If a mutation gives an organism a beneficial trait, they are more likely to survive, reproduce and pass on this trait to their offspring through the process of natural selection. For example, some bacteria have developed resistance to certain antibiotics due to mutations. Over generations, as the number of individuals with this trait increases, the population can become so different that it develops into a new species.

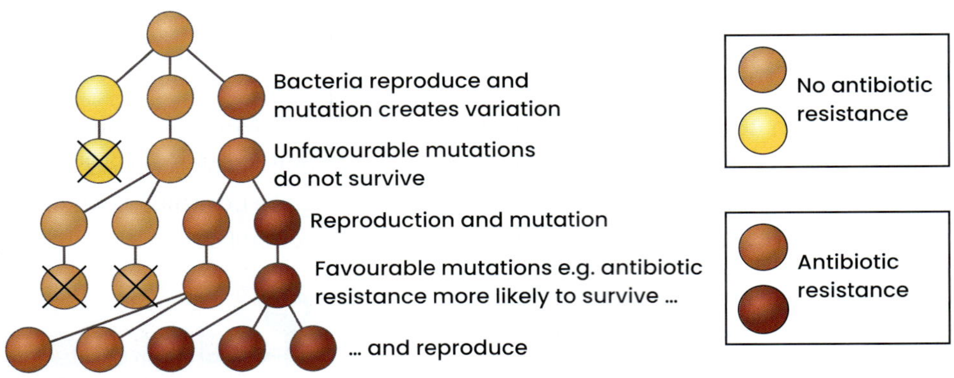

Figure 9.11: Mutations and variation in a population.

A **mutation** is a spontaneous, inheritable change in DNA (genes or number of chromosomes).

There are two main types of mutation: gene mutation and chromosome mutation. We will learn more about these types of mutation and how they occur in Chapter 24.

PORTFOLIO LINK **Go to your Learning Portfolio and complete Learning Check 2 to assess your knowledge and understanding of the topic so far.**

Section 3: Evidence for Evolution

At the end of this section, you should be able to:

- Outline the concept of a theory.
- Describe how embryology, fossil records, phylogeny, comparative anatomy, antibiotic resistance and speciation act as evidence for evolution.
- Explain how the theory of evolution can be used to predict common origins and future directions for highly specialised organisms.
- Explain the importance of the theory of evolution in understanding biology.

Unpacked from LOs 1.5.1, 1.5.3

CONCEPT OF A THEORY

A scientific **theory** is a well-established explanation for a natural phenomenon.

A scientific **theory** is a well-established explanation for a natural phenomenon that is supported by scientific evidence over a long period of time.

While it is not possible to prove a theory, a theory is testable. Should any new evidence not agree with a theory, the theory is refined or rejected. The theory of evolution is one of the most widely accepted and tested theories in science. We will outline evidence below that indicates that the theory of evolution presents the best available explanation for the progression and diversity of life (see Figure 9.12). However, it is important to remember that life on Earth continues to evolve to this day.

EVIDENCE FOR EVOLUTION

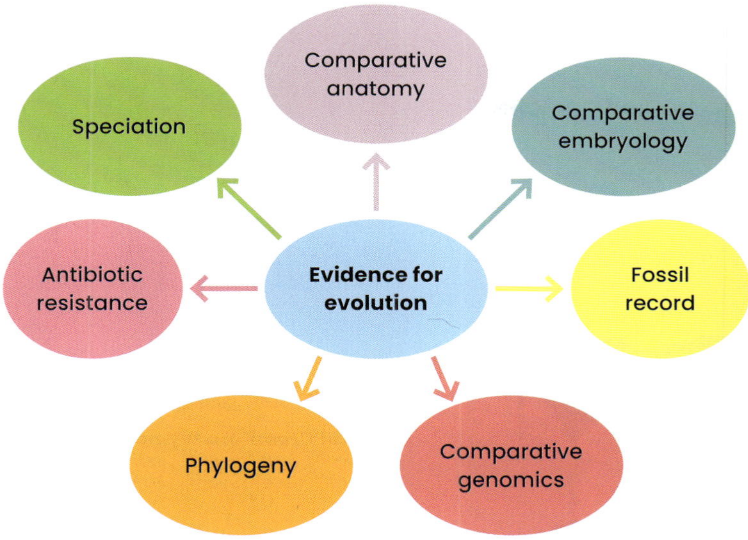

Figure 9.12: Evidence for evolution.

Comparative **G**enomics
Fossil **R**ecord
Comparative **A**natomy
Antibiotic resistance
Phylogeny
Comparative **E**mbryology
Speciation

Comparative anatomy

Many different vertebrate species share a similar or **homologous** bone structure feature. An example of this is the pentadactyl limb (five-fingered limb). Although they look different externally, humans, whales, bats, horses, birds and dogs all possess a pentadactyl limb. This indicates that these species share a common ancestor. During evolution, the pentadactyl limb in each species was modified as they adapted to their environment: the bat wing is adapted to flying, the whale fin is adapted to swimming and so on.

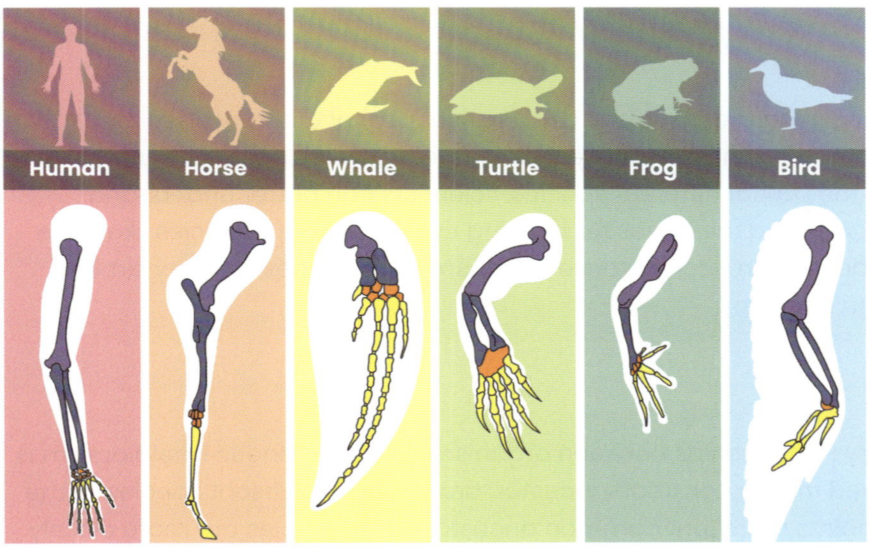

Figure 9.13: The pentadactyl limb is a homologous feature in many species.

Comparative embryology

All vertebrate embryos including those of humans, fish, birds and reptiles follow similar stages of development. Early in their development, these vertebrate embryos exhibit gill slits and a tail, reflecting the existence of a common ancestor. The embryos become more differentiated as they develop. In humans, the tail becomes the coccyx and the gill slits become the jaw and middle ear.

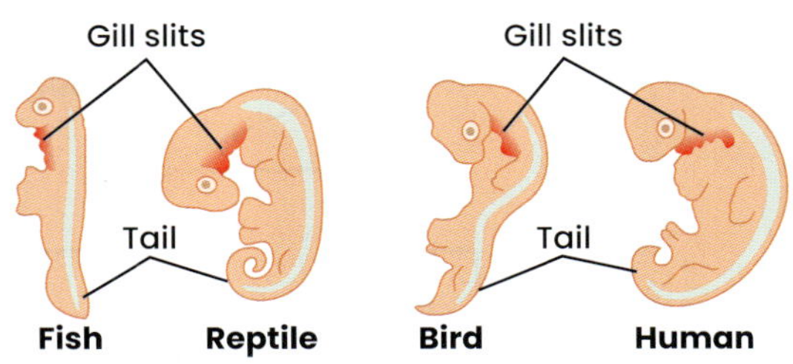

Figure 9.14: Comparative embryology of different species.

Fossil record

Fossils are the preserved remains in rocks of organisms that lived millions of years ago. Fossils of extinct organisms have been found allowing evolutionary timelines to be constructed. Gaps exist in the fossil record, as many early organisms were soft-bodied and did not form fossils.

Figure 9.15: Fossil of a seymouria (extinct).

Comparative genomics

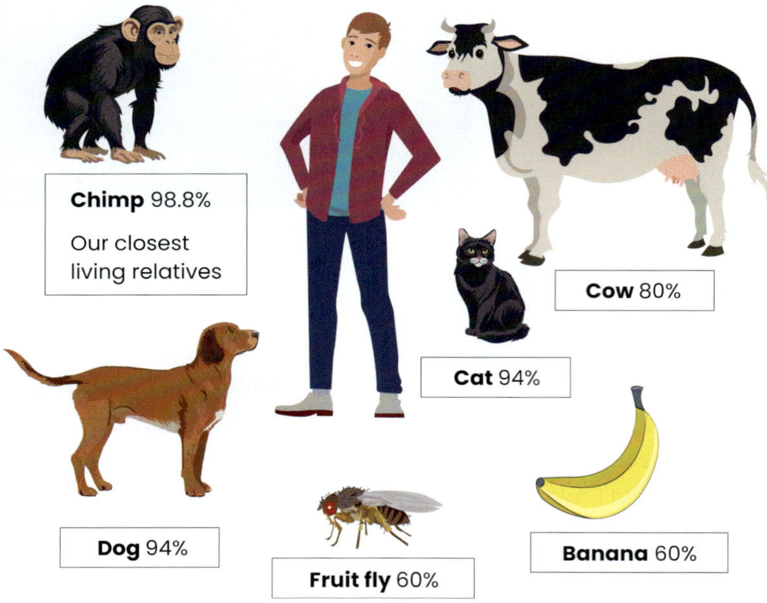

Figure 9.16: Comparative genomics.

Comparative genomics involves comparing the genomes (DNA sequences) of different species, e.g. humans and chimpanzees. The greater the similarity between genes in different species, the more closely they are related. By studying the DNA of different species, scientists can better understand their evolution, e.g. scientists have compared the genes for building eyes in different species like mice and sharks. Although different eyes develop in each of these organisms, their genes are very similar, indicating the presence of a common ancestor.

Phylogeny

Phylogeny is the study of evolutionary relationships between organisms. A phylogenetic tree diagram is used to represent how species evolved from a common ancestor; branch tips represent species at the end of their lineage and branching points represent common ancestors. Two species are more closely related if they have a more recent common ancestor.

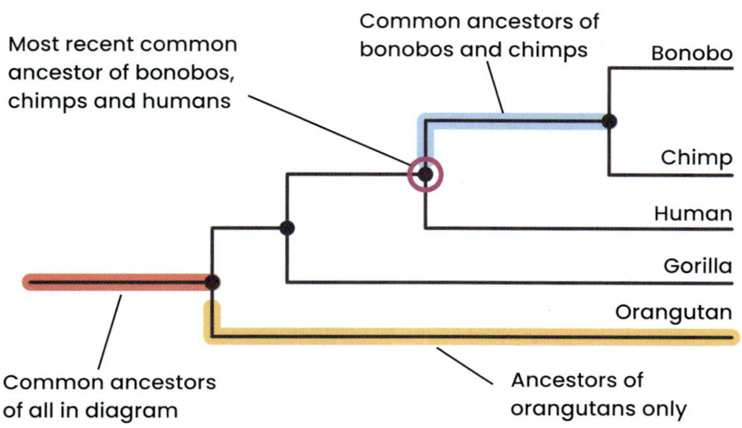

Figure 9.17: A phylogenetic tree.

Speciation

Speciation is the evolutionary process by which new species form from an original species.

Allopatric speciation occurs when species populations become geographically isolated from each other so they cannot transfer genetic material. The populations evolve separately and eventually become so genetically different that they can no longer interbreed, resulting in a new species.

Sympatric speciation is the evolution of a new species from an original species while both inhabit the same geographical location. Sympatric speciation can arise from reproductive isolation, e.g. a mutation occurs that only allows organisms with the mutation to successfully reproduce.

Go to YouTube.com and search for 'Evidence of Evolution' (5:31) by By: Rachel Taylor.

Figure 9.18: Allopatric and sympatric speciation provide evidence for evolution.

Antibiotic resistance

Both beneficial bacteria and harmful bacteria are killed by antibiotics. **Antibiotic resistance** is a growing problem as bacteria become immune to antibiotics that once killed them. When a person takes antibiotics, bacteria that are sensitive to the antibiotic are killed but resistant bacteria grow and multiply. Bacteria can acquire resistance through random genetic mutations or by acquiring genes for antibiotic resistance from another bacteria's plasmid. This is an example of natural selection – the antibiotic resistant bacteria survive and reproduce as they have a competitive advantage. Bacteria can reproduce every 20 minutes, allowing the number of bacteria with the antibiotic resistant mutation to grow rapidly. Thus, evolution can happen very quickly in bacterial populations. Misuses of antibiotics, e.g. using antibiotics to treat a viral infection or not completing a course of antibiotics, are some of the main causes of the spread of antibiotic resistance.

Antibiotic resistance occurs when bacteria are no longer affected by antibiotics which could once be used to kill them.

Go to YouTube.com and search for 'Evolution of Antibiotic Resistance in Bacteria' (2:41) by By: Rachel Taylor.

How does antibiotic resistance occur?

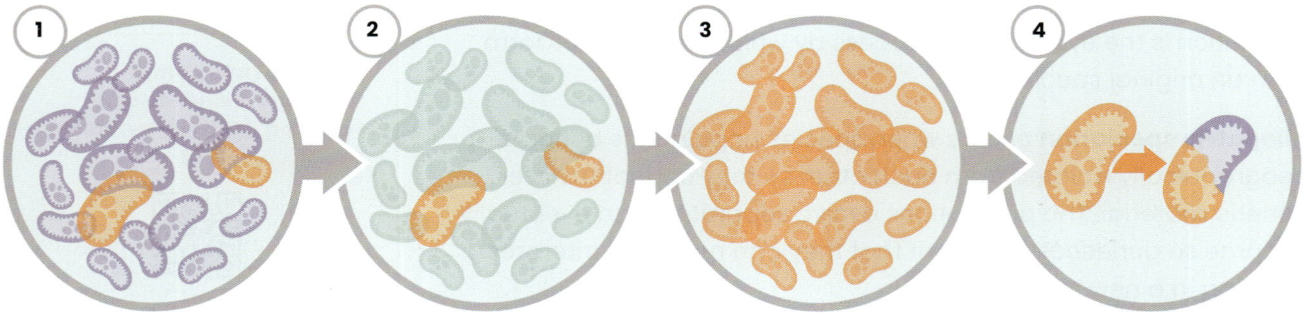

High number of bacteria. A few of them are resistant to antibiotics.

Antibiotics kill bacteria causing the illness, as well as good bacteria protecting the body from infection.

The resistant bacteria now have preferred conditions to grow and take over.

Bacteria can even transfer their drug resistance to other bacteria, causing more problems.

Figure 9.19: Development of antibiotic resistance by bacteria provides evidence for evolution.

FUTURE DIRECTIONS OF EVOLUTION

The Covid-19 pandemic reminded us that evolution can occur in weeks, days or even hours. Microbes such as bacteria reproduce quickly, going through several generations in a day, allowing scientists to watch evolution in action. Scientists look at past trends in evolution to make predictions about our future. Experiments on microbes have shown that they evolve in similar ways and that evolution is predictable in the short term, i.e. over the next few months or years, but it can't be predicted over long periods of time. If scientists can predict how viruses such as the flu or Covid-19 will evolve, they can work on ways to treat them.

THEORY OF EVOLUTION IN UNDERSTANDING MODERN BIOLOGY

Understanding evolution helps us to address biological problems that impact on society. We will look at the continued impact of evolution on agriculture, medicine and conservation.

Agriculture

In Ireland, we are aware how an overreliance on the potato for food resulted in famine in the 19th century. All the potatoes were of the same variety, so they were genetically identical and susceptible to the same potato blight which caused the potatoes to rot. We now know from evolution that lack of genetic variation makes a species more susceptible to disease.

Evolutionary theory also informs us on the speed at which insects can evolve to be resistant to pesticides. Using this knowledge, efforts can be made to slow the spread of pesticide resistant genes. Providing fields that are not sprayed with pesticides in between sprayed areas offers spaces where non-resistant insects can thrive. This knowledge is used to avoid crops being wiped out and helps ensure food supply.

THINK, PAIR, SHARE

How is evolution of importance in today's world?

SCIENCE IN SOCIETY: HEALTH

Medicine

Changes in our environment can occur much more quickly than the human genome can evolve. Our genetic makeup may not be adapted to deal with environmental factors that cause disease, e.g. a diet high in refined sugar and saturated fats. This mismatch partially explains how modern diseases such as diabetes or coeliac disease arise. We have also seen how evolution by natural selection explains how excessive use of antibiotics leads to drug resistance and provides a pathway to analyse the emergence of new diseases like Covid-19.

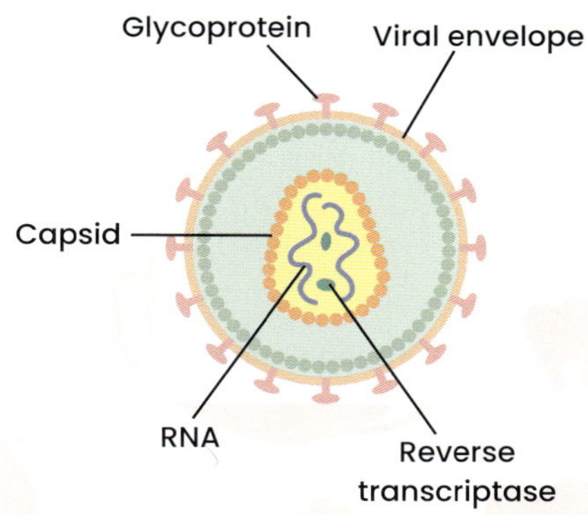

Figure 9.20: Studying the evolutionary history of viruses such as HIV may help us to treat against the disease.

SCIENCE IN SOCIETY: SUSTAINABILITY

Conservation

The number of endangered plant and animal species around the world is increasing. This may be caused by human activity. Invasive species or genetically modified organisms can also impact negatively on native populations in an ecosystem. If population sizes are allowed to become too low, inbreeding can occur and genetic variation is reduced. This can lower the population's ability to reproduce and survive. Evolutionary biology informs us of how to address conservation issues.

Figure 9.21: Scientists studied the evolutionary history of the endangered sea otter to inform conservation practices.

PORTFOLIO LINK **Go to your Learning Portfolio and complete Learning Check 3 to assess your knowledge and understanding of the topic so far.**

Section 4: Artificial Selection

At the end of this section, you should be able to:

- Define the term artificial selection.
- Distinguish between natural and artificial selection.

Unpacked from LO 1.5.2

Artificial selection – or 'selective breeding' – involves human intervention in plant or animal breeding. Parents with desirable traits are selected to breed together to produce offspring with these traits, rather than leaving the species to evolve gradually by natural selection. Artificial selection is often carried out for economic reasons, e.g. cattle that have been selectively bred to produce more lean meat, cows that produce more milk, chickens that produce larger eggs or crops with increased resistance to disease and nutritional content.

In **artificial selection**, humans intervene in animal or plant breeding by selecting parents with desirable traits to produce offspring with these traits.

Artificial vs natural selection

Natural selection and artificial selection both result in genetic changes in populations. Differences exist between the processes.

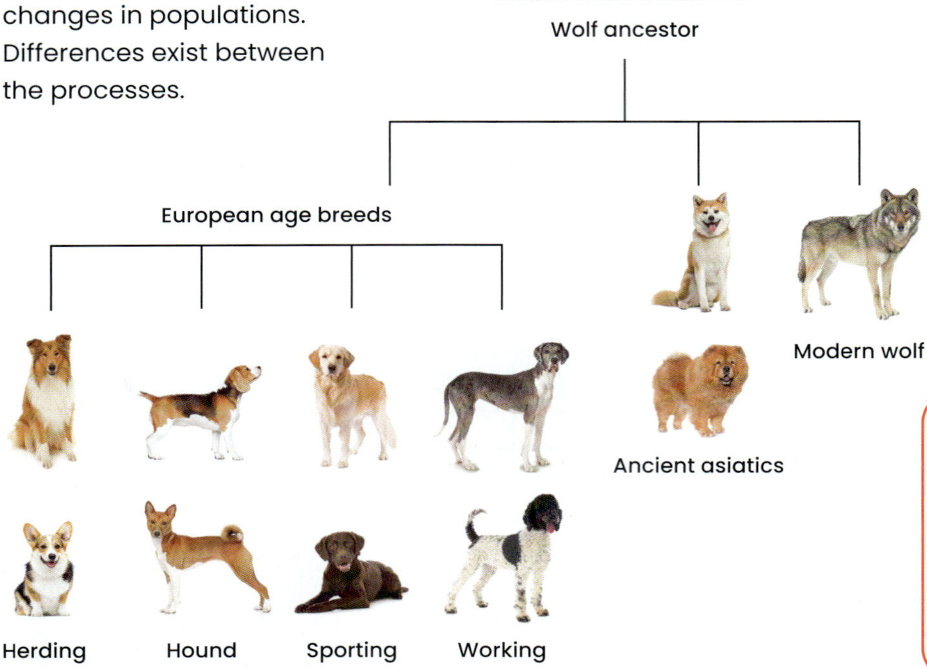

Figure 9.22: Artificial selection over thousands of years has resulted in many different domestic dog breeds.

Go to YouTube.com and search for 'GCSE Biology – Selective Breeding #77' (3:44) by Cognito.

Natural selection ...

- is driven by environmental factors, e.g. access to mates
- produces offspring with a diverse gene pool
- produces offspring better adapted to their environment that tend to survive and reproduce
- happens very slowly.

Artificial selection ...

- is driven by human intervention
- produces offspring with selected traits
- produces offspring that share similar genes, making them more susceptible to genetic diseases
- is driven by traits that do not necessarily enhance the species chance of survival; they are chosen for economic or aesthetic reasons
- occurs more quickly than natural selection.

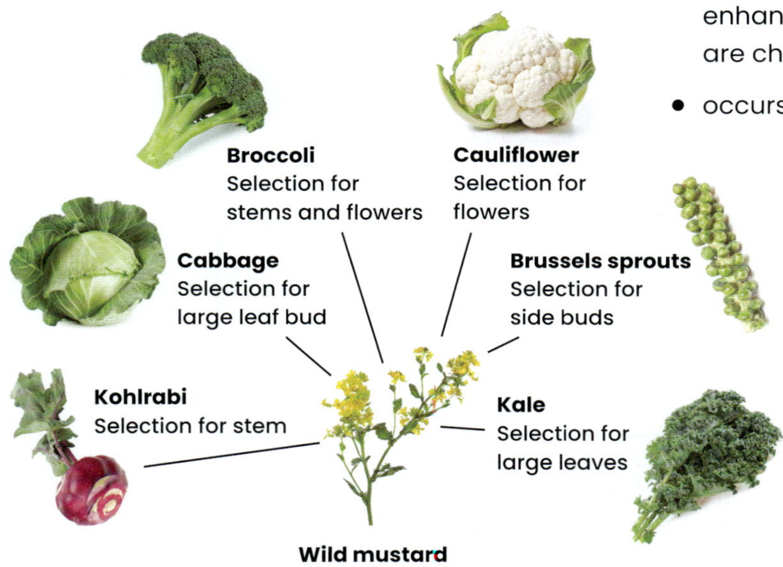

Figure 9.23: Humans have cultivated different types of crops from the wild mustard plant by artificially selecting for certain traits.

PORTFOLIO LINK Go to your Learning Portfolio and complete Learning Check 4 to assess your knowledge and understanding of the topic so far.

- Charles Darwin and Alfred Wallace proposed the theory of evolution by natural selection.
- **Natural selection** is the process by which members of a species with genetic characteristics making them better adapted to their environment tend to survive, have more offspring and pass on their genetic traits to the next generation.
- **Evolution** refers to inheritable changes in the characteristics of a species over generations due to natural selection and mutations, which may result in the formation of a new species.
- A **species** is a group of organisms with common characteristics that can mate to form fertile offspring.
- **Theory of evolution by natural selection:**
  - Variation exists between members of a species.
  - Members of a species compete for survival.
  - Individuals with features better suited to their environment more likely to survive and reproduce. 'Survival of the fittest'. Individuals more poorly adapted to their environment less likely to survive or reproduce.
  - Over successive generations, genes of the more successful individuals in a species become more common.
  - Given enough time, and due to the build-up of these genetic changes, a new species may evolve which is unable to breed with the original species.
- **Variation** describes differences in genetic characteristics among members of a species due to mutations and sexual reproduction (meiosis).
- **Mutations** are spontaneous, inheritable changes in DNA (gene or chromosome).
- A scientific **theory** is a well-established explanation for a natural phenomenon that is supported by scientific evidence over a long period of time.
- The following provide **evidence for evolution** of common ancestors between species: comparative genomics, fossil record, comparative anatomy, antibiotic resistance, phylogeny, comparative embryology, speciation.
- **Antibiotic resistance** occurs when bacteria are no longer affected by antibiotics which could once be used to kill them. Results from genetic mutations or acquiring antibiotic resistant genes from other bacteria. This is an example of natural selection.
- **Artificial** selection refers to human intervention in animal/plant breeding involving selecting parents with desirable traits to breed together to produce offspring with these traits.

NATURAL SELECTION	ARTIFICIAL SELECTION
Natural populations	Domestic populations
Driven by environmental factors	Driven by human intervention
Genetically diverse offspring	Genetically similar offspring
Slow process	Rapid process
Traits that enhance the chance of survival are passed on	Selected traits do not always enhance the chance of survival
Affects the entire population	Only affects the selected individuals
Diverse gene pool – less prone to genetic disease	Smaller gene pool – greater risk of genetic disease

END OF CHAPTER ASSESSMENT

REMEMBER

1. Define each of the following terms.
   (a) Evolution
   (b) Natural selection
   (c) Species
   (d) Mutation
2. Name the scientists who proposed the theory of natural selection.

UNDERSTAND

3. Summarise the theory of evolution by natural selection.
4. Discuss the variations that come from sexual reproduction.
5. Describe **two** pieces of evidence for evolution.

APPLY

6. Which of the following is an example of natural selection in action?
   (a) An increase in the number of antibiotic resistant bacteria in a population
   (b) A human increasing the size of the muscles in their arms by weight training
   (c) Breeding of thoroughbred racehorses
   Explain your answer.
7. A farmer sprayed an insecticide on crops in a greenhouse to kill insects. Most of the insects were killed, but a small number remained. A few months later, the insect numbers had increased, and he used the insecticide again. Again, most of the insects were killed but some remained. A few months later, the greenhouse was full of insects, and he sprayed the insecticide again. This time, only half the insects were killed. A few weeks later, he sprayed the insecticide again, but hardly any insects were killed.
   (a) Why do you think some of the insects were not killed by the insecticide?
   (b) Explain why more and more insects survived each time the insecticide was sprayed.
8. Grasshoppers, honeybees and butterflies use their mouths for different purposes. Grasshoppers bite and chew, honeybees bite and butterflies suck pollen. The mouths of all three are different but they share the same basic structure.
   (a) What does this information indicate about the origin of grasshoppers, honeybees and butterflies?
   (b) How did the differences in mouth structure of these insects arise?
9. Antibiotic resistance in bacteria is an example of evolution in response to an environmental change. Demonstrate, using another example, how an environmental change can lead to evolution.

ANALYSE

10. Compare and contrast natural selection and artificial selection.

EVALUATE

11. Evaluate the following statement in relation to evolution:
    A trait linked with survival in one population today may no longer be helpful in several generations or even in other modern-day populations.

CREATE

12. Hold a class debate on whether evolution is important to the modern world.
13. Create a model of a virus that shows the internal and the external structure.

UNIT 3

ECOSYSTEMS AND ECOLOGY

CHAPTERS

STRANDS

Unifying Strand: Nature of Science

Strand 3: Interactions of Life

10 Biodiversity and Ecosystems

LEARNING OUTCOMES: U3.1, U4.1, U4.2, 3.1.1, 3.1.2, 3.1.4, 3.1.5, 3.1.6, 3.1.7, 3.1.8, 3.1.9, 3.1.10, 3.1.11

EYE-OPENER

The bombardier beetle is a remarkable insect known for its unique defence mechanism. When threatened, it ejects a hot, noxious chemical spray from its abdomen. This defensive spray can deter predators, such as frogs and birds, by causing irritation or even pain. With over 500 species worldwide, bombardier beetles inhabit various ecosystems, from forests to grasslands. Their fascinating adaptations not only protect them from predators, but also highlight the incredible diversity and complexity of insect survival strategies.

By the end of this chapter, you should be able to:

- Describe the main features of ecosystems, understanding the organisation within them.
- Define the term abiotic factors, giving examples and describing how abiotic factors affect species distribution.
- Describe biodiversity, the various kinds and how it is calculated using the Simpson's species biodiversity index. HL
- Describe the various ecological relationships found in ecosystems, giving examples of each.
- Outline how food chains, food webs and ecological pyramids can be used to track the flow of energy and matter in an ecosystem.
- Describe how populations of species change over time and discuss the concept of carrying capacity and the factors affecting it.

LINKS TO:

Chapter 4	**Classification and Taxonomy**
Chapter 5	**The Animal Kingdom**
Chapter 6	**The Plant Kingdom**
Chapter 7	**Bacteria and Fungi**
Chapter 13	**Human Impact on the Environment**
Chapter 22	**Photosynthesis**

SUBJECT VOCABULARY

biosphere, ecology, ecosystem, habitat, community, population, adaptation, niche, biodiversity, species diversity, habitat diversity, conservation, producer, consumer, predation, competition, intraspecies competition, biomass, food chain, trophic level, food web, contest competition, scramble competition, symbiosis, pyramid of numbers, pyramid of biomass

Section 1: Ecology and Ecosystems

At the end of this section, you should be able to:

- Explain the term 'ecology'.
- Describe some of the interactions living things have with other living things and their surroundings.
- Describe what is meant by a community of organisms.
- Define 'population'.
- Describe ecosystems as the unit of ecology.
- Describe the parts of an ecosystem.
- Discuss the importance of adaptations in filling ecological niches.

Unpacked from LOs 3.1.1, 3.1.10

WHAT IS ECOLOGY?

As far as we know, Earth is the only planet in our solar system and perhaps the universe where life exists. As we discussed in Chapter 1, life exists in almost every corner of the planet and has adapted to some of the most extreme environments. Yet there are some parts of the planet that don't support life – deep below the surface of the Earth and high into the atmosphere. We call the part of the Earth where life can be found the **biosphere**.

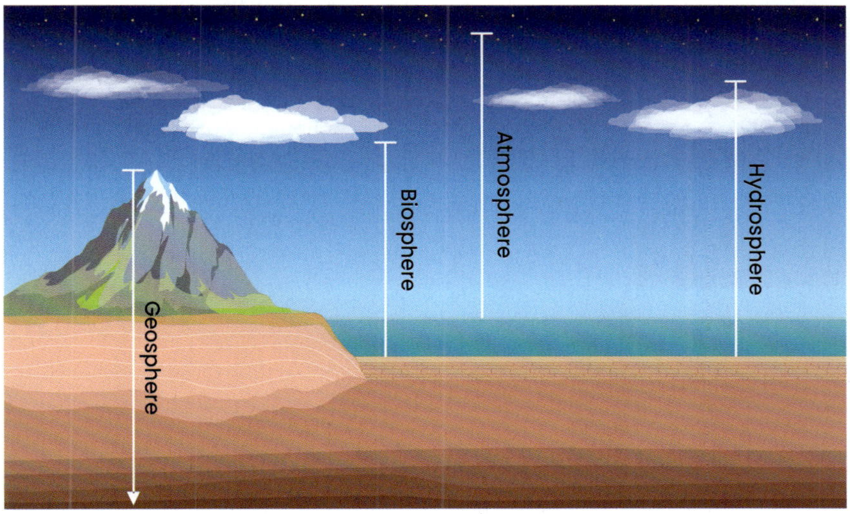

Figure 10.1: Earth's spheres.

- The **biosphere** is the part of the Earth where life exists.
- **Ecology** is the study of living things and their interactions with their environment.

Different parts of the world will contain different types of living things and **ecology** is all about understanding why this happens. For example, cacti live in dry areas while moss live in wetter environments. A living thing cannot survive on its own and will depend on other organisms for food, shelter or to make new organisms.

Living things interact with other living things all the time – they might eat them, be eaten by them, compete with them or, if they're the same species, mate with them. But a living thing will also interact with their environment – they will breathe the air around them, add nutrients to the soil when they urinate or perhaps set up home in a cave to avoid the weather conditions.

Ecology is the study of all these various interactions and it's a fascinating and interest area of biology. In many ways, it's about zooming right out and seeing the living world in all its glory. However, the ecological world is coming under enormous stress and strain, as pollution, global warming and the destruction of forests and other parts of the world are causing the extinction of many species and reducing biodiversity.

ECOSYSTEMS – THE BUILDING BLOCKS OF ECOLOGY

We talk about 'building blocks' a lot in biology; cells are the building blocks of all living things while biomolecules are the building blocks of cells. All molecules have building blocks themselves – atoms (at Junior Cycle Science, we learned that even atoms have their own building blocks).

Ecosystems are the building blocks of ecology and, since the entire living world would be impossible to study, ecosystems allow us examine relationships and interactions between organisms in a manageable way. We define an ecosystem as a group of organisms interacting with each other and their environment.

An ecosystem is made up of two main parts: a **habitat** and a **community**. The habitat is the place where the organisms live while the community is made up of all the various different species living there. A habitat can also be living: an oak tree, for example, can be home to many different species of animals (squirrels, insects, etc.) and plants (moss or ferns).

A community can be divided up into various **populations** of different species and, in many cases, these different populations are dependent on one another. For example, bees rely on plants for food, while the plants rely on the bees to act as pollinators. Populations fluctuate, and we will discuss this later in the chapter.

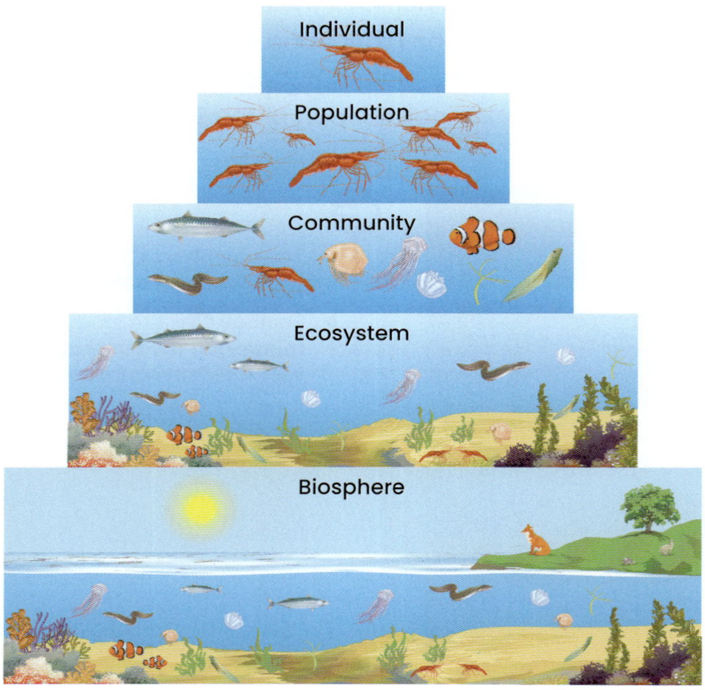

Figure 10.2: Ecological organisation.

- An **ecosystem** is the unit of ecology – a group of organisms interacting with each other and their environment.
- A **habitat** is a place where an organism lives.
- A **community** is a group of different species living in a particular ecosystem.
- A **population** is the number of each type of species in an ecosystem.

In an ecosystem, organisms interact with each other (mating, preying upon, digesting) and with their environment also (making homes, adding nutrients, breathing air). Studying and understanding these interactions is key to understanding how ecology works. As part of our studies, we will explore local ecosystems using some hands-on techniques (more on this in the next chapter) and engaging with current news and media.

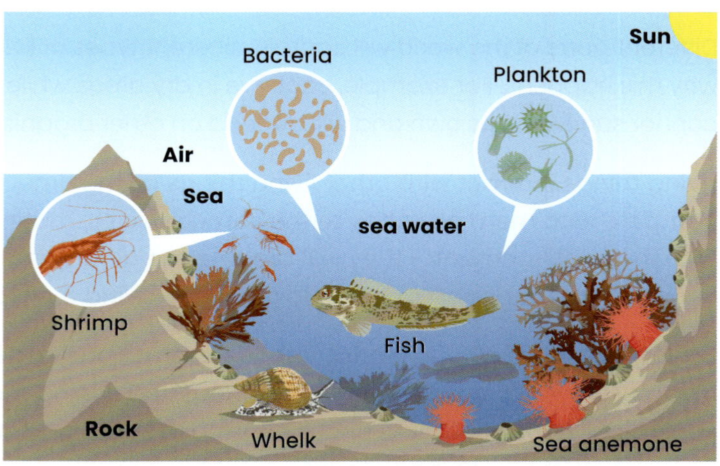

Figure 10.3: A rock pool ecosystem.

ADAPTATIONS AND NICHES

Organisms seem perfectly suited to the habitat they live in – think of a squirrel's bushy tail which gives it perfect balance as it climbs trees or its sharp teeth to cut through the hard shell of nuts or a hawk's razor-sharp eyesight and talons making it the perfect predator. We call these features **adaptations** – physical or behavioural features that allow each organism to secure the resources they need, evade predators and reproduce successfully.

Each organism has a particular role or **niche** in their ecosystem – some may make food, others might be predators while some might recycle nutrients. Adaptations allow organisms carry out their role and ensure survival in their ecosystem.

It is important to note that an organism's adaptations mainly develop through evolutionary processes like natural selection, which allow them to fill a particular niche in an ecosystem.

- An **adaptation** is a feature of an organism that allows it thrive and survive in its ecosystem.
- A **niche** is the role of an organism in its ecosystem.

Go to YouTube.com and search for 'What Is Adaptation? | Ecology & Environment | Biology' (1:22) by FuseSchool.

Figure 10.4: A camel's adaptations allow it to fill a particular niche in a desert ecosystem.

Figure 10.5: Water lilies have adaptations to fill a niche in a pond ecosystem.

THINK, PAIR, SHARE

Ecosystems are delicate and a change to the habitat or the community can destabilise it. There are many reasons why an ecosystem could become unstable or even collapse entirely but, sadly, human activity is highest on the list. In the news and media, we hear of stories of how human activity or population is having widespread impacts on ecosystems.

With a partner, discuss the ways an ecosystem might become unstable, apart from human interference. Share your thoughts with the class.

PORTFOLIO LINK Go to your Learning Portfolio and complete Learning Check 1 to assess your knowledge and understanding of the topic so far.

Section 2: Abiotic Factors

At the end of this section, you should be able to:

- Define the term abiotic factors, categorising them as air, ground, aquatic or edaphic factors.
- Give examples of abiotic factors and briefly describe their effects.
- Describe how abiotic factors affect species distribution.

Unpacked from LOs 3.1.5, 3.1.11

WHAT ARE ABIOTIC FACTORS?

So far, we have talked about the living things in an ecosystem: their diversity and adaptations. However, ecosystems are not just about the animals, plants and micro-organisms within them; remember, an ecosystem is defined as a community of organisms interacting with each other **and** their environment. That environment is not living – it is made up of the air, water, rocks and soil – but these features have a major influence on the living things they host.

Abiotic factors are the non-living parts of an ecosystem. They can be broadly divided into four categories:

Abiotic factors are the non-living components of an ecosystem.

1. **Air:** These are prevailing atmospheric conditions in the ecosystem, often called **climatic factors**. They include air temperature, humidity, wind speed and direction, and rainfall.
2. **Ground:** These refer to the physical characteristics of the land surface, particularly topographical elements. They include the slope, altitude, elevation and aspect (which direction the land faces).
3. **Aquatic:** These are specific to water environments and include temperature, oxygen concentration, pH, currents and salinity (how salty it is).
4. **Edaphic factors:** These factors relate to the soil in the ecosystem: its fertility, pH, composition and temperature.

Together, these factors create the conditions for life to thrive and are often considered the architects of ecosystems. Different abiotic factors, in different ecosystems, will mean different plants and animals might live there.

EXAMPLES OF ABIOTIC FACTORS

Air

ABIOTIC FACTOR	EFFECT ON ORGANISMS IN AN ECOSYSTEM
Temperature 	Temperature influences the metabolic rate, reproduction and overall physiological functions of organisms. Different species have specific temperature ranges within which they thrive, and extreme temperatures can stress or harm them.

Light intensity	Sunlight is crucial for photosynthesis in plants, which forms the basis of the food chain. The amount and duration of light exposure influence plant growth and the behaviours of animals that rely on light for visibility and navigation.
Rainfall	Water availability is vital for the survival of all organisms. It affects hydration, nutrient transport and various physiological processes. Aquatic ecosystems are particularly dependent on water quality and quantity.
Air	The composition of the atmosphere, including oxygen, carbon dioxide and nitrogen levels, is critical for the respiration and metabolism of organisms. Changes in atmospheric conditions can affect climate and weather patterns.
Humidity	Humidity levels in the air influence water loss and gain in organisms, affecting their ability to adapt to their ecosystem.
Wind	Wind speed and direction can influence temperature and water loss in organisms. It also plays a role in seed dispersal, pollination and the distribution of airborne organisms.

Ground

ABIOTIC FACTOR	EFFECT ON ORGANISMS IN AN ECOSYSTEM
Slope	Slope affects water and nutrient runoff, soil depth and microclimate. Steeper slopes typically have thinner soils and drier conditions, influencing the types of plants and animals that can survive there. Slope also impacts erosion rates and habitat stability.
Aspect	Aspect refers to the direction a slope faces. It affects the amount of solar radiation received, influencing temperature, moisture levels and vegetation types. In the Northern Hemisphere, south-facing slopes are generally warmer and drier than north-facing slopes.
Altitude	As altitude increases, air pressure and temperature decrease while UV radiation increases. This creates distinct vegetation zones and affects species distribution. Many organisms have adaptations for specific altitude ranges.
Landforms	The physical features of the land, such as mountains and valleys, can affect water drainage, exposure to sunlight, wave action, wind and temperature gradients.

Aquatic

ABIOTIC FACTOR	EFFECT ON ORGANISMS IN AN ECOSYSTEM
Salinity 	Salinity levels in aquatic environments affect the types of organisms that can thrive. Many species have specific tolerances to salt concentrations, influencing the composition of marine and freshwater ecosystems.
Currents 	Currents affect nutrient distribution, oxygen levels and the dispersal of organisms. Strong currents can shape physical habitats and influence the feeding strategies of aquatic species.
Water temperature 	Temperature impacts metabolic rates, dissolved oxygen levels and the distribution of aquatic species. Many aquatic organisms have specific temperature ranges they can tolerate.
Dissolved oxygen 	Critical for aquatic life, dissolved oxygen levels affect species distribution and abundance. Low oxygen levels can lead to fish kills.

Edaphic

ABIOTIC FACTOR	EFFECT ON ORGANISMS IN AN ECOSYSTEM
pH/acidity 	The acidity or alkalinity of soil and water can impact nutrient availability and the chemical reactions occurring in living organisms. Some species are adapted to specific pH ranges.
Soil temperature 	Soil temperature influences seed germination, root growth, nutrient uptake and microbial activity in the soil. It affects plant growth rates and the timing of seasonal events like flowering.
Fertility 	Soil fertility determines the availability of nutrients for plants, affecting vegetation types and productivity. It influences the entire food web built upon primary producers.
Composition 	The physical properties of soil (e.g. texture and organic matter content) influence water retention, nutrient availability and the types of plants and soil organisms that can thrive in an area.

ABIOTIC FACTORS AND SPECIES DISTRIBUTION

Abiotic factors are often the reason why a particular species of plant or animal is found in an ecosystem. In ecosystems where an abiotic factor may vary (e.g. light intensity in a woodland ecosystem), we might see a change in the distribution of that species there. In a woodland, ferns will be found in greater abundance in the shaded areas while bluebells grow in sunnier areas.

On a rocky seashore, the distance from the shore is an abiotic factor. Some species are better adapted to live close to the shoreline, where they are always covered in water (e.g. kelp or sea anemones), while other species are better adapted for tidal zones, where they are only covered by the tide occasionally (e.g. limpets and periwinkles).

Figure 10.6: Ferns prefer shaded areas of woodland ecosystems.

Figure 10.7: Bluebells prefer areas with more sunlight.

Figure 10.8: Kelp is adapted to life underwater.

Figure 10.9: Limpets prefer life away from the lower shore, in the tidal zone.

When you study an ecosystem, you will need to assess how an abiotic factor affects the distribution of a species. We will learn more about this in Chapter 13.

PORTFOLIO LINK **Go to your Learning Portfolio and complete Learning Check 2 to assess your knowledge and understanding of the topic so far.**

Section 3: Biodiversity

At the end of this section, you should be able to:

- Define the term 'biodiversity'.
- Discuss the importance of biodiversity.
- Explain the different kinds of biodiversity: species diversity, ecosystem diversity and genetic diversity.
- Using Simpson's species diversity index, calculate the variety of species in a particular ecosystem. (HL)
- Distinguish between species richness and species evenness.
- Define the term 'conservation'.
- Discuss some approaches that scientists are taking to conserve species.

Unpacked from LOs U3.1, U4.1, U4.2, 3.1.1, 3.1.2, 3.1.4

WHAT IS BIODIVERSITY?

We learned in Chapter 4 that there is a wide variety of life across the Earth – perhaps as many as 14 million different species. This variety of life is called **biodiversity**. We can describe biodiversity in a broad sense or we can focus on a small ecosystem and the variety of life within it.

Biodiversity is the variety of life on Earth or in an ecosystem.

Ecosystems with a wide range of organisms are more balanced as they may have multiple species filling each of the various niches in the ecosystem. If an ecosystem has a small diversity of species, the loss of just one species (with an important niche) might result in the ecosystem collapsing. Imagine that the grasses in the African savannah vanished because of a disease. What impact might that have on the grazing animals there – and the animals that eat those grazing animals?

Biodiversity is important for humans, too. Earth's various species provide us with food, medicine, fuel and help regulate the Earth's climate and soils. It is in our best interest to maintain a large biodiversity in the ecosystems around us but sometimes human activity – deforestation or agriculture – can impact biodiversity negatively.

RESEARCH ACTIVITY

Keystone species are essential species within ecosystems. Do some quick research to find examples of keystone species in your local ecosystems, e.g. bogland, woodland, etc. Are these species under threat?

Go to YouTube.com and search for 'Why is biodiversity important – with Sir David Attenborough' (5:39) by The Royal Society.

TYPES OF BIODIVERSITY

We can describe biodiversity in a number of ways: species diversity, habitat diversity and genetic diversity.

Figure 10.10: Types of biodiversity.

Species diversity is the number of and relative abundance of different species in an ecosystem.

SPECIES DIVERSITY

As the name suggests, **species diversity** refers to the number of species in an ecosystem but also their relative abundance. If you think of a forest ecosystem, there might be hundreds of different species of plants but relatively small numbers of each. In contrast, on farm grassland there might be only a few species but larger numbers of each.

FOCUS ON ... SPECIES RICHNESS VS SPECIES EVENNESS

Two important measures of species diversity are species richness and species evenness.

We refer to the number of different species in an ecosystem as the species richness while the species evenness refers to the relative abundance of the species.

Figure 10.11 shows two woodland ecosystems, A and B. Woodlands A and B both have three different species of tree in the woodland, meaning they have the same species richness; however, woodland A has a greater species evenness with more even numbers of each type of tree.

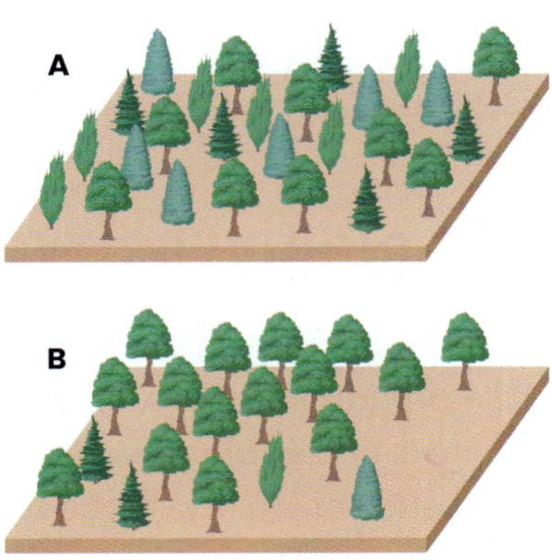

Figure 10.11: Two woodland ecosystems.

HABITAT DIVERSITY

Sometimes large ecosystems can contain more than one habitat. If you imagine a large woodland, numerous different habitats can exist. Squirrels, birds and insects will live in the trees, fungi and woodlice might live on a decaying tree trunk while rabbits might live in burrows under the soil. **Habitat diversity** describes the range of different habitats found within an ecosystem.

GENETIC DIVERSITY

Members of the same species will all contain the same genes but genes can have different versions called alleles (e.g. everyone has genes for their eye colour but some will have versions for brown, blue, etc.). **Genetic diversity** refers to the differences in the alleles amongst members of the same species in an ecosystem.

- **Habitat diversity** is a measure of the number of different habitats in an ecosystem.
- **Genetic diversity** is a measure of the variation of the alleles in a population in an ecosystem.

Genetic diversity gives rises to **variation**, an important factor in the evolution of a species through natural selection. Species with large genetic diversity are better able to adapt to environmental change, disease or other factors. Ecosystems with low genetic diversity are more likely to face challenges.

SCIENCE IN SOCIETY: SUSTAINABILITY

Modern farms, with their focus on uniformity and high yields, often struggle with the lack of species and genetic diversity. 'Monoculture' dominates, where large fields contain the same crop, limiting the variety of plants and animals. This lack of diversity makes farms vulnerable to pests and diseases, demanding heavy pesticide and fungicide use. Additionally, it weakens resilience to environmental changes and threatens food security. Many farmers are being encouraged to embrace more sustainable agricultural practices that incorporate more diverse crops and sustainable farming methods to overcome these challenges.

Figure 10.12: A monoculture crop.

RESEARCH ACTIVITY

Carry out some research on how farmers can improve the species and genetic diversity on their farms. Find out examples of sustainable ways to manage pests and diseases on farms.

MEASURING BIODIVERSITY

HL

It can be useful to quantify the biodiversity within an ecosystem. One of the most common ways of measuring biodiversity is by using the mathematical model called Simpson's **species diversity index**. This takes into account both the species richness and species evenness in an ecosystem to place a numerical value on the diversity: the value ranges from 0 (no diversity) to 1 (infinite diversity).

$$Ds = 1 - \left(\frac{\Sigma n(n-1)}{N(N-1)}\right)$$

Ds = Species diversity index
n = Number of organisms of a particular species (e.g. daisies)
N = Number of organisms of all species (in the ecosystem)
Σ = Sum of (in this case, we add all the values of n(n – 1) together)

Worked example: Species biodiversity index

A student carried out a survey of plants and small animals in a sand dune ecosystem, taking 10 random samples. The following species and their numbers were noted.

SPECIES	NUMBER FOUND (n)	SPECIES	NUMBER FOUND (n)
Marram grass	80	Lady's bedstraw	10
Hawkweed	24	Banded snail	32
Bird's foot trefoil	12	Ramshorn snail	20

Figure 10.13: A sand dune ecosystem.

To calculate the biodiversity index, it is best to use a table and have a calculator ready. We can calculate N by adding the number of each species together.

SPECIES	NUMBER FOUND (n)	n – 1	n (n – 1)
Marram grass	80	79	80 × 79 = 6320
Hawkweed	24	23	552
Bird's foot trefoil	12	11	132
Lady's bedstraw	10	9	90
Banded snail	32	31	992
Ramshorn snail	20	19	380
N	178		Σ n(n - 1) = 8466

HL

For the above example, the figures required by the formula are:

N = 178

N – 1 = 177

Σ n(n – 1) = 8466

$$\mathbf{Ds = 1 - \left(\frac{\Sigma n(n-1)}{N(N-1)}\right)}$$

$$Ds = 1 - \left(\frac{8466}{178 \times 177}\right)$$

$$Ds = 1 - \left(\frac{8466}{31506}\right)$$

$$Ds = 1 - 0.268$$

$$\mathbf{Ds = 0.732}$$

In this example, 0.732 indicates a relatively high diversity according to Simpson's index.

Note: You may be asked to carry out one of these calculations in an examination.

CONSERVATION

As mentioned previously, ecosystems are delicate and many are suffering from biodiversity loss. In many cases, the loss of biodiversity is a result of human agriculture, industry or construction. Any effort to protect ecosystems, including their biodiversity, is called **conservation**. It is a broad area and sometimes referred to as the 'wise management of Earth's resources'; Earth's various species are amongst its greatest resource. We will explore conservation in more detail in Chapter 13.

Conservation is the management of Earth's resources, including its biodiversity.

PORTFOLIO LINK **Go to your Learning Portfolio and complete Learning Check 3 to assess your knowledge and understanding of the topic so far.**

Section 4: Ecological Relationships

At the end of this section, you should be able to:

- Describe the main types of feeding relationships in ecosystems.
- Distinguish between producers and consumers, giving examples of each.
- Describe the different types of consumers in an ecosystem: herbivores, carnivores, omnivores, detritus feeders and decomposers.
- Define the term 'biotic factor'.
- Define the term 'predation'.
- Define the term 'competition' and describe the types of competition within ecosystems. HL
- Give examples of symbiosis in ecosystems, outlining the different types. HL

Unpacked from LO 3.1.11

All organisms encounter other organisms throughout their lives, interacting with them in various different ways. In this section, we will look at the various ecological relationships and interactions that could exist within ecosystems.

FEEDING RELATIONSHIPS

The most obvious form of interaction between two living things is a feeding relationship – where one organism eats another organism for food. Food is a source of energy and raw materials for new cells. We can track a feeding relationship within an ecosystem using food chains, which we will explore later in the chapter.

Producers (autotrophs) are organisms that can make their own food.

Consumers (heterotrophs) are organisms that cannot make their own food.

Within a feeding relationship, an organism can be described as a **producer** or a **consumer**. Producers or autotrophic organisms are able to make their own food. Plants are producers, making their own food by **photosynthesis**, a process through which light energy is used to make food. Some bacteria can make their food by a process called **chemosynthesis**, where they use energy from chemical reactions to make their food.

Consumers – or heterotrophic organisms – cannot make their own food. They obtain their food from other organisms, usually by eating them, their parts or their remains. There are various types of consumer.

Herbivores

Herbivores are referred to as primary consumers; they only feed on plant material (e.g. cows or sheep).

Figure 10.14: Deer are herbivores; they eat plant material.

Carnivores

Carnivores feed on the meat of animals (e.g. wolves, seals, cats). Some carnivores eat herbivores (e.g. a lion eating a zebra) and are referred to as secondary consumers while tertiary consumers feed on other carnivores.

Figure 10.15: Wolves are carnivores; they eat other animals.

Omnivores

Omnivores eat both animals and plants (e.g. humans or bears), so they may act as primary or secondary consumers in an ecosystem.

RESEARCH ACTIVITY

Carry out some research on how different animals are adapted to their role as herbivores, carnivores or omnivores.

Decomposers

Decomposers live off the remains of dead animals or plants (or their parts). Most are bacteria or fungi and they work by releasing digestive enzymes into their environment to break down the cells of the dead organisms, absorbing the nutrients after they have been digested.

Figure 10.16: Mushroom (fungi) are examples of a decomposer.

Detritus feeders

Detritus feeders, like decomposers, live off the remains of dead animals and plants (these remains are referred to as **detritus**). Unlike decomposers, detritus feeders **ingest** the dead material and **digest** it inside their bodies. Examples include woodlice, beetles and earthworms.

Figure 10.17: Earthworms are detritus feeders.

Decomposers and detritus feeders both play a critical role in the recycling of matter in ecosystems, which we will explore in greater detail in Chapter 12.

BIOTIC FACTORS

Feeding is one form of relationship between two organisms but there are others too, often called **biotic factors**. The relationships have an effect on the wider ecosystem, resulting in loss or gain of biodiversity or even changes in the physical features of the ecosystem. For example, earthworms alter the structure of the soil, improving aeration and drainage. This then has an effect on the types of plants the soil can support.

Another classic example occurred in Yellowstone National Park in the 1990s, when wolves were reintroduced. The wolves had a massive effect on the whole ecosystem there, reducing deer numbers which resulted in the course of the river changing.

Other examples of biotic factors are **predation**, **competition**, **symbiosis** and **interdependence**.

Humans – another major biotic factor – have significant impact on ecosystems which will be explored in a dedicated chapter later.

Go to YouTube.com and search for 'How Wolves Change Rivers' (4:33) by Sustainable Human.

Predation

Predation is a specific type of feeding relationship, with one organism (the predator) hunting, capturing and eating another organism for food (the prey). This interaction is a key aspect of food chains and plays a crucial role in regulating population sizes within ecosystems.

Predation is the killing of one organism by another for food.

Most predators are animals but there are some plants that can capture and consumer prey, like the Venus flytrap. Predators have specific adaptations which allows them to fill this important ecological niche.

Figure 10.18: Predators (L–R): Venus flytrap, great white shark, cheetah.

Competition

HL

Within an ecosystem, resources like food, water, places to live, mates and light are limited. This means there will often be a struggle between organisms for these vital resources. This struggle is called **competition** and it can occur between members of the same species (**intraspecific**) or between different species (**interspecific**). If different species are competing, they generally have the same niche in the ecosystem.

Figure 10.19: Examples of interspecific and intraspecific competition.

- **Competition** is the struggle between organisms for essential resources.
- **Intraspecific competition** occurs between members of the same species.
- **Interspecific competition** occurs between different species.

Competition within a species is more common. Animals compete for food, water, mates and places to live while plants compete for light, nutrients, water, space or pollinators.

Intraspecific competition can sometimes be confrontational, where a physical struggle can be observed e.g. two male deer (stags) competing for mating opportunities. This is called **contest competition** and generally involves one winner and one loser.

Scramble competition is a non-confrontational struggle between organisms for a resource. There is generally no winner or loser in scramble competition; instead, both get an equal share of the resource. An example of scramble competition is two herbivores competing for grass or drinking water from a stream.

Competition is another vital element in the evolution of species over time, ensuring the best adapted organisms survive (natural selection) and pass on their superior genetics to the next generation.

- **Contest competition** is a confrontational struggle between two organisms; the winner takes all.
- **Scramble competition** is a non-confrontational struggle between two organisms; each get an equal share.
- **Symbiosis** is a close relationship between two organisms where at least one benefits.
- A **parasite** is an organism that lives off another organism causing it harm.

SYMBIOSIS

Symbiotic relationships are another non-feeding ecological relationship that can be found in all ecosystems. **Symbiosis** involves two non-competing organisms living closely together and at least one benefitting from the relationship. There are three different types of symbiotic relationship, defined by how many of the organisms benefit and if one is harmed in the process.

HL

MUTUALISM	COMMENSALISM	PARASITISM
 Figure 10.20: Lichen.	 Figure 10.21: Barnacles on a whale.	 Figure 10.22: An insect infected by cordyceps.
Mutualism involves two organisms living closely together; both benefit and neither is harmed. A lichen is an example of a mutualistic relationship, where bacteria and fungi live together. The bacteria can photosynthesise, making food for both, while the fungi provide a safe place for the bacteria to live.	Commensalism occurs when two organisms live together and only one benefits, but the other is not harmed. A good example of commensalism is when barnacles live on a whale's back; the whale is not harmed but the barnacles have a safe place to live. Lichens growing on tree bark is another example of commensalism.	Parasitism involves one organism (the parasite) living on another (the host); the **parasite** benefits while the host is harmed. Cordyceps is a fungus that lives on various insect species, altering the insect's brain chemistry to help it spread its spores. Parasitism, like predation, plays an important role in controlling populations. Parasitism is responsible for most diseases.

DEPENDENCY AND INTERDEPENDENCE

When one organism relies on another to survive we call this **dependency**. The liver fluke is a parasitic flatworm that lives inside sheep and cattle but, oddly, it cannot reproduce inside its host. Instead, it relies on a species of mud snail to reproduce; it is dependent on it for survival. Farmers control liver fluke by ensuring their land is free from mud snails.

Interdependence occurs when two species rely on each other for survival, e.g. bees pollinate flowers and the flowers, in return, provide nectar as a food source for the bees. It can be seen as a form of mutualism, although the two species don't necessarily live close together.

THINK, PAIR, SHARE

In pairs, think of examples of mutualism, commensalism, parasitism, dependence and interdependence not mentioned above. Share your combined thoughts with the class.

Interdependence is a relationship where two species rely on each other for survival.

PORTFOLIO LINK

Go to your Learning Portfolio and complete Learning Check 4 to assess your knowledge and understanding of the topic so far.

Section 5: Energy Flow and Food Chains

At the end of this section, you should be able to:

- Describe how energy flows through an ecosystem.
- Outline the 10% rule.
- Define the term 'biomass'.
- Outline the features of a food chain, a food web and a pyramid of numbers.
- Outline the shapes and limitations of pyramids of numbers. HL
- Compare the use of pyramids of numbers and pyramids of biomass in the study of ecosystems. HL

Unpacked from LOs 3.1.7, 3.1.8, 3.1.9

ENERGY AND MATTER FLOW

You might remember learning about the conservation of energy and matter in Junior Cycle Science. To remind you:

Energy and matter cannot be created or destroyed but can only be changed from one form into another.

In the living world, most living things get their energy from the sun. Plants, that can make their own food by photosynthesis, get that energy directly from the sun and convert it into chemical energy stored in molecules like sugar.

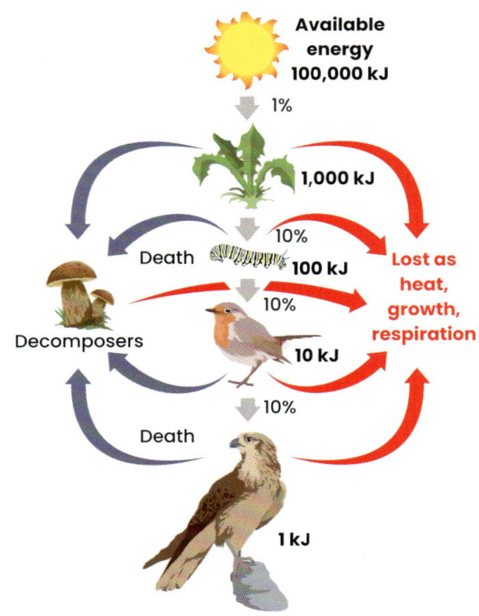

Figure 10.23: Energy flow in an ecosystem.

Animals get their energy from eating plant material (when it is alive or when it decays after dying) or by eating animals that eat plants. Other organisms (decomposers) feed off animals and plants after they die. In ecosystems, energy moves from one organism to another as they feed; we refer to the movement of energy from one organism to another as a **food chain** (see Figure 10.24).

Throughout the life of an organism, the energy they take in is converted to the other various stores of energy (kinetic, heat, sound, etc.) as the living thing goes about being a living thing (respiring, reproducing, moving, digesting, etc.). Some of the energy will be stored in their body as chemical energy. Animals store energy as fat while plants store energy as carbohydrates like starch or sugar. So, when an animal eats a plant or another animal, most of the energy has already been used up, around 90%. That means that just 10% of the energy an organism takes in is stored in its body and therefore available to another organism that might eat it.

Matter (atoms and molecules) moves through organisms as they feed in the same way as energy; **food is simply a way of moving matter and energy from one organism to another**. We refer to matter within living things as **biomass** – the total weight of all living things in a habitat. Like energy, only 10% of the biomass from one trophic level is made available to the next one.

- A **food chain** illustrates what eats what in an ecosystem.
- **Biomass** is the total weight of the community in an ecosystem.
- A **trophic level** is a step (feeding level) on a food chain.

FOOD CHAINS AND FOOD WEBS

Food chains are simple diagrams that track the flow of energy and biomass in an ecosystem, illustrating what eats what. Each step on a food chain is called a **trophic level** (see Figure 10.25).

Food chains always begin with a producer (first trophic level), usually a plant or another photosynthetic organism (bacteria or algae).

The plant is eaten by a herbivore (primary consumer) which, in turn, is eaten by a carnivore (secondary consumer, third trophic level). Finally, provided there is enough energy available, the secondary consumer may be eaten by a larger carnivore or predator (tertiary consumer).

Sometimes the largest carnivore in the food chain may have parasites, which live off it and cause it harm, e.g. a flea on a fox. It is unusual for food chains to be longer than four trophic levels because, as we've discussed, only 10% of the energy an organism collects is stored in its body and available to the animal that eats it.

Go to YouTube.com and search for 'Ecosystem Ecology: Links in the Chain' (10:09) by CrashCourse.

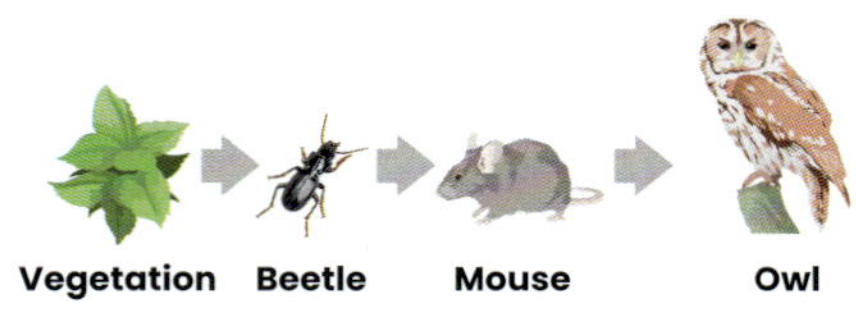

Figure 10.24: An example of a food chain.

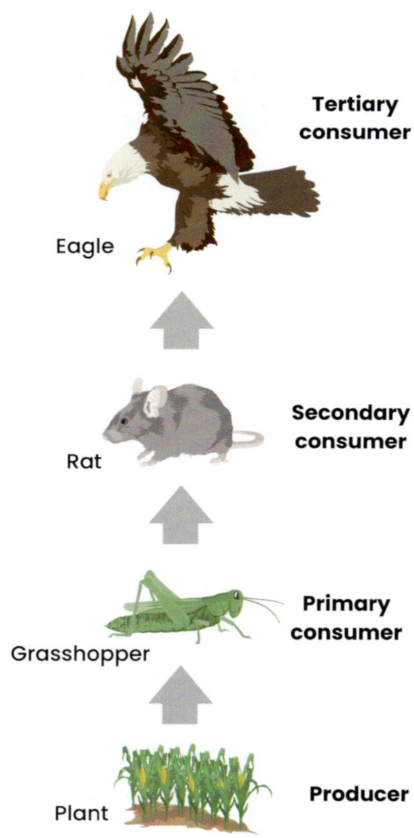

Figure 10.25: Trophic levels in a food chain.

There are two types of food chains. **Grazing food chains** begin with a living plant being eaten by a herbivore (see Figure 10.26). **Detritus food chains**, as the name suggests, begin with decaying plant or animal material, e.g. leaf litter or a rotting tree trunk (see Figure 10.27).

Figure 10.26: Grazing food chain.

Figure 10.27: Detritus food chain.

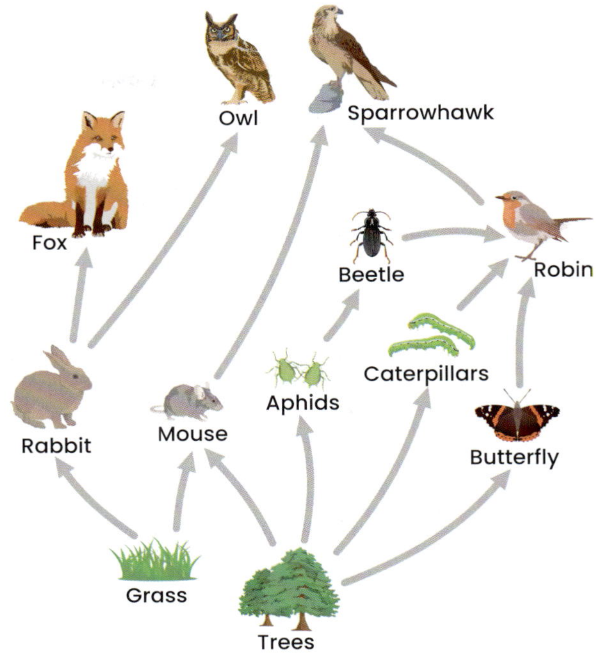

Figure 10.28: Food web in a woodland ecosystem.

Food chains can sometimes oversimplify the feeding relationships within an ecosystem. A **food web** can paint a more accurate picture of the feeding relationships and are essentially multiple food chains joined together (see Figure 10.28). They more accurately illustrate the sources of food on each trophic level and can show that some organisms, particularly omnivores, can appear on different levels in different food chains.

A **food web** is a set of interconnected food chains.

ECOLOGICAL PYRAMIDS

Both food chains and food webs provide useful **qualitative** information on feeding relationships but they don't provide **quantitative** data i.e. on the numbers of each organism within the ecosystem or their collective size or mass.

Ecological pyramids provide additional quantitative information on the organisms on each trophic level. **Pyramids of numbers** illustrate the relative numbers of organisms on each step of a food chain while **pyramids of biomass** illustrate the relative combined mass of the organisms on each trophic level.

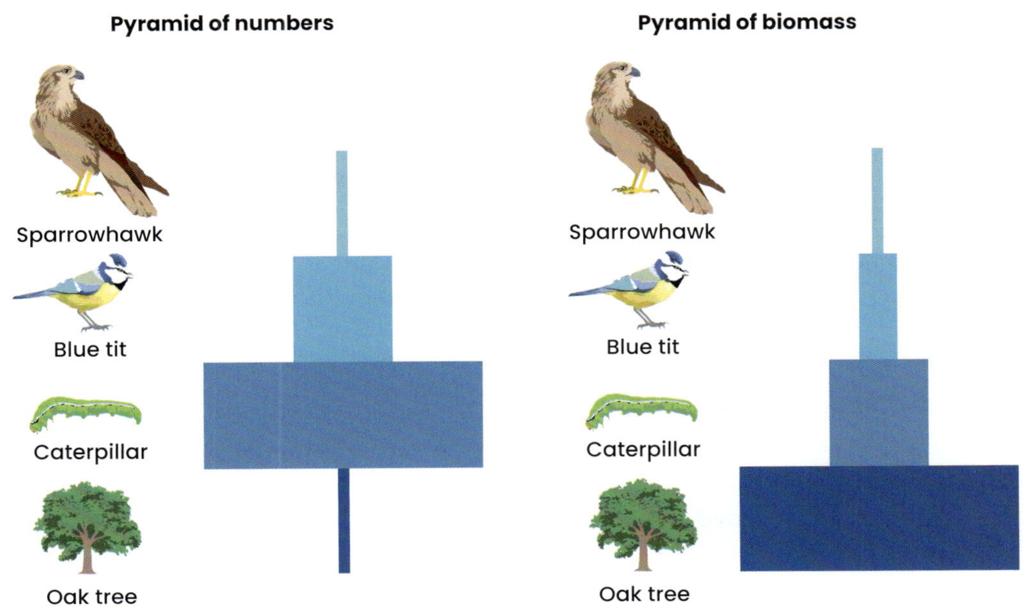

Figure 10.29: Pyramids of numbers and biomass.

PYRAMIDS OF NUMBERS

Most **pyramids of numbers** have a similar regular shape, where the numbers decrease as you go up through the tropic levels; the producer in the food chain (at the base) is the most abundant, followed by the herbivore and so on. This makes sense, as less energy is available as you progress through the food chain, so only smaller populations can be supported in higher trophic levels.

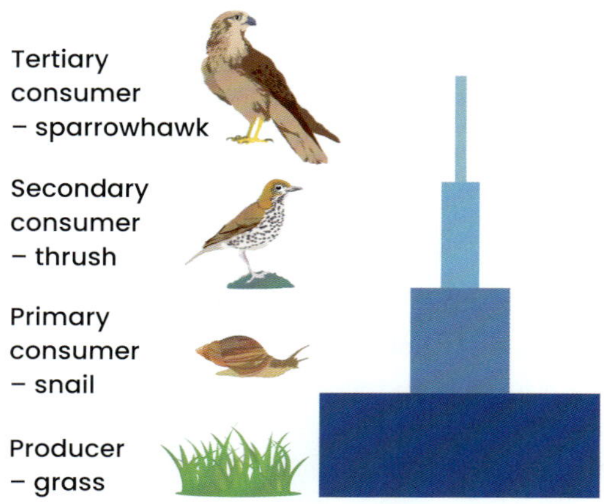

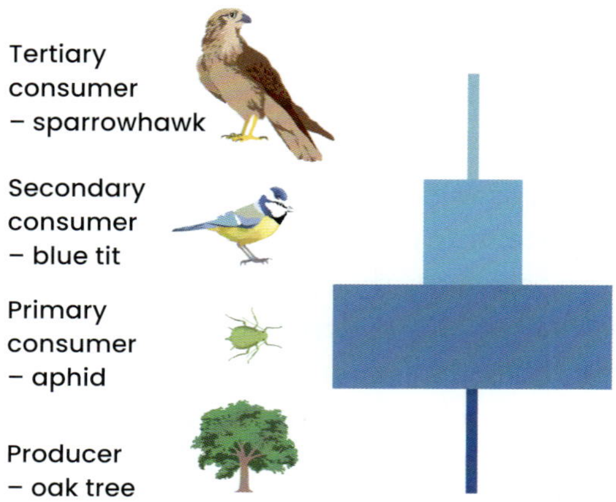

Figure 10.30: Regular and inverted pyramids of numbers.

Shapes of pyramids of numbers

However, this shape doesn't always apply. What if the producer is a large tree? Now the pyramid looks different, with a small base and much larger steps above it. What if the predator has internal parasites? Now the number on the top would be larger than the number of predators.

Limitations of pyramids of numbers

Pyramids of numbers are easier to create (because it is easy to count the number of organisms in an ecosystem), but they don't take the size of the organism into account so may not always resemble a pyramid.

HL

PYRAMIDS OF BIOMASS

Pyramids of biomass are arguably more useful than pyramids of numbers because they take into the account the size of the organism. Remember, biomass is the collective mass of the population in a food chain so the steps on the pyramid represent the combined mass of all members of that population.

In the case of a food chain where the producer is a tree, the size of the tree is more accurately represented with a large base on the pyramid and the biomass of the subsequent trophic levels reduces accordingly. This is more reflective of the '10% rule' of energy and matter flow.

The major disadvantage of pyramids of biomass is that biomass is difficult to measure. To measure the biomass, the organism is killed and then heated in an oven to remove any water, until a constant dry mass is obtained.

- A **pyramid of numbers** illustrates the number of organisms on each trophic level.
- A **pyramid of biomass** illustrates the collective mass of the organisms on each trophic level.

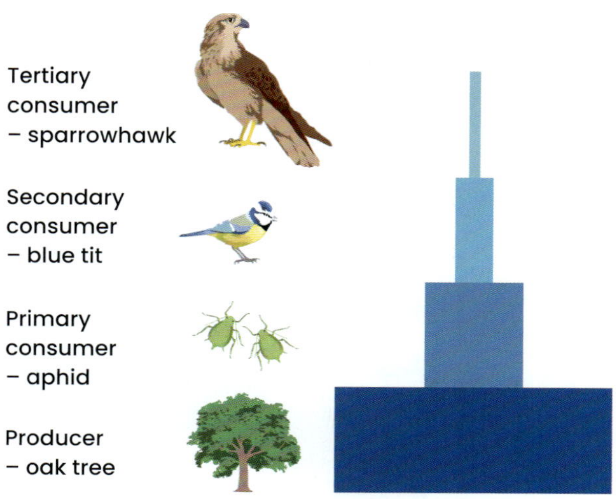

Figure 10.31: Pyramid of biomass in a woodland ecosystem.

PORTFOLIO LINK **Go to your Learning Portfolio and complete Learning Check 5 to assess your knowledge and understanding of the topic so far.**

Section 6: Carrying Capacity and Population

At the end of this section, you should be able to:

- Discuss patterns in species populations.
- Define the term carrying capacity.
- Discuss the factors that determine the carrying capacity for plants and animals in ecosystems.
- Distinguish between J and S population curves. HL
- Give examples from Irish ecosystems of both J and S population curves. HL
- Using population curves, discuss how carrying capacity might vary depending on limiting factors. HL

Unpacked from LOs 3.1.5, 3.1.6

We previously mentioned that a population is the number of a specific species in an ecosystem. A species' population doesn't stay constant though, and in reality, populations tend to fluctuate up and down, and ecosystems find a natural balance between population and the available resources.

The maximum number of a species that an ecosystem can support, without damaging that natural balance or running out of essential resources, is called the **carrying capacity** (sometimes the symbol K is used to represent carrying capacity). If the population exceeds the carrying capacity, it can lead to resource depletion, increased competition and potential harm to the ecosystem's overall health.

The **carrying capacity** is the maximum population of a species an ecosystem can support.

Figure 10.32: Aquarium owners need to consider the carrying capacity of their tank.

FOCUS ON ... FACTORS INFLUENCING CARRYING CAPACITY

The carrying capacity of a species in an ecosystem is affected by a number of factors, both abiotic and biotic. We sometimes call these 'limiting factors'.

ABIOTIC

- **Water:** The availability of fresh water is a key factor. Water is required by all living things and, if there's not enough water, the ecosystem can't support a large population of animals or plants.
- **Oxygen:** The vast majority of living things require oxygen; the more oxygen available, the greater the carrying capacity. This is particularly important in aquatic ecosystems.
- **Space:** The physical space in an environment limits how many organisms it can hold. Overcrowding can lead to stress and competition for resources.
- **Shelter:** The availability of suitable shelter, like nests or hiding places, affects carrying capacity. Without proper shelter, some species may struggle to survive.
- **Climate and weather:** Different organisms thrive in different climates. Extreme weather conditions can impact the ability of an ecosystem to support certain species.
- **Soil fertility:** The availability of nutrients in the soil affects plant carrying capacity: the more fertile the soil, the greater the carrying capacity.

BIOTIC

- **Food:** The more food available in the ecosystem, the greater the carrying capacity.
- **Predators and disease:** The presence of predators and the risk of diseases can control population sizes. If these factors increase, they can limit the number of organisms an ecosystem can sustain.
- **Reproductive rates:** The rate at which organisms reproduce influences carrying capacity. High reproductive rates can lead to rapid population growth, reaching the environment's limit sooner.
- **Human influence:** Human activities, such as deforestation, pollution and habitat destruction, can significantly alter an ecosystem's carrying capacity, often leading to a decline in support for native species.

Go to YouTube.com and search for 'Ecological Carrying Capacity-Biology' (3:26) by MooMooMath and Science.

POPULATION CURVES

Population curves are graphical representations of how populations of organisms change over time. They help to illustrate periods of population growth, decline or stability.

The two primary types of population curves are the J-curve and the S-curve, each reflecting different growth patterns and ecological dynamics.

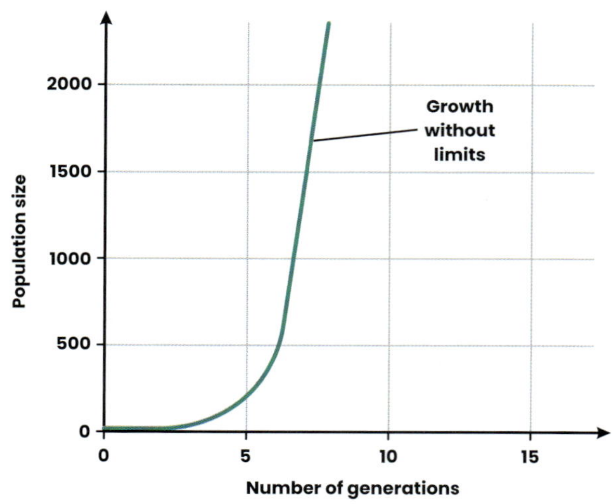

Figure 10.33: A 'J-curve' population graph.

J-curve (exponential growth)

HL

The J-curve represents exponential growth, where a population increases rapidly when resources are abundant and environmental conditions are favourable. This growth occurs in two main phases:

1. Lag phase**:** Initially, the population grows slowly as individuals adapt to their environment.
2. Exponential phase**:** Once conditions are optimal, the population experiences rapid growth, leading to a steep upward curve.

When a species' population growth resembles a J-curve, no carrying capacity is reached. The population continues to grow until all the resources are used up; then the population crashes. This pattern is often seen in newly established populations or when species are introduced to a new habitat with few competitors or predators.

SCIENCE IN SOCIETY: SUSTAINABILITY

Algal blooms

Algal blooms occur when the population of algae in lakes and ponds increases rapidly, causing serious environmental problems. These blooms often happen when there are too many nutrients, like nitrogen and phosphorus, entering the water from sources such as agricultural runoff and wastewater. The addition of nutrients to a water body is called eutrophication. Warm temperatures and plenty of sunlight also help algae grow quickly. Algal blooms usually follow a J-curve growth pattern, starting with slow growth and then exploding in numbers. Once the bloom peaks, the algae can die off quickly due to a lack of nutrients or other unfavourable conditions. This can lead to low oxygen levels in the water as the algae decay, the release of harmful toxins and fish kills.

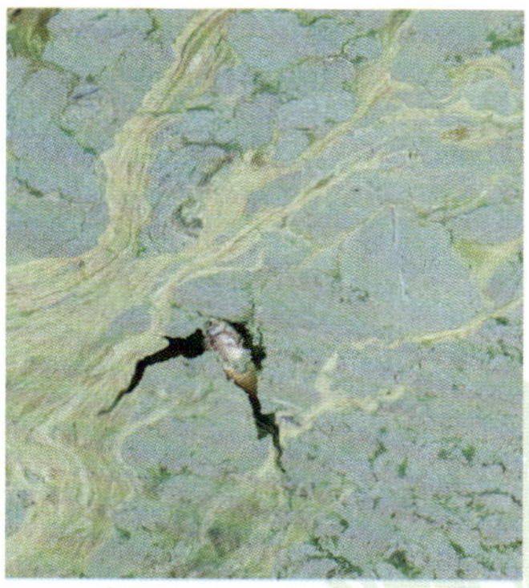

Figure 10.34: Algal blooms can cause fish kills.

S-curve (logistic growth)

The S-curve is evident when population growth slows as it approaches the carrying capacity of the environment. This curve includes four phases:

1. Lag phase: As with the J-curve, the population grows slowly at first.
2. Exponential phase: The population grows rapidly as resources are plentiful.
3. Deceleration phase: Growth slows as resources become limited, and competition increases.
4. Stationary phase: The population stabilises at or near the carrying capacity, where birth and death rates are roughly equal.

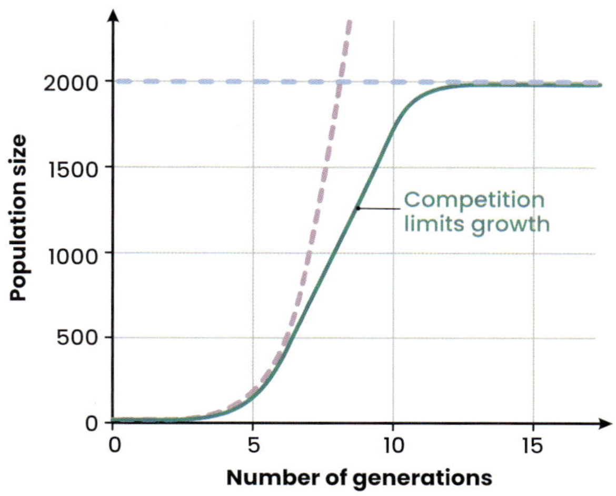

Figure 10.35: An 'S-curve' population graph.

HL

S-curves reflect stable ecosystems, with species reaching a carrying capacity. They indicate that resources are limited and populations cannot grow indefinitely.

Impact of limiting factors

Limiting factors (e.g. food and water availability, predation, disease and habitat space) can alter population curves:

- **Increased limiting factors:** If resources become scarce or competition increases, the growth rate may slow, shifting a J-curve towards an S-curve as the population approaches carrying capacity.
- **Decreased limiting factors:** Conversely, if a population experiences a sudden increase in resources (e.g. a new food source), it may shift towards a J-curve growth before stabilising again.

THINK, PAIR, SHARE

Do you think the human population is reaching carrying capacity? What evidence is there to support your view? Discuss in pairs.

PORTFOLIO LINK

Go to your Learning Portfolio and complete Learning Check 6 to assess your knowledge and understanding of the topic so far.

Chapter Summary

- The **biosphere** is the part of the Earth that can contain life.
- **Ecology** is the study of living things and their interactions with their environment.
- **Ecosystems** are the units of ecology.
- Ecosystems consist of a **community** of organisms (made up of multiple **populations**) interacting with each other and their environment.
- A **habitat** is the place where an organism lives.
- A **population** is the number of each type of species in an ecosystem.
- Organisms living in an ecosystem have particular **adaptations** that allow them thrive and survive there.
- Organisms assume roles or **niches** in ecosystems, e.g. predator or decomposer.
- **Abiotic factors** are non-living features of an ecosystem.
- They can be broadly classified as air (**climatic factors**), ground (**topography**), aquatic (water) and **edaphic** (soil).
- Abiotic factors can affect the distribution of a species in an ecosystem.
- **Biodiversity** is the variety of life in an ecosystem.
- There are three types of biodiversity: **genetic diversity**, **species diversity** and **habitat diversity**.
- HL Biodiversity can be calculated using Simpson's species biodiversity index.
- HL $Ds = 1 - \left(\frac{\Sigma n(n-1)}{N(N-1)}\right)$
- **Conservation** is the management of Earth's resources including its biodiversity.
- Feeding relationships describe how organisms obtain their food.
- **Producers** are organisms that can make their food while **consumers** need to eat other organisms to get food.
- Consumers can be **herbivores**, **carnivores** or **omnivores**.
- **Decomposers**, including bacteria and fungi, and **detritus feeders** live off the remains of dead animals and plants.
- **Biotic factors**, which include feeding relationships, are living factors in an ecosystem.
- Other biotic factors include: **predation**, **symbiosis**, **competition** and **interdependence**.
- A **predator** is an animal that kills another animal for food.
- Competition is the struggle between organisms for essential resources.
- **Symbiosis** is a close relationship between two organisms; at least one benefits. HL
- **Interdependence** occurs when two organisms rely on each other for survival.
- **Energy flow** describes the movement of matter and energy through food chains in an ecosystem.
- Only 10% of the energy and matter on one trophic level is available to the organisms on the next level.
- There are two types of food chain: **grazing food chains** and **detritus food chains**.
- A **food web** is an interconnected series of food chains.
- Ecological pyramids visually describe the number or collective mass of the organisms on each trophic level in a food chain.
- Pyramids of numbers illustrate the numbers on each trophic level while pyramids of biomass illustrate the collective mass of a species on a trophic level.
- The maximum population of a species in an ecosystem is called **carrying capacity**.
- Carrying capacity is influenced by the availability of food and water, space and shelter, climate, predators and disease, reproduction rates and humans.
- **Population curves** are graphical illustrations of population change over time. HL
- There are two main types: S-curve and J-curve graphs. HL
- **J-curves** illustrate exponential growth, where a species population doesn't reach a carrying capacity. HL
- **S-curves** reflect stable ecosystems where limiting factors cause populations to reach a carrying capacity. HL

END OF CHAPTER ASSESSMENT

REMEMBER

1. What is the *biosphere*?
2. Define *ecosystem* and identify its **two** main components.
3. Name the **four** categories of abiotic factors and provide examples of each.
4. What is meant by *biodiversity*? What are the three types of biodiversity?
5. Define a feeding relationship and explain its significance in the context of living organisms.
6. What is the primary source of energy for most living things, and how is it obtained in ecosystems?
7. HL Name the **three** types of symbiotic relationships and provide an example for each.
8. What is *predation* and why is it important?
9. What is meant by the carrying capacity of a species?

UNDERSTAND

10. Discuss how abiotic factors can shape ecosystems and influence species distribution.
11. How does biodiversity provide benefits to both ecosystems and human society?
12. Differentiate between species richness and species evenness in an ecosystem.
13. Briefly explain the role of decomposers in ecosystems and how they contribute to the recycling of matter.
14. Describe the difference between grazing food chains and detritus food chains.
15. How do ecological pyramids provide quantitative information about trophic levels in an ecosystem?
16. HL Discuss the concept of biomass and its significance in understanding energy and matter flow in ecosystems.

APPLY

17. How might changes in abiotic factors impact the distribution of species in a woodland ecosystem?
18. Identify **three** abiotic factors present in a local ecosystem.
19. HL Can you give an example of a mutualistic relationship not mentioned in the chapter?
20. HL Using Simpson's species biodiversity index, calculate the biodiversity of the habitat below.

SPECIES	NUMBER FOUND (n)
Daisy	12
Dandelion	13
Perennial ryegrass	80
Clover	30

21. HL Compare the use of pyramids of numbers and pyramids of biomass in ecosystem study. Which is a more useful tool?
22. HL Using examples, distinguish between S- and J-curve population graphs.

ANALYSE

23. How do abiotic factors act as architects of ecosystems, influencing the variety of plants that grow in them? Give examples.
24. An ecologist studied the population of algae in a lake over a 12-week period, after the lake was polluted with human waste.

WEEK	ALGAL POPULATION (MILLIONS)
1	0.1
2	0.2
3	0.5
4	1.0
5	2.5
6	5.0
7	10.0
8	20.0
9	35.0
10	40.0
11	15.0
12	0.1

(a) What term describes the addition of nutrients into a river or lake?

(b) Apart from the nutrients, what other limiting factors might affect the population of the algae?

(c) Briefly explain why the population of algae didn't increase immediately after the human waste entered the lake.

(d) Draw a graph of the data, with population on the y-axis.

HL (e) Does the data represent a J-curve or S-curve population graph?

(f) Can you use the data or graph to calculate carrying capacity?

(g) What effects might the growth and decline of the algae population have on other wildlife in the lake?

HL **25.** How do contest competition and scramble competition differ, and what are their implications for the organisms involved?

26. Examine the irregular pyramid of numbers shown and suggest an organism that might appear on each trophic level.

EVALUATE

27. Discuss the impact of human activities on biodiversity in a local ecosystem.

HL **28.** (a) Using Simpson's species diversity index, evaluate numerically the biodiversity in two woodland ecosystems below:

WOODLAND 1	
SPECIES	**NUMBER FOUND (n)**
Oak	12
Sycamore	8
Birch	2
Ash	2

WOODLAND 2	
SPECIES	**NUMBER FOUND (n)**
Oak	6
Ash	6
Horse chestnut	7
Scots pine	7

(b) Comment on the species evenness and species richness of both woodlands.

29. Assess the role of humans as a biotic factor in ecosystems.

30. Evaluate the effectiveness of using pyramids of numbers in ecosystems. HL

31. Angler fish live at depths of over 1000m. In clear water, sunlight does not usually reach more than 100m deep. Many angler fish have a transparent 'lure' containing a high concentration of bioluminescent bacteria. Bioluminescent bacteria produce light.

Suggest an advantage to the angler fish of having a lure containing bioluminescent bacteria.

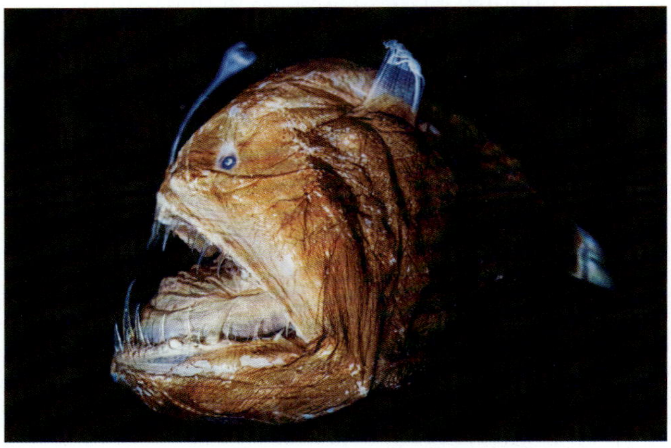

CREATE

32. Create a simple food chain for a given ecosystem, including a producer and three trophic levels.

33. Devise a food web for a specific ecosystem, illustrating the connections between different organisms and trophic levels.

34. Propose a research project to measure and analyse the biomass of different trophic levels in a given ecosystem. HL

11 Modelling an Ecosystem

LEARNING OUTCOMES: U1.1, U1.2, U2.1, U2.2, U2.3, U2.4, U2.5, U2.6, U4.1, 3.1.1, 3.1.2, 3.1.3, 3.1.11

EYE-OPENER

Rock pool ecosystems, found along rugged coastlines, come to life during low tide. These coastal micro-ecosystems provide a mix of colours and textures, with algae clinging to rocks, barnacles covering surfaces and seaweed swaying in the current. As the tide goes out, rock pools reveal a celebration of biodiversity, where small fish dart among the rocks, anemones unfold their tentacles and crabs scuttle across sandy bottoms. Look closer to discover periwinkles grazing, limpets gliding, sea stars clinging and sea urchins navigating.

At the end of this chapter, you should be able to:

- Discuss the importance of ecological surveys.
- Describe the equipment used during ecological studies, to collect organisms, measure abiotic factors and measure species distribution.
- Construct a model of a local ecosystem.
- Investigate the influence of abiotic factors on the distribution of a species.
- Analyse models of woodland, grassland and seashore ecosystems.

LINKS TO:

Chapter 2	**Scientific Knowledge and Research**
Chapter 3	**Experiments and Data Analysis**
Chapter 10	**Biodiversity and Ecosystems**

SUBJECT VOCABULARY

qualitative study · quantitative study · pictorial keys · dichotomous keys · topography · kick sampling · species abundance · percentage frequency · percentage cover · species population density · species distribution · random sampling · line transect · belt transect

Section 1: Introduction to Ecological Techniques

At the end of this section, you should be able to:

- Outline why ecosystem study is important.
- Distinguish between qualitative and quantitative ecosystem study.
- Describe the steps involved in ecosystem study.
- Name common specimen collection tools and describe how they work.
- Name common tools used to measure abiotic factors in an ecosystem.
- Outline how you can identify and classify organisms in an ecosystem.

Unpacked from LOs U1.2, U2.1, U2.2, U2.3, U2.4, U2.5, 3.1.11

WHY STUDY ECOSYSTEMS?

Ecologists are scientists who study ecosystems, the interactions between organisms and their environment. They investigate the relationships between living organisms (plants, animals, micro-organisms) and their physical surroundings, including the air, water, soil and climate.

Crucially, at a time when our climate is changing, ecologists aim to understand how we can protect ecosystems from natural and human-created changes (more on this in Chapter 13). In this chapter, you will become an ecologist. In your study of ecosystems, you are required to model a habitat, building a comprehensive picture of the species present, the relevant abiotic factors, the feeding relationships within and the distribution of species. Many of the skills you develop during this study may be important during your Biology in Practice investigation.

QUALITATIVE VS QUANTITATIVE STUDY

As we discovered in Chapter 3, scientists can collect qualitative or quantitative data when carrying out investigations. The same is true for ecological investigations and both are important in establishing a true picture of the ecosystem and its inhabitants.

Qualitative surveys

Qualitative studies in ecology aim to describe the characteristics of an ecosystem in a broad sense, without using numerical data. The data collected instead might include observations, interviews and open-ended surveys.

Examples of qualitative data in ecosystem study include identifying the animal and plant species in an ecosystem and evaluating their adaptations, recording the food chains or food webs present and assessing the ecological relationships (e.g. parasitism, symbiosis, etc.). Qualitative surveys may be followed by quantitative surveys later.

Qualitative		Quantitative
Answer 'why?' questions	**Purpose**	Answer 'how many/much?' questions
Observation, symbols, words, etc.	**Data type**	Number/statistical result
Observe and interpret	**Approach**	Measure and test
Grouping of common data/non-statistical analysis	**Analysis**	Statistical analysis

Figure 11.1: Comparing qualitative and quantitative surveys.

Qualitative studies do not collect numerical data during investigations.

Quantitative surveys

Quantitative studies in ecology involve collecting numerical and measurable data, like counting the number of species, measuring abiotic factors, ground cover and other measurable metrics in an ecosystem. Data is collected by structured surveys, sampling techniques, controlled experiments, ensuring the methods provide **numerical data**. After collecting the numbers, the data is analysed to look for patterns, relationships and trends.

Population surveys in ecology are examples of how we use numbers to estimate sizes, growth rates and how their distribution is affected by other factors (abiotic and biotic).

Quantitative studies collect numerical data during investigations.

HABITAT STUDY EQUIPMENT

When carrying out ecosystem surveys, specialist equipment is needed. The equipment will depend on the ecosystem and usually will have one of three purposes: collecting organisms, measuring abiotic factors or measuring species distribution.

Collecting organisms

NAME	PICTURE	FUNCTION
Beating tray		A beating tray is a simple tool which is placed under a tree or shrub. The tree or shrub is then shaken and small invertebrates fall onto the tray.
Pooter	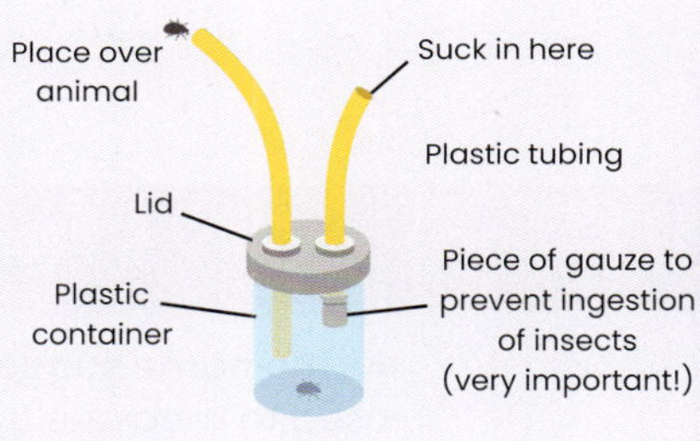	A pooter is an important tool when surveying habitats. It has a tube which, when inhaled through, can be used to suck up small invertebrates safely. A pooter can be used with a beating tray.
Net	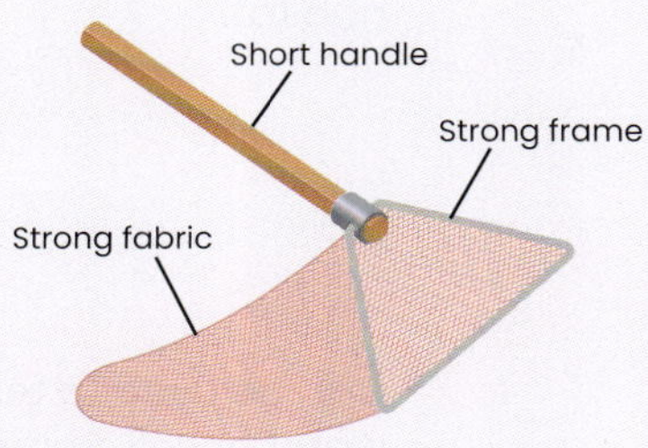	Nets have many uses in habitat surveys. Nets with large holes can be used to collect butterflies or large marine animals while finer nets are used to collect small animals, in grassland or aquatic environments.
Sieve	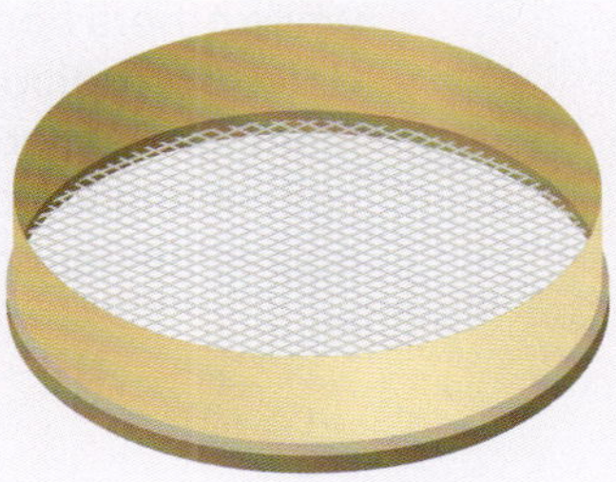	A laboratory sieve has multiple uses in habitat surveys. When leaf litter is shaken over a sieve, the small invertebrates within fall through the holes. Sieves can also be used to measure soil texture.

Tullgren funnel	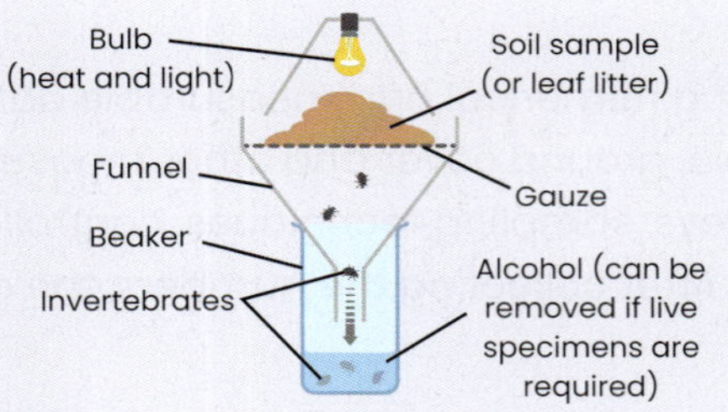	A Tullgren funnel uses a bright and hot lamp to extract small animals from a soil sample. The soil is placed on a gauze set over a funnel, then exposed to the lamp. The small animals move away from the light and heat, through the gauze and into a beaker positioned underneath.
Pitfall trap	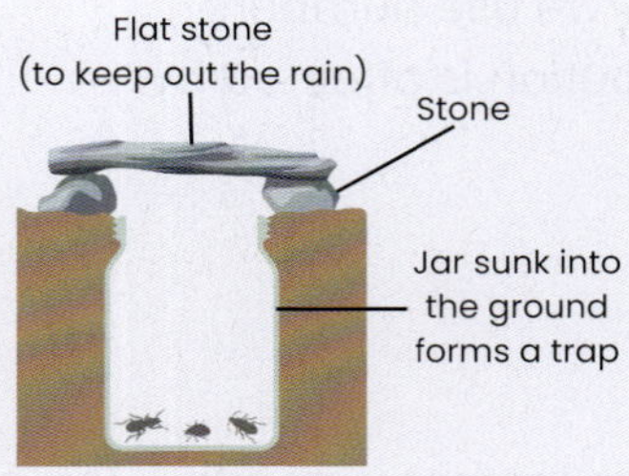	A pitfall trap is used to collect walking invertebrates on the forest floor. You can make a simple pitfall trap by digging a hole and placing a jar under the surface. A flat stone or plastic canopy is placed over the hole to prevent rain getting in.
Mammal trap	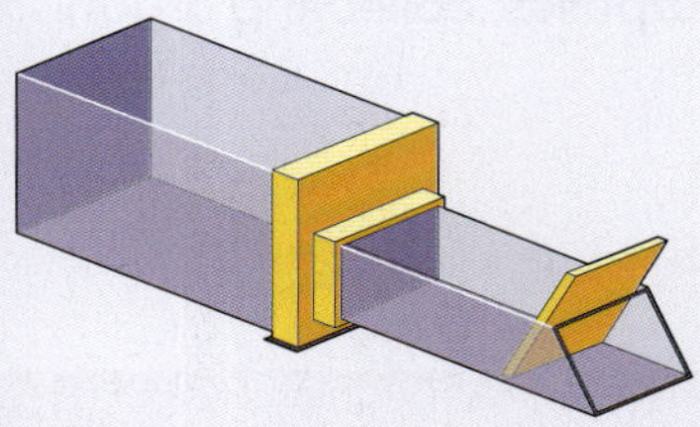	A mammal trap, as the name suggests, is used to collect small mammals like mice or shrews. Bait is placed inside, and the trap closes when a mammal enters. The animal is not harmed and can be easily released after being identified and counted.
Cryptozoic trap		A cryptozoic trap is a simple tool to collect small invertebrates. It consists of a flat object, usually a piece of wood or board, that is placed on the forest floor or over grass. Small invertebrates gather under the wood seeking shelter. A pooter can be used to collect the samples for identification.

Measuring abiotic factors

NAME	PICTURE	FUNCTION
Light (lux) meter		As the name suggests, a light meter is used to measure the light intensity in a habitat.
pH meter	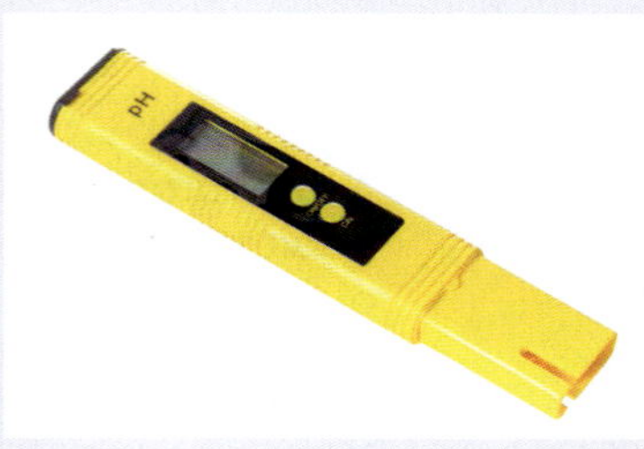	Can be used to test the pH of a sample of water or soil. Some can be used onsite while others can only be used in the lab. Universal indicator paper can also be used to measure pH. Many commercial soil meters will also measure soil moisture and temperature.

Air thermometer		Used to measure the temperature of the air in an ecosystem.
Soil thermometer		Soil thermometers are used to measure the temperature of the soil; they usually have a metal casing to protect a glass thermometer within. Digital thermometers are also available.
Anemometer		Used to measure windspeed, which can be a measure of how exposed an area is.
Rain gauge		Rain gauges measure the amount of precipitation in an area.
Soil testing kit	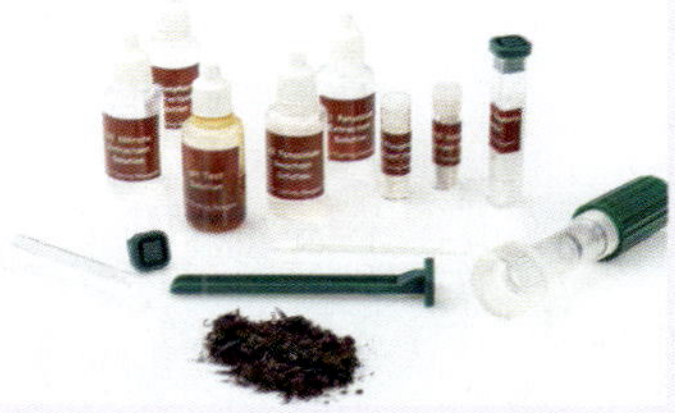	A soil testing kit can be used to measure various edaphic factors including soil pH, fertility (amount of nitrogen, phosphorous and potassium).
Hydrometer	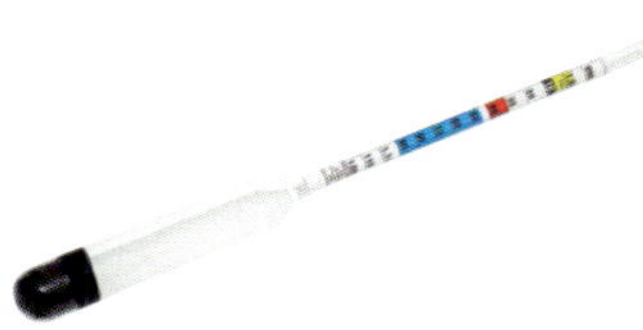	Hydrometers are used to measure the salinity of a liquid, useful for studying marine habitats.
Dissolved oxygen meter		A dissolved oxygen meter measures the amount of oxygen in a water sample; this is a useful tool for surveying aquatic ecosystems or ponds.

Compass	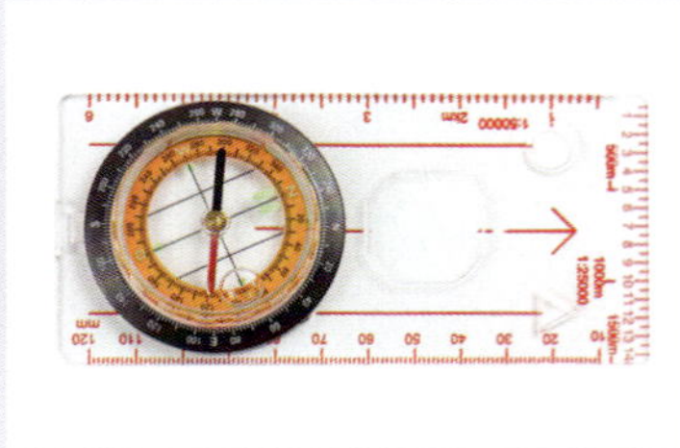	A compass can be used to determine aspect, whether a habitat is south or north facing. This can affect soil temperature and light exposure.
Datalogging tools	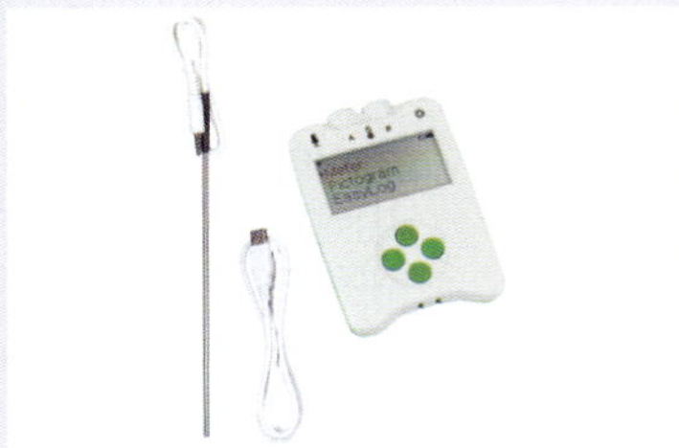	If your school has datalogging equipment, they may have sensors for measuring pH, humidity, salinity, temperature and more.

Investigating species distribution

NAME	PICTURE	FUNCTION
Quadrat		A quadrat is an essential tool for habitat surveys. It's a wooden or metal square that is used to sample parts of the habitat being surveyed. They can be different sizes (e.g. $0.25m^2$ or $1m^2$) and can be simple squares or contain grids.
Transect ropes		Transect ropes are used to assess how species might change as you move through a habitat in a straight line. They often have knots every 1–5 metres, depending on the habitat.
Tape measure		It is important to know the area of the habitat being surveyed and a long tape measure is useful for this. It can also be used as a transect.

IDENTIFYING ORGANISMS IN AN ECOSYSTEM (KEYS)

You are going to encounter plants, animals and micro-organisms during your ecosystem survey that you don't recognise. There are a number of tools that can be used to identify organisms in an ecosystem including pictorial keys, dichotomous keys and identification apps.

Pictorial keys

A **pictorial key** contains photos of common species within an ecosystem. Often prepared by ecologists, they allow students and others an opportunity to find and identify the most common plants and animals in an ecosystem easily.

Figure 11.2: An example of a pictorial key. Courtesy of the Tree Council of Ireland.

Dichotomous keys

A **dichotomous key** uses a series of paired, opposing statements on the characteristics of species in an ecosystem to identify unknown species. You select the statement that best fits the organism's physical features, moving to the next step, until you get to the name of the species. Sometimes images are also used. One of the most famous Irish keys is Webb's *An Irish Flora*, which can be used to identify every native Irish plant.

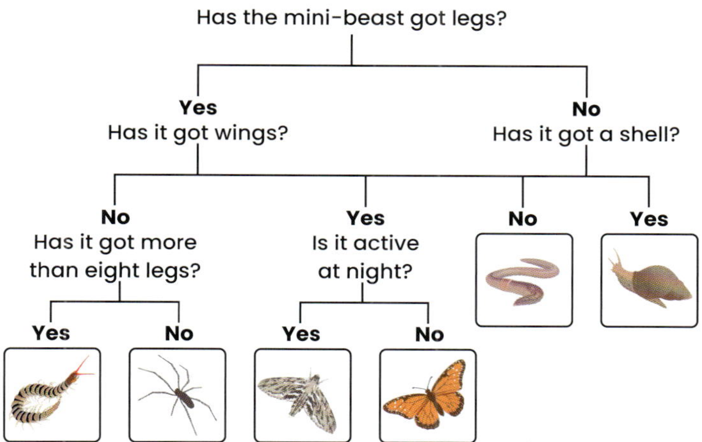

Figure 11.3: A simple dichotomous key.

Identification apps

There are many free smartphone apps that allow you to identify and even classify plants, animals and fungi. Google Lens can use photos to quickly find out what you've found while there are other specialist apps, such as Seek by iNaturalist (no fee or registration needed), which are specially designed to aid with identification.

- **Pictorial keys** use images to identify unknown species.
- **Dichotomous keys** use a series of questions to help identify unknown species.

SCIENCE IN SOCIETY: TECHNOLOGY

Ecology equipment in your pocket!

Mobile phones are equipped with many useful tools for ecological surveys. Google or Apple Maps can be great for mapping an area or getting a satellite image of the habitat. Most smartphones include environmental sensors including light intensity, humidity, pressure and temperature sensors; users may need to download an app to use them for habitat surveys, but most will be free. Most phones also have a compass app included.

Figure 11.4: Many phone apps are available to help identify plant and animal species.

PORTFOLIO LINK

Go to your Learning Portfolio and complete Learning Check 1 to assess your knowledge and understanding of the topic so far.

Specified Investigation 2

Model an ecosystem and investigate quantitatively the impact of variation in abiotic factors on the distribution and abundance of a species. **(LO 3.1.11)**

You can choose to study any ecosystem for Leaving Certificate Biology – a woodland, meadow, bogland, stream or seashore. You should always aim to leave an ecosystem as you found it, respecting nature and the species within it. Ideally, choose a location where you can visit throughout the year to get a sense of how it is affected by seasonal changes.

The following is a step-by-step guide on how to approach ecological study. Some steps take longer than others and may not apply to all ecosystems.

STEP 1: MAP THE ECOSYSTEM

The first step when carrying out an ecological study is to draw a simple but clear map of your chosen area. It is a good idea to set the boundaries of your area so that you can measure the total area being surveyed. This will allow you measure species density later. Don't try to survey too large an area, though.

On your map you should include the main features; add paths, walls, roads, rivers, streams and any wooded areas. Using the satellite images from Google or Apple Maps is very useful when mapping an ecosystem.

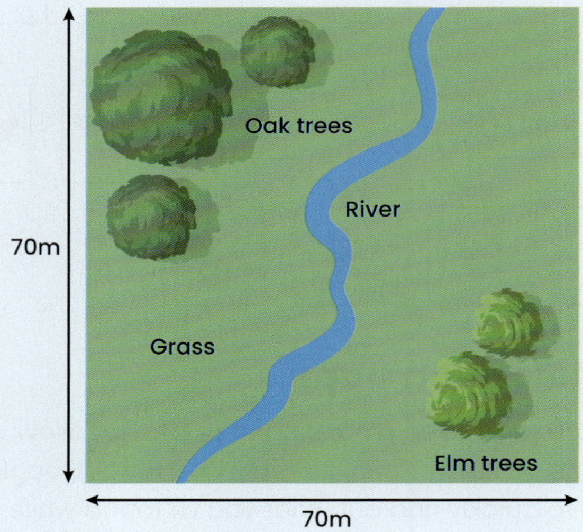

Figure 11.5: A simple ecosystem map.

STEP 2: QUALITATIVE SURVEY

Before gathering numerical data, you and your partners should aim to assess the habitat qualitatively. Here are some of the different types of information you can gather, which you can do while mapping the area.

Physical features

Assess and note the physical features of the habitat. What is the main type of vegetation i.e. trees, shrubs or grassland? Are there trees, paths, roads, rivers, streams or exposed soil? Are there areas of shade or open areas? Are there fallen trees, rockpiles or ponds that might be microhabitats within the larger area?

Topography

Is the area being assessed sloped or in a valley? Which way is north? How does that affect light availability?

Topography describes the lie of the land: slope and aspect.

Wildlife

Are there any outward signs of wildlife, e.g. burrows, tracks or nests present? Do you hear any bird song or animal sounds?

Soil
Does the soil appear dark or light? Is it rocky? Does it have a smooth or rough texture? Does it feel warm or cold?

Human influence
Are there signs of human interference in the area, i.e. pathways, litter or other forms of pollution?

Weather conditions
Record the weather conditions on the day of the survey.

Historical features
Are there historical features visible, i.e. forts, old houses, walls, archaeological sites or old roads?

STEP 3: IDENTIFY PLANTS PRESENT

Use pictorial or dichotomous keys, or plant identification apps, to gather a list of plants in the ecosystem. You should aim to identify at least five plants in the area being studied, although you will likely find many more. Plants are easier to spot on a walk through a habitat.

STEP 4: IDENTIFY ANIMALS PRESENT

Beating trays, nets and sieves can be used to find small animals in the area. Large animals might be hard to find but look for signs of their activity e.g. rabbit burrows, badger sets, bird nests and fox or deer droppings.

You might consider setting up cryptozoic, pitfall and mammal traps to find animal species in the area. Remember to record where you placed your traps on your map.

Take a sample of soil to use with a Tullgren funnel back in the science lab. In shallow rivers or streams, **kick sampling** can be used with nets to collect small animals.

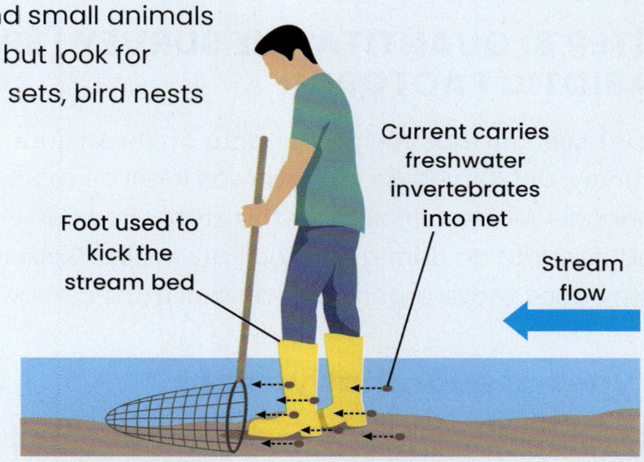

Figure 11.6: Kick sampling.

In some habitats, like a rocky seashore, the animals are sessile (meaning they don't move or move slowly) so are easier to find and identify; traps aren't usually needed.

Kick sampling is a technique used in stream studies, where the stream bed is kicked to disturb then collect invertebrates.

Later, use keys or identification apps to identify the species collected. Again, you should aim to identify at least five animals in your habitat.

STEP 5: CONSTRUCT A FOOD CHAIN AND FOOD WEB

When you have a list of the animals and plants present, you can build a few simple food chains. Ensure your food chains begin with a producer (a plant) and have at least three trophic levels. As you do, imagine energy and matter flowing from one organism to another. Sketch a simple food web too, if you can.

STEP 6: ADAPTATIONS AND NICHES

Consider the adaptations present in some of the plants and animals present in your ecosystem. These adaptations may be physical or behavioural.

Look for plants with bright colours, thorns for protection, poisonous berries or waxy leaves. Are the animals present camouflaged? Do they live underground, have horns to protect themselves or live in shells? All of these features are adaptations that allow organisms find a niche in their habitats.

Figure 11.7: Examples of adaptations.

STEP 7: OTHER BIOTIC FACTORS PRESENT

Assess other biotic factors present in the habitat you are studying. Do some organisms present have symbiotic relationships with other organisms? Is there evidence of disease in the ecosystem (e.g. ash dieback)? Are there predators in the area? Are decomposers evident? Are some species in competition? Are some species dependent on others in the habitat?

STEP 8: QUANTITATIVE SURVEY (SPECIES DISTRIBUTION AND ABIOTIC FACTORS)

It is essential that you gather data on the distribution of selected species during any habitat survey. Don't try to do too many as it will be more difficult to analyse later. Two plants or animals will be sufficient for this step – you can visually assess which plants and animals to gather data on during the qualitative step. Gathering data on plant populations is easier, as they don't move around, but animal traps can be used for counting invertebrates.

Types of quantitative data

Species abundance measures the number of a particular species in an ecosystem.

Species distribution records how spread out a plant or animal is and can be recorded by measuring species abundance data from multiple points in an ecosystem.

There are three main ways of measuring species distribution: **percentage cover**, **percentage frequency** and **species population density**.

Let's look at the three main methods that can be used to gather data and, later, three types of data that can be collected.

Later, we will compare the changes in the abiotic factors with changes in percentage cover, frequency or population density.

- **Percentage cover** is a measure of how much of the ground is covered by a species.
- **Percentage frequency** is a measure of how common a species is in an ecosystem.
- **Species population density** is a measure of the number of a species per m^2.

FOCUS ON ... MEASURING ABIOTIC FACTORS

In addition to species abundance, you also need to gather data on the abiotic factors present, which will allow you to analyse how abiotic factors affect the distribution of species in the ecosystem.

You should consider measuring a series of abiotic factors including temperature (air or soil), light intensity, wind speed, soil pH or distance from a feature (shoreline, pond, fallen tree, etc.). The factors you measure will depend on your habitat.

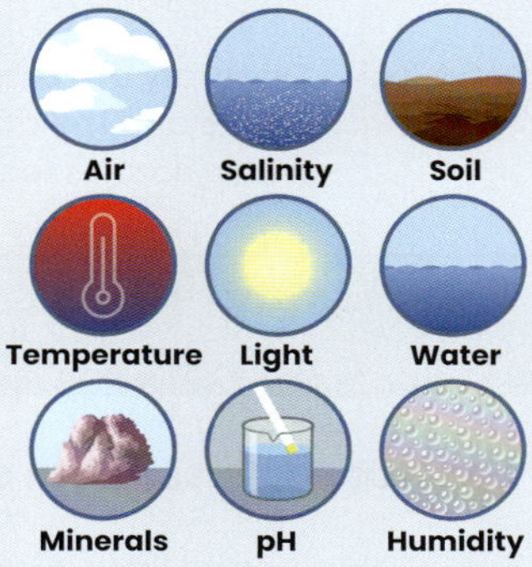

Method 1: Random sampling using quadrats or traps

1. Quadrats or traps (e.g. pitfall traps) are randomly positioned – this can be done by throwing a pen or pencil, then placing the quadrat/trap down where it lands.
2. You should aim for at least **ten** locations, randomly spread throughout the area being surveyed.
3. The presence or absence of each species in the quadrat or trap is recorded using a table.

Remember: You should also measure abiotic factors at each quadrat/trap location.

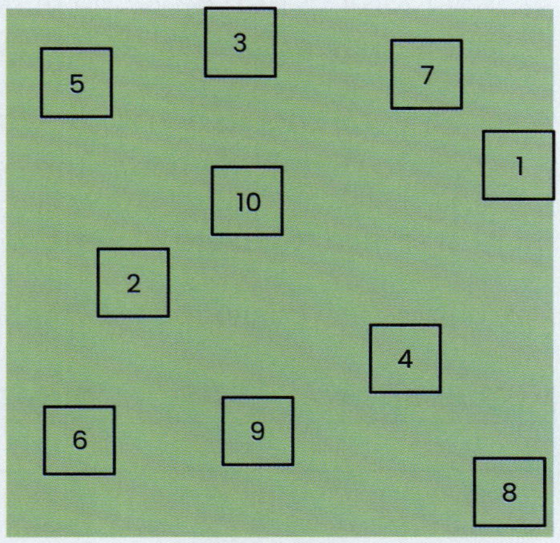

Figure 11.8: Random sampling in a grass habitat.

Random sampling in this context selects sites to place traps or quadrats randomly, without bias.

Method 2: Line transect

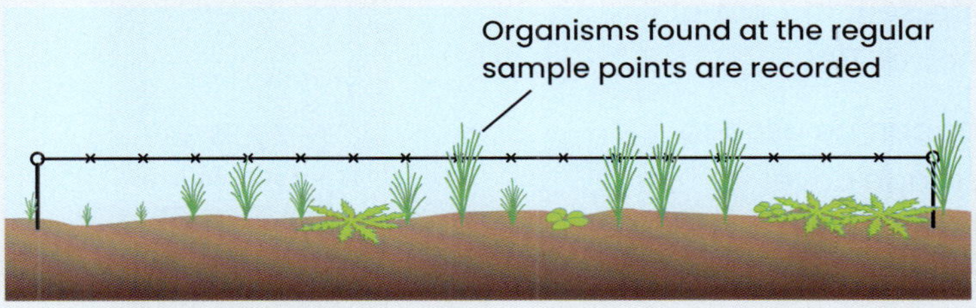

Figure 11.9: A line transect in a grass habitat.

1. A transect rope or long tape measure is extended across the habitat.
2. The transect rope usually has knots or marker tape at regular intervals, e.g. 2m.
3. The plant directly below the knot or marker is recorded along the transect.
4. It's good practice to take a measurement of some abiotic factors at each point too.

This can be a useful tool to see how species change in a habitat, in response to changes in environmental (abiotic) conditions.

Line transects use a long rope or tape to collect data in a straight-line sequence.

Method 3: Belt transect (recommended)

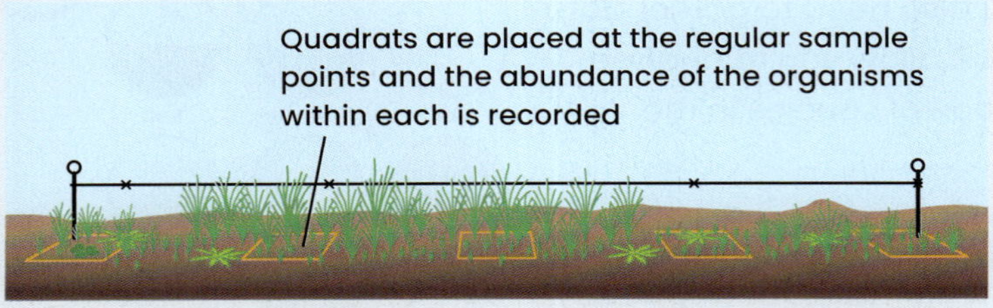

Figure 11.10: A belt transect in a grass habitat.

1. A **belt transect** uses both a transect rope and quadrats to gather data in a systematic way.
2. A transect rope or long tape measure is extended across the habitat.
3. A quadrat or trap is placed at regular intervals (e.g. every 2 or 5 metres) and the presence of the species, the number of them present or the percentage cover is recorded.
4. Again, you should aim for at least ten quadrats or traps, changing the distance between them to match the size of the habitat.
5. You should measure at least one abiotic factor at each quadrat/trap location.

Use a table like this to record your results:

QUADRAT NO.	SPECIES 1 NAME	SPECIES 2 NAME	ABIOTIC FACTOR
1			
2			

A **belt transect** gathers quadrat data in a straight-line sequence.

FOCUS ON ... GATHERING DATA WITH QUADRATS

Whatever method you choose, using quadrats will provide you with useful data for analysis.

Percentage frequency: Recording the presence or absence of a species in a quadrat will allow you calculate percentage frequency. If a species is present in 7 out of 10 quadrats, it has a percentage frequency of 70%.

Percentage cover: Quadrats with grids can be used to measure the percentage cover. Percentage cover is calculated by recording how many of the grids the species appears in (it is counted if it appears in over half of the grid). The average cover from at least 10 quadrats is then calculated.

Species population density: To calculate population density, the number of each species in each quadrat is recorded, again at least ten times. The average number per square metre (m^2) can then be recorded. The number per m^2 is then multiplied by the total habitat area to find the species population density.

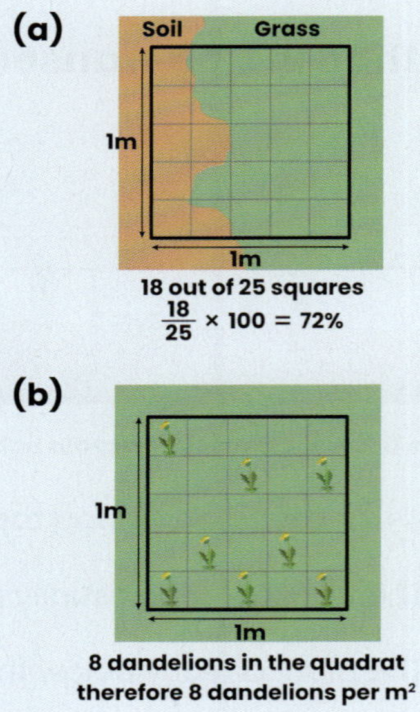

Figure 11.11: Determining (a) percentage cover and (b) species population density using a quadrat.

STEP 9: MEASURE SPECIES BIODIVERSITY INDEX

HL

As discussed in Chapter 10, the Simpson's species biodiversity index formula can be used to provide numerical data on biodiversity in your habitat. While not an essential step, it can be a useful tool in assessing changes in biodiversity over time.

$$Ds = 1 - \left(\frac{\Sigma n (n - 1)}{N (N - 1)}\right)$$

STEP 10: ANALYSE DATA AND LOOK FOR PATTERNS

It is important to present your data collected during the habitat survey. You should present the species abundance and abiotic factor data in both tables and graphs.

QUADRAT NO.	HERB-ROBERT (Species density: no./m^2)	DANDELION (Species density: no./m^2)	LIGHT INTENSITY (lux)
1	6	2	1500
2	4	2	1600
3	4	4	1800
4	3	5	2000
5	1	4	2200
6	1	6	2500
7	0	7	3600
8	0	8	3800
9	0	12	4000
10	0	13	4200

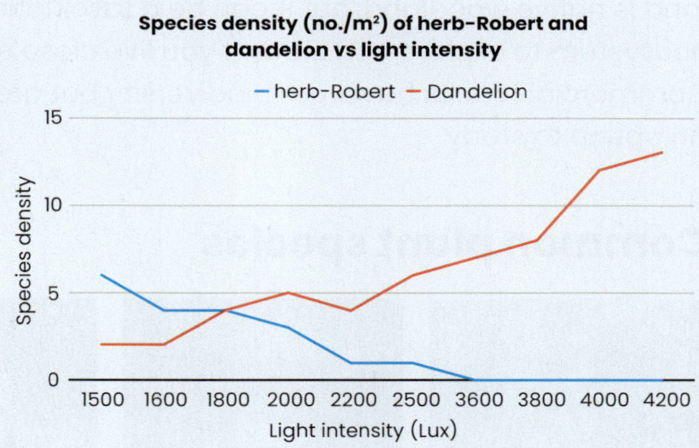

Look for patterns in the data. Do some species prefer higher temperatures, lower pH or higher light intensity? For example, in the sample data provided, a pattern emerges where the species population density of dandelions increases with increased light intensity while the population density of herb-Robert decreases. This would suggest that dandelions are sun-loving plants while herb-Robert is more tolerant of shade.

STEP 11: COMPLETE REPORT AND DRAW CONCLUSIONS

You should complete a report for your chosen ecosystem – in your Investigative Log you will find a blank report that you can use to record the results of your investigation. The final section of the report should include your overall conclusions.

INVESTIGATIVE LOG LINK

Go to your Investigative Log and complete your report on the ecosystem you studied.

Section 2: Sample Ecosystem Models

At the end of this section, you should be able to:

- Compare and contrast three varying ecosystem models.

Unpacked from LOs U1.1, U4.1, 3.1.11

The following is a simple guide to what to expect in three sample ecosystems.

1. WOODLAND ECOSYSTEM

Just over 10% of Ireland's land is woodland – the lowest coverage in Europe – though most of that is commercial forest. Only around 1% of Ireland's land is native woodland, but it can be a fascinating ecosystem to explore and study if you live close to one. Commercial forests have less biodiversity but are still an option to study.

Figure 11.12: Native Irish woodland.

Common plant species

Figure 11.13: (L–R) Nettle, herb-Robert, oak, wild garlic, bluebell.

Common animal species

Figure 11.14: (L–R) Woodlouse, ground beetle, devil's coach horse, ladybird, owl.

Sample food chain and food web

Figure 11.15: Woodland food chain.

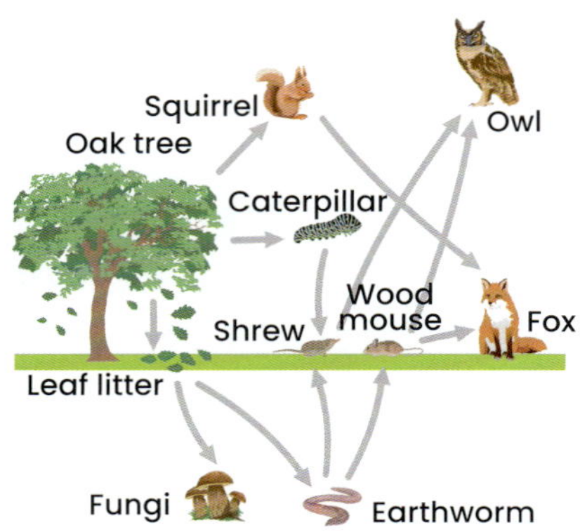

Figure 11.16: Woodland food web.

Adaptations, niches and biotic factors

A woodland ecosystem is home to multiple niches. Decomposers and detritus feeders are found on the forest floor, adapted to their role perfectly. Rabbits live in burrows and their keen sense of hearing allows them to avoid predators. Owls have amazing eyesight and can fly almost silently through the woods. Bats are nocturnal, using echolocations to navigate in the dark, eating midges and moths at night. Larger animals, like deer, are perfectly camouflaged and have antlers for protection.

Figure 11.17: Rabbits in a burrow – a behavioural adaptation.

Figure 11.18: Antlers on fallow deer – a physical adaptation.

Abiotic factors and species distribution

There are various abiotic factors that can be measured in a woodland ecosystem. Light intensity often changes as the density of the trees above change. Soil temperature and soil pH are also easy but useful factors to measure.

It is easy to observe and measure how these abiotic factors affect species distribution in a woodland. Some plants prefer shade while others prefer high light intensity. Some plants will grow directly under large trees, potentially forming symbiotic relationships with them.

Figure 11.19: Wild garlic prefers shaded areas of the woodland.

2. ROCKY SEASHORE ECOSYSTEM

The rocky seashore is a fascinating ecosystem to explore and study. Rocky seashores have a wide range of plant and animal species to explore, each perfectly adapted to life by the shore, and easy to measure abiotic factors.

Figure 11.20: A rocky seashore ecosystem.

Common plant species

Figure 11.21: (L–R) Sea lettuce, bladder wrack, kelp, serrated wrack, sea pink.

Common animal species

Figure 11.22: (L–R) Limpet, barnacle, dog whelk, sea anemone, sea gull.

Sample food chain

Figure 11.23: Seashore food chain.

RESEARCH ACTIVITY

Find out about other adaptations in plants and animals on the seashore. Record your research on a mini-whiteboard before sharing with your class.

Adaptations, niches and biotic factors

A rocky seashore is a dynamic environment. Some parts of the shore are always covered by the water, other parts are covered when the tide comes in while others are never covered by the water, but get splashed with salt water when the waves roll in. These areas are categorised as zones: the low tide zone, the middle tide zone, the high tide zone and the splash zone. Each zone provides unique niches and the plants and animals that live there are perfectly adapted to their roles.

Figure 11.24: Rocky seashore zones.

Bladder wrack has little air sacs that allow it float when the tide comes in, maximising photosynthesis; it also has a slimy layer that prevents it from drying out when the tide goes out. Limpets have a muscular foot that allow them cling tightly to the rocks and a rasping tongue to scrape algae off the rocks. Sea anemones have stinging tentacles to capture prey but can retract them when the tide goes out.

There are also lots of predators to look out for, including seagulls and crabs.

Abiotic factors and species distribution

The most significant abiotic factor on a rocky seashore is the distance from the shore and how much exposure to the tidal water a plant or animal will have. Tracking the species from the low tide zone to the splash zone with a transect will reveal dramatic changes in the plants and animals living there.

Pollution, especially plastic pollution, is a very significant abiotic factor in a rocky seashore. Salinity, water pH and temperature and wind exposure are also potential abiotic factors to measure.

3. GRASSLAND ECOSYSTEM

A wild grassland ecosystem is another important and interesting ecosystem to explore. Be mindful that most areas of grassland in Ireland are used for grazing animals and, therefore, may lack biodiversity. In fact, some may be monocultures – areas where only a single species grows.

Figure 11.25: A grassland ecosystem.

You may be able to find an area of grassland to study on the school grounds (sports fields are not suitable, as the constant topping prevents many plants from competing with the grass).

Common plant species

Figure 11.26: (L-R) Timothy grass, red clover, white clover, buttercup, bee orchid.

Common animal species

Figure 11.27: (L-R) Grasshopper, shrew, bee, admiral butterfly, swift.

Sample food chain

Figure 11.28: A grassland food chain.

Adaptations, niches and biotic factors

Grassland species have a number of adaptations. Grass itself has the ability to produce tillers (side shoots) when it is damaged from grazing or topping allowing it grow back more aggressively. Clover species form symbiotic relationships with nitrogen-fixing bacteria which provide them with natural nitrate fertiliser. The swift can fly at over 100 kilometres per hour, catching insects as it swoops over the grass, while the bee orchid uses mimicry to attract bees to pollinate it.

Abiotic factors and species distribution

Abiotic factors like soil pH, soil fertility, humidity, wind speed and exposure all have an impact on the grassland species. If there is a sheltered area in the grassland, near a hedgerow or wall, the species can vary as you move further away. Running a transect from the hedgerow or wall out to the middle of a grassland is a good technique for habitat study.

SCIENCE IN SOCIETY: TECHNOLOGY – SUSTAINABILITY

Technology in ecology

New and emerging technologies are changing how scientists monitor ecosystems. Here are some examples of technology use in ecosystem study:

Figure 11.29: Drones are being used to monitor ecosystems.

- **Satellite imaging** allows scientists to see how ecosystems change over time, e.g. monitoring deforestation (cutting down of trees) or desertification (expansion of deserts).
- **Drones** are being used to survey ecosystems from above, helping to map ecosystems in real time.
- **Remote sensors** are being used to continuously monitor abiotic factors in ecosystems, collecting data on temperature, humidity, soil moisture or light levels in real time.
- **GPS** monitors can be used to track animals helping to monitor migration patterns, behaviours and interactions. In addition, **cameras** and **acoustic monitors** can also help track animals in ecosystems.
- **DNA profiling** can be used to analyse samples of animals, plants or micro-organisms from habitats, helping to monitor genetic diversity, evolutionary history and even to find new species.

PORTFOLIO LINK **Go to your Learning Portfolio and complete Learning Check 2 to assess your knowledge and understanding of the topic so far.**

Chapter Summary

- Ecosystem and habitat studies are important to help us understand how and why living things interact with each other and their environment and, crucially, how they adapt to climate change and human interference.
- Ecological study can be **quantitative** and **qualitative**.
- Various pieces of equipment are used to collect or sample animal and plant species including: **beating trays**, **pooters**, **nets**, **sieves**, **Tullgren funnels** and **pitfall**, **cryptozoic** and **mammal traps**.
- Abiotic factors are measured in an ecosystem using tools like **light meters** (light intensity), **pH meters** (soil acidity), **thermometers** (soil, water or air temperature) and more.
- **Quadrats** and **transect ropes** are essential tools for investigating species distribution.
- **Keys** (pictorial and dichotomous) are used to help identify organisms in an ecosystem.
- Modelling an ecosystem is done in a series of steps:

1. Map the ecosystem.
2. Qualitative survey – assess the main features of the ecosystem.
3. Identify plants present.
4. Identify animals present – set some pitfall, cryptozoic or mammal traps.
5. Construct food chains and a food web.
6. Identify adaptations and niches.
7. Assess biotic factors.
8. Quantitative survey – using quadrats or transects (or both), collect data on the distribution of species and abiotic factors present in the ecosystem.
9. Measure biodiversity.
10. Analyse data and look for patterns.
11. Complete report and draw conclusions.

END OF CHAPTER ASSESSMENT

REMEMBER

1. Name **three** abiotic factors commonly measured in ecosystem studies and the equipment needed to measure them.
2. What does qualitative data aim to describe in ecology studies?
3. Name and briefly describe **three** tools used for collecting organisms during habitat surveys.
4. What square ecological tool is used to sample sections of an ecosystem?
5. Identify the main types of data collected in qualitative studies.

UNDERSTAND

6. How do beating trays work, and why are they used in habitat surveys?
7. Why is it important to adapt the approach during qualitative studies in ecology?
8. How do quantitative studies contribute to understanding species distribution in an ecosystem?
9. Differentiate between percentage cover and percentage frequency as methods of data collection in habitat surveys.

APPLY

10. Explain how a Tullgren funnel is used in collecting organisms during a habitat survey.
11. If you were conducting a habitat survey, how would you use a pitfall trap, and what type of organisms might it collect?
12. How would you use a quadrat to assess species abundance in a grassland ecosystem?
13. Apply the concept of a transect rope to assess how species change in a habitat in a straight line in a grassland ecosystem.
14. If you were conducting a quantitative study on a rocky seashore, what abiotic factors would you measure, and why?

ANALYSE

15. Compare and contrast qualitative and quantitative survey methods in terms of their goals and data collection processes.
16. What patterns or trends might you expect to find in species distribution data collected through quantitative surveys?
17. Analyse the importance of considering abiotic factors when studying species distribution in an ecosystem.
18. The table below shows the average number of two types of plant per m^2 in a grassland ecosystem. Study the data and answer the questions that follow.

SOIL pH	PERENNIAL RYEGRASS	COWSLIP
5.0	5	12
5.5	8	10
6.0	12	8
6.5	18	6
7.0	25	4
7.5	20	2
8.0	15	1

(a) How does the population of perennial ryegrass change as soil pH increases from acidic to alkaline conditions?

(b) At approximately which soil pH do perennial ryegrass and cowslip have equal populations?

(c) Which species appears to be more tolerant of acidic soil conditions (pH $<$ 7)?

(d) If the soil pH were to decrease to 4.5, predict how this might affect the populations of both species.

EVALUATE

19. Evaluate the impact of human interference on an ecosystem during a habitat survey.
20. Evaluate the potential biases that may arise in qualitative and quantitative studies and propose strategies to minimise them.
21. Evaluate the ethical considerations of collecting organisms during a habitat survey and propose ethical guidelines for habitat study for your classmates.
22. Assess the potential limitations of using identification apps in identifying organisms during an ecosystem study.

CREATE

23. Develop a hypothetical research question related to ecosystem study and propose a qualitative and quantitative approach to answer it.

12 Nutrient Recycling

LEARNING OUTCOMES: U4.1, U4.2, 3.2.4, 3.2.5, 3.2.6, 3.2.7, 3.2.8

EYE-OPENER

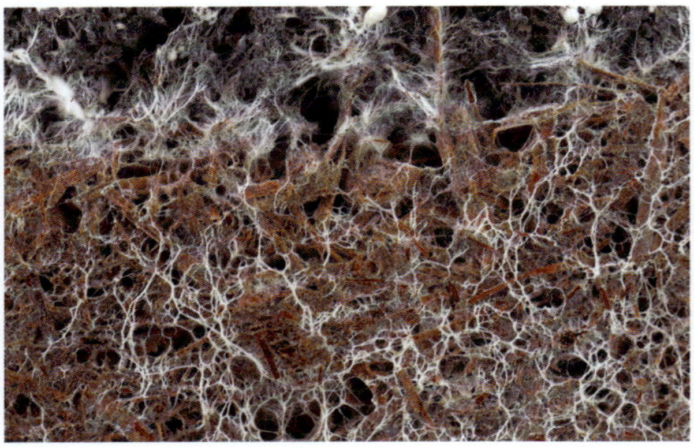

In the world beneath the soil, mycorrhizae play a vital role. These fungi form partnerships with plants, enhancing nutrient absorption and promoting healthier growth. As plants share sugars, mycorrhizae provide essential minerals, creating a sustainable and mutually beneficial relationship. This collaboration not only improves plant health but also reduces dependence on artificial fertilisers.

At the end of this chapter, you should be able to:

- Outline the role of micro-organisms in decomposition.
- Describe the importance of carbon and nitrogen to living things.
- Describe the main steps of the carbon cycle, including the role of carbon sinks.
- Discuss the link between carbon-based gases (carbon dioxide and methane) and climate change.
- Outline the main stages of the nitrogen cycle.

LINKS TO:

Chapter 7	**Bacteria and Fungi**
Chapter 10	**Biodiversity and Ecosystems**
Chapter 13	**Human Impact on the Environment**
Chapter 14	**Biomolecules**
Chapter 15	**Metabolism and Enzymes**

SUBJECT VOCABULARY

decomposition, organic matter, micro-organisms, greenhouse effect, carbon cycle, photosynthesis, respiration, combustion, carbon sink, nitrogen cycle, nitrogen fixation, assimilation, nitrification, denitrification

Section 1: Micro-organisms and Decomposition

At the end of this section, you should be able to:

- Define the term 'decomposition'.
- Describe what is meant by the soil microbiome.
- Outline the role of the soil microbiome in the recycling of nutrients.
- Discuss the factors that affect the rate of decomposition in ecosystems.

Unpacked from LO 3.2.4

DECOMPOSITION

As mentioned in Chapter 10, decomposers are organisms that break down the remains of dead animals or plants (often referred to as **organic matter**). **Decomposition** is a vital ecological process that plays a crucial role in returning nutrients to living organisms in the ecosystem. Most nutrients are returned to the soil, where plants can absorb them and bring them back into the food chain.

Decomposers are primarily **micro-organisms**, particularly bacteria and fungi, who live and carry out their important ecological niche in the soil. Collectively known as the **soil microbiome**, this community of bacteria and fungi have major roles in the recycling of carbon, nitrogen and phosphorous compounds.

ROLE OF THE SOIL MICROBIOME

Bacteria

Bacteria are the primary decomposers in ecosystems. They decompose by secreting enzymes directly onto the organic matter, which then begin to digest the complex molecules into simpler compounds. Decomposing bacteria are **saprophytes**.

Nitrogen-fixing bacteria play a pivotal role in converting atmospheric nitrogen into a form usable by plants. This nitrogen fixation is crucial for plant growth and, subsequently, sustains herbivores and carnivores within the ecosystem.

Fungi

Fungi, including yeasts, moulds and mushrooms, are another major group of decomposers. Fungi play a complementary role to bacteria, often targeting tougher organic materials like lignin and cellulose found in wood and plant cell walls. Many fungi have long thin cells called hyphae, which can penetrate and break down complex organic compounds. One particularly important group of fungi, called mycorrhizae, form symbiotic relationships with plants, allowing them to absorb more nutrients from their roots.

- **Organic matter** refers to the remains of dead living things or their by-products and wastes.
- **Decomposition** is the breakdown of the remains of dead animals and plants.
- **Soil microbiome** describes the community of bacteria and fungi in the soil.
- **Saprophytes** are organisms that release enzymes onto organic matter for decomposition.

Go to YouTube.com and search for 'What is Mycorrhizal Fungi?' (1:13) by PlantWorks.

FACTORS AFFECTING DECOMPOSITION

Bogs contain large quantities of organic matter because the rate of decomposition is very low. This is due to the waterlogged conditions which result in low oxygen levels, in addition to often acidic environments.

DID YOU KNOW?

Several factors influence the rate and efficiency of decomposition in ecosystems:

- **Temperature:** Bacteria and fungi need relatively warm conditions for their cells and, in particular, enzymes to function properly. At low temperatures, the rate of decay is lower as the collision rate for molecules is low. Climate change is affecting the temperature of soils and could have a large effect on the soil's micro-organisms.
- **Water:** Water is an essential nutrient for all living things, including micro-organisms. Water is needed as a medium to carry out enzymatic reactions. Too much water can lead to a lack of oxygen, making decomposition less efficient.
- **Oxygen availability:** Decomposition can occur in aerobic (with oxygen) or anaerobic (without oxygen) conditions but, as mentioned previously, aerobic decomposition is more efficient. Anaerobic conditions often result in the production of methane and other gases, which can be harmful to the environment.
- **Detritivores:** Detritivores, as discussed in Chapter 10, including insects, woodlice and earthworms, play a crucial role in breaking down organic matter physically. When detritivores are present, they increase microbial decomposition by increasing the surface area of the organic matter.
- **Type of organic matter:** The source and chemical nature of the organic matter influence the rate of decomposition. Woody plants contain a complex carbohydrate called lignin, which can take a long time to decompose. Animal bones, ligaments and tendons, which contain proteins, are slower to decompose than plant material.

Whale fall, where the colossal carcass of a dead whale sinks to the ocean floor, presents a significant challenge to ocean decomposers. The extreme cold, high pressure, low oxygen levels and complete darkness all contribute to a slower process than on the surface.

The vastness of a whale's body provides a sudden banquet for deep-sea decomposers. Specialised micro-organisms, often only found on the deep ocean floor, bone-eating worms and, later, octopi adapt to this challenging environment. Studying how these creatures survive in the deep sea helps us understand how life persists in extreme conditions.

Figure 12.1: Whale fall.

DID YOU KNOW?

PORTFOLIO LINK

Go to your Learning Portfolio and complete Learning Check 1 to assess your knowledge and understanding of the topic so far.

Section 2: The Carbon Cycle

At the end of this section, you should be able to:

- Outline why carbon is an important element for living things.
- Describe the main processes in the carbon cycle, including the role of photosynthesis, respiration, combustion and decomposition.
- Define a carbon sink and discuss their importance.
- Discuss the ethical and sustainability issues associated with the carbon cycle.

Unpacked from LOs 3.2.5, 3.2.7, 3.2.8

WHY IS CARBON IMPORTANT?

Carbon is a key element found in all biomolecules – the building blocks of cells. Carbon is found in carbohydrates, proteins, lipids, nucleic acids and energy carriers (e.g. ATP). In the air, carbon is found within molecules like carbon dioxide and methane, which contribute to the **greenhouse effect**.

The **greenhouse effect** is the natural process that warms the Earth's surface when gases, like CO_2, reabsorb heat back into the atmosphere.

When living things die, the carbon from their bodies is needed by other living things – remember, biomass is limited in an ecosystem – so natural mechanisms are in place to return the carbon to the food chain. We call this process the carbon cycle.

FOCUS ON ... THE GREENHOUSE EFFECT

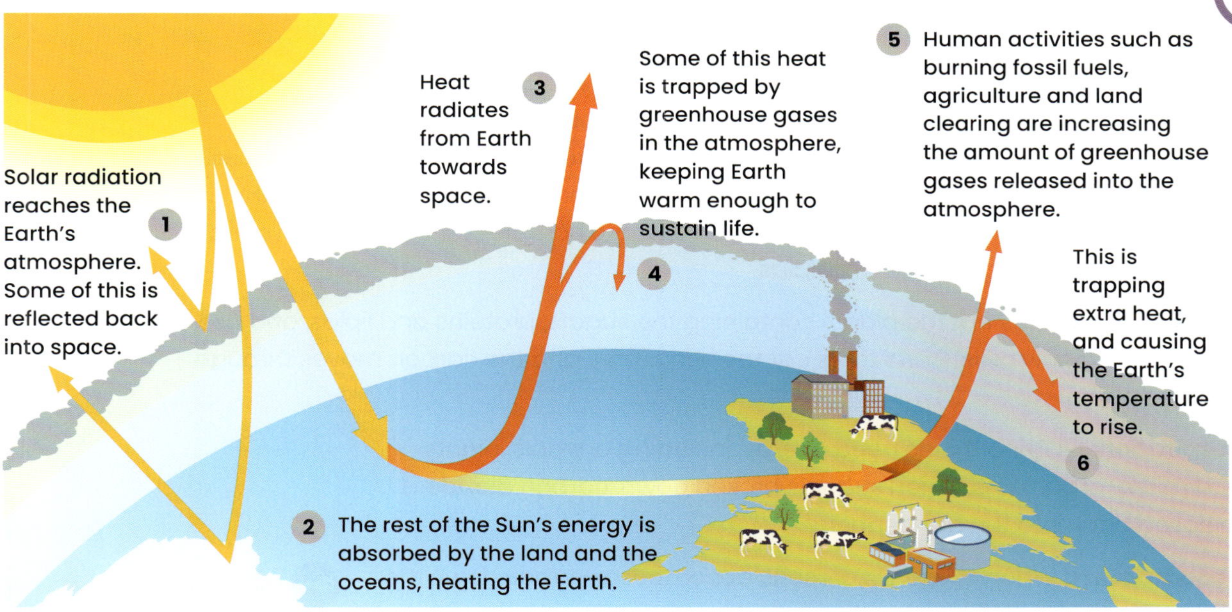

Figure 12.2: The greenhouse effect.

The greenhouse effect is a natural process that warms the Earth's surface. When the Sun's energy reaches the Earth, some of it is reflected back to space while the rest is absorbed and re-radiated by greenhouse gases in the atmosphere. The absorbed energy warms the atmosphere and the surface of the Earth, maintaining a moderate temperature that allows life to exist on our planet. However, human activities, such as the burning of fossil fuels, have intensified the greenhouse effect, leading to global warming and climate change.

Go to YouTube.com and search for 'What Is The Natural Greenhouse Effect?' (5:21) by FuseSchool – Global Education.

The carbon cycle

The movement of carbon from the environment to living organisms (and back again) is called the carbon cycle. It is an essential component of healthy ecosystems, returning this important building block to new living cells. Being a cycle, there is no beginning or end, but it's useful to begin in the atmosphere where carbon is found primarily as carbon dioxide, which makes up 0.04% of the volume of the air around you.

Go to YouTube.com and search for 'The carbon cycle – Nathaniel Manning' (3:54) by TED-Ed.

Stages of the carbon cycle

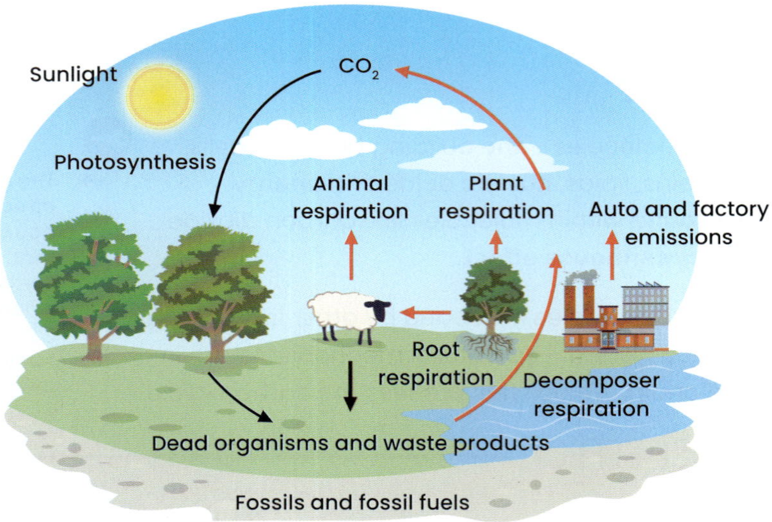

Figure 12.3: The carbon cycle.

- **Photosynthesis:** Plants absorb carbon dioxide from the atmosphere, through the stomata on the leaves, and use it to make food in the form of sugar. This vital process brings carbon from the abiotic (non-living part of the ecosystem) to the biotic (the living part). The plants can convert the sugar to more complex carbohydrates (starch, cellulose or lignin) or proteins and lipids.
- **Feeding:** Herbivores eat the plants containing the sugars, proteins and lipids, and use them to build new cells. Carnivores eat the herbivores and the carbon moves through the food chain.
- **Respiration:** All living things need energy, including animals, plants and micro-organisms. When sugar is broken down, carbon dioxide is released back into the atmosphere.
- **Decomposers:** When oxygen is present, decomposers break down the organic compounds from dead and decaying organisms, releasing more carbon dioxide into the atmosphere. Without oxygen, methane (CH_4) can be released. Methane is also a greenhouse gas.
- **Fossil fuels:** The burning of fossil fuels (**combustion**) releases large amounts of carbon dioxide into the atmosphere. Fossil fuels are the remains of organisms that lived over 300 million years ago and whose bodies didn't decay when they died. Instead, their carbon has been trapped under the surface of the Earth until humans began to mine them, burning them for electricity generation and heating. The release of this extra carbon dioxide is contributing to the enhanced greenhouse effect, global warming and climate change.

- **Photosynthesis** is the process by which plants make their food.
- **Respiration** is the release of energy from food.
- **Combustion** is the burning of fossil fuels, releasing heat and carbon dioxide into the atmosphere.

- **Carbon sinks:** Carbon isn't always cycling from plants to animals to decomposers and back to the atmosphere; sometimes, it is absorbed and stored by natural or artificial means. Anything that absorbs and stores carbon is known as a **carbon sink**.

A **carbon sink** is anything that absorbs more carbon than it releases. Forests are natural carbon sinks.

FOCUS ON ... CARBON SINKS

Carbon sinks are natural or artificial reservoirs that absorb and store more carbon than they release (a process called carbon sequestration). They play a crucial role in reducing carbon dioxide and methane levels in the atmosphere and therefore potentially reducing the impact of climate change. Forests, oceans, peat bogs and soil are natural carbon sinks. Forests and trees absorb carbon dioxide during photosynthesis, while oceans also act as important carbon sinks. The ocean is home to phytoplankton and algae, which remove carbon dioxide by photosynthesis; the ocean's water stores large amounts of dissolved carbon dioxide, keeping it out of the atmosphere. Soils store carbon from decomposed organic matter while peat bogs accumulate partially decomposed plant material over centuries, storing vast amounts of carbon in waterlogged conditions. Human-made carbon sinks include reforestation projects and carbon capture technologies. It is vital that humans preserve and enhance these sinks to maintain a balanced carbon cycle and reduce atmospheric carbon dioxide levels.

Figure 12.4: Trees act as carbon sinks.

PORTFOLIO LINK **Go to your Learning Portfolio and complete Learning Check 2 to assess your knowledge and understanding of the topic so far.**

Section 3: The Nitrogen Cycle

At the end of this section, you should be able to:

- Discuss the importance of nitrogen in living things.
- Outline the stages of the nitrogen cycle, including nitrogen fixation, assimilation, decomposition, nitrification and denitrification.
- Discuss the ethical and sustainability issues of the nitrogen cycle, including the uses of artificial fertilisers.

Unpacked from LOs 3.2.6, 3.2.7, 3.2.8

WHY IS NITROGEN IMPORTANT?

Like carbon, nitrogen is an essential element in living things. Nitrogen is found within important biomolecules like amino acids (which build proteins) and nucleic acids (DNA and RNA), found in all living things. It is also an essential component in chlorophyll, which plants need to photosynthesise, and also in haemoglobin, a protein which carries oxygen in the blood of many animals.

Like carbon, nitrogen is recycled from dead living things through natural processes, returning it to the atmosphere or living things; we call this carefully balanced, continuous recycling and movement the **nitrogen cycle**.

THE NITROGEN CYCLE

The **nitrogen cycle** maps the movement of nitrogen between the environment and the living world.

There is no beginning or end to the nitrogen cycle, but we will begin in the atmosphere. Nitrogen is the most abundant element in the atmosphere (78% of the air around you is nitrogen gas, N_2) but, unlike carbon dioxide, it cannot be absorbed and used by plants or animals directly (you can think of the atmosphere as a nitrogen sink). In nature, nitrogen is more readily available to plants as simple molecules called nitrates, nitrites and ammonia, which are absorbed in soil water.

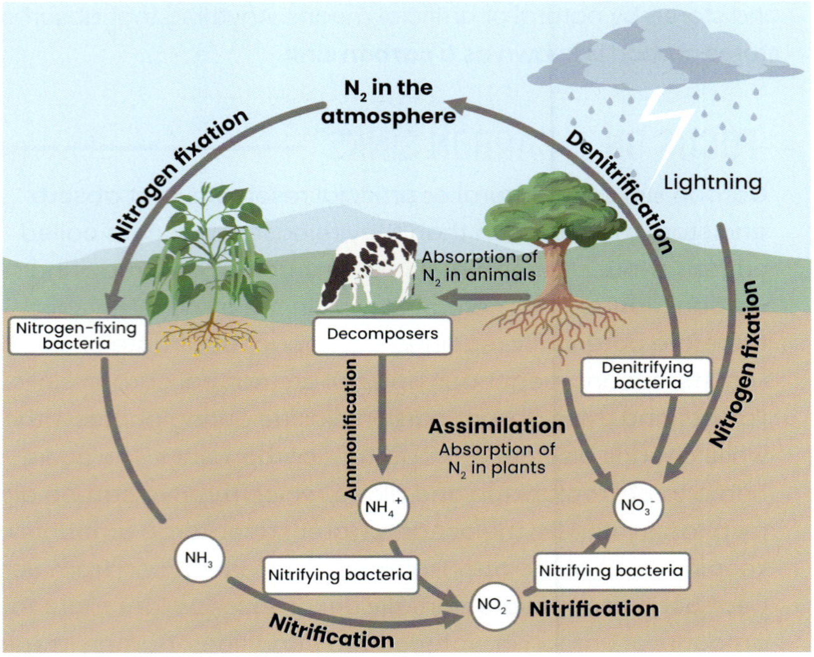

Figure 12.5: The nitrogen cycle.

Stages of the nitrogen cycle

- **Lightning:** This natural process provides the energy to react nitrogen and oxygen in the atmosphere to create **nitrates** (compounds containing the NO_3^- ion). The nitrates dissolve in atmospheric water, falling as precipitation and making their way into the soil. This process is called **nitrogen fixation**.
- **Nitrogen-fixing bacteria:** Lightning is not the only means by which nitrogen can be 'fixed'. Specialised soil bacteria are also able to convert atmospheric nitrogen into nitrates. These bacteria are called **nitrogen-fixing bacteria** and they often form symbiotic relationships with plants from the pea family (legumes), including clover, growing in small nodules on their roots.
- **Assimilation:** Now that the nitrogen is in a form that plants can use (nitrates), it is absorbed by the plant roots. The plant can use it to make new proteins and DNA in its cells. The use of a nutrient by a living thing or cell is called **assimilation**.
- **Feeding:** The nitrogen is now in the food chain; herbivores eat the plants and carnivores can eat the herbivores, thus bringing nitrogen into the animal kingdom.
- **Decomposers:** Plants and animals die and the nitrogen in their cells is broken by the decomposers, bacteria and fungi found mainly in the soil. The product of the breakdown of nitrogen compounds by decomposers is **ammonia** (**NH_3**) which quickly reacts with water to form ammonium ions (**NH_4^+**).

Go to YouTube.com and search for 'Nitrogen Cycle – Leaving Cert Biology 2022 – Updated for revision' (7:51) by Biology Bugbears.

Figure 12.6: Root nodules on a bean plant, containing nitrogen-fixing bacteria.

- **Nitrogen fixation** is the process by which atmospheric nitrogen is converted into nitrates or other compounds.
- **Assimilation** is the use of nutrients in a living cell.

- **Nitrification:** Ammonia, once again, is not easily absorbed by plants. Bacteria in the soil convert the ammonia to **nitrites** (compounds containing the NO_2^- ion) but these also can't be used by plants. So, finally, more bacteria convert the nitrites to nitrates, which can then be absorbed by the plants and used to make proteins, new cells and useful compounds like chlorophyll. These bacteria, which convert ammonia to nitrite and nitrite to nitrate, are called **nitrifying bacteria** and the process is called **nitrification**.

> **Nitrification** is the conversion of ammonia and nitrite compounds to nitrates by nitrifying bacteria in the soil.
>
> **Denitrification** is the conversion of nitrates to nitrogen gas by denitrifying bacteria in the soil.

- **Denitrification:** So far, all the processes have made nitrogen available to plants and therefore animals and decomposers. **Denitrification** does the opposite – it involves the conversion of nitrates back into nitrogen gas, returning it to the atmosphere (and finishing the cycle). Again, this is carried out by bacteria – called **denitrifying bacteria**. These bacteria thrive in waterlogged soils, with little oxygen, and are chemosynthetic.

Like ecosystems in general, the nitrogen and carbon cycles are fragile. They face a series of challenges as the widespread effects of human population growth, pollution, agriculture and industry continue to impact our environment. In the next chapter, we will discuss the impact our species is having on ecosystems including decomposition and the carbon and nitrogen cycles.

SCIENCE IN SOCIETY: TECHNOLOGY – SUSTAINABILITY

The Haber-Bosch process

The Haber-Bosch process is important for producing ammonia (NH_3). It works by combining nitrogen from the air with hydrogen under high pressure and temperature in the presence of a catalyst. This reaction allows for large-scale ammonia production, a key ingredient in the artificial fertilisers that have revolutionised agriculture.

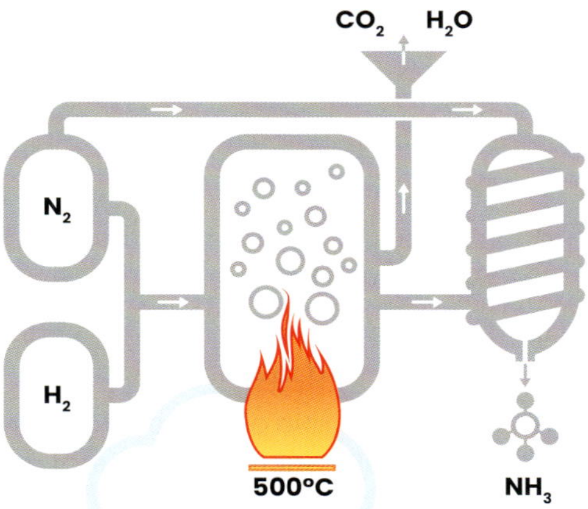

Figure 12.7: The Haber-Bosch process.

However, there are some ethical and sustainability issues associated with the process. It requires significant energy inputs, often obtained from burning fossil fuels, raising concerns about greenhouse gas emissions and climate change. Increased fertiliser use can lead to intensive farming practices that contribute to deforestation and habitat loss. Excess fertiliser runoff can pollute waterways, causing algal blooms and harming aquatic ecosystems. While the Haber-Bosch process has been crucial in feeding the world's population, its environmental impact means we need to explore more sustainable ways of producing artificial fertilisers in the future.

RESEARCH ACTIVITY

Carry out some research on more sustainable means of fertilising crops.

THINK, PAIR, SHARE

In pairs, discuss briefly how the human population is altering the natural balance of both the carbon and nitrogen cycles. Share your thoughts with the class.

PORTFOLIO LINK

Go to your Learning Portfolio and complete Learning Check 3 to assess your knowledge and understanding of the topic so far.

Chapter Summary

- **Decomposition** is a crucial ecological process that involves breaking down the remains of dead animals or plants, known as organic matter, to recycle nutrients in ecosystems.
- Micro-organisms, particularly **bacteria** and **fungi**, play a central role as decomposers in this process.
- Bacteria decompose organic matter by secreting enzymes directly onto it, initiating the digestion of complex molecules into simpler compounds.
- Fungi, including yeasts, moulds and mushrooms, complement bacteria by targeting tougher organic materials such as **lignin** and **cellulose** found in wood and plant cell walls.
- Several factors influence the **rate of decomposition**, including temperature, water availability, oxygen levels, detritivores (such as insects and earthworms) and the type of organic matter.
- The **carbon cycle**, essential for healthy ecosystems, encompasses processes like photosynthesis, respiration, combustion and decomposition.
- **Carbon sinks**, such as forests and oceans, play a crucial role in reducing atmospheric carbon dioxide levels by absorbing and storing more carbon than they release.
- The **nitrogen cycle** involves stages like nitrogen fixation, assimilation, decomposition, nitrification and denitrification.
- The nitrogen and carbon cycles face challenges due to widespread human activities, including population growth, pollution, agriculture and industry.

END OF CHAPTER ASSESSMENT

REMEMBER

1. Define *decomposers* and provide examples.
2. Identify the key nutrients recycled through decomposition.
3. Describe how nitrogen-fixing bacteria contribute to the nitrogen cycle.
4. Explain the role of micro-organisms – specifically bacteria and fungi – in decomposition.
5. Define a *carbon sink* and provide examples.
6. Summarise the processes involved in the carbon cycle.
7. Define *assimilation* in the context of the nitrogen cycle.

UNDERSTAND

8. Discuss the factors influencing the rate of decomposition in ecosystems.
9. Compare the roles of bacteria and fungi in decomposition.
10. Discuss the impact of burning fossil fuels on atmospheric carbon dioxide levels.
11. Explain the importance of nitrogen in living things.
12. Outline the stages of the nitrogen cycle.
13. Discuss the impact of denitrification on nitrogen availability in ecosystems.

APPLY

14. Develop a flowchart illustrating the stages of decomposition in an ecosystem.
15. Display the key processes of the carbon cycle on a diagram.
16. Analyse the impact of temperature on the efficiency of decomposition.
17. Evaluate the importance of water in decomposition and potential consequences of excess water.

END OF CHAPTER ASSESSMENT

ANALYSE

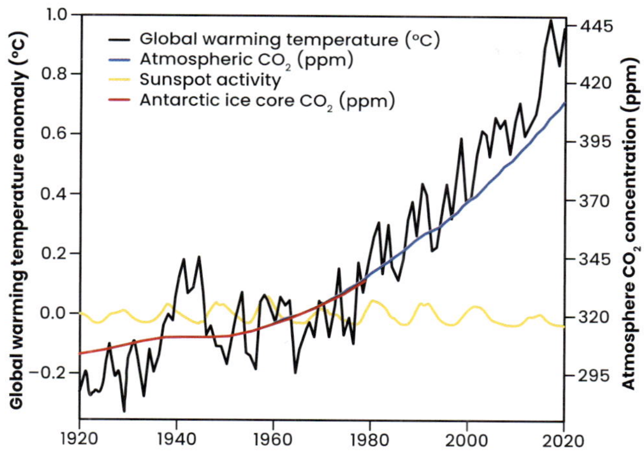

Source: ResearchGate.

18. The above graph shows the relationship between the average global warming temperature anomaly, atmospheric CO_2 concentration, sunspot activity in the period since 1920 (CO_2 concentration data before 1920 was obtained from ice cores in the Antarctic).

(a) Is there a relationship between CO_2 concentration in the atmosphere and global temperature anomaly according to the graph? Explain your answer.

(b) Why do you think global warming anomaly figures have risen quickly since the 1980s?

(c) What evidence is there to suggest that sunspot activity is **not** the cause of global warming in the period since 1920?

19. Examine the role of carbon sinks in climate change.

20. Nitrification is a crucial process in the nitrogen cycle where ammonium (NH_4^+) is converted to nitrate (NO_3^-) by nitrifying bacteria. A study was conducted to investigate how different factors affect the rate of nitrification in agricultural soils. The researchers measured the nitrate production (in mg/kg soil/day) under various conditions. Analyse the data and answer the following questions.

TEMP (°C)	SOIL pH	SOIL MOISTURE (%)	NITRATE PRODUCTION (mg/kg soil/day)
10	6.0	20	2.1
20	6.0	20	4.5
30	6.0	20	7.2
20	5.0	20	2.8
20	7.0	20	5.9
20	6.0	10	3.2
20	6.0	30	5.7

(a) How does temperature affect the rate of nitrification? Provide evidence from the data to support your answer.

(b) What appears to be the optimal soil pH for nitrification among the conditions tested? Explain your reasoning.

(c) Calculate the percentage increase in nitrate production when soil moisture increases from 10% to 30% at 20°C and pH 6.0.

(a) If a farmer wants to reduce nitrogen loss, which condition should they aim for to slow down nitrification? Justify your answer using the data.

EVALUATE

21. Evaluate the ethical implications of human activities contributing to the greenhouse effect.

22. Evaluate the significance of decomposition in maintaining ecosystem health.

23. Discuss the importance of biodiversity in supporting ecosystem stability.

24. Discuss the pros and cons of the Haber-Bosch process.

CREATE

25. Propose a hypothetical scenario where the absence of decomposers disrupts the nutrient cycle.

26. Create models of the nitrogen and carbon cycles.

13 Human Impact on the Environment

LEARNING OUTCOMES: U1.2, U3.1, U3.2, U4.2, 3.1.2, 3.1.3, 3.2.7, 3.2.8

EYE-OPENER

The Amazon rainforest is home to unparalleled biodiversity, hosting thousands of plant and animal species. However, deforestation in the Amazon poses a severe threat to one of the planet's most vital ecosystems, as humans aim to convert land for agriculture, acquire timber for construction, mine for precious metals or oil and build new roads, housing or other infrastructure projects. Thousands of species, many yet to be discovered, face habitat loss, pushing them towards extinction. The interconnected relationships between plants, animals and micro-organisms are disrupted, affecting ecosystem stability. Indigenous communities, deeply connected to the forest, also experience social and cultural upheaval. Deforestation exacerbates climate change, as trees act as carbon sinks. The release of stored carbon into the atmosphere intensifies global warming. Sustainable land use practices, conservation initiatives and international co-operation are imperative to safeguard this invaluable natural treasure.

At the end of this chapter, you should be able to:

- Discuss pollution, its various sources and effects on the air, water and soil.
- Outline mechanisms for controlling pollution.
- Discuss the causes and effects of biodiversity loss in our ecosystems.
- Describe conservation practices to help restore biodiversity and ecosystem health, including the use of technology in conservation.
- Define sustainability and discuss how biodiversity and conservation help form part of a sustainable Earth.

LINKS TO:

Chapter 2	**Scientific Knowledge and Research**
Chapter 3	**Experiments and Data Analysis**
Chapter 10	**Biodiversity and Ecosystems**
Chapter 12	**Nutrient Recycling**

SUBJECT VOCABULARY

pollution, eutrophication, greenhouse gas, afforestation, biodiversity, invasive species, food security, conservation, rewilding, sustainability

Section 1: Pollution and Climate Change

At the end of this section, you should be able to:

- Define the term 'pollution'.
- Give examples of pollutants from various sources.
- Name two 'greenhouse gases' and briefly describe how they contribute to climate change.
- Discuss methods to reduce pollution, including greenhouse gases.

Unpacked from LOs U3.1, 3.2.8

There are now more than 8 billion people on the planet, with more than half living in urban environments. Humans have, in their short time on the planet, colonised practically every corner of the globe. They have developed technologies which allowed their population to explode, including systematic agricultural practices.

Sadly, humans are having an enormous impact on life on our planet, as our lifestyle, industry and agriculture conflict with the natural world. In this chapter, we will explore how humans can influence ecosystems, both positively and negatively.

Figure 13.1: Humans have had a significant impact on the planet.

Go to YouTube.com and search for 'Human Population Through Time' (6:24) by American Museum of Natural History.

WHAT IS POLLUTION?

Pollution is an unwanted, harmful addition to an ecosystem by human activity. Pollutants can come from various sources including our homes (domestic), our transport (planes, cars, freight, etc.), our farms (agriculture) and our factories (industry).

Pollution is a harmful addition to an ecosystem by human activity.

DOMESTIC POLLUTION	AGRICULTURAL POLLUTION	INDUSTRIAL POLLUTION

• Plastic • Wastewater • Sewage • Waste detergents	• Fertiliser runoff • Herbicides/ fungicides/ pesticides • Animal waste • Excess nutrients	• Greenhouse gas emissions • Heavy metals • Waste runoff • Noise pollution

Pollution affects living things and processes in all ecosystems; it is often categorised into water, air and soil pollution.

Water pollution

Earth's oceans, rivers and lakes can be polluted with runoff from excess fertiliser, farm chemicals (e.g. pesticides), heavy metals from industry, sewage and, of course, plastic. These unwanted additions can lead to the poisoning of animals or sometimes to a process called **eutrophication**, where excess nutrients cause a rapid growth of plant material or algae before they quickly die, consuming the oxygen in the water as they degrade (displaying a typical J-curve population graph). This loss of oxygen causes the death of animal life in the water.

Figure 13.2: Eutrophication can cause algal blooms.

Eutrophication is the addition of nutrients to a waterway, which often results in an algal bloom.

A **greenhouse gas** is a gas, such as carbon dioxide or methane, that traps heat in the Earth's atmosphere, contributing to the greenhouse effect and influencing global climate patterns.

Go to YouTube.com and search for 'What are greenhouse gases and how do they contribute to climate change?' (1:52) by CBS News.

SCIENCE IN SOCIETY: SUSTAINABILITY

Human population growth

The population of humans has risen exponentially over the past thousand years, as our species has developed industry, agriculture, sanitation and medicine. The Earth's human population is now over 8 billion and it continues to rise, although the rate of growth has slowed in recent decades, particularly with better education and easier access to contraception. With over 8 billion mouths to feed, the production of food and the provision of clean water are under enormous pressure. There is considerable debate around human carrying capacity on Earth, with some estimating we've already reached that point at 8 billion.

Do you think the current rate of human population growth is sustainable?

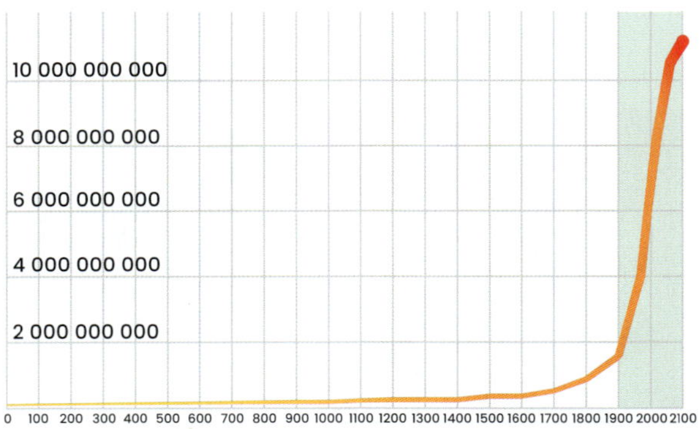

Figure 13.3: Human population growth over the last 2000 years.

Air pollution

Our homes, farms and factories can all add harmful chemicals to the air. The burning of fossil fuels like coal, peat or oil adds smoke, which contains tiny particles that can lead to lung damage in humans and other animals. Smoke can also contain toxic chemicals that can affect both animal and plant life. Most significantly, burning fossil fuels releases **greenhouse gases** like carbon dioxide, which can lead to increased global air temperatures and climate change. Methane, a gas released from bacteria and some farm animals, is also a greenhouse gas.

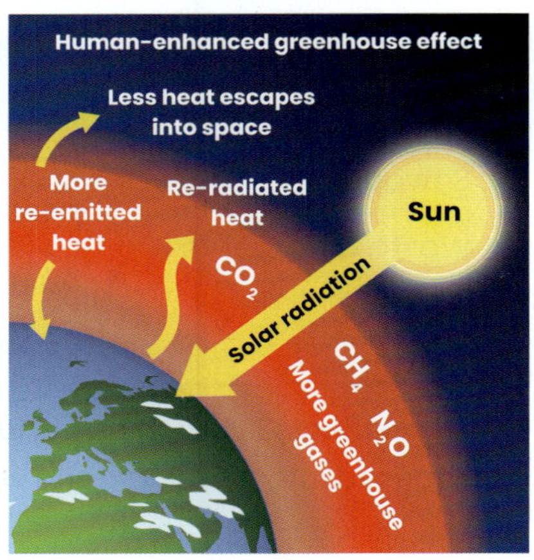

Figure 13.4: The effects of the additional greenhouse gases in our atmosphere.

Soil pollution

As waste from industry and agriculture enter waterways, they quickly make their way into the soil. In the soil, they can build up and affect the fertility (amount of nutrients) available to plants, perhaps limiting their growth. This in turn affects the population of the animal life in the ecosystem. Soil pollutants can also have a major effect on fungi and bacteria in the soil, essential to the recycling of matter and energy in the ecosystem.

CONTROLLING POLLUTION

Minimising and controlling pollution, from all sources, is a significant challenge. Everyone has a role to play – at home, in farms and factories, and in national and international politics. Below are some means of controlling pollution in the air, water and soil.

Afforestation is the planting of trees.

1 Alternative energy sources: Transition to renewable electricity production using solar, wind or hydroelectric, instead of burning fossil fuels.

2 Afforestation: Plant trees or green areas to absorb pollutants from the soil and remove CO_2 from the air.

3 Wastewater treatment: Treat wastewater from homes and businesses before returning it to rivers, lakes or the sea.

4 Increased regulation: Regulate fertiliser, herbicide and pesticide use on farms. Increase regulation on industrial emissions and vehicle exhausts.

5 Research: Invest in scientific research on pollution control and remediation, including the development of new technologies.

6 Minimise waste: Reduce household, farm and industrial waste. Ban single use plastics and excess packaging.

7 Bioremediation: Use bacteria or other organisms to remove plastics or oil slicks from the ocean.

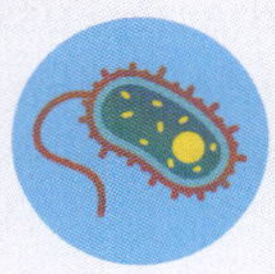

8 Education: Improve education on pollution and increase public awareness around clean energy.

9 Clean transport: Encourage the use of public transport, walking or cycling rather than car use.

10 Recycling: Encourage the recycling of plastic, paper, glass and aluminium. Encourage industry to use recycled plastics.

11 Policy: Encourage the government to implement policies that promote waste reduction at all levels, nationally and through international co-operation.

12 Public consultation: Encourage the public to get involved in the decisions around pollution and its control.

PORTFOLIO LINK **Go to your Learning Portfolio and complete Learning Check 1 to assess your knowledge and understanding of the topic so far.**

Section 2: Biodiversity Loss, Conservation and Sustainability

At the end of this section, you should be able to:

- Discuss the factors that contribute to biodiversity loss.
- Discuss the effects of biodiversity loss.
- Interpret data on the effect of human activity on biodiversity.
- Define conservation and give examples of conservation practices in a range of areas.
- Discuss how technology is shaping conservation practice.
- Define the term sustainability.
- Relate conservation to sustainable practices in agriculture, industry and at home.
- Evaluate the sustainability challenges facing the carbon and nitrogen cycles.

Unpacked from LOs U3.1, U3.2, 3.1.2, 3.1.3, 3.2.7

BIODIVERSITY LOSS

As discussed in Chapter 10, **biodiversity** is the variety of life on Earth or in a specific ecosystem. There are three categories of biodiversity: species diversity, genetic diversity and habitat diversity.

Sadly, both nationally and internationally, we are experiencing a rapid decline in biodiversity. It is estimated that we are experiencing loss of biodiversity at a rate 1000 times natural levels, driven mainly by human activity. Some groups of organisms are particularly at risk, with an estimated 40% of amphibian species facing the threat of extinction and insects also declining, including bees. Many insects are keystone species, essential for the stability of ecosystems. As well as loss of species diversity, a reduction in genetic diversity is also evident in many ecosystems.

Figure 13.5: Deforestation.

Biodiversity is the variety of life on Earth or in an ecosystem.

Causes of biodiversity loss

- **Natural biodiversity loss:** Most ecosystems undergo periods of reduced biodiversity, such as over the winter months when plant and insect populations decrease. Natural disasters like floods, wildfires or volcanic activity can lead to temporary biodiversity loss in an ecosystem.
- **Habitat destruction:** Deforestation and other land-clearing activities, for agriculture and urbanisation.
- **Climate change:** Changes in climate are affecting abiotic factors (e.g. rainfall, sea levels, etc.) at local level, leading to habitat changes.
- **Pollution:** The addition of harmful materials to the air, water and soil.

- **Invasive species:** The introduction of non-native species to ecosystems can lead to them outcompeting native species. The introduction of grey squirrels in Ireland had a massive impact on the native red squirrel.
- **Disease:** The spread of disease, often enhanced by changes in climate, is leading to the loss of species. Ash dieback is a fungal disease spreading throughout Ireland, killing native ash trees.
- **Overfishing:** The overexploitation of fishing areas and other natural food sources is leading to a decline in fish numbers and diversity worldwide.
- **Poaching:** The illegal hunting and killing of animals is affecting species diversity. The killing of elephants and rhinos for their ivory is one example of this; only two female northern white rhinos exist, as of 2025.
- **Herbicides and pesticides:** The overuse of plant, insect and fungi killing chemicals on farms is leading to many species going extinct; in many cases, these plants or insects are not the intended target of the chemicals.
- **Lack of policy:** A lack of policy, rules and regulations around conservation practices (more on this later) is also a contributing factor in biodiversity loss.

Figure 13.6: Japanese knotweed is an invasive species.

An **invasive species** is a non-native plant or animal that outcompetes native species.

Go to YouTube.com and search for 'Bees and Almonds' (1:15) by California Almonds.

Effects of biodiversity loss

The loss of species and genetic diversity has considerable consequences for ecosystems, humans and the under-threat species. We can categorise these effects into environmental, social, economic and cultural effects.

ENVIRONMENTAL EFFECTS

- **Reduced genetic diversity:** As numbers decrease, the variation in genes in a population also decrease. This can lead to inbreeding, which further reduces genetic diversity. A lack of genetic diversity means species are slow to adapt to a changing environment.
- **Species extinction:** A lack of genetic biodiversity can lead to full species extinction.
- **Ecosystem instability:** A loss of biodiversity can disrupt the balance in an ecosystem and cause it to collapse when environmental conditions change.
- **Nutrient recycling:** Biodiversity loss can lead to an imbalance or even a collapse of the carbon or nitrogen cycles.

SOCIAL EFFECTS

- **Displacement of communities:** The destruction of habitats can cause the displacement of local communities, especially indigenous groups.
- **Increased disease:** Biodiverse ecosystems provide a buffer zone for disease. A reduction in biodiversity increases the risk of plant, animal or human disease spreading.
- **Conflicts and war:** A potential increase in the likelihood of conflict over scarce natural resources.

ECONOMIC EFFECTS

- **Pharmaceuticals:** Many plant and fungus species provide or may provide vital medicines to treat human diseases.
- **Ecotourism:** A decline in natural spaces reduces ecotourism, having economic effects.
- **Food security:** Many food crops rely on insects to pollinate them. As insect populations decrease, crop yields might also decline. Overfishing and destruction of marine ecosystems could result in fish population collapse.

Food security means everyone has access to safe and nutritious food.

CULTURAL EFFECTS

- **Indigenous communities:** Loss of ecosystems and biodiversity can affect indigenous communities who have strong cultural connections with specific species or habitats e.g. under-threat coral reefs have strong cultural connections with Pacific Island communities.
- **Knowledge and cultural inspiration:** Biodiversity loss impacts artistic and creative expression in cultures that draw inspiration from nature. This affects everything from visual arts to storytelling traditions.

RESEARCH ACTIVITY

In small groups, carry out some quick research into one of the effects above which has relevance to your county or locality. Share your research with the class.

CONSERVATION

While human activity is the driving force behind biodiversity loss, our species can also be a force for good, protecting our environment through conservation practices. **Conservation** is a broad term describing any practice that aims to carefully manage Earth's resources, ecosystems and biodiversity.

Conservation is the management of Earth's resources, ecosystems and biodiversity.

Some examples of conservation practice in ecology include:

- **Protect habitats:** National parks and wildlife sanctuaries aim to preserve species and ecosystems, preventing habitat destruction, and maintain biodiversity.
- **Renewable energies:** As mentioned previously, renewable energies can help reduce air pollution but can also, in time, help decrease greenhouse gases and reduce the effects of climate change.
- **Protect endangered species:** Breeding programmes, habitat restoration and anti-poaching laws are helping to protect endangered species.
- **Organic farming:** Eliminating artificial fertilisers, herbicides and pesticides from food production is an effective way to improve biodiversity. Many farmers are sowing multiple species of plants in grazing land instead of monocultures, which contain just one grass species.

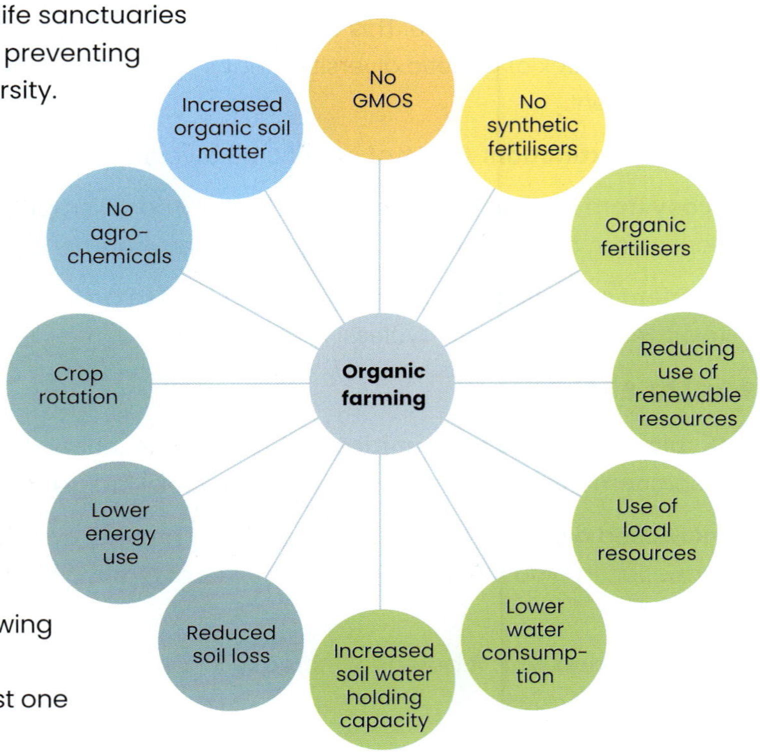

Figure 13.7: Principles of organic farming.

- **Fishery quotas:** Limiting the size of the fish catch and increasing the minimum mesh size in fishing nets, allowing smaller fish escape, is helping protect fish stocks.
- **Seed banks:** Seed banks save and store seeds from the world's important plant species, including crops.
- **Rewilding:** Farmland no longer used for food production is being allowed to return to its natural biodiverse state in a process called **rewilding**. Similarly, rewetting of harvested bogland returns this important ecosystem to its natural state.
- **Global co-operation:** Governments, international organisations (e.g. UN, UNESCO, WWF) and NGOs are collaborating on conservation projects worldwide. A great example of this is the UN's Sustainable Development Goals programme.

Rewilding allows unused farmland to return to its natural state.

SCIENCE IN SOCIETY: TECHNOLOGY

Technology in conservation

Technology is playing a major role in conservation practices worldwide. GPS (Global Positioning System) and GIS (Geographical Information System) are mapping ecosystems looking for changes in land cover (e.g. deforestation), biodiversity hotspots, endangered species habitats and even mapping migratory patterns. Remote video and tagging technologies are used to monitor the movements of endangered animals worldwide while drones are used to continuously monitor wildlife reserves and national parks for poaching. These important ecosystems are monitored by a range of AI-enabled sensors, providing real-time information on their environmental conditions. In addition, scientists carry out DNA analysis of species, monitoring their genetic diversity to ensure they can adapt to environmental change. In the future, gene editing technology could be used to improve genetic diversity of endangered species or return extinct species.

SUSTAINABILITY

Sustainability is a key concept in the battle to minimise the human impact on ecosystems. It focuses on how humans can meet the food, energy and water needs of our current population without compromising the needs of future generations.

Sustainability focuses on meeting the needs of the present without compromising the needs of future generations.

Sustainability involves finding a balance between our social, health, economic and environmental needs. Minimising pollution and biodiversity loss while implementing meaningful conservation practices are key concepts in sustainability; however, it is also about promoting education, providing clean energy, responsible infrastructure development, gender equality, eliminating poverty and ensuring economics fairness worldwide. At the core of sustainability is meaningful engagement with the public and building partnerships between governments, industry, scientists and NGOs.

Sustainable agriculture is extremely important to ensure our food is produced in a way that does not harm our environment or reduce biodiversity.

Go to YouTube.com and search for 'The Futuristic Farms That Will Feed the World' (6:19) by Freethink.

FOCUS ON ... SUSTAINABILITY AND THE CARBON & NITROGEN CYCLES

As mentioned in Chapter 12, the carbon and nitrogen cycles are fragile and the delicate balance between the amount of these nutrients in the living world and the environment is shifting, particularly in relation to carbon. Many of the issues we discussed already in this chapter are making the carbon and nitrogen cycles less sustainable.

Deforestation is removing vital carbon sinks; trees capture carbon dioxide during photosynthesis and store it, and mature trees store more carbon than newly planted ones. Deforestation also leads to soil erosion, which disrupts decomposition and the nutrient cycles. The burning of fossil fuels is adding more carbon dioxide to the atmosphere, contributing to global warming, while methane levels in the atmosphere are also on the rise. Methane is also released by farm animals, particularly cattle. There are also significant pockets of methane trapped in the permafrost and icecaps, which are being released as our planet warms. As the planet warms, the ocean temperature rises meaning it can store less carbon dioxide and other gases. The burning of fossil fuels releases nitrogen oxides (NO_x gases). NO_x contributes to air pollution and can form acid rain, which can alter soil pH. The manufacture and use of artificial fertilisers is disrupting the nitrogen cycle. When excess is applied, the compounds can make their way into waterways causing algal blooms and fish kills. The over-enrichment of the soil with nitrogen compounds, from fertilisers and even from animal waste, is disrupting the natural balance of the nitrogen cycle.

Many of these issues raise many ethical questions around the use of our land, our diet, our means of energy production and how we respect our soil.

DISCUSSION

- **Do we rely too heavily on artificial fertilisers?**
- **Are we eating too much meat?**
- **How do we combat deforestation?**
- **How do we move away from fossil fuels as our main energy source?**

PORTFOLIO LINK **Go to your Learning Portfolio and complete Learning Check 2 to assess your knowledge and understanding of the topic so far.**

Chapter Summary

- The Earth's population has surpassed 8 billion, with over half living in urban areas, and humans have colonised almost every corner of the planet.
- Human activities, particularly industry and agriculture, are impacting the natural world, leading to both positive and negative influences on ecosystems.
- **Pollution** is a harmful addition to an ecosystem.
- Pollution is categorised into **domestic**, **agricultural** and **industrial** types, affecting **water**, **air** and **soil**. Examples include plastic, sewage, greenhouse gas emissions and heavy metals.
- Water pollution results from runoff, leading to **eutrophication** and harming aquatic life.
- Air pollution from burning fossil fuels releases toxic chemicals and greenhouse gases, contributing to climate change.
- Soil pollution occurs as industrial and agricultural waste enters the soil, affecting fertility and disrupting the ecosystem.
- **Biodiversity loss** is occurring rapidly due to human activities like habitat destruction, climate change, pollution and overfishing.
- Consequences of biodiversity loss include reduced genetic diversity, species extinction, ecosystem instability and impacts on culture, disease, food security and economics.

- Conservation practices, including protecting habitats, using renewable energies and rewilding, aim to counteract biodiversity loss. Technology, such as GPS and GIS, plays a crucial role in monitoring and conserving ecosystems.
- **Sustainability** focuses on meeting the needs of the present without compromising the needs of future generations.
- Sustainable agriculture aims to produce food without harming the environment.
- The delicate balance of the carbon and nitrogen cycles is threatened by human activities like deforestation, burning fossil fuels and the use of artificial fertilisers, raising ethical questions about land use, diet and energy production.

END OF CHAPTER ASSESSMENT

REMEMBER

1. Define the term *pollution*.
2. Name **two** examples of pollutants from each of the following sources: domestic, agriculture, industry.
3. Give **two** examples of air pollution.
4. What is a greenhouse gas? Give **two** examples of greenhouse gases.
5. Summarise the impact of pollutants on ecosystems under air, water and soil pollution.
6. Define *biodiversity*.
7. List **three** causes of biodiversity loss.

UNDERSTAND

8. Briefly describe the process of eutrophication.
9. Describe **three** ways of controlling domestic pollution.
10. Discuss ways of reducing biodiversity loss.
11. Distinguish between conservation and sustainability.
12. Explain, in your own words, what is meant by the term sustainability.

APPLY

13. Propose methods to reduce pollution in the air.
14. Explain how wastewater treatment contributes to pollution control.
15. Outline the use of plants and bacteria in removing toxins from the soil.
16. Discuss the causes of biodiversity loss.
17. Discuss the economic and environmental effects of biodiversity loss.
18. What might be the effects of a loss in bee genetic diversity for human food production?
19. What are the advantages and possible disadvantages of organic farming?
20. Discuss the role biology might play in a post-fossil fuel world.

ANALYSE

21. The graph below shows the changes in the number of African elephants on the continent from 1913 to 2015. It is now estimated that there are 415,000 African elephants in the wild. Analyse the graph and answer the questions below.

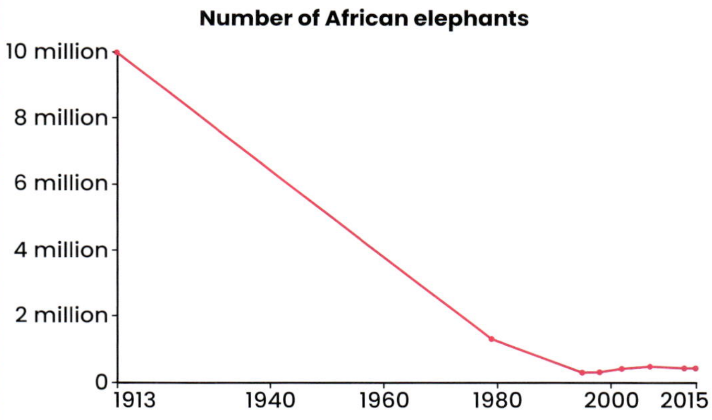

Source: African Elephant Specialist Group.

(a) What do you think is the main cause of the rapid decline in elephant population since 1913?

(b) There was a rise in the African elephant population at the turn of the century. Give **one** possible reason for this.

(c) What could be done to ensure the population of African elephants continue to recover?

22. The map below shows the net changes in forest expansion in 2015, either positively through afforestation and natural expansion or negatively through deforestation.

Annual change in forest area 2015

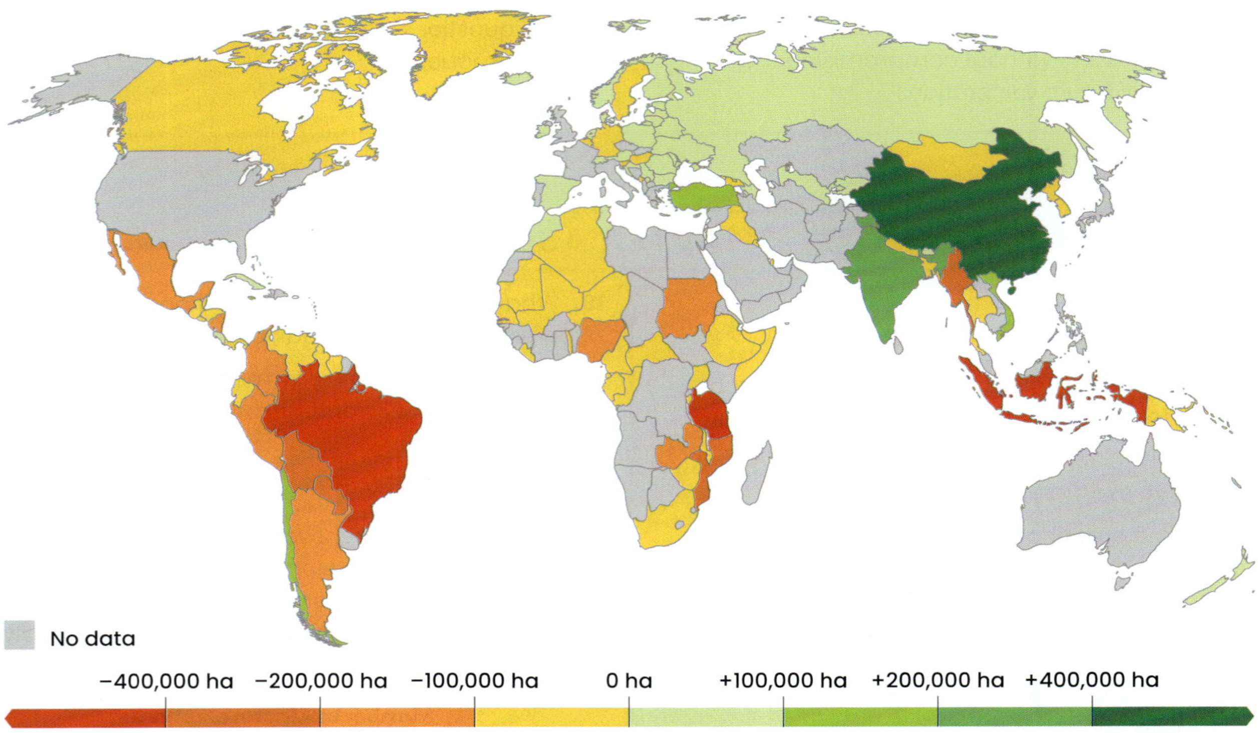

Source: UN Food and Agriculture Organization.

(a) In which countries is deforestation faster than forest expansion?

(b) In which countries is forest expansion faster than deforestation?

(c) Do you think a newly planted forest will have the same level of biodiversity as an established forest? Explain your answer.

(d) How might deforestation lead to the wider spread of disease?

EVALUATE

23. Discuss the factors that have contributed to the human species' rapid population growth.

24. In your opinion, which type of pollution has the greatest effect on climate change: air, water or soil? Give reasons for your answer.

25. Research the role of technology in modern conservation methods, including sustainable farming.

26. Gather evidence to help evaluate the sustainability issues associated with the carbon and nitrogen cycles.

27. What ethical questions are raised in relation to the sustainability of the carbon and nitrogen cycles?

CREATE

28. Propose conservation practices that can be implemented in Ireland to promote sustainable agriculture.

29. Design an awareness campaign for your school to educate about the importance of biodiversity and pollution control.

30. Develop a plan for a sustainable city. Consider pollution reduction and protecting biodiversity in the sustainable city.

UNIT 4

THE CHEMICALS OF LIFE

CHAPTERS

STRANDS

Unifying Strand: Nature of Science

Strand 1: Organisation of Life

14 Biomolecules

LEARNING OUTCOMES: U2.3, U2.5, 1.2.1, 1.2.2, 1.2.3, 1.2.4, 1.2.5, 1.2.6

EYE-OPENER

It was only at the start of the 20th century that scientists began investigating what parts of a diet were needed to prevent diseases. In 1912, Casimir Funk coined the term 'vitamine'. In 1913, vitamins A and D were the first to be extracted. The last vitamin to be extracted was folic acid in 1941.

At the end of this chapter, you should be able to:

- Describe the structure of carbohydrates, lipids and proteins.
- Give nutritional sources of carbohydrates, lipids, proteins, vitamins and minerals.
- Outline the structural and metabolic roles of a carbohydrate, lipid and protein.
- Describe the role of vitamins and minerals in biological processes.
- Explain the main roles of water in living organisms.
- Investigate qualitatively the presence of nutrients in a range of food samples.
- Investigate quantitatively the level of reducing sugars in a range of food samples.

SUBJECT VOCABULARY

biomolecule | balanced diet | carbohydrate | fat | oil | triglyceride | phospholipid | amino acid | peptide | polypeptide | vitamin | mineral | anabolic | catabolic

LINKS TO:

Chapter 15	**Metabolism and Enzymes**
Chapter 16	**Cell Structure and Diversity**
Chapter 18	**Movement through Cell Membranes**
Chapter 22	**Photosynthesis**
Chapter 23	**Respiration**
Chapter 24	**Information Flow in the Cell**
Chapter 30	**The Digestive System**
Chapter 32	**Infectious Disease and the Human Immune Response**

Section 1: Biomolecules – Nutrient Groups, Carbohydrates

At the end of this section, you should be able to:

- Define the term *biomolecule* and list different types of biomolecules.
- Explain why we need nutrients.
- Name the different nutrient groups.
- List the elements from which carbohydrates are formed and state their general formula.
- Describe mono-, di- and polysaccharide structures.
- Identify nutritional sources for each of the three main types of carbohydrate.
- State structural and metabolic roles (catabolic, anabolic, respiratory, photosynthetic) of carbohydrates.

Unpacked from LO 1.2.1

BIOMOLECULES

All living organisms are composed of elements (mainly carbon, hydrogen, oxygen, nitrogen, sulfur and phosphorus) which combine in different ratios to form **biomolecules**. Biomolecules are the building blocks of life and include:

A **biomolecule** is a chemical compound found in living organisms.

- **carbohydrates**
- **lipids**
- **proteins**
- **nucleic acids** formed from nucleotide units (you will study the structure and function of nucleotides and the nucleic acids DNA and RNA in Chapter 24)
- **energy transfer biomolecules** (you will learn about the structure and function of the energy transfer biomolecule **ATP** in Chapter 23)
- **electron transfer biomolecules** (you will learn about the electron transfer biomolecules **NAD$^+$** and **NADP$^+$** in Chapters 22 and 23).

It is important to note that the biomolecules can be rearranged into new molecules, e.g. the human body converts excess carbohydrate into lipids and excess protein can be converted to carbohydrates or lipids.

Figure 14.1: Food provides us with biomolecules needed for growth and to keep us healthy.

WHY WE NEED NUTRIENTS

Nutrients in food have many important roles in our bodies, providing us with energy for all our body activities, e.g. muscle movement, cell division. The energy stored in nutritional molecules is released during respiration, mainly in the form of heat. The elements needed for growth and repair of cells are obtained from nutrients in our food. Enzymes which control metabolic reactions in the body are made from protein. Vitamins and minerals help protect the body against disease.

NUTRIENT GROUPS

There are six main nutrient groups:

- Proteins
- Carbohydrates
- Fats
- Vitamins
- Minerals
- Water.

These should all be included in a **balanced diet**.

A **balanced diet** contains the correct amount of each of the different nutrient groups to keep a person healthy.

SCIENCE IN SOCIETY: HEALTH

Food Pyramid

The amounts of each nutrient group a person eats depends on their sex, age and how active their lifestyle is. Younger people such as teenagers need more food than older people, males need to consume more than females and those with active lifestyles need to take in more food than those who are less active.

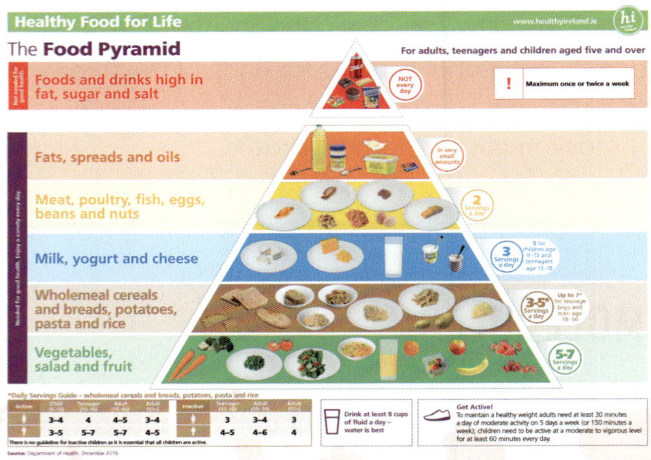

Figure 14.2: The food pyramid indicates the number of portions of each food group a person should eat daily.

CARBOHYDRATES

Carbohydrates are biomolecules formed from the elements carbon (C), hydrogen (H) and oxygen (O) in the ratio $C_x(H_2O)_y$.

A **carbohydrate** is a biomolecule formed from the elements carbon, hydrogen and oxygen in the ratio $C_x(H_2O)_y$.

There are three main types of carbohydrates: monosaccharides, disaccharides and polysaccharides.

The main function of monosaccharides and disaccharides, and the polysaccharide starch, is to provide energy. The polysaccharide cellulose (fibre) helps to prevent constipation.

- **Monosaccharides** are composed of single sugar units or monomers. **Glucose** and **fructose** have the same chemical formula – $C_6H_{12}O_6$ – but they have different structures.
- **Disaccharides** consist of two sugar units. **Maltose** is formed from two glucose subunits joined together and **sucrose** is formed from a glucose and fructose subunit joined together. Note that sucrose, found in table sugar, is not a reducing sugar like most other monosaccharides and disaccharides.
- **Polysaccharides** are formed from many sugar units, up to several thousand, joined together. Both **starch** and **cellulose** are formed from glucose subunits but there are many more crosslinks between the glucose units in cellulose.

General formula of carbohydrates:
$C_x(H_2O)_y$

Example: glucose
$x = 6, y = 6$
$C_6H_{12}O_6 = C_6(H_2O)_6$

MONOSACCHARIDES (SINGLE SUGAR UNIT)

Glucose

Source: fruit, honey, processed sweet food and drinks

Glucose molecule

Fructose

Source: fruit, honey

Fructose molecule

DISACCHARIDES (TWO SUGAR UNITS)

Maltose

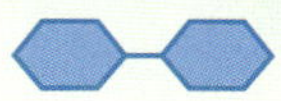

Source: cereal grains

Sucrose

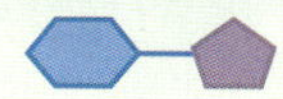

Source: sugarcane, table sugar

POLYSACCHARIDES (MANY SUGAR UNITS)

Starch

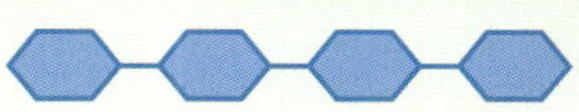

Source: bread, potatoes, pasta

Cellulose (fibre)

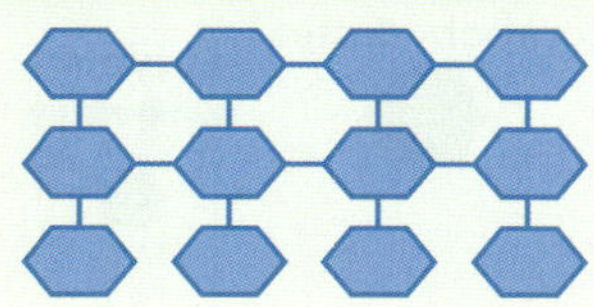

Source: wholegrain bread, skin of fruit and vegetables

Metabolic reactions

All the chemical reactions that occur in a living organism are referred to as metabolism. These metabolic reactions may be divided into two main types: anabolic or catabolic.

- In a **catabolic** reaction, large molecules are broken down into smaller ones, releasing energy.
- In an **anabolic** reaction, small molecules are converted into larger molecules, using energy.

CATABOLIC REACTIONS

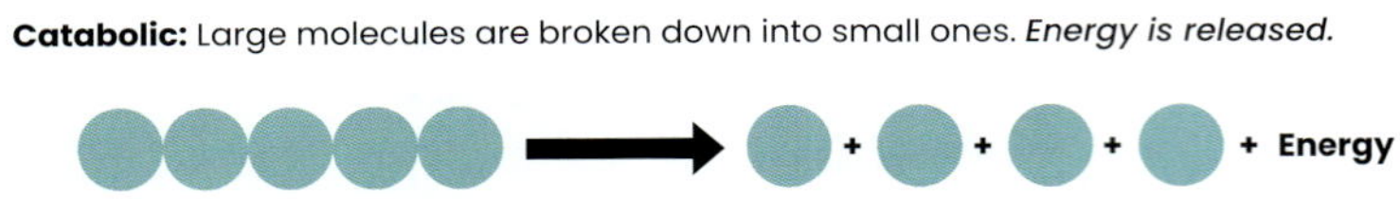

Figure 14.3: A catabolic reaction.

In a **catabolic** reaction, large molecules are broken down into smaller ones, releasing energy.

ANABOLIC REACTIONS

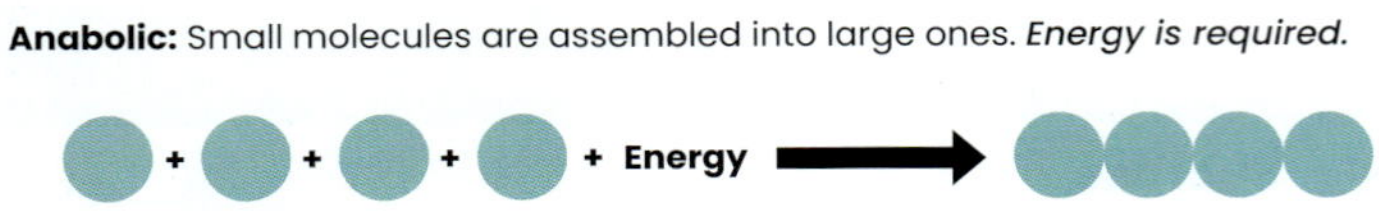

Figure 14.4: An anabolic reaction.

In an **anabolic** reaction, small molecules are converted into larger molecules using energy.

Metabolic role of carbohydrates

CATABOLIC REACTION: RESPIRATION

The chemical energy stored in carbohydrates such as glucose is released during the catabolic process of **respiration**. This energy is stored in the form of adenosine triphosphate (ATP). You will learn about ATP in Chapter 23.

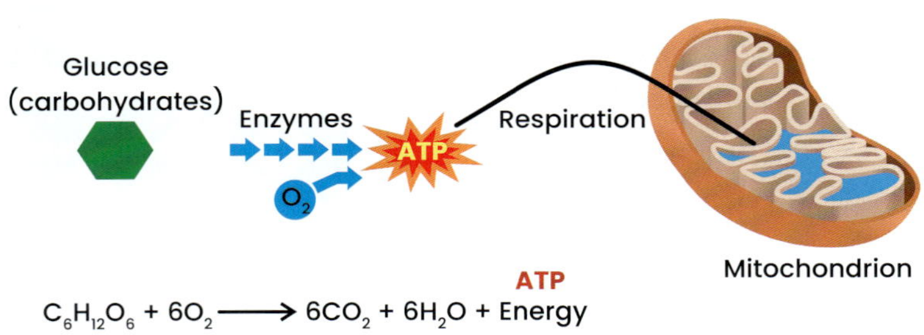

Figure 14.5: Catabolic role of glucose in respiration.

ANABOLIC REACTION: PHOTOSYNTHESIS

The carbohydrate glucose is formed during the anabolic reaction of **photosynthesis**.

Photosynthesis: $6CO_2 + 6H_2O \xrightarrow{\text{Light, chlorophyll}} 6C_6H_{12}O_6 + 6O_2$

Go to YouTube.com and search for 'Carbohydrates | Biological Molecules Simplified #1' (2:25) by 2 Minute Classroom.

PORTFOLIO LINK **Go to your Learning Portfolio and complete Learning Check 1 to assess your knowledge and understanding of the topic so far.**

Section 2: Biomolecules – Lipids and Proteins

At the end of this section, you should be able to:

- List the elements from which lipids are formed.
- Sketch the structure of a triglyceride and phospholipid.
- Distinguish between a fat and an oil.
- List the elements from which proteins are formed.
- Describe the structure of proteins (peptides and polypeptides).
- Identify nutritional sources for lipids and proteins.
- State metabolic roles of lipids and proteins.

Unpacked from LO 1.2.1

LIPIDS

Lipids are biomolecules formed from the elements carbon (C), hydrogen (H) and oxygen (O). The elements in a lipid are arranged differently to those in carbohydrates. Lipids that are solid at room temperature are known as **fats**. Lipids that are liquid at room temperature are called **oils**. The basic unit in a lipid is a **triglyceride** which is formed from a molecule of glycerol linked to three fatty acids. Other lipids called **phospholipids** are formed when one of the three fatty acids in the lipid structure is replaced by a phosphate group. Nutritional sources of lipids include butter, meat, cheese and olive oil.

- An **oil** is a lipid that is liquid at room temperature.
- A **fat** is a lipid that is solid at room temperature.
- A **triglyceride** is the basic unit in a lipid consisting of glycerol linked to three fatty acids.
- A **phospholipid** is a lipid unit consisting of glycerol, two fatty acids and one phosphate group.

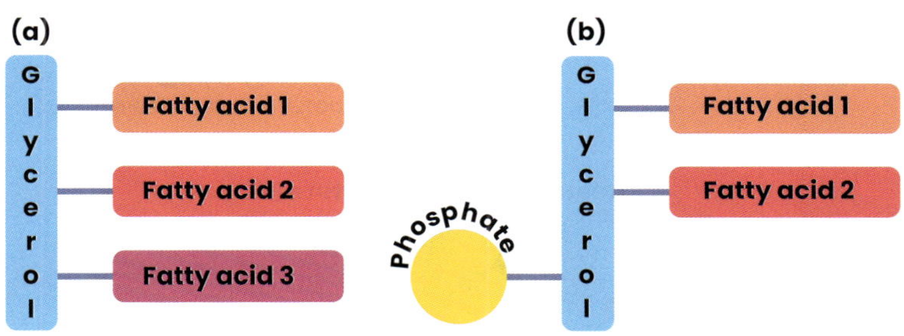

Figure 14.6: Structure of (a) a triglyceride (b) a phospholipid.

Figure 14.7: Sources of fats include butter, meat and cheese.

Figure 14.8: Sources of oils include olive oil and fish oil.

Metabolic role of lipids

1. CATABOLIC REACTIONS: RESPIRATION

Lipids are used for long-term energy storage. When carbohydrates are in low supply, lipids may be used to meet the body's energy needs. During respiration, the chemical energy stored in lipids may be released. Lipids provide 9 kcal of energy per gram, more than twice that of carbohydrates or proteins which provide only 4 kcal of energy per gram.

2. ANABOLIC REACTIONS: PHOTOSYNTHESIS

Lipids play an important role in photosynthesis, which is an anabolic reaction. In plant cells, light is trapped by chlorophyll held in a lipid bilayer called the thylakoid membrane, which we will learn more about in Chapter 22.

3. HORMONES

Hormones are chemical messengers carried around the body in the blood. Steroid hormones formed from lipids are important in regulating metabolism. We will learn more about hormones in Chapter 34.

Go to YouTube.com and search for 'Lipids | Fats, Steroids, and Phospholipids | Biological Molecules Simplified #4' (2:52) by 2 Minute Classroom.

PROTEINS

Proteins are biomolecules formed from the elements carbon (C), hydrogen (H), oxygen (O) and nitrogen (N). Some proteins also contain the elements phosphorus (P) and sulfur (S) or other elements.

The basic structure in a protein is an **amino acid**. Protein structure may be represented by CHONR, where R is a sidechain that varies for each amino acid.

An **amino acid** is the basic structure in a protein.

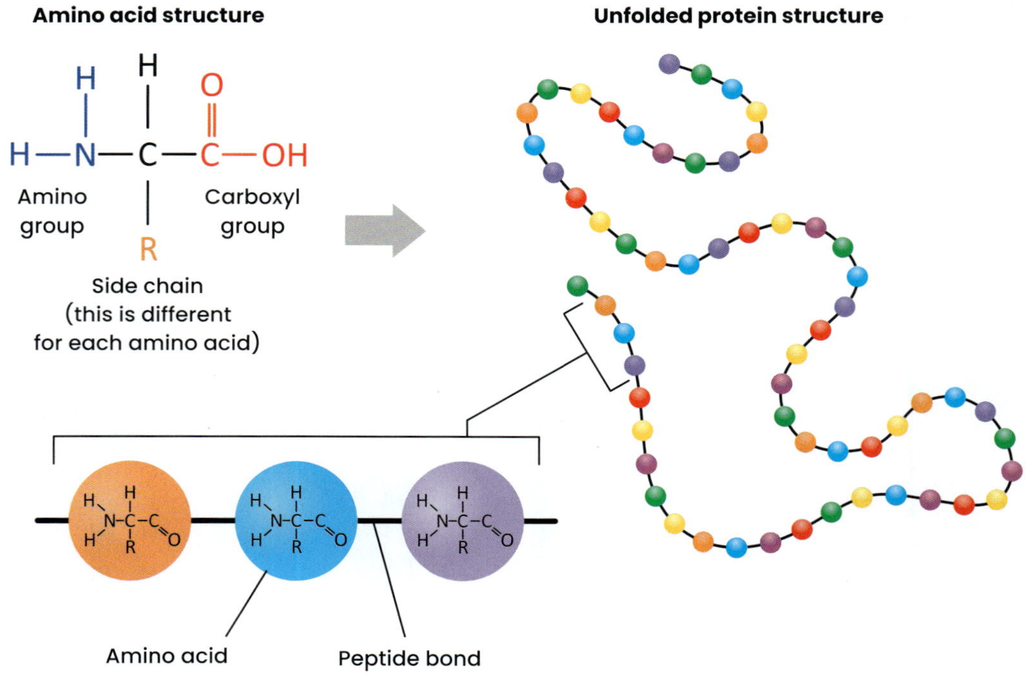

Figure 14.9: Amino acids forming an unfolded protein chain.

Proteins are formed from 20 different amino acids linked together by peptide bonds. Each protein chain has a different order of amino acids. A **peptide** is a short chain of amino acids (typically two to 50). A longer chain of linked amino acids (51 or more) is a **polypeptide**. Protein chains can contain hundreds to thousands of amino acids.

- A **peptide** is a short chain of amino acids linked by peptide bonds.
- A **polypeptide** is a long chain of amino acids linked by peptide bonds.

The protein chain forms a folded, three-dimensional, globular shape.

Nutritional sources of proteins include meat, fish, eggs, milk and cheese.

Figure 14.10: Sources of proteins include meat, fish, eggs, milk, cheese, nuts and pulses.

Structural role of proteins

Certain proteins form body structures:

- **Hair and nails:** Keratin is the structural protein found in hair and nails.
- **Muscle:** Myosin and actin are structural proteins found in muscle.

Figure 14.11: Keratin is a protein found in hair and nails.

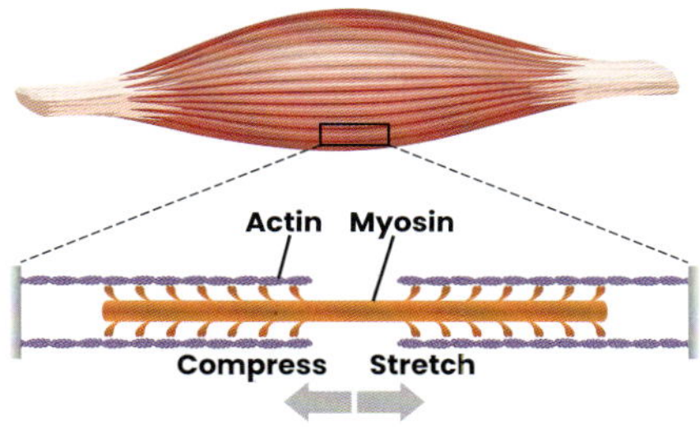

Figure 14.12: Myosin and actin are found in muscle tissue

Metabolic role of proteins

1. ENZYMES

Enzymes are biological catalysts that control the rate of cellular reactions. Amylase is a protein enzyme that breaks down starch to maltose during chemical digestion. We will study enzymes in detail in Chapter 15.

2. HORMONES

Some hormones are formed from protein. Hormones are regulators controlling metabolism, growth and development. For example, insulin is a protein hormone that controls blood sugar levels.

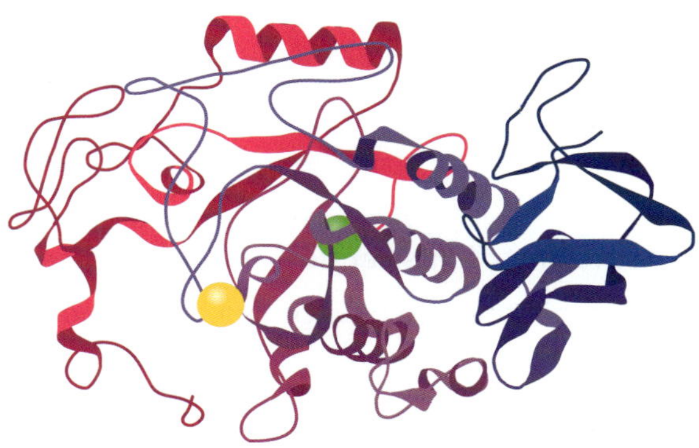

Figure 14.13: Structure of the salivary amylase enzyme.

3. MEMBRANE INTERACTIONS

In Chapter 16, we will discover that all cells are surrounded by a phospholipid bilayer cell membrane. The cell membrane has embedded and attached proteins that are used for communication with other cells and transport of chemicals and ions. Some enzymes are embedded in the cell membrane.

4. IMMUNITY

Proteins on cell surfaces allow the immune system to distinguish between self and foreign cells. Antibodies are proteins formed by white blood cells on detection of foreign bodies, e.g. bacteria. We will learn more about antibodies and the immune response in Chapter 32.

> The human body has the potential to make a quintillion – 1,000,000,000,000,000,000, or 1 million trillion – unique antibodies.
>
> **DID YOU KNOW?**

FOCUS ON ... PROTEINS AND MOLECULAR MECHANISMS

Proteins are involved in different processes in cells.

CHANNEL PROTEIN	TRANSPORTER PROTEIN	RECEPTOR PROTEIN	STRUCTURAL PROTEIN
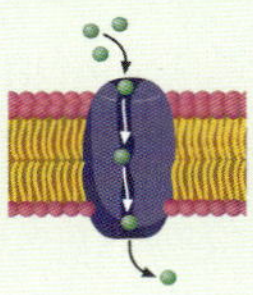	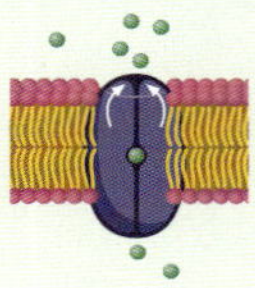		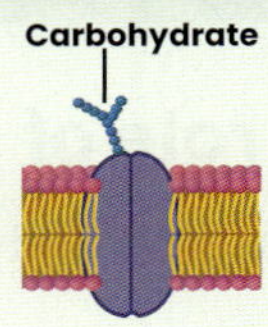
Channel proteins form openings across the cell membrane, allowing ions or molecules to diffuse in or out of the cell, down a concentration gradient.	Transport or carrier proteins carry ions or molecules, e.g. glucose, from one side of the cell membrane to the other by passive or active transport.	Certain molecules bind to receptor proteins, triggering cell responses, e.g. the release of hormones.	Some proteins allow cells to attach to each other, holding tissues together. As seen above, other proteins, e.g. keratin, actin and myosin, form structures such as hair, nails and muscle.

PORTFOLIO LINK **Go to your Learning Portfolio and complete Learning Check 2 to assess your knowledge and understanding of the topic so far.**

Go to YouTube.com and search for 'Protein functions in the human body' (4:57) by Joao's Lab.

Section 3: Vitamins, Minerals and Water

At the end of this section, you should be able to:

- Explain the role of vitamins and minerals in the diet.
- Distinguish between fat-soluble and water-soluble vitamins and give examples and a deficiency disease of each.
- Describe the main roles of water in organisms.

Unpacked from LOs 1.2.2, 1.2.3, 1.2.4

VITAMINS

Vitamins are substances produced by plants or animals. With the exception of vitamin D, vitamins cannot be formed by the body and sufficient amounts must be taken in from food to keep a person healthy. **Vitamins** are divided into two groups:

- Water-soluble vitamins, e.g. vitamins B and C
- Fat-soluble vitamins, e.g. vitamins A, D, E and K.

A **vitamin** is an organic molecule needed in small amounts by organisms to function properly.

Biological significance of vitamin solubility

Water-soluble vitamins dissolve in water, allowing them to be easily transported in blood and readily absorbed in the digestive system. Excess water-soluble vitamins are passed out of the body in urine.

Fat-soluble vitamins dissolve in fat and are stored in tissue, allowing the body access to them as needed. Excess fat-soluble vitamins are stored in the liver.

We will now compare water-soluble vitamin C and fat-soluble vitamin D.

	VITAMIN C	VITAMIN D
FAT- OR WATER-SOLUBLE?	Water-soluble	Fat-soluble
SOURCES	• Citrus fruits • Kiwis • Fresh vegetables 	• Oily fish • Liver • Egg yolks • The body can create vitamin D from direct sunlight on the skin 
FUNCTION	• Forms the connective tissue collagen to make skin, tendons and ligaments • Helps wounds to heal • Helps the absorption of iron from plants	• Helps the absorption of calcium
DEFICIENCY	A lack of vitamin C results in the disease scurvy. Some symptoms of scurvy include swollen and bleeding gums, slow wound healing and anaemia. 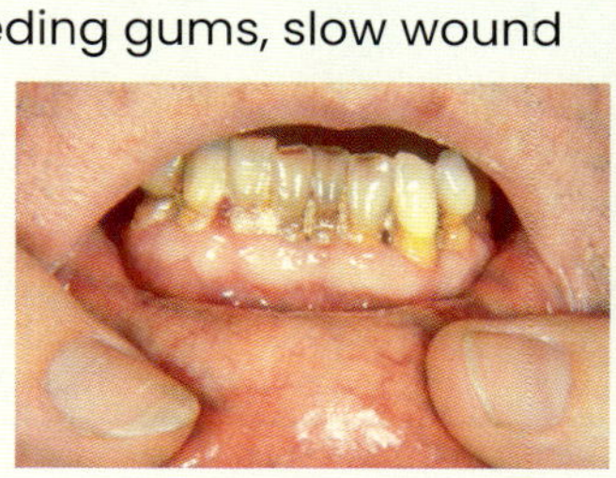	A lack of vitamin D in adults can result in osteomalacia or 'soft bones'. In children, a lack of vitamin D results in rickets and bowlegs. 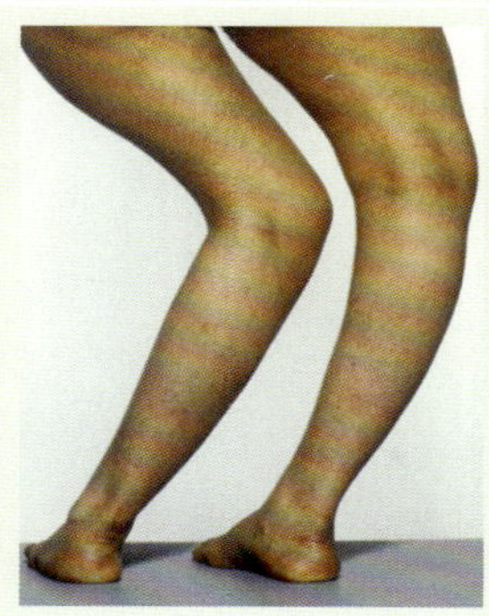

MINERALS

Minerals are chemical elements needed by the body. They are not necessarily built into biomolecules but found as mineral salts. Minerals are needed for:

- pH balance
- Regulation of enzyme systems
- Nerve impulse transfer
- Muscle contraction

> A **mineral** is a chemical element needed by organisms to carry out functions necessary for life.

Some **minerals** are needed in greater amounts, e.g. calcium (Ca), magnesium (Mg), potassium (K), sodium (Na) and chlorine (Cl). Other minerals needed in smaller amounts are called trace elements. These include iron (Fe), zinc (Zn) and copper (Cu).

Let's look at two minerals: calcium and iron.

MINERAL	CALCIUM	IRON
SOURCES	• Milk • Cheese • Green leafy vegetables such as broccoli and cabbage 	• Liver • Red meat • Nuts • Fortified breakfast cereals 
FUNCTION	• Builds strong bones and teeth • Controls muscle movement	• Used to make haemoglobin, a protein in red blood cells that carries oxygen around the body
DEFICIENCY	Weakened teeth and osteoporosis which causes brittle bones which break or fracture more easily 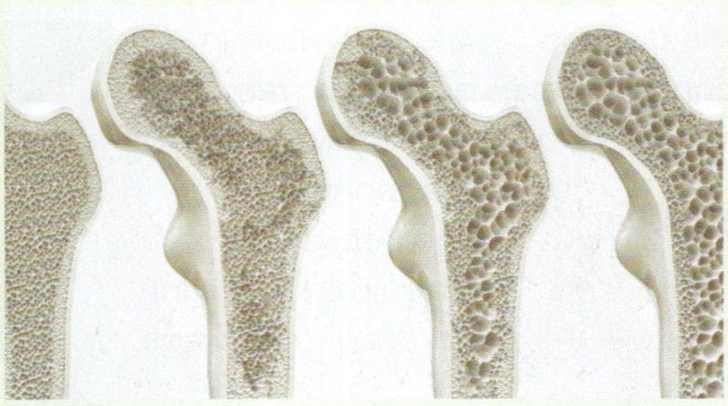	Anaemia, resulting in fatigue and shortness of breath 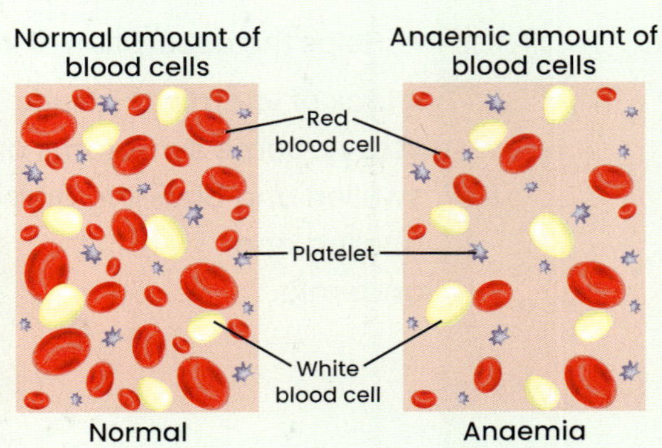

Role of water in living organisms

Hydrogen atoms in water have a partial positive charge and the oxygen atom has a partial negative charge. This is called polarity and results in a large network of bonds called 'hydrogen bonds' forming between water molecules. This gives water many special properties in living organisms e.g. high specific heat capacity.

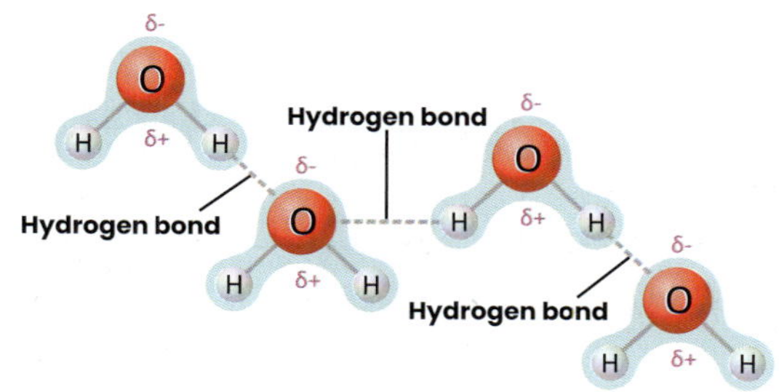

Figure 14.14: Hydrogen bonding between water molecules.

1. TRANSPORT SOLVENT

Water helps transport dissolved compounds in and out of cells and acts as a solvent for chemical reactions in the body. Water is called the universal solvent because of the range of compounds it dissolves including salts, sugars, amino acids, small proteins and nucleic acids. Lipids do not dissolve in water.

2. PARTICIPANT IN CHEMICAL REACTIONS

Water is a reactant or product in many chemical reactions, e.g. photosynthesis and respiration.

Aerobic respiration: $C_6H_{12}O_6 + 6O_2 \rightarrow 6CO_2 + 6H_2O$ + energy (water is a product)

Photosynthesis: $6CO_2 + 6H_2O \xrightarrow{\text{Light, chlorophyll}} C_6H_{12}O_6 + 6O_2$ (water is a reactant)

3. REGULATION OF BODY TEMPERATURE

Due to all the hydrogen bonds between water molecules, it takes a lot of energy to raise the temperature of water by one degree, giving water a high specific heat capacity. This helps to keep body temperature constant. Enzymes that control biochemical reactions are sensitive to temperature; high temperatures can cause enzymes to lose their function. The heat used to evaporate sweat prevents the body overheating. A similar process – transpiration – occurs in plants, where water evaporates mainly off the surface of leaves.

4. REGULATION OF pH

pH is a scale from 0–14, measuring how acidic or basic a solution is. Enzymes are sensitive to pH with those in the human body functioning best around pH 7. Water which makes up 70% of the cell's cytoplasm and 90% of blood is neutral and has a pH of 7.

5. OSMOREGULATION

Osmoregulation is the control of water balance in the blood, tissue or cytoplasm. If cells gain or lose too much water, they do not function properly. Water moves in and out of cells by a process called osmosis, which we learn more about in Chapter 18.

Go to YouTube.com and search for 'Biological properties of water – A Level Biology' (4:48) by Mr Exham Biology.

Figure 14.15: Role of water in living organisms.

PORTFOLIO LINK Go to your Learning Portfolio and complete Learning Check 3 to assess your knowledge and understanding of the topic so far.

Specified Investigation 3

Investigate qualitatively the presence of nutrients in a range of food samples. (LO 1.2.5)

Scan the QR code to watch a video demonstration of this investigation.

A. FAT

Hypothesis: If a food sample containing fat is rubbed on brown paper, a translucent spot will be observed.

Equipment: Food samples, brown paper, deionised water.

VARIABLES

Independent variable: Different food samples will be tested.
Dependent variable: The presence or absence of fat.
Control: Deionised water.

METHOD

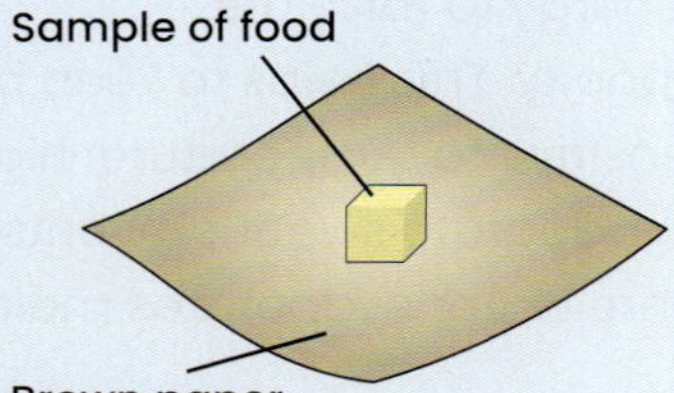

1. Rub a sample of the food on brown paper.
2. Repeat with a fresh piece of brown paper and water as a control.
3. Hold each piece of paper up to the light.

RESULTS

	RESULT	CONCLUSION
FOOD SAMPLE	Permanent stain – translucent spot	Fat present
DEIONISED WATER	No permanent stain	No fat present

Safety precautions: Do not eat or drink in the lab. Wear a lab coat and goggles.

B. REDUCING SUGAR (GLUCOSE)

Scan the QR code to watch a video demonstration of this investigation.

Hypothesis: If a solution of a food sample containing a reducing sugar is tested with Benedict's reagent and heated, the solution will turn brick red.

Equipment: Food samples, test tubes and rack, mortar and pestle, deionised water, Benedict's reagent, water bath, droppers.

VARIABLES

Independent variable: Different food samples will be tested.

Dependent variable: The presence or absence of a reducing sugar.

Control: Deionised water.

METHOD

1. Grind equal masses of food samples to be tested in a mortar using a pestle and dissolve in 10ml deionised water in test tubes.

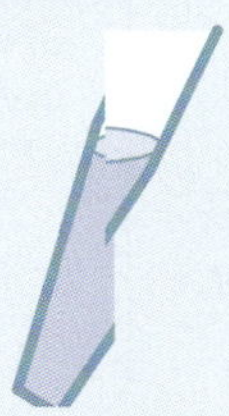

2. Set up control test tube containing 10ml deionised water (0% glucose).

3. Add 2ml Benedict's reagent to each of the test tubes and swirl to mix.

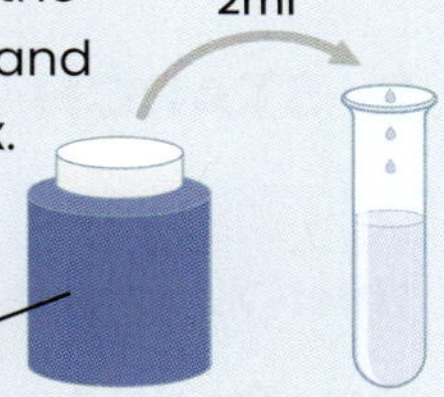

4. Place the test tubes in a hot water bath for 5 minutes. Observe and record any colour changes in a table.

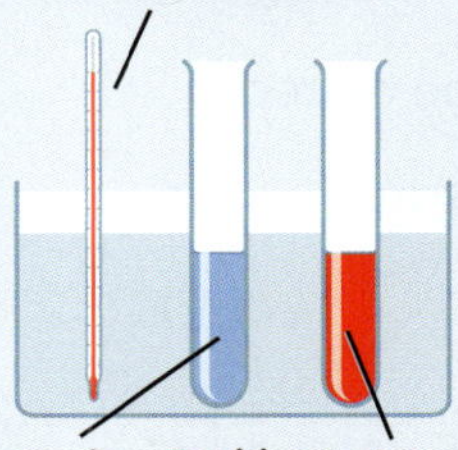

RESULTS:

	COLOUR CHANGE	CONCLUSION
FOOD SAMPLE	Blue to brick red	Reducing sugar present
WATER (CONTROL)	Remains blue	Reducing sugar absent

Safety precautions: Do not eat or drink in the lab. Wear a lab coat and goggles.

C. PROTEIN

Scan the QR code to watch a video demonstration of this investigation.

Hypothesis: If a solution of a food sample containing a protein is tested with Biuret reagent, the solution will turn purple.

Equipment: Food samples, test tubes and rack, mortar and pestle, deionised water, Biuret reagent, droppers.

METHOD

1. Grind equal masses of food samples to be tested in a mortar using a pestle and dissolve in 10ml deionised water in test tubes.
2. Set up control test tube containing 10ml deionised water (0% protein).
3. Add 2ml Biuret reagent to each of the test tubes and swirl.
4. Observe and record any colour changes in a table.

RESULTS

	COLOUR CHANGE	CONCLUSION
FOOD SAMPLE	Blue to purple	Protein present
WATER (CONTROL)	Stays blue	Protein absent

Safety precautions: Do not eat or drink in the lab. Wear a lab coat and goggles.

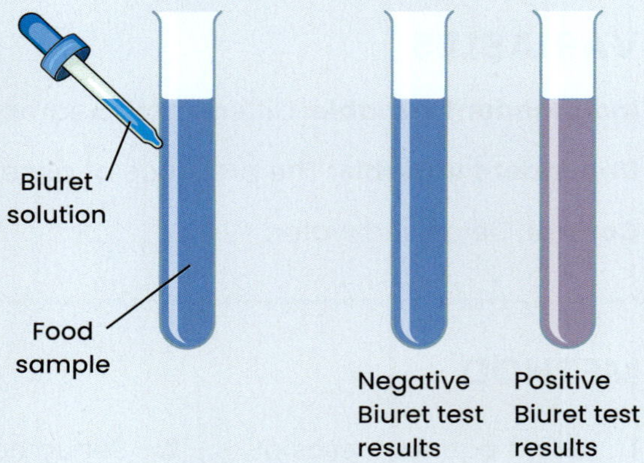

D. STARCH

Scan the QR code to watch a video demonstration of this investigation.

Title: Qualitative test for starch in food.

Hypothesis: If a solution of a food sample containing starch is tested with iodine, the solution will turn blue-black.

Equipment: Food samples, test tubes and rack, mortar and pestle, deionised water, iodine, droppers.

VARIABLES

Independent variable: Different food samples will be tested.

Dependent variable: The presence or absence of starch.

Control: Deionised water.

METHOD

1. Grind equal masses of food samples to be tested in a mortar using a pestle and dissolve in 10ml deionised water in test tubes.
2. Set up control test tube containing 10ml deionised water (0% starch).
3. Add 2ml iodine solution to each of the test tubes and swirl.
4. Observe and record any colour changes in a table.

RESULTS

	COLOUR CHANGE	CONCLUSION
FOOD SAMPLE	Red-brown to blue-black	Starch present
WATER (CONTROL)	Stays red-brown	Starch absent

Safety precautions: Do not eat or drink in the lab. Wear a lab coat and goggles.

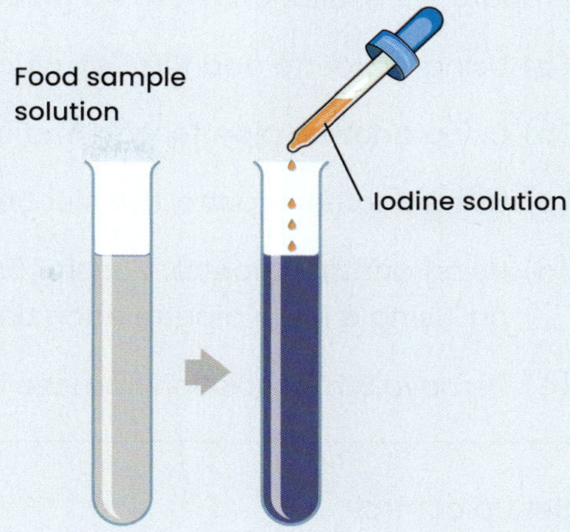

Specified Investigation 4

Investigate quantitatively the level of reducing sugars in a range of food samples. (LO 1.2.6)

Scan the QR code to watch a video demonstration of this investigation.

Hypothesis: Glucose solutions of known concentration can be used to determine the concentration of reducing sugar in a food sample.

Equipment: Test tubes, mortar and pestle, pipettes, water bath, Benedict's reagent, food samples, glucose, water.

VARIABLES

Independent variable: The reducing sugar concentration.

Dependent variable: The colour of each solution.

Control: Water.

METHOD

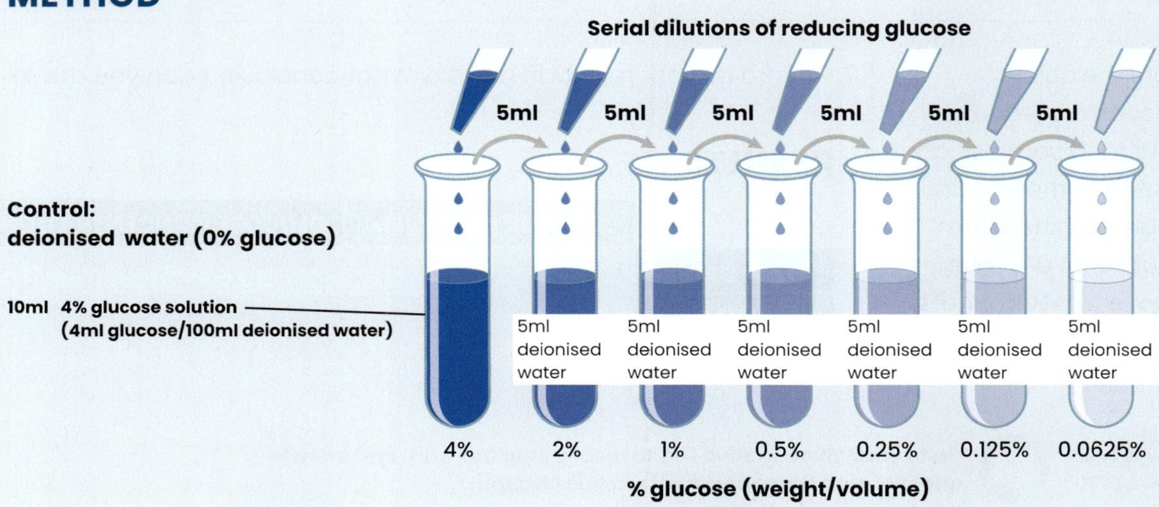

1. Prepare serial dilutions from 4% glucose stock solution as follows:
   (a) Using a pipette add 10ml 4% glucose stock solution to first test tube.
   (b) Using another pipette, add 5ml deionised water to six other test tubes.
   (c) Using another pipette, transfer 5ml from the stock solution in the first test tube to the next.
   (d) Using another pipette, transfer 5ml solution from the second test tube to the third and so on, using a fresh pipette each time. Swirl each solution.
   (e) Remove 5ml of solution from the last test tube so that each contains a total of 5ml of solution.

2. Set up control test tube containing 5ml deionised water (0% glucose).

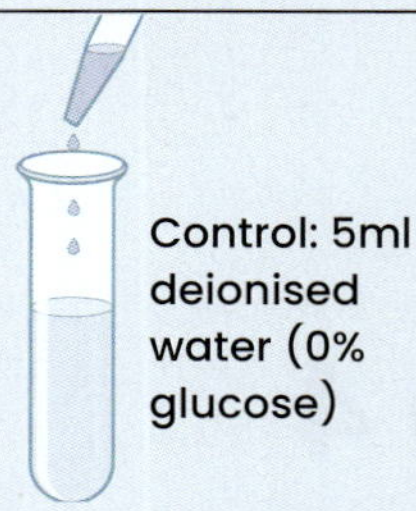

3. Add 2ml Benedict's reagent to each of the eight test tubes and swirl (seven glucose dilutions and one control test tube).

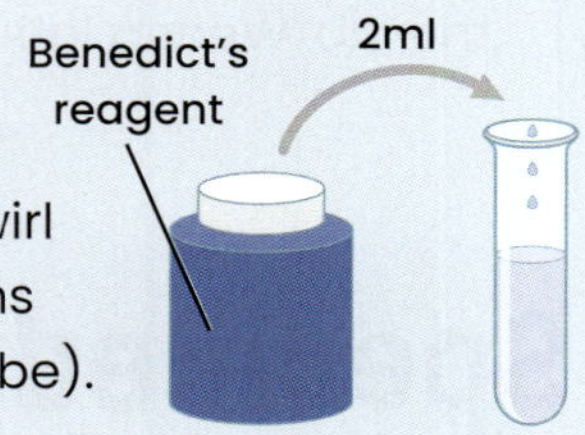

4. Place each of the eight test tubes in a hot water bath for 5 minutes. Observe and record any colour changes. Let any precipitates settle to the bottom of their test tube.

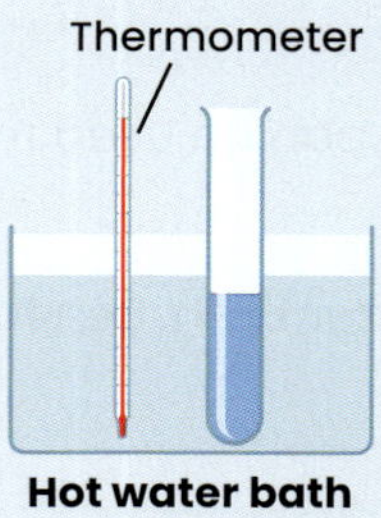

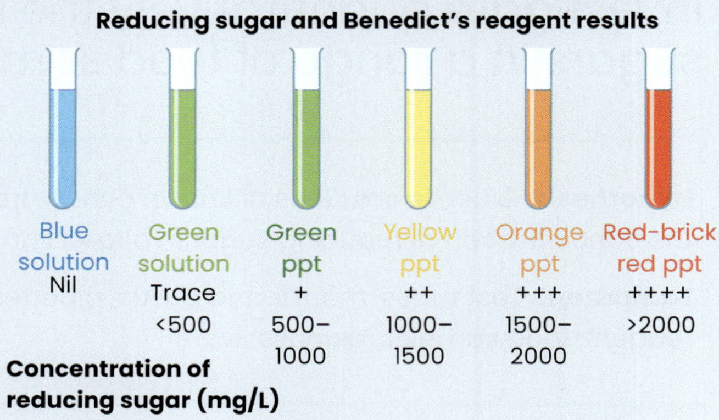

5. Grind equal masses of food samples to be tested in a mortar using a pestle and dissolve in 10ml deionised water in test tubes. Add 2ml Benedict's solution to each food sample and heat in water bath.

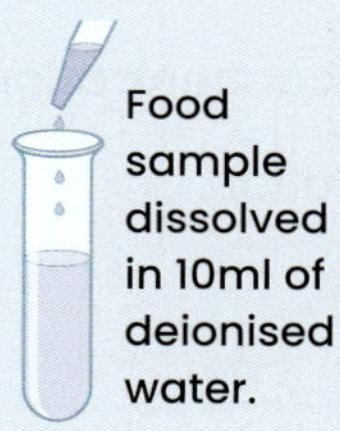

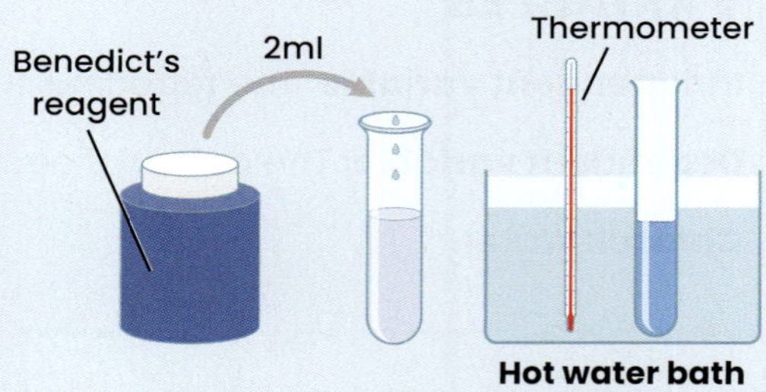

6. Note the colour observed in each of your food samples and match to the colour chart obtained from your serial dilutions above to determine its glucose concentration.

7. Record all your results in a table. What conclusion can you draw?

RESULTS

	COLOUR OBSERVED	GLUCOSE CONCENTRATION
FIZZY DRINK		
BREAD		

INVESTIGATIVE LOG LINK **Go to your Investigative Log to record your lab work and answer questions on the experiments in this chapter.**

Chapter Summary

- **Biomolecules** are chemical compounds found in living organisms.
- **Examples of biomolecules** include (i) carbohydrates (ii) lipids (iii) proteins (iv) nucleic acids formed from nucleotide units, e.g. DNA, RNA (v) energy transfer biomolecules, e.g. ATP (vi) electron transfer biomolecules, e.g. NAD^+ and $NADP^+$.
- **Monosaccharides** contain one sugar unit, e.g. glucose in fruit, honey.
- **Disaccharides** have two sugar units, e.g. sucrose in sugarcane, table sugar.
- **Polysaccharides** have many sugar units, e.g. starch in bread, potatoes, pasta.
- **Metabolic role of carbohydrates:** Energy stored in carbohydrates is released in respiration.
- **Lipids** are formed from the elements carbon, hydrogen and oxygen.
- Solid lipids are called fats; liquid lipids are called oils.
- The basic unit in a lipid is a **triglyceride** (glycerol linked to three fatty acids).
- **Phospholipids** contain glycerol, two fatty acids and one phosphate group.
- **Nutritional sources** of lipids include butter, meat, cheese and olive oil.
- The main **metabolic role of lipids** is as a source of energy. The energy stored in lipids is released during respiration.
- The elements found in **proteins** include carbon, hydrogen, oxygen and nitrogen. Some proteins contain phosphorus and sulfur.
- The basic structural unit in a protein is the **amino acid**.
- Proteins consist of long chains of amino acids linked by peptide bonds. Proteins have a folded shape.
- **Nutritional sources** of proteins include meat, fish, eggs, milk and cheese.
- **Biological structures** composed of protein include keratin in hair and nails, myosin and actin in muscle.
- Metabolic role of proteins include enzymes, membrane interactions and immunity.
- **Proteins necessary for molecular mechanisms:**
  - **Channel proteins** form openings across the cell membrane, allowing ions or molecules to diffuse in or out of the cell.
  - **Protein transporters** carry ions or molecules e.g. glucose from one side of the cell membrane to the other.
  - **Structural proteins** allow cells to attach to each other, holding tissues together.
  - **Receptors proteins** bind to certain molecules, triggering cell responses e.g. release of hormones.
- **Fat-soluble vitamins:** A, D, E, K.
- **Water-soluble vitamins:** B and C.
- **Foods rich in vitamin C** include citrus fruits and fresh vegetables.
- **Role of vitamin C:** Vitamin C helps to make skin, tendons and ligaments, helps wounds to heal, assists in the absorption of iron and allows white blood cells to function properly.
- **Deficiency of vitamin C** can cause scurvy, swollen and bleeding gums, slow wound healing and anaemia.
- **Nutritional sources of calcium** include milk, cheese and green leafy vegetables such as broccoli and cabbage.
- **Role of calcium:** Calcium is needed to build strong bones and teeth, clot blood and control muscle movement.
- **Deficiency of calcium** can cause weakened teeth, osteoporosis and muscle spasms.
- **Role of water in living organisms:** Water is needed as a medium for transport and chemical reactions and for regulation of temperature, pH and osmoregulation.

END OF CHAPTER ASSESSMENT

REMEMBER

1. What is a *biomolecule*? List the different types of biomolecule.
2. State a source of each of the following in a diet.
   (a) Carbohydrate
   (b) Protein
   (c) Lipid
3. Identify a metabolic role for each of the following.
   (a) Carbohydrate
   (b) Protein
   (c) Lipid
4. Name a vitamin.
   (a) What is the role of this vitamin in a diet?
   (b) What are the effects of a lack of this vitamin in a diet?
5. Name a mineral.
   (a) What is the role of this mineral in a diet?
   (b) What are the effects of a lack of this mineral in a diet?
6. Define the following terms.
   (a) Anabolic
   (b) Catabolic
7. Identify the chemicals used in each of the following food tests.
   (a) Protein
   (b) Reducing sugar
   (c) Starch

UNDERSTAND

8. Identify the biomolecule (food type) in each of the following.

(a)

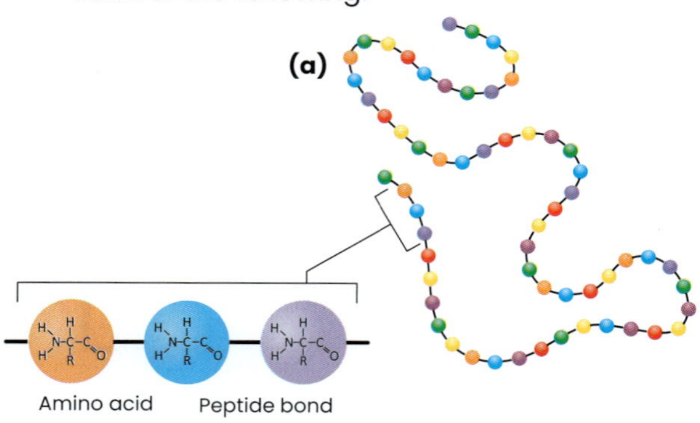

(b)

(c)

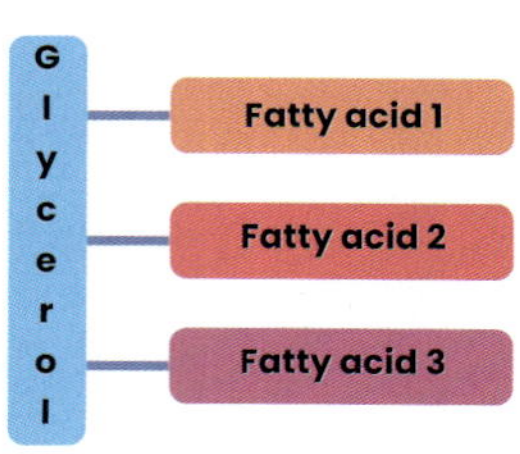

9. Explain the difference between a triglyceride and a phospholipid.
10. Give an example of each of the following.
    (a) Monosaccharide
    (b) Disaccharide
    (c) Polysaccharide
11. Distinguish between a peptide and a polypeptide.
12. Explain the main roles of water in living organisms.

APPLY

13. How would you test a food sample for each of the following?
    (a) Starch
    (b) Glucose
    (c) Protein
    (d) Fat
14. A student testing food samples for glucose using Benedict's reagent obtained the following results for three of their food samples: sample A – green precipitate, sample B – red precipitate, C – blue solution. What conclusion would you draw about each of the food samples?

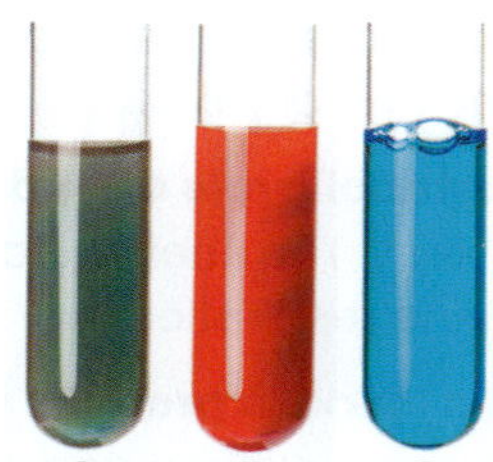

ANALYSE

15. Compare and contrast (i) carbohydrates (ii) proteins and (iii) lipids under the following headings:
    (a) Basic unit
    (b) Nutritional sources
    (c) Metabolic role

EVALUATE

16. Evaluate the effect of the deficiency of each of the following on a person's health.
    (a) A named vitamin
    (b) A named mineral

CREATE

17. Design an experiment to compare the glucose content of fresh fruit to frozen fruit.

15 Metabolism and Enzymes

LEARNING OUTCOMES: U1.2, U2.3, U2.5, U3.2, 2.1.1, 2.1.2, 2.1.3, 2.1.4

EYE-OPENER

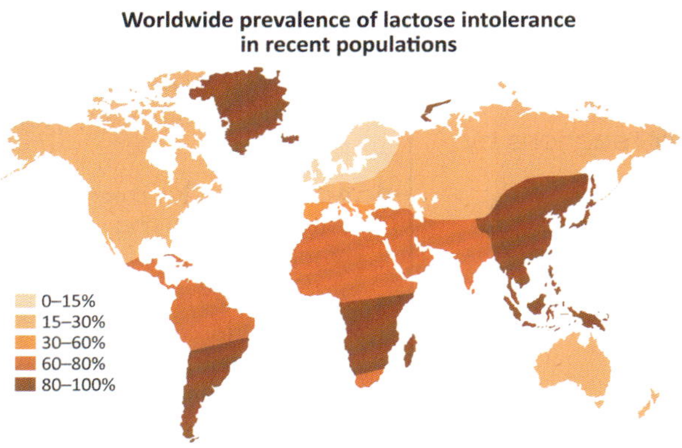

Enzymes are proteins that catalyse chemical reactions in organisms. The enzyme lactase helps the digestion of the sugar lactose in milk. People who do not produce enough lactase suffer from lactose intolerance. They cannot fully digest the lactose in milk and suffer from symptoms such as diarrhoea and bloating after consuming dairy products. Worldwide, most adults are lactose intolerant. However, in northern Europe, 80–90% of adults are lactose tolerant due to a gene mutation that arose approximately 7,000 years ago.

At the end of this chapter, you should be able to:

- Explain how enzymes work to catalyse chemical reactions that occur in living things.
- Explain enzyme specificity using the induced fit model.
- Investigate factors that affect rate of enzyme-catalysed reactions.
- Research the use of enzymes in industry.
- Understand the central role of enzymes including immobilised enzymes in industrial applications.

LINKS TO:

Chapter 14	**Biomolecules**
Chapter 22	**Photosynthesis**
Chapter 23	**Respiration**
Chapter 24	**Information Flow in the Cell**
Chapter 30	**The Digestive System**
Chapter 36	**Human Reproduction – From Fertilisation to Birth**

SUBJECT VOCABULARY

enzyme | catalyst | catalysis | substrate | active site | enzyme-substrate complex | product | specificity | metabolism | anabolic reactions | catabolic reactions | activation energy | optimum | denatured | induced fit | bioreactor | immobilised enzyme

Section 1: Biological Catalysts

At the end of this section, you should be able to:

- Define the terms 'enzyme', 'catalyst', 'catalysis' and 'metabolism'.
- Describe how enzymes are protein in nature.
- Describe the three-dimensional (3D) shape of enzymes.
- Explain the terms 'active site', 'substrate' and 'product' in relation to how enzymes work.
- Draw diagrams to illustrate enzyme action – formation of the enzyme-substrate complex and release of products.
- Explain the term 'enzyme specificity'.

Unpacked from LO 2.1.1

Approximately 37,000,000,000,000 (37 trillion) chemical reactions occur in the human body every second. These reactions are controlled by chemicals called enzymes, found in all cells of every organism. **Enzymes** are biological **catalysts** – they control the rate at which chemical reactions occur in organisms. The acceleration of chemical reactions by catalysts, including enzymes, is called **catalysis**.

- A **catalyst** is a chemical which generally speeds up the rate of a chemical reaction but is unchanged in the reaction.
- An **enzyme** is a selective catalyst controlling biochemical reactions.
- **Catalysis** is the acceleration of chemical reactions by catalysts, including enzymes.

ENZYME STRUCTURE AND SHAPE

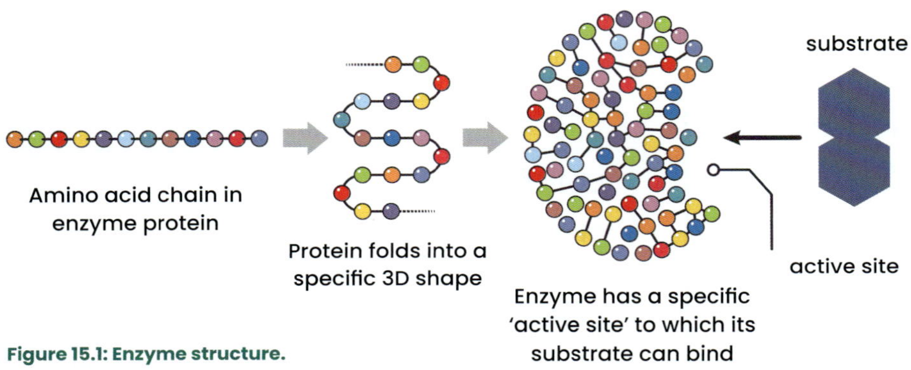

Figure 15.1: Enzyme structure.

Enzymes are made of protein and are formed in the cell's ribosomes. They consist of long chains (100–1,000) of amino acids in a unique sequence, linked by peptide bonds. This protein chain folds into a 3D structure; a ball-like shape. In Chapter 24, we will learn that this unique amino acid order is determined by the gene that codes for the enzyme.

HOW DO ENZYMES WORK?

Enzymes are selective – they act on a particular chemical called a **substrate**, e.g. the enzyme catalase breaks down its substrate hydrogen peroxide, in cells.

Each enzyme has a region to which its substrate binds, called the **active site**.

Go to YouTube.com and search for 'How Enzymes Work' (1:19) by RicochetScience.

- The **substrate** is the particular molecule an enzyme acts on.
- The **active site** is the region in an enzyme to which its particular substrate attaches and from which the product(s) are released.

Once the enzyme binds to its substrate, at the active site, it forms an **enzyme-substrate complex**. This complex undergoes an immediate chemical change: the substrate is converted to the new **product(s)**, which are released from the enzyme. The enzyme isn't altered in this process and can quickly bind with another substrate. Each enzyme will only bind to a specific substrate which fits the 3D shape of the active site. We call this **enzyme specificity**.

- The **enzyme-substrate complex** is the structure formed when an enzyme binds to its particular substrate.
- The **product** is the substance formed when an enzyme acts upon its substrate.
- **Enzyme specificity** describes how each enzyme will only act on one particular substrate.

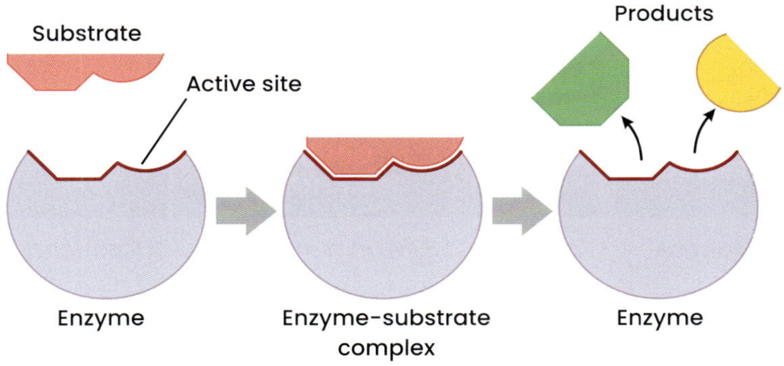

Figure 15.2: Enzyme action.

PORTFOLIO LINK **Go to your Learning Portfolio and complete Learning Check 1 to assess your knowledge and understanding of the topic so far.**

Section 2: Enzymes and Metabolism

At the end of this section, you should be able to:

- Explain the difference between anabolic and catabolic reactions.
- Give examples of enzymes, the pH they work at, their substrate and product.
- Define the term 'activation energy' and explain how enzymes affect activation energy in biochemical reactions.

Unpacked from LO 2.1.1

METABOLIC FUNCTIONS OF ENZYMES

The collective name for all the chemical reactions in an organism is **metabolism**. Metabolic reactions fall into two types – anabolic and catabolic. **Anabolic reactions** such as photosynthesis use small molecules to make larger ones, usually requiring energy. **Catabolic reactions** such as respiration break large molecules into smaller ones and release energy.

The main role of enzymes is to control the rate of an organism's chemical reactions or 'metabolism' – either speed them up or slow them down. Without enzymes, metabolic reactions would occur too slowly for living organisms to function.

- **Metabolism** describes all the chemical reactions that occur in an organism.
- In an **anabolic reaction**, small molecules are converted into larger molecules, using energy.
- In a **catabolic reaction**, large molecules are broken down into smaller ones, releasing energy.

Metabolic reactions need to take place under the normal temperature and pH conditions found in the organism's cells. In humans, the normal body temperature is 37°C and cell cytoplasm has a pH of around 7.2. Examples of reactions catalysed by enzymes include:

- photosynthesis
- respiration
- DNA replication
- pulling chromosomes apart during cell division
- generation of ATP.

Examples of enzymes

ENZYME	LOCATION	SUBSTRATE	PRODUCT	OPTIMUM pH	OPTIMUM TEMPERATURE	REACTION TYPE
Amylase	Mouth and pancreas	Starch	Maltose	7	37°C in humans	Catabolic
Catalase	Liver	Hydrogen peroxide	Water and oxygen	7	37°C in humans	Catabolic
DNA polymerase	Every cell	Nucleotides	DNA	9	75–80°C in bacteria	Anabolic

ACTIVATION ENERGY

A certain amount of energy – **activation energy** – is needed to start a chemical reaction. Enzymes lower a reaction's activation energy, allowing the reaction to occur at a lower temperature than would otherwise be needed. This enables reactions to occur at an organism's body temperature.

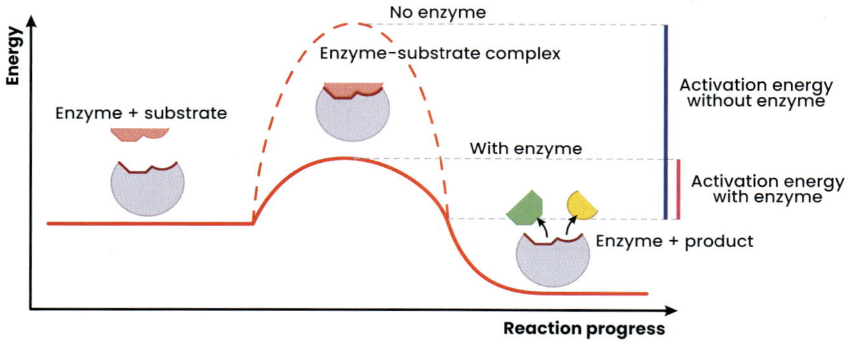

Figure 15.3: Energy profile diagram – effect of an enzyme on activation energy.

Activation energy is the minimum amount of energy needed for a reaction to occur.

Go to YouTube.com and search for 'Enzymes and Activation Energy' (1:56) by RicochetScience.

PORTFOLIO LINK **Go to your Learning Portfolio and complete Learning Check 2 to assess your knowledge and understanding of the topic so far.**

Section 3: Induced Fit Model

At the end of this section, you should be able to:

- Briefly outline the lock and key theory of enzyme action.
- Write a note to describe the induced fit model of enzyme action. HL
- Draw a labelled diagram to describe the induced fit model. HL

Unpacked from LO 2.1.2

LOCK AND KEY THEORY

The **lock and key theory** proposed that the enzyme's active site (the 'lock') matched the shape of its specific substrate (the 'key') exactly. Just like a car key will only open one car lock, each enzyme will only act on one particular substrate, e.g. the enzyme amylase will only break down the substrate starch. This is called **specificity**.

FOCUS ON ... THE INDUCED FIT MODEL (HL)

HL

The lock and key theory was replaced by the induced fit model of enzyme action, which states that:

- The substrate binds to its enzyme's active site and forms the enzyme-substrate complex.
- The shape of the enzyme's active site changes to fit its substrate more exactly upon binding with it.
- The biochemical reaction occurs, and the substrate is converted to its product(s).
- The product(s) are released.
- The enzyme's active site returns to its original shape.
- The enzyme can act again on another substrate molecule.

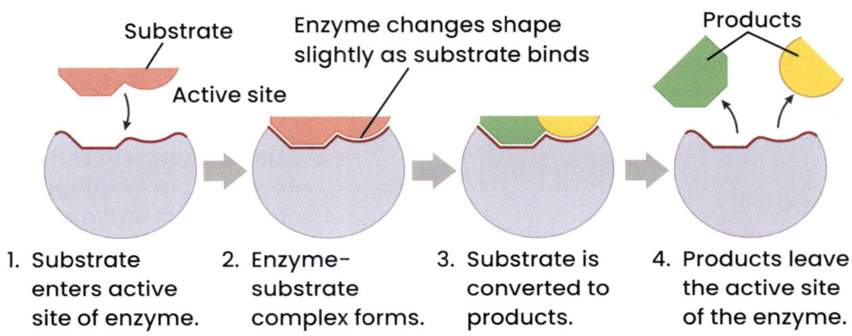

Figure 15.4: Induced fit model of enzyme action.

PORTFOLIO LINK

Go to your Learning Portfolio and complete Learning Check 3 to assess your knowledge and understanding of the topic so far.

Go to YouTube.com and search for 'Enzymes: The Induced Fit Model' (1:53) by RicochetScience.

Section 4: Factors That Affect Enzyme-Catalysed Reactions

At the end of this section, you should be able to:

- List the factors that affect enzyme action.
- Describe with the aid of a labelled graph, the effect of (i) temperature (ii) heat denaturation (iii) environmental pH (iv) substrate concentration (v) enzyme concentration on enzyme action.
- Explain the term 'denatured'.
- Explain the term 'optimum'.
- Give an example of an enzyme and its optimum pH and temperature.

Unpacked from LO 2.1.3

OPTIMUM CONDITIONS: WHAT MAKES ENZYMES WORK BETTER?

The unique shape of the enzyme can be affected by changes in temperature or pH (acidity/alkalinity), which can cause the protein structure to unfold and lose its shape. Each enzyme has an '**optimum**' temperature and a pH value at which it works best.

Effect of temperature on enzyme action

- An increase in temperature increases the rate of reaction up to a maximum or **optimum temperature** at which the enzyme works best.
- Raising the temperature above the enzyme's optimum changes the enzyme's shape so that the substrate can no longer bind to the active site. This lowers the enzyme's rate of reaction.
- If the **temperature is too high** the **enzyme** shape is lost permanently. The enzyme **no longer functions** and is said to be **denatured**.
- Enzymes found in the human body function best at body temperature 37°C.
- Plant enzymes function best at lower temperatures, e.g. those in onion cells have an optimum temperature of 25°C.
- Some enzymes can survive extremes of temperatures.

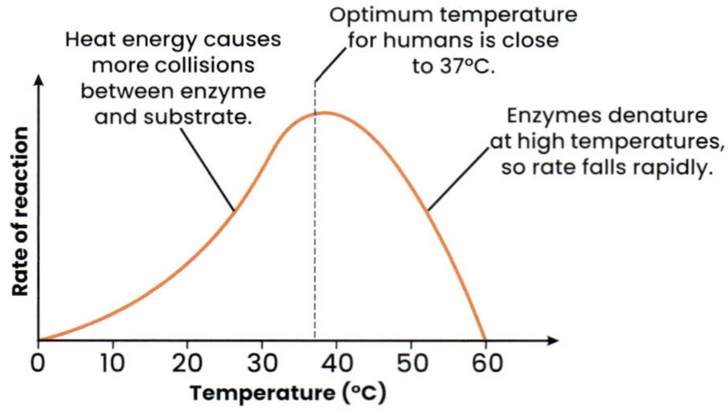

Figure 15.5: Effect of temperature on human enzyme action.

The **optimum temperature** is the temperature at which the enzyme works best.

A **denatured** enzyme is an enzyme that has permanently lost its shape and can no longer carry out its function.

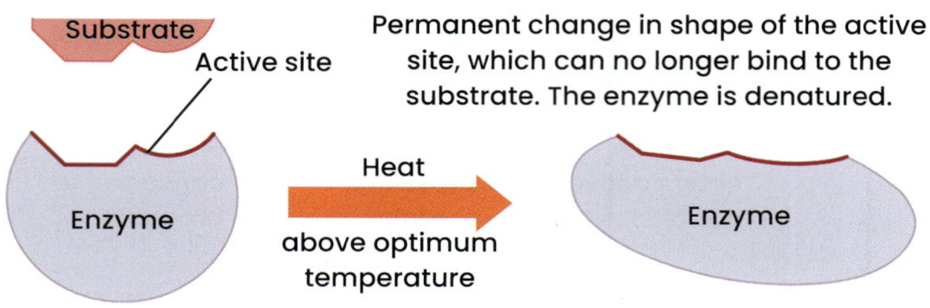

Figure 15.6: Enzyme denaturation.

Effect of environmental pH on enzyme action

The pH scale (0–14) measures how acidic or alkaline a substance is. Acids have pH values less than 7, alkalis are greater than 7 and neutral substances have a pH of 7. Enzyme shape is affected by its environment's pH. The enzyme pepsin which breaks down proteins in the stomach works best at pH 1. Amylase which breaks down starch in your mouth works best around pH 7.

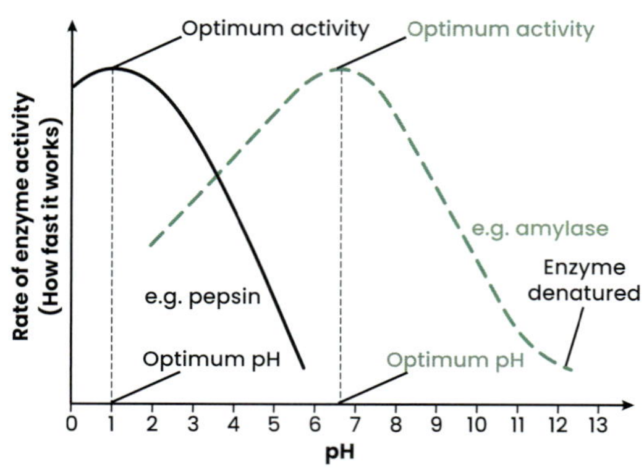

Figure 15.7: Effect of pH on enzyme action.

- An enzyme only has the correct shape to attach to its substrate at a particular pH.
- The **pH at which each enzyme works best** is called its **optimum pH**.
- Decreases or increases in pH from the optimum pH result in the enzyme changing shape and a decline in its ability to function.
- Extremes of pH **denature enzymes**, i.e. result in a **permanent loss of function as the enzyme shape is irreversibly changed**.

The **optimum pH** is the pH at which an enzyme works best.

Effect of substrate concentration on enzyme action

The rate of an enzyme reaction is affected by substrate concentration.

- At low substrate concentrations, the reaction rate is low.
- **The rate of the enzyme reaction increases as the substrate concentration increases:** this is because the enzyme has more substrate molecules to bind with.
- After a certain substrate concentration, all the enzyme molecules' active sites are occupied by substrates – the enzyme is said to be **saturated**. Increasing the substrate concentration further will not increase the reaction rate.

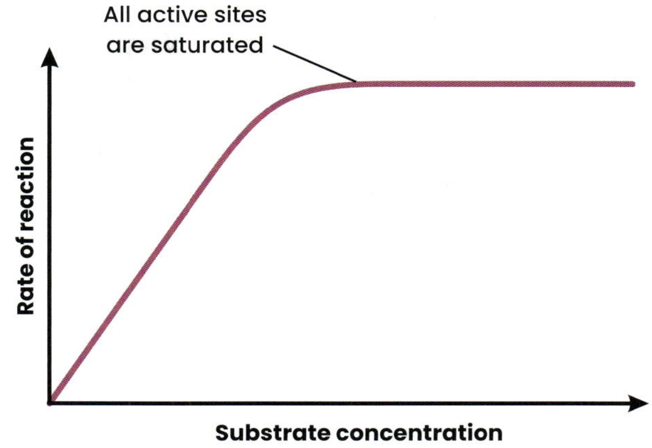

Figure 15.8: Effect of substrate concentration on enzyme action.

Effect of enzyme concentration on enzyme action

Enzyme concentration alters the rate of an enzyme-catalysed reaction.

- Increasing the enzyme concentration increases the rate of reaction: there are more enzyme molecule active sites for substrate molecules to bind with.
- If the enzyme concentration exceeds the number of substrate molecules, the reaction rate levels off. There are not enough substrate molecules to react with the enzyme.

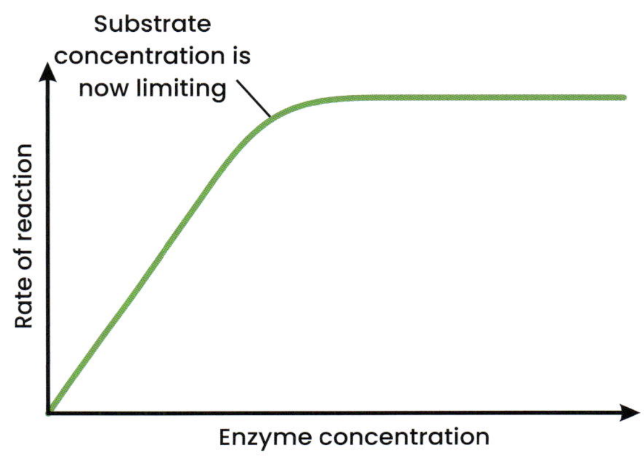

Figure 15.9: Effect of enzyme concentration on enzyme action.

Measuring rate of enzyme-catalysed reactions

Later in the chapter, we will investigate factors that affect enzyme activity. The time taken for the reaction to occur – i.e. for the substrate to be broken down or the product to form – can be measured experimentally.

- The less time taken for a reaction to occur, the greater the rate is.
- The longer the time taken for the reaction to occur, the lower the rate of reaction is.
- Rate of reaction is inversely proportional to the time taken and is measured as **1/time (1/t)**.

1/t is a measure of rate of enzyme activity, where t = time taken.

Go to YouTube.com and search for 'GCSE Biology – Enzymes – How Temperature and pH Affect Rate of Reaction' (3:36) by Cognito.

PORTFOLIO LINK **Go to your Learning Portfolio and complete Learning Check 4 to assess your knowledge and understanding of the topic so far.**

Section 5: Use of enzymes in industry

At the end of this section, you should be able to:

- Explain the terms bioreactor and immobilised in relation to enzymes.
- Give examples of the use of enzymes in the following industries: (i) pharmaceuticals (ii) food and beverages (iii) biofuels (iv) medicine.
- Explain the advantages and disadvantages of using a named enzyme in producing a particular product in industry.
- Research the use of enzymes in industries.
- Describe how to investigate the effect of different factors on the rate of enzyme activity.

Unpacked from LO U1.2, U3.2, 2.1.3, 2.1.4

Enzymes are used in industry to catalyse a wide variety of reactions, producing commercially valuable products. Enzymes are useful in industry due to their specificity. They can also speed up reactions at much lower temperatures than other catalysts and they can be reused.

Biochemical reactions are carried out in sterile, stainless-steel vessels called **bioreactors**. Enzymes can be used in these bioreactors either '**free**' in solution or '**immobilised**'. Immobilised enzymes are 'held in place' – the enzyme can be:

- adsorbed on an inert solid surface or
- trapped within a gel made from a substance such as alginate.

When an immobilised enzyme is used in a bioreactor, the substrate is passed over the immobilised enzyme and the product is collected.

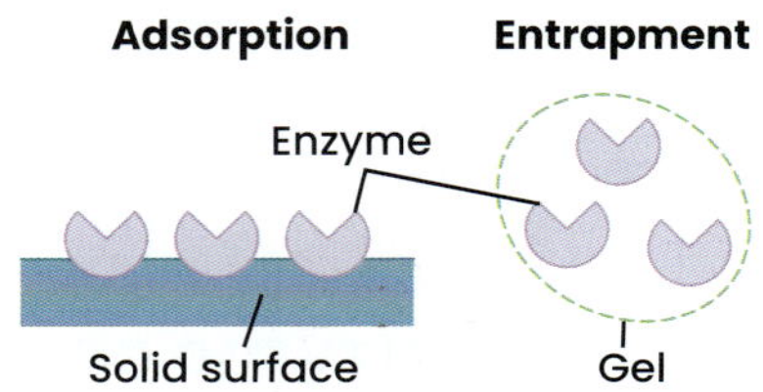

Figure 15.10: Enzyme immobilisation techniques.

A **bioreactor** is a sterile stainless-steel vessel in which biochemical reactions are carried out using enzymes.

An **immobilised** enzyme is an enzyme that is held in place on an inert surface or trapped in a gel.

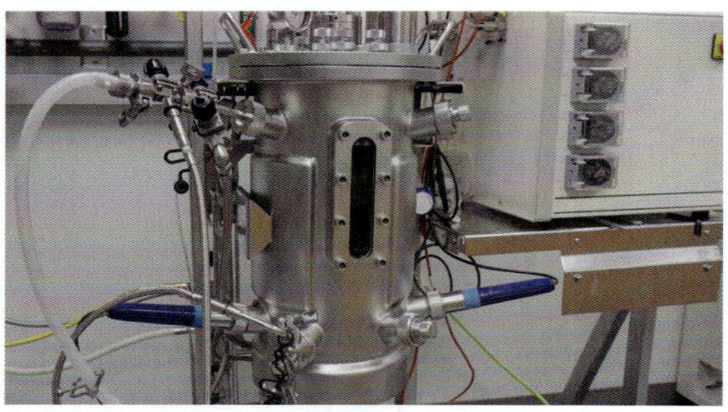

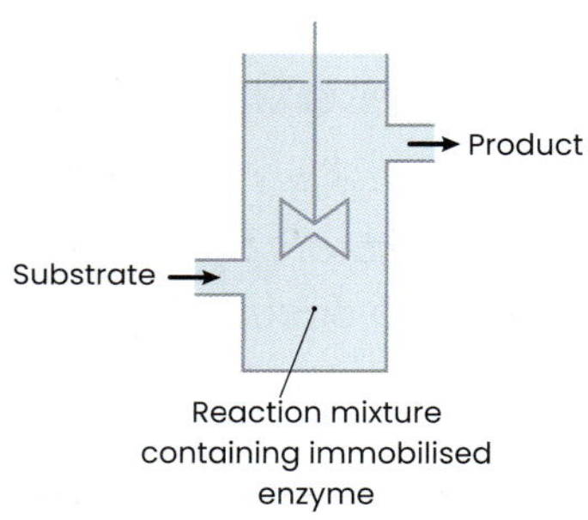

Figure 15.11: Bioreactor with immobilised enzymes.

Immobilised enzymes offer **advantages** over free enzymes:

1. The product is pure and does not need to be separated from the reaction solution.
2. Immobilised enzymes are easily retrieved for reuse.
3. Immobilised enzymes are more stable to temperature/pH changes.

Disadvantages associated with the use of immobilised enzymes include:

1. There is a lower reaction rate than with a free enzyme. When the enzyme is immobilised in a gel, the substrate must diffuse into the gel to react with the enzyme and the product must diffuse out.
2. There is the cost of immobilising the enzyme.

USES OF ENZYMES IN INDUSTRY

Enzymes are used in a wide variety of industries, including:

- pharmaceuticals
- food and beverages

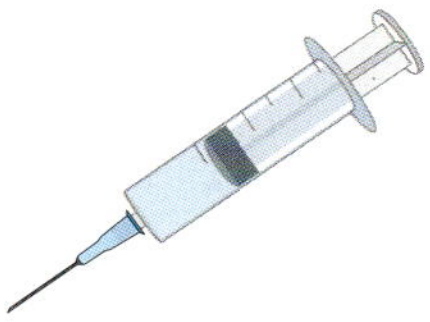

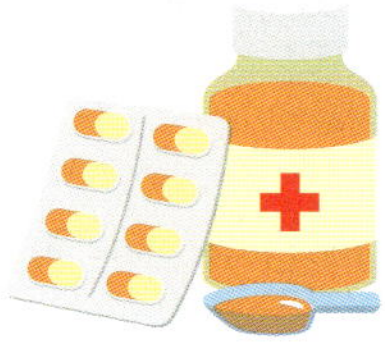

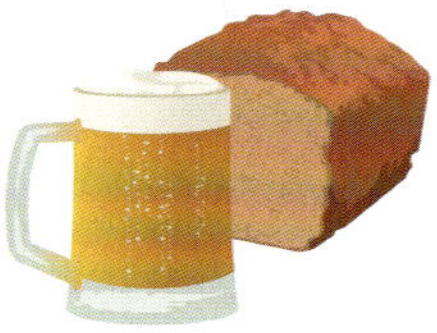

- medicine
- biofuels.

64% of industrial enzymes are produced from fungi, including yeast.

24% are from bacteria, and the rest are from plant and animal sources.

DID YOU KNOW?

RESEARCH ACTIVITY

LO2.1.4

Research the use of enzymes in industries (food and beverages/ biofuels/medicine/ pharmaceuticals).

Use of enzymes in pharmaceuticals

- Enzymes are used to **produce antibiotics** such as semi-synthetic penicillin.

Use of enzymes in food and beverages

- Enzymes are used to control sugar and alcohol levels in alcoholic drinks during **brewing**.
- In the **dairy** industry, enzymes are used to control the levels of proteins or sugars in milk.
- **Corn syrup** is produced by enzymes which convert starch to the sugar fructose.
- The enzyme rennet is used in making **cheese**.
- The enzymes maltase and zymase in yeast are used in **baking bread**. The yeast breaks down sugar releasing carbon dioxide, causing the bread to rise.

Figure 15.12: Rennet used in cheese production.

SCIENCE IN SOCIETY: SUSTAINABILITY

Use of enzymes in biofuels

Biofuels are fuels derived directly from plant or animal sources. Biofuels are renewable and are more environmentally friendly than fossil fuels. Examples of biofuels include ethanol, biodiesel and biogas.

- **Ethanol** is derived from starch in corn or sugarcane. Amylase is used to break down starch in these plants to glucose which is fermented by bacteria or yeast to produce ethanol.
- **Biodiesel** is produced from vegetable oils.
- **Biogas** is produced from methane derived from animal manure.

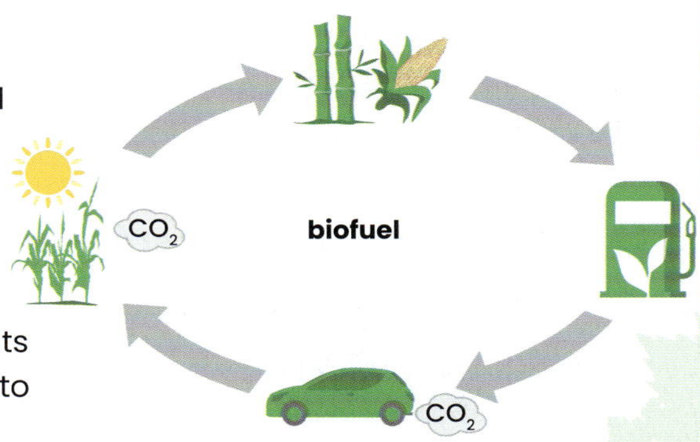

Figure 15.13: Biofuel production.

SCIENCE IN SOCIETY: HEALTH

Use of enzymes in medicine

- **Analytical tests:** Diabetics use glucose test strips, containing the enzyme glucose oxidase, to monitor blood glucose levels. Enzymes are also used in pregnancy tests.
- **Enzyme replacement therapy:** In the disease cystic fibrosis, a defective gene causes the production of a thick mucus which blocks the pancreas from releasing enzymes. Pancreatic enzymes, e.g. pancreatic amylase, pancreatic lipase and trypsin, are needed to break down starch, fats and proteins. People with cystic fibrosis take an enzyme supplement with their meals to aid food digestion.

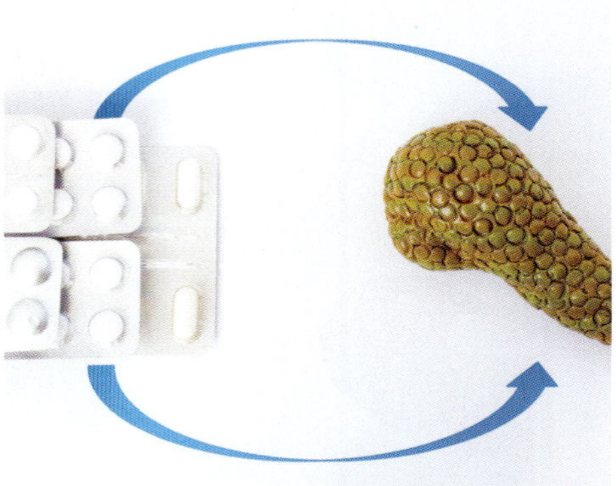

Figure 15.14: Pancreatic enzyme replacement therapy.

- **Assisting drug delivery:** The enzyme hyaluronidase is frequently injected with drug treatments into muscle tissue to speed up the drug's absorption and dispersion to its target.
- **Medical device cleaning:** Reusable medical equipment is cleaned using detergents containing enzymes, e.g. proteases that break down protein-rich blood stains and lipases that break down adipose fat tissue.

SCIENCE IN SOCIETY: HEALTH

Lactose-free milk

Lactose-free milk is produced by treating the milk with the enzyme lactase.

Lactase enzyme from yeast or bacteria is immobilised, e.g. by trapping in alginate gel.

Milk is passed through a bioreactor containing the immobilised lactase.

As the milk passes over the immobilised lactase, lactose is converted to glucose and galactose, which are both easily absorbed by lactose intolerant individuals.

Lactose-free milk is collected from the bioreactor and can be used to produce other lactose-free dairy products e.g. yoghurt or cheese.

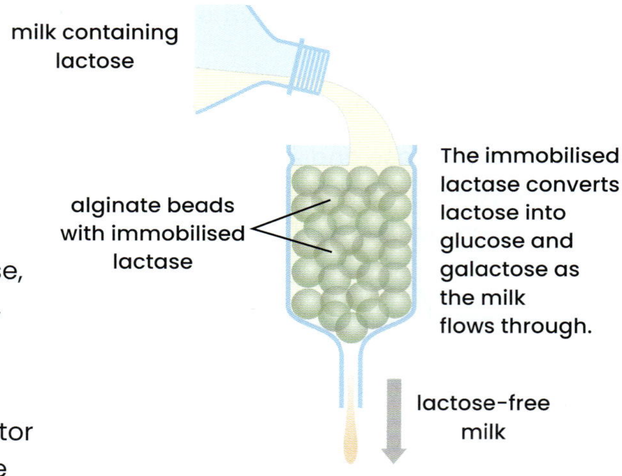

Figure 15.15: Use of immobilised lactase to convert lactose to glucose and galactose.

Lactose-free milk has a number of advantages:

- Source of dairy products for lactose intolerant individuals.
- Cheese and yoghurt may be produced more quickly using lactose-free milk as bacteria can ferment the monosaccharides glucose and galactose more quickly.
- Ice cream produced using lactose-free milk freezes at a lower temperature forming softer ice cream.

Specified Investigation 5

Investigate factors that affect the rate of enzyme-catalysed reactions. (LO2.1.3)

Scan the QR code to watch a video demonstration of this investigation.

A. pH

Hypothesis: If a starch solution is reacted with amylase enzyme at different pH values, the amylase will break down the starch most quickly at its optimum pH.

Equipment: Starch solution, amylase, iodine, pH buffers (4, 7, 9), dropping tiles, syringes, droppers, stopwatch, test tubes, water bath.

VARIABLES

Independent variable: Different pH values (4, 7, 9) will be tested.

Dependent variable: The time taken for starch to be broken down by amylase will be measured.

Control variables: The temperature will be kept at 37°C, the same starch solution (4% concentration), same amylase solution concentration.

METHOD

1. Place single drops of iodine solution in wells in a dropping tile. Note the solution's orange-brown colour.

2. Using different syringes, place $2cm^3$ amylase and $1cm^3$ pH 7 buffer solutions in a test tube.

   Add 4% starch solution to a second test tube.

   Place both test tubes in a water bath at 37°C. Wait 5 minutes to allow solutions to adjust to the temperature.

   Add $2cm^3$ starch solution to the amylase-buffer solution. Start the stopwatch immediately and mix the solution.

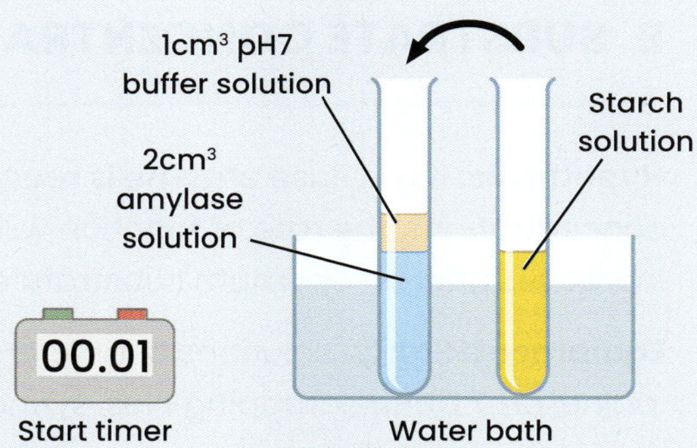

3. After 10s, use a dropper to place one drop of the starch-amylase mixture on the first drop of iodine. This should turn blue-black, indicating starch is present.

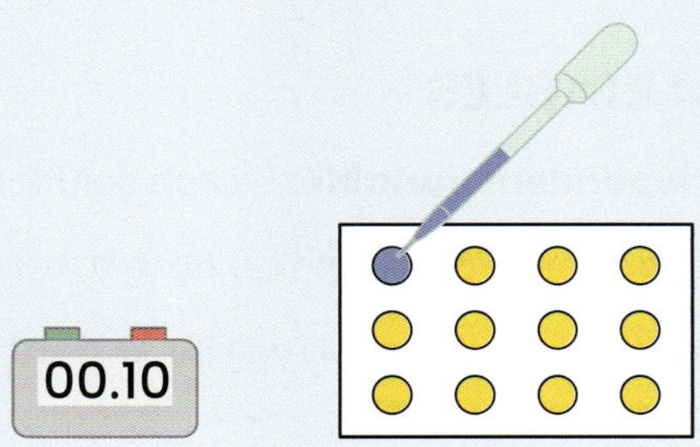

4. Wait 10s and place another drop of the starch-amylase mixture in the next dropping well.

   Repeat every 10s until the iodine solution remains orange-brown indicating amylase has broken down all the starch.

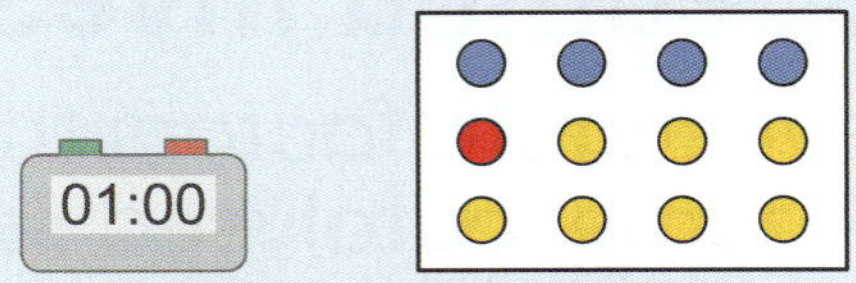

pH7 = 60 seconds for reaction to complete

---

5. Repeat experiment at different pH values using different pH buffers (pH4, pH9).

   Record all the times in a table and calculate the reaction rate (1/time) at each pH. Graph rate of enzyme activity against pH.

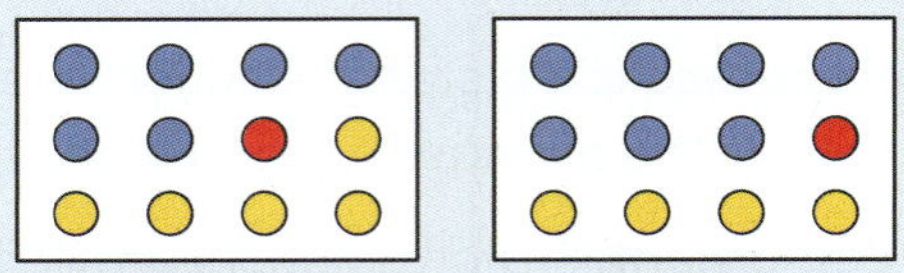

pH4 pH9

RESULTS

pH	TIME FOR STARCH TO BE BROKEN DOWN (s)	RATE (1/time) s^{-1}
4		
7		
9		

CONCLUSION

Amylase works best at pH 7. The less time the iodine solution takes to remain orange-brown, the quicker all the starch has been digested and the better the enzyme works at that pH.

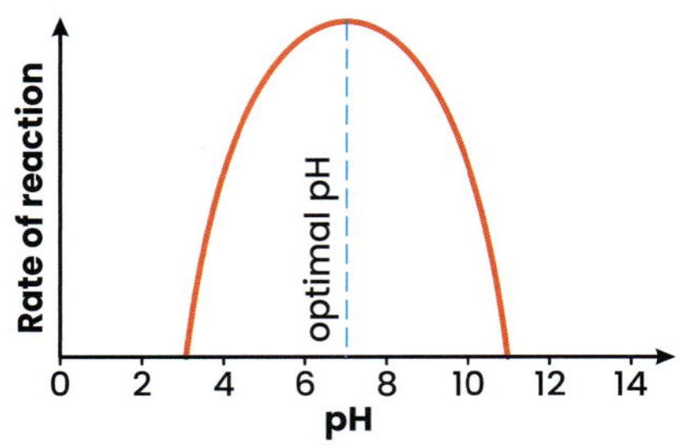

B. SUBSTRATE CONCENTRATION

Hypothesis: If amylase enzyme is reacted with starch solutions of different concentrations, the rate of reaction will increase as the substrate concentration increases until an optimum substrate concentration is reached.

Equipment: Starch solutions of different concentration (1%, 2%, 3%, 4%, 5%), amylase solution, iodine, pH 7 buffer, dropping tiles, syringes, droppers, stopwatch, test tubes, water bath.

VARIABLES

Independent variable: Starch solutions of different concentrations will be tested (1%, 2%, 3%, 4%, 5%).

Dependent variable: The time taken for starch to be broken down by amylase will be measured.

Control variables: pH will be kept at pH7 using a buffer, the temperature will be kept at 37°C using a water bath.

METHOD

1. Repeat Steps 1–4 of Investigation 5A using the different starch solution concentrations (1%, 2%, 3%, 4%, 5%).

   Record all the times in a table and calculate the rate of reaction (1/time) at each substrate concentration. Graph rate of enzyme activity against starch concentration (substrate concentration).

(L–R) 1%, 2%, 3%, 4%, 5% starch solution

RESULTS

STARCH CONCENTRATION	TIME FOR STARCH TO BE BROKEN DOWN (s)	RATE (1/t)
1%		
2%		
3%		
4%		
5%		

CONCLUSION

The 4% starch solution gives the optimum reaction rate for amylase. The less time the iodine solution takes to remain orange-brown, the quicker all the starch has been digested and the better the enzyme works.

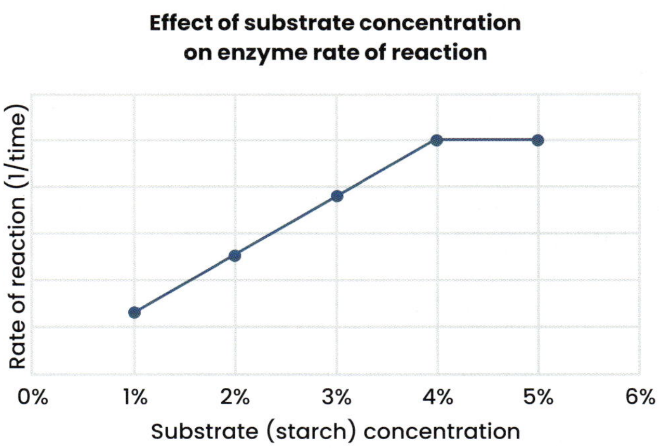

C. TEMPERATURE

Hypothesis: If a starch solution is reacted with amylase enzyme at different temperatures, the amylase will break down the starch most quickly at its optimum temperature (37°C).

Equipment: Starch solution, amylase solution, iodine, pH 7 buffer, dropping tiles, syringes, droppers, stopwatch, test tubes, water bath.

VARIABLES

Independent variable: Different temperature values will be tested (10°C, 20°C, 37°C, 60°C).

Dependent variable: The time taken for starch to be broken down by amylase will be measured.

Control variables: pH will be kept at pH7 using a buffer, the same starch solution (4% concentration), same amylase solution concentration.

METHOD

1. Repeat Steps 1–4 of Investigation 5A.
2. After 1 minute, place one drop of the starch-amylase mixture on the first drop of iodine. This should turn blue-black, indicating starch is present.

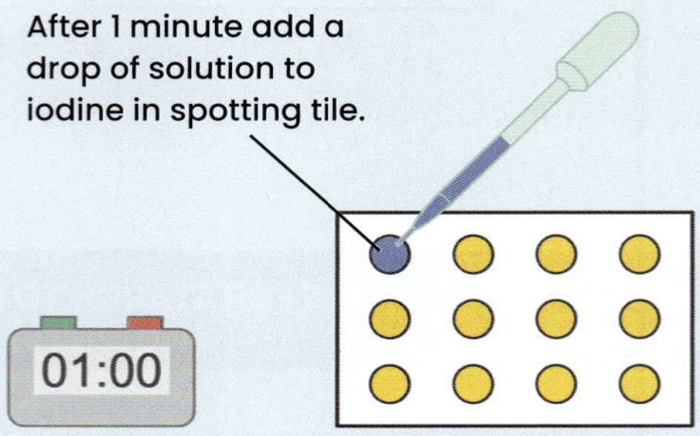

3. Wait another minute and place another drop of the starch-amylase mixture in the next dropping well.
4. Repeat every minute until iodine solution remains orange-brown indicating amylase has broken down all the starch.

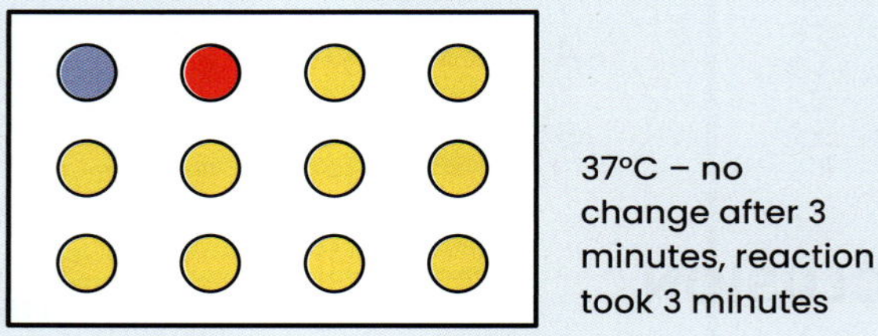

5. Repeat experiment at different temperature values using water baths (10°C, 20°C, 60°C).
6. Record all the times in a table and calculate the reaction rate (1/time) at each temperature. Graph rate of enzyme activity against temperature.

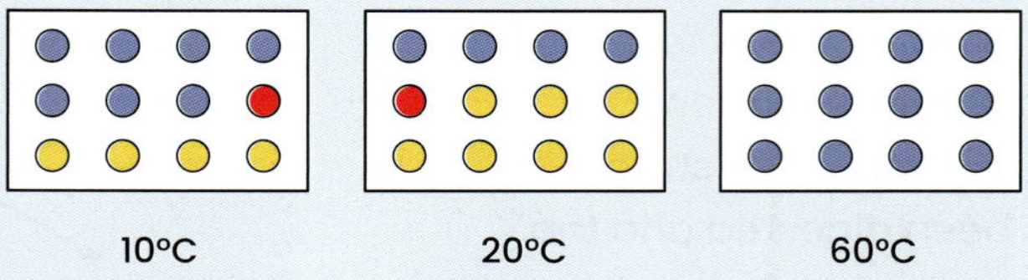

Note: no change in colour at 60°C indicates that the amylase did not break down the starch.

RESULTS

TEMPERATURE (°C)	TIME FOR STARCH TO BE BROKEN DOWN (s)	RATE (1/t)
10		
20		
37		
60		

CONCLUSION

Human amylase works best at 37°C.
The less time the iodine solution takes to remain orange-brown, the quicker all the starch has been digested and the better the enzyme works at that temperature.

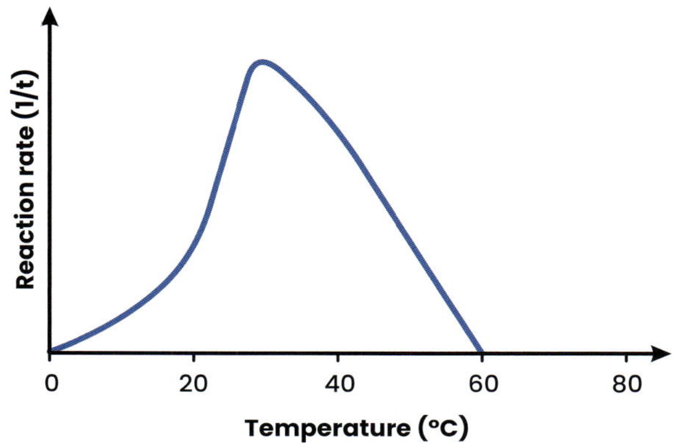

D. ENZYME CONCENTRATION

Scan the QR code to watch a video demonstration of this investigation.

The enzyme catalase breaks down hydrogen peroxide (substrate) to water and oxygen (products). In this experiment, you will trap oxygen released by catalase using washing-up liquid to form foam. The greater the rate of catalase activity, the greater the rate at which oxygen and foam is formed.

Hypothesis: The more concentrated a catalase enzyme sample the greater the rate at which it will break down hydrogen peroxide to oxygen and water.

Equipment: Different amounts of celery (2g, 3g, 4g, 5g, 6g), pH 9 buffer, hydrogen peroxide (20% or less), graduated cylinders, droppers, stopwatch, boiling tubes, balance, water bath.

VARIABLES

Independent variable: Different amounts of celery – source of different catalase concentrations.

Dependent variable: The amount of foam produced in 2 minutes.

Control variables: pH will be kept at pH 9 using a buffer, the temperature will be kept at 25°C.

METHOD

1. Finely chop celery.
2. Add 20ml of pH 9 buffer, one drop of washing-up liquid and 2g of celery into a graduated cylinder.
3. Add 2ml of hydrogen peroxide to a boiling tube.
4. Place the graduated cylinder and boiling tube in a water bath at 25°C.
5. Add the hydrogen peroxide to the graduated cylinder and record the initial volume immediately.
6. Start a stopwatch and record the final volume after 2 minutes.
7. Repeat Steps 2–6 with the following masses of chopped celery: 3g, 4g, 5g, 6g.
8. Calculate the volume of foam produced in 2 minutes at each of the masses of celery (enzyme concentration).

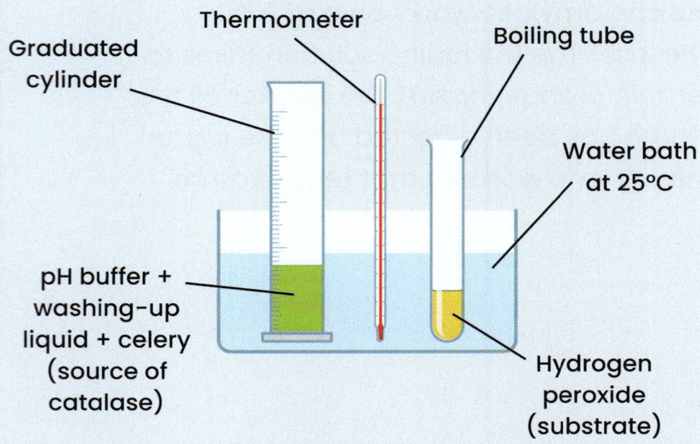

RESULTS

MASS OF CELERY (CATALASE CONCENTRATION)	ENZYME RATE OF REACTION = VOLUME OF FOAM IN 2 MINUTES (FINAL VOLUME – INITIAL VOLUME)
2g	
3g	
4g	
5g	
6g	

CONCLUSION

The greater the enzyme catalase concentration, the greater the rate of enzyme reaction.

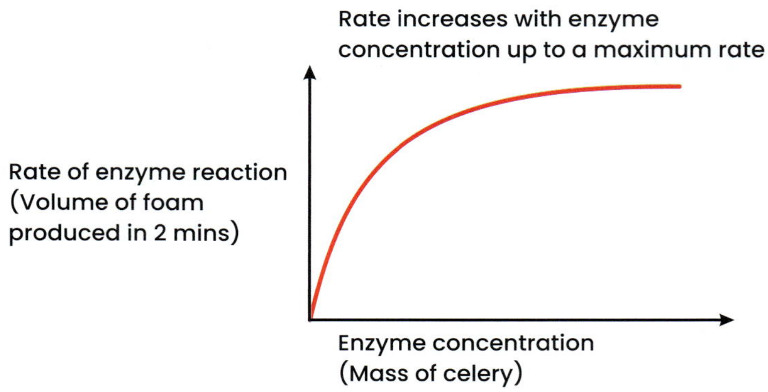

E. HEAT DENATURATION

Hypothesis: If a hydrogen peroxide solution is reacted with denatured catalase enzyme, the catalase will be unable to break down hydrogen peroxide to water and oxygen.

Equipment: Hydrogen peroxide solution, celery as a source of catalase enzyme, pH 9 buffer, measuring cylinders, washing-up liquid, balance, boiling tubes, stopwatch, water baths.

VARIABLES

Independent variable: 2g celery will be boiled prior to use to denature the enzyme catalase.

Dependent variable: The presence or absence of foam will be noted.

Control variables: pH will be kept at pH9, the same hydrogen peroxide solution (substrate), same enzyme concentration (amount of celery). An unboiled sample of celery will be used as a control.

METHOD

Repeat Steps 1–6 of Investigation 5D using (a) 2g of fresh celery and (b) 2g of celery that was boiled for 5 minutes prior to use to denature its catalase. Note the presence or absence of foam.

RESULTS

	FORMATION OF FOAM
UNBOILED CATALASE	Foam formed
BOILED DENATURED CATALASE	No foam formed

CONCLUSION

No foam was formed with boiled celery as the enzyme catalase was denatured.

INVESTIGATIVE LOG LINK Go to your Investigative Log to record your lab work and answer questions on the experiments in this chapter.

Chapter Summary

- **Metabolism** describes all the chemical reactions that occur in an organism.
- **Enzymes** are selective catalysts controlling biochemical reactions, formed from protein. They are not altered by the reaction and may be reused.
- Enzymes speed up reactions by lowering the 'activation energy'. This allows the reaction to occur at temperatures at which the cell functions.
- **Enzymes** consist of a chain of amino acids linked by peptide bonds. Enzymes have a three-dimensional globular shape.
- The **substrate** is the particular molecule the enzyme acts on.
- The **active site** is the region in an enzyme to which its particular substrate attaches and from which the product(s) are released.
- The **product(s)** are the substance(s) formed when the enzyme acts on its substrate.
- The **optimum pH** is the pH at which an enzyme works best.
- The **optimum temperature** is the temperature at which an enzyme works best.
- A **denatured enzyme** is an enzyme whose shape is irreversibly changed due to extreme changes in pH or temperature causing permanent loss of enzyme function.
- **Enzyme specificity** means that each enzyme will only act on a particular substrate.
- HL The **induced fit model** states that the shape of the enzyme's active site changes to fit its substrate more exactly upon binding with it.
- Enzyme activity increases with increasing **substrate concentration** up to a maximum where no more enzyme molecules are available to react with.
- Enzyme activity increases with increasing **enzyme concentration** up to a maximum where no more substrates are available to bind with.
- **Rate of enzyme activity** is measured as 1/t, where t = time.
- An **immobilised enzyme** is an enzyme that is held in place on an inert surface or trapped in a gel.
- A **bioreactor** is a sterile stainless-steel vessel in which biochemical reactions are carried out using enzymes.
- **Advantages of enzyme use in industry** include (i) their specificity, (ii) they speed up reactions and (iii) they allow reactions to occur at lower temperatures.
- **Advantages of immobilised enzymes** are that the product is pure, the enzyme is easily separated and can be reused, the enzyme is more stable.
- **Disadvantages of immobilised enzymes** are that the reaction rate may be slower than that with free enzymes and the cost of immobilising the enzyme.
- **Uses of enzymes in industry:**

INDUSTRY	USE OF ENZYMES
Food and beverages	Control sugar levels in alcoholic drinks, enzymes in yeast ferment sugars releasing CO_2 causing bread to rise, production of lactose-free milk
Biofuels	Enzymes are used to produce fuels such as ethanol from sugarcane
Medicine	Testing glucose levels in blood, pregnancy tests
Pharmaceuticals	Production of antibiotics, e.g. semi-synthetic penicillin

- **Lactose-free milk** is produced as follows: (i) milk is passed through a bioreactor containing immobilised lactase (ii) lactose in the milk is converted to glucose and galactose (iii) lactose-free milk is collected from the bioreactor.
- **Lactose-free milk has a number of advantages:** (i) It provides a source of dairy products for people who are lactose intolerant; (ii) Cheese and yoghurt may be produced more quickly using lactose-free milk; (iii) Ice cream produced using lactose-free milk will freeze at a lower temperature, resulting in softer ice cream.

END OF CHAPTER ASSESSMENT

REMEMBER

1. Define the following terms.

(a) Enzyme
(b) Catalysis
(c) Activation energy

2. List **four** factors that affect enzyme activity.

3. Copy and label the following diagram illustrating enzyme action.

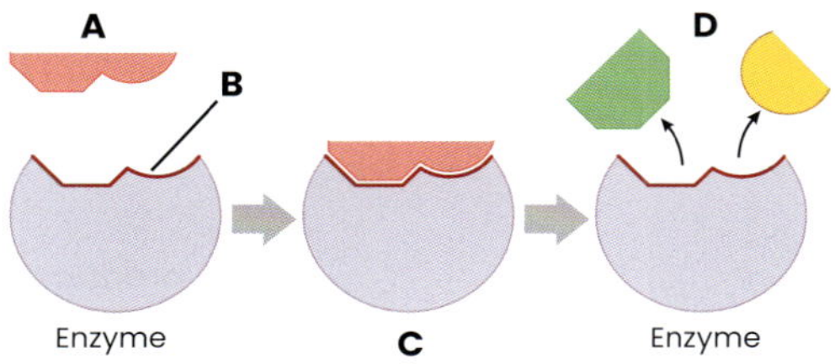

4. Match the key words below with their correct definition.

(a) Anabolic	(i) Area in an enzyme to which its particular substrate attaches
(b) Catabolic	(ii) Molecule the enzyme acts on
(c) Substrate	(iii) Reaction in which small molecules are converted to larger ones, usually requiring energy
(d) Active Site	(iv) Substance(s) formed when enzyme acts on its substrate
(e) Product	(v) Reaction in which large molecules are converted to smaller ones, usually releasing energy

HL **5.** Identify the model of enzyme action referred to in the following diagram.

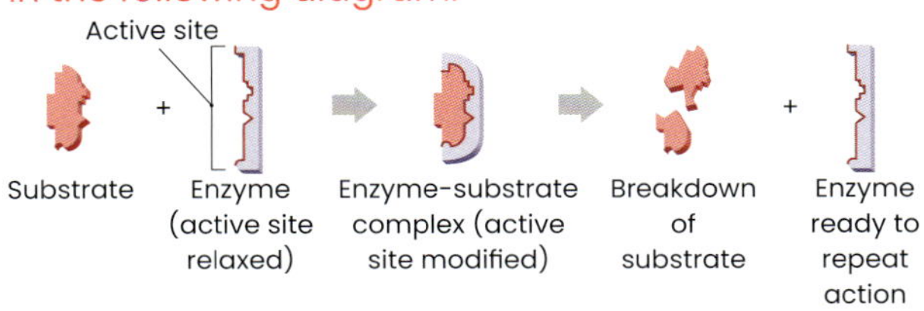

6. Define the term *denatured enzyme*.

7. Select from the following list the temperature at which enzymes in humans work best.

(a) 25°C (b) 37°C (c) 50°C

8. Name an enzyme and state the pH at which it works best.

9. Draw and label the following graph illustrating the effect of pH on enzyme action.

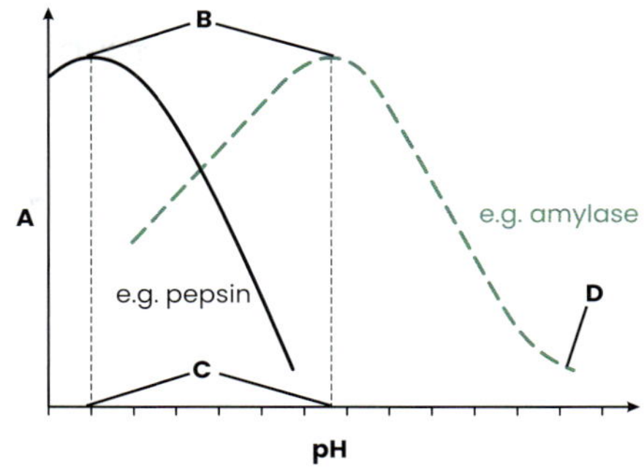

10. Choose the correct substrate for the following enzyme from the selection below:

Enzyme:

Substrates:

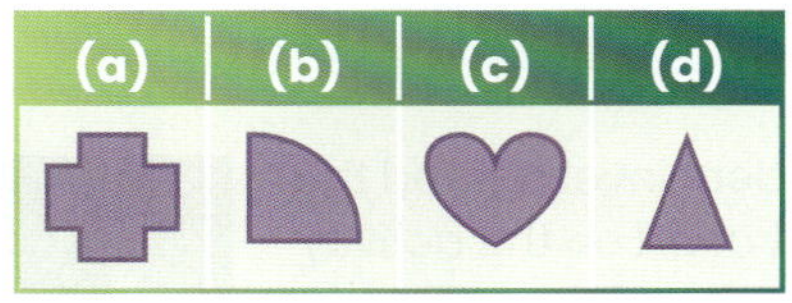

11. Define each of the following terms.

(a) Bioreactor (b) Immobilised enzyme

UNDERSTAND

12. Summarise how enzymes work. Include a labelled diagram in your answer.

13. Explain what is happening at stages A, B and C of the graph below.

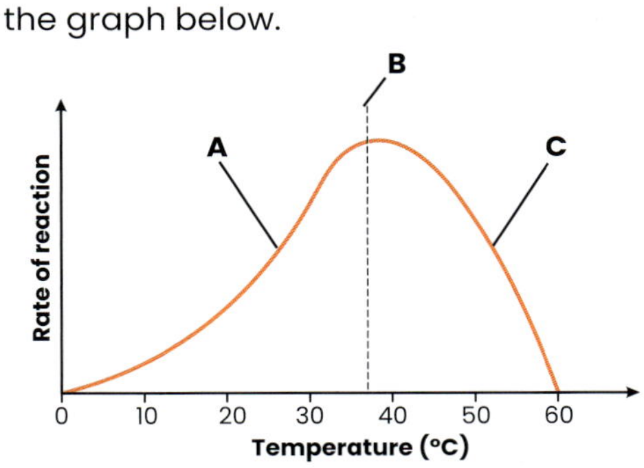

14. Describe what has happened the enzyme in the diagram below.

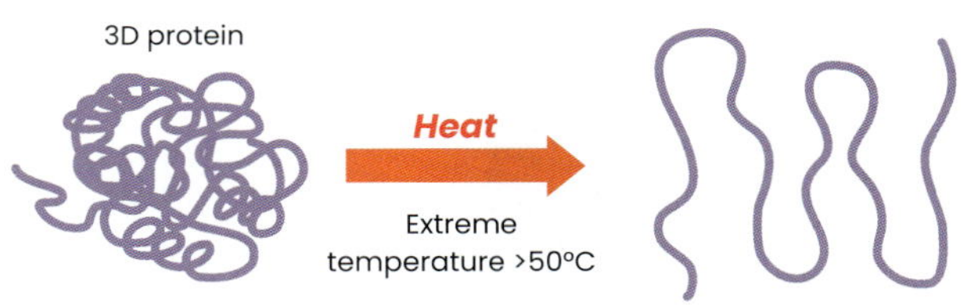

15. Explain the differences between A and B on the graph below in relation to the effect of enzyme concentration on enzyme activity. Refer to substrate concentration in your answer.

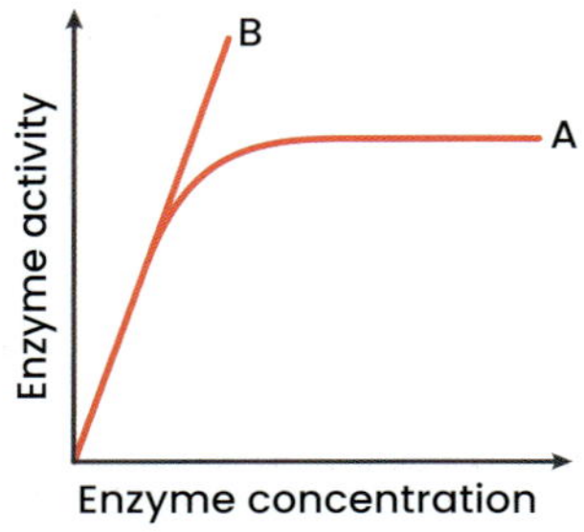

16. Discuss the effect of substrate concentration on rate of enzyme activity.

APPLY

17. A student investigated the effect of pH on the activity of the enzyme catalase with hydrogen peroxide. Catalase breaks down hydrogen peroxide to produce water and oxygen. The oxygen produced is used to form a foam.

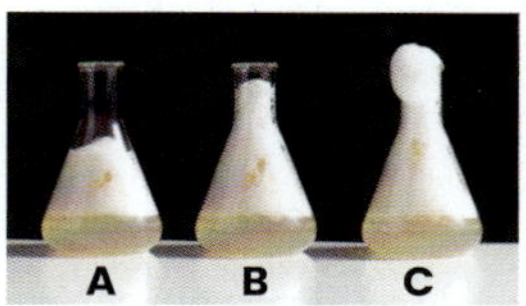

The photo above shows their results.

(a) In which flask was the enzyme most active?

(b) What pH do you think was employed in this flask? Explain your answer.

(c) How may the effect of temperature on the rate of catalase activity be measured?

18. The following results were obtained by students investigating the effect of substrate concentration on enzyme activity. They reacted starch and amylase and monitored the length of time taken for the starch to be broken down. Calculate the rate of reaction at each substrate concentration and plot a graph to illustrate the effect of substrate concentration on enzyme activity.

SUBSTRATE CONCENTRATION (% starch)	1%	2%	3%	4%	5%
TIME FOR STARCH TO BREAK DOWN	90s	70s	50s	30s	50s
RATE OF REACTION (s^{-1})					

19. What would happen if a student forgot to use a pH buffer when investigating the effect of temperature on enzyme action?

ANALYSE

20. Compare and contrast the lock and key theory of enzyme action with the induced fit theory of enzyme action. HL

21. The graph illustrates how a biochemical reaction proceeds in the absence of an enzyme.

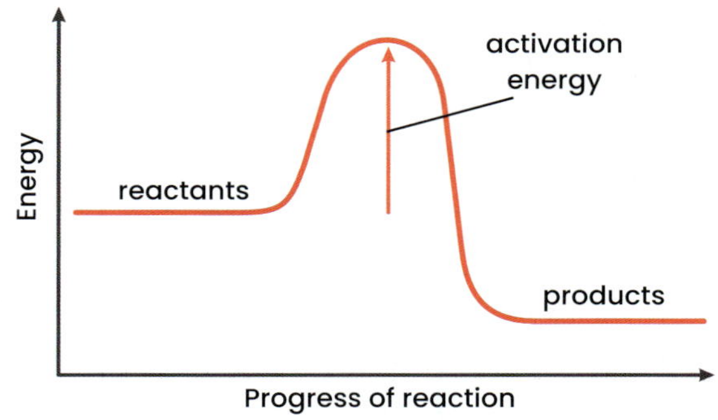

(a) How would the presence of an enzyme affect this reaction?

(b) What impact does this have on reactions occurring in living cells?

22. How would the conditions required for the breakdown of hydrogen peroxide by the enzyme catalase differ in a plant cell to an animal cell? Explain your answer.

EVALUATE

23. Evaluate the importance of enzymes in a living organism.

24. *All enzymes are made of protein, but all proteins are not enzymes.*
Evaluate this statement.

25. What are the advantages and disadvantages of using enzymes in industry?

26. Research the impact of 'fuel crops', e.g. corn or sugarcane, on each of the following.

(a) Food supply

(b) The environment

CREATE

27. Design an experiment to immobilise an enzyme and use it to produce a product.

28. Design an experiment to investigate the effect of enzyme concentration on rate of enzyme action.

29. Design a poster to illustrate the industrial applications of enzymes using lactose-free milk as a case study.

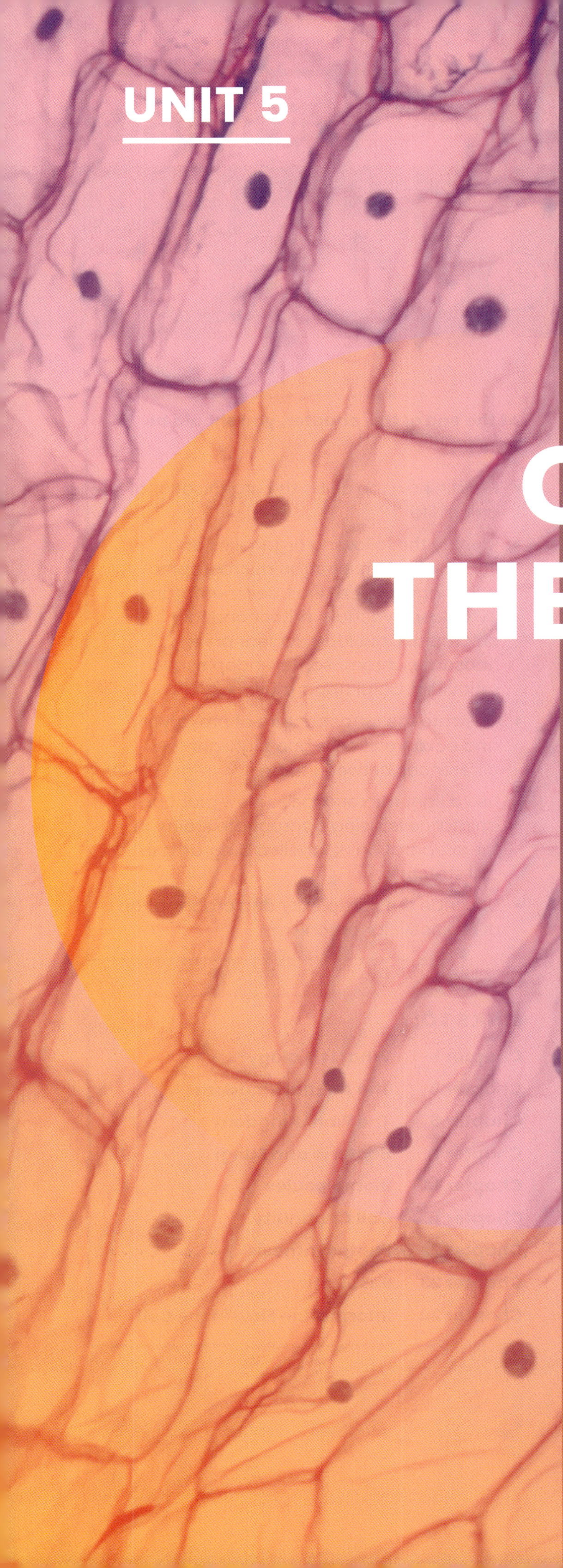

UNIT 5

CELLS – THE UNITS OF LIFE

CHAPTERS

STRANDS

Unifying Strand: Nature of Science

Strand 1: Organisation of Life

Strand 2: Structures and Processes of Life

Strand 3: Interactions of Life

16 Cell Structure and Diversity

LEARNING OUTCOMES: U2.3, U2.5, U3.2, U4.1, 1.3.1, 1.3.2, 1.3.3, 3.3.7

EYE-OPENER

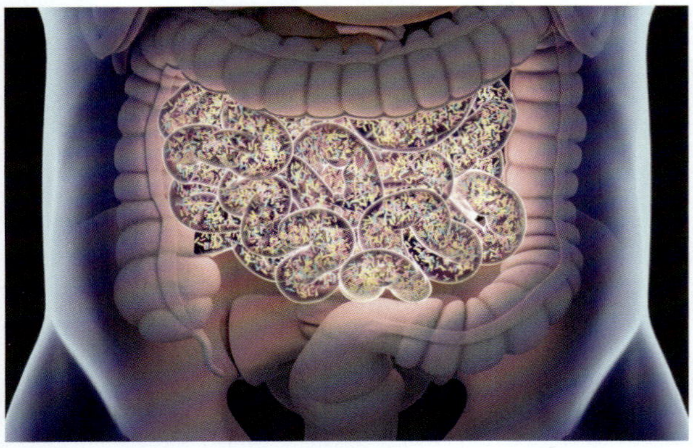

Humans are covered in bacteria. They can be found all over the body and most of them live inside us, particularly in the digestive tract. Recent research indicates that humans contain as many bacterial cells in their body as their own cells. Bacterial cells, however, are tiny compared to human cells, and therefore take up far less space.

At the end of this chapter, you should be able to:

- Explain how the cell is the organisational unit of life.
- Identify the following organelles and structures from electron micrographs and diagrams: cell wall, cell membrane, cytoplasm, vacuole, mitochondrion, chloroplast, nucleus, nuclear pore, ribosome, chromosome, endoplasmic reticulum, Golgi apparatus.
- Illustrate the ultrastructure of prokaryotic and eukaryotic cells and highlight the differences between them.
- Use a light microscope to view plant and animal cells, using appropriate stains, in the preparation of labelled diagrams of cell structure.
- Describe the complexity and organisation of multicellular organisms.

SUBJECT VOCABULARY

unicellular, multicellular, eukaryotic, organelles, prokaryotic, cytoplasm, cytosol, semi-permeable, selectively permeable, nucleus, chromatin, chromosome, gene, mitochondria, ribosomes, endoplasmic reticulum, Golgi apparatus, cell wall, fully permeable, middle lamella, chloroplast, vacuole, turgor, specialised, tissue, organ, organ system, stem cells

LINKS TO:

Chapter 1	**What is Biology?**
Chapter 5	**The Animal Kingdom**
Chapter 6	**The Plant Kingdom**
Chapter 7	**Bacteria and Fungi**
Chapter 14	**Biomolecules**
Chapter 17	**Cell Continuity**
Chapter 22	**Photosynthesis**
Chapter 23	**Respiration**
Chapter 24	**Information Flow in the Cell**

Section 1: Cells – The Building Blocks of Life

At the end of this section, you should be able to:

- Recognise cells as a characteristic of all living things.
- Distinguish between unicellular and multicellular organisms.
- Compare eukaryotic and prokaryotic cells.
- Compare the basic structure of plant and animal cells.
- Distinguish between cytoplasm and cytosol.
- Explain what is meant by 'cell ultrastructure'.
- Discuss the importance of electron microscopes in the study of cells.

Unpacked from LOs 1.3.1, 1.3.2

One of the key characteristics of all living things is that they are organised into tiny structures called cells. Cells are the building blocks of all organisms, the units of life. They are extremely small and cannot be seen by the naked eye. Therefore, a microscope is required to view cells. Some organisms, such as bacteria, consist of simply one cell and are called **unicellular**. Many organisms, however, are made up of lots of cells. These are referred to as **multicellular**.

Unicellular describes an organism consisting of just one cell.

Multicellular organisms consist of lots of cells.

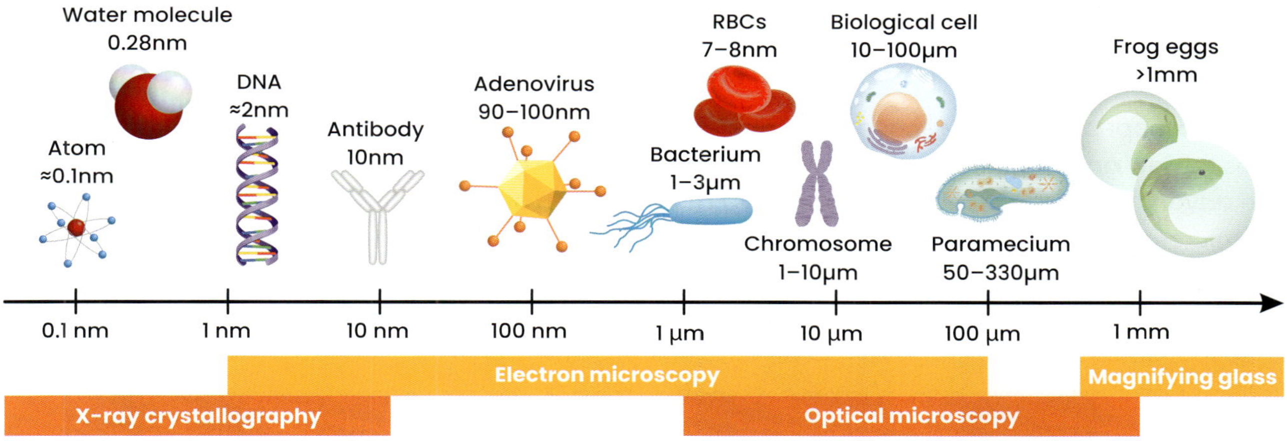

Figure 16.1: The scale of cells.

EUKARYOTIC CELLS AND PROKARYOTIC CELLS

Cells can be divided into two main categories. **Eukaryotic** cells contain a nucleus and other **membrane-bound organelles**. Organelles are structures within cells which have a particular function, similar to organs within the body. Membrane-bound organelles are surrounded by a **phospholipid** membrane – examples include the nucleus, mitochondria and chloroplasts. Ribosomes, the cell's protein factories, are not bound by a membrane, but are found in both eukaryotic and prokaryotic cells.

The nucleus is home to the cell's genetic material, DNA, found in structures called **chromosomes**. The cells of all animals, plants and fungi are eukaryotic.

Eukaryotic cells contain a nucleus and membrane-bound organelles.

Organelles are structures within cells which carry out a particular role.

Prokaryotic cells do not contain a nucleus or membrane-bound organelles; instead, their genetic material is found in one circular chromosome called a **nucleoid**. Bacteria and Archaea are prokaryotic. Prokaryotic cells are explored in more detail in Chapter 7.

Prokaryotic cells do not contain a nucleus or membrane-bound organelles.

Eukaryotic cells are much bigger and more complex than prokaryotic cells. Organisms which are multicellular tend to have eukaryotic cells, whereas a lot of unicellular organisms are prokaryotic. The first ever cell that originated on Earth (LUCA – last universal common ancestor) would have resembled a prokaryotic cell. Eukaryotic cells later evolved from the prokaryotic cells.

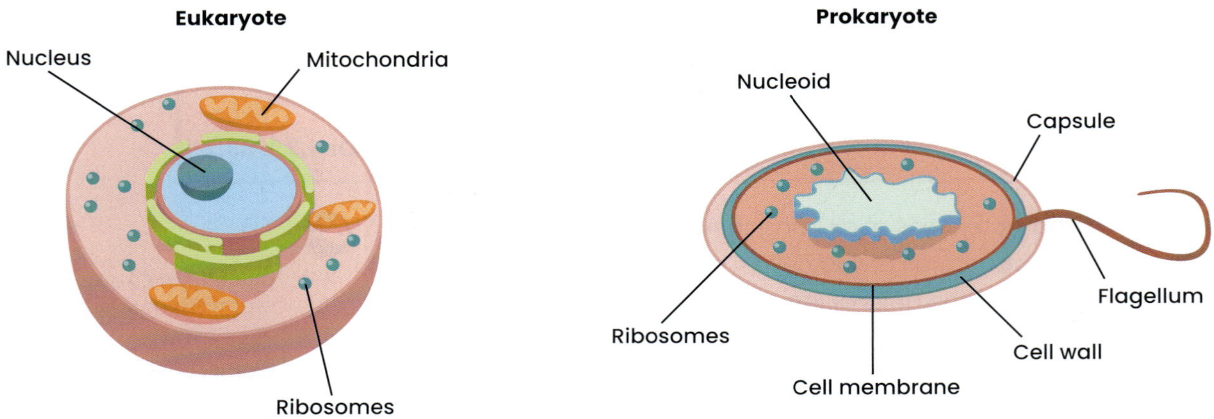

Figure 16.2: Eukaryotic and prokaryotic cells.

BASIC STRUCTURE OF PLANT AND ANIMAL CELLS

The microscopes used in the school laboratories are called **light microscopes**. They use light and lenses to view and magnify the specimen and typically have a magnification of up to 400 times. More advanced light microscopes can have a magnification of up to 2,000 times. We will study the light microscope in more detail in Section 4.

When we view cells under the microscopes we can see the basic structure of the cell, as shown in the diagrams.

	ANIMAL CELL	PLANT CELL
ILLUSTRATION	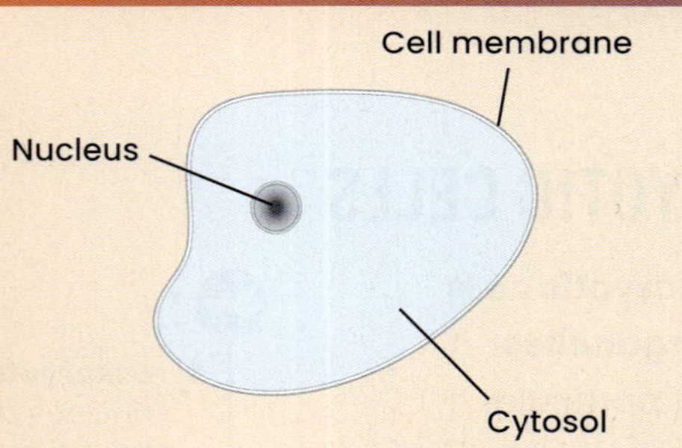	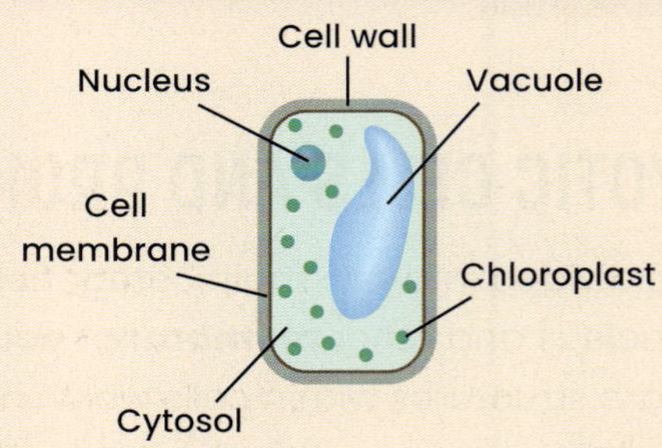
AS SEEN UNDER LIGHT MICROSCOPE	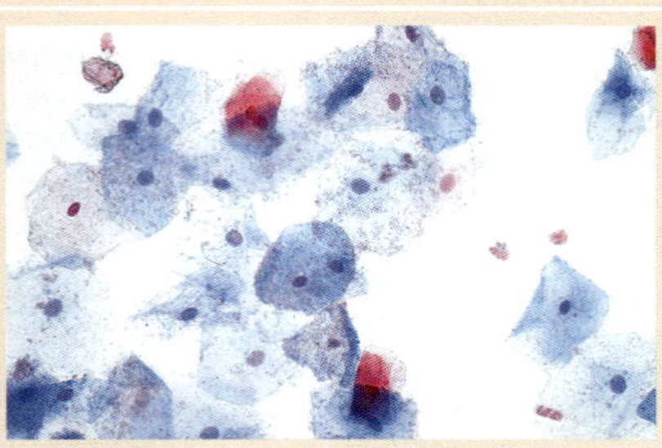	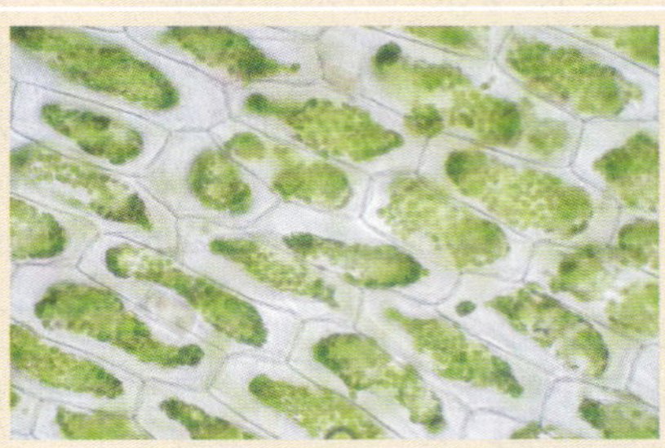

Animal cells consist of a nucleus, containing the genetic material of the cell, embedded in a gel-like substance called the **cytoplasm**. The entire cell is surrounded by a **cell membrane**.

The cytoplasm consists of the **cytosol**, along with all the organelles embedded in it, apart from the nucleus. The cytoplasm also contains several other substances, such as sugars, amino acids, lipids and proteins. Since the cytoplasm contains a liquid cytosol, which is predominantly water, its main function is to act as a medium for chemical reactions to take place inside the cell.

Plant cells also contain a nucleus, cytoplasm and a cell membrane, but also contain a **large vacuole**, a **cell wall** and some plant cells contain **chloroplasts**, in which photosynthesis takes place.

- The **cytoplasm** describes the contents of the cells, except the nucleus.
- The **cytosol** is the liquid part of the cytoplasm.

DIFFERENCES BETWEEN ANIMAL AND PLANT CELLS	
ANIMAL CELLS	**PLANT CELLS**
Do **not** have a cell wall	Do have a cell wall
Do **not** have chloroplasts	Do have chloroplasts
Contain small, non-permanent vacuoles	Contain large, permanent vacuoles

Figure 16.3: The main differences between animal cells and plant cells.

FOCUS ON ... ELECTRON MICROSCOPES

In order to view inside cells in much more detail, an electron microscope can be used. The electron microscope is a type of microscope that uses a beam of electrons to create an image of the specimen; it can magnify cell structures up to 1 million times. The structure of a cell as seen under an electron microscope is called the **cell ultrastructure**.

There are two types of electron microscope. A **transmission electron microscope** (TEM) shows the internal structure of a specimen, whereas a **scanning electron microscope** shows the surface of a specimen. An electron micrograph is a digital image taken through an electron microscope.

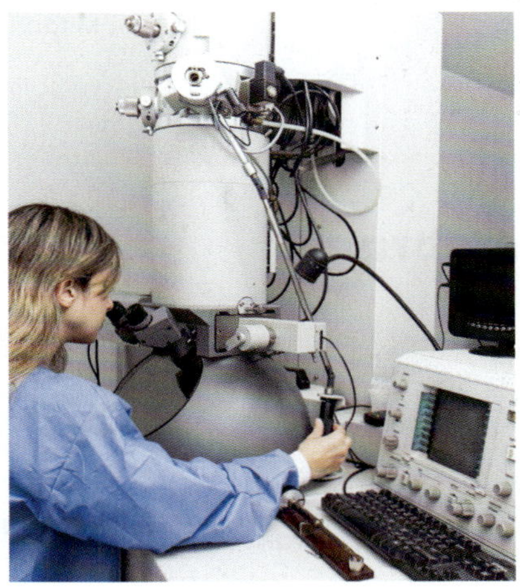

Figure 16.4: An electron microscope.

RESEARCH ACTIVITY

Carry out further research on the next generation of microscopes.

PORTFOLIO LINK

Go to your Learning Portfolio and complete Learning Check 1 to assess your knowledge and understanding of the topic so far.

Go to YouTube.com and search for 'GCSE Biology – Cell Types and Cell Structure #2' (6:48) by Cognito.

Section 2: Animal Cell Ultrastructure

At the end of this section, you should be able to:

- Describe the structure and function of the cell membrane.
- Outline the structure of the nucleus and the chromosomes found within.
- Describe the structure and function of mitochondria.
- Discuss the importance of ribosomes to the cell.
- Compare the structure and function of the endoplasmic reticulum and Golgi apparatus in cells. HL

Unpacked from LOs 1.3.2, 1.3.3

ANIMAL CELLS

Animal cells are so called because the cells of all animals share these common features. When viewed with an electron microscope, the ultrastructure of an animal cell reveals a complex cityscape of organelles including mitochondria and ribosomes. The inside of the nucleus can also be viewed in more detail using an electron microscope.

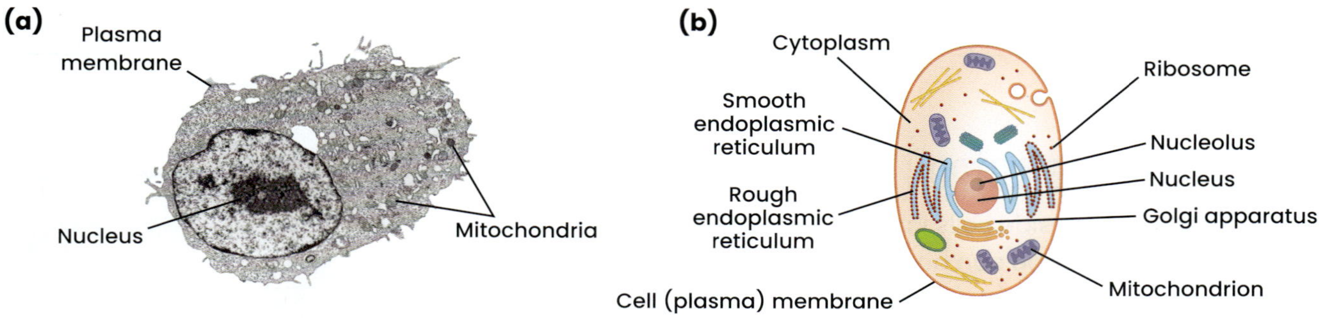

Figure 16.5: (a) Electron micrograph and (b) illustration of an animal cell ultrastructure.

Cell membrane

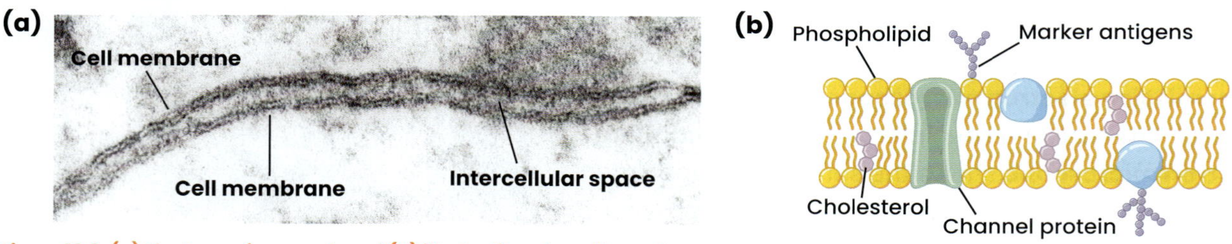

Figure 16.6: (a) Electron micrograph and (b) illustration of a cell membrane.

The cell membrane is made up of two layers of **phospholipids**. This is known as a lipid bilayer. Proteins are embedded in this bilayer, providing support and acting as special channels to control what comes in and out of the cell. Some of these proteins on the surface of the cell allow the cell to be identified as part of the body, so that it is not attacked by the body's immune system. These are called marker antigens.

The cell membrane is **semi-permeable**, meaning it acts like a sieve around the cell. Small molecules (e.g. water or oxygen) can pass through the cell, but larger molecules cannot. The cell membrane is also described as a **selectively permeable** membrane. This means it can actively control which of those larger molecules can enter and exit the cell. These larger molecules pass through one of the channel proteins embedded in the membrane.

- **Semi-permeable** describes how a membrane allows some small molecules pass through it.
- **Selectively permeable** describes how a cell membrane actively controls the entry and exit of materials in a cell.

Nucleus

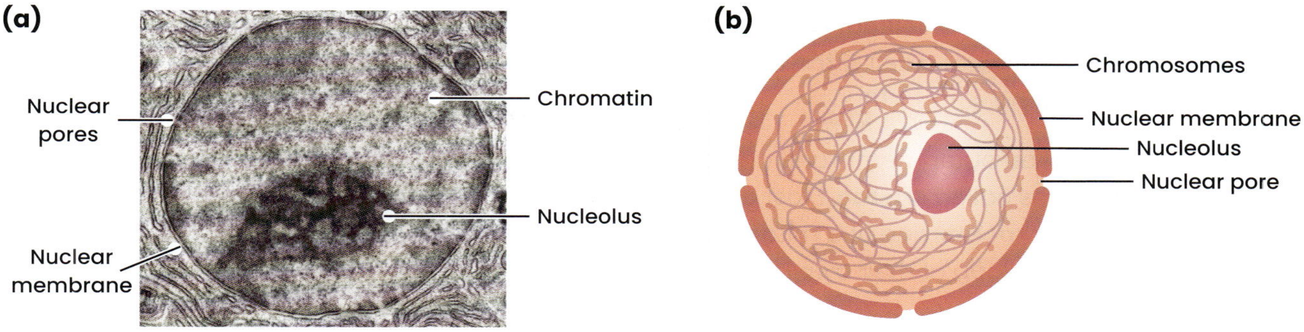

Figure 16.7: (a) Electron micrograph and (b) illustration of a nucleus.

The **nucleus** is the control centre of the cell and home to the cell's genetic material. It is wrapped in a phospholipid membrane, which has small openings called **nuclear pores**. These allow materials, such as RNA, to pass in and out of the nucleus.

The nucleus contains a structure called the **nucleolus**. The function of the nucleolus is to make a type of RNA. We will discuss the role of RNA in Chapter 24.

The nucleus also contains **DNA** found in structures called **chromosomes**. The DNA is wrapped around proteins called **histones**. When the cell is not dividing (a period called interphase), this DNA is sprawled out inside the nucleus and is called **chromatin**. The cell is busy during interphase, copying the DNA and preparing to divide. Just before the cell is about to divide, the DNA coils up into a distinctive shape and is referred to as **chromosomes**.

The chromosomes contain **genes** – sections of DNA which carry instructions for making proteins. We will explore the role of genes in greater detail in Chapter 24.

- The **nucleus** is the control centre of the cell. It contains DNA found in structures called chromosomes.
- **Chromatin** consists of DNA wrapped around proteins called histones.
- **Chromosomes** are condensed, wound-up chromatin.
- **Genes** are sections of DNA in chromosomes that carry instructions for making proteins.

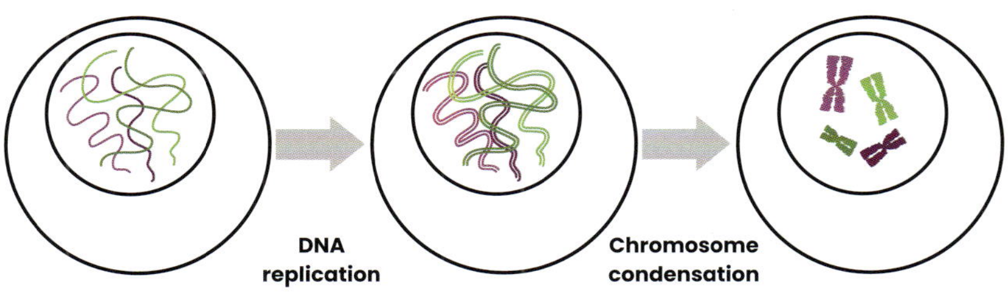

Figure 16.8: Chromosome formation in a cell.

Mitochondria

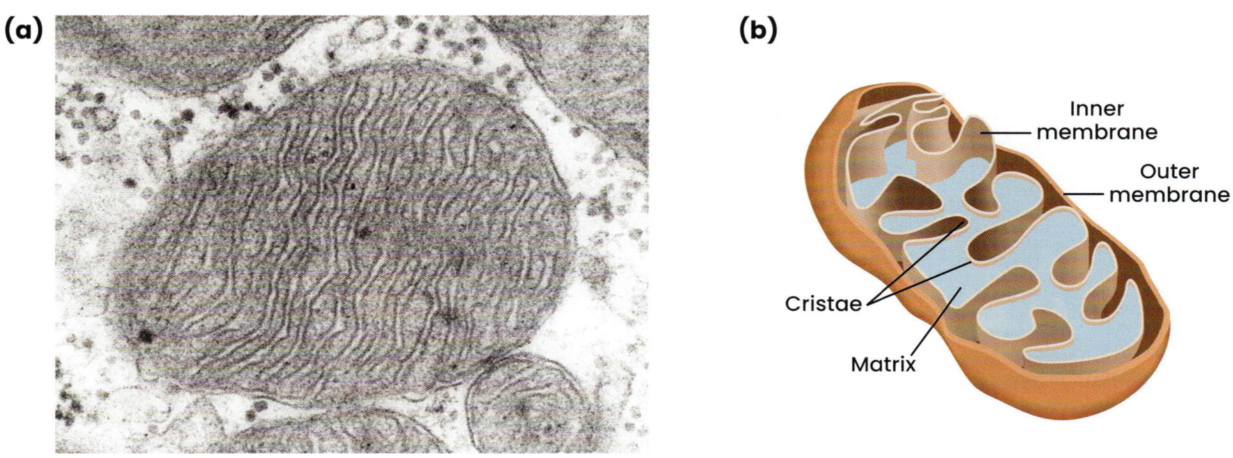

Figure 16.9: (a) Electron micrograph and (b) illustration of a mitochondrion.

The **mitochondria** (singular: mitochondrion) are like the battery packs of the cell, converting food into an energy-rich molecule called ATP in a process called **cellular respiration**. ATP is often called the energy currency of a cell; if the cell needs to make something, it must spend some ATP.

Mitochondria are organelles responsible for respiration, the release of energy from food.

Mitochondria contain an outer membrane and a much-folded inner membrane. The inner membrane is folded into structures called **cristae**. Most cellular respiration occurs in the mitochondria. The more folds in the inner membrane of a mitochondrion, the more energy it can release. Interestingly, mitochondria contain a loop of their own DNA, separate to that of the nucleus.

We will explore cellular respiration in more detail in Chapter 23.

Ribosomes

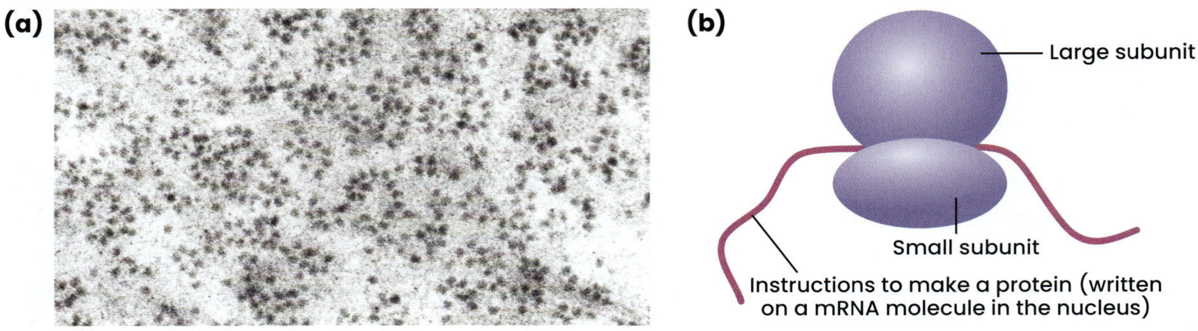

Figure 16.10: (a) Electron micrograph and (b) illustration of a ribosome.

Ribosomes are some of the smallest organelles in a cell, but arguably amongst the most important. They are the protein factories of the cell, building the various proteins needed to regulate the cell's metabolism or build an organism's tissues. Ribosomes are made of **ribosomal RNA** and **proteins**; they are not surrounded by a membrane.

Ribosomes are the organelles responsible for protein synthesis.

If a cell becomes damaged or undergoes some type of infection, it may self-destruct. This process is called apoptosis. This is an important process in the body because, if a cell becomes damaged and apoptosis does not occur, this could result in the development of cancer.

DID YOU KNOW?

Go to YouTube.com and search for 'Eukaryopolis – The City of Animal Cells' (11:34) by CrashCourse.

HL

Endoplasmic reticulum

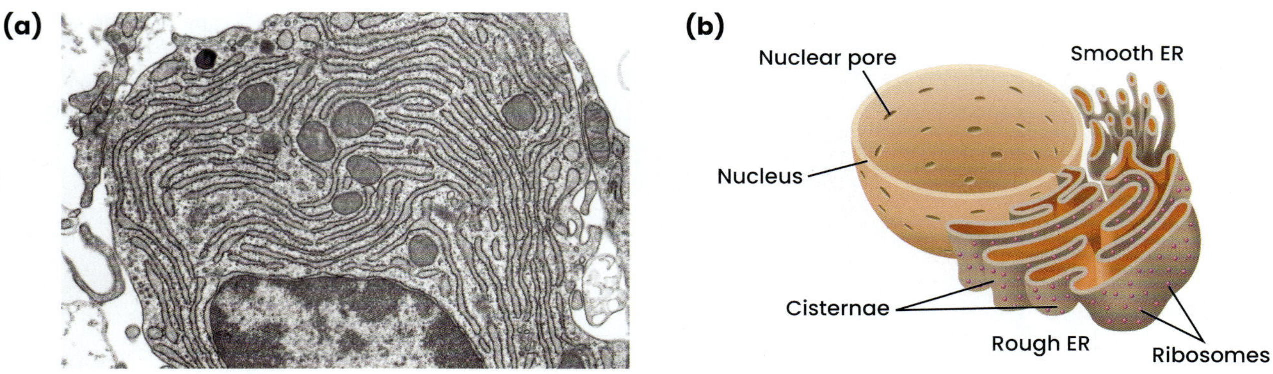

Figure 16.11: (a) Electron micrograph and (b) illustration of endoplasmic reticulum.

Endoplasmic reticulum (sometimes called ER) is a membrane-bound organelle with many functions in both animal and plant cells. ER consists of a set of interconnected tubes called **cisternae** and is a production site for important cellular fats, like phospholipids and cholesterol, and also steroids and hormones. ER also stores and releases calcium. Some ER is covered in ribosomes (called **rough endoplasmic reticulum**), so rough ER is a site of protein synthesis also. ER without ribosomes is called **smooth endoplasmic reticulum**. Both types of ER also play a role in transporting materials around the cell.

Endoplasmic reticulum is an organelle responsible for protein and lipid synthesis.

HL

Golgi apparatus

(a)

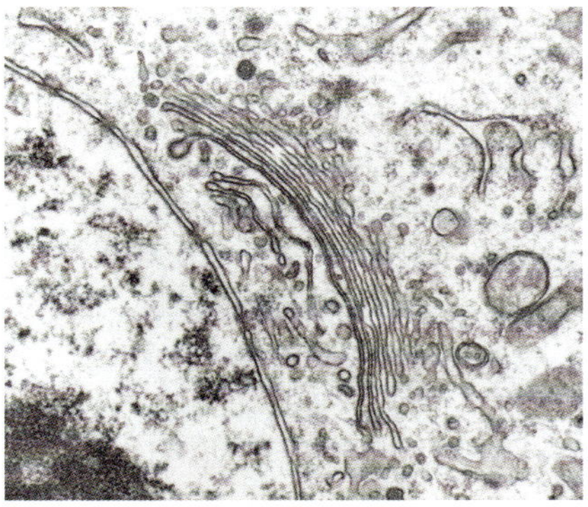

(b)

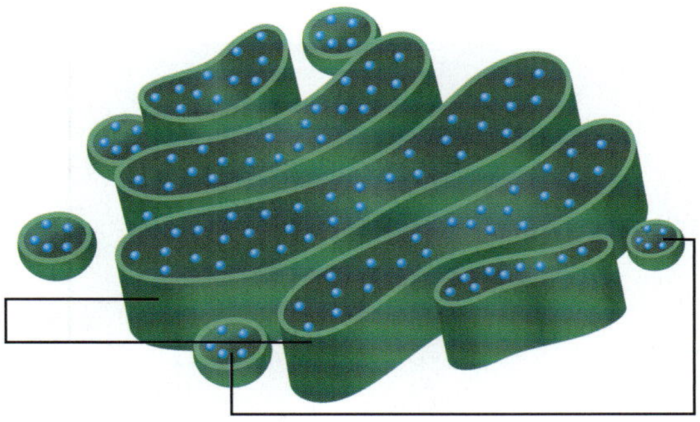

Figure 16.12: (a) Electron micrograph and (b) illustration of a Gogli apparatus.

The **Golgi apparatus** (plural: apparata) works closely with the endoplasmic reticulum, packaging the proteins and lipids manufactured in the ER and moving them to other parts of the cell or even outside the cell. It consists of stacks of membrane-bound sacs, again called cisternae, but this time they are not interconnected. Golgi apparata exist in both animal and plant cells. In plant cells, they produce the carbohydrates needed to build cell walls.

The **Golgi apparatus** is responsible for packaging protein and lipids manufactured in the ER.

PORTFOLIO LINK **Go to your Learning Portfolio and complete Learning Check 2 to assess your knowledge and understanding of the topic so far.**

Section 3: Plant Cells

At the end of this section, you should be able to:

- Compare plant cell structure to animal cells.
- Describe the function of the cell wall.
- Relate the structure of the chloroplast to its function in a plant cell.
- Describe the functions of the large vacuole in plant cells.

Unpacked from LOs 1.3.2, 1.3.3

Just like animal cells, plant cells also contain a cell membrane, nucleus, mitochondria, ribosomes, endoplasmic reticulum and Golgi apparata. However, they also contain a cell wall, chloroplasts and large permanent vacuoles.

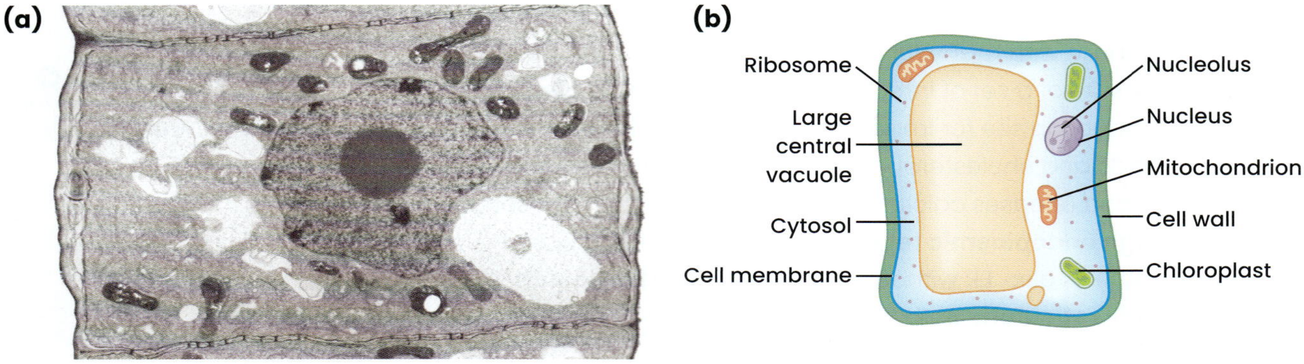

Figure 16.13: (a) Electron micrograph and (b) illustration of plant cell ultrastructure.

Cell wall

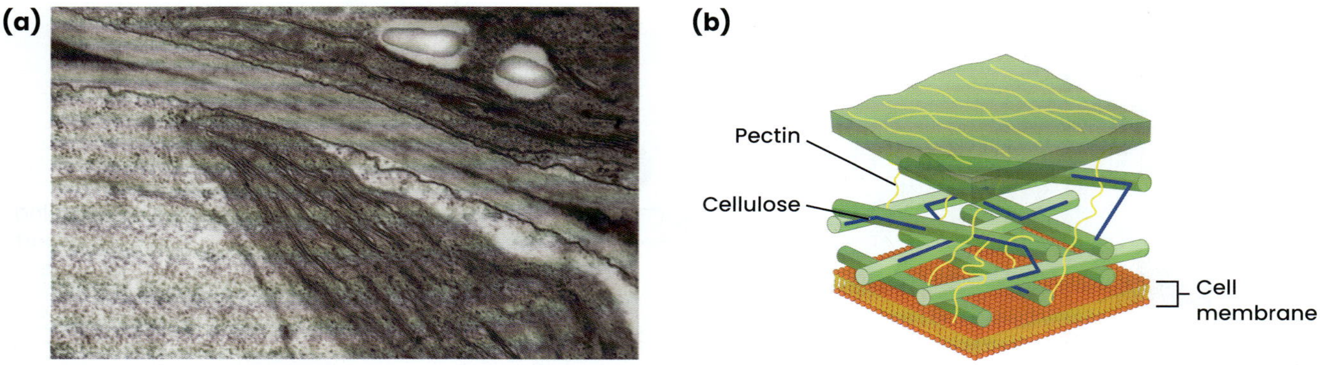

Figure 16.14: (a) Electron micrograph and (b) illustration of a plant cell wall.

Plant **cell walls** are **fully permeable**, which means they allow all substances to pass through them. The role of the cell wall is to provide strength and support to the cell. The wall also prevents the cell from bursting if it absorbs too much water. The cell wall is made from a complex carbohydrate (polysaccharide) called **cellulose**.

The **middle lamella** is the region between two adjacent cell walls. It contains a substance called pectin which holds the cells together.

Although animals do not contain cell walls, some organisms such as fungi do. However, their cell walls are composed of a different carbohydrate called chitin.

- The **cell wall** provides strength and support to the cell.
- **Fully permeable** means all substances pass through it.
- The **middle lamella** is the space between the cell walls of two cells.

Chloroplast

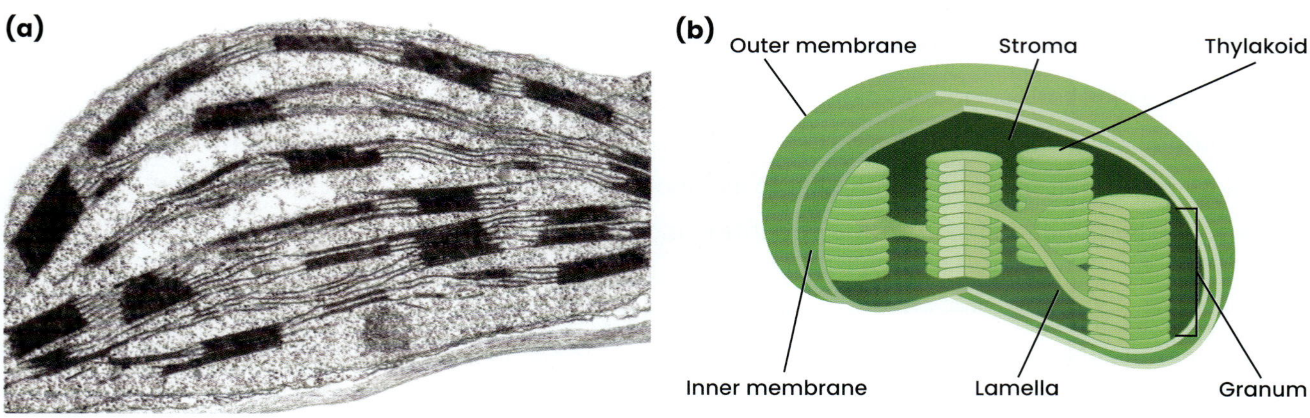

Figure 16.15: (a) Electron micrograph and (b) illustration of a chloroplast.

The function of **chloroplasts** is to make food for plant cells during the process of **photosynthesis**.

Similar to mitochondria, chloroplasts also contain a double membrane, although the inner membrane is not folded, and they also contain a loop of DNA.

Chloroplasts are the organelles responsible for photosynthesis; they contain the pigment chlorophyll.

Inside, there is a liquid called the **stroma** and stacks of green pancake-like structures called **thylakoids**; a stack of thylakoids is called a **granum** (plural: grana).

Chloroplasts are green in colour due to the green chemical pigment **chlorophyll** within the grana. Most plant cells don't contain chloroplasts; the cells found in the leaves and other green parts tend to have the greater concentration of chloroplasts. Therefore, the leaves are the part of the plant responsible for most photosynthesis. We will explore photosynthesis in more detail in Chapter 22.

RESEARCH ACTIVITY

Research endosymbiosis and the origins of double-membrane organelles like mitochondria and chloroplasts in eukaryotic cells.

Vacuoles

(a) (b)

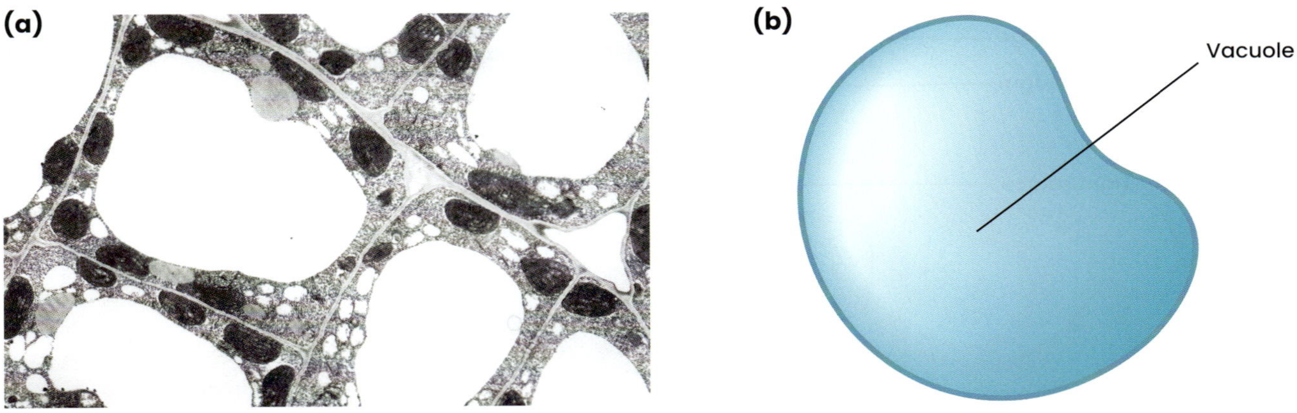

Figure 16.16: (a) Electron micrograph and (b) illustration of a vacuole.

Vacuoles are filled with a fluid called **cell sap**. The sap contains water, dissolved sugars and salts and waste materials. The vacuoles found in animal cells tend to be small and are not permanent while plant cells contain large permanent vacuoles.

Vacuoles give the plant cell strength also, pushing against the cell wall from the inside. The pressure exerted on the wall by the vacuole is called **turgor**.

Turgor is the internal pressure exerted by the large vacuole on the cell walls.

PORTFOLIO LINK

Go to your Learning Portfolio and complete Learning Check 3 to assess your knowledge and understanding of the topic so far.

Go to YouTube.com and search for 'Plant Cells' (10:27) by CrashCourse.

Section 4: The Light Microscope

At the end of this section, you should be able to:

- Label the parts of the light microscope, outlining the functions of each part.
- Describe the preparation of plant and animal cells for viewing using a light microscope.

Unpacked from LOs U2.3, 1.3.3

The light microscope can be used to see the basic structure of plant and animal cells. It has a number of different parts, shown below.

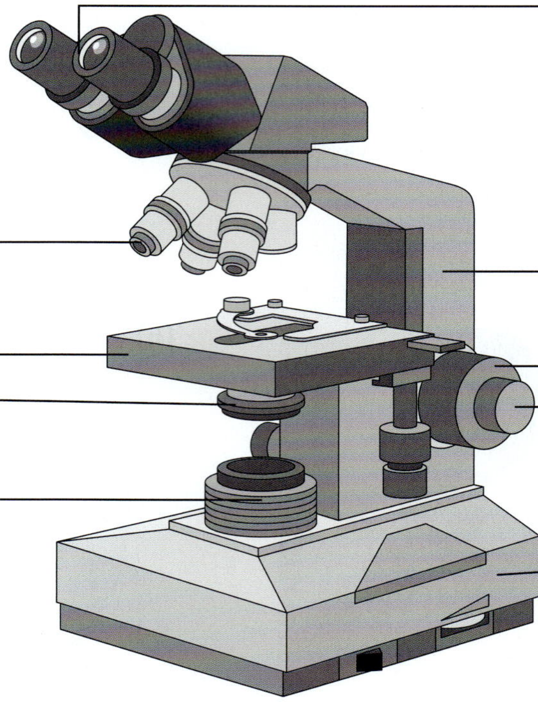

Figure 16.17: The light microscope.

PART	FUNCTION
Eyepiece lens	Magnifies the specimen The specimen is viewed through here
Objective lenses	Magnifies the specimen: low (shortest), medium and high power (longest)
Stage	Holds the slide
Coarse focus wheel	Brings the specimen into rough focus
Fine focus wheel	Brings the specimen into sharp focus
Light source	Provides light to view the slide
Iris diaphragm	Controls the amount of light shining onto the slide
Slide	Holds the specimen
Coverslip	Prevents the cells from drying out Holds the specimen in place Protects the lens from the stain

Figure 16.18: Parts of the light microscope and their functions.

Specified Investigation 6

Investigate the structures and organelles of animal and plant cells and relate them to their functions. **(LO 1.3.3)**

Scan the QR code to watch a video demonstration of this investigation.

A. ANIMAL CELLS

Equipment: Light microscope, cotton bud, dropper, methylene blue stain, pointing needle, tissue paper, glass slide, coverslip.

METHOD

1. Place a drop of water on a slide.

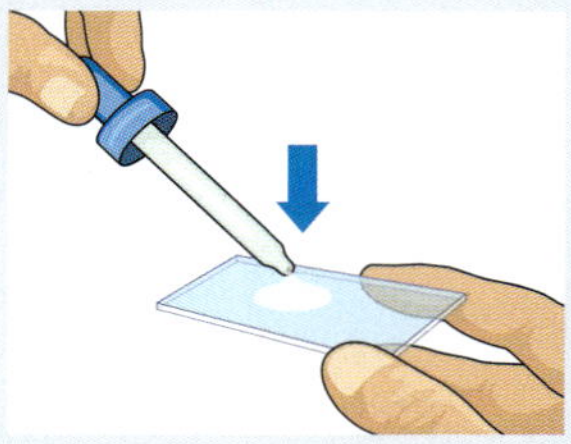

2. Using a gloved finger or a cotton bud, gently scrape some cells from the cheek.

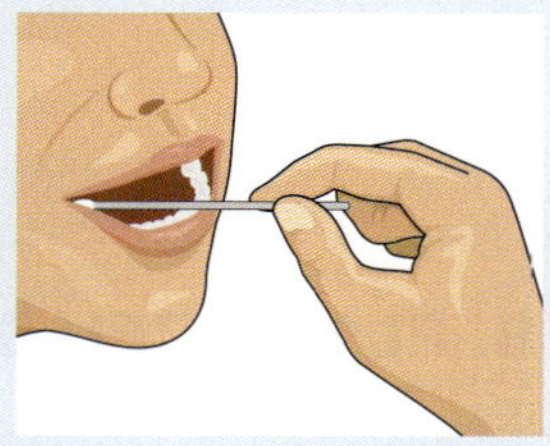

3. Smear the cells on the drop of water on the slide.

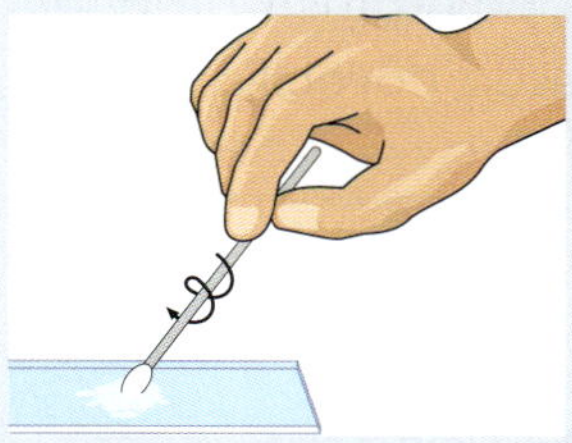

4. Using a dropper, add 1–2 drops of methylene blue to stain the cells.

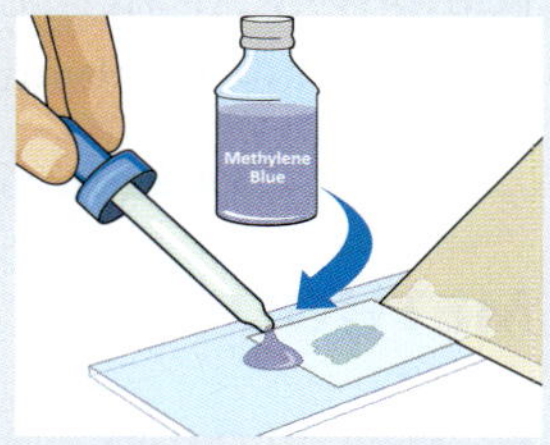

5. Allow dye to dry and then blot off any excess dye.

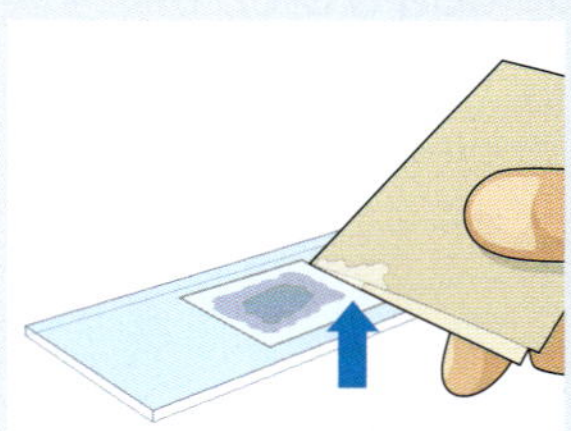

6. Add a coverslip at a 45° angle and lower slowly using mounted needle. This prevents trapping air bubbles on the slide.

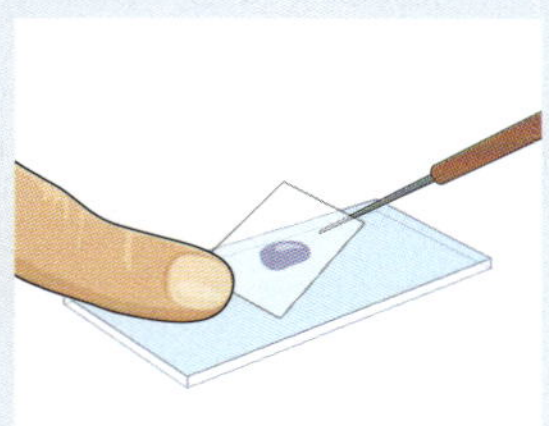

7. Alternatively, apply the stain by placing a drop of methylene blue on one side of the coverslip and using a piece of tissue or filter paper on the other side of the coverslip to draw the stain across the cells.

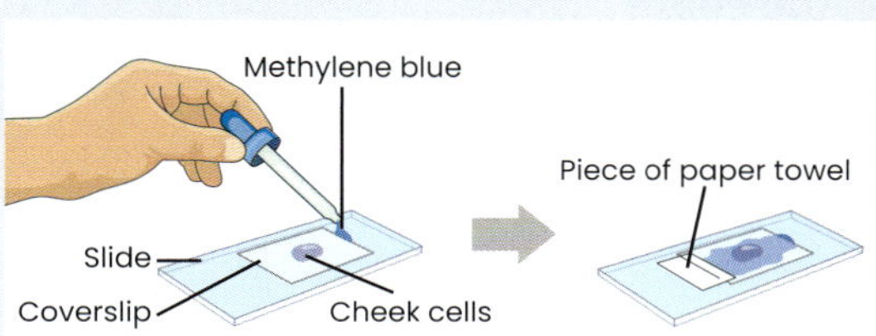

8. Place the slide on the stage of the microscope and observe the cells at low power first, then increase the magnification.

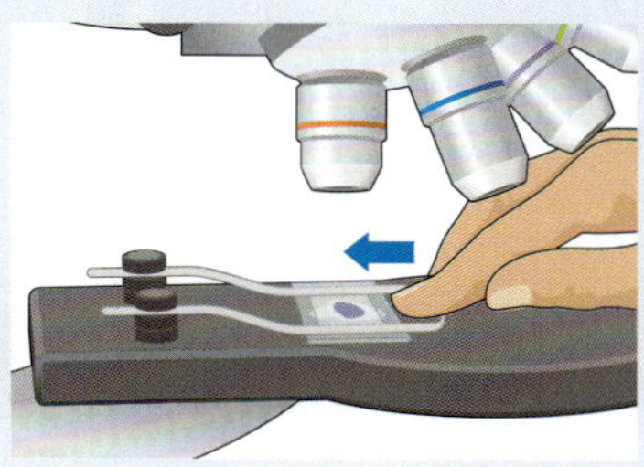

RESULTS

Due to the methylene blue stain, the nucleus appears darker than cytoplasm and can be seen clearly.

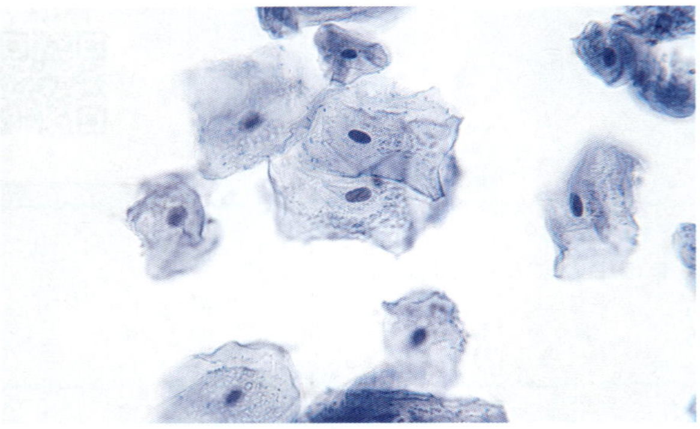

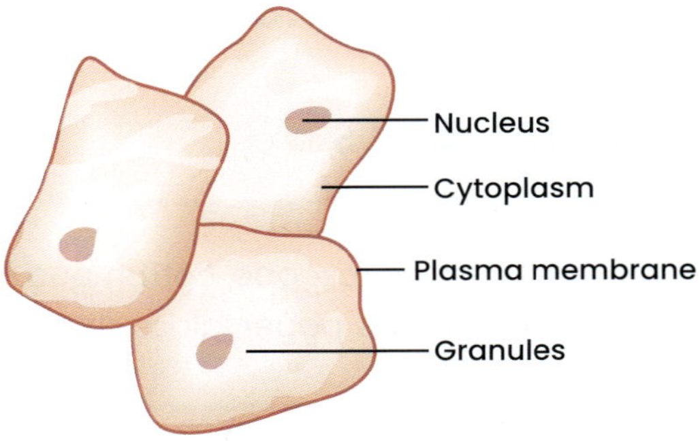

B. PLANT CELLS

Equipment: Light microscope, onion, knife or scalpel, paintbrush, dropper, iodine solution, pointing needle, tissue paper, glass slide, coverslip.

METHOD

1. Peel off a thin layer (one cell thick) of onion tissue from the onion using a forceps or tweezers. Place in a dish of water.

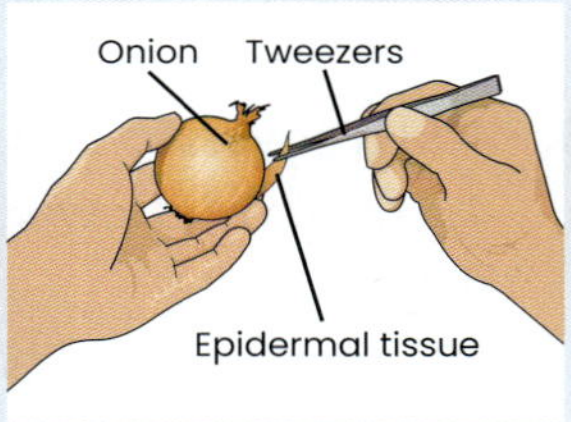

2. Using a dropper, add a drop of water onto a clean slide.

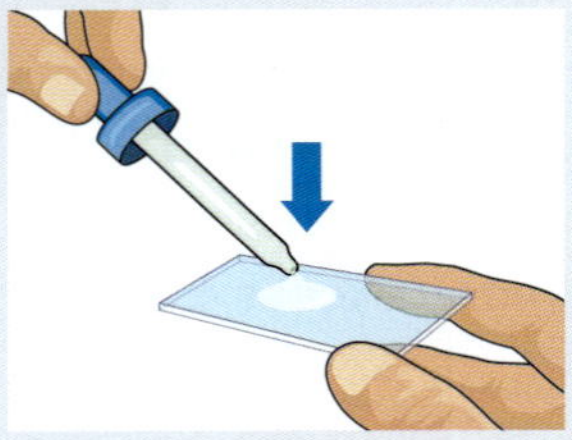

3. Using a paintbrush, place the onion tissue on the drop of water and spread it out.

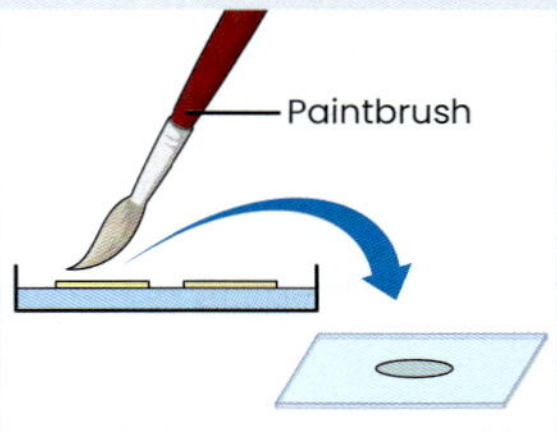

4. Using a dropper, add 2–3 drops of iodine solution to stain the cells.

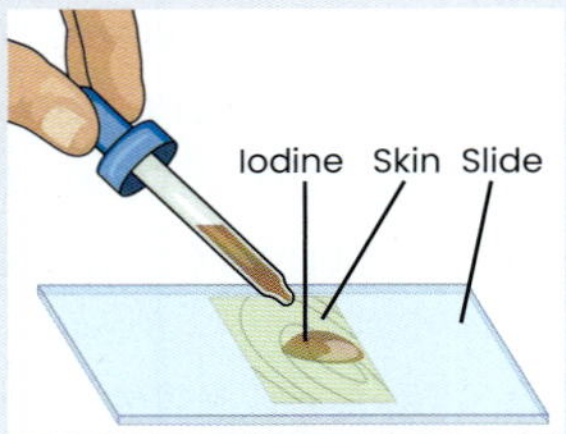

5. Add a coverslip at a 45° angle and lower slowly using mounted needle to avoid trapping air bubbles.

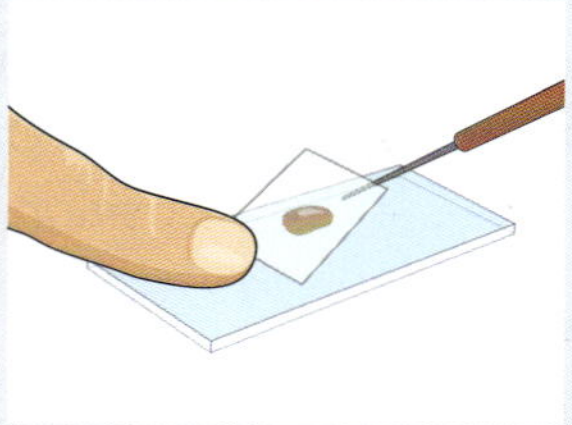

6. Blot off any excess iodine solution.

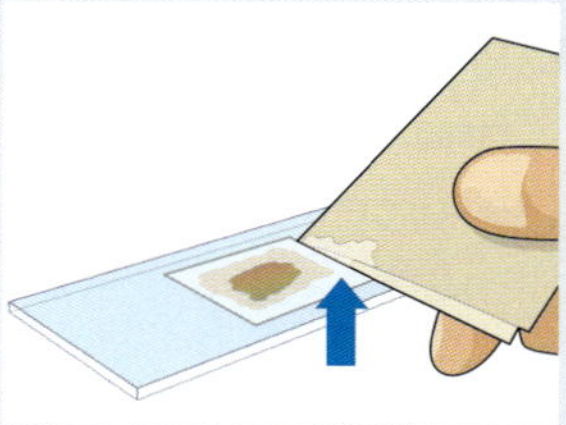

7. Alternatively apply the stain by placing a drop of iodine solution on one side of the coverslip and using a piece of tissue or filter paper on the other side of the coverslip to draw the stain across the tissue.

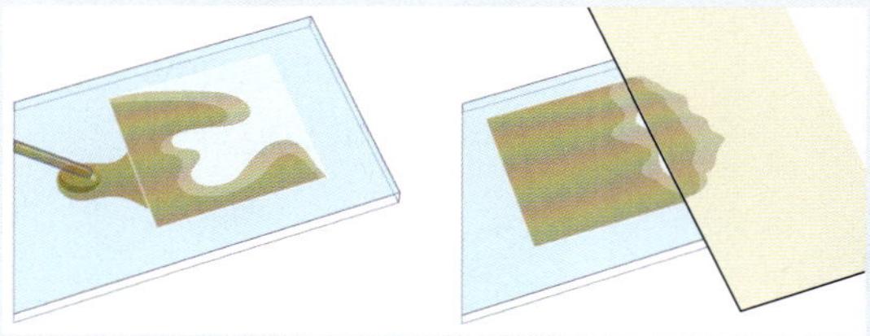

8. Place the slide on the stage of the microscope and observe the cells using the method described in the **Focus on** feature on the next page..

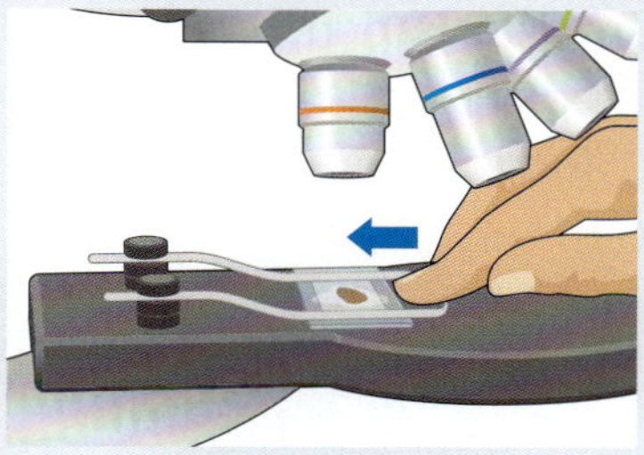

RESULTS

Due to the iodine stain, the nucleus and cell wall appear darker than the cytoplasm and can be seen clearly.

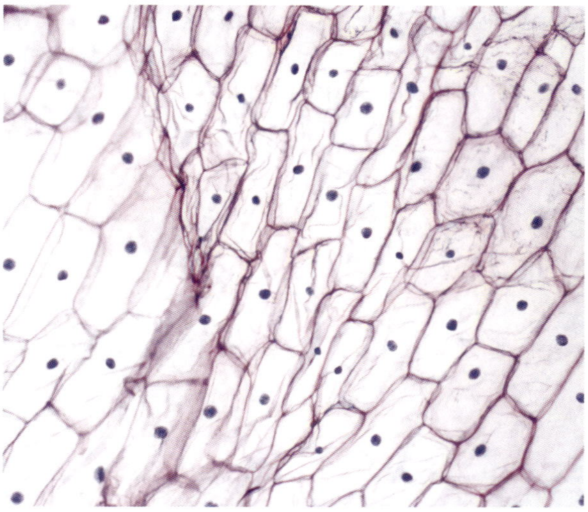

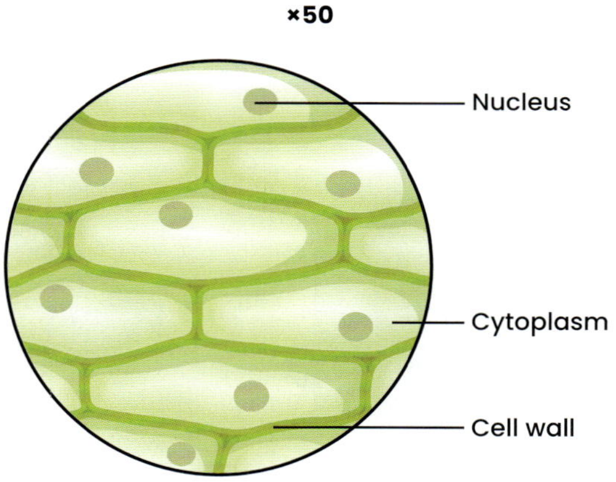

FOCUS ON ... VIEWING PLANT AND ANIMAL CELLS UNDER THE LIGHT MICROSCOPE

1. Plug in the microscope and turn on the light source.

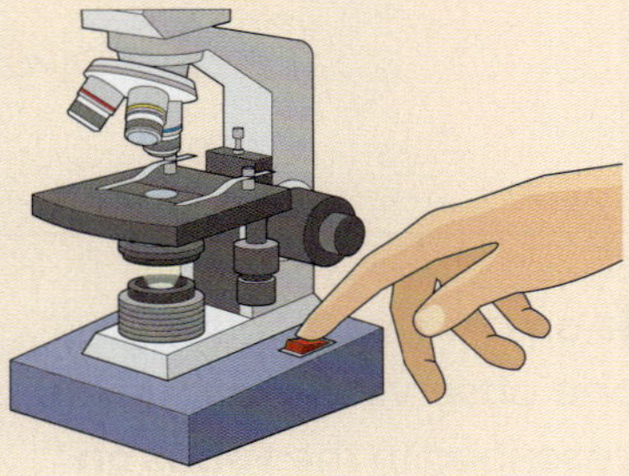

2. Make sure the low-power objective lens is in place.

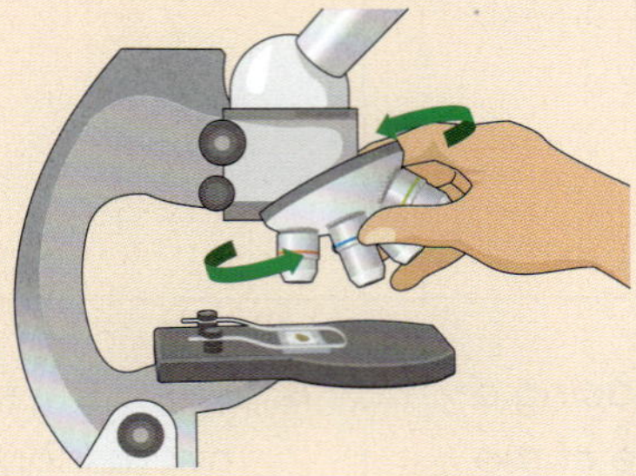

3. Place the slide on the stage. Clip the slide into position.

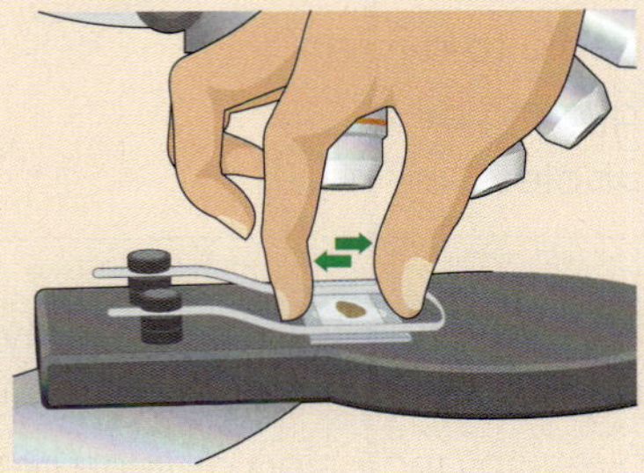

4. Use the coarse focus wheel to bring the specimen into rough focus.

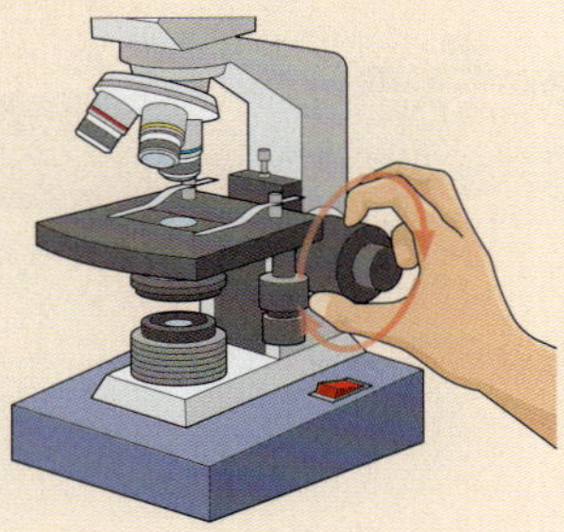

5. Use the fine focus wheel to bring the specimen into sharp focus.

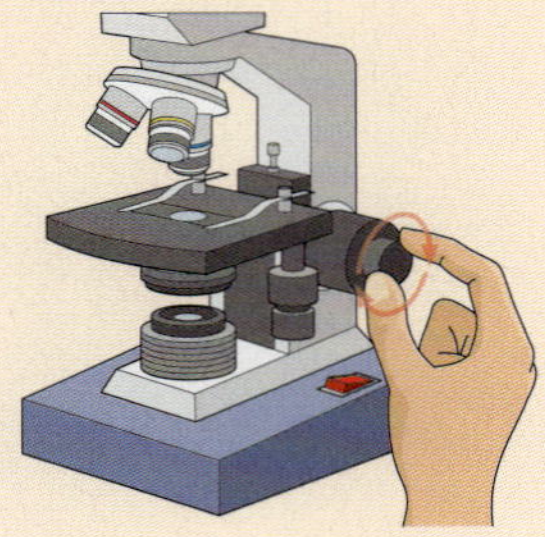

6. Use the iris diaphragm to adjust the amount of light on the specimen so it can be seen most clearly. Sketch what you see.

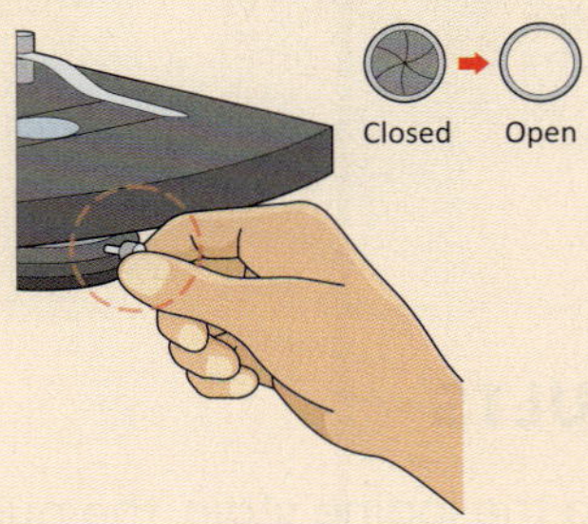

7. Rotate to select the medium-power objective lens in place. Sketch what you see.

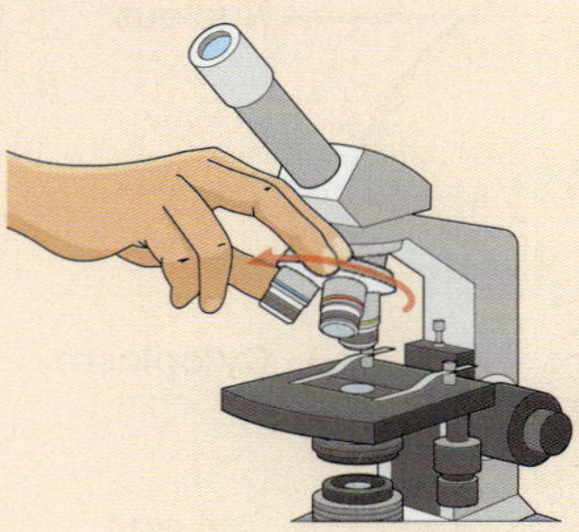

8. Use the fine focus wheel only to bring the specimen into focus.

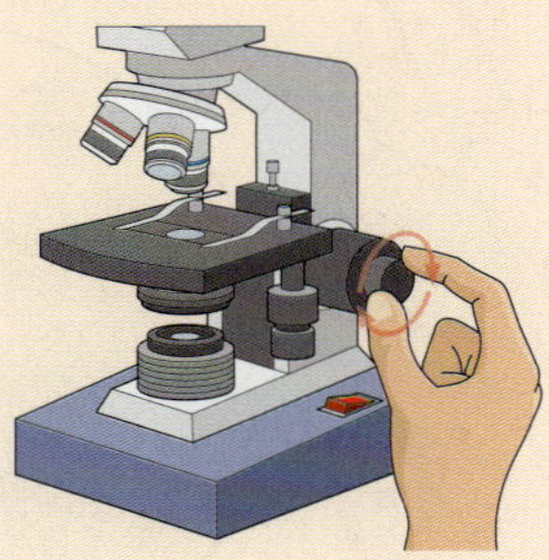

9. Rotate to select the high-power objective lens, if available. Again, sketch what you see.

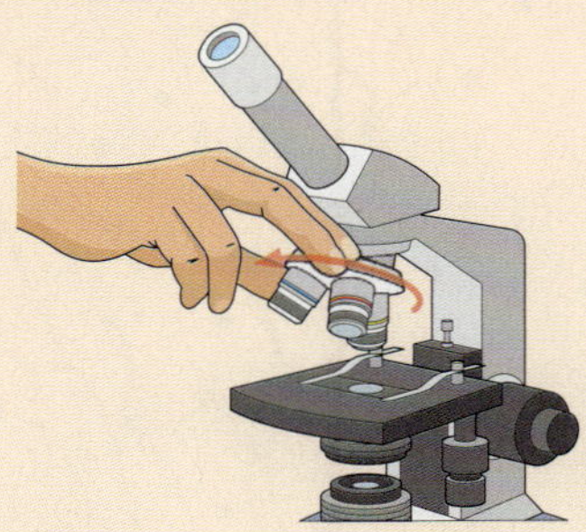

Calculating magnification

If the magnification of the eyepiece lens is 10× and the magnification of the objective lens is 4×, the total magnification is 4 × 10 = 40×. This means the object will appear 40 times bigger.

PORTFOLIO LINK **Go to your Learning Portfolio and complete Learning Check 4 to assess your knowledge and understanding of the topic so far.**

INVESTIGATIVE LOG LINK **Go to your Investigative Log to record your lab work and answer questions on the experiments in this chapter.**

Section 5: Cell Diversity and Organisation

At the end of this section, you should be able to:

- Describe the complexity of multicellular organisms.
- Outline the organisation of cells into tissues, organs and organ systems.
- Outline the function of stem cells in the body and discuss the potential use of stem cells for the treatment of disease.
- Describe the different types of stem cells.

Unpacked from LOs 1.3.1, 3.3.7

Bacteria and protists (like amoeba) are **unicellular** organisms and their whole lives are spent as a single cell. The majority of species on Earth are **multicellular** and consist of many different types of cells. The greater the number of different cells, the greater the level of complexity.

One of the first multicellular animals was the sponge, which contains just a few different types of cells while humans, being rather complex, have over 200 types of cells.

Cells are **specialised** for particular functions within an organism, each adapted perfectly for their role. For example, in humans, we have cells specialised for carrying oxygen (red blood cells), for making insulin to control our sugar levels (beta cells in the pancreas), to fight infection (white blood cells) or to transmit messages (nerve cells). In plants, there are specialised cells for photosynthesis (called parenchyma) and water transport (xylem).

Similar cells co-operate forming larger structures known as **tissues**. A tissue is a group of similar cells carrying out a specific function. Tissues also co-operate forming **organs**. Organs are groups of tissues designed to carry out a particular function. Examples of organs in animals include the heart, lungs, liver or skin. In plants, the leaves, roots or stem are considered organs as they contain different types of tissue working together.

A group of organs can also work together creating an **organ system**. You will be familiar with many of the organ systems in humans including the circulatory system, the nervous system or the endocrine (hormone) system. We will learn more about these in Chapter 27.

Go to YouTube.com and search for 'GCSE Biology – Levels of Organisation – Cells, Tissues, Organs and Organ Systems' (4:24) by Cognito.

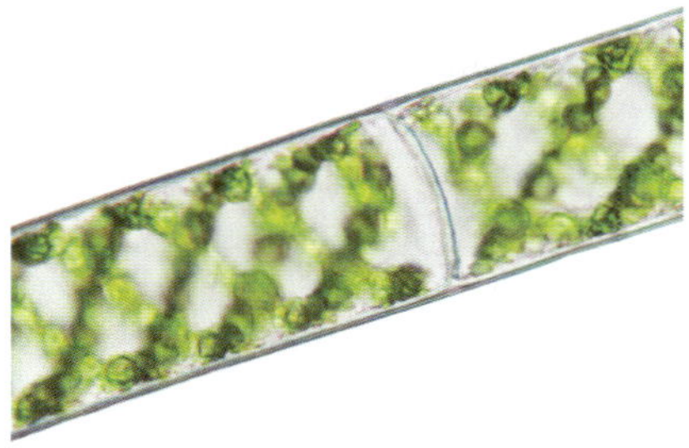

Figure 16.19: Spirogyra is a simple filamentous multicellular organism but with only one type of cell.

Figure 16.20: A sponge is a simple animal with only a small number of specialised cells.

- **Specialised** cells have altered structures to match their specific function.
- A **tissue** is a group of similar specialised cells, carrying out a specific function.
- An **organ** contains a group of different tissues working together to perform a specific function.
- An **organ system** consists of multiple organs co-operating to carry out a specific function.

STEM CELLS

Not all cells in our bodies are specialised. **Stem cells** are unspecialised cells in our bodies which, under the right conditions, have the potential to transform into any type of cell. Biologists are very excited by the potential of stem cells as they could provide potential cures for many human diseases. Plants also contain stem cells, within the **meristem** of their roots, stems and leaves.

Stem cells are unspecialised cells in the body.

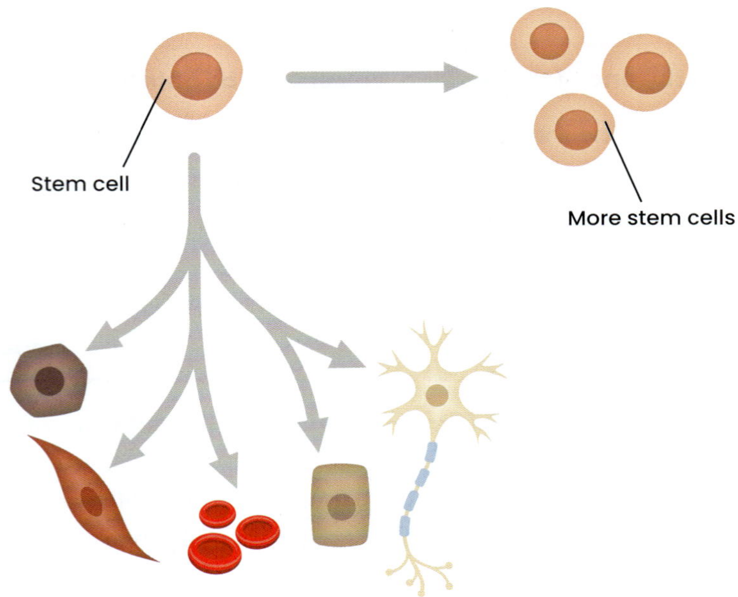

Figure 16.21: Stem cells have the ability to turn into various types of cell in the body.

FOCUS ON ... TYPES OF STEM CELLS

Stem cells are unique cells with the remarkable ability to develop into various specialised cell types. There are a number of different types, categorised based on their potential for specialisation.

Pluripotent stem cells

These cells are capable of giving rise to all the cell types that make up a human body. Embryonic stem cells, found in early animal embryos, are an example of pluripotent stem cells. They hold immense promise for the treatment of various diseases.

Multipotent stem cells

These cells can develop into a limited number of cell types within a specific tissue or organ. Haematopoietic stem cells are an example of multipotent stem cells; they can only produce blood cells (red blood cells, white blood cells, etc.). Mesenchymal stem cells are another example and they can become bone, cartilage or fat cells. They are used in therapies like bone marrow transplants and are being explored for treating various tissue injuries.

Cloning stem cells

Cloning stem cells do not naturally occur in the body. Instead, they are artificial copies of cells. Cloning stem cells are produced using a process called somatic cell nuclear transfer (SCNT), where the nucleus of a somatic cell (a typical body cell) is transferred into an egg cell whose own nucleus has been removed. This creates an embryo that is a genetic clone of the donor.

Induced pluripotent stem cells

These cells begin as a somatic cell (e.g. muscle or blood cell) but are converted into pluripotent stem cells by scientists. They can then be turned into any kind of cell, giving them the same properties as embryonic stem cells.

Therapeutic use of stem cells

No matter the type of stem cells, there are multiple therapeutic uses for stem cells in medicine. The process usually involves obtaining stem cells and differentiating them into specific cell types, and then transplanting them into a patient to repair damaged tissue. Pluripotent and induced pluripotent stem cells have the greatest potential for therapy, as they can be turned into any kind of body cell. They can be used to treat a wide range of diseases, including Parkinson's disease and diabetes, as well as spinal cord injuries.

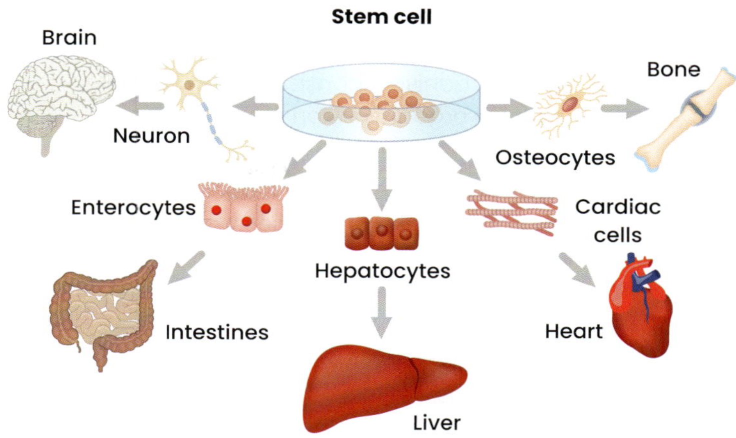

Figure 16.22: Stem cell therapy has lots of potential in medicine.

Ethical considerations

While stem cells may provide opportunities for treating disease, there are ethical issues to discuss. In particular, the use of embryonic (pluripotent) stem cells for research has been controversial. This used to require the fertilisation of eggs in vitro (outside of the body), but now most embryonic stem cells are obtained from abortion clinics. The use of cloning stem cells has already led to clones of sheep, horses and other animals being produced, and leaves open the possibility of human cloning in the future.

DISCUSSION

Have a debate in your class about the therapeutic use of stem cells.

SCIENCE IN SOCIETY: TECHNOLOGY

Stem cells used to produce egg cells

In 2012, a team of scientists in Japan reported the successful production of egg cells from stem cells created from a male adult mouse.

In this experiment, the scientists used blood cells to first create **induced pluripotent stem cells**. They removed the Y chromosome found in the male cells, duplicated the X chromosome and then incubated the cells in artificial ovaries. After four months, the stem cells had developed into egg cells. The eggs were extracted, fertilised by sperm from another male mouse and implanted into a surrogate mouse mother. The fertilised eggs successfully formed baby mice, which themselves were able to reproduce.

In 2023, the same group of scientists reported that human egg cells had been produced using the same method, potentially allowing for those with fertility problems, same sex couples, or others unable to conceive, to create eggs from their own cells. These cells could then be fertilised by sperm cells in vitro, forming an embryo.

Go to YouTube.com and search for 'What are stem cells? – Craig A. Kohn' (4:10) by TED-Ed.

PORTFOLIO LINK

Go to your Learning Portfolio and complete Learning Check 5 to assess your knowledge and understanding of the topic so far.

Chapter Summary

- All living things contain cells.
- **Unicellular** organisms consist of one cell while **multicellular** organisms consist of many specialised cells.
- **Prokaryotic** cells do not have a nucleus or membrane-bound organelles while **eukaryotic** cells have a nucleus and membrane-enclosed organelles.
- **Animal cells** contain a cell membrane, nucleus, mitochondria, cytosol, ribosomes and other organelles.
- **Plant cells** contain the same organelles and structures as an animal cell but also a cell wall, chloroplasts and large permanent vacuoles.
- The **cytoplasm** describes the contents of the cell, excluding the nucleus.
- The **cytosol** is the liquid part of the cytoplasm.
- A **light microscope** uses lights and lenses to magnify specimens and can have a magnification of up to 2,000 times.
- An **electron microscope** uses a beam of electrons to create an image of the specimen and can have magnifications up to 10,000,000 times.
- Electron microscopes allow the study of **cell ultrastructure**.
- An **electron micrograph** is a photograph or digital image taken through an electron microscope.
- The **cell membrane** is composed of a phospholipid bilayer, with embedded proteins.
- The cell membrane is both **semi-permeable**, allowing small molecules pass but not large ones, and **selectively permeable**, actively controlling what enters and exits the cell.
- The **nucleus** is the control centre of the cell; it contains most of the cell's DNA. The nucleus is surrounded by a **nuclear membrane** containing **nuclear pores**.
- DNA is found within structures called **chromosomes** in the nucleus; chromosomes are composed of DNA and proteins.
- When the cell is not dividing, the chromosomes are unwound and are called **chromatin**.
- **Genes** are sections of a chromosome which carry instructions for making protein.
- **Mitochondria** are the organelles responsible for cellular respiration – the release of energy from food.
- **Ribosomes** are the organelles responsible for building protein.
- **Endoplasmic reticulum** (ER) is a membrane-bound organelle responsible for lipid production. HL
- There are two types of endoplasmic reticulum – smooth ER and rough ER, which is covered in ribosomes. HL
- The **Golgi apparatus** is the organelle responsible for lipid and protein packaging and transport in both plant and animal cells. HL
- Plants and fungi contain **cell walls**; they give the plant cell **strength** and **support**.
- **Chloroplasts** are the organelles responsible for photosynthesis.
- **Vacuoles** store **cell sap**, which contains water, food and waste materials.
- A **tissue** is a group of similar cells working together to carry out a particular function.
- Examples of tissues in animals include epidermis, connective tissue, blood.
- An **organ** is a group of tissues working together to carry out a particular function.
- Examples of organs in humans include the heart, lungs, liver while the root, stem and leaf are examples of plant organs.
- An **organ system** is a group of organs working together to carry out a particular function.
- Examples of organ systems include the circulatory system, the nervous system and the reproductive system.
- **Stem cells** are unspecialised cells capable of becoming any type of cell, under the right conditions.
- There are different kinds of stem cell: **pluripotent**, **multipotent**, **cloning** and **induced pluripotent** stem cells.
- Stem cells have great **therapeutic** potential, but there are ethical considerations to their use in medicine.

END OF CHAPTER ASSESSMENT

REMEMBER

1. Name the type of microscope used in the school laboratory.
2. Name the type of microscope you would use to view the ultrastructure of a cell.
3. Identify the cell type shown below.

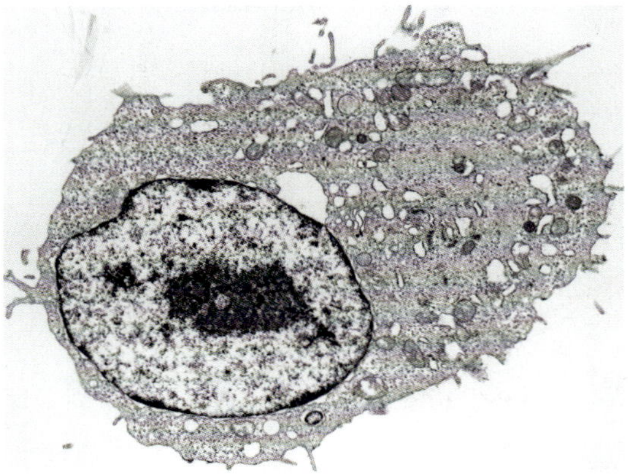

4. Give the function of the cell membrane.
5. Name the materials that chromosomes are composed of.
6. State the function of the following parts of the cell:
   (a) Mitochondria
   (b) Ribosomes
   (c) Cytoplasm
   (d) Cell wall
   (e) Chloroplast
   (f) Vacuole
7. Name the chemical you would use to stain each of the following.
   (a) An animal cell
   (b) A plant cell
8. State the function of the following parts of the light microscope.
   (a) Eyepiece
   (b) Objective lens
   (c) Stage
   (d) Coarse focus wheel
   (e) Fine focus wheel
9. What are stem cells?

UNDERSTAND

10. Explain the difference between chromatin and chromosomes.
11. Describe where in the cell ribosomes may be found.
12. HL Distinguish between smooth endoplasmic reticulum and rough endoplasmic reticulum.
13. When viewing plant cells under the light microscope, describe how you obtained a thin piece of plant tissue.
14. Describe how you prepared the slide of plant tissue to view under the microscope.
15. Explain why a coverslip is used when preparing a slide of cells.
16. When preparing a slide of cells, a coverslip is applied at an angle. Explain why the coverslip is applied in this manner.
17. Distinguish between tissues, organs and organ systems.
18. What is the function of stem cells in the body?

APPLY

19. If the eyepiece lens on a light microscope had a magnification of 10X and an objective lens had a magnification of 40X, calculate the total magnification of the specimen.
20. Illustrate the structure of the cell membrane, labelling each of its parts.
21. Illustrate the structure of a mitochondrion, labelling each of its parts.
22. Compare the structure and role of the endoplasmic reticulum and the Golgi apparatus. HL
23. Using a labelled diagram, demonstrate how you would prepare a slide of animal cells to view under a light microscope.
24. Suggest a reason why someone might get a stem cell transplant (injection of stem cells).

ANALYSE

25. Compare and contrast a light microscope with an electron microscope.
26. Compare and contrast the structure of an animal cell with that of a plant cell.
27. Contrast the structure of a prokaryotic cell with that of a eukaryotic cell, giving examples of each cell type.
28. Classify each of these cells as either prokaryotic or eukaryotic.
   (a) Human
   (b) Plant
   (c) Bacteria
   (d) Fungus
   (e) Amoeba
29. Identify how each of the following cell structures and organelles are adapted to their functions.
   (a) Cell wall
   (b) Cell membrane
   (c) Mitochondrion
   (d) Chloroplast

EVALUATE

30. Place the following structures and cells in order of size, starting with the smallest:

Plant cell   Human egg cell   Virus   Atom   Bacterium

31. Conduct research and investigate why both mitochondria and chloroplasts have their own DNA.

32. Compare the therapeutic potential of pluripotent, multipotent and induced pluripotent stem cells.

33. Decide whether the cell below is a eukaryotic or prokaryotic cell. Give reasons for your answer.

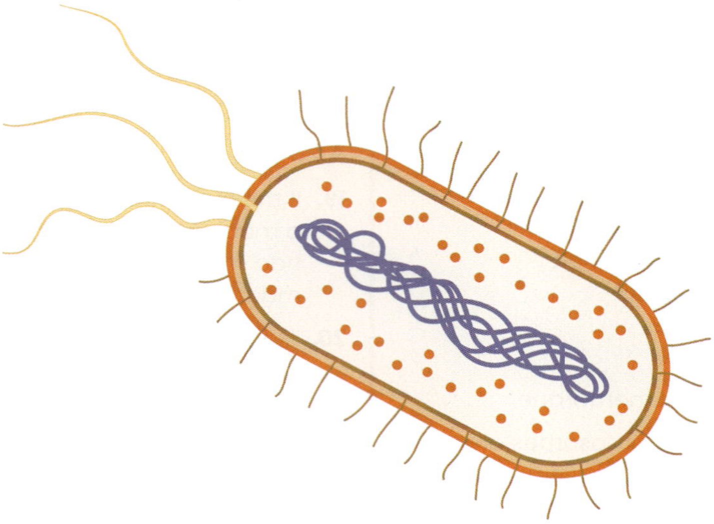

CREATE

34. Design a 3D model of a plant or animal cell, labelling all the parts in the model.

35. Design a new type of specialised cell for the human body. Detail the function and features to allow it carry out this function.

36. Design a poster showing the structure of the light microscope and labelling each of its parts.

17 Cell Continuity

LEARNING OUTCOMES: U1.2, U3.1, U3.2, 2.3.1, 2.3.2, 2.3.3, 2.3.7, 2.3.8

EYE-OPENER

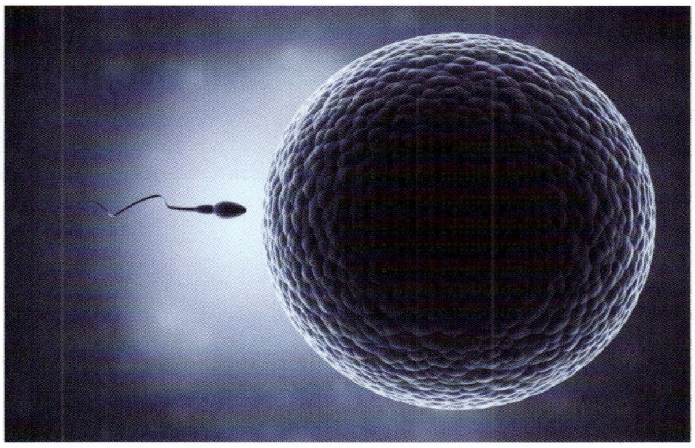

Each one of us began life as a single cell – a fertilised egg – but we now comprise trillions of cells, each with a specific role and function in our bodies. We make millions of new cells every single second as older cells are replaced, yet we rarely notice this at all. Our red blood cells are replaced every four months or so, our skin epidermis every month and our stomach lining every week. Every second, approximately 4 million new cells are produced in each of our bodies.

At the end of this chapter, you should be able to:

- Outline the cell cycle and relate it to DNA replication and mitosis in the cell cycle.
- Describe simply the process and role of mitosis with the aid of labelled diagrams.
- Compare the roles of mitosis and meiosis.
- Explain the factors that can lead to development of cancers and how cancer forms and is treated.

LINKS TO:

Chapter 7	**Bacteria and Fungi**
Chapter 16	**Cell Structure and Diversity**
Chapter 24	**Information Flow in the Cell**
Chapter 25	**Genetics and Inheritance**

SUBJECT VOCABULARY

cell cycle, interphase, mitosis, meiosis, asexual reproduction, binary fission, daughter cells, diploid, haploid, cancer, metastasis, carcinogens

Section 1: Cell Continuity and the Cell Cycle

At the end of this section, you should be able to:

- Outline what is meant by cell continuity.
- Describe the cell cycle.
- Describe the three types of cell division: binary fission, mitosis and meiosis.
- Explain when DNA replication takes place during the life of the cell.
- Describe the role of mitosis in unicellular and multicellular organisms.
- Compare the roles of mitosis and meiosis in living things.

Unpacked from LOs 2.3.1, 2.3.2, 2.3.3

Each new cell produced in a human body comes from an existing cell, and we can follow this chain back to the first single cell produced during fertilisation. This is known as **cell continuity**. It's also important to remember that, since all cells come from other cells, all living things derive from other living things. This means we can trace back every living thing on the planet to the first cell – our last universal common ancestor (LUCA) – which was probably formed over 4 billion years ago.

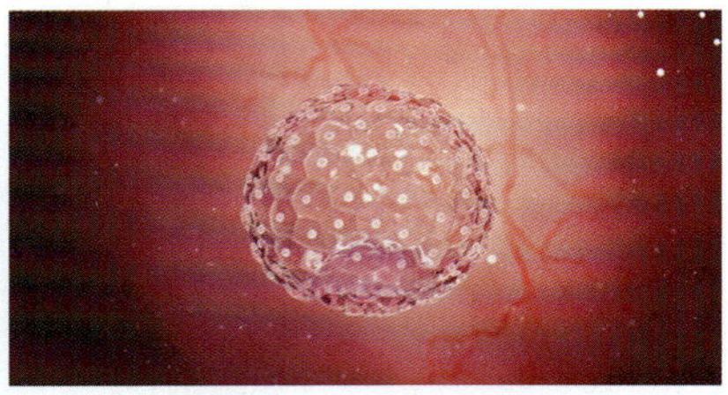

Figure 17.1: A blastocyst is a ball of cells formed from the division of a fertilised egg.

Cells don't live forever – in fact, millions die in your body every second – but new cells are made to replace those that died. When the rate of cell production is higher than the rate at which they die, the organism is said to be growing.

If new cells are being made, the organelles within the cell, including the DNA in the nucleus, need to be copied before the process of cell division takes place. In this chapter we will study the life of a typical cell, examine what occurs within the cell before, during and after it divides and find out what happens if things go wrong.

THE CELL CYCLE

As we discovered in Chapter 16, there are over 200 types of specialised cells in the human body. Each cell is adapted to its specific role and how long they live depends on their structure and function. Red blood cells, for example, live for around 120 days, skin cells live for 30 days and some white blood cells only live for a couple of hours. However long the cells live for, they all go through the same series of stages. These stages are collectively known as the **cell cycle**.

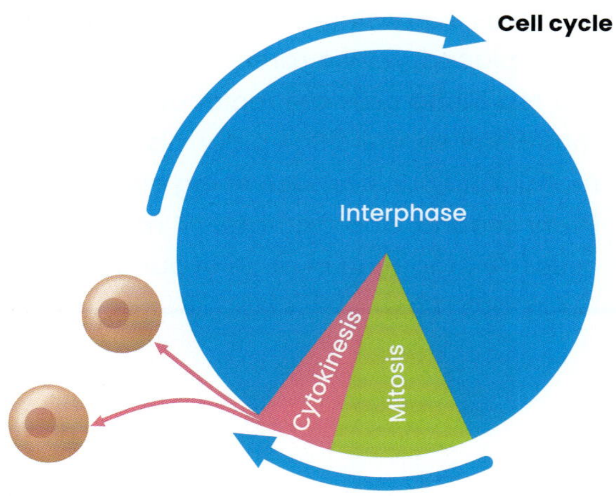

Figure 17.2: The cell cycle.

The cell cycle can be broken down into a number of stages:

- Interphase: a period of growth and preparation for the next division. DNA is copied during this interphase.
- Mitosis: the division of the nucleus.
- Cytokinesis: the division of the cell into two cells.

The **cell cycle** refers to the changes that occur over the life of a cell. It involves growth, DNA synthesis and division.

INTERPHASE

Interphase is the time in a cell's life when it isn't dividing. Instead it is carrying out its function. That will, of course, depend on the type of cell it is. If it's a muscle cell then it will be contracting and contributing to joint movement. If it's a white blood cell, then it will be fighting infection. If it's a neuron, it will be creating electrical impulses and sending messages around the body. Or if it's a pancreas cell it may be producing the hormone insulin, which regulates sugar levels in the blood.

However, during interphase the cell is also preparing for its next division. This involves:

- the cell growing, creating the space for the new cell to form.
- replicating the cell's DNA (which are in the form of **chromatin** in the cell).
- copying the cell organelles.

After interphase, the cell may undergo **mitosis**, creating an exact copy of its nucleus and organelles before splitting to form two identical cells. Other cells may undergo a different type of cell division called **meiosis**.

HL

Many of our cells complete the cycle, forming new cells once they have copied their organelles and DNA. However, when some cells grow and mature during interphase, they often differentiate for a specialised function (e.g. muscle cells, nervous tissue or sperm cells) and never divide. Other undifferentiated cells, called **stem cells**, are tasked with replacing these cells after they die.

Interphase is the period of a cell's life when it is not dividing.

Mitosis is the copying and division of the nucleus of a cell.

Meiosis is the type of nuclear division associated with the production of sex cells or gametes.

Go to YouTube.com and search for 'Mitosis and the Cell Cycle: How Cells Clone Themselves' (11:10) by CrashCourse.

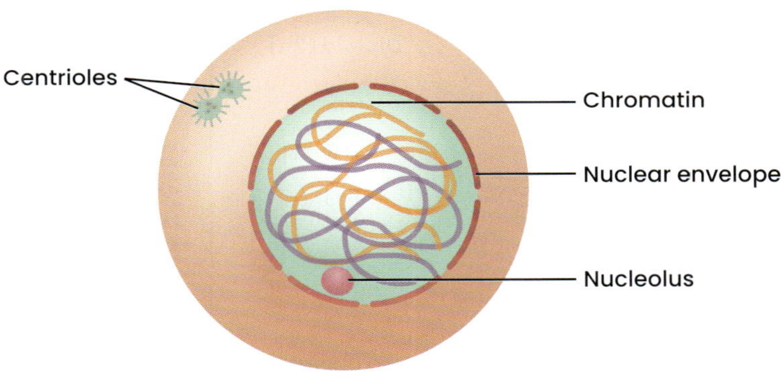

Figure 17.3: A cell during interphase.

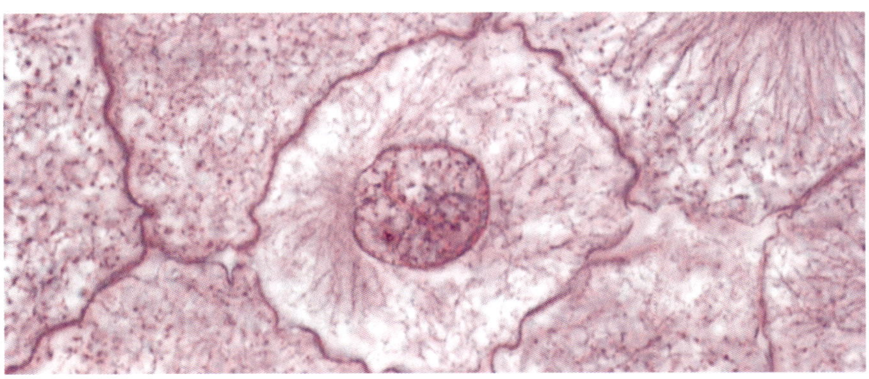

Figure 17.4: A cell during interphase as shown under a light microscope.

TYPES OF CELL DIVISION

There are three types of cell division. These depend on the cell type, the organisms and the function of the cells being produced.

Binary fission

Binary fission refers to the division of bacterial cells and is a form of **asexual reproduction**. In bacteria, the DNA is found within circular chromosomes that are not enclosed within a nuclear membrane. **Binary fission** takes place in a number of stages and bacteria can complete it in less than 20 minutes (a human cell takes around two hours to divide).

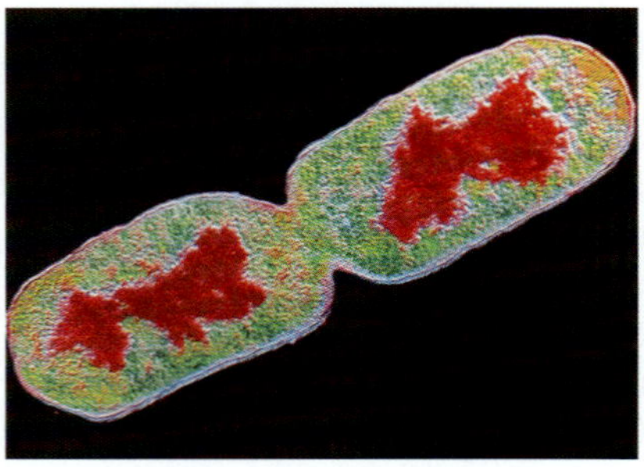

Figure 17.5: Bacteria dividing by binary fission, as seen under a light microscope.

- **Asexual reproduction** results in the formation of genetically identical offspring.
- **Binary fission** is the division of bacterial cells in a form of asexual reproduction.

Mitosis

Mitosis occurs in all **eukaryotic cells** (i.e. cells with membrane-bound nuclei). Mitosis is technically the division of a cell's nucleus and, since bacteria do not have a nucleus, it can only occur in eukaryotic cells (hence why bacterial division is known as binary fission).

Mitosis occurs in distinct stages (prophase, metaphase, anaphase and telophase, which we will explore in more detail later) and results in the production of two **daughter cells**. These daughter cells are **genetically identical**, meaning they have the same chromosomes, DNA and genes.

If a cell was originally **diploid** (two sets of homologous chromosomes) then the resulting daughter cells will also be diploid. If the original cell was haploid (only one set of chromosomes) then the resulting daughter cells would be **haploid**.

- **Daughter cells** are the cells produced from the result of mitosis or meiosis.
- A **diploid** cell contains two sets of chromosomes, e.g. a cheek cell.
- A **haploid** cell contains just one set of chromosomes, e.g. a sperm cell.

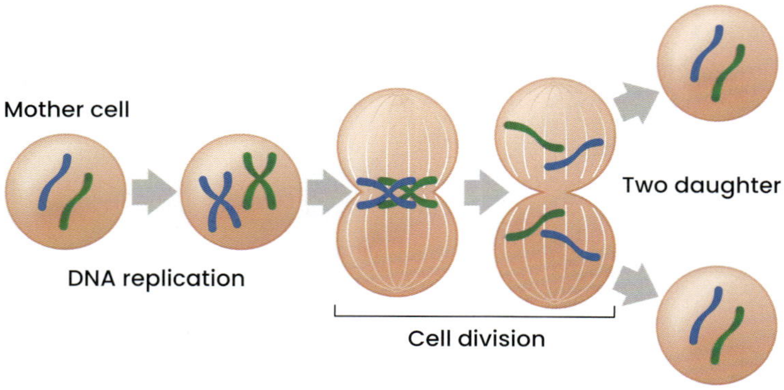

Figure 17.6: The stages of mitosis.

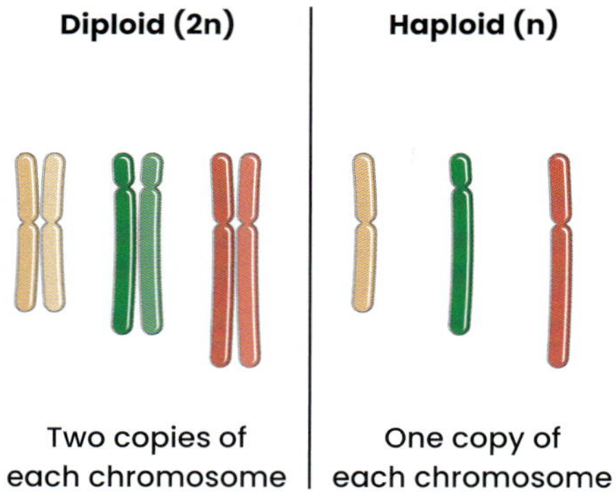

Figure 17.7: Haploid and diploid cells.

Mitosis has two distinct roles:

1. In single-celled organisms, like amoeba, mitosis is involved in **reproduction**.
2. In multicellular organisms, mitosis is responsible for **growth** (of muscle, bone and other types of tissue), **repair** of damaged tissues and organs and the **replacement** of dying cells.

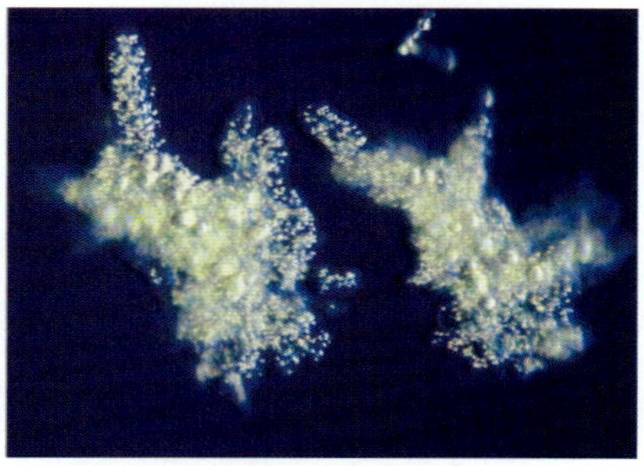

Figure 17.8: Amoeba just after dividing.

Meiosis

Meiosis is a special type of cell division which is responsible only for the production of cells called **gametes**. Gametes are sex cells and include sperm and egg cells in animals and pollen and eggs in plants. Meiosis only occurs in diploid cells and results in the formation of four daughter cells. These daughter cells are **haploid** and are **genetically different** from their parent cell.

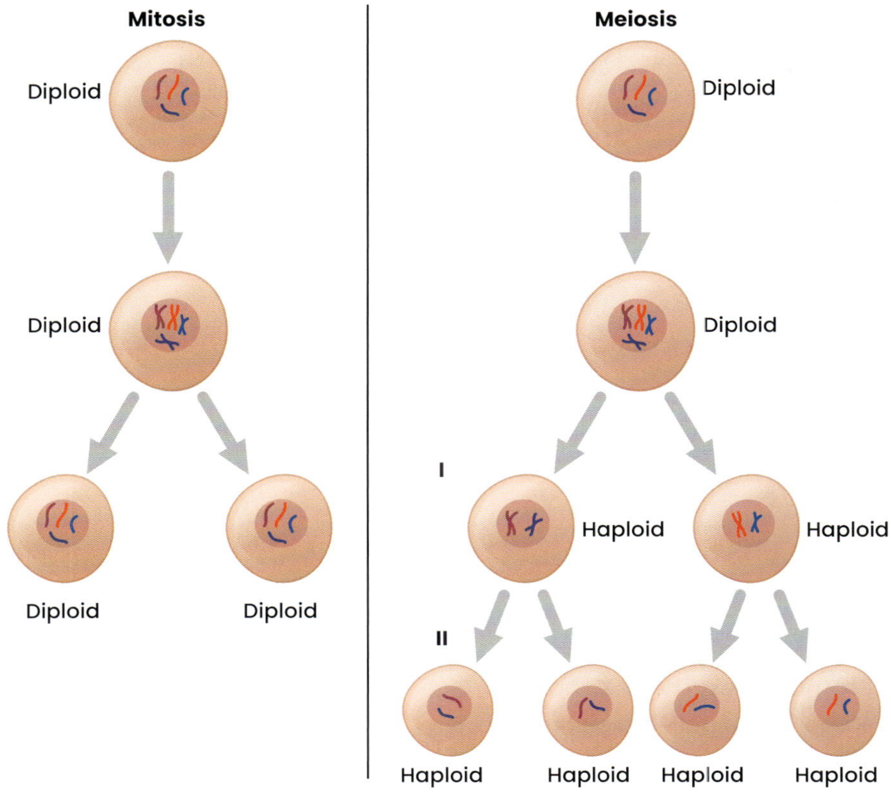

Figure 17.9: Meiosis results in the formation of four genetically different daughter cells.

In humans, only certain types of cells – known as **oogonia** in females and **spermatogonia** in males – undergo meiosis. In females, the first series of stages occurs before birth and meiosis is completed at the onset of puberty and once a month thereafter. In males, meiosis occurs constantly after the onset of puberty, with the average teenage boy producing nearly 1 billion sperm cells a day.

DID YOU KNOW?

Meiosis has several roles in living things:

1. It is responsible for the production of both male and female **gametes** in animals (**sperm** and **eggs**).
2. It is responsible for the production of the **female gamete** (ovum) in plants and of the **pollen grain**.
3. It creates **variation** in the offspring after sexual reproduction. Variation is the driving force behind evolution.
4. It ensures the diploid number is maintained after sexual reproduction.

PORTFOLIO LINK **Go to your Learning Portfolio and complete Learning Check 1 to assess your knowledge and understanding of the topic so far.**

Section 2: Mitosis – A Detailed Study

At the end of this section, you should be able to:

- Outline the changes within cells during the four stages of mitosis, using simple diagrams.
- Distinguish between chromatin, chromosomes, chromatids.
- Discuss the role of centrioles and spindle fibres in mitosis.
- Outline what is meant by cytokinesis and describe how it occurs in plant and animal cells.

Unpacked from LO 2.3.1

THE STAGES OF MITOSIS

Mitosis occurs in a series of distinct stages and can take approximately two hours to complete. Below is an account of the changes that occur within the cell during each of the stages. You should familiarise yourself with the diagrams and practise drawing them regularly.

Stage 1 – Prophase

HL

- The DNA, in the form of chromatin during interphase, condenses by coiling around proteins called histones to form distinct double-stranded X-shaped chromosomes. These X shapes are two copies of the chromosome (each called a **sister chromatid**) stuck together at a point known as a centromere.
- Organelles known as **centrioles** begin to move towards opposite ends of the cell. These organelles produced tiny fibres known as **spindle fibres**.
- The nuclear membrane, which encloses the DNA/chromosomes in the nucleus, begins to break apart.

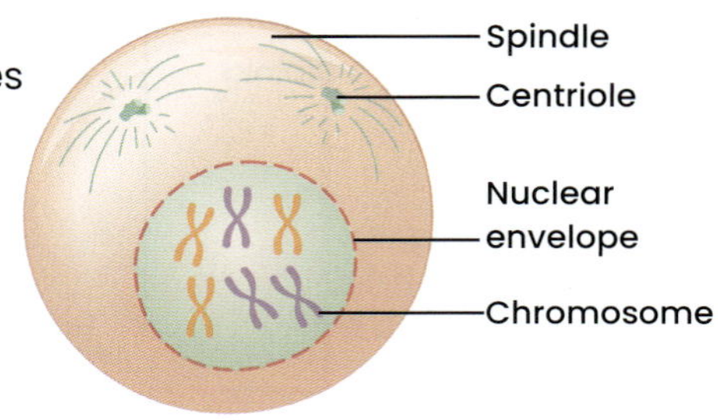

Figure 17.10: Prophase of mitosis.

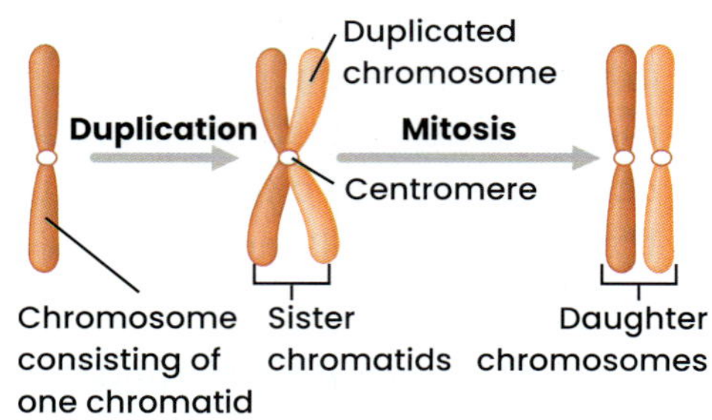

Figure 17.11: Chromosomes and chromatids.

Stage 2 – Metaphase

- Metaphase involves moving the chromosomes around the cell in preparation for their separation.
- The nuclear membrane has now completely disintegrated, and the DNA is suspended in the cytoplasm.
- The spindle fibres attach to each chromosome pair at the centromere, while specialised motor proteins begin to align the chromosomes along the middle (or equator) of the cell.
- Other motor proteins tighten the fibres connecting the centrioles with the centromere of the chromosome pair.

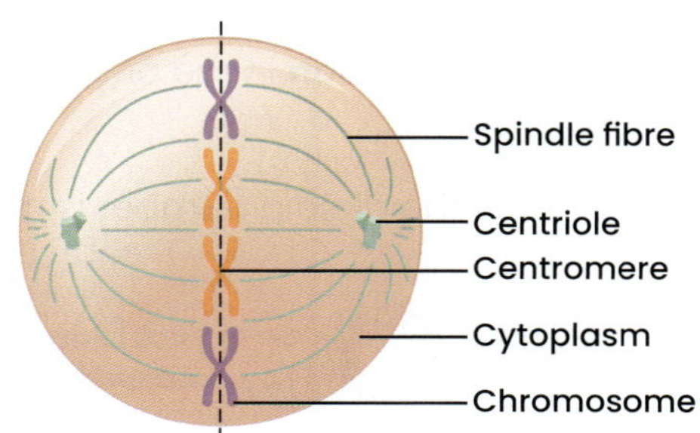

Figure 17.12: Metaphase of mitosis.

Stage 3 – Anaphase

- Anaphase is the shortest stage.
- The spindle fibres are pulled towards the centrioles, causing the pairs of chromatids to separate at the centromere.
- The strands, once again called chromosomes, continue to be pulled towards opposite ends of the cell.

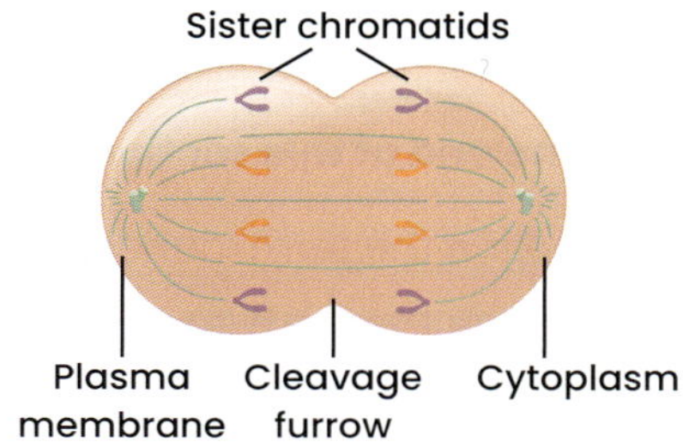

Figure 17.13: Anaphase of mitosis.

Stage 4 – Telophase

- At the beginning of telophase, there are two sets of chromosomes at either end of the cell, which begin to unwind back to the indistinct chromatin.
- New nuclear membranes form around the chromatin, while the cell begins to stretch.
- Mitosis is now complete – the nucleus has divided – but the cell now has two nuclei.
- A **cleavage furrow** begins to form, where the cell will eventually split in two.

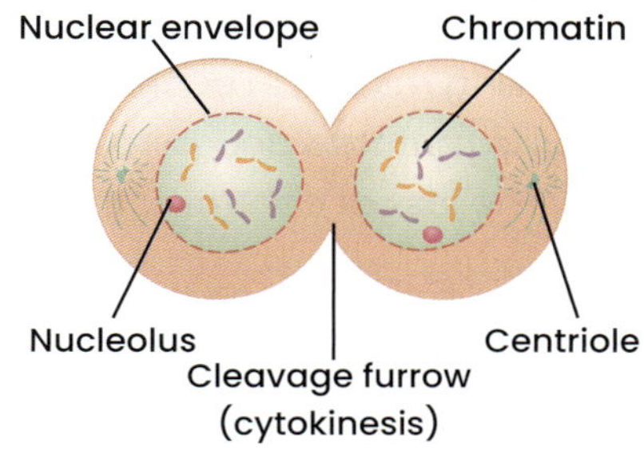

Figure 17.14: Telophase of mitosis.

HL

Stage 5 – Cytokinesis

- Cytokinesis completes cell division and begins when proteins (actin and myosin – the same contractile proteins found in your muscles) form a contractile ring, causing the cell membrane to pinch in.
- The cleavage furrow deepens, until the cells separate, and cell division is complete.
- The process is slightly different in plant cells: a cell plate begins to form during telophase, which builds a new cell membrane and cell wall for the new daughter cells.

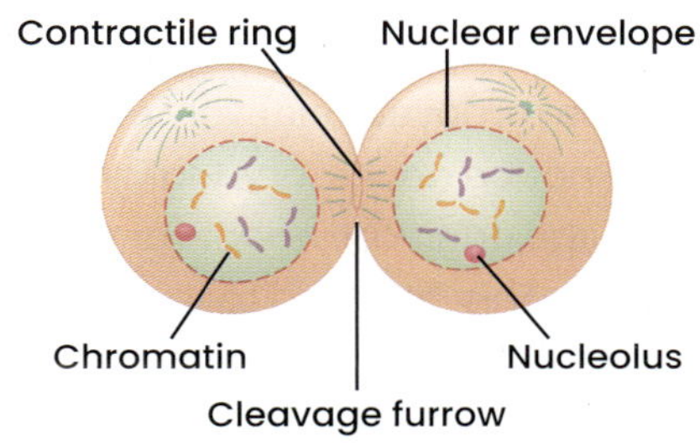

Figure 17.15: Cytokinesis phase of mitosis.

Go to YouTube and search for 'Mitosis – Stages of Mitosis' (3:03) by FuseSchool – Global Education.

PORTFOLIO LINK

Go to your Learning Portfolio and complete Learning Check 2 to assess your knowledge and understanding of the topic so far.

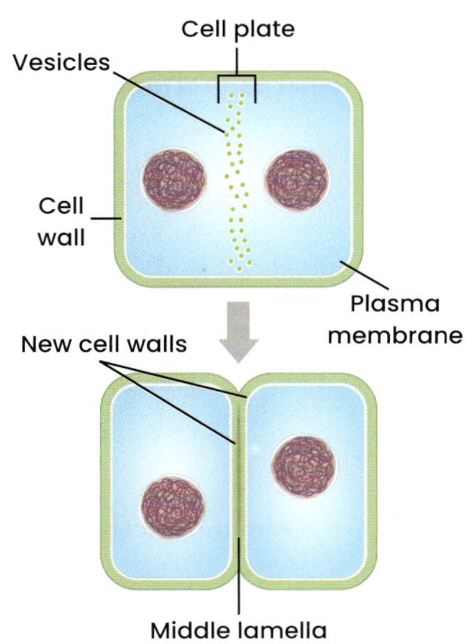

Figure 17.16: Cell plate forms a new cell wall in plant cells.

Section 3: Cancer – Causes and Treatments

At the end of this section, you should be able to:

- Define the term 'cancer' and relate it to cell division.
- Define the term 'tumour'.
- Briefly outline how cancer spreads in the body.
- Discuss the causes of cancer in living things.
- Outline methods of treating cancer in humans.

Unpacked from LOs 2.3.7, 2.3.8

Cell division is a carefully controlled process with over 600 genes controlling mitosis in each cell. However, sometimes these genes can become mutated and mitosis can lose control, producing **abnormal cells**. The loss of control of the rate of mitosis is called **cancer**.

These mutations don't happen very often and the vast majority of cells divide without any issues. When cancer cells do form, our immune system often detects and destroys them – but, unfortunately, not in all cases.

The abnormal cells continue to divide and can produce a mass of tissue called a **tumour**. These tumours can be categorised as either benign or malignant. Benign tumours pose a less serious health risk and the mass of cells stays in the same location. This means the cells can be removed relatively easily by surgery and further treatment may not be necessary. Malignant tumours are more serious and cells from the original tumour (or **primary tumour**) can spread to other tissues. This can lead to the formation of **secondary tumours** in other tissues in the body. The spreading of cancer cells is called **metastasis**.

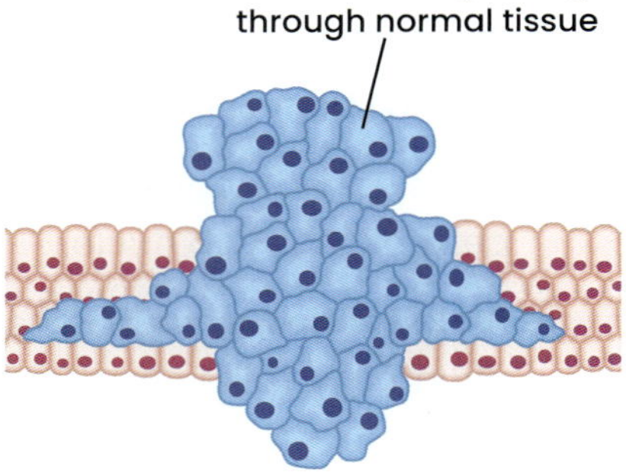

Figure 17.17: Cancer forms abnormal cells which divide uncontrollably.

Cancer is a group of disorders characterised by the loss of control of mitosis.

Metastasis is the spreading of cancer from one part of the body to another.

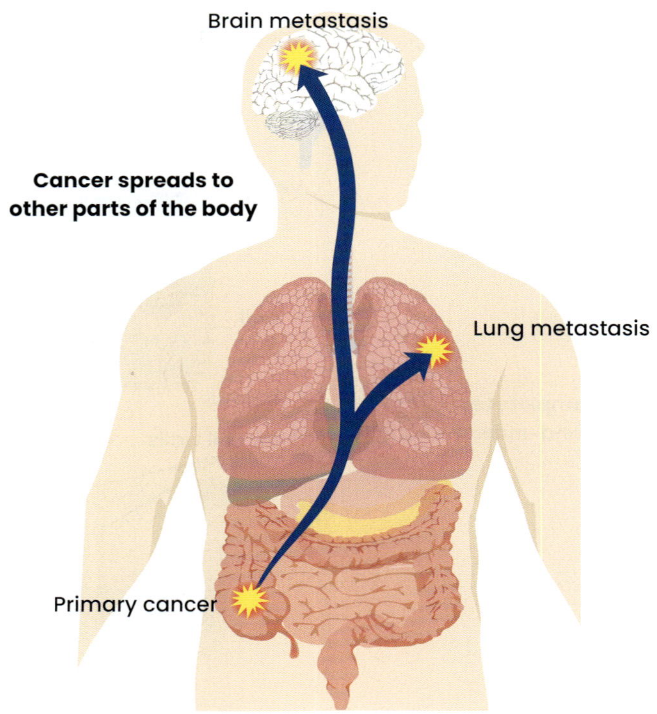

Figure 17.18: Metastasis.

Go to YouTube.com and search for 'How does cancer spread through the body? – Ivan Seah Yu Jun' (4:43) by TED-Ed.

RESEARCH ACTIVITY

Carry out some research on how cancer cells can spread in the body. Share your results with the class.

WHAT CAUSES CANCER?

Cancer, as mentioned previously, is caused by mutations in the genes that control mitosis. The following are factors that can increase the risk of these mutations, and therefore cancer, occurring.

- **Smoking cigarettes** is the most significant preventable cause of cancer worldwide. Cigarette smoke contains over 70 known **carcinogens**.

> A **carcinogen** is a chemical known to cause cancer.

- Being **overweight** or **obese** can increase the risk of developing cancer. A healthy diet, high in fibre, can help reduce the risk. Similarly, regular exercise can help prevent cancer.
- Exposure to harmful **UV radiation** from the sun or sunbeds can increase the risk of developing cancer.
- Other types of **ionising radiation**, like x-rays or nuclear radiation, can contribute to cancer.
- Some viruses, including the **human papilloma virus** (HPV), can cause cancers including cervical cancer in girls and testicular cancer in boys. A vaccine against HPV is available to all boys and girls in first year of secondary school.
- Drinking alcohol excessively can also increase the risk of cancer.
- Air pollution and exposure to high levels of radon gas can lead to an increased risk of developing cancer.
- Some people may **genetically inherit mutated genes** which increase the risk of developing cancer. The BRCA1 (breast cancer gene) is an example of a mutated gene which can increase the risk of breast and ovarian cancer.

Cancer misconceptions

There are many untruths and misconceptions about the causes of cancer. Some suggest Wi-Fi, 5G and the use of mobile phones can increase the risk but these produce **non-ionising** radiation which is unable to cause genetic mutations. There is also **no** convincing evidence to suggest plastic water bottles, cosmetics, toiletries, air travel or stress can cause cancer.

DISCUSSION

Do you or did you have any misconceptions about the causes of cancer?

DETECTING CANCER

One of the most important factors in increased survival rates is early detection and diagnosis. Regularly checking for lumps on breasts or testicles can help detect those common cancers early. Being aware of common cancer symptoms is important; the Irish Cancer Society has a helpful 'Spot cancer early' section on their website that shows common early signs which might indicate something is wrong. There are also three national cancer screening services (BreastCheck, BowelScreen and CervicalCheck) as well as genetic testing that can be carried out to determine if a faulty gene is present.

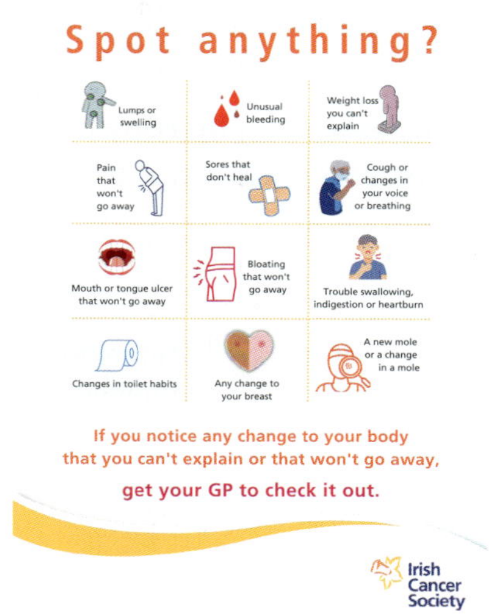

Figure 17.19: A leaflet from the Irish Cancer Society on detecting cancer symptoms early.

HOW IS CANCER TREATED?

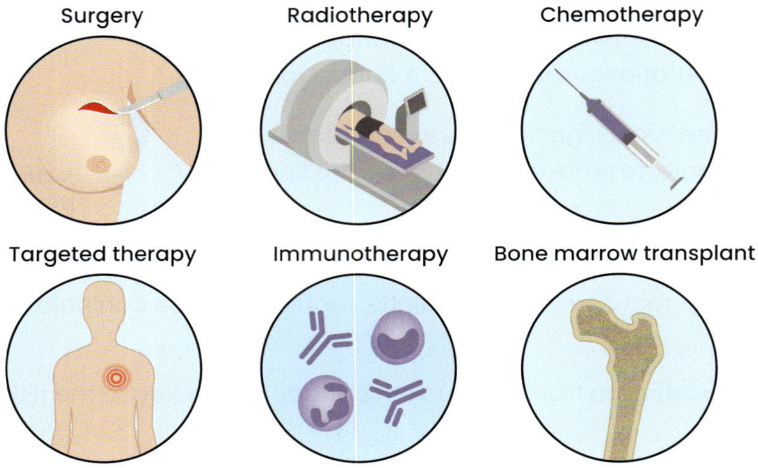

Figure 17.20: Types of cancer treatment.

Advances in modern medicine now mean most cancers are very treatable. There is no one cause of cancer, so therefore the treatments often vary. Cancer patients are usually treated in **oncology** departments in hospitals. Here are some of the more common treatment options:

- If the tumour is benign and confined to a specific tissue, then **surgery** may be an option. The tumour is removed by excision (cutting out), usually under general anaesthetic.
- **Chemotherapy** is often used, with or without surgery, as a treatment for cancer. A cocktail of chemicals is administered to the patients, usually over a period of weeks, killing actively dividing cells (including cancer cells). There are some side effects to this treatment, as the chemicals also attack other non-cancerous cells, including follicular (hair producing) cells and cells that line the stomach. This can cause the patient to lose their hair and develop nausea.
- For some blood cancers, including leukaemia, **bone marrow transplants** are often used. The bone marrow contains **stem cells** which can form into new blood cells to replace those lost due to chemotherapy or other treatments.
- The use of radiation to treat cancer is called **radiotherapy** and involves exposing the tumour to high energy radiation (usually gamma radiation). The exposure is targeted to avoid any damage to healthy tissue.
- **Immunotherapy** is a new cancer treatment that utilises the body's own immune system to destroy cancer cells. Immune cells destroy cancer cells when they identify them. However, some cancer cells evade them because the immune cells don't recognise the cancer cells as abnormal, so they ignore them. Immunotherapy helps the immune cells to detect and subsequently destroy these cancer cells.
- **Targeted drug therapy** is another emerging treatment option for cancer, specifically melanoma and leukaemia. The drugs target the cancer cells directly, slowing their growth or destroying them.

PORTFOLIO LINK

Go to your Learning Portfolio and complete Learning Check 3 to assess your knowledge and understanding of the topic so far.

Go to YouTube.com and search for 'How does chemotherapy work? – Hyunsoo Joshua No' (5:25) by TED-Ed.

Chapter Summary

- All cells are formed from existing cells. This is called **cell continuity**.
- The **cell cycle** is a series of stages in the life of a cell.
- HL The stages of the cell cycle are: **interphase**, **mitosis** and **cytokinesis**.
- HL Interphase is the period in a cell's life when it is not dividing.
- HL During **interphase** cells become larger, and the cell's DNA and organelles are copied.
- There are three main types of cell division: binary fission, mitosis and meiosis.
- **Binary fission** is the division of bacterial (prokaryotic) cells that don't contain a nucleus.
- **Mitosis** is the copying and division of the nucleus of a (eukaryotic) cell.
- **Meiosis** is a type of cell division associated with the production of (haploid) sex cells or gametes.
- The role of mitosis in single-celled organisms is **reproduction**.
- The role of mitosis in multicellular organisms is **growth**, **repair** and **replacement** of dying cells.
- The roles of meiosis are in the production of **sex cells** or **gametes**, creating **variation** amongst species and **maintaining the diploid number** of chromosomes after reproduction.
- The stages of mitosis are: **prophase**, **metaphase**, **anaphase** and **telophase**.
- HL **Cytokinesis** is the division of the cell after mitosis (division of the nucleus) is complete.
- Plant cells produce a **cell plate** to complete cell division.
- **Cancer** is a group of disorders characterised by a loss of control of mitosis.
- A **tumour** is a mass of cancerous (abnormal) tissue.
- **Metastasis** is the spreading of cancer around the body.
- Causes of cancer include: **smoking** cigarettes, exposure to **UV radiation**, exposure to **ionising radiation** (gamma or x-rays), some **viruses**, excess **alcohol** and **mutated genes**.
- Cancer screening involves self-checking, genetic testing and national screening services to help detect and diagnose cancer early.
- Cancer can be treated using one or more of the following: **surgery**, **chemotherapy**, **radiotherapy** and **immunotherapy**.

END OF CHAPTER ASSESSMENT

REMEMBER

1. What is meant by the term *cell continuity*?
2. What is a zygote? From what does it form?
3. Name the two types of cell division.
4. What are stem cells?
5. HL What is interphase?
6. HL Name the four stages of mitosis.
7. What is the role of mitosis in single-celled organisms?
8. What is cancer?
9. Name **three** possible causes of cancer.
10. Explain the following terms.
    (a) Haploid
    (b) Diploid

11. HL Copy the diagram of the cell undergoing prophase of mitosis and label it in as much detail as possible.

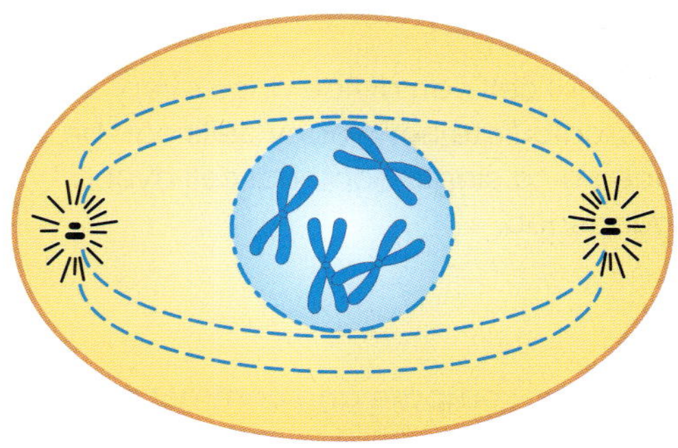

UNDERSTAND

12. Briefly describe the stages of the cell cycle.
13. *All cells complete the cell cycle.* Discuss this statement.

HL **14.** Outline the processes that take place during interphase of a cell.

15. Distinguish between binary fission and mitosis.

HL **16.** Which stage of mitosis is represented in this photograph? How do you know?

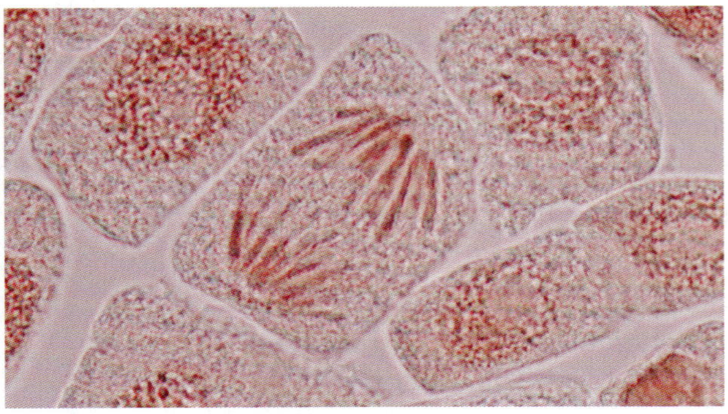

17. Where in a plant would you expect mitosis to occur most often?

18. Where in an animal would you expect meiosis to occur?

HL **19.** Briefly outline the events that occur during telophase of mitosis.

20. *Cancer is caused by cells losing control of mitosis.* Discuss this statement.

21. Outline **two** methods of treating cancer.

APPLY

22. How does cell division in a plant cell differ from cell division in an animal cell?

HL **23.** Sketch labelled diagrams of each stage of mitosis.

HL **24.** Using diagrams distinguish between a chromatid and a chromosome.

HL **25.** Outline the role of spindle fibres in the process of cell division.

26. Why is smoking considered one of the leading causes of cancer?

27. If a bacterial cell can divide once every 20 minutes, calculate how many cells can be produced by a single bacterial cell over a 24-hour period.

ANALYSE

28. Using a table, compare and contrast mitosis and meiosis.

29. Suggest why binary fission is a good form of reproduction for bacterial cells.

30. Research the stages of meiosis and compare and contrast the process with mitosis.

31. The Irish Cancer Society claim that 1 in 2 people will develop cancer during their lifetime. Why do you think the incidence of cancer is increasing amongst the Irish population?

32. The table below shows the duration of each phase of mitosis in onion root tip cells. Study the data and answer the questions that follow. HL

PHASE OF MITOSIS	DURATION (minutes)
Prophase	60
Metaphase	20
Anaphase	2
Telophase	40

(a) What is the total duration of mitosis in onion root tip cells? HL

(b) Which phase of mitosis takes the longest time? What percentage of the total mitosis duration does this phase occupy? HL

(c) Anaphase is the shortest phase of mitosis. Suggest a biological reason for its brief duration compared to the other phases. HL

(d) The data shows that prophase takes three times as long as telophase. Propose an explanation for this difference, considering the events that occur during each of these phases. HL

EVALUATE

33. Have a class debate on the motion: the house supports the administration of the HPV vaccine to all teenagers in Ireland.

34. Smoking is a leading cause of cancer, with half of all smokers dying from a smoking related illness. Carry out some research on the other health issues associated with smoking.

35. Carry out research on immunotherapy and its potential as a means of fighting cancer.

CREATE

36. Create a model that illustrates the main stages of mitosis. HL

37. Design a poster promoting the administering of the HPV vaccine to Irish teenagers.

38. Plan a 'No Smoking' campaign for your school, highlighting the health benefits of quitting.

18 Movement through Cell Membranes

LEARNING OUTCOMES: U2.3, U2.5, 2.6.1, 2.6.2

EYE-OPENER

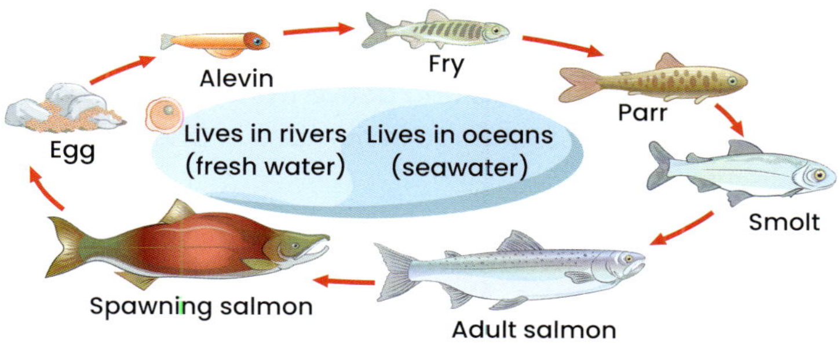

Sea fish are adapted to live in salty water and cannot usually survive in fresh water. In fresh water, too much water enters their cells by a method called osmosis, causing them to swell up and die. However, salmon have adaptations that allow them to live in both fresh water and sea water environments at different stages of their life cycle. Salmon use proteins in their cell membranes to move salt in or out of their cells by a method called active transport to keep the salt and water concentration in their cells constant.

At the end of this chapter, you should be able to:

- Distinguish between diffusion, osmosis and active transport.
- Investigate factors affecting rate of osmosis across a semi-permeable membrane.
- Examine the role of osmosis in food preservation and plant health.

LINKS TO:

Chapter 7	Bacteria and Fungi
Chapter 14	Biomolecules
Chapter 16	Cell Structure and Diversity
Chapter 20	Plant Transport and Responses
Chapter 28	The Breathing System
Chapter 29	The Urinary System
Chapter 30	The Digestive System

SUBJECT VOCABULARY

diffusion, passive, concentration gradient, active transport, osmosis, semi-permeable, isotonic, hypotonic, hypertonic, turgor, plasmolysis

Section 1: Diffusion and Active Transport

At the end of this section, you should be able to:

- Define the term 'diffusion'.
- Define the term 'active transport'.
- Describe the factors that affect the rate of diffusion.

Unpacked from LOs 2.6.1, 2.6.2

In Chapter 16, we learned that the cell membrane is **selectively permeable**, allowing certain substances to pass through, but not others. Living organisms are made of cells which must take in nutrients, carry out respiration and get rid of waste; all these processes require the movement of substances such as water, carbon dioxide and oxygen across the cell membrane. Cells must keep internal conditions as constant as possible to function properly. This is known as **homeostasis**. Homeostasis is maintained by controlling the movement of substances in and out of cells. Different substances travel across the cell membrane in different ways. In this chapter, we will learn about three methods by which substances move across cell membranes: diffusion, osmosis and active transport. We will learn more about the transport of substances in plants and gas exchange in humans in Chapters 20 and 28.

DIFFUSION

If you have ever sprayed perfume, you will have observed how the perfume particles spread throughout the room. This is an example of **diffusion** in everyday life. There was a higher concentration of perfume particles in the area in which the perfume was sprayed than in the rest of the room. As a result, the perfume particles moved throughout the room until the perfume was evenly distributed. No energy was required for the perfume to move from the area of **higher concentration** (where it was sprayed) to the area of **lower concentration** (the rest of the room).

Figure 18.1: Perfume spraying is a good example of the concept of diffusion.

In biology, we describe a process that occurs without the need of energy as **passive**, so we can refer to diffusion as a type of **passive transport**.

- **Diffusion** is the passive movement of a substance (solute) from an area of high concentration to an area of low concentration.
- **Passive transport** does not require energy for the movement of a substance.

Diffusion is very important in living organisms, e.g. exchange of oxygen and carbon dioxide in the alveoli of the lungs. Only small molecules such as oxygen and carbon dioxide move by diffusion. Larger molecules such as proteins require energy to move.

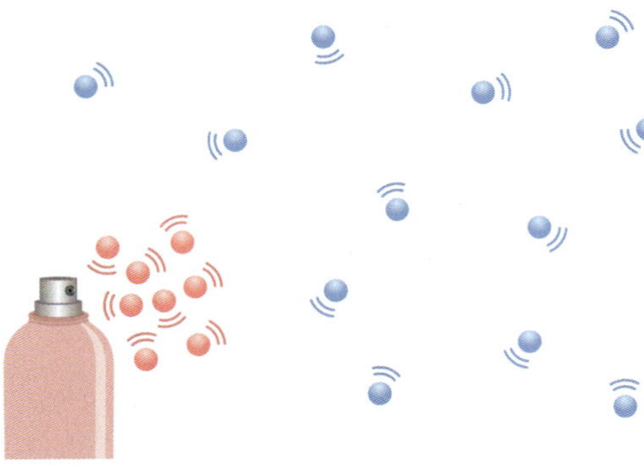

Figure 18.2: Diffusion.

Go to YouTube.com and search for 'What is Diffusion? How Does it Work? What Factors Affect it?' (5:17) by Cognito.

FACTORS AFFECTING DIFFUSION

The rate at which a substance diffuses is affected by a number of factors.

1. Concentration gradient

Concentration describes how much of a substance is dissolved in a certain volume of solution, e.g. 5g of sugar in 100cm³ water is a lower concentration of sugar than 90g of sugar in 100cm³ water. The difference between the concentrations of a solution in two areas is known as the **concentration gradient**.

Concentration gradient describes the difference in the amount of a substance per unit volume between two different areas.

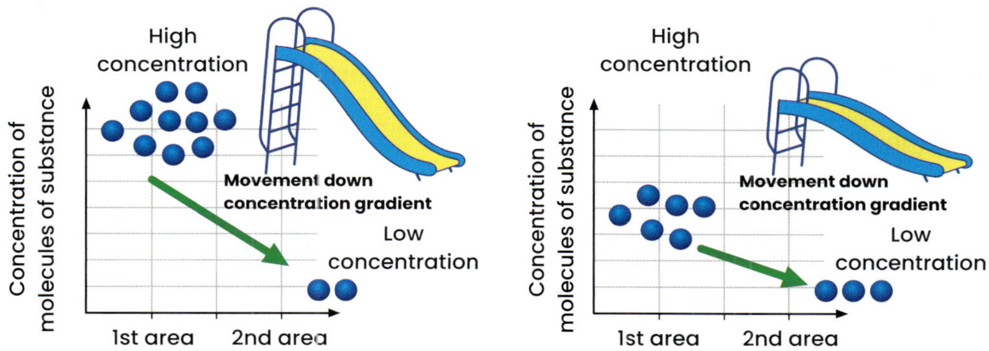

(a) Steeper concentration gradient: substance diffuses more quickly

(b) Shallower concentration gradient: substance diffuses more slowly

Figure 13.3: Comparing concentration gradients in different solutions. We can think of the concentration gradient as being like a slide at a playground: the steeper the gradient, the faster the diffusion.

In diffusion, molecules move from the area of high concentration to low concentration, or 'down the concentration gradient'. The bigger the difference in concentration in the two areas, the larger the concentration gradient and the more quickly the substance molecules will diffuse. In Figure 18.3, there is a greater difference in the concentration of the molecules between the two areas in graph (a) than in graph (b), meaning that the substance will diffuse more quickly in situation (a) than in (b). As the concentration of the substance in both areas becomes more equal, the rate of diffusion slows down.

2. Temperature

Molecules in a substance are in constant random motion. The temperature of the substance affects the molecules' energy, hence their rate of diffusion. The higher the temperature, the more energy the molecules have and the more quickly they diffuse. As the temperature lowers, the energy the molecules have decreases and they will diffuse more slowly.

3. Surface area of cell membrane between the two areas

The greater a cell's surface area, the greater the rate of diffusion across its cell membrane. As a cell increases in size, its volume increases at a greater rate compared to its surface area. The surface of a large cell would not be enough to allow materials to diffuse to all parts of the cell. Smaller cells have a higher surface area-to-volume ratio, allowing materials to diffuse throughout the cell quickly.

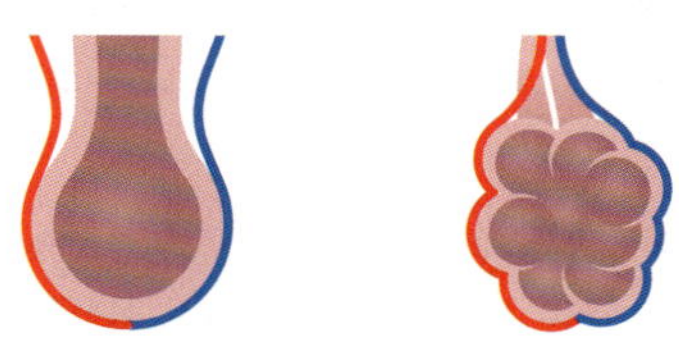

Figure 18.4: Comparing surface area and rate of diffusion in cells.

ACTIVE TRANSPORT

In diffusion, substances move down a concentration gradient from an area of high concentration to an area of low concentration. In other cases, such as in the reabsorption of glucose in the kidney or the uptake of mineral ions from the soil by roots in plants, substances must move from an area of low concentration to an area of high concentration – against the concentration gradient.

The movement of these substances is necessary to maintain their correct balance in cells, known as homeostasis. This movement of molecules from a region of lower concentration to a region of higher concentration requires energy in the form of ATP (adenosine triphosphate) and is known as **active transport**.

Note that active transport occurs across a semi-permeable membrane, whereas diffusion does not require the presence of a membrane.

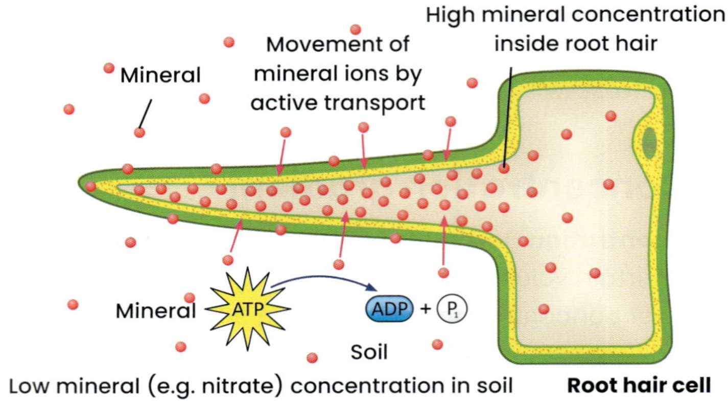

Figure 18.5: Movement of nitrate ions from soil into root hairs by active transport.

Active transport is the movement of a substance from an area of low concentration to an area of high concentration across a semi-permeable membrane, using energy.

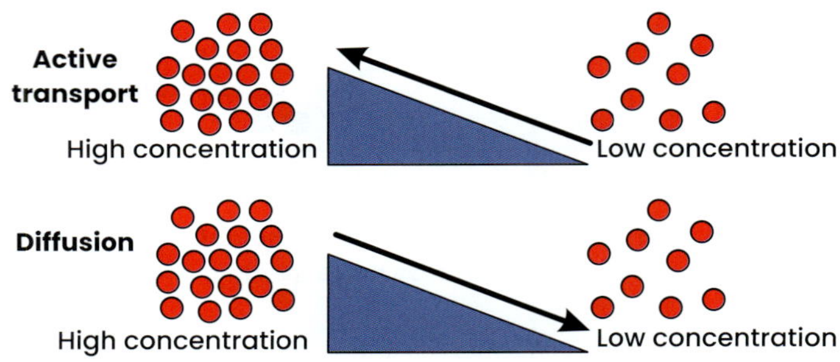

Figure 18.6: Differences between active transport and diffusion.

SCIENCE IN SOCIETY: HEALTH AND TECHNOLOGY

Dialysis

People awaiting a kidney transplant need to spend hours every week attached to a dialysis machine. This is necessary to remove toxic urea and to maintain the correct balance of salts and water in the blood. The patient's blood is passed through a dialysis filter. Excess salts and urea in the blood diffuse through the filter into a dialysis fluid as they move from the higher concentration in the blood to the lower concentration in the dialysis fluid. The filtered blood is returned to the patient.

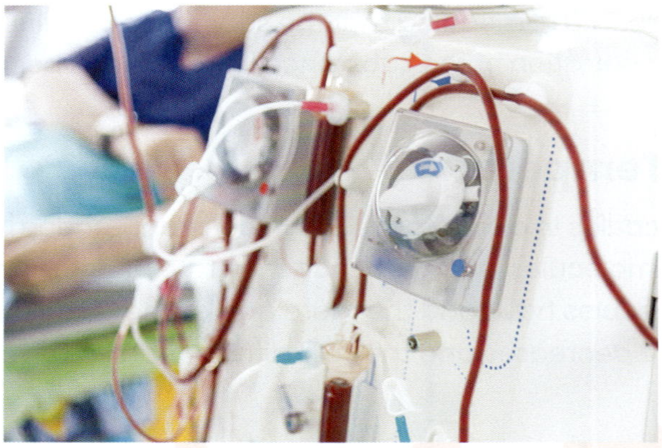

Figure 18.7: A patient undergoing dialysis treatment.

PORTFOLIO LINK Go to your Learning Portfolio and complete Learning Check 1 to assess your knowledge and understanding of the topic so far.

Section 2: Osmosis

At the end of this section, you should be able to:

- Define the term 'osmosis'.
- Describe the role of osmosis in food preservation and plant health.

Unpacked from LOs 2.6.1, 2.6.2

Osmosis is a type of diffusion that refers to the movement of water from an area of high water concentration to an area of low water concentration across a **semi-permeable membrane**. Note that normal diffusion can occur with or without a semi-permeable membrane. Like diffusion, osmosis is **passive** and does not require energy to occur. Examples of osmosis in living organisms include the absorption of water by plant roots, absorption of water through the walls of the colon and reabsorption of water in the nephron in the kidney.

- **Osmosis** is the passive movement of water from an area of high water concentration to an area of low water concentration across a semi-permeable membrane.
- A **semi-permeable membrane** is a barrier layer that allows certain substances to pass through but not others.

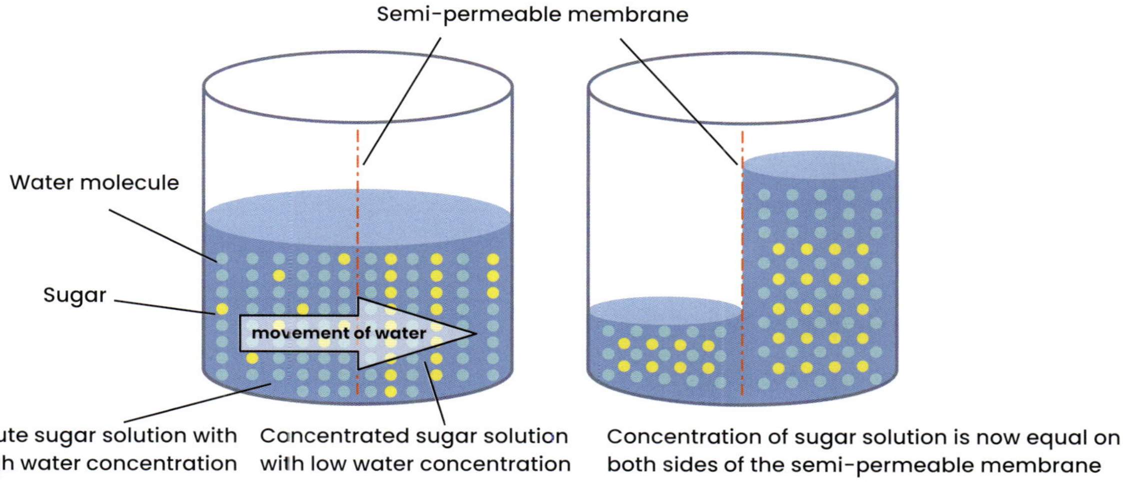

Figure 18.8: Water moves from an area of high water concentration to an area of low water concentration across a semi-permeable membrane by osmosis.

Figure 18.8 shows a beaker containing two solutions separated by a **semi-permeable membrane**. In the beaker on the left, before osmosis occurs, the concentration of water on each side of the semi-permeable membrane is different. Water will move from the region of high water concentration (dilute sugar solution) on the left across the semi-permeable membrane to the region of low water concentration (concentrated sugar solution) on the right. The movement of water by osmosis across the semi-permeable membrane results in the concentration of the sugar solution on both sides of the semi-permeable membrane being equalised. Only water moves by osmosis and not any other solutes or substances dissolved in the water.

Go to YouTube.com and search for 'GCSE Biology – Osmosis' (4:23) by Cognito.

EFFECT OF OSMOSIS ON CELLS

If a cell such as a red blood cell is placed in a solution, what happens depends on the concentration of water inside the red blood cell compared to that in the external solution.

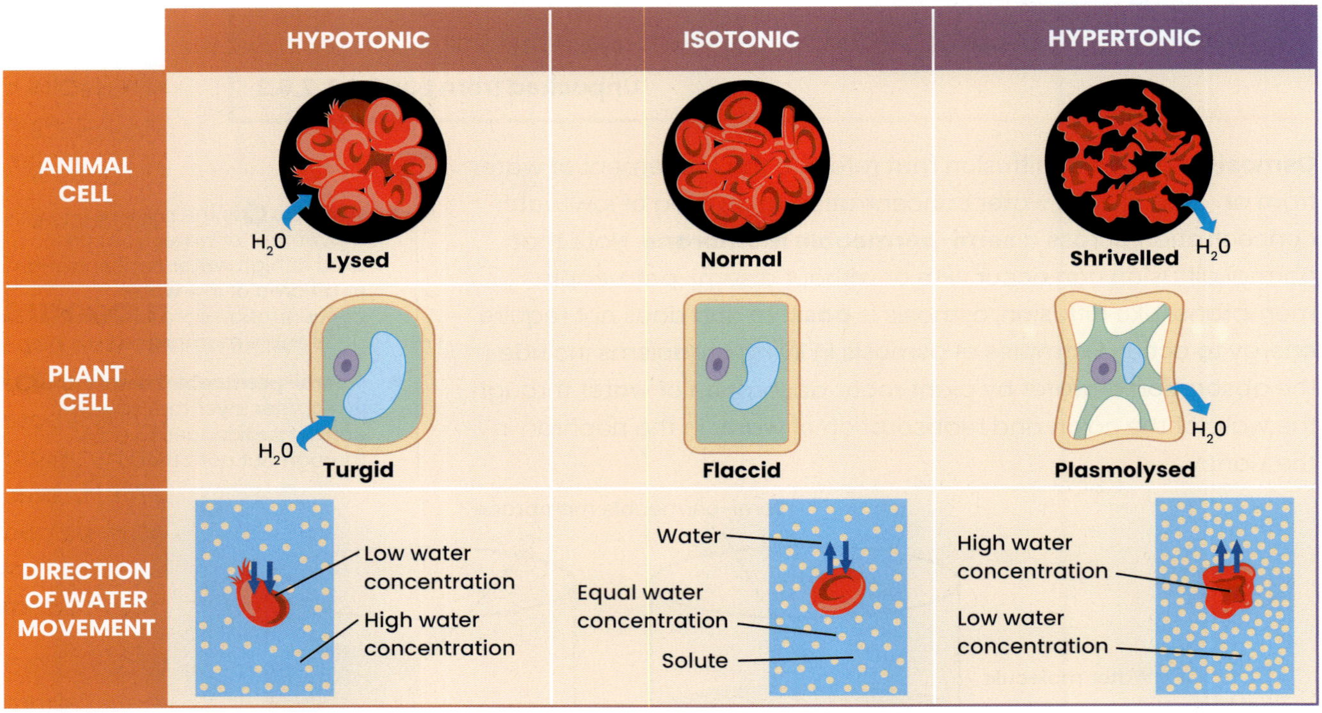

Figure 18.9: Hypotonic, isotonic and hypertonic solutions.

1. Hypotonic solution

If the water concentration in the external solution is higher than that inside the cell, water will move by osmosis into the cell. This causes the cell to swell in size. If too much water moves into an animal cell by osmosis, the cell could burst or 'lyse'. Swollen plant cells are described as 'turgid' and experience an outward pressure or **turgor** of the cytoplasm and cell membrane against the cell wall. Turgid plant cells do not lyse, as the cell wall prevents more water entering. A healthy herbaceous (non-woody) plant with turgid cells will stand upright. Plant health is best in **hypotonic** solutions.

2. Isotonic solution

If the water concentration in the external solution is the same as that inside the cell, no net movement of water will occur. This is the best situation for animal cell health. Plant cells in an **isotonic** solution will become flaccid, causing the plant to be 'floppy' and limp.

3. Hypertonic solution

If the water concentration in the external solution is lower than the water concentration inside the cell, water will move out of the cell by osmosis. This causes the cell to shrink. In a **hypertonic** solution, animal cells will shrivel, and plant cells will undergo **plasmolysis**. This means that the cytoplasm shrinks as it loses water, but the cell wall maintains its shape. As the cell membrane shrinks and pulls away from the cell wall, it leaves a gap. If a herbaceous plant loses too much water, it will wilt. Plasmolysed cells are unlikely to survive.

- An external solution is **hypotonic** when its water concentration is higher than that inside a cell.
- **Turgor** is the outward pressure of the contents of a plant cell against the cell wall.
- An external solution is **isotonic** when its water concentration is equal to that inside a cell.
- An external solution is **hypertonic** when its water concentration is lower than that inside a cell.
- **Plasmolysis** is the shrinking of the contents of a plant cell away from the cell wall, due to loss of water.

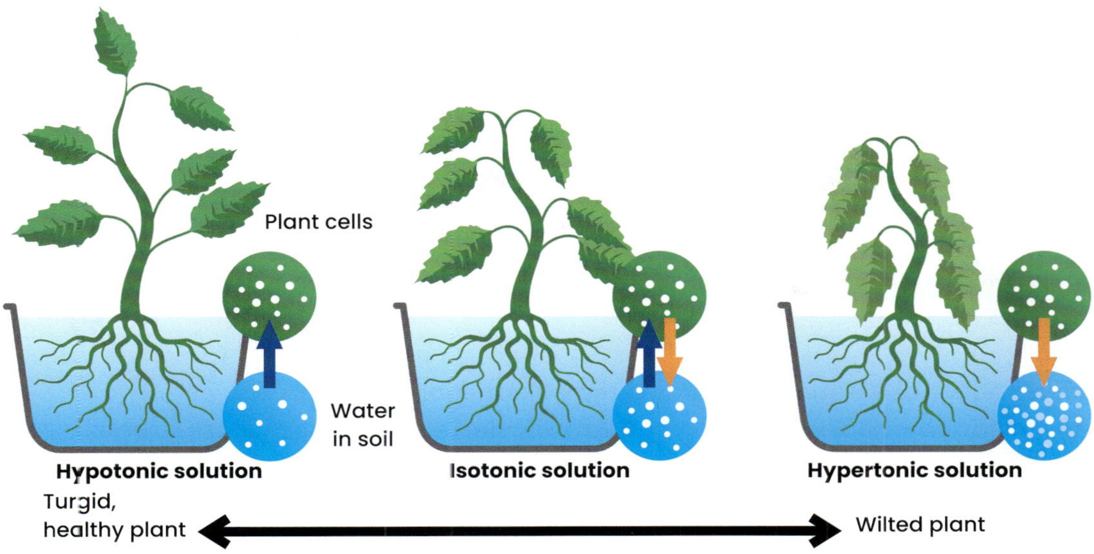

Figure 13.10: Effect of turgor on plant health.

OSMOSIS AND FOOD PRESERVATION

Micro-organisms, e.g. bacteria and fungi, cause food spoilage. In Chapter 7, we learned that bacteria need water to survive. This fact is the basis of many food preservation techniques. Meat and fish are often preserved by 'salting' or 'curing' – removing the water from the food. Fruit is preserved using a sugar solution to form jam. In both cases, the food is placed in a hypertonic solution with a low water concentration. The microbial cells have a high water concentration compared to the surrounding salt or sugar solution. Water moves by osmosis from the microbial cells to the surrounding solution, causing the microbial cells to shrivel and die. Some bacterial cells may survive by forming endospores.

Go to YouTube.com and search for 'Effects of Osmosis in Plant and Animal Cells' (5:43) by LearnStalk.

Figure 18.11: Jam and cured (salted) ham – preserving food by osmosis.

PORTFOLIO LINK Go to your Learning Portfolio and complete Learning Check 2 to assess your knowledge and understanding of the topic so far.

Specified Investigation 7

Investigate factors affecting rates of osmosis across semi-permeable membranes. (LO 2.6.2)

Scan the QR code to watch a video demonstration of this investigation.

A. CONCENTRATION GRADIENT

In this investigation, you will use Visking tubing as a semi-permeable membrane. Visking tubing has microscopic holes that allow small molecules, e.g. water or glucose to pass through but prevent larger molecules, e.g. sucrose or starch passing through.

Hypothesis: If sucrose solutions of different concentration in Visking tubing are placed in deionised water, the greater the concentration gradient, the more water will move into the tubing by osmosis and the greater its increase in mass.

Equipment: Visking tubing, beakers, balance, funnel, sucrose, deionised water, tissue paper, stopwatch.

VARIABLES

Independent variable: Sucrose solutions of different concentration (0%, 5%, 10%, 20%, 40%). The greater the concentration, the greater the concentration gradient.

Dependent variable: The change in mass of each Visking tubing over a set time period.

Control variables: Temperature, time and initial volume of solution will be kept constant.

METHOD

1. Prepare sucrose solutions of the following concentrations: 0%, 5%, 10%, 20% and 40%. (Measure the masses of sucrose in the table and make up to 100cm³ with deionised water in separate beakers.)

	SUCROSE SOLUTION CONCENTRATION (%)	MASS OF SUCROSE (grams)/ 100cm³ DEIONISED WATER
A	0 (deionised water)	0 (highest water concentration)
B	5	5
C	10	10
D	20	20
E	40	40 (lowest water concentration)

2. Cut five 40cm lengths of Visking tubing (A, B, C, D, E) and soften each by dipping in a beaker of deionised water.
3. Tie a knot at one end of the length of Visking tubing A. Using a funnel, half-fill the tubing with solution A (0%). Eliminate as much air as possible from the tubing and tie a knot at the other end. Dry the outside of the tube with tissue paper.
4. Record its mass on a balance. Repeat this step for solutions B–E. Note the turgidity of each tube.
5. Place each of the five different sucrose solution Visking tubings (A–E) into separate beakers containing deionised water and leave for 15 minutes.

6. Remove the tubing from each beaker, pat dry and reweigh on a balance. Note any changes in turgidity.

7. Calculate any change in the mass of each tube over the 15 minutes. Plot a graph of change in mass/15 minutes against the sucrose concentration.

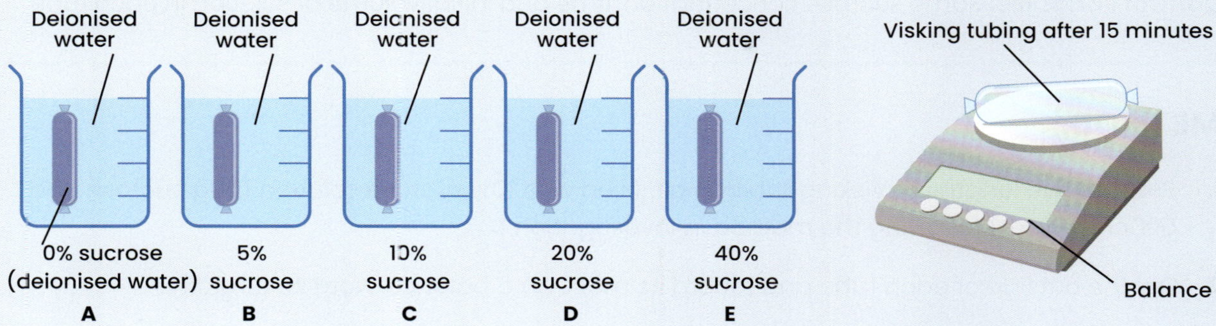

RESULTS

SUCROSE SOLUTION CONCENTRATION (%)	A: 0%	B: 5%	C: 10%	D: 20%	E: 40%
Initial mass of tubing					
Final mass of tubing					
Change in mass in 15 minutes					

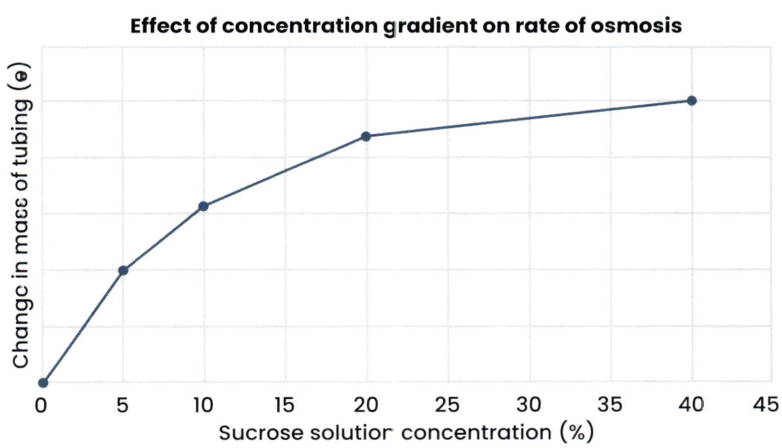

CONCLUSION

No change in mass occurred with the 0% sucrose solution as the water concentration inside the tube and outside in the beaker was the same. The more concentrated the sucrose solution, the more water moved into the tubing by osmosis and the greater its increase in mass. The greater the concentration gradient, the greater the rate of osmosis.

B. TEMPERATURE

Hypothesis: If sucrose solutions in Visking tubing are placed in deionised water at different temperatures, the higher the temperature, the more water will move into the tubing by osmosis and the greater its increase in mass.

Equipment: Visking tubing, beakers, balance, funnel, tissue paper, stopwatch, water baths, sucrose, deionised water, ice.

VARIABLES

Independent variable: Sucrose solutions in water baths at different temperatures.

Dependent variable: The change in mass of each Visking tubing over a set time period.

Control variables: Same sucrose concentration, time and initial volume of solution in each tube.

METHOD

1. Prepare five lengths of Visking tubing half-filled with 10% sucrose solution (50g sucrose and 500cm³ water) following the method in Investigation 1.
2. Dry the outside of each tube and record its mass on a balance. Note its turgidity.
3. Place the Visking tubing into beakers containing deionised water at the following temperatures: 0°C, 20°C, 30°C, 40°C and 50°C. Leave for 15 minutes.
4. Remove each tubing, pat dry and reweigh on a balance. Note any changes in turgidity.
5. Calculate any change in the mass of each tube over the 15 minutes. Plot a graph of change in mass over 15 minutes against the temperature.

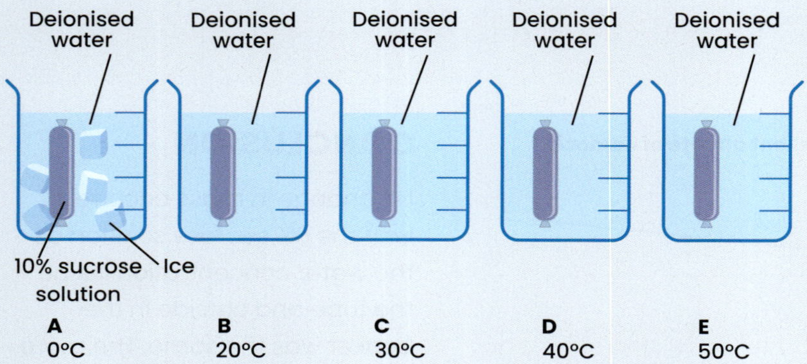

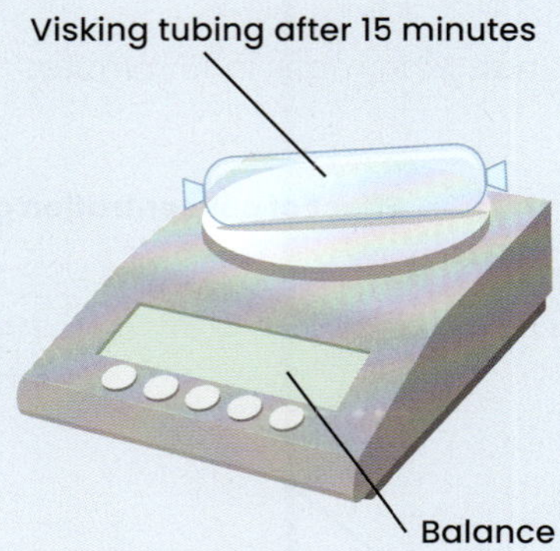

RESULTS

TEMPERATURE	A: 0 °C	B: 20°C	C: 30°C	D: 40°C	E: 50°C
Initial mass of tubing (g)					
Final mass of tubing (g)					
Change in mass in 15 minutes (g)					

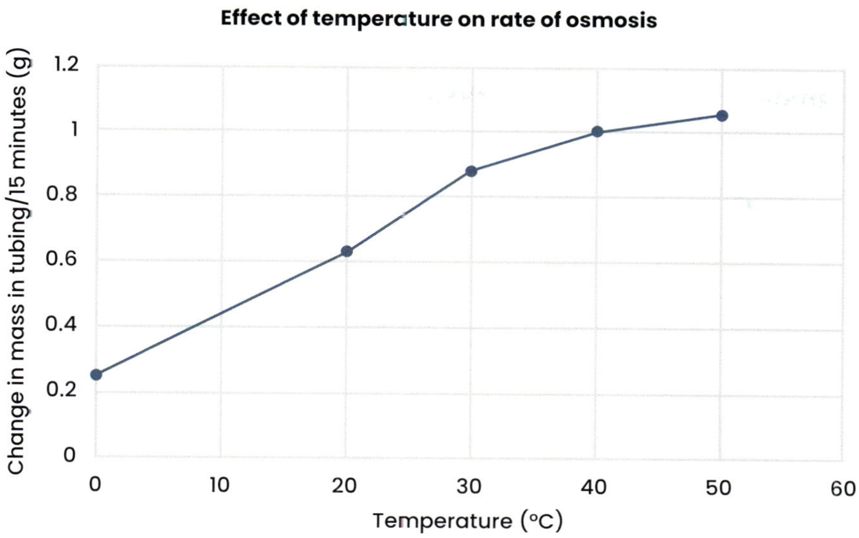

CONCLUSION

The greater the temperature, the greater the increase in mass and the greater the rate of osmosis.

C. SURFACE AREA

Hypothesis: If potato cubes with different surface area are placed in water, the greater the total surface areas, the more water the potato will absorb by osmosis and the greater its increase in mass.

Equipment: Potatoes, beakers, balance, knife, tissue paper, stopwatch, deionised water.

Scan the QR code to watch a video demonstration of this investigation.

VARIABLES

Independent variable: Potato cubes of different surface area.

Dependent variable: To measure the rate of diffusion of water, you will record the change in mass of potato cubes with different surface areas over a set time period.

Control variables: Time, temperature and initial total volume of the potato will be kept constant.

METHOD

1. Cut two cubes of side length 2cm from a potato.

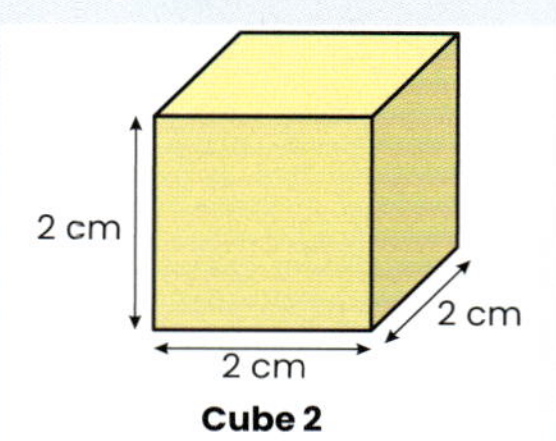

Cube 1 (2cm × 2cm × 2cm)

2. Cut one of these cubes into eight equal-sized cubes of side length 1cm. This will provide a potato sample of equal mass but a larger surface area.

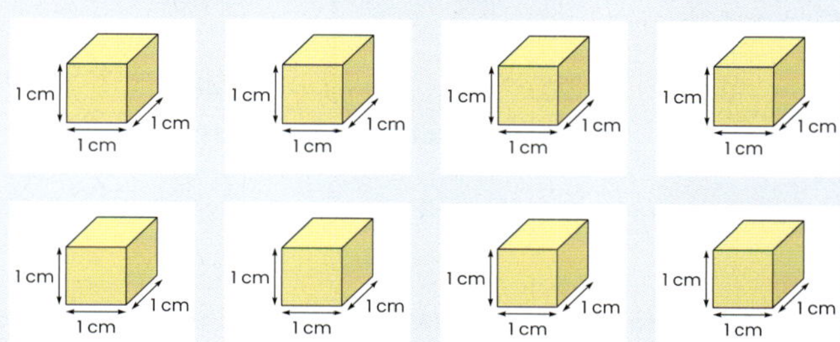

Cube 2 (divided into eight equal cubes 1cm × 1cm × 1cm)

3. Dry cube 1 with tissue paper and weigh on a balance. Dry the eight smaller cubes and measure their total mass.
4. Place cube 1 (2cm × 2cm × 2cm) in a beaker containing 100ml water for 15 minutes. Place the eight small cubes in a separate beaker containing 100ml water for 15 minutes.
5. Remove the cubes from each beaker, pat dry and reweigh on a balance. Note any changes in total mass.
6. Repeat the test three times and calculate the overall average change in mass. Plot a graph of change in mass over 15 minutes against the total surface area for each potato sample.

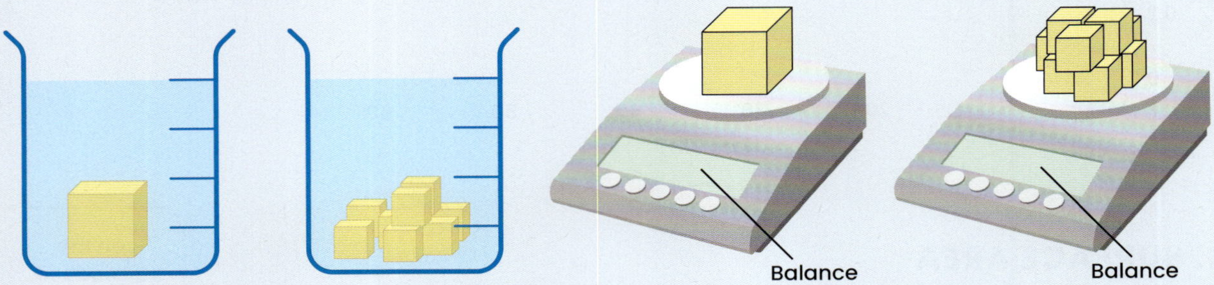

RESULTS

	TOTAL SURFACE AREA	TOTAL INITIAL MASS (g)	TOTAL FINAL MASS (g)	CHANGE IN MASS IN 15 MINUTES (g)
CUBE 1	6 x (2cm × 2cm) = 24cm^2			
CUBE 2 DIVIDED INTO EIGHT EQUAL-SIZED CUBES	8 × (6 × 1cm × 1cm) = 48cm^2			

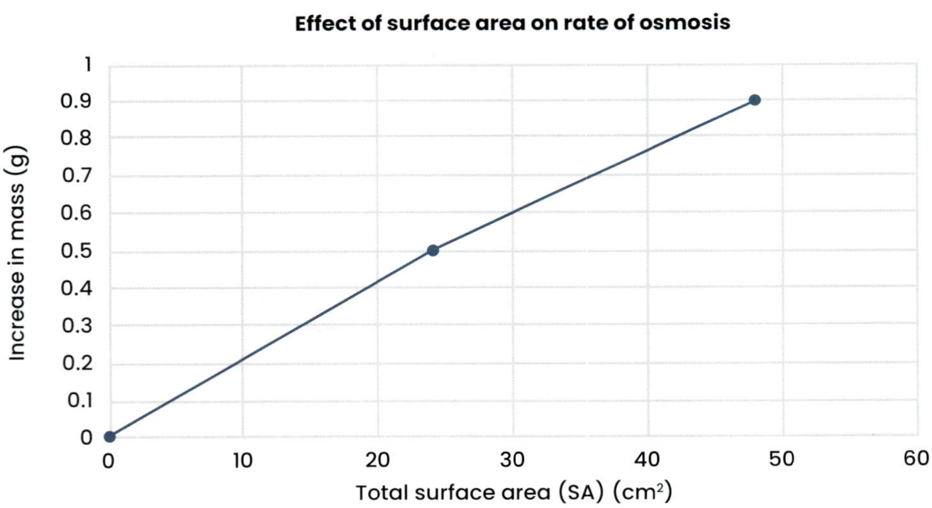

CONCLUSION

The greater the surface area, the greater the increase in mass and the greater the rate of osmosis.

INVESTIGATIVE LOG LINK **Go to your Investigative Log to record your lab work and answer questions on the experiments in this chapter.**

Chapter Summary

- **Diffusion** is the passive movement of a solute from an area of high concentration to an area of low concentration until the concentration of the solute is equal in both areas.
- **Rate of diffusion** increases with increasing solute concentration gradient, temperature and surface area.
- **Osmosis** is the movement of water from an area of high water concentration to an area of low water concentration across a semi-permeable membrane until the solution concentration is equal in both.
- If a cell is placed in a solution, what happens depends on the concentration of the solutions inside and outside the cell:

	ISOTONIC	HYPOTONIC	HYPERTONIC
CONCENTRATION OF WATER IN EXTERNAL SOLUTION COMPARED TO INSIDE THE CELL	Same	Higher	Lower
MOVEMENT OF WATER	• No movement • Animal cell: normal – best for animal cell health • Plant cell: flaccid	• Into cell Animal cell: swelling/lysis • Plant cell: turgid – best for plant health	• Out of cell • Animal cell: shrivels • Plant cell: plasmolysis – unlikely to survive

- Meat and fish are **preserved** by salting and fruit by making jam using sugar. Food is placed in a salt/sugar solution that is hypertonic, i.e. has a low water concentration.
- **Active transport** is the movement of substances from an area of low concentration to an area of high concentration with the use of energy.

	DIFFUSION	OSMOSIS	ACTIVE TRANSPORT
MOVEMENT OF SUBSTANCES	High concentration to low concentration (down a concentration gradient)	Movement of water from an area of high to low concentration	Low concentration to high concentration (against a concentration gradient)
ENERGY REQUIRED	No	No	Yes
EXAMPLE	Diffusion of oxygen from the alveoli into the blood	Uptake of water from the soil by roots in plants	Uptake of mineral ions from the soil by roots in plants

END OF CHAPTER ASSESSMENT

REMEMBER

1. Define each of the following terms.
   (a) *Diffusion* (b) *Osmosis*
2. State **two** factors which would increase the rate of diffusion of a solute.
3. Which of the following solutions is more concentrated?
   Solution A: containing 35g sugar/100ml water; or
   Solution B: containing 60g sugar/200ml water.
4. At which of the following temperatures would diffusion occur the fastest? Explain your answer.
   (a) 5°C (b) 15°C (c) 35°C
5. The diagram below shows sugar molecules in a beaker of water.

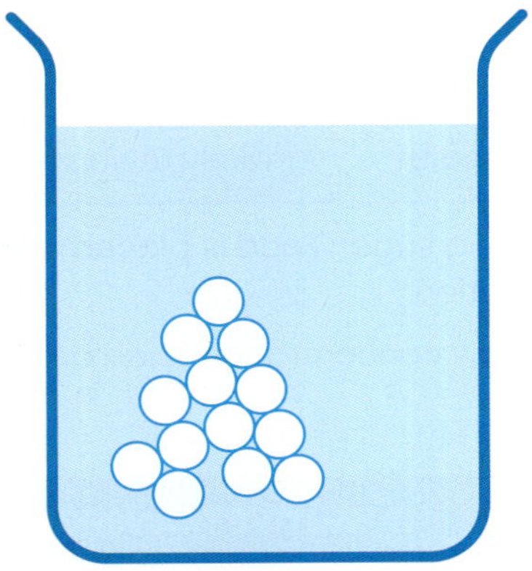

   (a) Draw a diagram to indicate the position of the molecules of sugar in the beaker of water after a period of time.
   (b) Explain why the sugar molecules have moved in this way.

UNDERSTAND

6. Discuss the role of osmosis in food preservation.
7. The diagram shows an epithelial cell from the villi of the small intestine. What structural feature of this cell assists diffusion of nutrients across its membrane?

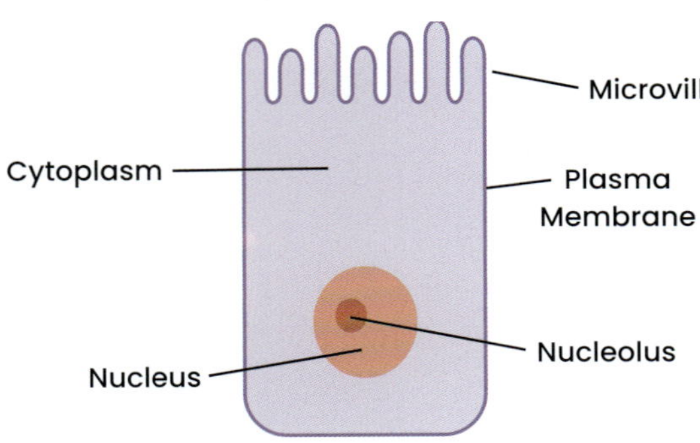

8. A student carried out an investigation into the effect of concentration of sucrose solutions on the rate of osmosis.
   (a) Describe how the student controlled the temperature of each trial.
   (b) What measurements did the student record during the investigation?
   (c) What controls did the student employ to ensure a fair test?
   (d) What was the dependent variable?
   (e) What was the independent variable?
   (f) Sketch a graph of the expected results for this investigation.
   (g) How do the results of this investigation explain the effect of concentration on the rate of osmosis across a plasma membrane?

APPLY

9. Diagrams A and B illustrate carbon dioxide molecules in a space of fixed size.
   (a) Copy both diagrams and indicate the net direction of diffusion in each case
   (b) Sketch graphs to illustrate the concentration gradient for each case
   (c) In which situation – A or B – would diffusion of the carbon dioxide occur more quickly? Assume that all other factors are kept constant. Explain your answer.

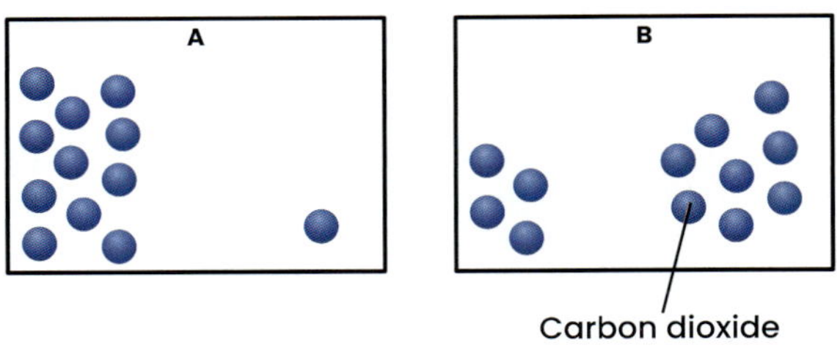

10. The diagram below shows two solutions, A and B, of differing salt concentrations. A cell of a certain internal salt concentration is found in each solution. Copy the diagrams and indicate using arrows the direction of movement of solution in each case.

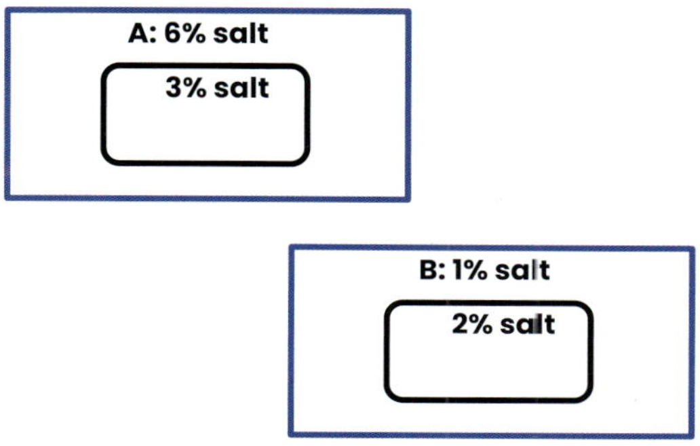

ANALYSE

11. A student cut potato cubes of equal size and placed them in sucrose solutions of different concentration for 24 hours. The % change in mass of each potato cube was recorded as follows.

SUCROSE SOLUTION CONCENTRATION (MOLAR)	% CHANGE IN MASS
0	+28%
0.2	+5%
0.4	−8%
0.6	−18%
0.8	−27%
1.0	−30%

(a) Draw a graph of the results.

(b) Explain the shape of the graph.

(c) What is the approximate concentration of solution in a potato? Explain your answer.

12. Compare and contrast diffusion, osmosis and active transport under the following headings.

(a) Substances transported

(b) Energy use

(c) Concentration gradient

13. In an experiment, a student cut a piece of red onion epidermis skin, placed it on a microscope slide and added a coverslip. The student observed the sample under a microscope. A drop of concentrated salt solution was placed at the side of the coverslip and allowed to draw across the sample.

(a) What observation would the student make about the onion cells after a few minutes? What conclusion would the student make from this observation?

(b) If the student soaked up the salt solution using tissue paper and added a drop of water to the sample, what would the student observe under the microscope? What conclusion would the student make from this observation?

14. Explain why minerals such as nitrates or magnesium are transported from the soil into a plant's root hair cells by active transport rather than by diffusion.

15. Active transport works by using transporter or carrier proteins embedded in the cell membrane to pick up specific molecules e.g. glucose and take them through the cell membrane. Why do you think transporter proteins are required for active transport?

EVALUATE

16. What are the (a) advantages and (b) disadvantages of osmosis to living cells?

CREATE

17. Design an investigation to examine the effect of the surface area on the rate of diffusion across a semi-permeable membrane in a potato.

18. Research one of the following subjects.

(a) How gardeners use the principle of osmosis to kill slugs.

(b) Why some athletes take salt tablets.

UNIT 6

PLANTS – GREEN MACHINES

CHAPTERS

STRANDS

Unifying Strand: Nature of Science

Strand 2: Structures and Processes of Life

19 Flowering Plants – Structure and Function

LEARNING OUTCOMES: 2.5.7, 2.6.14, 2.6.15

EYE-OPENER

The fly orchid is a plant found in several locations around Ireland but is most common in the Burren region of Co. Clare. The flowers of the fly orchid closely resemble a fly, hence the name, and they attract flies and occasionally wasps. The orchid also produces a scent which mimics chemicals produced by flies and wasps. The flies and wasps attempt to mate with the flower and unknowingly receive a 'packet' of orchid pollen, which they then may deliver to another fly orchid plant, fertilising the plant and beginning its reproduction. This is not the only Irish orchid species that uses insects in this way – the bee orchid and the butterfly orchid are also commonly found in the Burren.

At the end of this chapter, you should be able to:

- Describe the parts of a typical flowering plant.
- Describe the types of plant tissue and relate them to their function.
- Investigate and describe the structure of the leaf, stem, roots and flower.
- Discuss how the structure of the leaf, stem, root and flower relates to their function in the plant.
- Discuss plant responses and the role of plant growth regulators.

LINKS TO:

Chapter 4	**Classification and Taxonomy**
Chapter 16	**Cell Structure and Diversity**
Chapter 17	**Cell Continuity**
Chapter 20	**Plant Transport and Responses**
Chapter 21	**Plant Reproduction**
Chapter 22	**Photosynthesis**

SUBJECT VOCABULARY

angiosperm, dermal tissue, vascular tissue, xylem, phloem, ground tissue, stomata, buds, vascular bundles, lenticels, stamen, carpel, pollination, translocation, root hairs, cortex, plant growth regulator

Section 1: The Flowering Plant and Plant Tissues

At the end of this section, you should be able to:

- Name the parts of a typical flowering plant.
- Describe the typical features of flowering plants (angiosperms).
- Name the types of plant tissue (dermal, vascular, ground) and their location in roots, stems and leaves.
- Outline how dermal cells are adapted for their specific role in the plant.
- Draw a labelled diagram of xylem and phloem.
- Give three examples of ground tissue and their functions.
- Understand that meristems are undifferentiated cells in plants.

Unpacked from LO 2.6.14

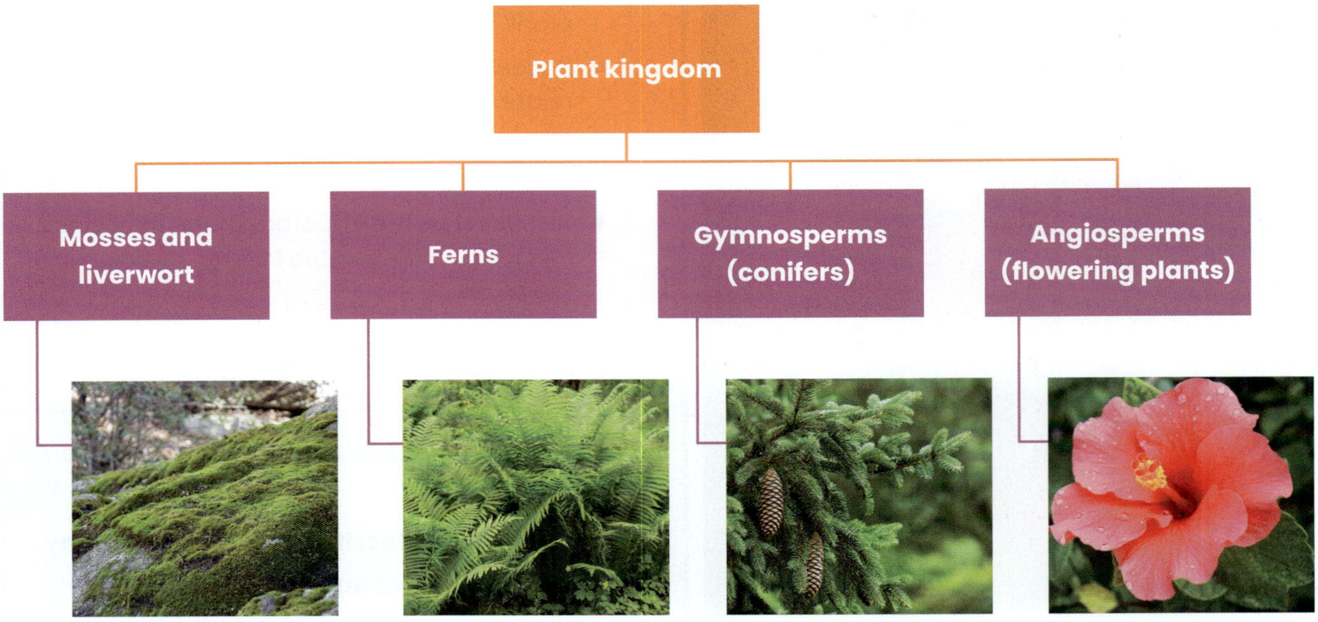

Figure 19.1: Classification of plants.

As we learned in Chapter 6, the Plant Kingdom is home to a wide variety of species categorised into four main groups: mosses, ferns, gymnosperms and angiosperms. **Angiosperms**, also known as flowering plants, are the most abundant and diverse type of plant, with over 370,000 known species. Yet, these flowering plants share many of the same structures and characteristics. In this chapter, we will look at a typical flowering plant and examine its parts and consider how each part is perfectly adapted to carry out its role.

Go to YouTube.com and search for 'Flowering and Non-flowering Plants' (3:38) by Next Generation Science.

Angiosperms are plants that produce flowers during their life cycle.

COMMON CHARACTERISTICS IN ANGIOSPERMS

All flowering plants have the following shared characteristics:

- At some point in their life cycle, angiosperms have **flowers**. Flowers are the reproductive parts of the plant and contain male and female structures responsible for producing gametes.
- Angiosperms also have a root system, to anchor the plant and draw up water and nutrients, and a shoot system that consists of the stem and leaves.
- Flowers contain a series of **petals** and **sepals**. These structures protect the internal structures and can help attract pollinators.
- Angiosperms produce seeds that are surrounded by a fruit.

Flowering plants also have similarities in how they arrange their plant cells and tissues within the roots, stem and leaves. Before we look at these structures in detail, let's talk about how plants arrange their cells into tissues and organs.

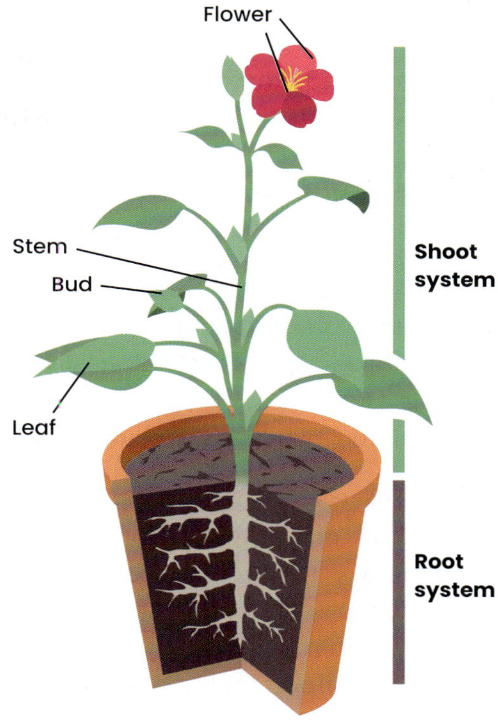

Figure 19.2: A typical flowering plant.

PLANTS CELLS AND TISSUES, INCLUDING XYLEM AND PHLOEM

We have already learned, in Chapter 16, that plant cells have some typical features. They have a **cell wall** made of cellulose, **large vacuoles** for storage which help maintain the shape of the cell and organelles called **chloroplasts** that are responsible for photosynthesis – how plants make their food. Plants contain many diverse types of cell, each adapted for their particular role, which work together as plant **tissues**.

Plant cells fall into one of three types of plant tissue: **dermal**, **vascular** and **ground**.

Dermal tissue

Dermal tissue is **skin-like tissues** found in plants which act as protection for the plant. Just like our skin, they are found on the outside of plant organs like the roots, stems and leaves.

Types of dermal cells include **dermal tissue cells**, **guard cells** (crescent-shaped cells which form stomata) and **trichomes** (hair-like cells). Typically, dermal cells have a waxy outer layer called a **cuticle**, which helps the plant prevent water loss. Dermal tissue is normally only one cell thick and referred to as **epidermis** but sometimes it can form a thicker layer called **cork**, typically on the bark of older woody plants.

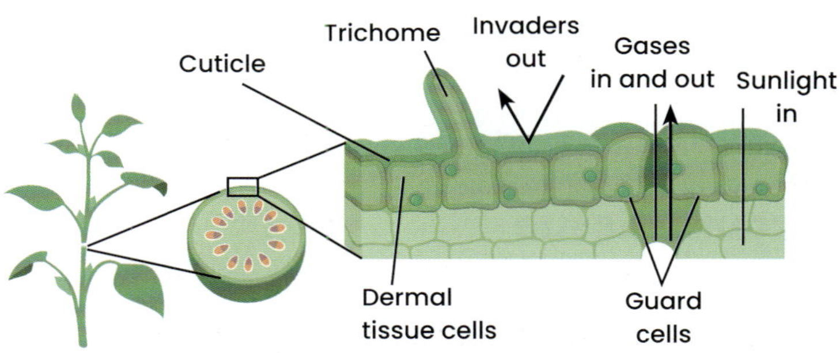

Figure 19.3: Types and position of dermal cells.

Dermal tissue is formed from protective skin-like cells in plants.

Vascular tissue

Vascular tissue is responsible for transporting materials around a plant. Plants transport water, food, minerals, growth regulators and other materials from their various organs to aid their growth. There are two main types of vascular tissue: xylem and phloem. These tissues are usually packaged together in **vascular bundles**, which can be seen in the leaves and stems.

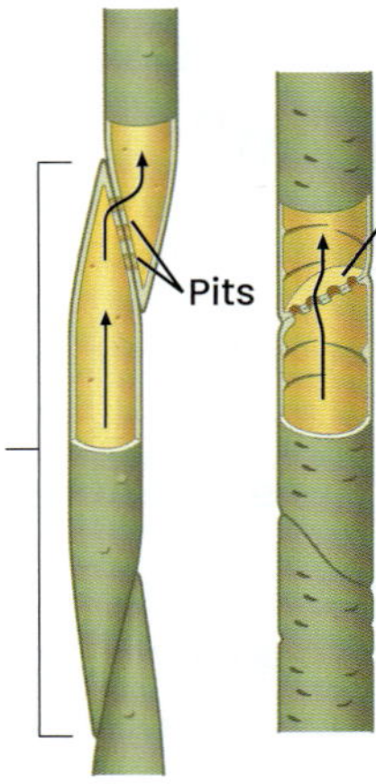

Figure 19.4: Xylem is a form of vascular tissue.

Xylem is a plant tissue responsible for transporting water in the plant. Water transport goes in one direction – from the root to the leaves – and is a passive process, i.e. the plant can do it without using up any energy. There are two types of xylem tissue: **tracheids** and **vessels**.

Xylem vessels are long, continuous tubes, formed from open-ended cells called vessel elements (a bit like the cardboard tube inside a toilet roll), while tracheids have tapered end walls. Xylem cells don't contain a nucleus or cytoplasm, so are 'dead' tissue. As well as transporting water, xylem provides structural support for the plant. The walls of the xylem are strengthened with a substance called lignin, which gives them strength and support.

- **Vascular tissues** are groups of cells specialised for the transport of materials in a plant.
- **Xylem** (pronounced *zai-lum*) is a plant vascular tissue that transports water from root to leaf.
- **Phloem** (pronounced *flow-em*) is a plant tissue that transports dissolved food in a plant.
- **Translocation** is the movement of food from one part of a plant to another.

Phloem is a plant tissue responsible for transporting food in the plant, usually in the form of the sugar sucrose. The food is usually made in the leaves during photosynthesis and then transported down the plant for storage in the stems or roots. This stored food can be broken down later and moved back up the plant if required. The movement of food through the plant is called **translocation**. Phloem tissue is a 'living' tissue as it contains cytoplasm and nuclei. It is made up of two types of cells: sieve tubes and companion cells.

We will discuss the role of xylem and phloem in plant transport in the next chapter.

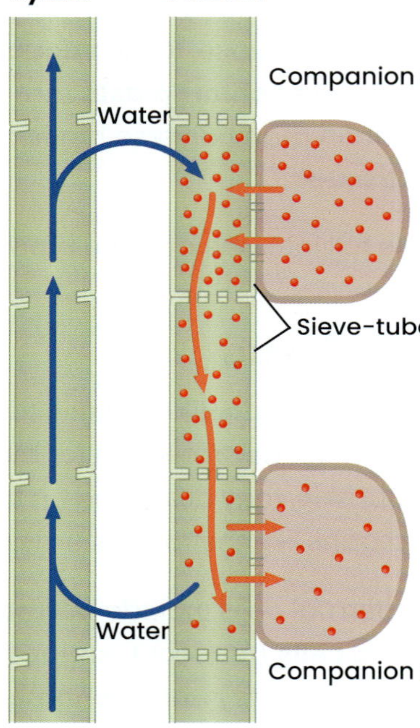

Figure 19.5: Xylem and phloem.

Ground tissue

In plants, **ground tissue** has lots of roles and functions. They are responsible for making food (photosynthesis), strengthening and supporting the plant, food and water storage, repair and regeneration of tissues and protection. The cells that make up ground tissue are generally large and don't have a regular shape; they make up the bulk of the plant. Any cell that isn't dermal or vascular is considered ground tissue.

- **Ground tissue** are groups of cells in plants that are not categorised as dermal or vascular – they have many functions.

FOCUS ON ... MERISTEM

Meristems or meristematic tissue are areas of a plant where there is active cell division (mitosis). These cells are usually very small and are unspecialised, meaning they can later become dermal, vascular or ground tissue as the plant grows. Unspecialised cells are often referred to as stem cells.

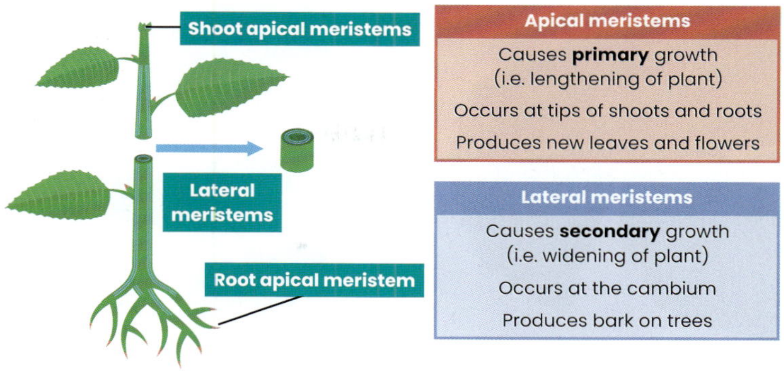

PORTFOLIO LINK **Go to your Learning Portfolio and complete Learning Check 1 to assess your knowledge and understanding of the topic so far.**

Section 2: Plant Organs – Leaves and Stems

At the end of this section, you should be able to:

- Describe the structure and function of the leaf and explain how the structure is adapted to carry out those functions.
- Distinguish between longitudinal cross-sections (LS) and transverse cross-sections (TS) when preparing samples for microscope study.
- Draw a labelled diagram of a leaf in LS.
- Describe the structure and function of the stem and explain how the structure is adapted to carry out those functions.
- Draw a labelled diagram of a stem in TS.

Unpacked from LOs 2.6.14, 2.6.15

As discussed in Chapter 17, tissues can form organs. Examples of plant organs include leaves, stem, flowers and roots.

THE LEAF

The leaf is the part of the plant responsible for making food by **photosynthesis**. The leaf is perfectly adapted for this role:

- Leaves are typically thin structures with a large surface area for capturing light energy.
- Xylem vessels deliver the water necessary for photosynthesis; xylem is found within the veins of the leaves.
- Small pores called **stomata** allow carbon dioxide to diffuse into the leaf. Oxygen, produced during photosynthesis, can escape via the pores.

Stomata are tiny leaf pores that allow for gas exchange and transpiration.

- Despite being flat, the leaf has lots of air spaces inside, allowing the gases move into and out of the photosynthesising cells.
- As it makes food, phloem vessels in the veins transport it elsewhere in the plant.

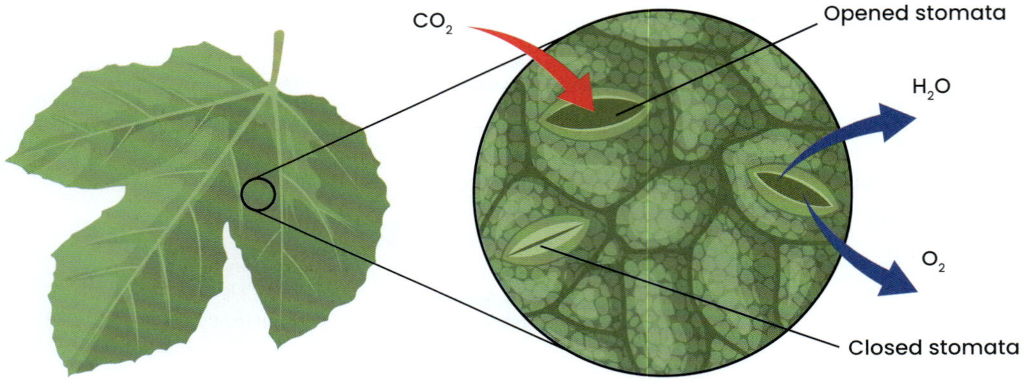

Figure 19.6: Stomata are small poles on the underside of leaves that allow gases (O_2 and CO_2) to be exchanged and for water vapour to leave.

Functions of the leaf

The main functions of the leaf are as follows:

- **Photosynthesis:** The leaves are mainly responsible for the production of the plant's food. Many of the cells in the leaves contain chloroplasts, the cell organelles responsible for photosynthesis.
- **Gas exchange:** The stomata open and close to allow oxygen and carbon dioxide diffuse in and out of the leaves.
- **Water transport:** The leaves are the site of **transpiration** – the loss of water from the plant through the stomata. This helps in the **uptake and transport** of water (and dissolved minerals) in the xylem, as well as **cooling** the plant.

Figure 19.7: Structure of the leaf.

- **Food storage:** Some plants can store food (sugars) in their leaves. The best example of this is the onion, which has layers of swollen leaves full of sugars and starch. Other plants like lettuce and cabbage also store food in their leaves.
- **Waste removal:** The leaf can be considered an **excretory organ** as it helps remove some of the plant's waste. Examples of plant wastes include oxygen gas, produced during photosynthesis, and waste products of metabolism stored by leaves over the summer and excreted when they fall in the winter.
- **Vegetative propagation:** This is a form of asexual reproduction in plants. Some plants can produce clones of themselves using their leaves. It's rare but one example is a common house plant, the snake plant, which produces small 'plantlets' on its leaves. These plantlets, when they fall off, grow into genetic clones of the parent plant.

Figure 19.8: Snake plant (*Sansevieria trifasciata*) plantlets on a leaf.

Go to YouTube.com and search for 'GCSE Biology – Structure of a Leaf and Stomata' (4:44) by Cognito.

THINK, PAIR, SHARE

In pairs, discuss how the various parts of the leaf are adapted to help the leaf carry out its functions. Summarise your thoughts on a mini-whiteboard.

FOCUS ON ... LOOKING AT PLANT TISSUES AND ORGANS – TS vs LS

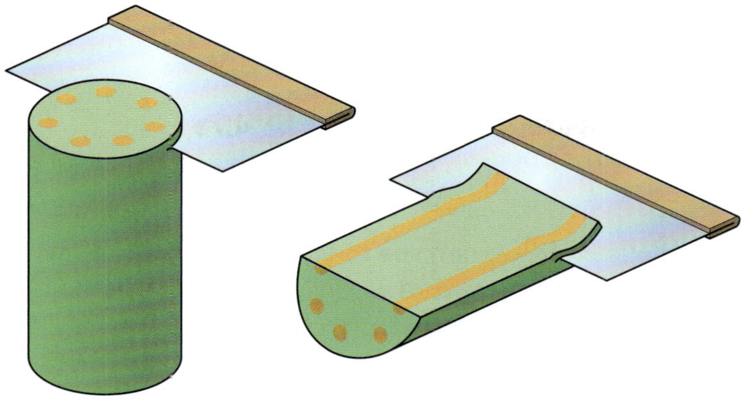

Leaves, flowers, stems and roots can be considered plant organs as they are made up of various tissues that work together to carry out specific functions. These tissues are arranged in specific ways to ensure the organ works effectively. When we view these organs under the microscope we can look at them in a number of ways. If we cut across them and look down we call this a transverse section (TS) or a cross-section. If we cut lengthways and look from the side we call this a longitudinal section (LS). Both are useful for studying how the tissues are arranged in a plant organ.

When you examine a **LS** of a leaf under the microscope, you see it is made up of various layers of tissues. Dermal tissue is found on the top and bottom of the leaf (**upper** and **lower epidermis**) with the top layer coated in a **waxy cuticle** to prevent water loss during warm weather. The lower epidermis has specialised dermal cells called **guard cells** that allow exchange of gases with the environment through tiny pores called **stomata**. These cells can control how the stomata open and close during the day.

Vascular tissue (**xylem** and **phloem**) is found within vascular bundles, within the leaf veins. Ground tissue makes up the bulk of the leaf with photosynthetic cells, densely packed below the upper epidermis. More photosynthetic cells are found within the aptly named **spongy layer**. This layer has lots of air spaces, allowing oxygen, carbon dioxide and water vapour to freely move around the inner section of the leaf.

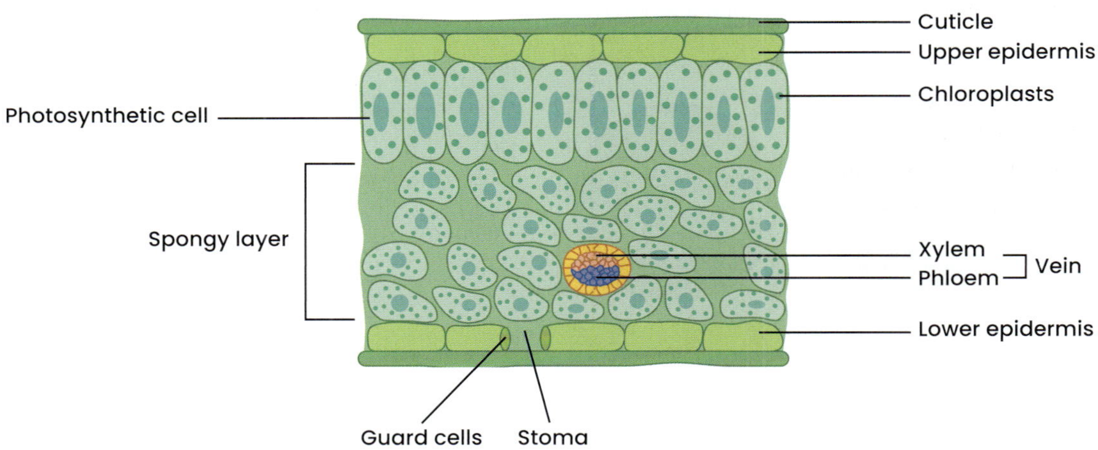

Figure 19.9: LS of a leaf.

THE STEM

The stem of a plant helps to **support** the leaves and flowers, positioning them to maximise photosynthesis and reproduction.

Along the stem, at the **nodes**, are a series of **buds**; these are sites of new stem, leaf or flower growth. The spaces between the nodes (and buds) are called **internodes**.

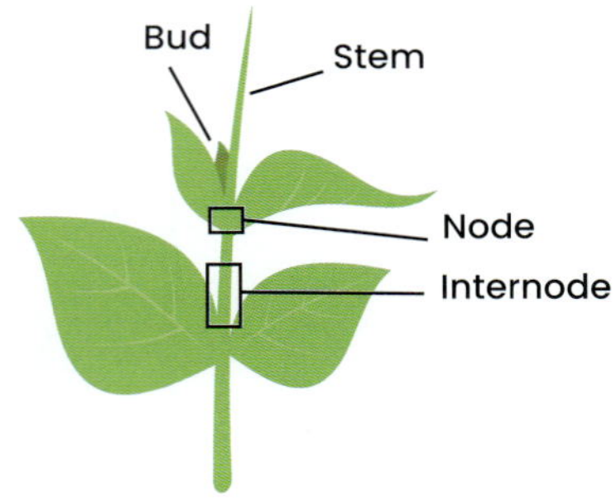

Figure 19.10: External structure of a stem.

If we view a cross-section (TS) of a stem, we can see the three types of plant tissue arranged in order. The **dermal** tissue is located on the outside of the stem as epidermis. This protects the plant. Some plants have pores called **lenticels** on the outside of their stems. These have a similar role to stomata, allowing the exchange of gases between the plant and atmosphere.

- A **bud** is a site of new stem, leaf or flower growth.
- **Lenticels** are pores on the surface of stems which allow for gas exchange.
- **Vascular bundles** contain xylem and phloem in the stem and leaves.
- **Cortex** is the tissue found between the dermal and vascular tissue in a plant's stem or roots.

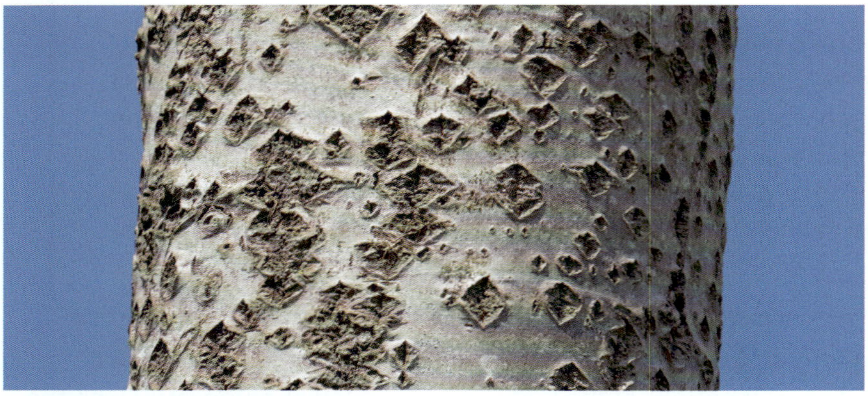

Figure 19.11: Lenticels on the surface of a tree.

Vascular tissue (xylem and phloem) is found within **vascular bundles**, arranged in a ring in dicot plants but scattered randomly in monocots.

Ground tissue again makes up the bulk of the cells and is found in the **cortex**. If the stem is green, some of these cells will contain chloroplasts for photosynthesis.

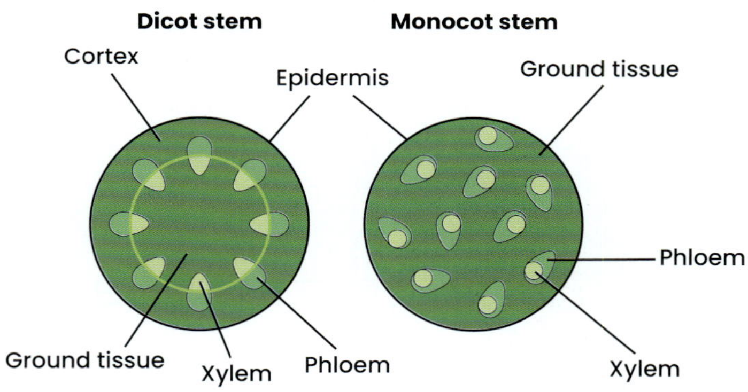

Figure 19.12: TS of a stem, with vascular bundles.

PORTFOLIO LINK

Go to your Learning Portfolio and complete Learning Check 2 to assess your knowledge and understanding of the topic so far.

Section 3: Plant Organs – Flowers and Roots

At the end of this section, you should be able to:

- Describe the structure and function of the flower and explain how the structure is adapted to carry out those functions.
- Investigate the structure of insect-pollinated and wind-pollinated flowers.
- Describe the structure and function of the root and explain how the structure is adapted to carry out those functions.

Unpacked from LOs 2.5.7, 2.6.14

THE FLOWER

The flower is the part of the plant responsible for reproduction, specifically sexual reproduction. Sexual reproduction involves the fusion of specialised sex cells called **gametes**.

The **stamen** is the male part of the plant and produces pollen, which produces the male gamete. The stamen has two parts: the anther and the filament. The **carpel** is the female part of the plant and produces the female gamete, the egg. The carpel has three parts: the stigma, style and ovary.

- The **stamen** is the male part of the flower.
- The **carpel** is the female part of the flower.
- **Pollination** is the transfer of pollen from the anther to the stigma.

Sexual reproduction in flowering plants involves a number of complex stages and we will explore the process in total in Chapter 21; however, the first stage is called **pollination** and involves the transfer of pollen from the anther of the stamen to the stigma of the carpel.

Some plants utilise insects and other animals to pollinate their flowers. These plants tend to have bright, colourful petals, a scent and sometimes nectar – a sticky sugar liquid – to attract and reward their pollinators. They also have sticky stigmas to catch the pollen.

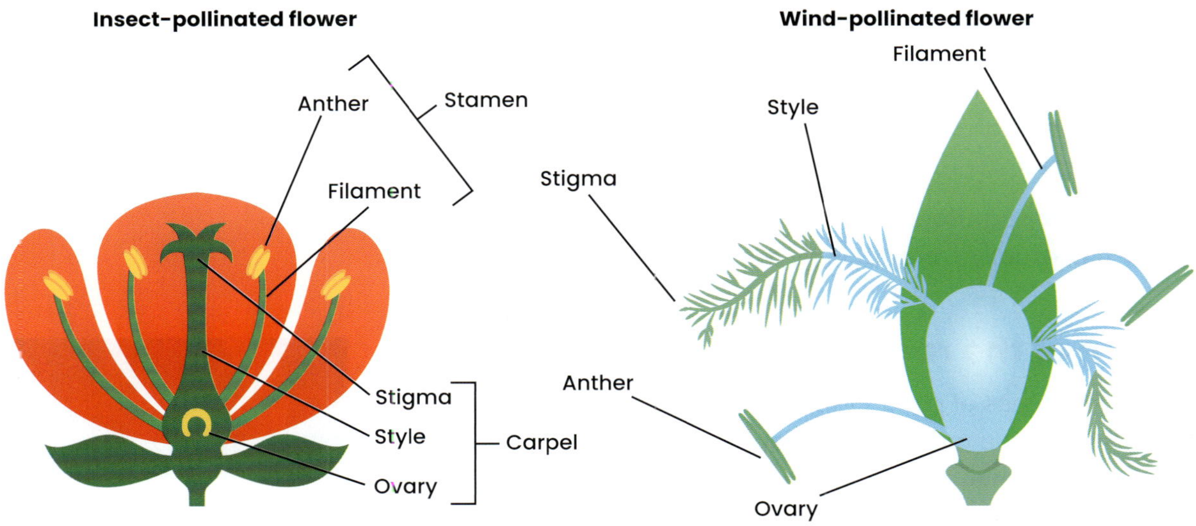

Figure 19.13: Insect- vs wind-pollinated flowers.

Wind-pollinated plants don't need colourful petals or a scent. They have evolved to produce large amounts of small, smooth pollen as the wind is less reliable than insects or animals. They have large feathery stigmas to trap the pollen.

The differences between insect- and wind-pollinated plants will be investigated further in Chapter 21.

Figure 19.14: The cherry flower is pollinated by insects like bees.

Figure 19.15: The flowers of plantain are pollinated by the wind.

THE ROOT

The root is the part of the plant typically found underground. The root has a number of functions in plants including:

- **Uptake of water:** The root is adapted to absorb water from the soil and pass it to xylem vessels in the stem (which transport it to the leaves for photosynthesis).
- **Uptake of nutrients:** The root is also adapted for the uptake of minerals from the soil. These minerals are needed in small amounts but essential for growth. We will explore the details of how plants absorb and transport water and minerals in Chapter 20.
- **Anchorage:** The roots provide stability for the plant, keeping the plant in the soil and preventing it from being blown over.
- **Food storage:** Many plants store their excess sugars or proteins in their roots during the growing season. Phloem vessels transport these excess foods from the leaves to the roots. The plant can use this food to regrow in the spring. We will explore this in more detail in Chapter 20.
- **Vegetative propagation:** Some plants can use their roots to reproduce asexually – producing clones of the parents.

THINK, PAIR, SHARE

Using your mini-whiteboards and working in pairs, discuss how the root is adapted to its functions.

Structure of the root

Again, the three types of plant tissue (dermal, vascular, ground) can all be found within the root. A layer of epidermal cells is found on the outside (epidermis), protecting the inner cells of the root from infection. Some of these dermal cells have **root hairs** – long narrow extensions of the cells – that are perfectly adapted for absorbing water and nutrient minerals from the soil.

Figure 19.16: Root hairs on a pea root.

Root hairs are narrow extensions of dermal cells which absorb water and minerals.

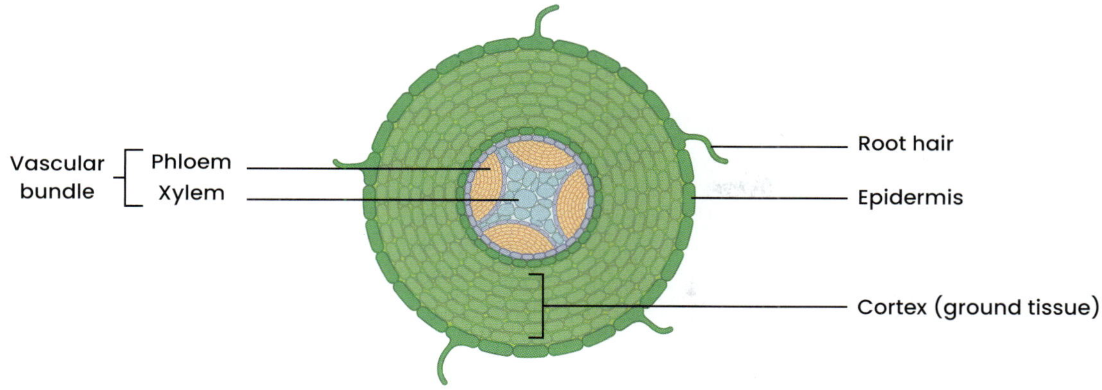

Figure 19.17: Cross-section (TS) of a root tip.

Deeper inside the root is the vascular tissue – xylem and phloem. The xylem transports water, absorbed initially by the root hairs, up through the stem to the leaves where it is needed for photosynthesis. The phloem transports food from the leaves to the roots for storage.

Ground tissue is found in the space between the epidermis and the vascular bundle, called the **cortex**, and provides support to the plant and stores excess food. There are large spaces between these cells allowing for water to move from the root hairs into the xylem. How these structures are arranged can be seen in TS of the root shown in Figure 19.17.

FOCUS ON ... MONOCOTS AND DICOTS

Flowering plants can be further classified as 'monocots' and 'dicots'. These are actually abbreviations for 'monocotyledons' and 'dicotyledons'. A cotyledon is a seed leaf; monocots have one seed leaf while dicots have two.

Monocots and dicots have the same plant tissues and cell types, but they are arranged differently in their roots, stems and leaves. Can you use the images below to identify some monocots and dicots around your school or home?

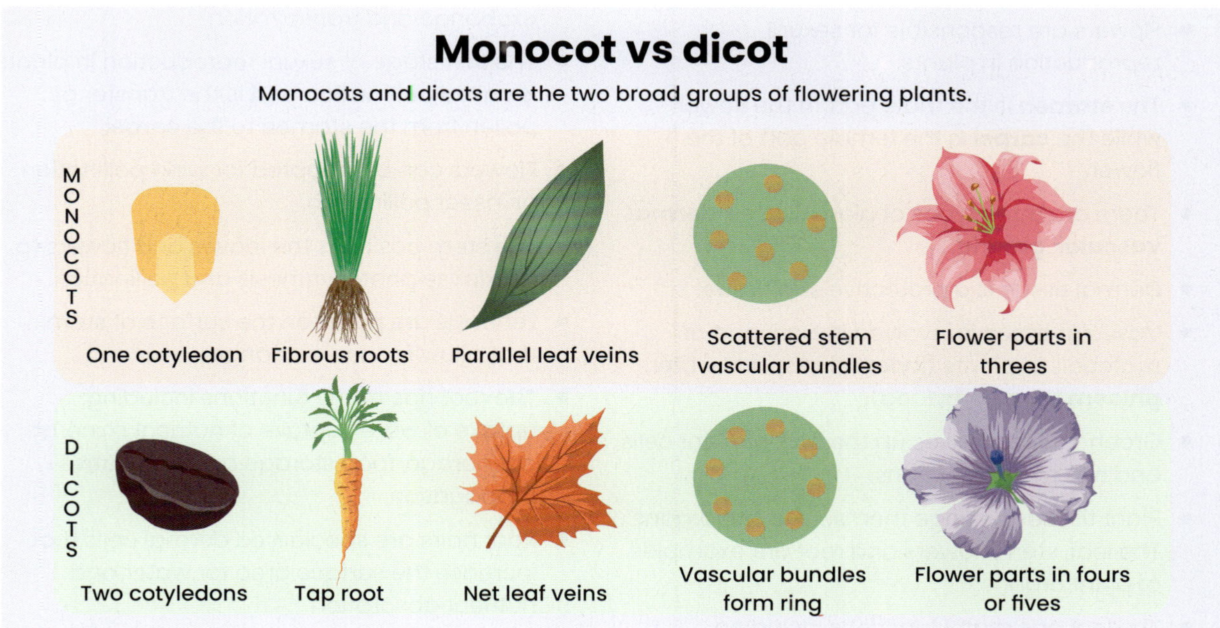

Figure 19.18: Features of monocots and dicots.

SCIENCE IN SOCIETY: SUSTAINABILITY

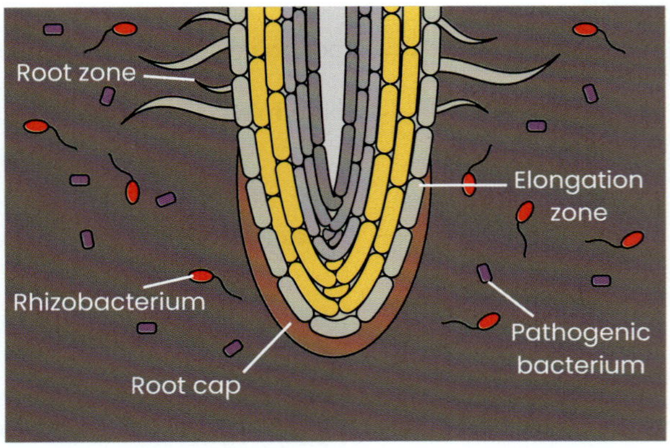

Figure 19.19: The rhizosphere is home to beneficial rhizobacteria.

Plant growth-promoting rhizobacteria (PGPR)

Rhizobacteria live in an area known as the rhizosphere – the part of the soil directly influenced by the roots. These bacteria can have a large effect on plant growth, helping to absorb nutrients for the plant and producing chemicals to promote their growth. Some rhizobacteria can fix nitrogen, converting nitrogen gas (N_2) into nitrate (NO_3^-), while others can produce more easily absorbed forms of iron and phosphate. Rhizobacteria can also protect the plant from disease by producing antibiotic and antifungal chemicals.

PORTFOLIO LINK

Go to your Learning Portfolio and complete Learning Check 3 to assess your knowledge and understanding of the topic so far.

RESEARCH ACTIVITY

Carry out further research on PGPR and their proposed use in agriculture and horticulture.

Chapter Summary

- Angiosperms are plants that produce flowers.
- Flowers are responsible for sexual reproduction in plants.
- The **stamen** is the male part of the flower while the **carpel** is the female part of the flower.
- There are three types of plant tissues: **dermal**, **vascular**, **ground**.
- Dermal tissue is a protective skin tissue.
- Vascular tissue is involved in transport of materials in plants (**xylem** transports water, **phloem** transports food).
- Ground tissue makes up the bulk of plant cells and has many functions.
- Plant tissues arrange themselves into organs. The leaf, stem, flowers and root are examples of plant organs.
- The leaf has many functions including photosynthesis, water transport, gas exchange, food storage, excretion and vegetative propagation.
- Leaves contain **stomata** – tiny pores for gas exchange and transpiration.
- The first stage of sexual reproduction in plants is called **pollination** and is the transfer of pollen from the stamen to the carpel.
- Flowers can be adapted for wind pollination or insect pollination.
- The stem positions the leaves and flowers to maximise photosynthesis and pollination.
- Lenticels are pores on the surface of stems that allow for gas exchange.
- The root has many functions including: uptake of water, uptake of nutrient minerals, anchorage, food storage and vegetative propagation.
- Root hairs are specialised dermal cells that increase the surface area for water and nutrient absorption.

END OF CHAPTER ASSESSMENT

REMEMBER

1. What is the primary function of flowers in plants?
2. Name the male part of a flower and state its function.
3. What is the term used to describe the fusion of an egg with a sperm nucleus in flowering plants?
4. What surrounds the seeds in flowering plants?
5. How many types of plant tissue are there? Name them.
6. What are the **two** main types of vascular tissue?
7. What is the function of xylem tissue in plants?
8. Stomata are present in leaves, name the corresponding structure in stems.
9. Name two gases that exit or enter the leaf.
10. Name a plant in which the leaves are modified for food storage.
11. (a) What is the meristem?
    (b) Give two locations in plants where meristematic tissue may be found.
    (c) What is the function of meristematic tissue?

UNDERSTAND

12. Explain why flowers are considered the reproductive parts of a plant.
13. Describe the difference between xylem and phloem tissue in terms of their function.
14. How do flowers attract pollinators, and why is this important for plant reproduction?
15. Why do wind-pollinated plants produce large amounts of small, smooth pollen grains?
16. How does the structure of a leaf support its function in photosynthesis?
17. Compare and contrast the functions of dermal, vascular and ground tissue in plants.
18. Explain how root hairs contribute to the uptake of water and nutrients in plants.

APPLY

19. You are designing a garden to attract pollinators. How would you arrange the types of flowering plants to ensure continuous flowering throughout the year?
20. You are a botanist studying a new species of plant. How would you determine whether it is wind-pollinated or insect-pollinated based on its flower structure?

ANALYSE

21. Analyse the adaptations of a leaf that make it suitable for its primary functions.
22. Examine the differences in the arrangement of tissues in the stem and root of a plant, and discuss how these differences relate to their functions.
23. Evaluate the advantages and disadvantages of sexual reproduction compared to asexual reproduction in flowering plants.
24. Indicate how a root is adapted for the uptake of water.
25. Distinguish between monocots and dicots under the following headings:
    (a) Leaf venation
    (b) Number of cotyledons
    (c) Arrangement of vascular bundles
    (d) Flower parts

EVALUATE

26. Discuss the effectiveness of different methods of pollination (insect vs wind) in ensuring successful plant reproduction.
27. Discuss the impact of environmental factors (light, water, temperature, etc.) on the growth and development of plant organs such as leaves, stems, flowers and roots.
28. Evaluate the ethical considerations surrounding the use of genetically modified plants for agricultural purposes.

CREATE

29. Design an experiment to investigate the effect of different types of soil on root growth in plants.
30. Design an investigation on the effect of rhizobacteria in the soil on plant growth.
31. Propose a research project to explore the potential benefits of introducing new pollinators to a specific ecosystem.

20 Plant Transport and Responses

LEARNING OUTCOMES: U2.3, U2.5, 2.4.1, 2.5.8, 2.6.8, 2.6.14, 2.6.15, 2.6.16

EYE-OPENER

Xerophytes are plants adapted to survive in dry, arid environments where water is scarce. Cacti are an example of a xerophyte. To minimise water loss, cacti have adapted their leaves into spines, significantly reducing the surface area for water loss. Their stems are covered with a thick, waxy cuticle that prevents water loss through transpiration, and the few stomata they possess are sunken, reducing exposure to dry air. Cacti also have an expandable stem that can store water when available, along with extensive root systems that reach both shallow and deep soil layers.

At the end of this chapter, you should be able to:

- Name materials transported within plants and describe the reason for their transport.
- Define the term 'transpiration' and investigate the factors affecting it.
- Describe the mechanism of water uptake and transport in the plant.
- Describe the uptake of nutrient minerals in the plant.
- Define the term 'gas exchange' and illustrate how gases, like carbon dioxide and oxygen, are taken in by the plant.
- Describe the transport of the products of photosynthesis through the plant.
- Explain how plant tropisms are controlled by plant growth regulators.

LINKS TO:

Chapter 16	**Cell Structure and Diversity**
Chapter 17	**Cell Continuity**
Chapter 19	**Flowering Plants – Structure and Function**
Chapter 21	**Plant Reproduction**
Chapter 22	**Photosynthesis**

SUBJECT VOCABULARY

transpiration, transpiration stream, stomata, root hairs, osmosis, passive process, active process, osmotic pressure, adhesion, cohesion, capillary action, transpiration pull, cohesion-tension model, guard cells, turgid, flaccid, gas exchange, translocation, phloem, stimulus, tropism, plant growth regulator, alkaloid

Section 1: Transpiration and Stomata

At the end of this section you should be able to:

- Discuss the terms 'transpiration' and the 'transpiration stream'.
- Describe the factors affecting the rate of transpiration in plants.
- Describe how the structure of stomata allows for controlling the rate of water loss from the plant.
- Outline how CO_2 levels affect guard cell turgidity and thus the opening and closing of stomata. HL

Unpacked from LOs 2.6.8, 2.6.14, 2.6.15

THE TRANSPIRATION STREAM

Water is constantly flowing through living plants. The roots absorb water from the soil, which then travels up through the plant (in xylem vessels) and arrives at the leaf. The leaf will use some of that water for photosynthesis but the vast majority of it will escape through tiny pores called **stomata** on the underside of the leaf.

This continuous flow of water, from the roots to the leaves (and out of the plant), is called the **transpiration stream**.

- **Stomata** are tiny leaf pores that allow for gas exchange and transpiration.
- The **transpiration stream** is the continuous flow of water from the roots to the leaves.
- **Transpiration** is the loss of water from the leaves, mainly through tiny pores called stomata.

Figure 20.1: The transpiration stream.

The leaf has lots of air spaces (in the spongy layer) which allow for the movement of water from the xylem (in the vascular bundles or veins) out of the plant via the stomata. The loss of water from the leaves is called **transpiration**.

Transpiration is vitally important for the plant. Just like sweating, the loss of water from the leaves helps to **cool the plant** and it allows more water to be absorbed by the roots. The 'new' water absorbed will have more mineral nutrients dissolved in it.

Go to YouTube.com and search for 'Water transport in a plant' (3:01) by OVD THIMO.

FACTORS AFFECTING TRANSPIRATION

The rate at which water is lost through the leaves isn't constant and can change due to the time of the day or environmental conditions. Later we will investigate how these factors affect the rate of transpiration.

The rate of transpiration in plants is affected by several key factors:

1. **Temperature:** As temperature increases, the rate of transpiration increases. Higher temperatures cause water molecules to have more kinetic energy, increasing evaporation from leaf surfaces.
2. **Light intensity:** Increased light intensity leads to higher transpiration rates. This is because light causes stomata to open wider, allowing more water vapour to escape.
3. **Humidity:** Transpiration rate decreases as humidity increases. Higher humidity reduces the water vapour concentration gradient between the leaf and the surrounding air, slowing water loss.
4. **Wind speed:** Increased air movement raises the transpiration rate by removing water vapour from around the leaf surface.
5. **CO_2 concentration:** At night, CO_2 levels increase when photosynthesis stops, causing the stomata to close.
6. **Leaf surface area:** Larger leaf surfaces allow for more transpiration compared to smaller leaf surfaces.
7. **Number and distribution of stomata:** More stomata and their arrangement on the leaf surface can increase transpiration rates.
8. **Cuticle thickness:** A thicker waxy cuticle on the leaf surface reduces water loss through transpiration.

During the day, stomata are generally **open**, allowing water to leave and carbon dioxide to enter the photosynthetic cells in the leaf. At night, when photosynthesis isn't occurring, plants don't need water in the leaves (or carbon dioxide) so the stomata **close**. The transpiration stream is switched off, reducing water loss.

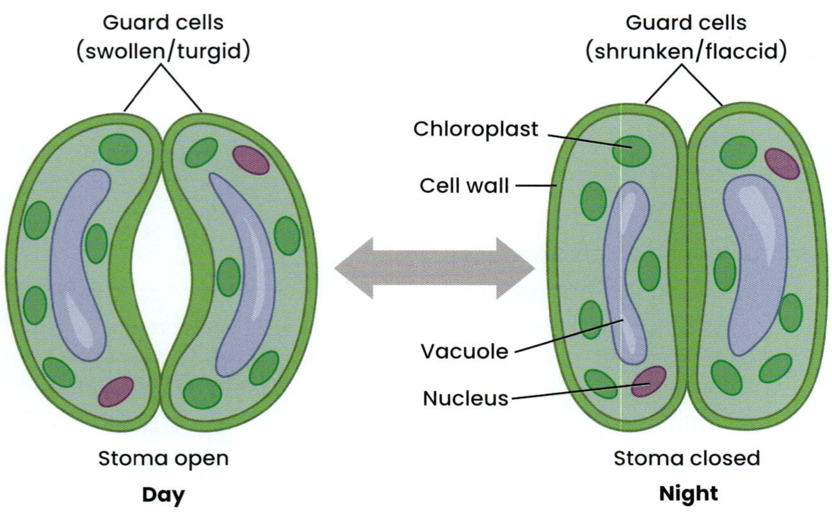

Figure 20.3: Stomata generally open during the day and close at night.

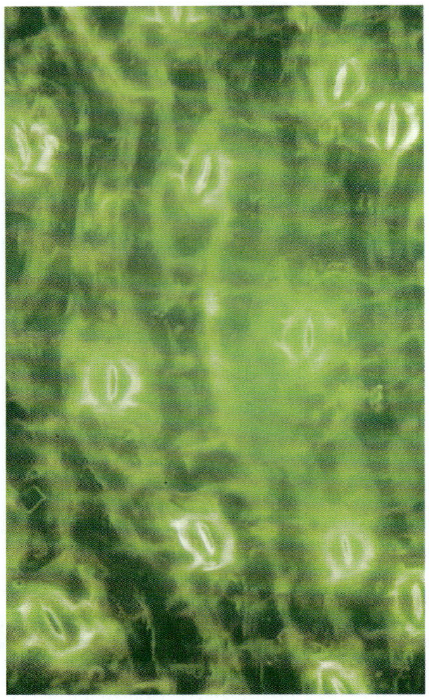

Figure 20.2: Stomata under a microscope.

The stomata can sometimes close during the day, if required. In very hot weather, or if there is a drought, the plant is in danger of losing too much water. In this case, the stomata close, preventing the cells of the plant from wilting.

Figure 20.4: Guttation sees water droplets form on leaves in the morning.

Even when the stomata are closed and the transpiration stream is switched off, water can still enter the plant via osmosis. This can sometimes result in water building up in the leaves in the early morning, forming drops at the edges of some plants. This phenomenon is called guttation.

DID YOU KNOW?

STOMATA – EXTENDED STUDY

Stomata have two main functions in plants: to facilitate the exchange of gases with the plant's environment **and** to control the rate of transpiration. A typical leaf has thousands of these tiny pores, mainly on the lower surface.

Stomata are surrounded by guard cells. **Guard cells** are crescent-shaped and two guard cells are found around a stoma. These guard cells can shrink or enlarge to control the opening and closing of the pore. Guard cells usually have chloroplasts too for photosynthesis.

- **Guard cells** are crescent-shaped cells that actively control the size of stomata.
- **Turgid** describes cells that contain lots of water and are firm.
- **Flaccid** describes cells that contain little water and are not firm.

As mentioned previously, stomata are generally open during the day. The opening of the stomata is caused by the guard cells filling with water and becoming **turgid**. When they close, water moves out of the guard cells and the cells become **flaccid**. This change in the shape of the cells isn't an accident – it is controlled by the guard cells themselves.

HL

During the day, guard cells pump potassium ions (K^+) into their cells. This creates a more concentrated cytoplasm (lower water concentration) in the guard cells which, in turn, results in water moving into the guard cells by osmosis. The guard cells become turgid and the stomata open.

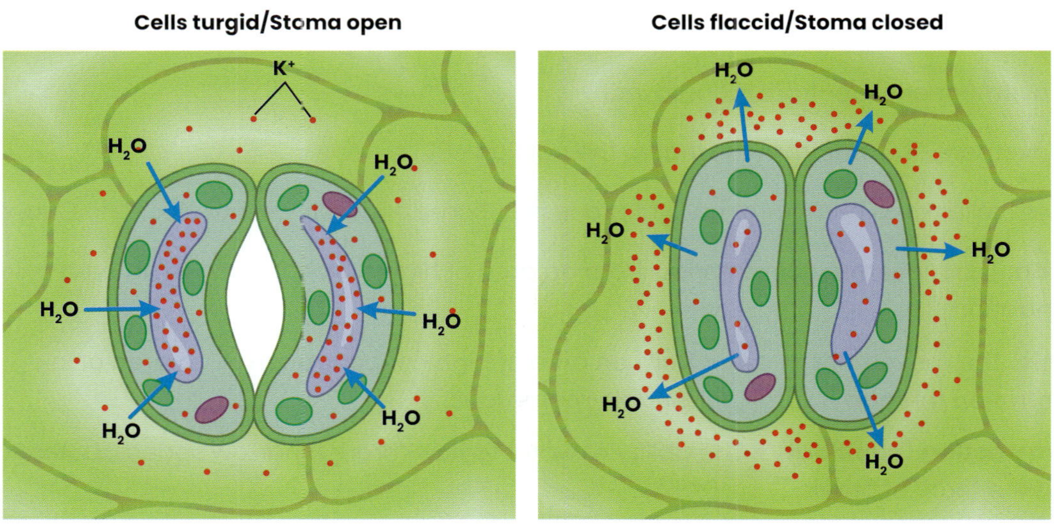

Figure 20.5: The movement of potassium ions (K^+) facilitates the opening and closing of stomata.

At night, or when the plant needs to conserve water, the guard cells pump the potassium ions out of their cells, creating more concentrated cytoplasm within the surrounding cells. Water then moves out of the guard cells and into the surrounding cells and the guard cells turn flaccid. The stomata then close.

HL There are a number of factors that cause the closing of the stomata. High carbon dioxide (CO_2) levels (typically at nighttime, as photosynthesis isn't occurring but respiration is) is the most significant controlling factor in stomatal closing but low humidity levels around the leaf (in very dry or windy conditions) can also cause the stomata to close.

PORTFOLIO LINK **Go to your Learning Portfolio and complete Learning Check 1 to assess your knowledge and understanding of the topic so far.**

Section 2: Water Uptake and Transport in Plants

At the end of this section, you should be able to:

- Describe the mechanism of water uptake in the plant.
- Describe the mechanism of water transport up through the plant, i.e. the cohesion-tension model of water transport. HL
- Relate structure to function in xylem.

Unpacked from LOs 2.6.15, 2.6.16

UPTAKE OF WATER BY PLANTS

Let's look more closely at how plants absorb water at the roots and transport it to the leaves.

There are various stages involved in the process:

1. Movement of water from the soil to the root epidermis.
2. Movement of water from the root epidermis to the xylem.
3. Movement of water through the stem.
4. Movement of water from stem to leaf.
5. Movement of water from the leaf to the outside.

The process relies on a number of natural processes, some of which we introduced in Chapter 18.

- **Diffusion:** The movement of materials from an area of high concentration to an area of low concentration.
- **Osmosis:** The movement of water from an area of high concentration to an area of low concentration through a semi-permeable membrane.
- **Capillary action (capillarity):** The ability of water to flow through narrow spaces, potentially against the force of gravity.
- **Cohesion:** The attraction of water to other water molecules.
- **Adhesion:** The attraction of water to other substances or surfaces.

1. Movement of water from the soil to the root epidermis

- Water moves from the soil into the dermal cells of the root by **osmosis**.
- The epidermis of the root contains **root hairs** – thin extensions of the cell that don't have a waxy cuticle.
- The vast number of root hairs **increases the surface area** for absorption of water.
- The cytoplasm of the dermal cells is more concentrated (contains more dissolved solids) than the soil water, meaning water moves from the soil into the root hair cells to balance the concentration of water (i.e. water moves from its greater concentration to an area of lower concentration).
- Osmosis is a **passive process** – no energy is used up in the process.

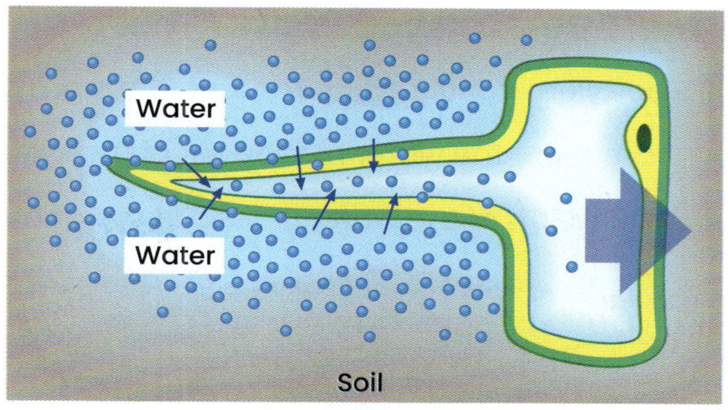

Figure 20.6: Uptake of water by the root hair cell.

- **Osmosis** is the movement of water from an area of high concentration to an area of low concentration through a semi-permeable membrane.
- **Osmotic pressure** (or root pressure) is the upward movement of water caused by the movement of water (by osmosis) into xylem vessels.
- **Capillary action** is the movement of water upwards in thin tubes against the force of gravity.
- **Adhesion** is the ability of water to stick to surfaces.

2. Movement of water from the root epidermis to the xylem

There are two ways in which water can travel from the dermal cells to the xylem within the root.

- Path 1: water moves by osmosis from cell to cell, via the cytoplasm or the vacuoles, from the root hair to the xylem.
- Path 2: water moves from cell wall to cell wall, by diffusion (it doesn't cross a membrane), from the root hair to the xylem.

HL

When water enters the xylem in these ways, it 'pushes' water already in the vessel further up the tube. This pushing force is known as root pressure or **osmotic pressure**.

3. Movement of water through the stem

- Water moves up through the stem in xylem, found within the vascular bundles, by **capillary action**.
- The thinner the capillary, the higher the water will travel, and xylem is extremely thin.
- Capillary action is caused by water's great **adhesion**.
- There is a continuous column of water due to water's cohesion – the molecules are attracted to each other.
- Capillary action is another **passive process**.

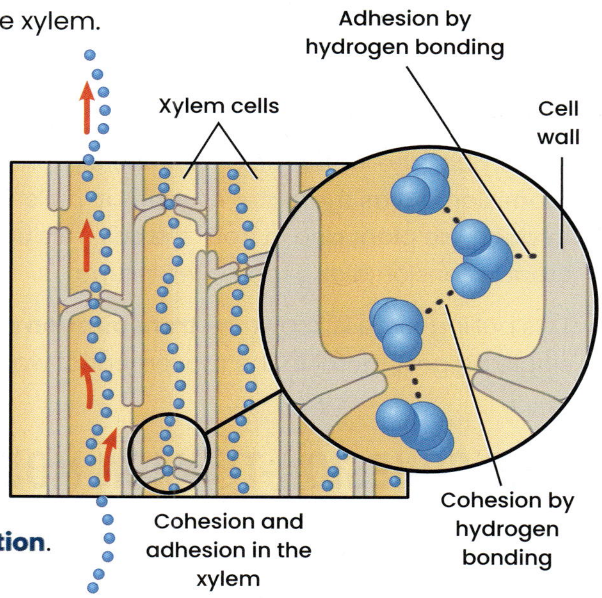

Figure 20.7: Cohesion and adhesion.

Go to YouTube.com and search from 'Armchair Animation | Capillary Action' (0:20) by Armchair.

4. Movement of water from stem to leaf

- The xylem continues up through the stem and into the leaf via the leaf veins.
- From there the water moves out of the xylem and into the leaf cells.
- Around 10% of the water is used in the photosynthetic cells to make sugars.

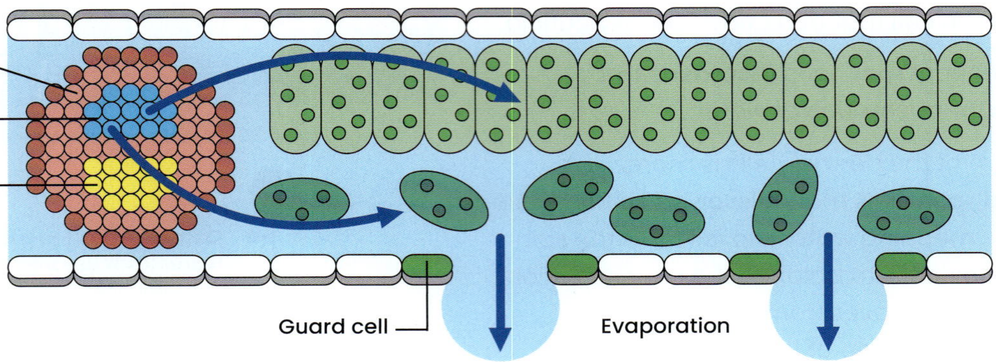

Figure 20.8: Movement of water in the leaf.

5. Movement of water from the leaf to the outside

- Water evaporates from the leaf cells into the air spaces in the spongy layer.
- The water vapour moves through the air spaces and out through the stomata on the undersurface of the leaf. The loss of water from the leaf is called transpiration.
- 90% of the water that reaches the leaf is lost through transpiration.
- Remember, the water molecules are all connected (cohesion) so as the water escapes it 'pulls' water up through the plant and in from the soil. This upward pull creates tension along the water column.
- This pulling force is known as **transpiration pull** and is the single biggest factor in the movement of water in plants.

Transpiration pull is a pulling force, caused by water's great cohesion, that draws water up through a plant.

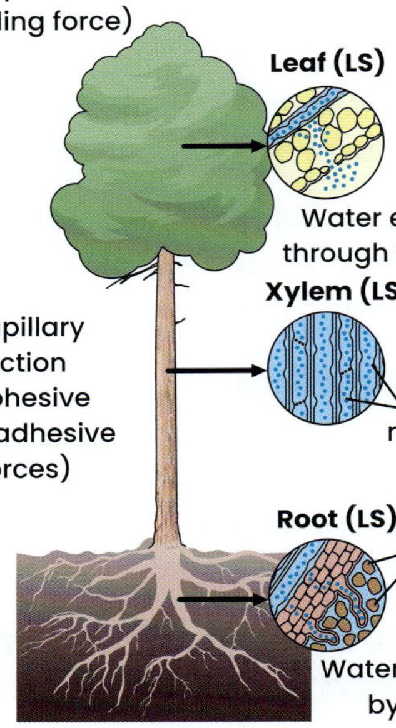

Figure 20.9: The cohesion-tension model.

THE COHESION-TENSION MODEL

HL

The overall mechanism for water transport in plants is referred to as the **cohesion-tension model**. This model was developed by two Irish scientists, **John Joly** and **Henry Dixon**. They proposed that three natural forces account for a plant's ability to transport water against the force of gravity. These are osmotic pressure, capillary action and transpiration pull.

Plants, even the tallest trees, use no energy to transport water through their roots, stems and leaves. It is said to be a **passive process**.

Go to YouTube.com and search for 'Transpiration and Cohesion-Tension Theory. Cohesion and adhesion in the transport of water in plants' (10:38) by Miss Estruch.

FOCUS ON ... THE WATER MOLECULE

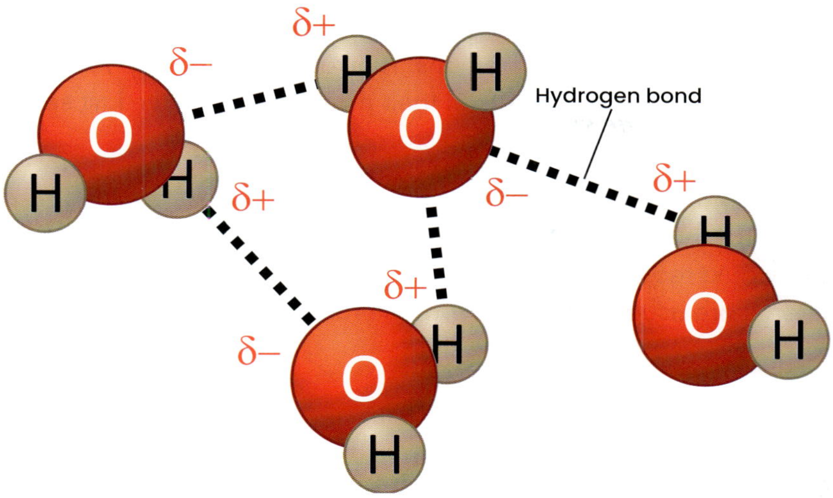

Figure 20.10: Hydrogen bonds between water molecules.

Water is a compound made from two atoms of hydrogen and one atom of oxygen. Its chemical formula is H_2O and its molecule forms a V shape.

This V shape gives water some of its unique chemical and physical properties. It results in the molecule's electrical charge being unevenly distributed, with the hydrogen atoms having a slight positive charge and the oxygen having a slight negative charge.

This means that, in its liquid form, the hydrogen in one water molecule is weakly attracted to the oxygen in another nearby water molecule (like a north pole of a magnet would be attracted to the south pole of another magnet). In fact, in a body of water, all the water molecules are attracted to each other.

These weak connections between the water molecules are called **hydrogen bonds** and are vital for understanding how water moves through plants. Hydrogen bonds also explain why water sticks to surfaces, like glass, and forms beads on waxy surfaces.

PORTFOLIO LINK **Go to your Learning Portfolio and complete Learning Check 2 to assess your knowledge and understanding of the topic so far.**

Hydrogen bonds are weak bonds between molecules.

Section 3: Mineral Nutrients, Gas Exchange and the Products of Photosynthesis

At the end of this section, you should be able to:

- Describe the uptake of nutrient minerals in the plant.
- Describe the role of transport proteins (and active transport) in the uptake of nutrient minerals from the soil.
- Describe how nutrient minerals are transported through the plant and outline the role of diffusion in their movement to plant cells.
- Define the term gas exchange and illustrate how gases, like carbon dioxide and oxygen, move within a plant.
- Describe the transport of the products of photosynthesis through the plant.

Unpacked from LOs 2.6.14, 2.6.15

UPTAKE OF MINERAL NUTRIENTS

Plants need a steady supply of nitrogen (in the form of nitrate), phosphorous (in the form of phosphate), potassium (as potassium ions) and many other ions. They use these to help form proteins, pigments (like chlorophyll) and to control various aspects of their metabolism (we've just seen how potassium ions play an important role in opening and closing stomata).

While the uptake of water by the plant is a completely passive process, requiring no energy, mineral absorption is generally an **active process**. Mineral nutrients can diffuse from soil water into the dermal cells of the root but this would require a higher concentration of the nutrient in the soil, which rarely happens. Mineral nutrients are usually 'pumped' into the root hair cells in young roots, and this process requires energy.

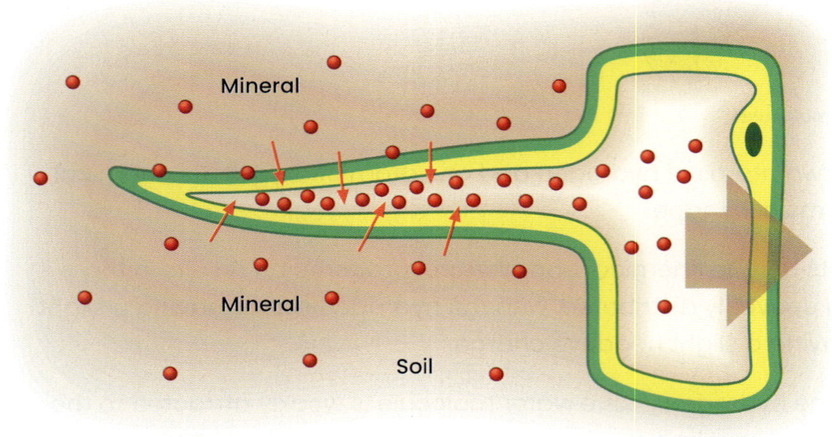

Figure 20.11: Uptake of mineral nutrients by the root hair cell.

Transport proteins move materials across biological membranes and are involved in active transport.

The membranes of these root hair cells contain **transport proteins** that can pump nutrients from the soil water into the root hair cells. These cells also have lots of **mitochondria** powering the process. The process can result in the concentration of some nutrients being over a thousand times higher in the root dermal cells than in the soil (in turn, this increased concentration results in the osmotic movement of water into the root hair, as discussed in the previous section).

Once in the root hair cells, the minerals then **diffuse** from cell to cell until they reach the **xylem**, by which they then travel up through the plant. The minerals can diffuse from the xylem into the cells that need them, e.g. a photosynthetic cell may need more magnesium ions to form new molecules of chlorophyll.

RESEARCH ACTIVITY

The Venus flytrap is an insectivorous or carnivorous plant that generally lives in soils that are low in mineral nutrients. The consumption of insects provides the additional nitrogen and other mineral nutrients the plant needs. Carry out some research on three other carnivorous plants (find at least one native Irish one) and present your research to the class.

Go to YouTube.com and search for 'Biology: Cell Transport' (2:02) by Nucleus Medical Media.

SCIENCE IN SOCIETY: SUSTAINABILITY

Mycorrhizae

Mycorrhizae (meaning fungus root) are types of fungi that form symbiotic (mutualistic) relationships with plants. The fungi live within and around the roots of the plants and greatly improve their uptake of water and nutrients. In fact, a plant with mycorrhizae around its roots can absorb over 700 times more mineral nutrients than a plant without them. They also improve soil structure, increase resistance to drought and help the roots find better soil. They can also inhibit harmful and disease-causing microbes. In turn, the plants provide sugars and amino acids for the fungi.

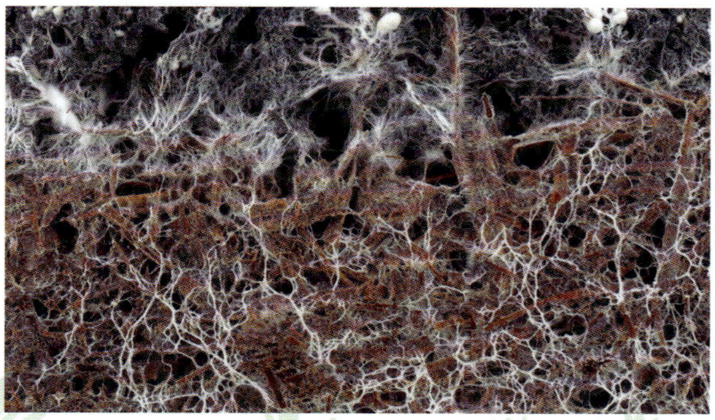

Figure 20.12: Mycorrhizae can be beneficial to plants.

Plant growers can promote the growth of mycorrhizae (and subsequently their plants) by adding organic matter to the soil, such as manure, by mulching and by minimising soil tillage (digging up or ploughing). Research is continuing into the use of mycorrhizae by plants to 'communicate' chemically with other plants nearby. The mycorrhizae help to transmit plant growth regulators from one plant to another.

GAS EXCHANGE IN PLANTS

Photosynthesis requires a constant supply of carbon dioxide to the photosynthetic cells in the leaf, mainly found on the upper layer of the leaf. There are two main sources of this carbon dioxide: the air surrounding the leaf (atmospheric CO_2) or from the cells themselves (CO_2 released by cellular respiration). Atmospheric CO_2 is the most abundant source and its uptake is again a **passive process**, requiring no energy input by the plant. Again, the plant relies on **diffusion**.

When levels of CO_2 are low, the stomata open allowing more CO_2 to diffuse from the atmosphere (high concentration) into the air spaces of the spongy layer of the leaf (low concentration). If the plant is photosynthesising, the CO_2 will be quickly used up and more CO_2 will diffuse in. At the same time the plant is producing oxygen (O_2) as a by-product of photosynthesis. Some of this will be used by the cells in leaves for respiration but most will **diffuse** out of the leaf (high concentration) into the air around the leaf (low concentration).

Plant gas exchange is the diffusion of carbon dioxide and oxygen through stomata or lenticels.

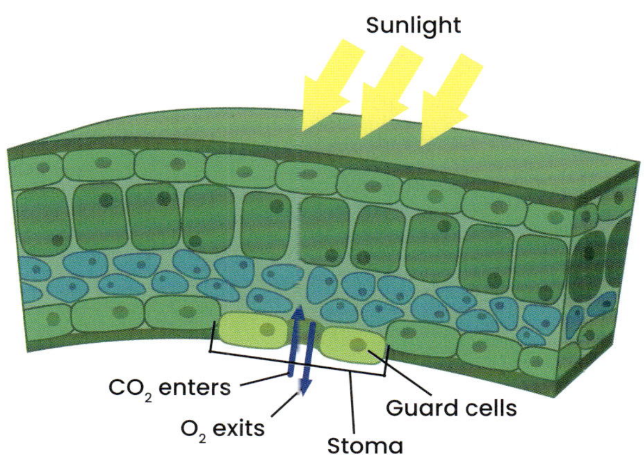

Figure 20.13: Movement of oxygen and carbon dioxide in the leaf during photosynthesis.

The stomata close at nighttime as a result of high CO_2 levels, limiting the diffusion of gases.

Gas exchange can also take place through small pores called **lenticels**. These can be found on stems (e.g. on the bark of trees, as shown in Figure 20.14) or in tubers (e.g. potatoes), storage roots (e.g. carrots) and even fruits (e.g. pears).

Figure 20.14: Lenticels on the stem are pores that facilitate gas exchange.

PHOTOSYNTHESIS: REACTANTS AND PRODUCTS

The process of photosynthesis is presented in detail in Chapter 22. In summary, the process requires water, carbon dioxide, light and the green pigment chlorophyll. The reaction can be summarised in the equation shown below.

Sun's energy

$$6CO_2 + 6H_2O \rightarrow 6C_6H_{12}O_6 + 6O_2$$

Carbon dioxide | Water | Chlorophyll | Sugars | Oxygen

The two **reactants** are carbon dioxide (CO_2) and water (H_2O). Water is transported from the roots to the leaves through a combination of adhesion, cohesion and capillary action. We also learned that CO_2 diffuses into the leaf spaces through the stomata or is produced by respiring cells in the plant. Chlorophyll is required to trap light energy and we learned that magnesium is a key nutrient mineral required for its production. Minerals are absorbed at the roots by active transport and move through the plant via the xylem.

There are two **products** of the reaction: glucose ($C_6H_{12}O_6$) and oxygen (O_2). We have already learned about the fate of oxygen – it diffuses out of the leaf through the stomata and into the atmosphere.

Some glucose will be used by the cells in the leaf for respiration or converted to **starch** or **sucrose** and stored in the leaf. A large proportion of the glucose produced during photosynthesis will need to be transported to other parts of the plant – either for respiration or for storage.

Reactants are substances required and used up during a chemical reaction.

Products are the substances formed during a chemical reaction.

Go to YouTube.com and search for 'Introduction to Photosynthesis' (1:40) by Teacher's Pet.

PHLOEM AND FOOD TRANSPORT

Translocation is the process by which plants transport food molecules, primarily sucrose and amino acids (the building blocks of proteins), via tubes called **phloem**. These materials are usually made by leaf cells during photosynthesis but need to be transported to other parts of the plant for storage, for energy or to make new cells. These leaf cells are often referred to as **sources** – they make more sugars and proteins than they need. The cells that do not make their own sugars but still require them are known as **sinks**.

- **Translocation** is the process by which food molecules are transported from one part of a plant to another.
- **Sources** are the cells in a plant where excess foods are produced.
- **Sinks** are the cells in a plant where excess foods are moved and stored.

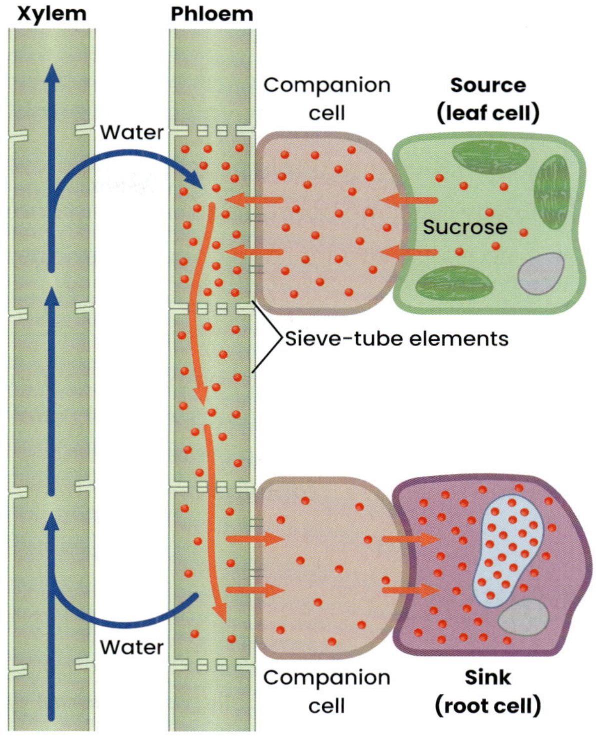

Figure 20.15: Phloem is responsible for translocation.

Translocation is an **active process**, requiring energy in the form of ATP produced during respiration (the breakdown of sugars for energy). Translocation can occur in both directions, e.g. in early spring food stored in a bulb or tuber can act as the source and newly formed cells in the shoot can be the sink.

Go to YouTube.com and search for 'Translocation – GCSE Biology' (2:15) by Launchpad Learning.

FOOD STORAGE

Plants produce most of their food during the summer, when warm temperatures and long days allow for maximum levels of photosynthesis. A lot of plants store the excess food (sugars, proteins) in modified organs like roots, stems or buds. These are often used for winter survival (the green parts of the plant may die at the end of the summer) and for asexual reproduction.

Examples of plants that use modified roots include carrots, turnips and beetroot. These roots store the excess sugars produced by the plant during the summer. An onion bulb is an excellent example of a modified bud – each layer of an onion is a young, non-photosynthetic leaf. Plants like rhubarb and potatoes use modified stems for food storage. The potato is an example of a tuber – a swelling of the stem.

Figure 20.16: L–R: Carrot, a modified root; onion, a modified bud; rhubarb, a modified stem.

PORTFOLIO LINK Go to your Learning Portfolio and complete Learning Check 3 to assess your knowledge and understanding of the topic so far.

Section 4: Plant Responses

At the end of this section, you should be able to:

- Define the term 'tropism' and give examples in plants.
- Briefly describe how tropisms are regulated by plants.
- Discuss anatomical and chemical adaptations of plants.

Unpacked from LOs 2.4.1, 2.5.8

Like all living things, plants can respond to changing environmental conditions. They are sensitive to external and internal **stimuli** (**stimulus**, singular). Plants have various mechanisms for detecting the stimuli and for responding, ultimately allowing them to get the resources they need and to survive. **Tropisms** are growth responses of plants to stimuli.

TROPISMS

Tropisms allow plants to respond to changing environmental conditions. They are often called **directional growth responses**, as they cause the plant to either grow in the direction of the stimulus (e.g. towards light) or away from the stimulus (e.g. against the force of gravity).

A **stimulus** is an internal or external detectable change which prompts a response in a living thing.

A **tropism** is a directional growth response in a plant to a stimulus.

Examples of tropisms

Figure 20.17: Examples of tropisms.

- **Phototropism:** The growth response of a plant towards a light source. Leaves will exhibit positive phototropism as they grow towards a light source.
- **Thigmotropism:** The growth response to touch or physical contact. This is evident in climbing plants that wrap around supports.
- **Geotropism:** The growth response to gravity. Roots will exhibit **positive geotropism**, growing down with gravity, while the shoots will exhibit **negative geotropism**, growing up against the force of gravity.

PLANT GROWTH REGULATORS

Tropisms, and other plant responses, are controlled by chemical messengers called **plant growth regulators** (PGRs). They are produced in small amounts by various plant tissues but have a large effect on the growth and development of the plant.

A **plant growth regulator** is a chemical messenger produced in small amounts which affects a plant's growth development.

Auxins are PGRs that control tropisms by causing some cells to elongate more than others (e.g. auxin causes cells on the shaded side of a stem to elongate more, bending the stem towards the light). Auxins, along with other PGRs called **gibberellins**, are produced by seeds as they germinate. **Ethylene** is a gas produced by ripening fruits.

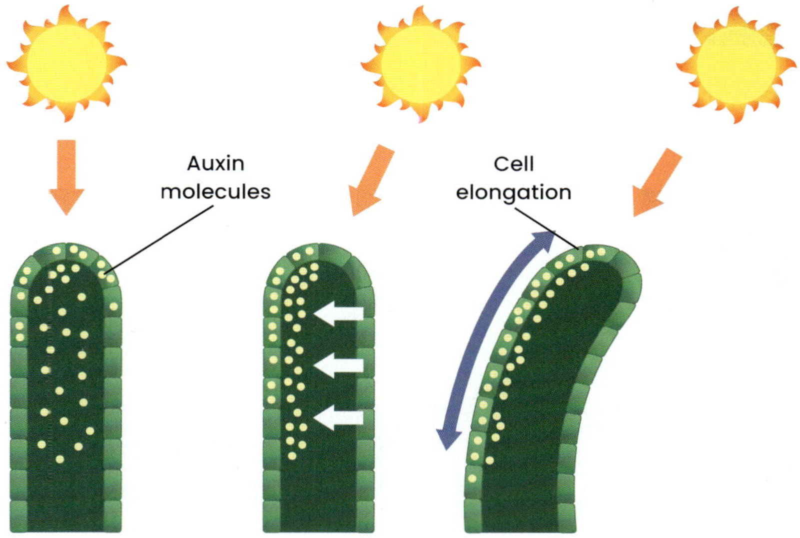

Figure 20.18: Auxins cause cells on the shaded side of a stem to elongate more.

PGRs can act as either growth promoters or inhibitors depending on their type and concentration. In agriculture and horticulture, both natural and artificial PGRs can be used to control plant growth, enhance crop yields and improve plant quality. For example, ethylene can be sprayed on green bananas to ripen them, and synthetic gibberellins can be used to increase fruit size.

PHYSICAL AND CHEMICAL ADAPTATIONS OF PLANTS

Plants have, over millions of years of evolution, developed a number of physical and chemical adaptations to increase their chance of survival. Some physical adaptations include thorns, spines, thick waxy cuticles or bark. Plants produce many chemicals which can maximise their growth or deter herbivores from eating them. Many of these chemicals are also plant growth regulators or belong to a family of chemicals called **alkaloids**. Some of these alkaloids have been extracted from plants for use as medicines including morphine, codeine and atropine.

Alkaloids are chemicals produced by plants to protect them from predators and to control their growth.

PHYSICAL ADAPTATIONS	CHEMICAL ADAPTATIONS

Some plants, like roses, have thorns for protection.	Poppies produce powerful alkaloids including morphine and codeine, both used as strong painkillers.

Some plants have thick waxy cuticles on their leaves, preventing drying out and protecting them from disease.	Nettles contain specialised stinging dermal cells (also a physical adaptation) that contain a cocktail of chemicals including some neurotransmitters to cause irritation to any animal that might eat them.

Some climbing plants can produce 'tendrils' which support them as they climb to great heights.	The bark of the cinchona tree contains *quinine*, a bitter and toxic substance that deters insects from attacking it. Quinine is used as an antimalaria drug and is also found in tonic water.

Trees have developed thick bark to protect themselves against infection and pathogens.

Deadly nightshade produces an alkaloid called *atropine* which has many medicinal uses including to treat forms of poisoning.

Sundew, a native Irish insectivorous plant, produces a sticky substance at the end of its leaves to trap insects.

The coffee plant produces the alkaloid *caffeine* which is a natural pesticide. In humans, caffeine is a stimulant.

RESEARCH ACTIVITY

Carry out some further research on alkaloids. Identify one not mentioned and research its use by the plant and any potential medicinal uses it has.

PORTFOLIO LINK

Go to your Learning Portfolio and complete Learning Check 4 to assess your knowledge and understanding of the topic so far.

Specified Investigation 8

Investigate factors affecting the rate of transpiration in plants. (LO 2.6.16)

Scan the QR code to watch a video demonstration of this investigation.

A. AIR MOVEMENT

Hypothesis: If air movement around the leaf increases, then the rate of transpiration will increase.

Equipment: Potometer, plant shoot, fan, timer, Vaseline, sharp blade, water basin.

VARIABLES

Independent variable: Wind speed, varied using a fan.

Dependent variable: Amount of water taken up by the plant (proportional to the water lost by the plant).

Controlled variables: Temperature, type of plant.

METHOD

A bubble potometer

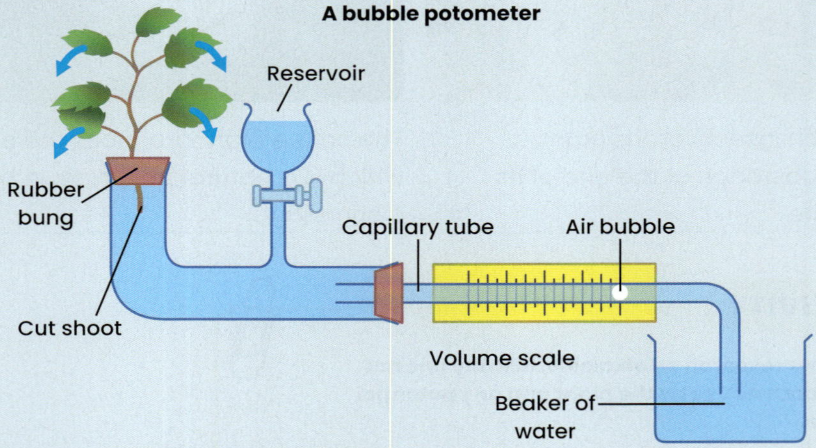

A potometer is a laboratory instrument which can be used to measure the rate of transpiration in a plant.

1. Fill the potometer with water, ensuring no air bubbles are present.
2. Cut a leafy shoot underwater to prevent air entering the xylem vessels.
3. Insert the cut stem into the potometer (under water), sealing any gaps with Vaseline. Ensure the system is airtight and fully filled with water.
4. Create an air bubble in the capillary tube by briefly removing it from the water reservoir.
5. Allow the plant to acclimatise for five minutes in still air conditions.
6. Record the initial position of the air bubble on the scale.
7. Start a timer and leave the setup for five minutes in still air.

8. Record the final position of the air bubble and calculate the distance moved.
9. Reset the bubble to its starting position by opening the tap of the reservoir.
10. Set up a fan at a fixed distance from the plant, on its lowest setting.
11. After five minutes, record the position of the air bubble and calculate distance moved.
12. Reset the bubble using the reservoir.
13. Increase the fan speed to medium and repeat Steps 11 and 12.
14. Increase the fan speed to maximum and repeat Steps 11 and 12.
15. Calculate the rate of transpiration for each wind speed by dividing the distance the bubble moved by the time (five minutes).
16. Record results.

RESULTS

FAN SPEED	DISTANCE TRAVELLED (mm)	RATE OF TRANSPIRATION (distance/time)
Still		
Low fan speed		
Medium fan speed		
High fan speed		

B. LIGHT INTENSITY

Hypothesis: If the light intensity increases on the leaf, then the rate of transpiration will increase.

Equipment: Conical flask (or beaker), plant shoot, lamp, ruler, timer, vegetable oil, sharp blade, electronic balance.

Scan the QR code to watch a video demonstration of this investigation.

VARIABLES

Independent variable: Light intensity, varied using a lamp at different distances from the plant.

Dependent variable: Amount of water taken up by the plant (proportional to the water lost by the plant).

Controlled variables: Temperature, type of plant.

METHOD

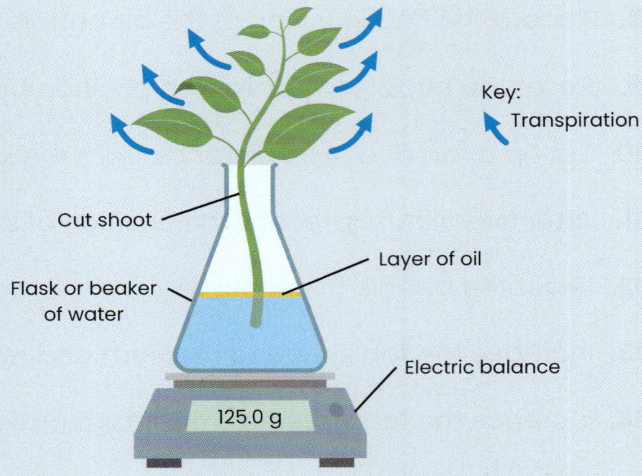

1. Set up the apparatus as shown in the diagram:
   - Fill a conical flask or beaker with water.
   - Place the cut end of the shoot into the water.
   - Cover the water surface with a layer of oil to prevent evaporation.
   - Ensure the leaves are dry and not touching the container.
2. Place the apparatus (plant and container) on a sensitive electronic balance.
3. Place a lamp at 10cm away from the plant and allow the plant to acclimatise for five minutes.
4. Record the initial mass of the setup.
5. Start a timer and leave the setup for 10 minutes.
6. Record the final mass after 10 minutes. Calculate the mass of water lost (initial mass – final mass).
7. Place the lamp 20cm from the plant and allow to adjust for five minutes.
8. Repeat Steps 5 to 7.
9. Repeat at 30cm and 40cm, recording the mass lost each time.
10. Calculate the rate of transpiration for each wind speed by dividing the mass of water lost by the time (10 minutes).

RESULTS

DISTANCE OF LAMP	LIGHT INTENSITY (1/d × 100)	CHANGE IN MASS	RATE OF TRANSPIRATION (mass/time)
10cm			
20cm			
30cm			
40cm			

DESIGN YOUR OWN EXPERIMENT

This investigation can be altered by changing the independent variable being investigated. Using either of the two methods shown, design another experiment to investigate the effect of one of the following factors on the rate of transpiration:

C. Temperature **D.** Surface area **E.** Presence or absence of a cuticle **F.** CO_2 concentration

Note for teachers: The full experiments for factors C–F are available in your *Teacher's Resource Book.*

INVESTIGATIVE LOG LINK

Go to your Investigative Log to record your lab work and answer questions on the experiments in this chapter.

Chapter Summary

- Plants have a network of vascular tubes (xylem and phloem) which are specialised for transporting water and food.
- The **transpiration stream** is the continuous flow of water in the plant, from the root to the stem.
- **Transpiration** is the loss of water from a plant.
- Transpiration cools the plant and allows for the uptake of more water.
- The rate of transpiration is controlled by the opening and closing of **stomata**
  - The movement of **potassium ions** (K^+) causes changes in the turgidity of guard cells, opening and closing the stomata.
  - HL High levels of **carbon dioxide** can cause the stomata to close but so too can high temperatures and humidity.
- There are five stages in the uptake and movement of water through a plant (called the cohesion-tension model):
  - **Osmosis** allows or the movement of water from the soil to the root hairs of root dermal cells.
  - Water moves from the dermal cells to the xylem in two ways: osmosis and diffusion.
  - HL Water moves through the xylem vessels (in the stem) by osmotic pressure, capillary action and transpiration pull (water molecules have great **adhesion** and **cohesion**).
  - HL Osmotive (root) pressure is a force that drives water up xylem vessels.
  - Xylem continues from the stem through to the leaf, via the leaf veins.
  - Water diffuses from the leaf to the air surrounding the plant, leaving via the stomata (transpiration). **Tension (transpiration pull)** is a pulling force that draws water up through the plant.
- Plants require mineral nutrients (e.g. nitrogen, phosphorous and potassium) in various amounts to build proteins, produce pigments and for metabolic reactions.
  - Mineral nutrients are absorbed by root dermal cells by **active transport**.
  - Mineral nutrients are transported through the plant in xylem and diffuse into cells that need them.
- Plants take in carbon dioxide through stomata in the leaves and **lenticels** in the stem by diffusion. Oxygen can also diffuse out of these pores.
- Sugars, produced during photosynthesis, are transported around the plant in a process called **translocation**.
- Translocation is carried out by specialised tissue called **phloem** and is an **active process**.
- Translocation describes the movement of sugars from source to sink.
  - A source is a cell that produces food by photosynthesis.
  - A sink is a cell where excess foods are stored.
  - Translocation is an active process, requiring ATP.
- Plant tissues can be adapted for food storage:
  - Onions store food in modified buds.
  - Carrots store food in modified roots.
  - Rhubarb store food in modified stems.
- Plants have a number of mechanisms of response, including **tropisms** – which improve their chance of accessing resources and survival.
- Tropisms are directional growth responses in plants to stimuli.
- Examples of tropisms include **phototropism** (growth response to light), **thigmotropism** (growth response to touch) and **geotropism** (growth response to gravity).
- Tropisms are controlled by **plant growth regulators**.
- Examples of plant growth regulators include **auxins**, **gibberellins** and **ethylene**.
- Auxins and gibberellins are produced by seeds to promote fruit development.
- Plants have evolved a number of physical and chemical adaptations to increase their chance of survival.
- **Alkaloids** are chemicals produced by plants to protect them from predators and to control their growth.

END OF CHAPTER ASSESSMENT

REMEMBER

1. Define the terms *transpiration* and *transpiration stream*.
2. List the factors affecting the rate of transpiration in plants.
3. Name the pores found on the underside of leaves.
4. Name the pores on stems, where gas exchange can also take place.
5. Identify the primary gases involved in gas exchange during photosynthesis.
6. Osmosis is a key process in water uptake. Identify the main process involved in gas exchange.
7. Name the primary nutrient minerals required by plants and briefly outline their functions.
8. What term describes the transport of sugars and proteins in the plant?
9. What is a tropism?
10. Which chemical messengers do plants use to control tropisms?

UNDERSTAND

11. Explain the process of water uptake in plants, detailing the stages involved.
12. What is the difference between active transport and diffusion?
13. Distinguish between a source and a sink, in relation to food transport in a plant.
14. Summarise the significance of stomata in regulating water loss and gas exchange in plants.
15. Discuss the structural adaptations of xylem for efficient water transport in plants.
16. Briefly describe how tropisms improve a plant's chance of survival.

APPLY

17. Create a diagram illustrating the process of water transport from soil to leaf through a plant's vascular system.
18. Devise a plan to conserve water in plants during periods of drought, considering both biological and environmental factors.
19. Why would food growers use ethylene gas after harvesting fruit?

ANALYSE

20. Analyse the cohesion-tension model of water transport in plants, identifying the key forces involved. HL
21. Evaluate the impact of environmental conditions on stomatal opening.
22. Assess the role of mycorrhizae in enhancing nutrient uptake in plants, considering both benefits and limitations.
23. Compare and contrast the mechanisms of water and nutrient uptake in plants, highlighting their similarities and differences.
24. Discuss the various methods used by plants to store excess food.
25. In an experiment, scientists tested the effects of CO_2 concentration on the transpiration rate (measured in ml of water lost per hour, i.e. ml/h) in tomato plants. The results of the experiment are shown below. HL

CO_2 CONCENTRATION (ppm)	TRANSPIRATION RATE (ml/h)	LEAF TEMPERATURE (°C)
200	35	24.5
400	30	25.0
600	25	25.5
1000	20	26.0

(a) Draw a graph of CO_2 concentration against transpiration rate. HL

(b) Describe the relationship, as described in the data, between CO_2 concentration and transpiration rate in tomato plants. HL

(c) Suggest why leaf temperature increases as CO_2 concentration increases. HL

(d) Discuss the potential real-world implications of these findings in the context of global climate change and its impact on agriculture. HL

26. In an experiment, scientists measured how wind speed affected the rate of transpiration in sunflowers, under controlled temperature and light intensity.

WIND SPEED (m/s)	TRANSPIRATION RATE (ml/hour)	TRANSPIRATION RATE INCREASE (%)
0 (Control)	1.7	n/a
1	2.2	29
2	2.7	59
3	3.2	88
4	3.7	118

(a) Draw a graph showing wind speed (m/s) against transpiration rate (ml/hour).

(b) Suggest a reason for the increase in transpiration rate at higher wind speeds.

(c) What other factors could also influence the rate of transpiration, apart from those mentioned?

(d) Explore the potential impact of climate change on wind patterns and its potential impact on plant water use. For example, would increased wind speeds exacerbate drought stress in certain regions?

27. In an experiment, scientists wanted to ascertain how pH affected the absorption of minerals in lettuce plants, grown under similar conditions. At harvest, samples of the plants were dried and the calcium and magnesium contents were calculated. The results are shown below.

pH	CALCIUM CONTENT (mg/g)	MAGNESIUM CONTENT (mg/g)
5.5	2.5	1.8
7.0	3.0	2.2
8.5	1.5	1.0

(a) Explain the relationship between mineral absorption and pH, according to the data presented.

(b) Propose a scientific explanation for why neutral pH (7.0) might promote better mineral uptake compared to acidic or basic conditions.

(c) Discuss the potential limitations of this experiment and suggest improvements for future research.

EVALUATE

28. Assess the ethical implications of genetically modifying plants to enhance their nutrient uptake capabilities, weighing potential benefits against risks.

29. Evaluate the ecological importance of plants' gas exchange processes, considering their role in carbon sequestration and atmospheric oxygen levels.

30. Critique the sustainability of current agricultural practices in terms of water usage and nutrient management, proposing alternative approaches.

31. Research the role of hormones on plant transport mechanisms.

32. Discuss the importance of alkaloids in medicine.

33. Considering the value of alkaloids to medicine, suggest **two** reasons why it is important to maintain plant biodiversity in rainforests.

CREATE

34. Develop a model illustrating the relationship between stomatal size, transpiration rate and environmental factors.

35. Research a specific plant known for its unique adaptations to extreme environments (e.g. mangroves in saltwater, cacti in deserts). Explain how these adaptations help the plant overcome challenges related to water and mineral acquisition.

36. Create a multimedia presentation explaining the journey of a water molecule from soil to leaf, incorporating animations and narration.

21 Plant Reproduction

LEARNING OUTCOMES: U2.3, U2.5, 2.5.7, 2.5.8

EYE-OPENER

Some plants have evolved very strange reproductive methods. Take the corpse flower (*Amorphophallus titanum*), for instance. Its enormous flower emits a putrid odour like rotting flesh, and attracts pollinators like carrion beetles and flesh flies, which mistake it for a decaying animal carcass. Once inside, these insects get coated in pollen before moving on to another bloom. This bizarre adaptation highlights nature's ingenuity in ensuring the survival of even the most peculiar plant species.

At the end of this chapter, you should be able to:

- Distinguish between sexual and asexual reproduction.
- Investigate the structure of the flower, its parts and their functions.
- Discuss the forms of pollination in plant sexual reproduction.
- Describe fertilisation in plant sexual reproduction.
- Outline the structure and function of seeds and fruits; discuss various methods of seed and fruit dispersal.
- Describe dormancy and the conditions needed for germination.
- Outline various methods of vegetative reproduction in plants.

LINKS TO:

Chapter 16	**Cell Structure and Diversity**
Chapter 17	**Cell Continuity**
Chapter 19	**Flowering Plants – Structure and Function**

SUBJECT VOCABULARY

reproduction, sexual reproduction, variation, asexual reproduction, vegetative reproduction, stamen, carpel, pollination, fertilisation, seed, fruit, seed dispersal, seed dormancy, geophytes

Section 1: The Flower

At the end of this section, you should be able to:

- Compare and contrast sexual and asexual reproduction in plants.
- Describe the basic structure of a flower, outlining the functions of each part.
- Investigate the differences between wind-pollinated and insect-pollinated flowers.

Unpacked from LO 2.5.7

WHAT IS REPRODUCTION?

We learned right back in Chapter 1 that all living things are able to form new living things of the same species. Plants are no different and each of the 391,000 types of plant makes new ones by one of two ways. Some plants **reproduce** by **sexual reproduction** and others by **asexual reproduction**.

Sexual reproduction involves the formation of specialised haploid sex cells called **gametes**. Gametes are produced by a type of cell division called meiosis (see Chapter 17), which halves the number of chromosomes in the cell. There are typically male and female gametes, which **fuse** together in a process called **fertilisation**. Sexual reproduction results in **variation** – the offspring are not genetically the same as their parents. Variation plays a major role in evolution by natural selection (see Chapter 9).

Asexual reproduction in plants involves the formation of new individuals without producing gametes. Plants have evolved many different mechanisms for reproducing in this way but all result in new plants that are **genetically identical** to their parents. They are often referred to as **clones**. Asexual reproduction in plants is also known as **vegetative reproduction**.

- **Reproduction** is the formation of individuals of the same species.
- **Sexual reproduction** is the formation of new individuals by the fusion of haploid sex cells called gametes.
- **Asexual reproduction** is the formation of new individuals by means other than gametes.
- **Variation** describes the differences between members of the same species.
- **Vegetative reproduction** is asexual reproduction in plants which results in the formation of clones.

Some plants can reproduce both sexually and asexually.

DID YOU KNOW?

THE FLOWER

The vast majority of plants (94% of all plant species) produce a specialised organ for sexual reproduction – **the flower** – which contains structures for making the male and female gametes. The flower provides a safe place for fertilisation to take place while also nourishing and protecting the new plant **embryo** when it forms.

In plants, the male gamete is found within structures called **pollen grains**. The female gamete is called the **egg** and it stays within the flower, waiting for the male gamete to fertilise it.

Figure 21.1: Sexual reproduction is carried out by specialised structures called flowers.

Figure 21.2: Asexual reproduction occurs when other parts of the plant are modified to form new clones, e.g. potato tubers.

We generally assume that all flowers are bright, colourful and have an attractive scent but this is not always the case. Many flowers are actually rather dull and have no bright colours or scent. It generally depends on how the plant has evolved to transfer its pollen from the male part of the flower to the female part – **pollination**. Brightly coloured and heavily scented flowers attract insects to aid the transfer, while plants that use wind to transfer their pollen have no need for these characteristics.

Go to YouTube.com and search for 'The hidden beauty of pollination | Louie Schwartzberg' (7:40) by TED.

Figure 21.3: Sedge is a plant that uses the wind to transfer its pollen.

Figure 21.4: Some plants have bright petals to attract bees and other insects to transfer pollen.

RESEARCH ACTIVITY

In some plants, sexual reproduction can occur without flowers. Mosses, ferns and gymnosperms (which includes coniferous trees) all reproduce sexually but without flowers. Carry out some further research into sexual reproduction in one non-flowering plant and present it on your mini-whiteboard.

Flower structure

To fully understand sexual reproduction in plants, we need to be very familiar with the basic structure of the flower. Insect- and wind-pollinated flowers generally have the same parts that perform the same roles, but their structure might be different. In this section, we will study the main parts of the flower, discuss their functions and outline how they might differ in wind- and insect-pollinated plants.

Most flowers are **hermaphrodites**, meaning they contain both the male and female parts, but some only contain one or the other.

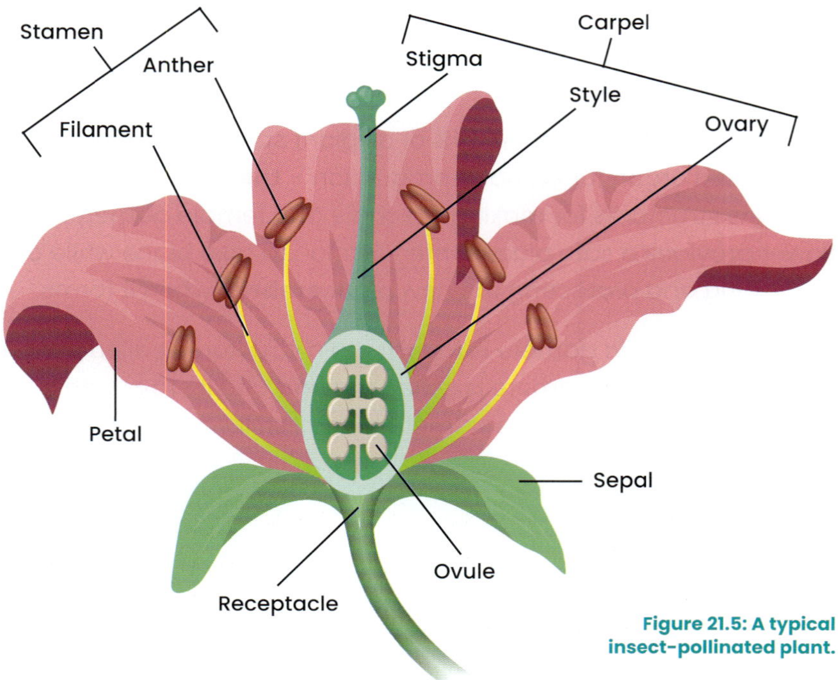

Figure 21.5: A typical insect-pollinated plant.

STAMEN

The **stamen** is the male, pollen-producing part of the flower, which forms and houses the male gamete. The stamen has two parts – the **anther** and the **filament**. Pollen is produced through meiosis and stored in the anther.

> The **stamen** is the male part of the flower. It consists of the anther and the filament.

The filament is the stalk that positions the anther where pollen can be best transferred. It contains vascular tissues (xylem and phloem) that provide the anther with water and nutrients. Some flowers may have just one stamen, but most have many.

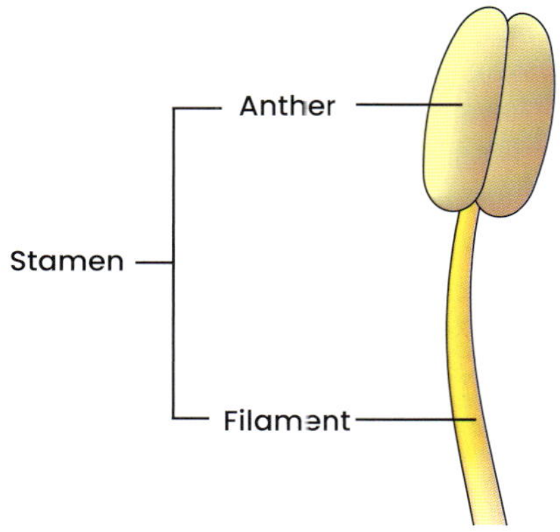

Figure 21.6: Structure of the stamen.

Figure 21.7: A close-up of pollen grains on the outside of the anther.

CARPEL

The female part of the plant is called the **carpel**. Most flowers have a single carpel although some flowers can have more than one. It comprises three parts: the stigma, style and ovary.

> The **carpel** is the female part of the flower. It consists of the stigma, style and ovary.

- The **stigma** is adapted for pollen collection. In wind-pollinated plants, the stigma is feathery while in insect-pollinated plants stigmas are sticky.
- The **style** is the stalk that holds the stigma in place, positioning it in the best place to catch as much pollen as possible.
- The **ovary** is home to one or more ovules which each contain an egg.

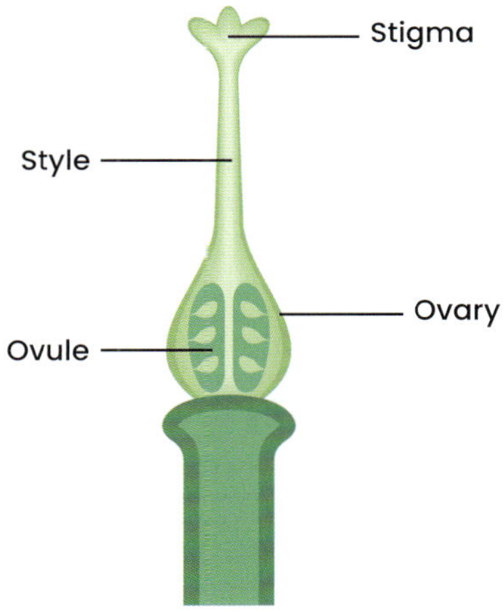

Figure 21.8: Structure of the carpel.

PETALS

Petals are used by most plants to attract pollinators – usually insects. They are often bright, colourful and scented. Many insects can see light in the ultraviolet spectrum and so flowers can appear differently to insects than they do to humans. In wind-pollinated plants, petals tend to be dull or may not even be present at all.

Figure 21.9: Petals are used to attract pollinators.

SEPALS

The **sepals** protect the flower, either before it emerges or when it closes at night. Sepals are usually green but can be colourful.

Figure 21.10: The sepals can be seen clearly before the flower emerges.

RECEPTACLE

The **receptacle** supports the flower. It is the thickened part of the stem directly under the sepals.

NECTARY

Some flowers contain a **nectary**, a small organ that produces a sugary reward for insect pollinators. Depending on the species, the nectary can be found throughout the flower. Bees can use nectar to make honey.

PORTFOLIO LINK **Go to your Learning Portfolio and complete Learning Check 1 to assess your knowledge and understanding of the topic so far.**

Section 2: Sexual Reproduction in Plants: Pollination and Fertilisation

At the end of this section, you should be able to:

- Discuss the importance of pollination in plant sexual reproduction.
- Distinguish between self and cross pollination.
- Discuss the role of insects in pollination.
- Outline the basic structure of pollen.
- Describe how fertilisation occurs in plants.

Unpacked from LO 2.5.7

There are a number of stages of sexual reproduction in plants.

STAGE 1 – POLLINATION

The first stage of sexual reproduction is the transfer of pollen from the anther to the stigma – this is known as **pollination**. There are two types of pollination:

Pollination is the transfer of pollen from the anther to the stigma.

1. **Self-pollination** occurs when pollen from an anther is transferred to the stigma of the same flower (or a flower on the same plant).
2. **Cross pollination** occurs when the pollen from an anther is transferred to the stigma of a flower of a different plant (of the same species).

Figure 21.11: Self-pollination and cross pollination.

Cross pollination is more desirable that self-pollination as it creates greater **variation** – differences between members of the same species – and means the species will be more capable of adapting to changing environmental conditions. In many plants, the carpels mature before the anthers to ensure they catch pollen from a different plant first.

The method of pollination can be broadly divided into two categories:

1. **Insect (animal) pollination**

   Most flowers use insects (bees, butterflies, moths) to transfer their pollen from anthers to stigmas. Many plants have bright, colourful and scented flowers to attract insects who then unintentionally pick up pollen and bring it to another flower. Some flowers give insects a sugary reward – **nectar** – to further encourage them to act as pollinators. Some flowers have adapted to look like insects, luring pollinators with a chance to mate rather than with food. The bee orchid, a native Irish plant, closely resembles a bee and uses this method to attract pollinators. Insects are not the only animal pollinators though – bats, birds and even rodents can be used to transfer pollen.

Go to YouTube.com and search for 'The corpse flower may keep you up at night – Daniel Nickrent' (5:27) by TED-Ed.

Insects and flowering plants have a strong **mutualistic** relationship, meaning they both benefit from their relationship. This relationship is a good example of co-evolution – where two groups of organisms evolved at the same time to create strong interdependence.

DID YOU KNOW?

2. **Wind pollination**

Not all flowers are colourful, with many preferring to transfer their pollen from anther to stigma using the wind. This method is a little less reliable so these plants typically produce larger amounts of pollen. Most wind-pollinated plants belong to the grass family (which includes many crops like barley and wheat) and the large volumes of pollen produced by grass in the summer contribute to hay fever – an allergic reaction to pollen.

	INSECT-POLLINATED	WIND-POLLINATED
PETALS	Large and brightly coloured to attract insects	Small, often dull in colour
SCENT AND NECTAR	Usually scented and with nectar to attract insects	No scent or nectar
NUMBER OF POLLEN GRAINS	Moderate, insects transfer pollen grains effectively	Large amounts of pollen as most grains are not transferred to another flower
POLLEN GRAINS	Sticky or spiky to adhere to insects	Smooth and light, easily carried by the wind without clumping together
ANTHERS	Inside flower, stiff and firmly attached to brush against insects	Outside flower, loose on long filaments to release pollen grains easily
STIGMA	Inside flower, sticky so that pollen grains stick to it when an insect brushes past	Outside flower, feathery to form a network to catch drifting pollen grains

Figure 21.12: A comparison between insect-pollinated plants and wind-pollinated plants.

STAGE 2 – FERTILISATION

Once pollination is successfully complete, i.e. a pollen grain lands on a stigma of a flower of the same species, **fertilisation** can occur.

Fertilisation is the fusion of two haploid gamete cells forming a diploid zygote.

The pollen 'germinates' and begins to form a tube which will bring the male gamete down the style and into the ovary, where the egg awaits. Fertilisation occurs when the male gamete fuses with the egg (the female gamete) forming a **diploid zygote**.

PORTFOLIO LINK **Go to your Learning Portfolio and complete Learning Check 2 to assess your knowledge and understanding of the topic so far.**

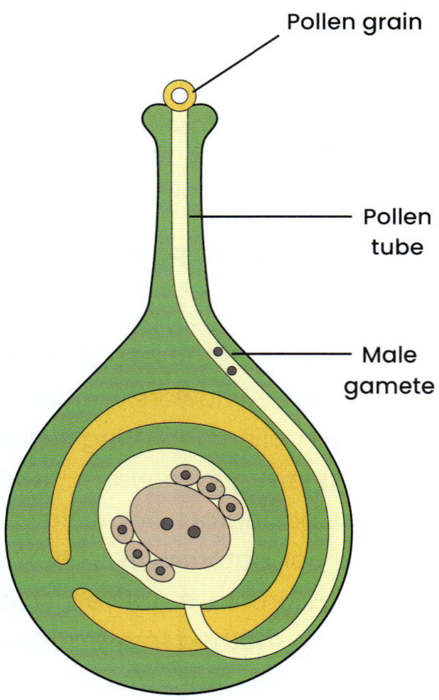

Figure 21.13: A pollen grain and the male gamete.

Section 3: Sexual Reproduction in Plants: Seeds and Fruits

At the end of this section, you should be able to:

- Outline the basic structure of a seed.
- Discuss the functions of a fruit.
- Outline the origins of the seed and fruit in the embryo sac.
- Discuss the importance of seed and fruit dispersal.
- Describe the methods of seed and fruit dispersal.

Unpacked from LO 2.5.8

STAGE 3 – FORMATION OF SEEDS AND FRUITS

Seeds

Once fertilisation is complete **a seed forms from the ovule**. Plants with multiple ovules will form multiple seeds.

- A **seed** contains a young plant embryo, a food store and a protective outer coat called the testa.
- The **endosperm** is a specialised tissue which acts as a food store in seeds.

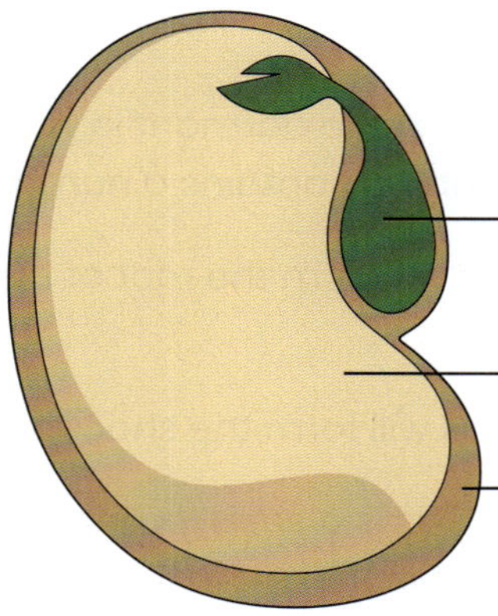

Figure 21.14: Basic structure of a seed.

On a basic level, a seed has three main parts:

1. The plant **embryo** – an immature, dormant plant formed from the diploid zygote.
2. A **food store** – the young embryo needs nutrients and the food store nourishes the young plant as it grows.
3. The **testa** – a tough outer coating, made from the ovule wall, that protects the embryo and the food reserve.

Seeds generally use one of two structures as their food store. Some seeds use a specialised tissue called the **endosperm**, while other seeds store their food in **cotyledons**, sometimes called seed leaves.

As the seed reaches maturity it begins to lose its water content and stops growing – the seed is said to be **dormant**.

Seeds of monocot plants differ from those of dicot plants. In fact, the names monocots and dicots derive from this difference. Monocot seeds have just one 'seed leaf' while dicots have two 'seed leaves'.

FOCUS ON ... SEEDS

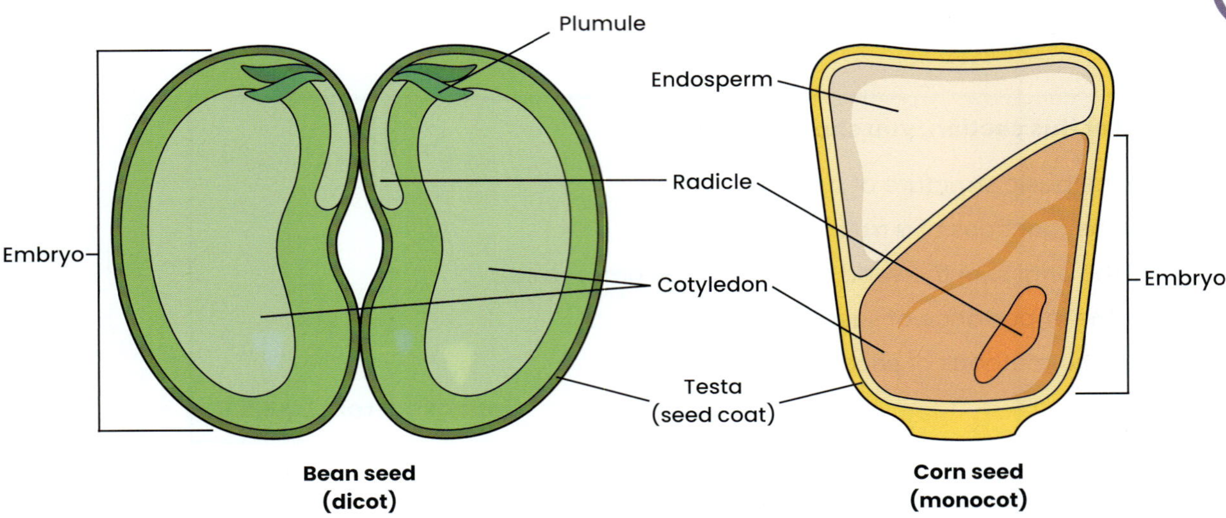

Figure 21.15: Monocot and dicot seeds.

Let's look at the three main parts in a little more detail.

Embryo

The seed embryo is the dormant immature plant – once it reaches a certain stage it stops growing. The embryo contains a number of parts:

- **Radicle:** This will form the root of the new plant. It is the first part to emerge from the seed at germination.
- **Plumule:** This will form the shoot of the new plant. It emerges just after the radicle.
- **Cotyledon:** This is a seed leaf, the first leaves of the new plant and the ones seen during germination. Monocots have just one seed leaf while dicot plants have two seed leaves.

Food reserve

After fertilisation, the endosperm begins to supply the young plant with nutrients and continues to nourish the embryo until dormancy in most monocot plants. In most dicot plants, the cotyledons take over the role as the main food reserve and the endosperm is absorbed. Different seeds store different nutrients – some will contain high levels of carbohydrate (e.g. cereals) while others will be high in proteins (e.g. soya beans) or fats (e.g. sunflower seeds). Seeds contain digestive enzymes, which break down these food reserves when they begin to germinate.

Testa

The outer coating of the seed is formed by the wall of the ovule. It is very tough and protects the embryo from adverse conditions, such as extreme heat or cold. The testa helps control seed germination – the regrowth of the seed in favourable conditions following a period of dormancy. The testa has a small opening called the micropyle which allows water and oxygen to enter during germination.

Fruits

After fertilisation, a fruit begins to develop around the seed. The **fruit** forms from the ovary. Fruits aid in the dispersal of the plant seeds (helping them travel away from their parents, to reduce competition) while also protecting the seeds from disease and from drying out.

A **fruit** is a structure that develops from an ovary and contains a seed or seeds.

Many fruits contain high levels of sugar, usually as a reward for herbivores that might unknowingly aid their dispersal, while others do not. Tomatoes, cucumbers, pumpkins, avocados and peppers are all fruits but often thought to be vegetables because of low sugar. Not all fruits are edible though and some may contain toxins or poisons.

Figure 21.16: A selection of edible fruits.

THE ROLE OF GROWTH REGULATORS ON FRUIT FORMATION AND DEVELOPMENT

After fertilisation, the seed starts producing **plant growth regulators**, such as auxins and gibberellins (see Chapter 20). These chemicals stimulate the growth and development of the ovary, which eventually becomes the **fruit**. They also influence the enlargement of the cells in the ovary, causing the fruit to swell. Later, another growth regulator, **ethylene** (a gas), is produced to ripen the fruit. Ethylene gas can be easily manufactured, allowing fruit farmers to ripen fruits early. This can be used to produce seedless fruits.

STAGE 4 – SEED AND FRUIT DISPERSAL

The new plant, now formed within the seed, will have a greater chance of surviving if it can avoid competition or colonise new areas. The fruit often plays a big role in dispersing the seed or seeds within. **Seed dispersal** can occur in a number of ways:

Seed dispersal is the movement of seeds away from the parent plant.

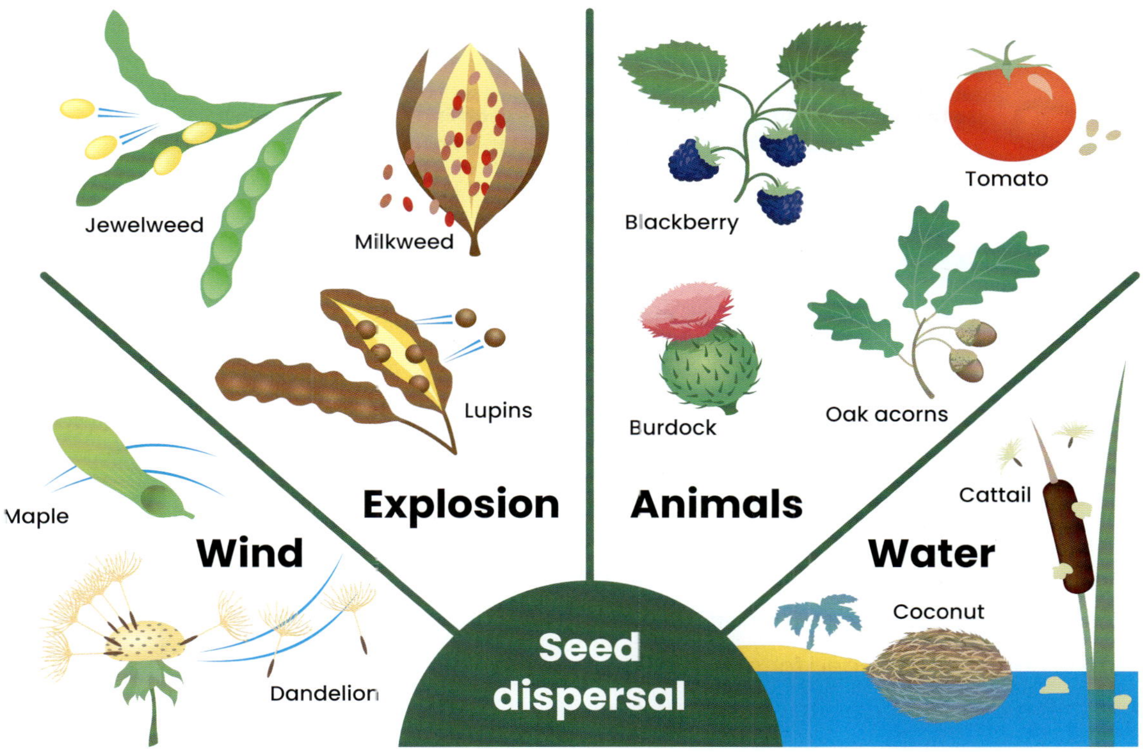

Figure 21.17: Methods of seed dispersal.

Wind dispersal

Many plants, including dandelions and sycamore trees, use the wind to disperse their seeds. The seeds are often light or contain a parachute or wing.

Animal dispersal

Animals are also used to unwittingly disperse fruits and seeds. Plants like Burdock and Goosegrass produce sticky fruits which attach to animal fur or even human clothing, carrying them far from their parent plant.

Animals, including many birds, swallow edible fruits containing seeds. In many cases, the seed is unharmed and later dropped in the bird's faeces. In fact, seeds may not germinate (continue growing) until they have passed through the acidic environment in an animal's stomach.

Figure 21.18: Burdock seeds have hooks to attach to animal fur.

Go to YouTube.com and search for 'Sir David Attenborough Gives a Lesson on Seeds | The Green Planet' (3:16) by BBC Earth.

Explosive (self) dispersal

Many plants use explosive methods to disperse their seeds, including gorse, Himalayan balsam and squirting cucumbers. These plants usually store their seeds in pods, building pressure inside using air or water before exploding as they dry out.

Water dispersal

Many aquatic plants, including coconuts and cattails, drop their fruits and seeds into water. The water carries the fruits and seeds along the current, carrying them away to potential new areas free from competition.

PORTFOLIO LINK **Go to your Learning Portfolio and complete Learning Check 3 to assess your knowledge and understanding of the topic so far.**

Section 4: Sexual Reproduction in Plants: Seed Dormancy and Germination

At the end of this section, you should be able to:

- Define the term 'seed dormancy'.
- Discuss the benefits of dormancy to the new plant.
- Describe the stages of seed germination.
- Outline the conditions needed for germination.
- Discuss the role of plant growth regulators in seeds.

Unpacked from LO 2.5.8

STAGE 5 – SEED DORMANCY

Many plants produce their seeds towards the end of the summer or even into the autumn. If the seeds were to germinate (regrow) as soon as they are formed, they might not survive the harsh winter months. For this reason, most seeds undergo a period of **dormancy** – a time when the plant within doesn't grow even if the conditions might be suitable.

Dormancy is an important stage in a seed's life. It allows time for the seed to disperse from its parent plant or find new areas to colonise, while also ensuring it doesn't begin to grow before the winter weather.

For farmers or horticulturists, dormancy might mean a plant doesn't germinate immediately after planting, which isn't convenient. Farmers have adopted a number of techniques to overcome dormancy, including soaking seeds in acids, which mimics passing seeds through an animal's stomach.

RESEARCH ACTIVITY

Carry out some research on other methods used by farmers and horticulturists to overcome seed dormancy.

Seed dormancy is a period when a seed does not grow, even though conditions might be favourable.

STAGE 6 – SEED GERMINATION

Germination is the final stage of plant reproduction, when the newly formed plant embryo begins to regrow and emerge from the seed. The embryo will use the food reserve to power its early growth, before its first leaves emerge and they start to photosynthesise.

Seeds need a number of conditions before they begin to germinate:

1. **Water:** Seeds need water to start growing; water is a medium for important metabolic reactions in seeds.
2. **Oxygen:** Growth requires energy, and oxygen is necessary for the release of the energy from the seed's food reserve, i.e. cellular respiration.
3. **Suitable temperature:** While every seed is different, seeds require some warmth to increase their metabolic rate and therefore respiration and growth.

Go to YouTube.com and search for 'Bean Time-Lapse – 25 days | Soil cross section' (3:09) by GPhase.

Germination occurs when the embryo begins to regrow after a period of dormancy.

Figure 21.19: Stages of seed germination.

Germination takes place in a number of stages:

1. **Water absorption:** The seed absorbs water, causing it to swell. The water also softens the testa (seed coat), allowing the new plant break through the tough coating more easily.
2. **Metabolic activation:** Water is a medium for many metabolic reactions. It triggers the seed's internal metabolism, including the digestion of the food store by enzymes.
3. **Root emergence:** The radicle (the embryonic root) begins to grow first, anchoring the plant and absorbing more water.
4. **Leaf emergence:** The embryo's first leaves (cotyledons) emerge next and begin to photosynthesise, providing more nutrients to the growing plant.
5. **True leaf development:** The new plant forms 'true leaves', which resemble those of the adult plant, and the new plant (now known as a seedling) continues to develop. Germination is now over.

PORTFOLIO LINK **Go to your Learning Portfolio and complete Learning Check 4 to assess your knowledge and understanding of the topic so far.**

Specified Investigation 9

Investigate and compare the structures of insect- and wind-pollinated plants. (LO 2.5.7)

Equipment: Samples of flowers: at least one wind-pollinated and one insect pollinated flower.

METHOD

1. Collect various samples of flowers from the locality, including a mix of wind- and insect-pollinated species.
2. Note differences in colour, scent and shape of the various flowers.
3. Compare the sepals and petals of each flower. Note which ones are colourful.
4. Check for nectar. Is there a sweet food reward for insects inside the flower?
5. Investigate the structure of the stamens and carpels in each flower. Are the carpels sticky, feathery, inside the flower or outside?
6. Collect pollen from each flower. Note differences in size, colour and texture. Consider viewing each under the microscope for further detail.
7. Look for specific adaptations. Insect-pollinated flowers may have landing platforms, intricate shapes or guides directing insects towards reproductive structures.
8. Pollen transfer and collection – look for signs on how the plant transfers and collects its pollen.

RESULTS

Complete the report in your Investigative Log to record the results of your investigation.

INVESTIGATIVE LOG LINK **Go to your Investigative Log to record your lab work and answer questions on the experiments in this chapter.**

Chapter Summary

- Living organisms reproduce to create new individuals of the same species.
- Plants can reproduce using both sexual and asexual reproduction.
- Sexual reproduction involves the formation of specialised sex cells called **gametes** through meiosis, leading to variation in offspring.
- Asexual reproduction produces genetically identical offspring, or **clones**, through various mechanisms like runners, bulbs or tubers, known as vegetative reproduction.
- Flowers are specialised organs for sexual reproduction in 94% of plant species, housing male (pollen) and female (egg) gametes.
- The **stamen** is the male part of the flower, consisting of the anther and filament.
- The **carpel** is the female part of the flower, consisting of the stigma, style and ovary.
- The **testa** is the seed coat.
- The **plumule** is an immature shoot in a seed.
- The **radicle** is an immature root in a seed
- The **embryo** is an immature plant in a seed consisting of a radicle, plumule and cotyledons.
- The **endosperm** is a food reserve in a seed.
- Sexual reproduction occurs in a number of stages:
  - **Pollination** is the transfer of pollen to the stigma via animals or wind. Animal-pollinated flowers have brightly coloured petals and are scented, with the anther and sticky stigma inside the flower. Wind-pollinated flowers have dull petals and no scent, with the anther and feathery stigma outside the flower.
  - **Fertilisation**, the fusion of the male and female gamete forming a zygote.
  - **Seed formation** follows fertilisation in the ovary, leading to the development of fruits surrounding seeds.
  - **Seed and fruit dispersal** aids in spreading offspring, utilising methods like wind, animals, self-dispersal or water.
  - **Dormancy** in seeds ensures survival during unfavourable conditions before germination.
  - **Germination** requires water, oxygen and suitable temperature conditions, leading to the emergence of a new plant.
- Seeds produce plant growth regulators that play a role in fruit formation; **ethylene** is a gas that ripens fruits.

END OF CHAPTER ASSESSMENT

REMEMBER

1. Name and briefly describe the **two** primary methods of plant reproduction.
2. What is the structure that houses the male gamete in a flower?
3. Name the female part of the flower.
4. What is *pollination*?
5. Draw a diagram of a typical insect-pollinated flower.
6. What term describes the fusion of a male and female gamete?
7. Name the **three** parts of a seed.
8. What is the term used to describe the process of a seed beginning to grow?
9. What is the purpose of dormancy in seeds?
10. Name **two** plant growth regulators involved in fruit formation and development.

UNDERSTAND

11. Explain the difference between sexual and asexual reproduction in plants.
12. Why is cross-pollination considered more desirable than self-pollination?
13. Distinguish between the endosperm and the cotyledons.
14. From what structure in the flower does the fruit form?
15. Discuss the various forms of fruit and seed dispersal.
16. Describe the process of germination in seeds.

APPLY

17. If you were a farmer, how would you ensure that your crop seeds overcome dormancy quickly and uniformly?

18. The photo shows the flower of a maize plant. Do you think this flower is insect- or wind-pollinated? Justify your answer.

19. Imagine you are a botanist studying a new species of plant. How would you determine whether it reproduces sexually or asexually?

20. What are the advantages of seed dormancy?

21. Suggest a reason why bananas are picked when they're green rather than yellow.

22. Suggest how green unripe bananas might be ripened before sale.

ANALYSE

23. Discuss the role of flowers in sexual reproduction.

24. Compare and contrast, using diagrams if appropriate, the differences between insect-pollinated and wind-pollinated flowers.

25. Discuss seed dispersal mechanisms and their effect on plant distribution and diversity.

26. The following data shows changes in the mass of a sunflower seed in the days following germination. The mass of the root, shoot and total mass are also shown. Study the data and answer the following questions.

TIME (days)	SEED MASS (mg)	ROOT MASS (mg)	SHOOT MASS (mg)	TOTAL PLANT BIOMASS (mg)
0	200	0	0	200
2	190	10	0	200
4	180	20	2	202
6	170	30	10	210
8	160	40	20	220
10	150	50	30	230
12	140	60	40	240
14	130	70	50	250
16	120	80	60	260
18	110	90	70	270

(a) Account for the loss of mass in the seed over the 18 days.

(b) According to the data, which forms first, the root or the shoot? Explain your answer.

(c) Explain why the total biomass starts to increase after day 4.

(d) Draw a graph, using the data above, showing the changes in seed mass over the 18-day period.

27. A student carried out an experiment on mustard seeds to see how treating them with different priming solutions affected the seeds' germination speed and success. At the end of seven days, the student took a series of measurements.

Treatments:

- Control: Unprimed seeds soaked in water for 24 hours
- Saltwater: Soaked in 1% saline solution for 24 hours
- Potassium nitrate (KNO_3): Soaked in 0.1% KNO_3 solution for 24 hours
- Gibberellic acid (GA3): Soaked in 50ppm GA3 solution for six hours

Measurements:

- Germination rate: Percentage of seeds germinated within seven days
- Mean germination time: Average time taken for 50% of seeds to germinate
- Shoot length: Average length of the longest shoot for germinated seedlings after seven days

The results obtained are shown below. Study the data and answer the questions that follow.

TREATMENT	GERMINATION RATE (%)	MEAN GERMINATION TIME (days)	SHOOT LENGTH (cm)
Control	75	5.0	3.5
Saltwater	85	4.5	4.0
Potassium nitrate	90	4.0	4.5
Gibberellic acid	95	3.5	5.0

(a) Describe the observed trends in germination rate, mean germination time and shoot length across different treatments.

(b) Which treatment resulted in the fastest germination? How much faster was it compared to the control?

(c) Based on the results, which priming solution do you think would be most beneficial for farmers who prioritize early harvest? Justify your answer.

(d) Propose a scientific explanation for why priming solutions might improve germination speed and success.

(e) Discuss the limitations of this experiment and suggest improvements for future research.

EVALUATE

28. Evaluate the role of plant growth regulators in fruit formation and development.

29. Research various mechanisms for producing seedless fruits. Which, in your opinion, is the most effective?

30. Critically analyse the impact of human activities on plant reproduction and biodiversity.

CREATE

31. Design a follow-up experiment to the one in Question 27 to investigate the long-term effects of these priming solutions on plant growth and yield.

32. Design an experiment to investigate the effect of different environmental factors (e.g. temperature, light) on seed germination.

33. Create a flowchart illustrating the process of sexual reproduction in flowering plants, including pollination, fertilisation and seed formation.

UNIT 7

PHOTOSYNTHESIS AND RESPIRATION

CHAPTERS

STRANDS

Unifying Strand: Nature of Science

Strand 1: Organisation of Life

Strand 2: Structures and Processes of Life

22 Photosynthesis

LEARNING OUTCOMES: U2.2, U2.3, U2.5, 1.2.8, 2.2.1, 2.2.2, 2.2.4, 2.2.5, 2.6.14

EYE-OPENER

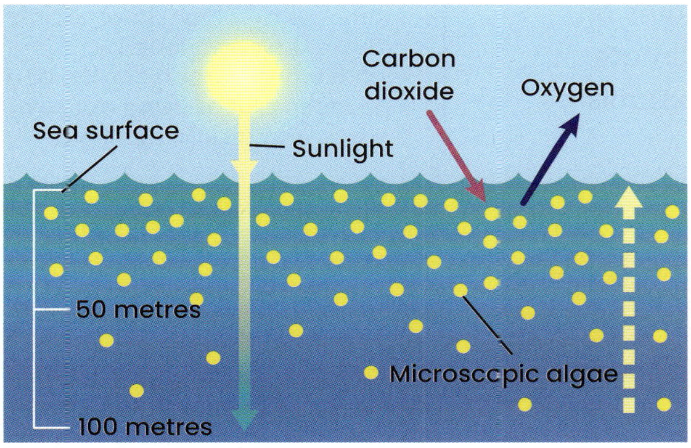

It is estimated that 50% of Earth's oxygen is produced by microscopic algae called phytoplankton during photosynthesis.

At the end of this chapter, you should be able to:

- Write a balanced chemical equation for photosynthesis.
- Understand the role of photosynthesis as a carbon sink.
- Outline the role of ATP in photosynthesis.
- Investigate the factors that affect the rate of photosynthesis.
- Relate leaf structure to its role in photosynthesis.
- Describe the internal structure of chloroplasts and how it facilitates photosynthesis. HL
- Outline the role of $NADP^+$ in photosynthesis. HL
- Describe the light-dependent stage of photosynthesis with reference to transport molecules. HL
- Describe the light-independent stage of photosynthesis with reference to transport molecules. HL

SUBJECT VOCABULARY

photosynthesis | chlorophyll | chloroplast | carbon sink | ATP | light-dependent stage | light-independent stage | $NADP^+$ | electron transport chain | concentration gradient

LINKS TO:

Chapter 10	**Biodiversity and Ecosystems**
Chapter 12	**Nutrient Recycling**
Chapter 14	**Biomolecules**
Chapter 16	**Cell Structure and Diversity**
Chapter 19	**The Plant – Structure and Function**
Chapter 20	**Plant Transport and Responses**
Chapter 23	**Respiration**
Chapter 30	**The Digestive System**

Section 1: What Is Photosynthesis?

At the end of this section, you should be able to:

- Define photosynthesis and write a balanced equation for photosynthesis.
- Explain how leaf structure is adapted to photosynthesis.
- Understand the role of photosynthesis as a carbon sink.
- Outline the role of ATP in photosynthesis.

Unpacked from LOs 1.2.8, 2.2.1, 2.6.14

WHAT IS PHOTOSYNTHESIS?

Nearly all life on Earth depends on the process of **photosynthesis**. While studying ecology, we learn that almost all food chains on Earth begin with a green plant and that the Sun is the ultimate source of our planet's energy. Plants, unlike most other organisms, can make their own food using energy from sunlight.

Photosynthesis is the process by which green plants make their own food by converting carbon dioxide and water to glucose and oxygen, using energy from sunlight.

Photosynthesis word equation:

Reactants → **Products**

carbon dioxide + water $\xrightarrow{\text{light, chlorophyll}}$ **glucose + oxygen**

Photosynthesis chemical equation:

$$6CO_2 + 6H_2O \xrightarrow{\text{light, chlorophyll}} C_6H_{12}O_6 + 6O_2$$

The green pigment chlorophyll found in chloroplasts in plant cells allows plants to absorb sunlight. Photosynthesis is an **anabolic reaction** where small molecules are converted to larger molecules using energy. In photosynthesis, carbon dioxide from the air is absorbed by the aerial parts of the plant and water is taken in from the soil by the roots. A series of chemical reactions occur in the chloroplasts which use the energy from sunlight to convert the reactants carbon dioxide (CO_2) and water (H_2O) to the following products: glucose ($C_6H_{12}O_6$) and oxygen (O_2). The glucose is converted to sucrose for transport around the plant through the phloem or stored as the complex carbohydrate starch in the leaves. The by-product oxygen is released to the atmosphere through the leaves' stomata or used by the plant in respiration.

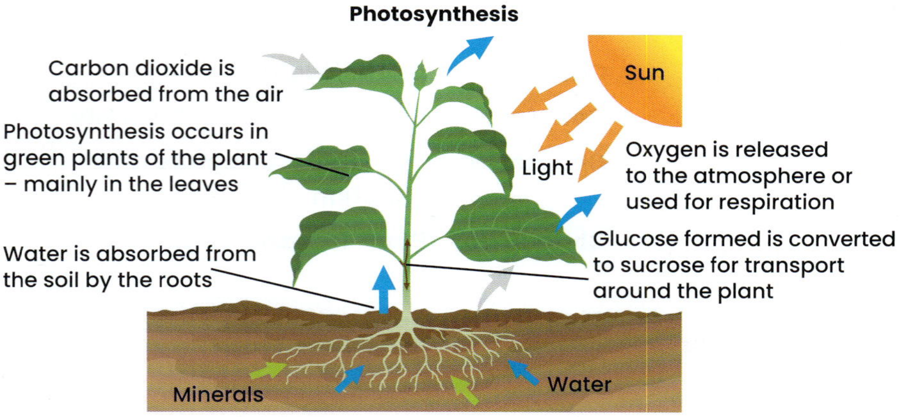

Figure 22.1: Green plant absorbing light, water and carbon dioxide for photosynthesis.

Go to YouTube.com and search for 'Definitions in the Field: Photosynthesis' (0:47) by National Geographic Education.

The sources of each of the requirements for photosynthesis are outlined below:

REQUIREMENTS FOR PHOTOSYNTHESIS	SOURCE
Carbon dioxide (CO_2)	• Air – carbon dioxide is taken in from the air through the stomata in leaves and lenticels in stem. • Produced during respiration. • Produced during combustion of fuels.
Water (H_2O)	Water is absorbed from the soil by the roots. It is transported through the xylem up the stem to the leaves.
Chlorophyll	Green pigment found in chloroplasts in plant cells that allows plants to absorb energy from sunlight.
Light	The Sun.

The destinations for each of the overall products of photosynthesis are outlined below:

PRODUCTS OF PHOTOSYNTHESIS	DESTINATION
Glucose ($C_6H_{12}O_6$)	Glucose is converted to sucrose for transport in the phloem or stored in the form of the complex carbohydrate starch in leaves. Note that other complex carbohydrates such as cellulose found in plant cell walls can also be formed from glucose.
Oxygen (O_2)	• Released to the air through the stomata in leaves or lenticels in the stem. • Used by organisms in respiration.

HOW LEAF STRUCTURE IS ADAPTED FOR PHOTOSYNTHESIS

Leaves are structurally adapted for their role in photosynthesis.

- They have a **large surface area** to absorb as much light as possible.
- They are **thin**, so carbon dioxide has a short distance to diffuse across into cells.
- **Chloroplasts** and the pigment chlorophyll allow leaves to absorb energy from sunlight.
- **Stomata** allow carbon dioxide to diffuse in and oxygen to diffuse out.
- A **network of veins** containing vascular tissue facilitates the transport of water through the xylem and sucrose in the phloem.

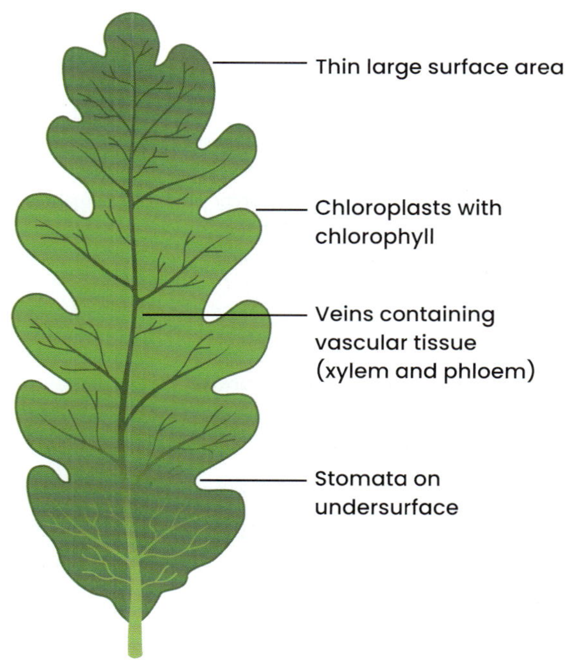

Figure 22.2: The structure of a leaf.

CHLOROPLASTS

- Green plant cells contain the **chloroplast** organelle, which contains the green pigment **chlorophyll**, allowing the plant to absorb light for photosynthesis.

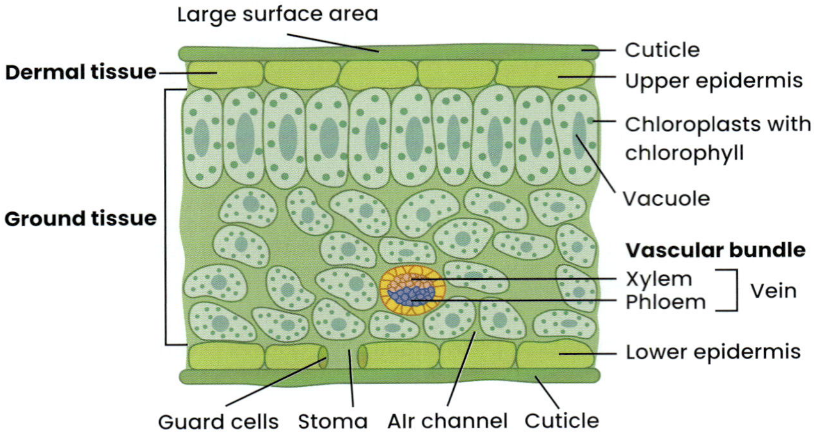

Figure 22.3: Transverse section of a leaf.

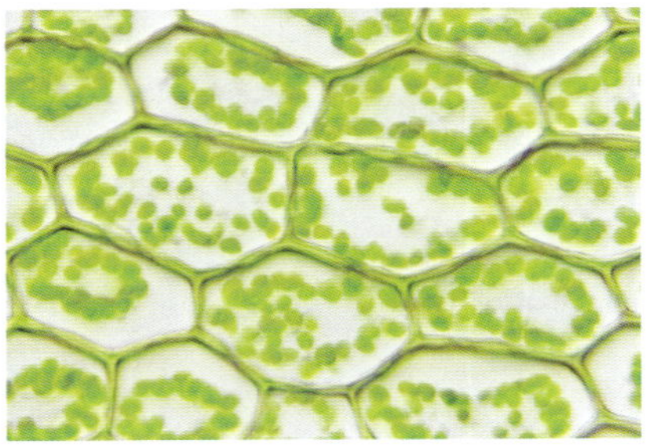

Figure 22.4: Chloroplasts.

Role of photosynthesis as a carbon sink

In Chapter 12, we learned about **carbon sinks**. Plants carrying out photosynthesis continually take carbon (in the form of carbon dioxide) out of the atmosphere and convert it to glucose and then to sucrose, or store it as the complex carbohydrate starch. The carbon is now in a form stored away from the atmosphere. Photosynthesis is described as a carbon sink, as it takes more carbon out of the atmosphere than is released.

- The **chloroplast** is an organelle found in plant cells that contains the pigment chlorophyll. It is the site of photosynthesis.
- **Chlorophyll** is the green pigment found in the chloroplast in plant cells that absorbs light to provide energy for photosynthesis.
- A **carbon sink** is anything that absorbs more carbon than it releases. Forests are natural carbon sinks.
- The molecule **ATP** is used to store and transfer energy in cells.

TRANSFER MOLECULES

A molecule called adenosine triphosphate, or **ATP**, is used to store and transfer energy in cells.

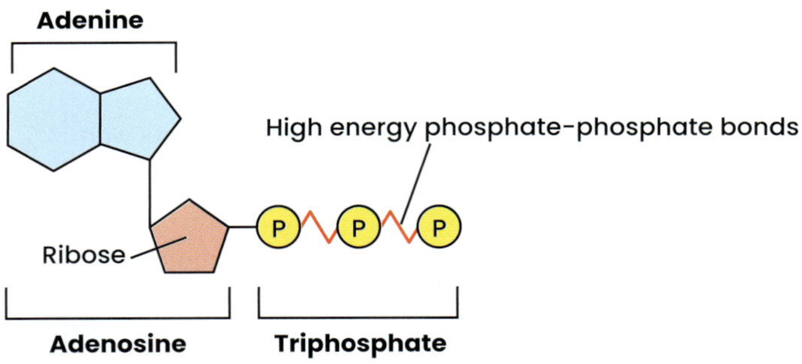

Figure 22.5: The structure of ATP.

ATP is formed from a nitrogenous base adenine, a five-carbon sugar ribose and three phosphate groups linked together. The adenine and ribose molecules bonded together are called adenosine. Energy is stored in the bonds between the phosphate groups.

The role of ATP in photosynthesis

Energy from sunlight is used to attach a phosphate group to a molecule adenosine diphosphate (ADP), which has two phosphate groups, to form adenosine triphosphate.

ADP + P + energy ⟶ ATP **adenosine diphosphate + phosphate + energy ⟶ adenosine triphosphate**

When a cell needs energy, the last phosphate group is lost from ATP, releasing energy and reforming ADP.

ATP ⟶ ADP + P + energy

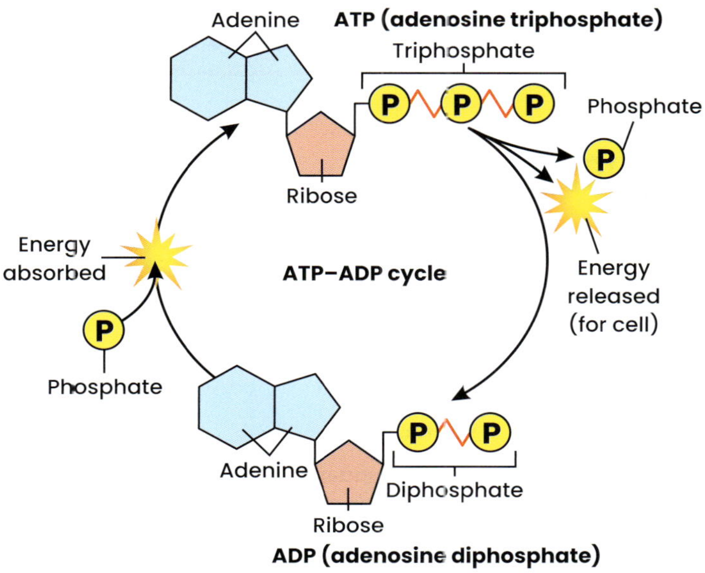

Figure 22.6: Storage and release of energy from ATP.

Go to YouTube.com and search for 'ATP (ATP cycle and ADP) for Leaving Cert Biology' (4:07) by Biology Bugbears.

We will learn more about ATP in Chapter 23.

FOCUS ON ... CHLOROPHYLL

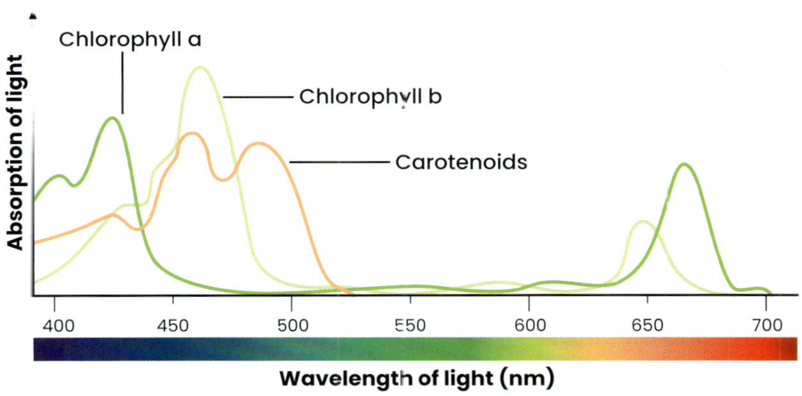

Figure 22.7: Absorption of light by chlorophyll molecules.

Photosynthetic pigments in the leaf's chloroplasts capture visible light energy. There are two main types of photosynthetic pigments: chlorophylls and carotenoids. Chlorophylls and carotenoids absorb light of different wavelengths, allowing more light to be captured. Chlorophylls absorb red and blue light but reflect green light, giving plants their green colour. In autumn, chlorophyll breaks down and the colours reflected by the carotenoids – yellow, orange and red – become visible.

PORTFOLIO LINK **Go to your Learning Portfolio and complete Learning Check 1 to assess your knowledge and understanding of the topic so far.**

Section 2: Two-Stage Process of Photosynthesis

At the end of this section, you should be able to:

- Describe the internal structure of the chloroplast and how it relates to its role in photosynthesis. HL
- Outline the role of $NADP^+$ in photosynthesis. HL
- Describe the light-dependent and light-independent reactions of photosynthesis. HL

Unpacked from LO 1.2.8, 2.2.4, 2.2.5

CHLOROPLAST INTERNAL STRUCTURE

HL

The chloroplast has an outer double membrane.

- Chloroplasts contain long, flattened membranes called thylakoid membranes.
- A stack of thylakoid membranes is called a granum (plural: grana). The thylakoid membranes contain chlorophylls, e.g. chlorophyll a and chlorophyll b, which absorb different wavelengths of light.
- The thylakoid membranes contain electron transport chains, which are involved in photosynthetic reactions.
- The aqueous liquid between the grana and the inner membrane is called the stroma. (Be careful not to confuse it with stoma!)

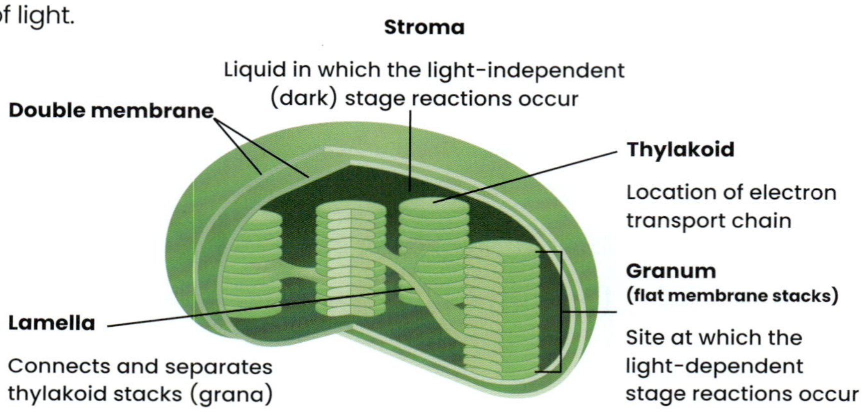

Figure 22.8: The internal structure of a chloroplast.

FOCUS ON ... TRANSFER MOLECULES HL

Before we look at the two stages involved in photosynthesis, it is important that we look at the transport molecule $NADP^+$.

In photosynthesis, the biomolecule nicotinamide adenine dinucleotide phosphate ($NADP^+$) acts as a proton (a hydrogen ion, H^+) and electron (e^-) carrier.

$NADP^+$ picks up two high-energy electrons and a proton forming the high-energy molecule NADPH: $NADP^+ + 2e^- + H^+ \rightarrow NADPH$.

NADPH can release these electrons and protons when needed during photosynthetic reactions, reforming the low energy molecule $NADP^+$: $NADPH \rightarrow NADP^+ + 2e^- + H^+$.

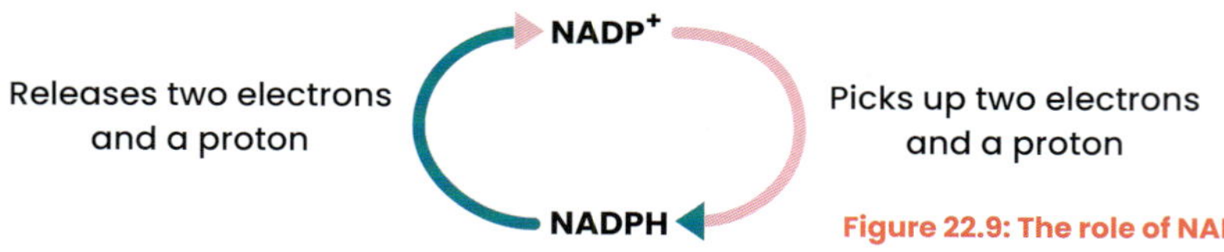

Figure 22.9: The role of $NADP^+$ in photosynthesis.

Photosynthesis occurs within the chloroplast in two stages:

1. the light-dependent stage, or 'light' stage
2. the light-independent stage, or 'dark' stage.

The light-dependent reactions occur in the thylakoid membrane, whereas the light-independent reactions occur in the stroma.

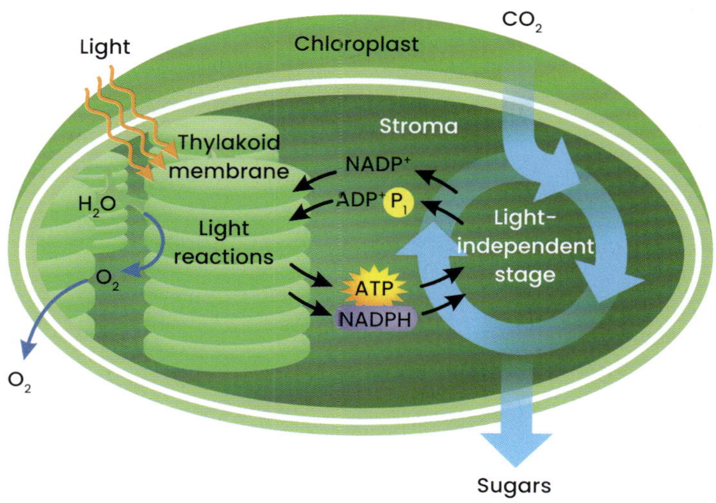

Figure 22.10: Light-dependent and light-independent stages of photosynthesis.

STAGE 1: LIGHT-DEPENDENT STAGE

HL

Cyclic pathway of light-dependent stage – electrons return to the chlorophyll

The **light-dependent or 'light' stage** will only occur when light is present.

- This stage occurs in the thylakoid membranes in the chloroplast, where a photon of light is absorbed by chlorophyll pigments.
- This light energy is passed from one chlorophyll pigment molecule to the next until it reaches a reaction centre in a chlorophyll molecule.

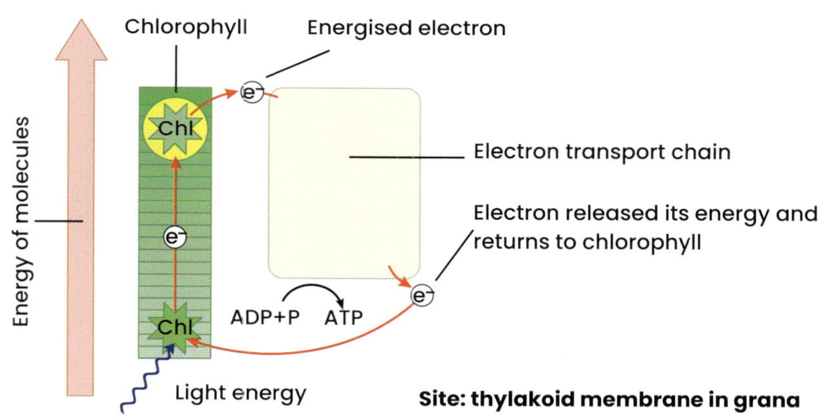

Figure 22.11: Cyclic pathway of light-dependent stage of photosynthesis.

- Energy from the light is transferred to electrons within the chlorophyll, energising them. These high energy electrons can enter one of two pathways: cyclic **or** non-cyclic pathways.
- Some of the energised electrons travel along an **electron transport chain** (ETC) in the thylakoid membrane.
- As the energised electrons travel along this chain, they release their surplus energy and return to the chlorophyll.
- This energy is used to form ATP.

- The **light-dependent stage**, the first stage of photosynthesis, involves the absorption of light energy by chlorophyll and the production of NADPH, ATP and oxygen.
- The **electron transport chain** is a series of protein complexes and mobile electron carriers embedded in the thylakoid membrane along which high energy electrons are passed, releasing their energy.

Non-cyclic pathway of light-dependent stage

- Energised electrons are picked up by an electron carrier, $NADP^+$, and do not return to chlorophyll.
- Each $NADP^+$ picks two electrons, forming $NADP^-$.

$$NADP^+ + e^- \rightarrow NADP$$

$$NADP + e^- \rightarrow NADP^-$$

$$\text{Overall: } NADP^+ + 2e^- \rightarrow NADP^-$$

Nicotinamide adenine dinucleotide phosphate ($NADP^+$) is a low-energy molecule that acts as an electron and proton (H^+) carrier in cells, forming high-energy NADPH.

- The loss of electrons to $NADP^+$ leaves chlorophyll positively charged.
- The positively charged chlorophyll attracts electrons (e^-) from water to replace those lost to the $NADP^+$.
- This causes **photolysis**, or splitting of water using energy from light. Water splits into protons (H^+), electrons (e^-) and oxygen (O_2):

$$H_2O \xrightarrow{\text{light}} ½O_2 + 2H^+ + 2e^-$$

- **Oxygen** is released to the air or used for respiration by the plant.

HL

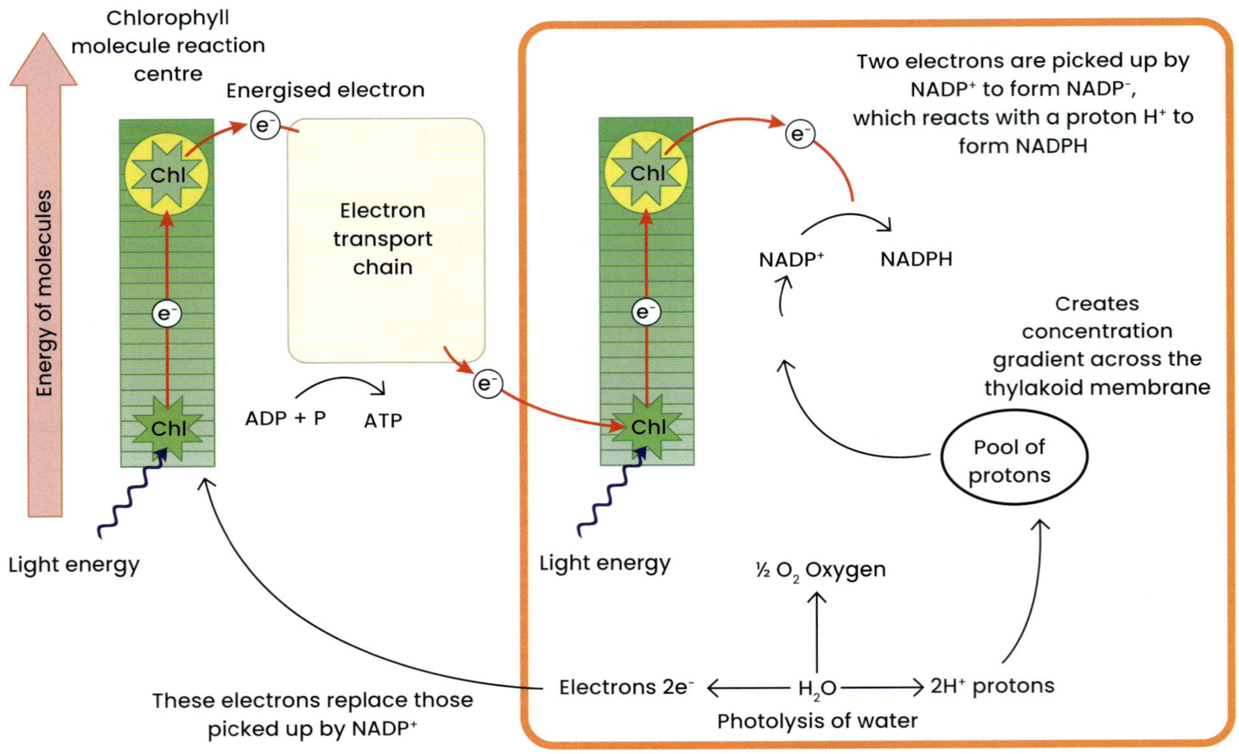

Figure 22.12: Non-cyclic pathway light-dependent stage photosynthesis.

ELECTRONS AND PROTONS

- The movement of electrons to protein complexes in the thylakoid membrane's electron transport chain causes some of the protein complexes to become negatively charged, creating a charge difference.
- This attracts protons (H^+) from donor molecules such as water in the stroma to the protein complex.

- These protons move inside the thylakoid lumen (space inside the thylakoid membrane). This results in a high proton (H^+) concentration inside the thylakoid lumen and a low proton (H^+) concentration outside in the stroma, creating a proton **concentration gradient**.
- Protons flow back down the concentration gradient, from high to low proton concentration, across the thylakoid membrane, through the enzyme ATP synthase.
- ATP synthase uses the energy from this flow of protons to attach a third phosphate group to ADP to form ATP.

ADP + P + energy ⟶ ATP

- $NADP^-$ picks up a proton (H^+) from the proton pool in the thylakoid lumen to form NADPH.

$NADP^- + H^+ \longrightarrow$ NADPH

HL

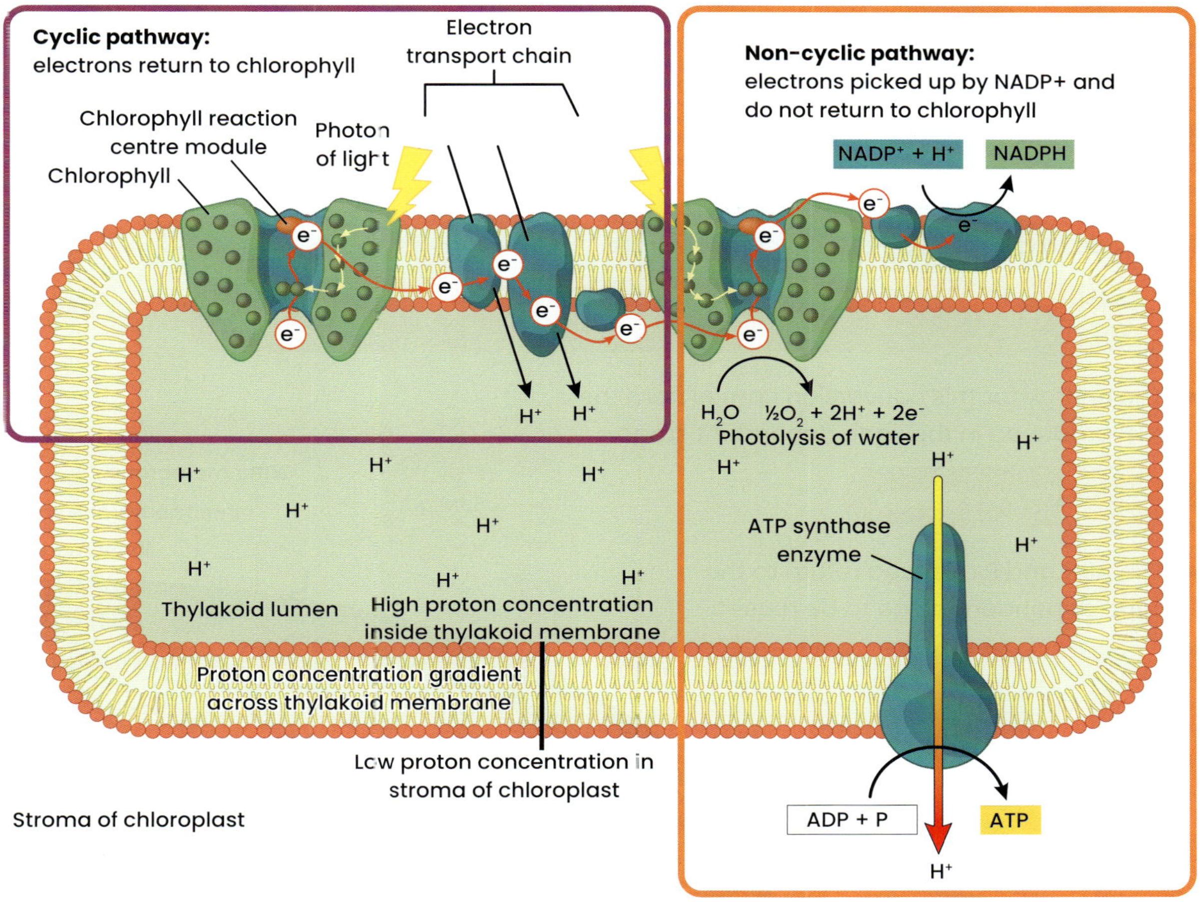

Figure 22.13: Cyclic and non-cyclic pathway of electron transfer of the light-dependent stage 1 of photosynthesis.

PRODUCTS OF LIGHT-DEPENDENT STAGE	FATE
Oxygen	Released to air or used in respiration.
ATP	Transfers energy to the light-independent stage (dark stage) to convert carbon dioxide to carbohydrate.
NADPH	Transfers electrons and protons to the light-independent stage (dark stage) for the conversion of carbon dioxide to carbohydrate.

STAGE 2: LIGHT-INDEPENDENT (DARK) STAGE

- This stage occurs in the liquid stroma in the chloroplast.
- ATP and NADPH produced in the light-dependent stage feed into the **light-independent stage**.
- NADPH from the light-dependent stage releases protons and electrons.

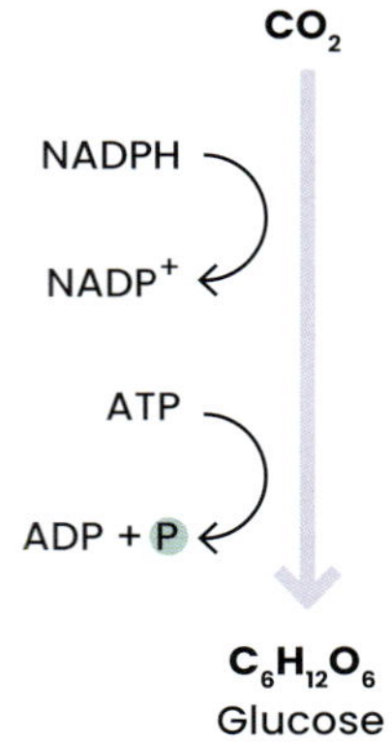

$NADPH \rightarrow NADP^+ + 2e^- + H^+$

- The $NADP^+$ then returns to the light-dependent stage to be reused.
- The protons and electrons released by NADPH react with carbon dioxide. A series of reactions convert carbon dioxide to glucose ($C_6H_{12}O_6$), a carbohydrate.

carbon dioxide (CO_2) + protons (H^+) + electrons (e^-) $\xrightarrow{\text{energy}}$ glucose ($C_6H_{12}O_6$)

- The energy for this conversion comes from the ATP produced in the light-dependent stage.

HL

ATP → ADP + P + energy

- The ADP and P released return to the light-dependent stage to be reused.

The **light-independent stage**, the second stage of photosynthesis, involves the conversion of carbon dioxide to carbohydrate using protons, electrons and ATP from the light-dependent stage.

Go to YouTube.com and search for 'Photosynthesis Intro and Light-Dependent Reactions' (3:54) by BioMan Biology.

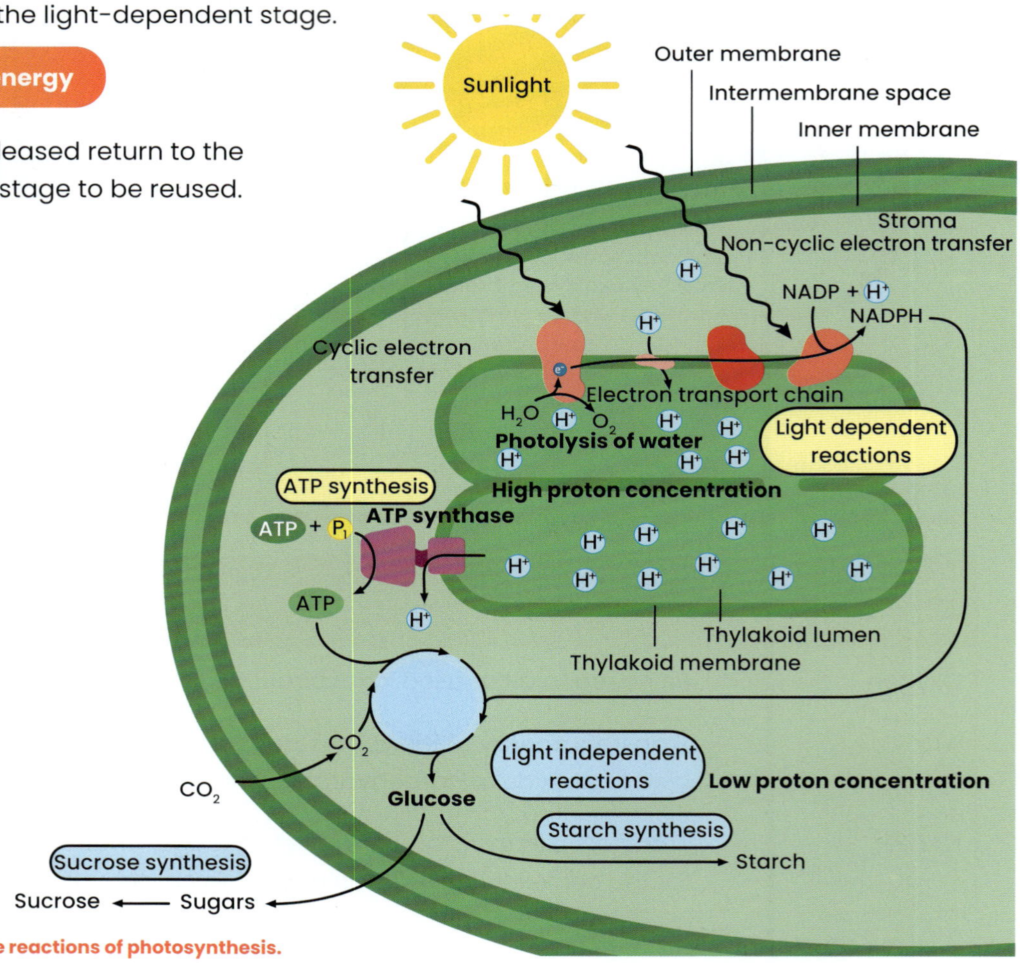

Figure 22.14: Overview of the reactions of photosynthesis.

PORTFOLIO LINK **Go to your Learning Portfolio and complete Learning Check 2 to assess your knowledge and understanding of the topic so far.**

Section 3: Factors Affecting Rate of Photosynthesis

At the end of this section, you should be able to:

- Understand the effect of temperature, light intensity and carbon dioxide levels on the rate of photosynthesis.

Unpacked from LO 2.2.2

FACTORS AFFECTING RATE OF PHOTOSYNTHESIS

Three main factors affect the rate of photosynthesis in plants: light intensity, carbon dioxide concentration and temperature.

LIGHT INTENSITY

Figure 22.15: Artificial lighting in an industrial greenhouse..

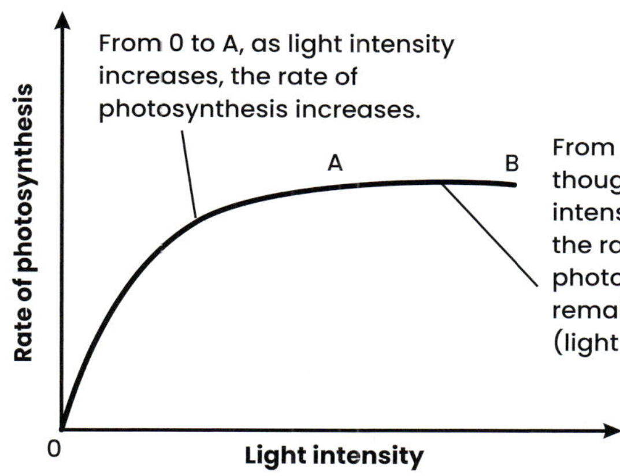

Figure 22.16: Effect of light on rate of photosynthesis.

The rate of photosynthesis increases with increasing light intensity up to a certain point, after which it levels off. The plant is 'light saturated' at this point – increasing the light intensity alone will not increase the rate of photosynthesis. The rate of photosynthesis is also limited by carbon dioxide concentration or temperature. Increasing the carbon dioxide concentration or temperature may increase the rate of photosynthesis beyond the light-saturation point. Horticulturists use artificial lighting to increase light intensity and allow photosynthesis to continue after daylight.

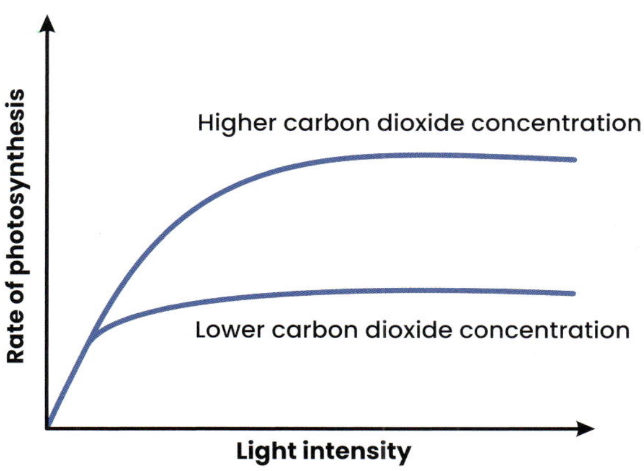

Figure 22.17: Increasing carbon dioxide concentration enhancing rate of photosynthesis.

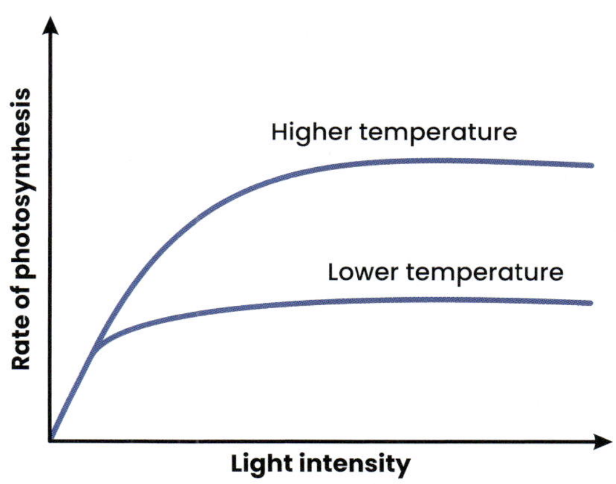

Figure 22.18: Increasing temperature enhancing rate of photosynthesis.

CARBON DIOXIDE CONCENTRATION

Increasing the carbon dioxide concentration will also increase the rate of photosynthesis up to a certain point, after which it levels off. As with light, the plant becomes 'saturated', and further increasing the levels of carbon dioxide will not increase the rate of photosynthesis. Either the light intensity or the temperature needs to be raised to increase the rate of photosynthesis further.

Horticulturists can make good use of gas heaters, which release carbon dioxide when they burn fuel. This increases both carbon dioxide concentration and temperature (see below).

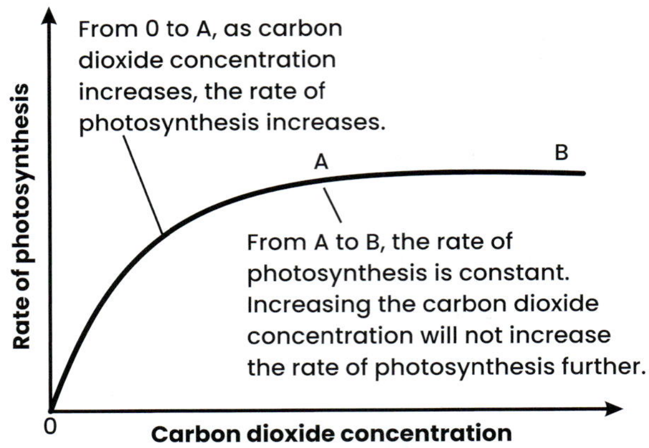

Figure 22.19: Effect of carbon dioxide concentration on rate of photosynthesis.

Figure 22.20: Gas heating in a greenhouse.

TEMPERATURE

Increasing the temperature increases the rate of photosynthesis up to a certain point – an optimal temperature. Increasing the temperature beyond this optimum can **denature** the enzymes that control photosynthetic reactions. Maintaining the optimum temperature is important to ensure the maximum rate of photosynthesis. Horticulturists use gas heaters in greenhouses to increase temperatures. Different plants have different optimum temperatures, but for most it is between 20–25°C.

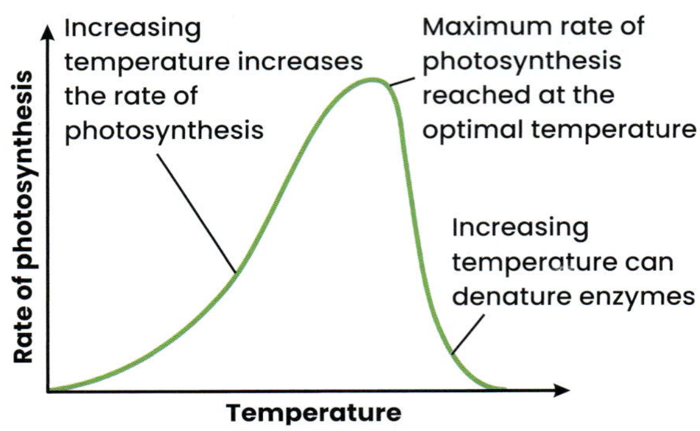

Figure 22.21 : The effect of temperature on the rate of photosynthesis.

Go to YouTube.com and search for 'Harnessing plant power' (0:55) by European Research Council.

SCIENCE IN SOCIETY: SUSTAINABILITY

As the world's population continues to rise, we need more food to feed a growing population. However, less space is available for growing crops. Boosting plant productivity benefits food security, so a group of scientists from Germany, the UK and Czechia are researching ways to enable plants to use a greater portion of the energy available from the Sun. Watch the video above to find out more.

PORTFOLIO LINK

Go to your Learning Portfolio and complete Learning Check 3 to assess your knowledge and understanding of the topic so far.

Specified Investigation 10

Investigate factors affecting the rate of photosynthesis (LO 2.2.2)

Scan the QR code to watch a video demonstration of this investigation.

A. LIGHT INTENSITY

In the following investigations, discs of smooth spinach or ivy leaves are placed in a partial vacuum to remove any initial oxygen present in the leaves. The leaf discs are exposed to light and undergo photosynthesis, releasing oxygen which causes the leaves to float in solution. The more the plants undergo photosynthesis, the more oxygen they produce.

Hypothesis: Increasing the light intensity will increase the rate of photosynthesis, up to a maximum, when the plant is light saturated.

Equipment: Beakers, cork borer, lamp, stirring rod, forceps, straw, syringe, ivy or spinach leaves, sodium bicarbonate, washing-up liquid, water.

VARIABLES

Independent variable: The distance of a lamp from the plant is varied.

Dependent variable: The number of leaf discs floating every minute.

Control variables: Temperature is kept constant at 25°C using a water bath. Carbon dioxide concentrations are kept constant using a solution of sodium bicarbonate.

METHOD

1. Using a cork borer, cut 10 leaf discs from spinach leaves.

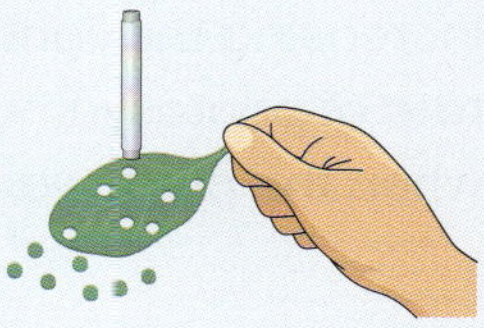

2. Dissolve 0.5g sodium bicarbonate in 500ml deionised water in a beaker to prepare a 0.1% solution. This acts as a source of dissolved carbon dioxide for photosynthesis.

3. Add a few drops of liquid dish soap to the 0.1% bicarbonate solution and stir gently. Avoid making any bubbles. The soap breaks down molecules on the leaves' surfaces allowing them to take in the bicarbonate solution.

4. Remove the plunger from a large syringe and, using a forceps, place the leaf discs into the syringe, near its tip. A straw can be used to gently blow the discs to the bottom of the syringe.

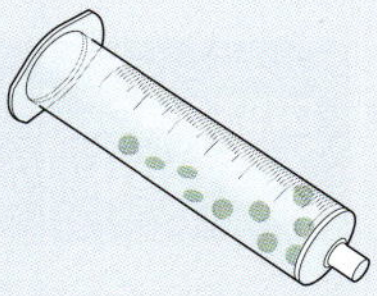

5. Replace the plunger into the syringe, being careful not to damage the leaf discs.

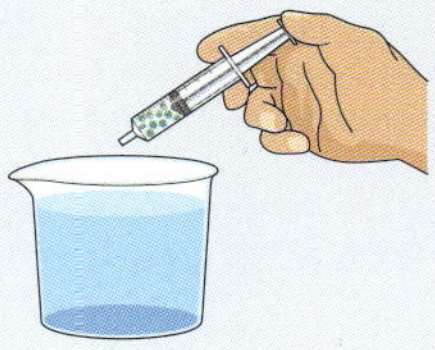

6. Insert the syringe tip into the 0.1% sodium bicarbonate solution and draw 6-8 ml into the syringe. The leaf discs should float in the solution.

7. Hold the syringe with the tip upward and expel the air by gently pushing the plunger, without releasing any of the solution.

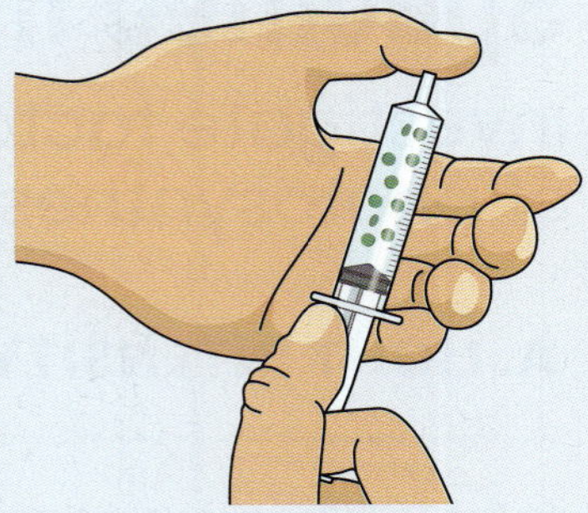

8. Place a finger on the syringe tip and gently pull the syringe plunger to create a partial vacuum. Tiny bubbles of air will be released from the discs. The leaf discs take in the bicarbonate solution and sink to the bottom. Repeat until all the leaf discs sink.

9. Carefully remove the plunger and pour the discs and solution into the beaker containing the sodium bicarbonate solution at 25°C. The leaf discs should settle to the bottom of the beaker. Remove any floating discs.

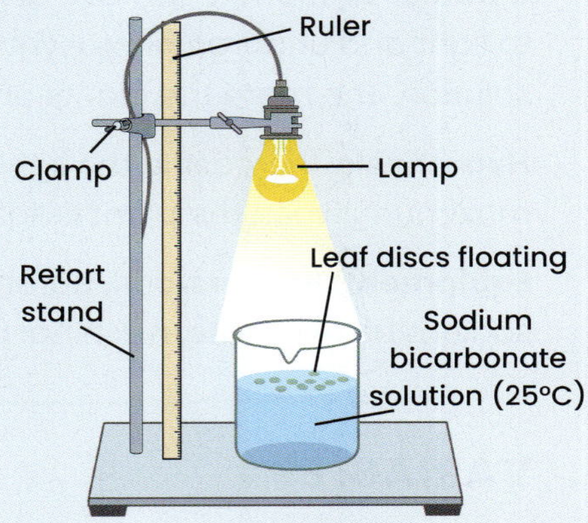

10. Place a lamp 10cm above the beaker.

11. Turn on the lamp and start a timer. Count the time taken for half of the leaf discs to float. As the discs photosynthesise, they release bubbles of oxygen that cause the discs to float.

12. Repeat the above steps with the lamp at the following heights above the beaker: 15cm, 20cm, 25cm, 30cm, 35cm.

13. Record your results in a table and plot a graph of rate of photosynthesis vs light intensity. Light intensity is calculated as follows, where d is the distance:
$\frac{1}{d^2}$
This value may be multiplied by 10^4 to obtain a value that is easier to graph.

RESULTS

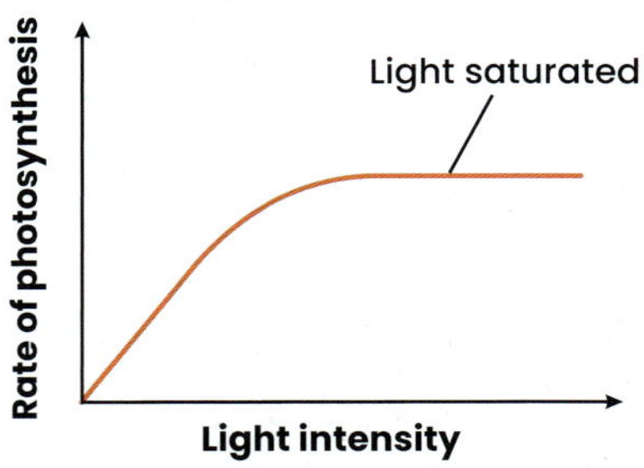

DISTANCE (d) (cm)	LIGHT INTENSITY = $\frac{1}{d^2} \times 10^4$	TIME TAKEN FOR FIVE DISCS TO FLOAT (seconds)	RATE OF PHOTOSYNTHESIS (number of discs floating/second)
35cm			
30cm			
25cm			
20cm			
15cm			
10cm			

CONCLUSION

Increasing the light intensity increased the rate of photosynthesis, up to a maximum, when the plant was light saturated.

B. CO_2 LEVELS

Hypothesis: Increasing the carbon dioxide concentration will increase the rate of photosynthesis, up to a maximum, when the plant is saturated with CO_2.

Equipment: Beakers, cork borer, lamp, stirring rod, forceps, straw, syringe, ivy or spinach leaves, sodium bicarbonate, washing-up liquid, water.

VARIABLES

Independent variable: Carbon dioxide concentrations are varied using sodium bicarbonate ($NaHCO_3$) solutions of different concentrations: 0.2%, 0.4%, 0.6%, 0.8% and 1.0%.

Dependent variable: The number of leaf discs floating each minute due to the production of oxygen.

Control variables: Temperature is kept constant at 25°C using a water bath. Light intensity is kept constant using a lamp of a fixed wattage at a set distance from the plant.

METHOD

1. Using a cork-borer, cut 10 leaf discs from spinach leaves.
2. Dissolve 0.2g sodium bicarbonate in 100ml of deionised water in a beaker to prepare a 0.2% solution.
3. Repeat Steps 3–10 of Investigation 10A.
4. Repeat the above steps with each of the following sodium bicarbonate solutions: 0.4%, 0.6%, 0.8% and 1.0% solution.

SODIUM BICARBONATE CONCENTRATION (W/V) (%)	0.2	0.4	0.6	0.8	1.0
MASS OF SODIUM BICARBONATE IN 100cm³ WATER (g)	0.2	0.4	0.6	0.8	1.0

5. Record your results in a table and plot a graph of rate of photosynthesis (time taken for five leaf discs to float) against carbon dioxide concentration.

RESULTS

CO_2 CONCENTRATION	TIME TAKEN FOR FIVE LEAF DISCS TO FLOAT (seconds)	RATE OF PHOTOSYNTHESIS (number of discs floating/second)
0.2%		
0.4%		
0.6%		
0.8%		
1.0%		

CONCLUSION

Increasing the carbon dioxide concentration increases the rate of photosynthesis up to a maximum when the rate levels off. Increasing carbon dioxide concentrations beyond this level will not increase the rate of photosynthesis any further.

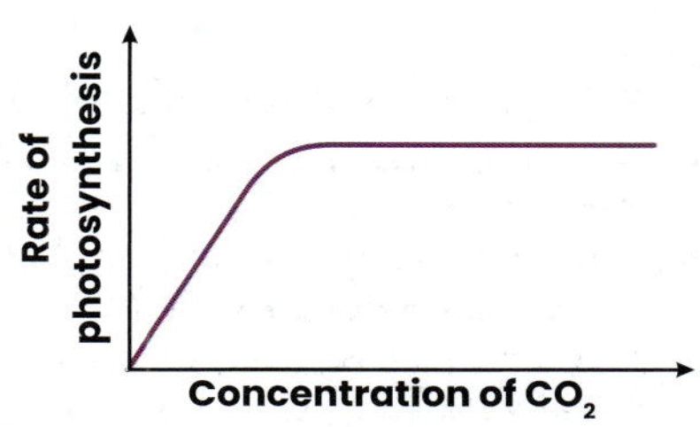

C. TEMPERATURE

Scan the QR code to watch a video demonstration of this investigation.

In the following investigation, a water plant (elodea) is used, as it allows the bubbles of oxygen gas released during photosynthesis to be seen.

Hypothesis: Increasing the temperature will increase the rate of photosynthesis up to an optimum, above which the rate declines.

Equipment: Elodea (Canadian pondweed), beaker, boiling tube, hot plate, lamp, metre stick, stopwatch, thermometer, ice, sodium bicarbonate, water.

VARIABLES

Independent variable: Temperatures are varied using water baths at different temperatures.

Dependent variable: The number of bubbles of oxygen released in a minute indicates the rate of photosynthesis.

Control variables: Carbon dioxide concentration is kept constant using a solution of sodium bicarbonate. Light intensity is kept constant using a lamp of a fixed wattage at a set distance.

METHOD

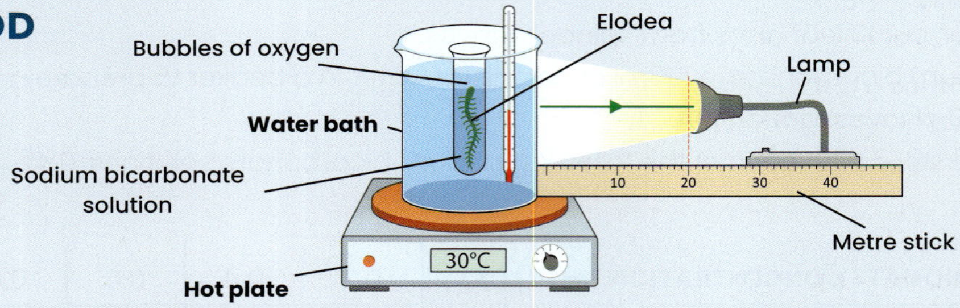

1. Cut a stem of Elodea (Canadian pondweed) at an angle and remove several leaves from its cut end.
2. Place the Elodea, cut end up, into a boiling tube containing 0.2% w/v sodium bicarbonate solution. It may be held down using a small weight.
3. Place the boiling tube, containing the Elodea, into an ice bath at 0°C.
4. Place the water bath containing the Elodea 20cm from a lamp. Turn on the lamp and wait five minutes for the Elodea to adjust to the light intensity.
5. Count the number of bubbles released from the cut end of the stem in a minute. Repeat twice and calculate the average.
6. Repeat the above steps using water baths at the following temperatures: 20°C, 30°C, 40°C and 60°C.
7. Record your results in a table and plot a graph of rate of photosynthesis (number of bubbles of oxygen/min) vs temperature.

RESULTS

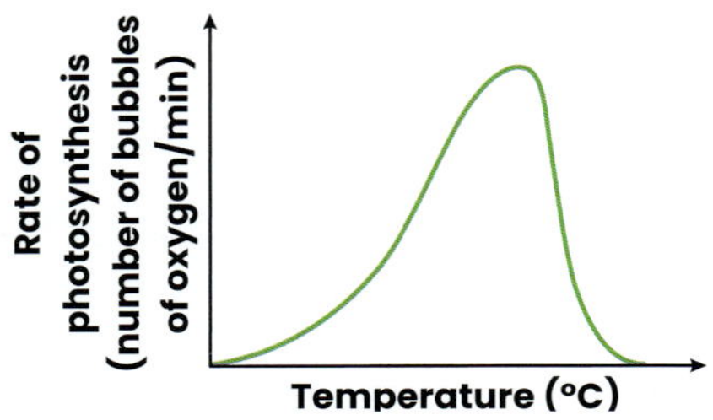

TEMPERATURE (°C)	RATE OF PHOTOSYNTHESIS
0	
20	
30	
40	
60	

CONCLUSION

Increasing the temperature increases the rate of photosynthesis up to an optimum when the rate levels off. If temperatures are increased above this, the rate of photosynthesis decreases.

INVESTIGATIVE LOG LINK

Go to your Investigative Log to record your lab work and answer questions on the experiments in this chapter.

Chapter Summary

- Chemical equation for *photosynthesis*:
  $6CO_2 + 6H_2O \xrightarrow{\text{light, chlorophyll}} C_6H_{12}O_6 + 6O_2$
- **Photosynthesis** is the process by which green plants make their own food by converting carbon dioxide and water to glucose and oxygen, using energy from sunlight.
- Photosynthesis is described as a **carbon sink**, as it takes more carbon out of the atmosphere than is released.
- **Reactants:** Carbon dioxide is taken in through the stomata in the leaves, water is absorbed through the roots, while the Sun is the main source of light.
- **Products:** Glucose is converted to sucrose for transport or stored as complex carbohydrates, e.g. starch. Glucose is used for respiration and oxygen is released or used in respiration.
- **The site of photosynthesis** is the chloroplast – organelle found in plant cells that contains the pigment chlorophyll.
- **Chlorophyll** is the green pigment in the chloroplast in plant cells that absorbs light to provide energy for photosynthesis.
- HL Light-dependent stage is the first stage of photosynthesis and only occurs in the light.
- HL Light is absorbed by **chlorophyll** pigment in thylakoid membrane in chloroplast.
- HL Energy from light is transferred to **electrons** in chlorophyll, energising them.
- HL High energy electrons enter one of two pathways: cyclic or non-cyclic.
- HL Pathway 1: Cyclic electron movement: Energised electrons travel along an electron transport chain, release their energy to form ATP and return to the chlorophyll:
  ADP + P + energy ⟶ ATP
- HL Pathway 2: Non-cyclic electron movement: Energised electrons are picked up by an electron carrier, $NADP^+$.
- HL Each $NADP^+$ picks two electrons, forming $NADP^-$.
- HL The loss of electrons to $NADP^+$ leaves chlorophyll positively charged.
- HL The positively charged chlorophyll attracts electrons from water to replace those lost to the $NADP^+$. This causes the 'photolysis', or splitting of water using energy from light.
- HL Water splits into protons (H^+), electrons (e^-) and oxygen (O_2).
- HL **Photolysis of water:** $H_2O \xrightarrow{\text{light}} \frac{1}{2}O_2 + 2H^+ + 2e^-$.
- HL Oxygen is released to the air or used for respiration by the plant.
- HL The movement of electrons along the thylakoid membrane's electron transport chain causes a charge difference. This attracts protons inside the thylakoid membrane, which results in a proton concentration gradient across the thylakoid membrane.
- HL Protons flow back down the concentration gradient, across the inner membrane of the thylakoid membrane, through the enzyme ATP synthase. ATP synthase uses the energy from this flow of protons to drive the **formation of ATP**:
  ADP + P + energy ⟶ ATP
- HL $NADP^-$ picks up a proton (H^+) from the pool of protons to form NADPH:
  $NADP^- + H^+ \longrightarrow NADPH$
- HL **Nicotinamide adenine dinucleotide phosphate ($NADP^+$)** acts as an electron and proton (hydrogen ion) carrier in cells.
- HL Light-independent stage is the second stage of photosynthesis and can occur with or without light. It occurs in the liquid stroma in the chloroplast.
- HL The ATP and NADPH produced during the light-dependent stage feed into the light-independent stage.
- HL NADPH releases protons and electrons:
  $NADPH \longrightarrow NADP^+ + 2e^- + H^+$ ($NADP^+$ reused in light dependent stage)
- HL Protons and electrons released by NADPH react with carbon dioxide. Series of reactions occur which result in the conversion of carbon dioxide to the carbohydrate glucose ($C_6H_{12}O_6$).
  Carbon dioxide (CO_2) + protons (H^+) + electrons (e^-) $\xrightarrow{\text{energy}}$ glucose ($C_6H_{12}O_6$)
- HL ATP produced in the light-dependent stage provides energy for this reaction:
  ATP ⟶ ADP + P + energy
- HL ADP and P return to the light-dependent stage to be reused.

END OF CHAPTER ASSESSMENT

REMEMBER

1. Define *photosynthesis*.
2. Write a balanced chemical equation for photosynthesis.
3. List the **four** things needed by a plant for photosynthesis to occur and state the source of each.
4. Name the **two** products of photosynthesis and give the fate of each.
5. Where in a plant cell does photosynthesis occur?
6. Write a word equation to illustrate the effect of light on water during photosynthesis.
7. HL Define each of the following terms.
   (a) HL Light-dependent stage of photosynthesis
   (b) HL Light-independent stage of photosynthesis
8. HL Where do each of the stages in Question 7 occur?

UNDERSTAND

9. Describe how leaves are adapted for photosynthesis.
10. HL Select the products of the light-dependent stage of photosynthesis from the word bank below.

ADP ATP carbon dioxide
electrons glucose $NADP^+$
NADPH oxygen protons

11. HL Describe the reaction that occurs in the light-independent stage of photosynthesis.

APPLY

12. Students investigating the effect of light intensity on rate of photosynthesis using algal gel beads obtained the following results.

DISTANCE FROM LAMP (cm)	LIGHT INTENSITY ($1/d^2$) ($\times 10^{-5}$)	ABSORBANCE AT 550nm
250	1.60	0.70
350	0.81	0.65
500	0.40	0.55
780	0.16	0.48
1250	0.06	0.41
dark	0.00	0.31

Draw a graph of absorbance against light intensity. Explain the shape of your graph.

ANALYSE

13. The light absorbed by solutions of two plant pigments – chlorophyll a and chlorophyll b – was measured using a spectrophotometer. The results are presented in the graph below. Study the graph and answer the following.
   (a) What colour of light is not absorbed by chlorophyll a?
   (b) At what wavelength is the maximum absorption of light by chlorophyll b?

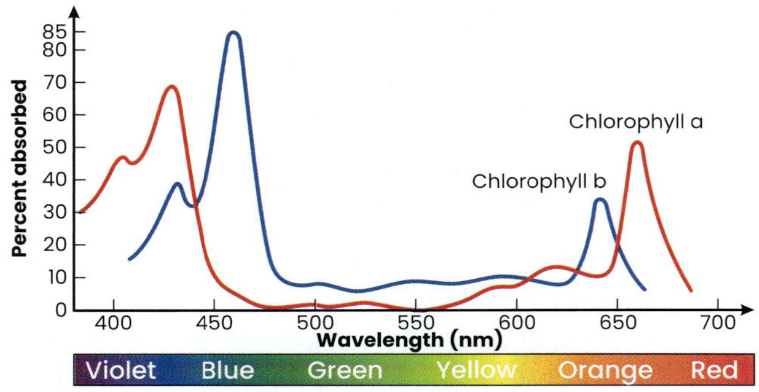

Absorption spectra

EVALUATE

14. What is the biological importance of photosynthesis to life on Earth?

CREATE

15. Design another investigation that you could carry out to determine the effect of temperature on rate of photosynthesis using spinach leaf discs.

23 Respiration

LEARNING OUTCOMES: U2.3, U2.5, 1.2.8, 2.2.1, 2.2.3, 2.2.4, 2.2.5

At the end of this chapter, you should be able to:

- Outline the processes of anaerobic respiration and aerobic respiration.
- Write a balanced chemical equation for aerobic respiration.
- Investigate the conditions necessary for fermentation.
- Outline the role of transfer molecule ATP in respiration.
- Outline the role of transfer molecule NAD^+ in respiration. HL
- Describe stage 1 and stage 2 of aerobic respiration, with reference to transfer molecules. HL
- Relate mitochondrial structure to respiration. HL

LINKS TO:

Chapter 1	What Is Biology?
Chapter 12	Nutrient Recycling
Chapter 14	Biomolecules
Chapter 16	Cell Structure and Diversity
Chapter 20	Plant Transport and Responses
Chapter 22	Photosynthesis
Chapter 28	The Breathing System
Chapter 30	The Digestive System
Chapter 31	The Circulatory System

SUBJECT VOCABULARY

respiration | aerobic respiration | anaerobic respiration | fermentation | ATP | matrix | cristae | NAD^+ | glycolysis | citric acid cycle | electron transport chain | concentration gradient

EYE-OPENER

Meat such as beef provides valuable protein to the human diet. However, livestock production has many negative impacts on the environment including deforestation and greenhouse gas emissions. Research published in the peer-reviewed journal *Nature* in 2022 indicates that replacing 20% of meat eaten globally with fermentation-derived microbial protein (such as Quorn) by 2050 would cut annual deforestation and related carbon dioxide emissions roughly in half.

Section 1: What Is Respiration?

At the end of this section, you should be able to:

- Define respiration and write a balanced equation for respiration.
- Distinguish between aerobic and anaerobic respiration.
- Outline the role of ATP in respiration.

Unpacked from LOs 1.2.8, 2.2.1

ROLE OF RESPIRATION IN CELLULAR ACTIVITY

We have already learned that respiration is one of the characteristics of living things. **Respiration**, carried out by all living cells, is a **catabolic process** as it converts large molecules into smaller ones with the release of energy. The energy released during respiration is stored in the form of ATP (adenosine triphosphate).

Respiration is the enzyme-controlled release of energy from food.

There are two main forms of respiration, depending on whether oxygen is present or not:

1. aerobic respiration
2. anaerobic respiration.

Breathing and respiration are two different processes. Respiration is a chemical process in cells that releases energy from the food you eat, whereas breathing is a physical process involving the inhalation and exhalation of air in and out of your lungs.

DID YOU KNOW?

Aerobic respiration

Aerobic respiration occurs in the presence of oxygen, breaking down glucose to form carbon dioxide and water with the release of energy in the form of ATP.

Aerobic respiration is the enzyme-controlled release of energy from food using oxygen.

Aerobic respiration is summarised by the following chemical equation.

Reactants		Products
$C_6H_{12}O_6 + 6O_2$	→	$6CO_2 + 6H_2O$ + energy (ATP)
Glucose + oxygen	→	carbon dioxide + water + energy

Cells in living organisms need energy to carry out different functions, e.g. for muscle contraction, cell replication or protein formation. Living things obtain this energy from their food – mainly glucose. Most of the energy released during respiration is in the form of heat.

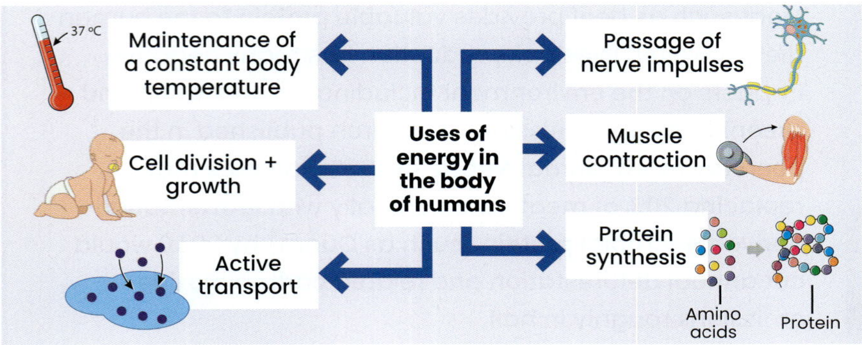

Figure 23.1: Uses of energy from respiration for cellular activity.

SITE OF RESPIRATION

We will see later that respiration occurs in two separate stages. In eukaryotic cells, e.g. in plants, animals and fungi, the first stage occurs in the cell's cytosol and the second stage occurs in the mitochondria. Bacteria do not have mitochondria and carry out respiration only in the cytoplasm.

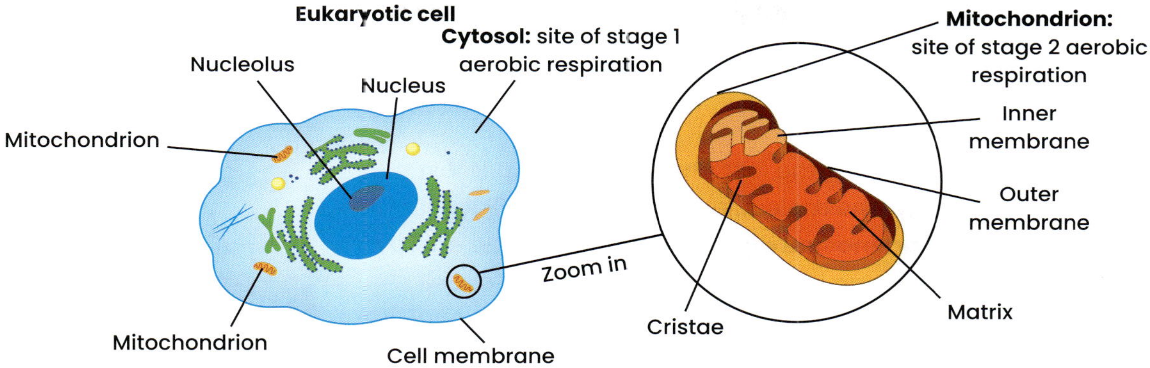

Figure 23.2: Eukaryotic cell showing cytosol (site of stage 1 respiration) and mitochondrion (site of stage 2 respiration).

ROLE OF RESPIRATION IN CARBON RELEASE

It is important not to confuse respiration with breathing which is the physical process of exchanging the gases oxygen and carbon dioxide. It should also be remembered that plants undergo respiration both day and night. Respiration results in the release of the carbon containing compound carbon dioxide to the atmosphere, from where it may be taken in by plants for photosynthesis. In Chapter 12, we learned about the importance of respiration and photosynthesis in the carbon cycle.

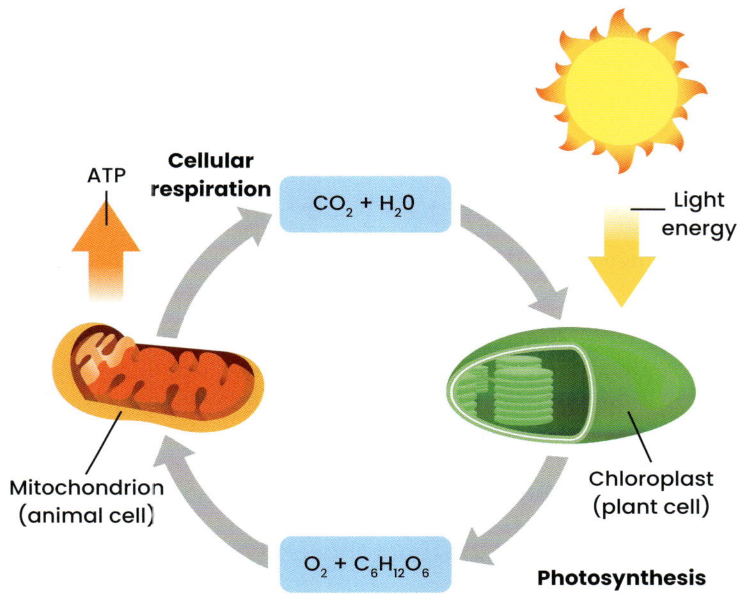

Figure 23.3: Role of respiration in carbon release.

Anaerobic respiration (fermentation)

Anaerobic respiration, also called fermentation, takes place in the absence of oxygen. **Alcohol fermentation** occurs in fungi (e.g. yeast), some bacteria and plant cells lacking oxygen.

glucose → ethanol + carbon dioxide + small amount energy (ATP)

$$C_6H_{12}O_6 \rightarrow 2C_2H_5OH + 2CO_2 + \text{energy}$$

This process is important in the production of alcoholic beverages. It is also used in baking bread with yeast – the carbon dioxide formed during anaerobic respiration by the yeast causes the bread to rise and the ethanol evaporates due to the heat of the oven. We will see later that lactic acid may be formed during fermentation in other organisms.

Go to YouTube.com and search for 'What Is Aerobic Respiration? | Physiology | Biology' (2:52) by FuseSchool – Global Education.

Anaerobic respiration (fermentation) is the controlled release of energy from food in the absence of oxygen.

THE ROLE OF ATP IN RESPIRATION

Organisms such as humans get their energy from the food they eat. In Chapter 22, we learned that chemical energy is stored in living organisms in a molecule called **adenosine triphosphate (ATP)**. We saw that ATP is formed from a nitrogenous base adenine, a five-carbon (5C) sugar ribose and three phosphate groups linked together. Energy is stored in the bonds between the phosphate groups.

Adenosine triphosphate (ATP) is a molecule used to store and transfer energy in cells.

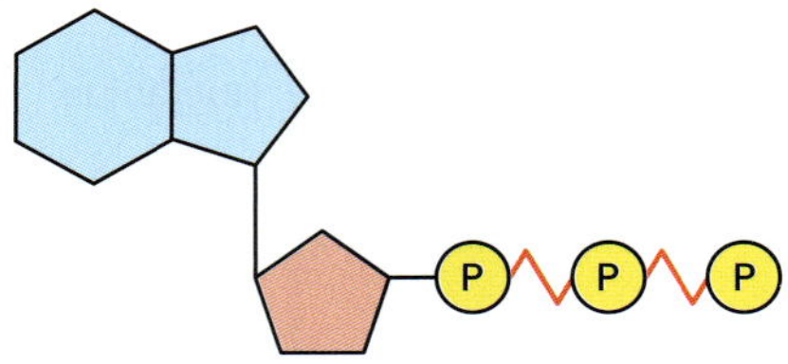

Figure 23.4: Structure of ATP.

When energy is needed in a cell, the last phosphate group is lost from ATP, releasing energy. The molecule formed when ATP loses one phosphate is called **adenosine diphosphate (ADP)**. The energy released in these reactions can be used by the organism for cell replication, enzyme formation, muscle action, etc.

ATP → ADP + P + energy

When energy is not required, ATP can be reformed in the cell by attaching phosphate groups to ADP using the energy obtained from food. This ATP formation occurs during the process of respiration.

ADP + P + energy → ATP

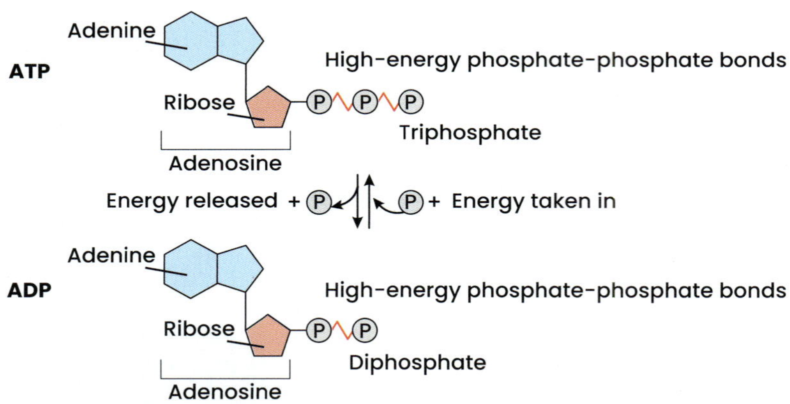

Figure 23.5: ATP storing and releasing energy.

Go to YouTube.com and search for 'ATP (ATP cycle and ADP) for Leaving Cert Biology' (4:07) by Biology Bugbears.

All the stages in respiration are controlled by different enzymes in cells.

The overall source of the reactants and products of respiration are outlined in the tables below:

REACTANTS FOR RESPIRATION	SOURCE
Glucose	• Formed by plants during photosynthesis • Product of breakdown of carbohydrates in the digestive system
Oxygen	• Formed by plants during photosynthesis • Taken in from the air by organisms

PRODUCTS OF AEROBIC RESPIRATION	SOURCE
Carbon dioxide (CO_2)	Carbon dioxide produced from the breakdown of glucose is released into the air
Water (H_2O)	Oxygen is converted to water in a series of reactions in respiration
ATP	ADP + P + energy → ATP

PRODUCT OF ANAEROBIC RESPIRATION	SOURCE
Lactic acid/ethanol	Lactic acid or ethanol are produced from the breakdown of glucose
Carbon dioxide	Carbon dioxide produced from the breakdown of glucose is released into the air
ATP	ADP + P + energy ⟶ ATP

SCIENCE IN SOCIETY: TECHNOLOGY

Precision fermentation

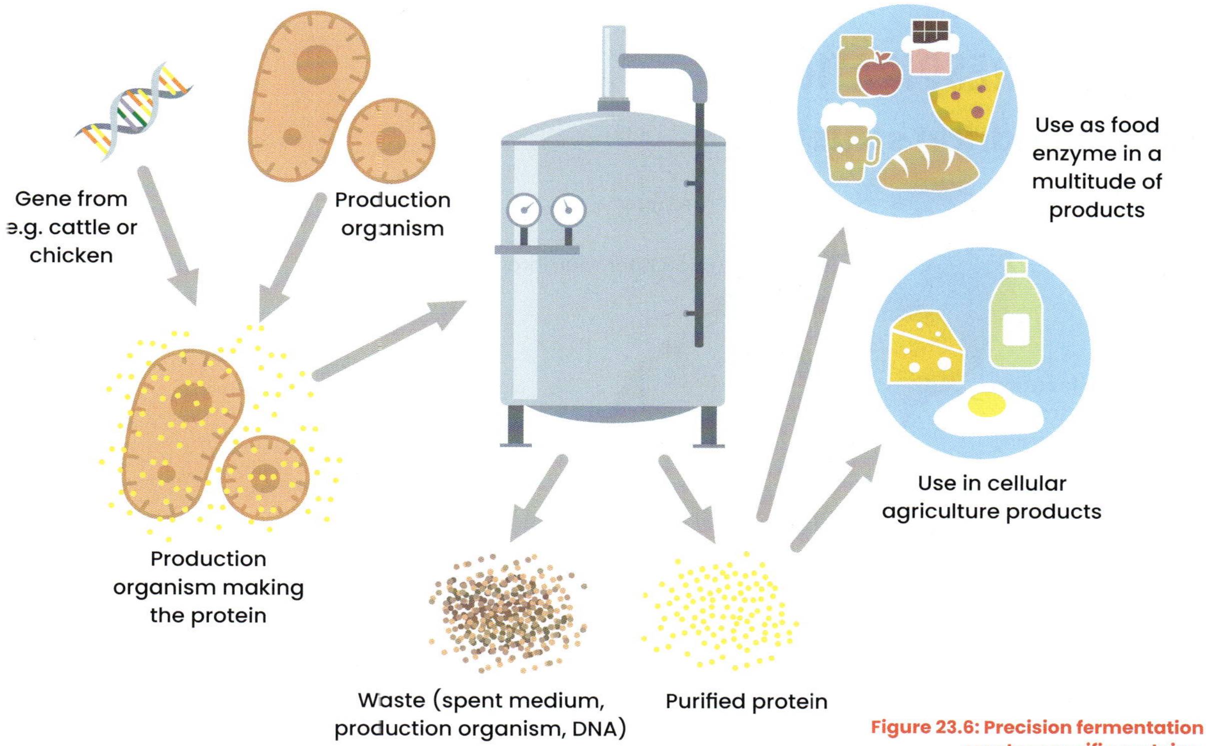

Figure 23.6: Precision fermentation creates specific proteins.

Precision fermentation involves using genetically modifying micro-organisms to create specific proteins found in food. It aims to produce high-quality proteins at a lower cost and with minimal environmental impact compared to traditional food sources. The genetically modified micro-organisms are introduced to a fermentation vessel containing water, nutrients and a carbohydrate substrate, which form the desired protein. The protein is then purified and dried into a powdered form for use as an ingredient in other products, such as plant-based protein or cultivated meat. Precision fermentation of haemoglobin has been used to give plant-based burgers the look and smell of red meat. Chicken-free egg products have also been produced with precision fermentation technology.

PORTFOLIO LINK **Go to your Learning Portfolio and complete Learning Check 1 to assess your knowledge and understanding of the topic so far.**

Section 2: Two-Stage Process of Respiration

At the end of this section, you should be able to:

- Describe stage 1 (glycolysis) of respiration. HL
- Describe stage 2 (citric acid cycle and electron transport chain) of respiration. HL
- Explain the role of transfer molecules in respiration. HL
- Identify the cellular location of the two stages in aerobic respiration. HL
- Relate mitochondrial structure to respiration. HL
- Investigate the conditions necessary for fermentation.

Unpacked from LOs 1.2.8, 2.2.3, 2.2.4, 2.2.5

AEROBIC RESPIRATION

HL

Mitochondrion structure

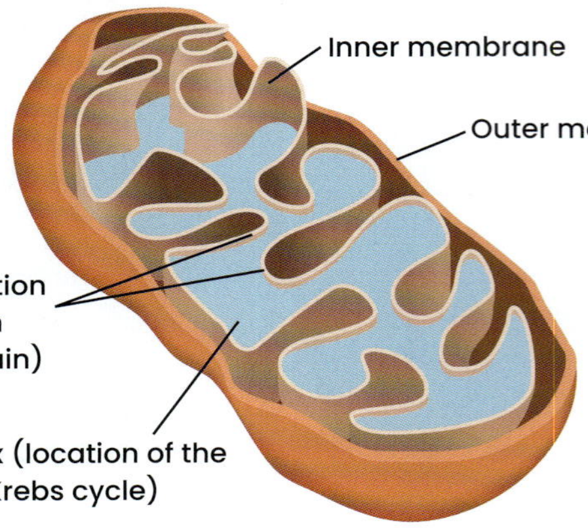

Figure 23.7: Structure of a mitochondrion showing the cristae and matrix.

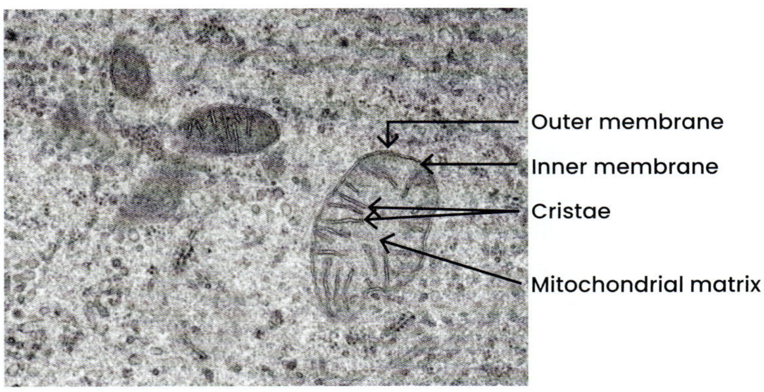

Figure 23.8: Mitochondrion under an electron microscope.

The mitochondrion is the site of aerobic respiration in cells. Mitochondria are small organelles with a double membrane. The inner membrane has folds called **cristae**. The greater surface area of the folded cristae maximises the production of ATP. Inside the inner membrane is a fluid called the **matrix**, which contains ribosomes, enzymes and mitochondrial DNA. Cells that need a lot of energy and respire more have larger numbers of mitochondria, e.g. muscle cells and sperm cells.

- The **matrix** is the fluid-filled inner space in the mitochondrion.
- The **cristae** are folds on the inner membrane of the mitochondrion.

Process of respiration

Before we examine the two-stage process of respiration in detail, we will look at the transfer molecule **nicotinamide adenine dinucleotide (NAD^+)**.

Nicotinamide adenine dinucleotide (NAD^+) is a low energy molecule that acts as an electron and proton (H^+ hydrogen ion) carrier in metabolic reactions to form a high-energy molecule NADH.

FOCUS ON ... NAD^+: A TRANSFER MOLECULE (HL)

NAD^+ is a biomolecule that acts as a proton (H^+) and electron (e^-) carrier in metabolic reactions. When NAD^+ picks up two high-energy electrons and a proton (H^+) in metabolic reactions such as respiration, it forms the compound NADH:

$$NAD^+ + 2e^- + 2H^+ \rightarrow NADH + H^+$$

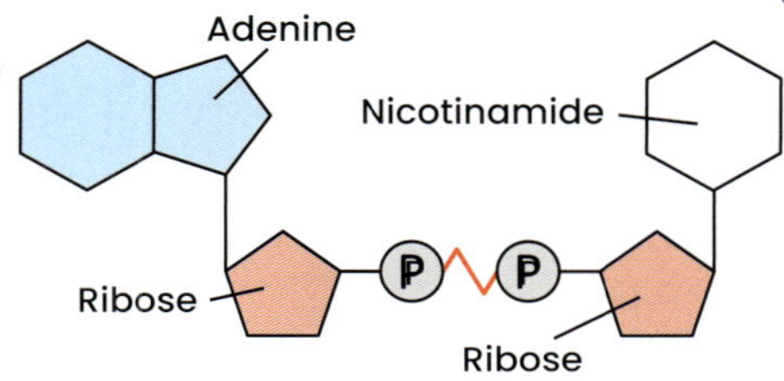

Figure 23.9: Structure of NAD^+.

NADH can then donate these protons and electrons when needed in respiration, releasing energy, which is used to form ATP:

$$NADH \rightarrow NAD^+ + 2e^- + H^+$$

Aerobic respiration stage 1: Glycolysis

- Aerobic respiration is a two-stage process. The first stage of respiration is called **glycolysis**, which means 'splitting sugar'.

HL

- Glycolysis occurs in the cytosol – the liquid filling the cell.
- This process does not require oxygen – it can occur in the presence or absence of oxygen.
- In glycolysis, the six-carbon glucose molecule is broken down to two molecules of three-carbon (3C) pyruvate (also known as pyruvic acid).
- Stage 1 releases a small amount of energy (2 ATP).
- Two protons (hydrogen ions, H^+) and four electrons (e^-) removed from glucose join with NAD^+ to form two molecules of NADH.

Go to YouTube.com and search for 'NAD^+ and NADH Leaving Cert Biology' (2:31) by Biology Bugbears.

Glycolysis is the breakdown of six-carbon glucose to two molecules of three-carbon pyruvate.

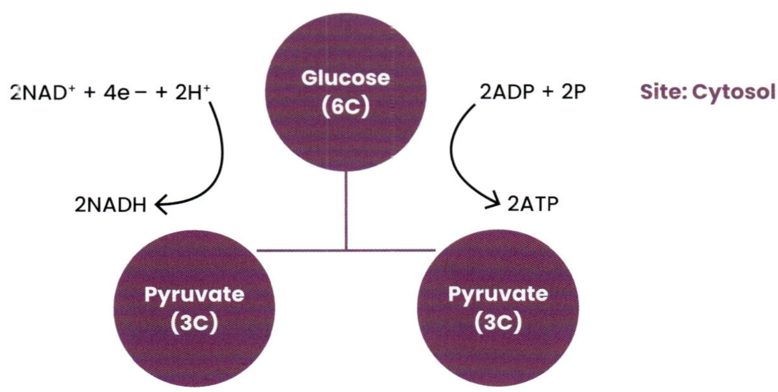

Figure 23.10: Stage 1 of respiration: Glycolysis.

The fate of the pyruvate molecules formed during glycolysis depends on whether oxygen is present or not.

Aerobic respiration stage 2

- This stage is aerobic and only occurs if oxygen is present.
- If oxygen is present, the pyruvate diffuses into the matrix of the mitochondrion.
- Here, each pyruvate (3C) breaks down to form carbon dioxide (1C) and a two-carbon (2C) molecule, acetyl coenzyme A (Acetyl-CoA). CO_2 is breathed out.
- A proton (H^+) and two electrons (e^-) removed from pyruvate are picked up by NAD^+ to form NADH.

$$NAD^+ + 2e^- + H^+ \rightarrow NADH$$

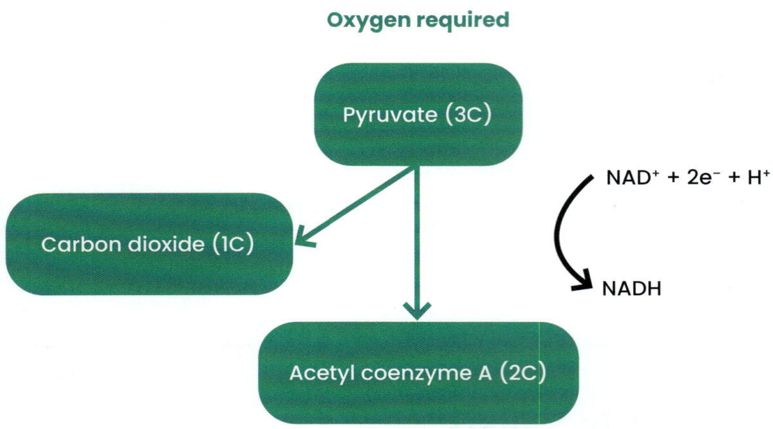

Site:
matrix of
mitochondrion

Products per glucose:
2 CO_2, 2 Acetyl-CoA, 2 NADH

Products per pyruvate:
1 CO_2, 1 Acetyl-CoA, 1 NADH

Figure 23.11: Stage 2 of respiration: Preparation for the citric acid (Krebs) cycle.

THE CITRIC ACID CYCLE

HL

- If oxygen is present, each acetyl coenzyme A molecule will enter a series of reactions called the **citric acid cycle** (also called the Krebs cycle).
- The citric acid cycle takes place in the fluid-filled space, the matrix.
- The citric acid cycle produces carbon dioxide (CO_2).
- Electrons from the cycle are transferred through an electron transport chain; energy from this transfer is used in the production of ATP molecules which can be used for other biochemical reactions within cells.
- Protons and electrons are transferred to NAD^+ to form NADH.

> The **citric acid cycle** is a series of reactions in the matrix of the mitochondrion, breaking down Acetyl-CoA to release carbon dioxide, ATP and NADH.

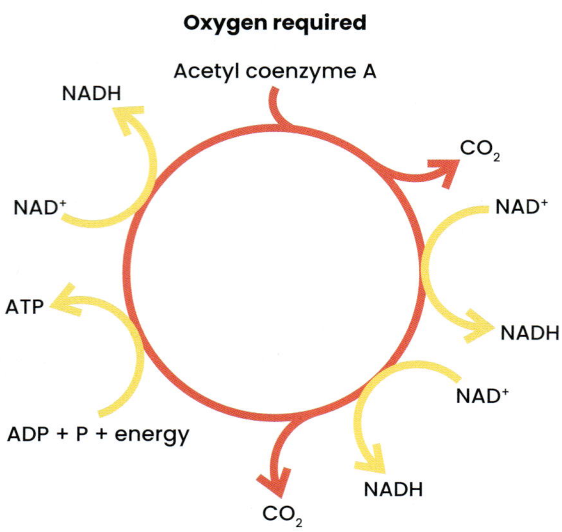

Site:
matrix of
mitochondrion

Products per Acetyl-CoA:
3 NADH, 1 ATP, 2 CO_2

Products per glucose:
6 NADH, 2 ATP, 4 CO_2

Figure 23.12: Stage 2 of respiration: Citric acid cycle. (Remember, each glucose molecule gives rise to two Acetyl-CoA molecules.)

The electron transport chain (ETC)

- Most of the energy produced during aerobic respiration occurs during the **electron transport chain**.
- The electron transport chain is formed from a series of protein complexes and mobile electron carriers in the cristae.
- NADH from stage 1 (glycolysis) and stage 2 (citric acid cycle) transfer high-energy electrons to the electron transport chain. NAD^+ from this reaction can be reused in stages 1 and 2.

The electron transport chain (ETC) is a group of protein complexes and mobile electron carriers embedded in the inner mitochondrion membrane along which electrons move – releasing their energy to drive the formation of ATP.

$NADH \rightarrow NAD^+ + 2e^- + H^+$

- Electrons are transferred from one protein complex to the next, moving from a higher energy level to lower energy level, gradually releasing their energy.
- This causes a charge difference which attracts protons to the complex from donor molecules such as NADH.
- The energy released by the electrons is used to pump protons (H^+) from the matrix into the intermembrane space, where they build up.

HL

- This creates a **concentration gradient** with a low proton (H^+) concentration in the matrix and a high proton (H^+) concentration in the intermembrane space between the inner and outer mitochondrial membranes.

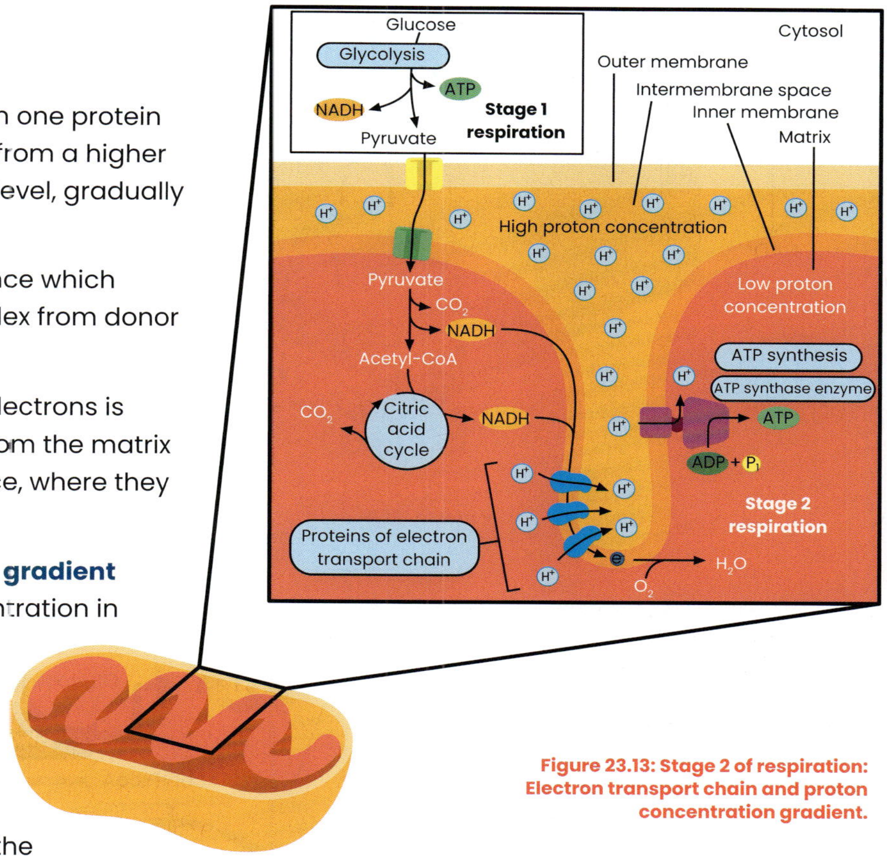

Figure 23.13: Stage 2 of respiration: Electron transport chain and proton concentration gradient.

- Protons (H^+) flow back down the gradient across the inner membrane, through the enzyme ATP synthase embedded in the inner mitochondrial membrane.
- This flow of protons releases energy which is used to make ATP.

ADP + P + energy $\rightarrow$ ATP (34 ATP per glucose)

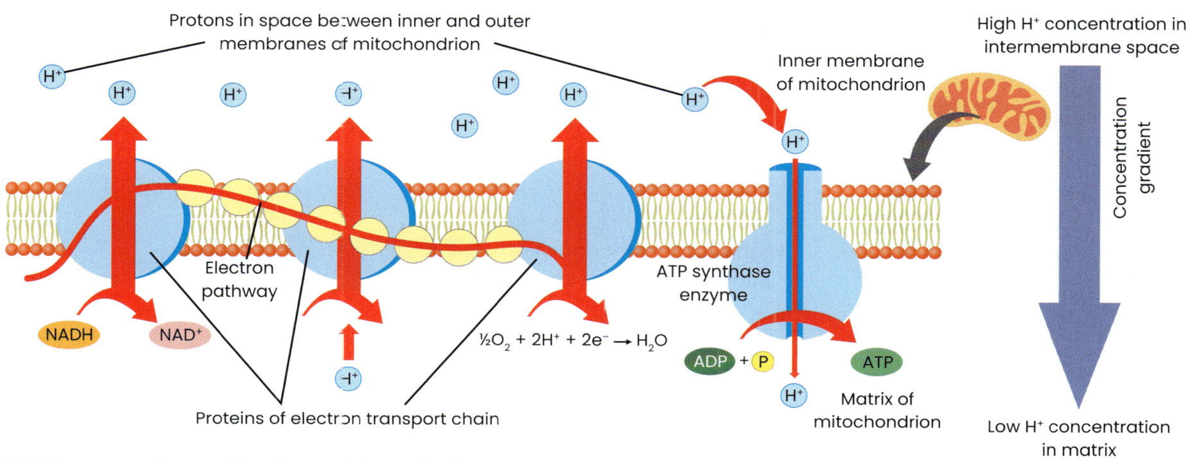

Figure 23.14: Concentration gradient in aerobic respiration.

The **concentration gradient** in respiration is the difference in the distribution of protons between the intermembrane space and the inner matrix of the mitochondrion.

- At the end of the chain, electrons are transferred to oxygen which reacts with protons (H^+) from the mitochondrion matrix to form water (H_2O) ($6H_2O$ per glucose).

$\frac{1}{2}O_2 + 2H^+ + 2e^- \rightarrow H_2O$

Go to YouTube.com and search for 'Aerobic Respiration – Updated to help with revision – Leaving Cert Biology Respiration' (8:45) by Biology Bugbears.

HL

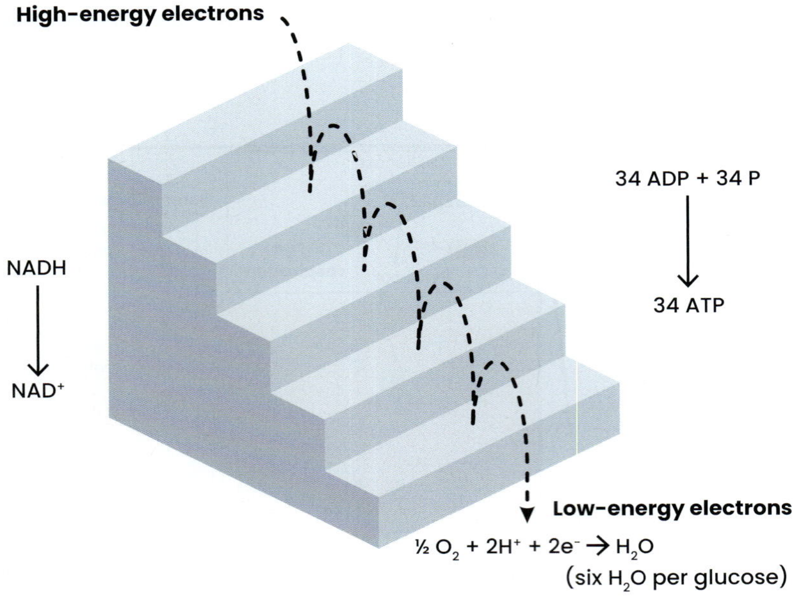

Figure 23.15: Electrons losing energy along electron transport chain in aerobic respiration.

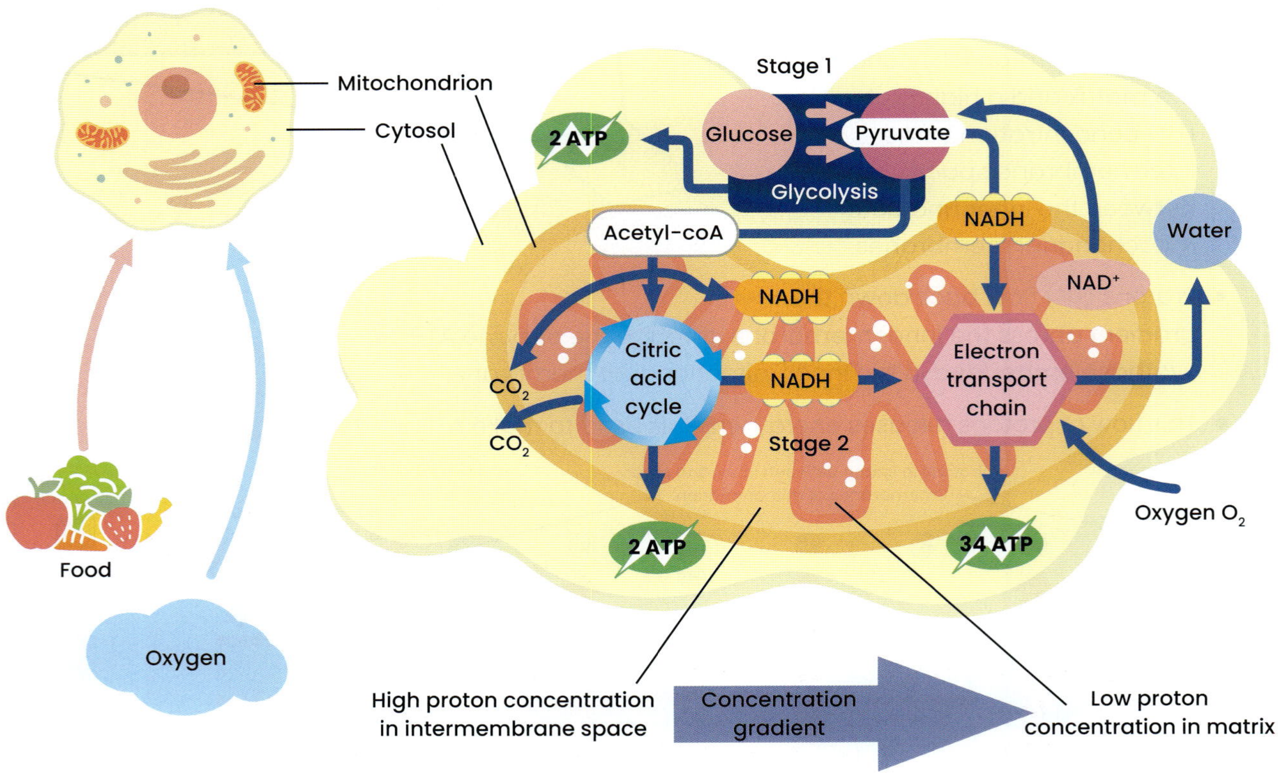

Figure 23.16: Overview of aerobic respiration.

ANAEROBIC RESPIRATION: FERMENTATION

If oxygen is not present, pyruvic acid is converted to one of two products: **ethanol** or **lactic acid**.

Alcohol fermentation occurs in fungi e.g. yeast, some bacteria and plant cells lacking oxygen.

glucose → ethanol + carbon dioxide + small amount of energy (ATP)

$C_6H_{12}O_6 \rightarrow 2C_2H_5OH + 2CO_2 + \text{energy}$

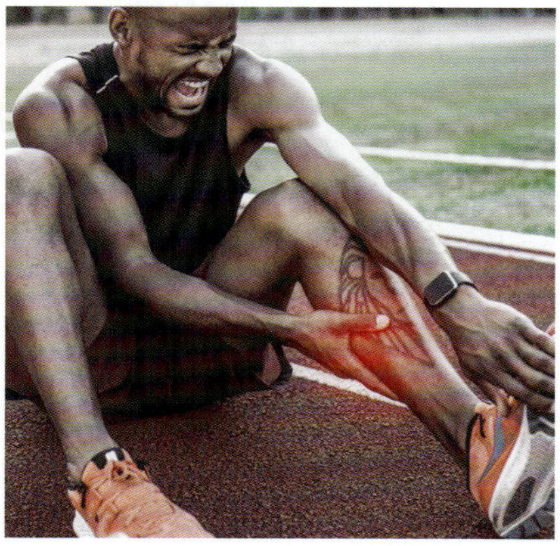

Figure 23.17: An athlete suffering from leg cramp (anaerobic lactic acid fermentation in muscle cells).

Lactic acid fermentation occurs in some bacteria, fungi and animal cells lacking oxygen, e.g. muscle cells after strenuous exercise.

glucose → lactic acid + small amount energy (ATP)

Lactic acid formation in animal cells, e.g. muscle cells, can cause painful cramping. To break down the lactic acid, more oxygen must be taken in, which requires more rapid and deeper breathing. The lactic acid is converted to carbon dioxide and water which are breathed out.

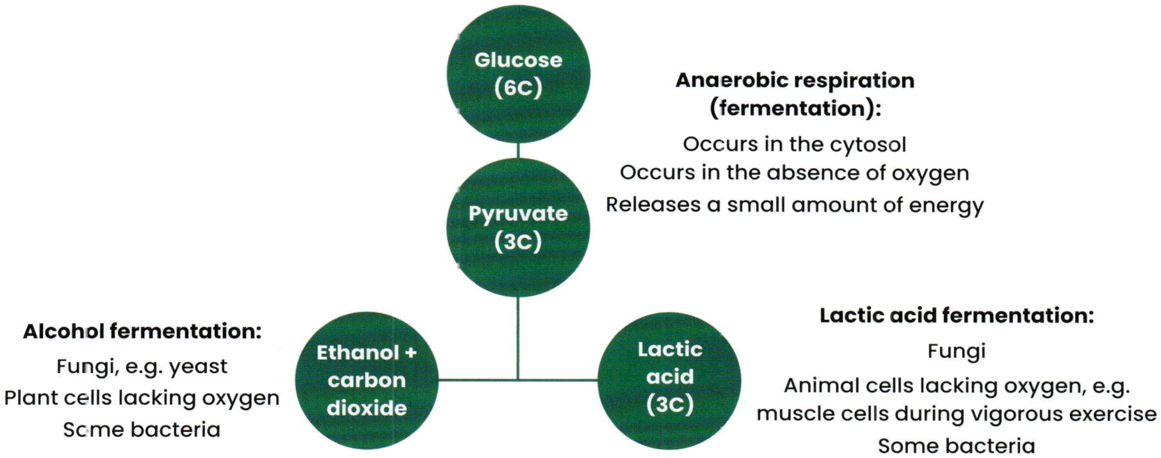

Figure 23.18: Outline of anaerobic respiration (fermentation).

FOCUS ON ... PRODUCTION OF BIOETHANOL FROM FERMENTATION

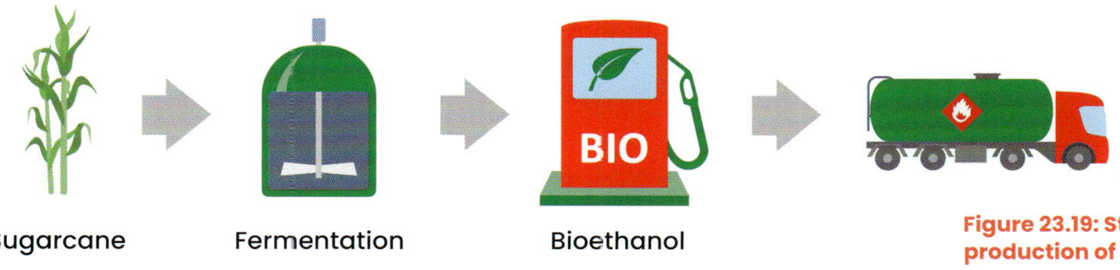

Figure 23.19: Stages in the production of bioethanol.

Ethanol produced by the fermentation of crops such as sugarcane, sugar beet, maize or wheat may be used as a fuel for vehicles. This biofuel or bioethanol is a renewable source of energy as it is produced from plants that may be readily grown. It is less toxic and more environmentally friendly than fossil fuels. Most vehicles can use a blend of petrol with up to 10% bioethanol without any modifications to their engine. In 2023, E10 (10% bioethanol) was introduced as the standard grade of petrol in Ireland. This forms part of Ireland's Climate Action Plan 2021 (CAP21), which aims to achieve a 51% reduction in carbon emissions by 2030.

PORTFOLIO LINK

Go to your Learning Portfolio and complete Learning Check 2 to assess your knowledge and understanding of the topic so far.

Specified Investigation 11

Investigate the conditions necessary for fermentation – presence of sugars and micro-organisms (yeast), anaerobic environment and suitable temperature. (LO 2.2.3)

Scan the QR code to watch a video demonstration of this investigation.

Hypothesis: In anaerobic conditions, yeast will ferment a glucose solution, forming the alcohol ethanol and releasing carbon dioxide.

Equipment: Five conical flasks, four fermentation locks, yeast, glucose solution, water bath, test tubes, funnel, limewater, potassium iodide solution, sodium hypochlorite solution, graduated droppers.

VARIABLES

Independent variable: Presence or absence of yeast, presence or absence of fermentation lock, water baths at 4°C, 30°C and 60°C.

Dependent variable: The production of the alcohol ethanol.

Control variables: The same volume and concentration of glucose solution was used in each case.

METHOD

1. Add 250cm³ 10% w/v glucose solution to each of five conical flasks.
2. Add 5g of yeast to flasks A, C, D and E.
3. Attach a fermentation lock (half-filled with limewater) to flasks A, B, D and E.
4. Place flasks A, B and C in a water bath at 30°C overnight. Place flask D in a water bath at 60°C overnight. Place flask E in a fridge at 4°C overnight.

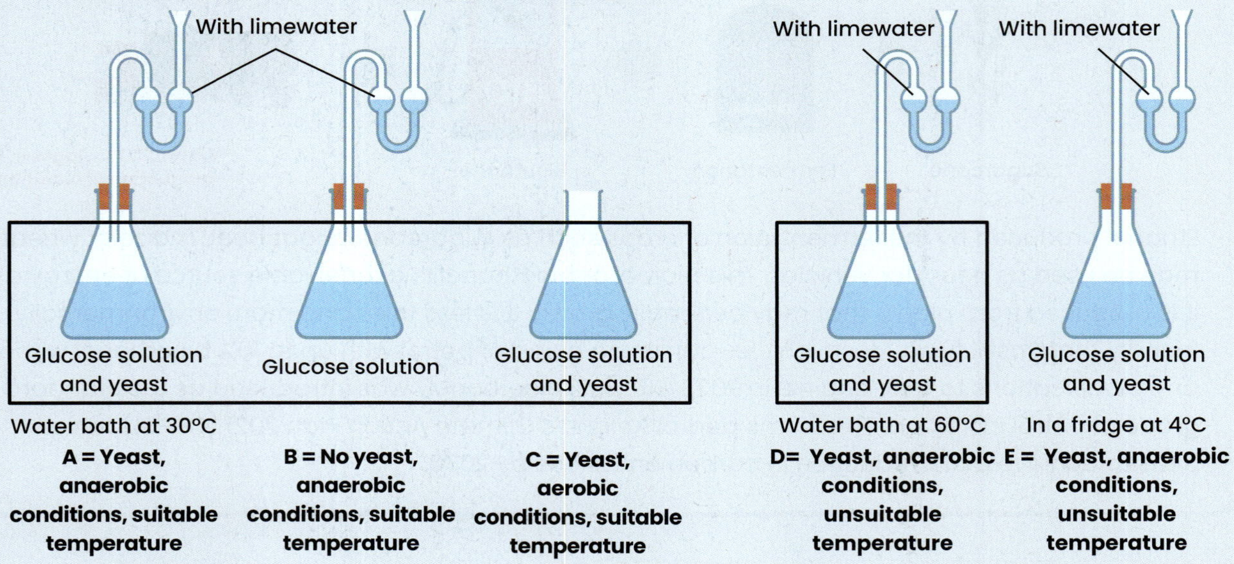

5. Filter the solution in each conical flask into separate beakers.

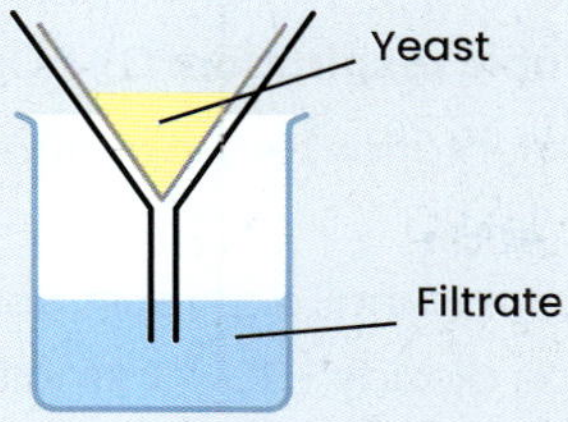

6. Place 3cm³ of each filtrate into separate test tubes using separate graduated droppers.

7. Add 3cm³ potassium iodide solution and 5cm³ of the sodium hypochlorite solution (iodoform solution) to each test tube.

8. Place both test tubes in a water bath at 60°C for 5 minutes.

9. Observe any colour change and record your results in a table.

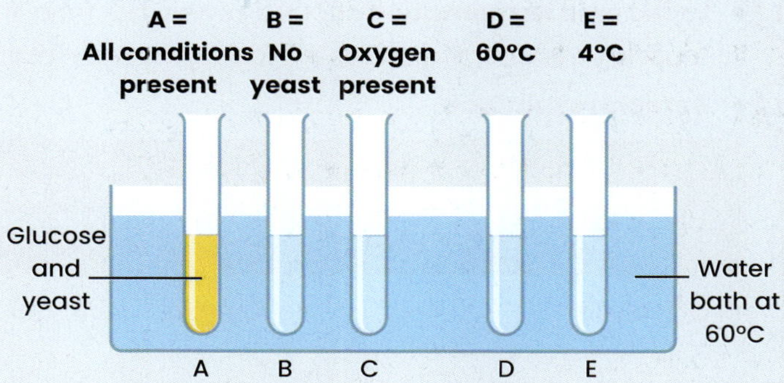

Add 3cm³ potassium iodide and 5cm³ sodium hypochlorite to each test tube

RESULTS

	CONDITION ABSENT	APPEARANCE OF LIMEWATER	APPEARANCE IN IODOFORM	CONCLUSION
A: Yeast and glucose solution, anaerobic, 30°C	All conditions present	Milky	Yellow crystals	Ethanol (alcohol) formed
B: Glucose solution, anaerobic, 30°C	No yeast	No change	No change	No ethanol present
C: Glucose and yeast in aerobic conditions, 30°C	Aerobic rather than anaerobic	Not used	No change	No ethanol present
D: Yeast and glucose solution, anaerobic, 60°C	Unsuitable temperature	No change	No change	No ethanol present. High temperature kills the yeast
E: Yeast and glucose solution, anaerobic, 4°C	Unsuitable temperature	No change	No change	No ethanol present. Low temperature slows down yeast's metabolism

CONCLUSION

Formation of solid yellow crystals of iodoform indicates the formation of alcohol (ethanol). Fermentation requires the presence of a sugar (glucose), micro-organism (yeast), anaerobic conditions and a suitable temperature to occur.

INVESTIGATIVE LOG LINK

Go to your Investigative Log to record your lab work and answer questions on the experiment in this chapter.

Chapter Summary

- **Aerobic respiration** is the controlled release of energy from food using oxygen.
- **Chemical equation for aerobic respiration:** $C_6H_{12}O_6 + 6O_2 \rightarrow 6CO_2 + 6H_2O$ + energy (ATP).
- **Anaerobic respiration** (fermentation) is the controlled release of energy from food in the absence of oxygen.
- **Alcohol fermentation** occurs in fungi, e.g. yeast, some bacteria and plant cells lacking oxygen: glucose → ethanol + carbon dioxide (CO_2) + small amount energy (ATP).
- HL **Lactic acid fermentation** occurs in some bacteria, fungi and animal cells lacking oxygen, e.g. muscle cells after strenuous exercise: glucose → lactic acid + small amount energy.
- HL **Aerobic respiration**

STAGE	LOCATION	OXYGEN REQUIREMENT	ENERGY YIELD	PRODUCTS (PER GLUCOSE MOLECULE)
1 Glycolysis	Cytosol	Anaerobic	2 ATP	2 pyruvate, 2ATP, 2NADH
2 (a) Formation of Acetyl-CoA	Matrix	Aerobic		2 CO_2, 2NADH
2 (b) Citric acid cycle	Matrix	Aerobic	2 ATP	6 NADH, 2 ATP, $4CO_2$
2 (c) Electron transport chain	Cristae	Aerobic	34 ATP	$6H_2O$, 34 ATP

- HL Stage 1 occurs in the cytosol and does not require oxygen.
- HL Six-carbon glucose molecule is broken down to two molecules of three-carbon (3C) pyruvate.
- HL Stage 1 releases a small amount of energy (2 ATP). Two protons (H^+) and four electrons (e^-) removed from glucose join with NAD^+ to form two molecules of NADH.
- HL If oxygen is present, respiration can proceed to the second stage of the process.
- HL Pyruvate diffuses into the mitochondrial matrix. Each pyruvate (3C) breaks down to form carbon dioxide (1C) and a (2C) molecule of Acetyl-CoA.
- HL A proton (H^+) and two electrons (e^-) from pyruvate are picked up by NAD^+ to form NADH: $NAD^+ + 2e^- + H^+ \rightarrow NADH$.
- HL If oxygen is present, each Acetyl-CoA molecule enters the citric acid cycle in the mitochondrial 'matrix'.
- HL The citric acid cycle produces carbon dioxide (CO_2).
- HL Electrons from the cycle are transferred through an electron transport chain; energy from this transfer is used in the production of ATP molecules which can be used for other biochemical reactions within the cells.
- HL The electron transport chain is formed from a series of protein complexes and mobile electron carriers in the cristae of the mitochondrion.
- HL Electrons transfer from one protein complex in the ETC to the next, moving from a higher energy to lower energy level, gradually releasing energy.
- HL This causes a charge difference which attracts protons to the complex from donor molecules, e.g. NADH.
- HL Energy released by electrons in the ETC is used to pump protons (H^+) from the matrix into the intermembrane space, creating a concentration gradient.
- HL Protons (H^+) flow back down the gradient through the enzyme ATP synthase, releasing energy which is used to form ATP: ADP + P + energy → ATP.
- HL At the end of the chain, electrons are transferred to oxygen which reacts with protons (H^+) from the mitochondrion matrix to form water (H_2O): $\frac{1}{2}O_2 + 2H^+ + 2e^- \rightarrow H_2O$.

END OF CHAPTER ASSESSMENT

REMEMBER

1. Define each of the following terms.

(a) *Aerobic respiration*

(b) *Anaerobic respiration*

2. Write a balanced chemical equation for respiration.

3. Give **two** examples of how the energy released in respiration is used by human cells.

4. Identify a commercially valuable product of anaerobic respiration.

5. (a) What does ATP stand for?

(b) Describe the role of ATP in cells.

HL **6.** Name the products of anaerobic respiration in the following cells.

HL (a) Muscle cells

HL (b) Yeast cells

HL **7.** Define each of the following terms.

HL (a) Glycolysis

HL (b) Citric acid cycle

HL (c) Electron transport chain

UNDERSTAND

8. Distinguish between breathing and respiration.

HL **9.** Summarise the process of anaerobic respiration.

HL **10.** Give an example of when anaerobic respiration occurs in animal cells.

HL **11.** Summarise the following three stages of aerobic respiration.

HL (a) Glycolysis

HL (b) Citric acid cycle

HL (c) Electron transport chain

HL **12.** (a) Identify the organelle in the diagram below.

HL (b) Identify parts A and B.

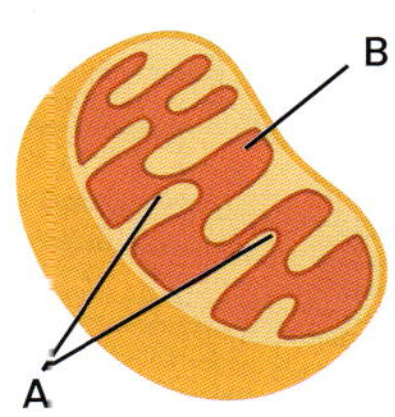

HL **13.** Give the full name of NAD^+ and explain its purpose in cells.

APPLY

HL **14.** The blood lactic acid concentrations in a runner were analysed before, during and after exercise and plotted as in the graph.

HL (a) Identify the type of respiration (aerobic or anaerobic) occurring at each of the following stages of the race, giving a reason for your answer in each case.

(i) A–B (ii) B–C (iii) D–E HL

Blood lactic acid concentrations in a runner during a race

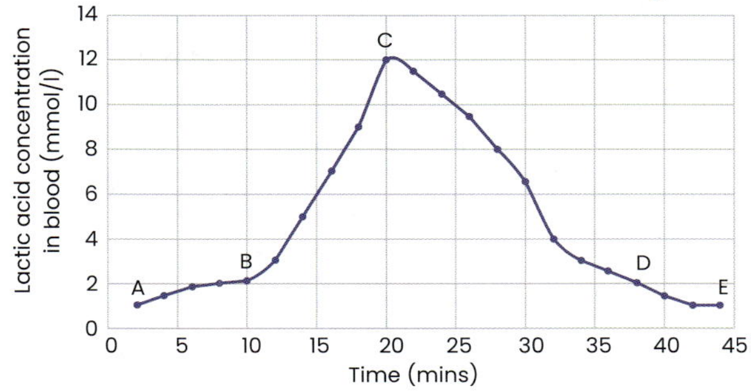

(b) What advice would you give to runners based on the data in this graph? HL

ANALYSE

15. Compare and contrast aerobic and anaerobic respiration under the following headings. HL

(a) Cell location HL

(b) Products HL

(c) Oxygen requirements HL

(d) Energy release HL

16. Compare and contrast each of the three stages of aerobic respiration (glycolysis, citric acid cycle, electron transport chain) under the following headings. HL

(a) Cell location HL

(b) Products HL

(c) Oxygen requirements HL

(d) Energy release HL

17. Research the importance of anaerobic respiration in industry.

EVALUATE

18. Compare the value of aerobic and anaerobic respiration to living organisms.

19. (a) How does the slow, gradual release of energy during the process of respiration, rather than a single release of energy, benefit living organisms?

(b) How would living organisms be affected if this was not the case?

CREATE

20. Research and design an experiment to demonstrate the effect of different sugars on anaerobic respiration by yeast.

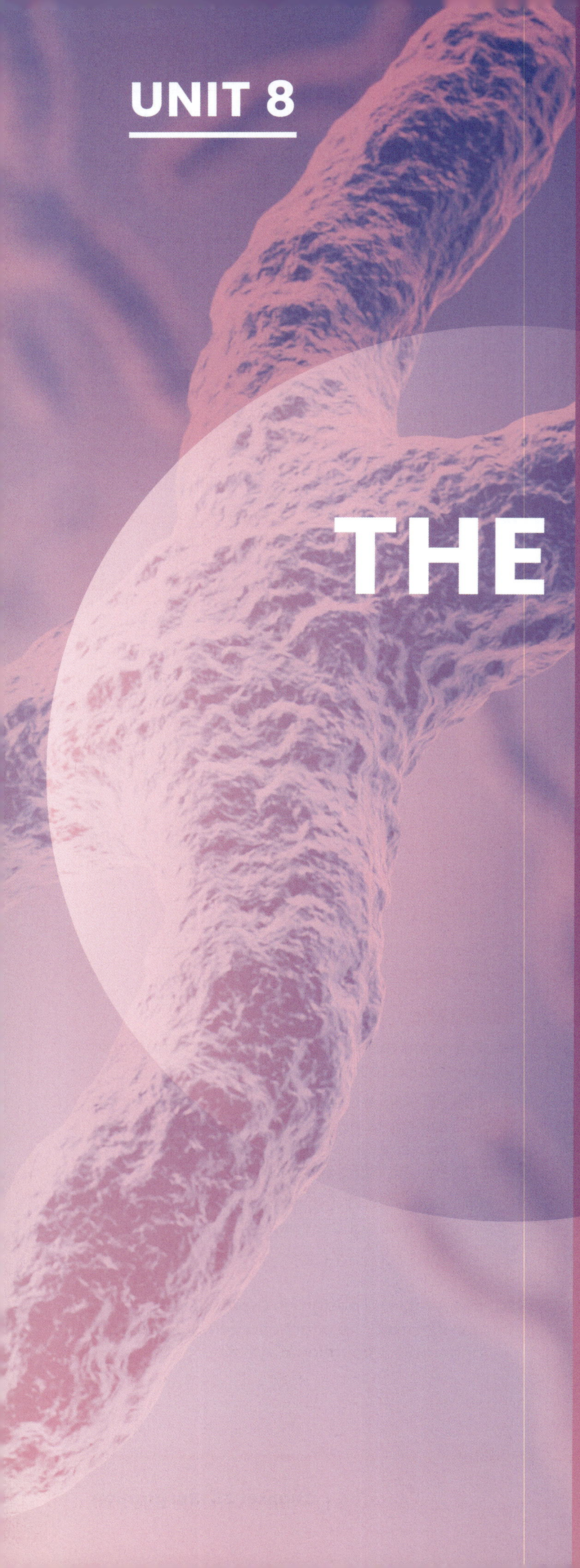

UNIT 8

THE LIVING CODE

CHAPTERS

STRANDS

Unifying Strand: Nature of Science

Strand 1: Organisation of Life

Strand 2: Structures and Processes of Life

Strand 3: Interactions of Life

24 Information Flow in the Cell

LEARNING OUTCOMES: 1.2.7, 1.2.9, 2.3.3, 2.3.4, 2.3.5, 2.3.6

EYE-OPENER

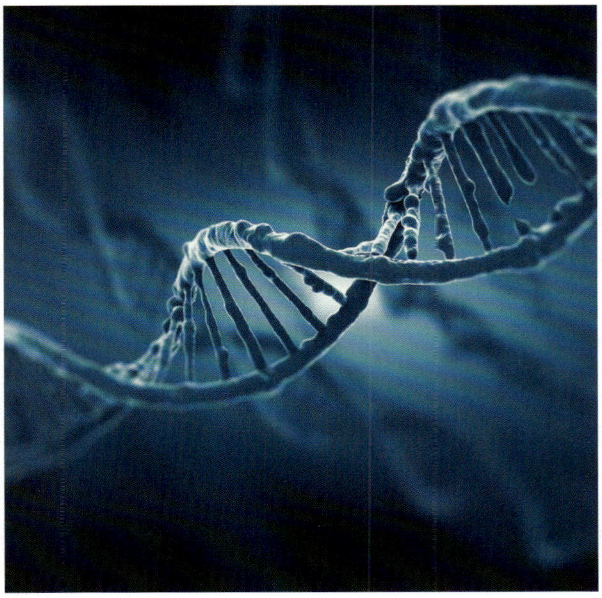

Around 99.9% of a human's genetic material is the same as that for all other humans. Only 0.1% of our DNA is unique to us and determines our physical characteristics such as eye colour, hair type and skin tone.

At the end of this chapter, you should be able to:

- Describe the structure and function of DNA nucleotide.
- Describe the structure and function of RNA nucleotide.
- Describe DNA and RNA as nucleic acids.
- Relate genes, proteins and traits in organisms.
- Outline the concept of the genetic code, referring to codons and anticodons. (HL)
- Explain the process of DNA replication.
- Describe how DNA is replicated and the flow of information from DNA through mRNA to protein. (HL)
- Describe the processes of transcription and translation. (HL)
- Relate the structure of tRNA to the codon nature of the genetic code. (HL)
- Describe how point mutations and chromosomal mutations occur, giving examples of each. (HL)

SUBJECT VOCABULARY

chromosome, gene, deoxyribonucleic acid (DNA), nucleotide, ribonucleic acid (RNA), gene expression, DNA replication, mutation, genetic code, transcription, translation

LINKS TO:

Chapter 9	**Evolution and Natural Selection**
Chapter 16	**Cell Structure and Diversity**
Chapter 17	**Cell Continuity**
Chapter 25	**Genetics and Inheritance**
Chapter 26	**Genetic Technologies and Genomics**
Chapter 35	**Human Reproduction Anatomy**
Chapter 36	**Human Reproduction – From Fertilisation to Birth**

Section 1: DNA, RNA and the Genetic Code

At the end of this section, you should be able to:

- Describe the structures of DNA and RNA.
- Describe the functions of DNA and RNA.
- Compare DNA and RNA.
- Distinguish between chromosomes, DNA and genes.
- Relate genes, proteins and traits in organisms.

Unpacked from LO 1.2.7, 1.2.9

STRUCTURE OF CHROMOSOMES

We learned in Chapter 16 that **chromosomes** are thread-like structures of DNA wrapped around proteins. Different species of organisms have different numbers of chromosomes in their nucleus. All human cells, except for red blood cells, contain DNA. In humans, non-sex cells have 46 chromosomes, or 23 homologous pairs. To fit in the nucleus of a cell, DNA must be tightly wound around the proteins which are then tightly packed together. About 40% of chromosomes are made from DNA with the other 60% being made from protein. The appearance and number of chromosomes in a typical body cell of an organism is called a **karyotype**.

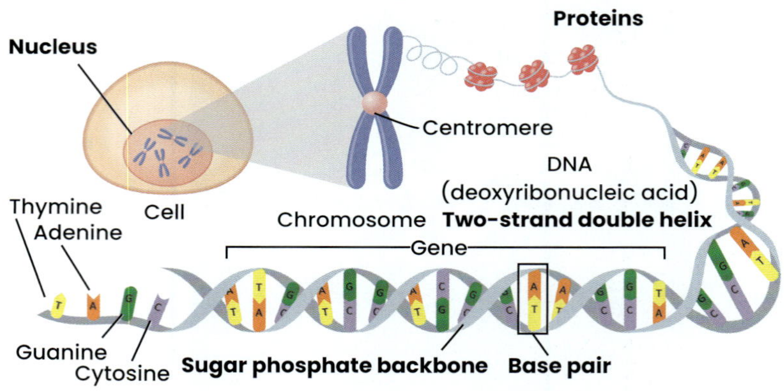

Figure 24.1: Chromosome structure of DNA and protein.

A **chromosome** is a thread-like structure of DNA and protein that becomes visible in the nucleus during cell division.

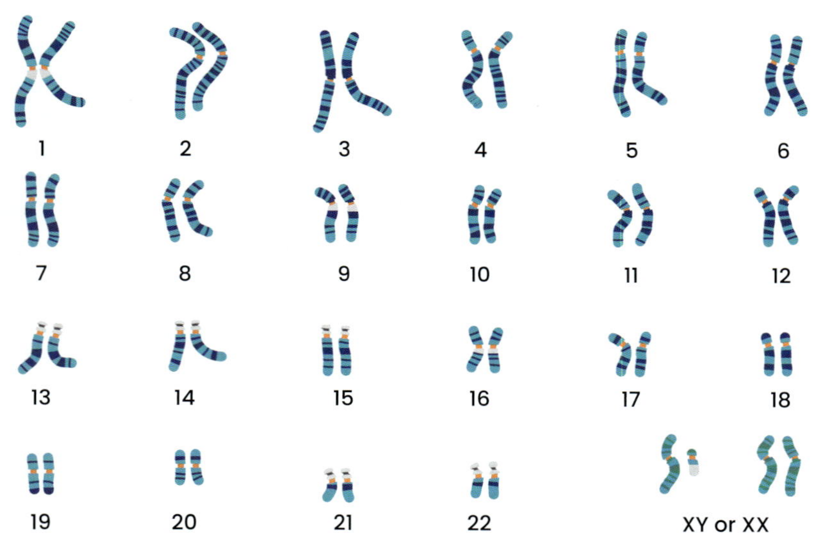

Figure 24.2: A human karyotype.

DNA is an extremely long biomolecule – the DNA in one of the 37 trillion cells in the human body is approximately two metres in length!

DID YOU KNOW?

GENES AND NON-CODING DNA

If you were to unwind a chromosome and remove all the proteins, you would find that DNA has sections called genes. **Genes** contain the genetic instructions for building proteins. Only about 1–2% of DNA is made up of protein-coding genes. The other 99% is referred to as **non-coding**, or 'junk', **DNA**.

- A **gene** is a section of DNA that codes for a protein.
- **Non-coding DNA** does not provide instructions for making proteins but rather contains regulatory sequences that determine when and where genes are turned on and off in a cell.

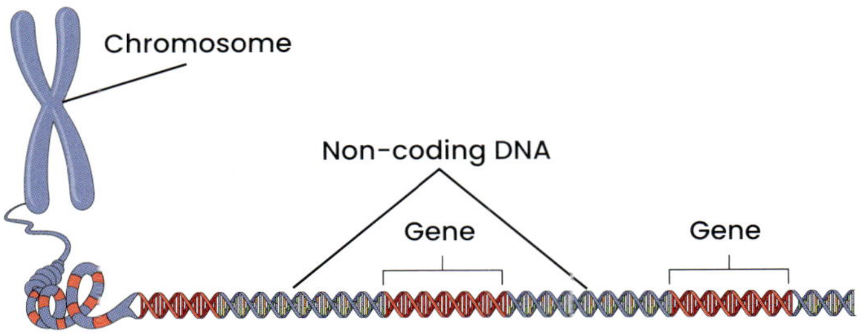

Figure 24.3: Genes and non-coding DNA.

STRUCTURE OF DEOXYRIBONUCLEIC ACID (DNA)

DNA, or **deoxyribonucleic acid**, has a two-stranded double helix structure. The structure of DNA is often compared to a ladder with the sides or backbone made of alternating deoxyribose sugar and phosphate units. Each rung or step in the ladder is made of a pair of complementary nitrogenous bases.

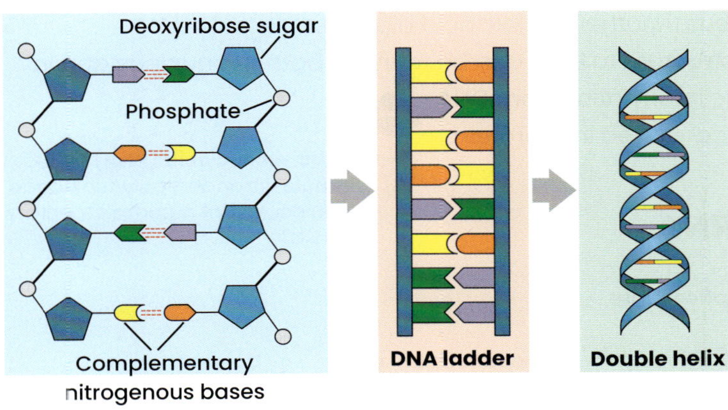

Figure 24.4: Structure of DNA.

- **Deoxyribonucleic acid (DNA)** is a two-stranded double helix molecule containing genetic information for the development and functioning of an organism.
- A **nucleotide** is a repeat unit in a nucleic acid (DNA or RNA) consisting of a phosphate molecule, a sugar and a nitrogenous base.

Nucleotide structure

DNA is an example of a nucleic acid or polynucleotide as it is made up of repeating building blocks called nucleotides. Each **nucleotide** consists of a phosphate molecule, a five-carbon (5C) deoxyribose sugar and a nitrogenous (nitrogen-containing) base. There are four different types of bases: adenine (A), thymine (T), cytosine (C) and guanine (G).

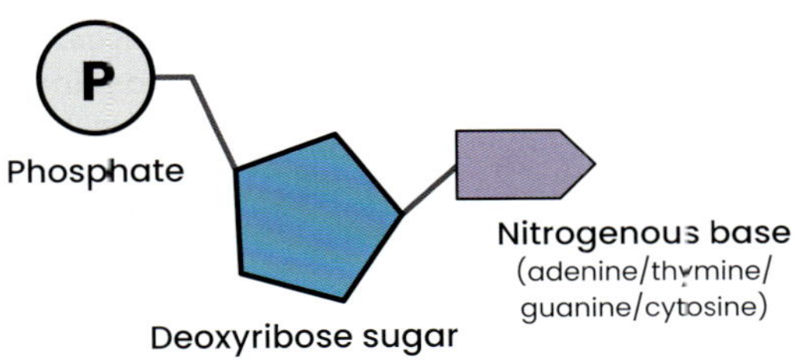

Figure 24.5: DNA nucleotide structure.

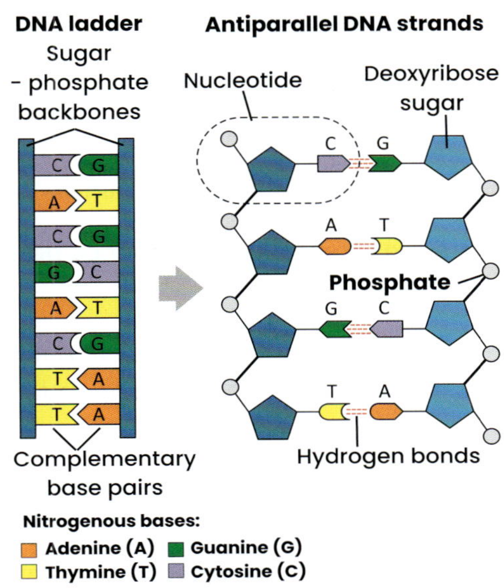

Figure 24.6: Complementary base pairs in DNA.

COMPLEMENTARY BASE PAIRS

The base **adenine always bonds with thymine**. The base **cytosine always bonds with guanine**. This is referred to as **complementary base pairing**. The two strands of DNA are linked by weak hydrogen bonds between the bases, two between each adenine and thymine (A and T) and three between each guanine and cytosine (G and C). All four bases are composed of carbon, nitrogen, oxygen and hydrogen, but differ in their structure. The two DNA strands run in opposite directions (antiparallel) to allow the base pairs to bond together correctly. Adenine and guanine have a double-ring structure and are called **purines**, whereas thymine and cytosine contain a single-ring structure and are called **pyrimidines**.

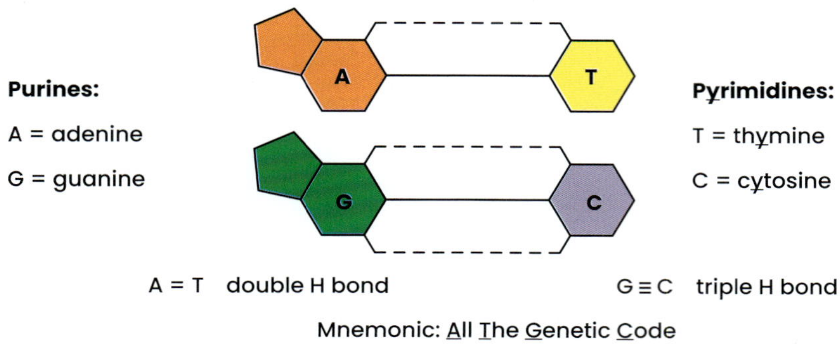

Figure 24.7: Purine and pyrimidine bases in DNA.

Go to YouTube.com and search for 'DNA Structure' (4:21) by Teacher's Pet.

STRUCTURE OF RIBONUCLEIC ACID (RNA)

In a eukaryotic cell, DNA is found inside the nucleus. Another nucleic acid biomolecule, called **ribonucleic acid (RNA)**, which is very similar in structure to DNA, can be found in both the nucleus and in the cytoplasm of the cell. RNA carries the genetic information from DNA in the nucleus to the ribosome where it may be converted into protein.

Ribonucleic acid (RNA) is a single-stranded polynucleotide molecule that is complementary to DNA.

Differences between DNA and RNA

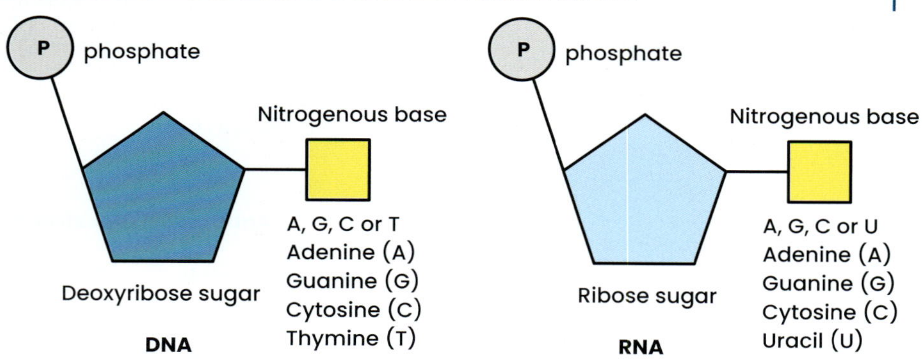

Figure 24.8: DNA and RNA nucleotide structures.

There are three major differences between DNA and RNA:

- The base thymine (T) in a DNA nucleotide is replaced by uracil (U) in an RNA nucleotide.
- The sugar in a DNA nucleotide is deoxyribose, whereas in RNA it is ribose.
- DNA is double-stranded, whereas RNA is single-stranded.

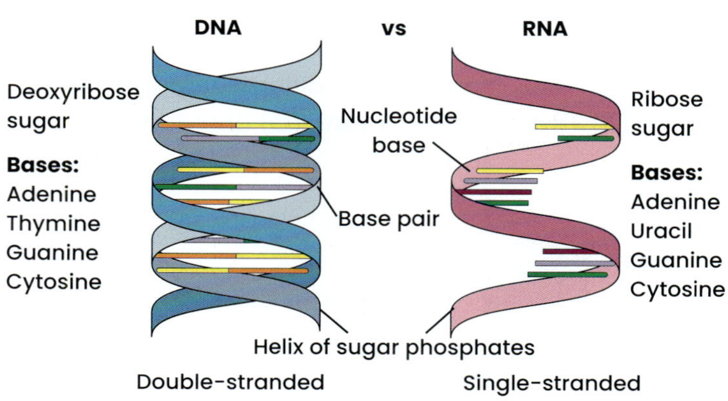

Figure 24.9: DNA and RNA structures.

Complementary structure of RNA to DNA

The structure of RNA is said to be complementary to DNA which that:

- where a strand of DNA contains **adenine**, its complementary RNA strand will contain **uracil**
- where a strand of DNA contains **thymine**, its complementary RNA strand will contain **adenine**
- where a strand of DNA contains **guanine**, its complementary RNA strand will contain **cytosine**
- where a strand of DNA contains **cytosine**, its complementary RNA strand will contain **guanine**.

Types of RNA

There are three types of RNA: messenger RNA (mRNA), ribosomal RNA (rRNA) and transfer RNA (tRNA). We will look at the function of each in more detail later in the chapter.

FOCUS ON ... COMPARING DNA AND RNA

	DNA	RNA
Sugar	Deoxyribose	Ribose
Nitrogenous bases	Adenine, thymine, guanine, cytosine	Adenine, uracil, guanine, cytosine
Strands	Double-stranded helix	Single-stranded
Location	Nucleus (some in mitochondria and chloroplast)	Nucleus, cytoplasm, ribosome

GENES, PROTEINS AND TRAITS IN ORGANISMS

Proteins, along with other chemicals, form cells that combine to build an organism. For a cell to function properly, proteins that are needed must be made at the correct time. Each gene in DNA contains the instructions for linking amino acids together in a particular order to form a specific protein. The types of proteins formed in certain cells dictate the traits which appear in an organism e.g. skin colour, eye colour and height.

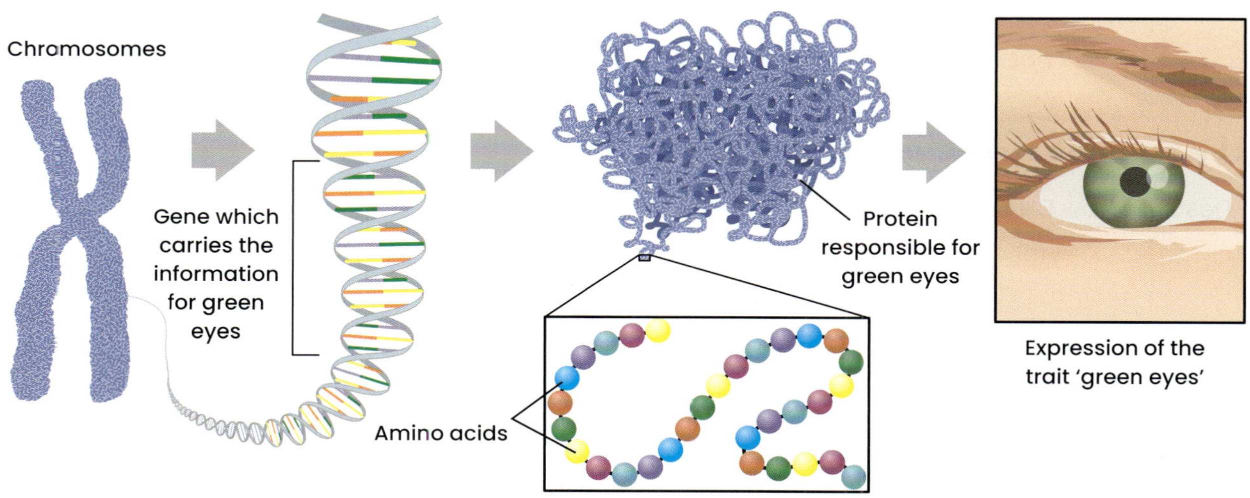

Figure 24.10: Genes, proteins and traits.

SCIENCE IN SOCIETY: HEALTH

Gene therapy

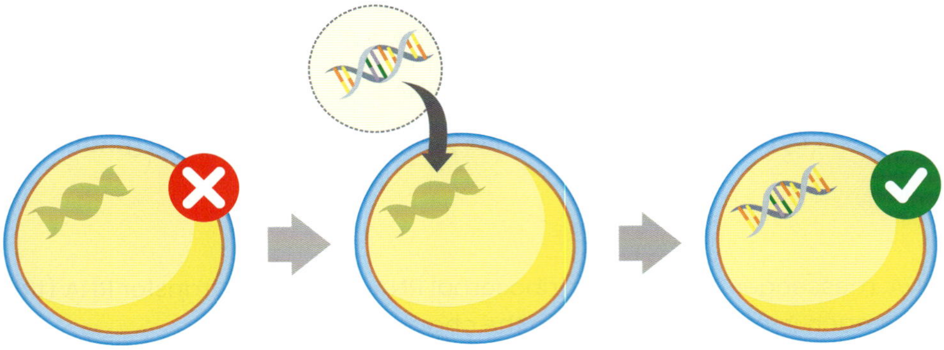

1. Cell with non-functioning gene
2. Adding DNA containing a functional version of the lost gene
3. Cell functioning normally

Figure 24.11: Process of gene therapy

Gene therapy is a technique that modifies a person's genes to treat disease. It involves introducing a new or modified gene into the body, inactivating a disease-causing gene that is not functioning properly or replacing a defective or missing gene in a patient's cells with a healthy version of that gene.

Both inherited genetic diseases (e.g. haemophilia and sickle cell anaemia) and acquired disorders (e.g. leukaemia) have been treated with gene therapy. Some forms of gene therapy use bacterial or viral vectors (carriers) to transport therapeutic genes into human cells. In other cases, cells are removed from the patient, genetically modified and then returned to the patient. Researchers all over the world are studying different aspects of gene therapy, understanding how genes work and how changes in genes can affect our health.

PORTFOLIO LINK **Go to your Learning Portfolio and complete Learning Check 1 to assess your knowledge and understanding of the topic so far.**

Section 2: DNA Replication and Translation into Protein

At the end of this section, you should be able to:

- Describe how DNA is replicated.
- Outline the concept of the genetic code which is read by cellular machinery to produce proteins. (HL)
- Describe the flow of information from DNA through mRNA to protein. (HL)
- Describe the processes of transcription and translation. (HL)
- Relate the structure of tRNA to the codon nature of the genetic code, referring to anticodons. (HL)
- Describe how point mutations and chromosomal mutations occur, giving examples of each. (HL)

Unpacked from LOs 1.2.9, 2.3.3, 2.3.4, 2.3.5, 2.3.6

DNA REPLICATION

As we learned in Chapter 17, for a cell to divide and produce new cells during mitosis, the DNA in each chromosome must make an exact copy or replicate of itself. **DNA replication** is the process by which the DNA in each chromosome makes an exact copy of itself. This process takes place in the nucleus, during interphase of the cell cycle, prior to mitosis. The chromosomes that enter the daughter cells during mitosis are identical to the chromosomes in the parent cell. During DNA replication, the double strand of DNA separates into two single strands, each of which serves as the template for the synthesis of a complementary strand of DNA.

DNA replication is facilitated by enzymes.

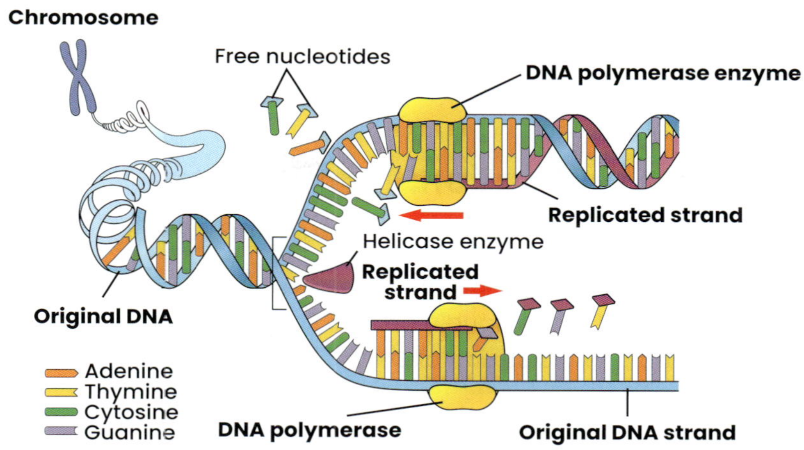

Figure 24.12: DNA replication.

DNA replication is the process by which DNA in each chromosome makes an exact copy of itself.

DNA REPLICATION – IN DEPTH

HL

Figure 24.13 illustrates the process of DNA replication. DNA replication can be thought of as a three-step process: initiation, elongation and termination.

DNA replication: Initiation

- The DNA double helix unwinds.
- The enzyme DNA helicase breaks the hydrogen bonds between the complementary base pairs, separating the two DNA strands.

DNA replication: Elongation

- Each old strand acts as a template for a new complementary strand.
- Free nucleotides (A, T, G, C) enter the nucleus from the cytoplasm and attach to their complementary bases on the exposed strands. (These nucleotides came from the breakdown of older RNA molecules.)
- The nucleotides join up to form the sugar-phosphate backbone. This is controlled by the enzyme DNA polymerase.

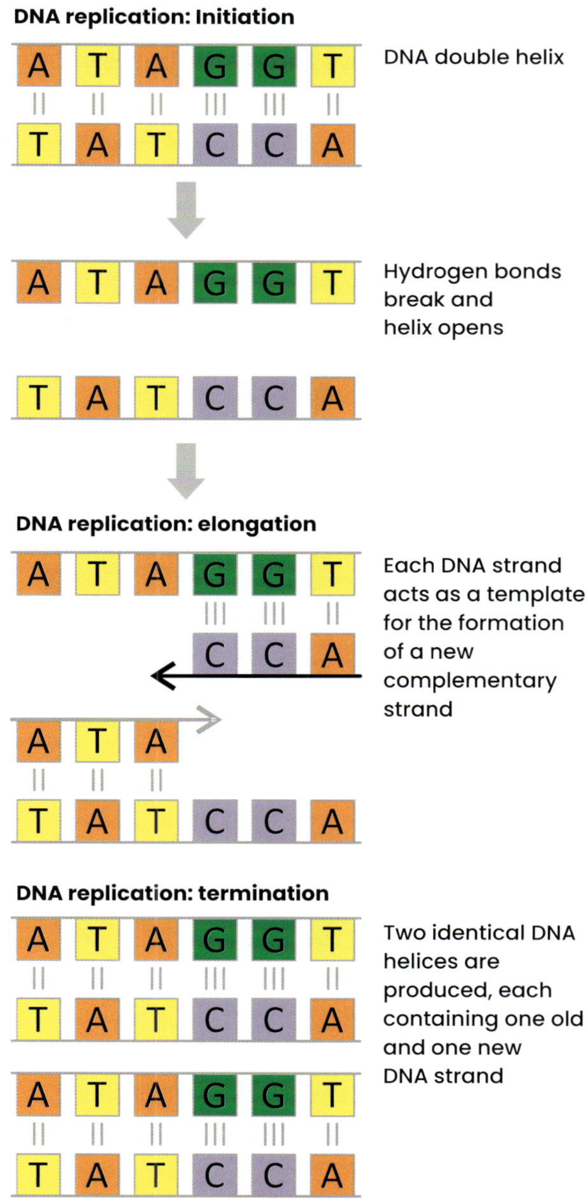

Figure 24.13: The stages of DNA replication.

- The enzyme **DNA polymerase** also controls the complementary bases in the old and new DNA strands forming hydrogen bonds.

DNA polymerase is an enzyme that joins DNA nucleotides together to form a strand of DNA.

DNA replication: Termination

- The double strands coil back up into a double helix structure.
- DNA replication produces two identical DNA double helices. Each of these double helices contains one new strand of DNA and one old strand of DNA.

Go to YouTube.com and search for 'DNA replication – 3D' (3:27) by yourgenome.

THE FLOW OF INFORMATION FROM DNA THROUGH mRNA TO PROTEIN

Genetic code

The DNA in a gene contains a **genetic code**, which can be read by cellular machinery to build a specific protein. A **codon**, or set of three consecutive nucleotide bases in DNA or RNA, codes for one of the 20 amino acids found in the human body. Each codon codes for only one amino acid, e.g. the codon UCC codes only for the amino acid serine. Sometimes, different codons will code for the same amino acid, e.g. UCC, UCA and UCG all code for the amino acid serine.

The **genetic code** is the sequence of triplets of nucleotides (codons) in DNA or RNA that determines the amino acid chain in a protein.

A **codon** is a sequence of three consecutive bases of DNA or RNA that codes for a specific amino acid.

HL

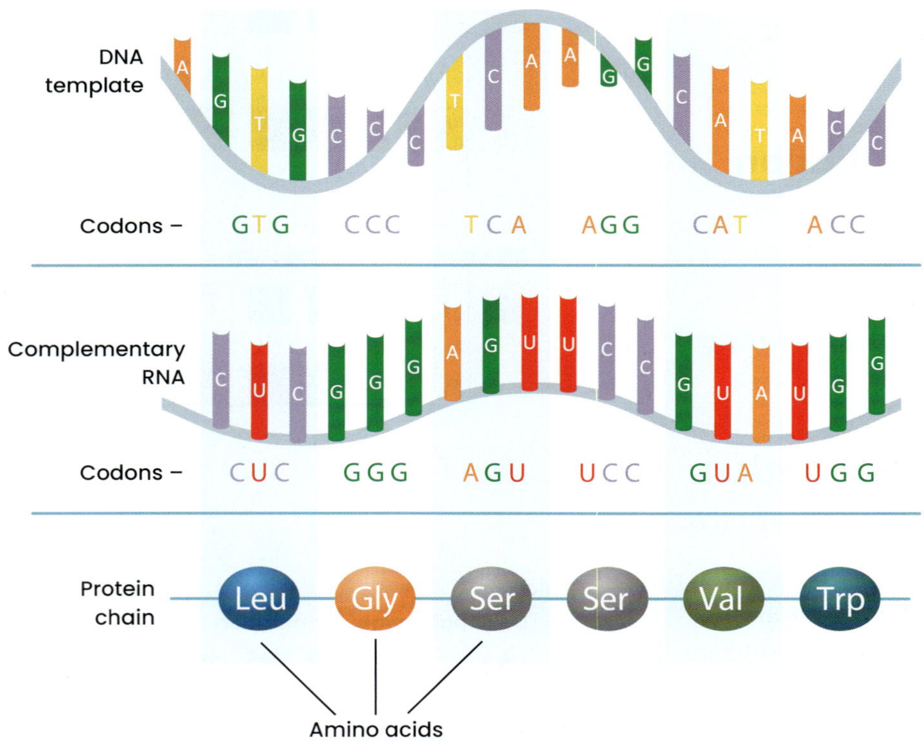

Figure 24.14: Codons in DNA and RNA.

DNA to mRNA

For the sequence of codons to be converted to protein, the genetic code must travel to a ribosome within the cell. DNA does not leave the nucleus; it is copied into a complementary copy of **messenger RNA (mRNA)**.

Messenger RNA (mRNA) is a single-stranded molecule complementary to DNA that carries the genetic code from the nucleus to the ribosome.

mRNA to protein

The mRNA travels from the nucleus to a ribosome in the cytoplasm. In the cytoplasm, amino acids are found attached to another type of RNA molecule called transfer RNA (tRNA). These tRNA molecules carry amino acids to the ribosome. The ribosome 'reads' the sequence of codons in the mRNA and attaches the matching amino acid for each codon, assembling a protein.

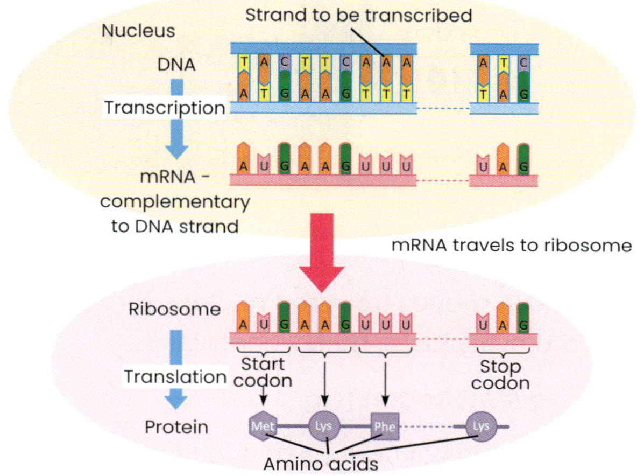

Figure 24.15: Information flow from DNA to mRNA to protein.

PROTEIN SYNTHESIS: THE PROCESSES OF TRANSCRIPTION AND TRANSLATION OF THE GENETIC CODE INTO AMINO ACIDS

We will now look at the two stages involved in **protein synthesis** in more detail:

HL

- Stage 1: Transcription of DNA to mRNA
- Stage 2: Translation of mRNA to amino acids.

Protein synthesis stage 1: Transcription

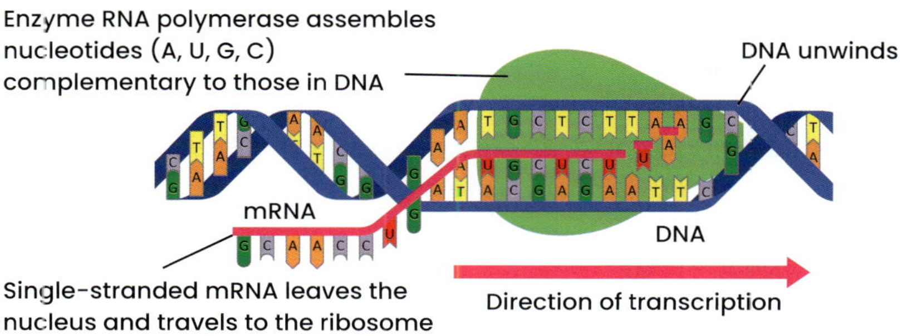

Figure 24.16: Transcription of DNA into mRNA.

Transcription is the production of mRNA using the sequence of DNA bases as a template.

Protein synthesis is the two-step process of converting the triplet nitrogenous base code from DNA to mRNA to amino acids to form a protein chain.

RNA polymerase is an enzyme that joins RNA nucleotides together to form a strand of mRNA.

Transcription involves the production of a single strand of messenger RNA (mRNA) that is complementary to its DNA template.

- The enzyme DNA helicase separates the DNA strands at the site of a particular gene.
- RNA nucleotides (A, U, G, C) bind to their complementary DNA bases on one strand of the DNA. Adenine binds with thymine in DNA, uracil with adenine and guanine with cytosine.
- The enzyme **RNA polymerase** joins the RNA nucleotides together to form a strand of mRNA.
- When transcription is complete, the mRNA strand is released and enters the cytoplasm.
- The two DNA strands reform their hydrogen bonds and coil up again.

Protein synthesis stage 2: Translation

The second stage of protein synthesis is called **translation**.

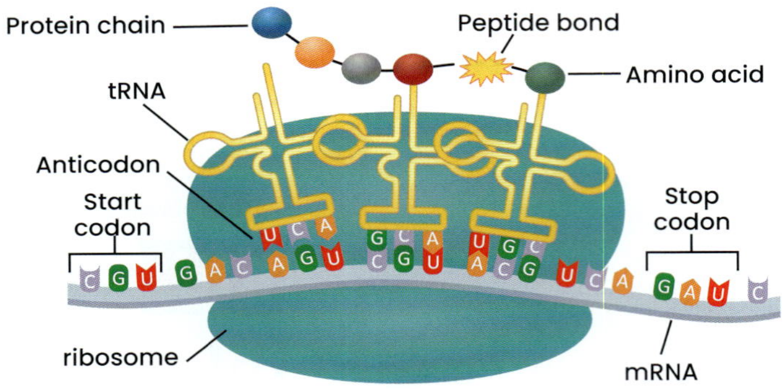

Figure 24.17: Translation of mRNA codons into amino acids to form protein.

During this process:

- mRNA moves from the nucleus through the nuclear pores into the cytoplasm. mRNA contains:
  - a start codon (e.g. AUG)
  - a series of codons to specify particular amino acids
  - a stop codon (e.g. UAA).
- The mRNA strand forms weak bonds with the **ribosomal RNA (rRNA)**.
- **Transfer RNA (tRNA)** molecules with an **anticodon** complementary to a specific mRNA codon are found in the cytoplasm. Each tRNA carries an amino acid corresponding to its anticodon.

HL

- The first tRNA anticodon binds to an mRNA start codon. This brings its particular amino acid to the ribosome. The ribosome moves along the mRNA to the next codon. The tRNA with the anticodon complementary to this codon brings in its amino acid. As this process is repeated, adjacent amino acids form peptide bonds with each other and break away from their tRNA molecules. The growing chain of amino acids forms a new protein. This process continues until the ribosome reaches a stop codon.
- The amino acid chain is released from the ribosome and folds into a specific three-dimensional, globular shape to form the completed functional protein. Protein folding occurs in the endoplasmic reticulum and the Golgi apparatus.

- **Translation** involves linking amino acids together to make a protein using the mRNA code.
- **Ribosomal RNA (rRNA)** forms part of the ribosome.
- **Transfer RNA (tRNA)** carries amino acids to the ribosomes to be assembled into protein.
- An **anticodon** is a group of three nucleotide bases on one end of a tRNA molecule that is complementary to a codon in mRNA.

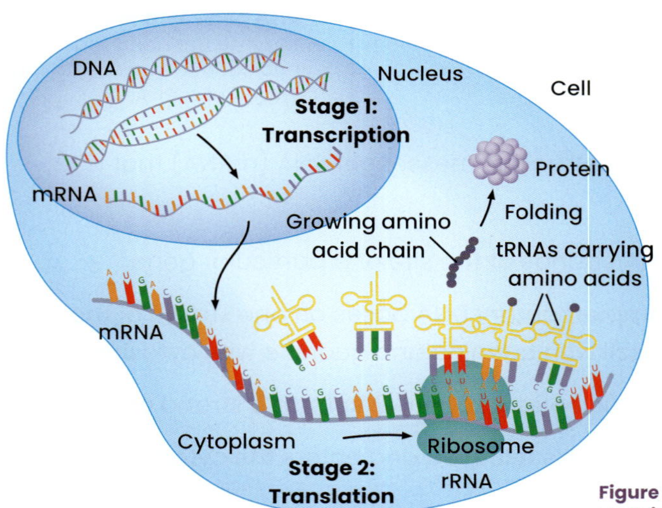

Figure 24.18: Overview of protein synthesis.

Go to YouTube.com and search for 'Protein Synthesis – Leaving Cert Biology Basic Summary' (4:48) by Biology Bugbears.

FOCUS ON ... tRNA STRUCTURE

The structure of tRNA is very important to its role. It folds into a three-leaf clover structure with an anticodon consisting of three nucleotides at the base of the molecule. This anticodon is complementary to a specific codon in the mRNA, allowing the tRNA to bind to the mRNA at this position. On its opposite end, the tRNA carries the amino acid specific to the codon to which its anticodon attaches.

Figure 24.19: tRNA structure.

SCIENCE IN SOCIETY: TECHNOLOGY

Fast artificial protein synthesis

Many proteins are useful as drugs for treating diseases such as cancer and diabetes. However, the artificial synthesis of these proteins is a time-consuming process requiring genetically engineered microbes to produce the required protein.

Scientists at Massachusetts Institute of Technology have developed a tabletop automated flow synthesis machine that dramatically reduces the amount of time needed to generate synthetic proteins. This technology can string together hundreds of amino acids within hours. It is hoped that this technology could speed up the manufacture and development of new drugs and also allow scientists to design new artificial proteins.

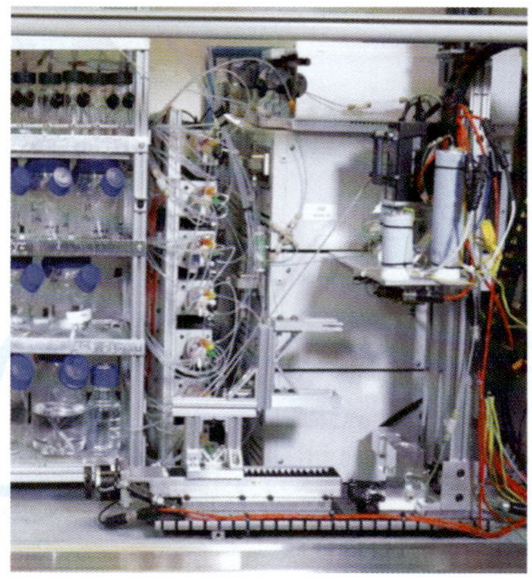

Figure 24.20: The MIT automated flow synthesis machine. Courtesy of the researchers/MIT.

HOW MUTATIONS OCCUR WHEN DNA IS BEING REPLICATED

HL

During DNA replication, an incorrect base might randomly be inserted into the new DNA strand, e.g. thymine might incorrectly bond with guanine. These incorrect base pairings occur on average once in every 100,000 base additions in a human. Luckily for us, immediately after DNA replication, specialised enzymes scan the DNA and check for errors. Most incorrect bases are detected, and the correct nucleotide is inserted into the new DNA strand. Therefore, the actual incidence rate of an incorrect base being inserted into a DNA strand is approximately one in every billion bases. These mistakes in DNA are known as **mutations**.

A **mutation** is a spontaneous, random change in the nucleotide sequence of an organism's DNA.

There are two main types of mutation: point mutations and chromosomal mutations.

Point mutation

A **point mutation** is a type of gene mutation that occurs due to incorrect base pairing during replication. Point mutations result from changes to a single base in the DNA sequence. These base changes may involve a substitution, insertion or deletion:

A **point mutation** is a mutation that occurs due to incorrect base pairing during replication resulting in a change in nucleotide sequence.

- **Substitution** refers to a nucleotide being replaced by an incorrect nucleotide.
- **Insertion** means an extra nucleotide is put in the DNA by mistake.
- **Deletion** occurs when a nucleotide is removed.

A point mutation results in a permanent change in the nucleotide sequence in DNA.

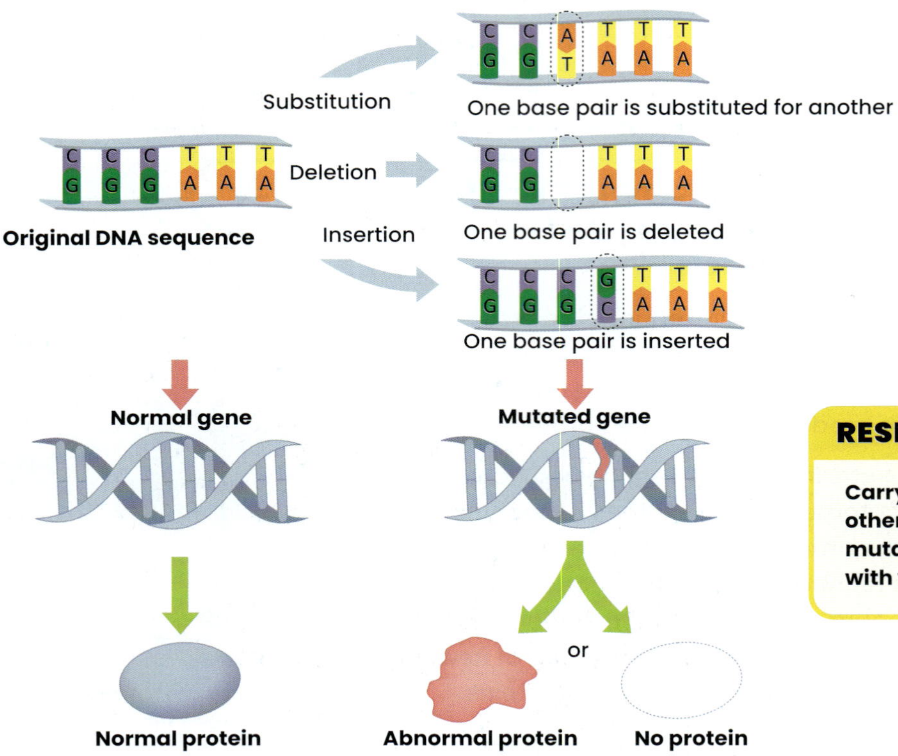

Figure 24.21: Point mutations.

RESEARCH ACTIVITY

Carry out some research on other diseases caused by point mutations. Share your results with your class.

If the mutations occur in non-coding DNA, this will most likely have no impact on the individual, but if the error occurs within a gene, it may result in a mutated protein being produced in the cell, which could lead to a disease. For example, a single base change in the gene that codes for the protein haemoglobin causes the disease sickle cell anaemia. This gene mutation causes the structure of haemoglobin to change and become less efficient at carrying oxygen. The red blood cell changes shape as a result. Another genetic disorder caused by gene mutation is cystic fibrosis.

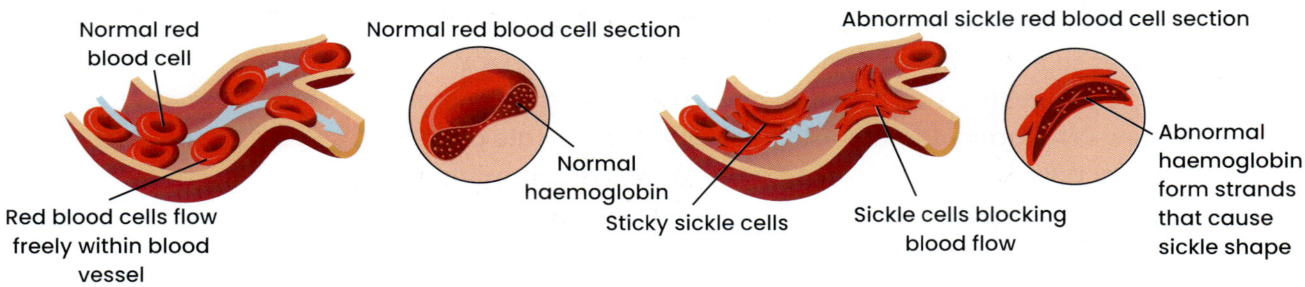

Figure 24.22: Point mutation: sickle cell anaemia.

Chromosome mutation

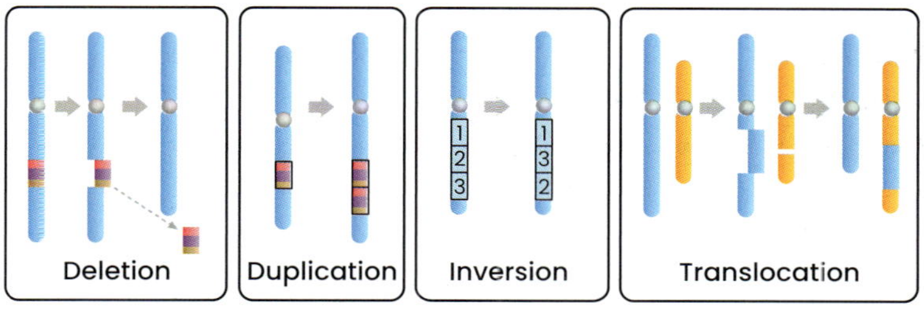

Figure 24.23: Types of chromosomal mutation.

During cell division, mutations may occur in chromosomes. **Chromosomal mutations** may change the number of chromosomes or their structure by deletion, duplication, inversion or translocation.

A **chromosome mutation** is a change in the number or structure of chromosomes.

1. Deletion is where part of a chromosome is lost.
2. Duplication is where a gene or a whole chromosome is copied.
3. Inversion occurs when a section of a chromosome breaks off and reattaches upside down to the same chromosome.
4. Translocation occurs when a chromosome (or a part thereof) breaks off and attaches to a different chromosome.

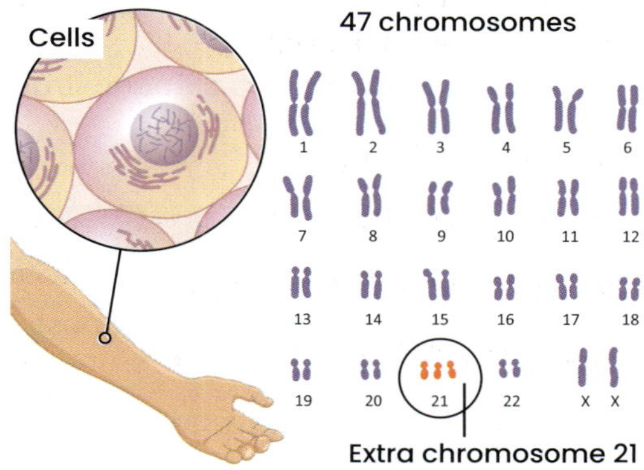

Figure 24.24: Chromosome mutation: Down syndrome karyotype.

HL

Mutations that change the structure of chromosomes are often lethal. Some chromosome mutations result in genetic disorders. For example, in Down syndrome, an error occurs during meiosis, resulting in an extra copy of chromosome 21 going into either the egg or sperm. A zygote that forms from one of these sex cells will have an extra third copy of chromosome 21 and a total of 47 chromosomes instead of 46 in each of its body cells.

Causes of mutations

As seen above, mutations can occur naturally during DNA replication or meiosis. External factors or 'mutagens' such as cigarette smoke, UV light and viruses can also cause damage to DNA, resulting in mutations that can lead to the development of cancers.

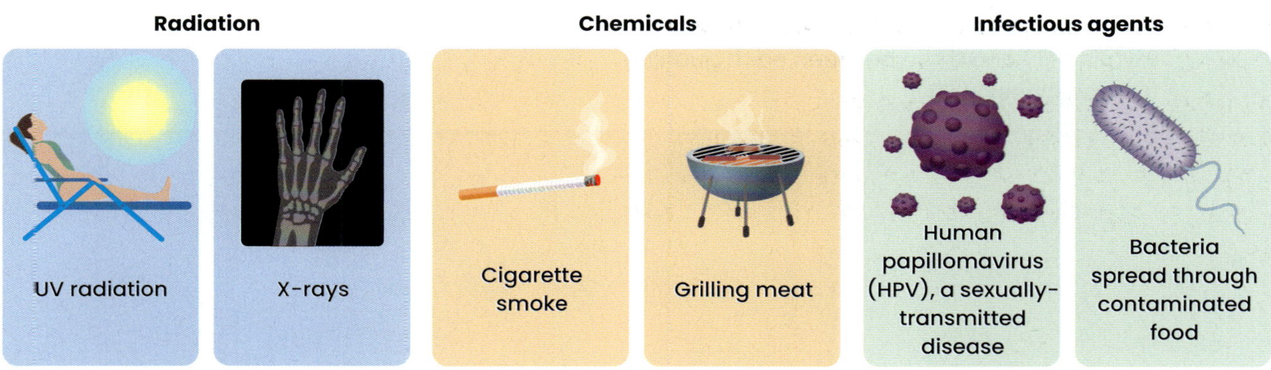

Figure 24.25: Classes and examples of mutagens.

PORTFOLIO LINK **Go to your Learning Portfolio and complete Learning Check 2 to assess your knowledge and understanding of the topic so far.**

Chapter Summary

- A **chromosome** is a thread of DNA and protein that becomes visible in the nucleus during cell division.
- A **gene** is a section of DNA that codes for a protein.
- **Non-coding DNA** is DNA that does not code for a protein.
- **DNA (deoxyribonucleic acid)** is a two-stranded double helix molecule containing genetic information for the development and functioning of an organism.
- A **nucleotide** is a repeat unit in DNA consisting of a phosphate molecule, deoxyribose sugar and a nitrogenous base.
- **RNA (ribonucleic acid)** is a single-stranded polynucleotide molecule that is complementary to DNA.

	DNA	RNA
SUGAR	Deoxyribose	Ribose
NITROGENOUS BASES	Adenine, thymine, guanine, cytosine	Adenine, uracil, guanine, cytosine
STRANDS	Double-stranded helix	Single - stranded
LOCATION	Nucleus (some in mitochondria and chloroplast)	Nucleus, cytoplasm, ribosome

- **Purines:** adenine, guanine.
- **Pyrimidines:** thymine, cytosine.
- Adenine always binds with thymine; guanine always binds with cytosine. Two hydrogen bonds form between each adenine (A) and thymine (T) and three between each guanine (G) and cytosine (C).
- **DNA replication** is the process by which DNA in each chromosome makes an exact copy of itself.
- A **mutation** is a change in the nucleotide sequence of an organism's DNA.
- **Point mutation** occurs due to incorrect base pairing during replication resulting in a change in nucleotide sequence, e.g. cystic fibrosis.
- A **chromosome mutation** is a change in the number or structure of chromosomes due to deletion, duplication, inversion or translocation, e.g. Down syndrome.
- **Genetic code** is the sequence of triplets of nucleotides (codons) in DNA or RNA that determines the amino acid chain in a protein.
- A **codon** is a sequence of three consecutive bases of DNA or RNA that codes for a specific amino acid.
- **Protein synthesis** is the two-step process of converting the triplet base code from DNA to mRNA to amino acids to form a protein chain. HL
- **Transcription** is the production of mRNA using the sequence of DNA bases as a template. HL
- **Translation** involves linking amino acids together to make a protein using the RNA code. HL
- An **anticodon** is a group of three nucleotide bases on one end of a tRNA molecule that is complementary to a corresponding codon in mRNA. HL
- **mRNA**, or messenger RNA, is a single-stranded molecule complementary to DNA that carries the genetic code from the nucleus to the ribosome.
- **rRNA**, or ribosomal RNA, forms part of the ribosome. It forms weak bonds with mRNA during translation. HL
- **tRNA**, or transfer RNA, carries amino acids to the ribosomes to be assembled into protein.

END OF CHAPTER ASSESSMENT

REMEMBER

1. Define each of the following terms.
   (a) Gene
   (b) DNA
   (c) Non-coding DNA
   (d) Nucleotide
   (e) RNA
2. Identify the complementary base pairs in DNA.
3. Label molecules A, B, C, D and structures E, F, G in the diagram below, with reference to protein synthesis.

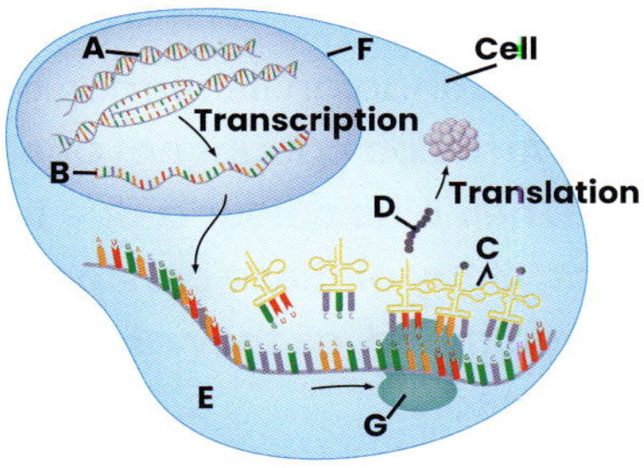

UNDERSTAND

4. With the aid of a labelled diagram, describe the process of DNA replication.
5. Compare DNA and RNA under the following headings: (i) strands (ii) nitrogenous bases (iii) location (iv) sugar.
6. HL Classify mutations under the headings (i) point (ii) chromosomal, giving examples in each case.
7. HL Identify **three** different forms of mutagens and give an example in each case.
8. HL What is meant by the 'genetic code'?
9. HL Identify the **three** forms of RNA present in cells and state the function of each.
10. HL Name the **three** types of codon present in mRNA and explain the function of each.
11. HL Compare transcription and translation, the two stages in protein synthesis, under the following headings: (i) location (ii) molecules involved (iii) products formed.

APPLY

12. Draw a chain of DNA consisting of three complementary base pairs.
13. Identify the sequence of bases present in a section of RNA that is complementary to the following strand of DNA: A T G C T A A C G A.
14. HL Study the following human karyotype and identify the evidence that the person has the genetic disorder Down syndrome.

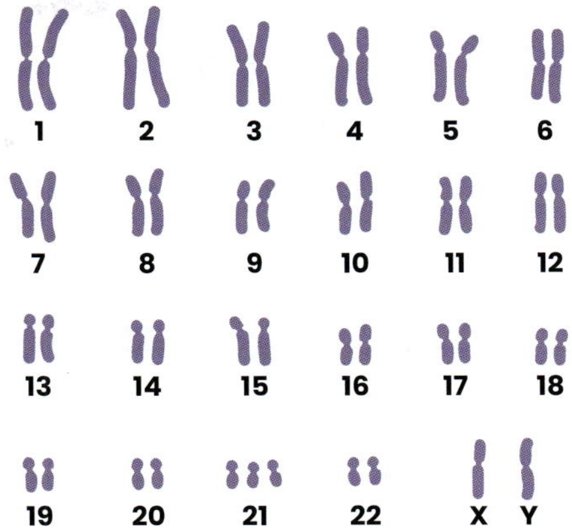

ANALYSE

15. An analysis of a DNA molecule showed it contained 30% adenine (A). What percentage of the DNA molecule would be cytosine (C)?
16. How many nitrogenous bases are in a codon?

EVALUATE

17. Why is it essential that DNA is replicated carefully in cells?
18. Evaluate the importance of protein synthesis to living things.
19. Why is non-coding DNA of importance to cells?

CREATE

20. Research an investigation to extract DNA from a plant cell.
21. Create a model to illustrate the structure of the following:
    (a) DNA
    (b) RNA
22. HL Ireland has the highest incidence of the disease cystic fibrosis in the world. Research the cause and symptoms of the disease and how it may be treated. Present a poster of your findings to your class.

25 Genetics and Inheritance

LEARNING OUTCOMES: U1.2, U3.2, 1.4.1, 1.4.2, 1.4.3, 1.4.4, 1.4.5, 1.4.6, 1.4.7, 1.4.8, 1.4.9, 1.4.10, 1.4.11

EYE-OPENER

In May 2016, two Canadian fishermen off the coast of North America caught not one, but two blue lobsters. Their blue colour is due to a rare genetic defect. Many of these crustaceans don't survive until adulthood, as they are easily spotted by predators.

At the end of this chapter, you should be able to:

- Describe the structure of a chromosome and the role of a gene.
- Explain nuclear inheritance.
- Compare nuclear inheritance and non-nuclear inheritance. (HL)
- Compare genetic and epigenetic mechanisms.
- Research one example of epigenetic inheritance in nature. (HL)
- Predict inheritance to the first generation of a single unlinked trait.
- Predict a cross involving incomplete dominance. (HL)
- Illustrate and state Mendel's laws of segregation and independent assortment. (HL)
- Predict inheritance to the second generation of two unlinked traits. (HL)
- Explain how linkage affects Mendel's law of independent assortment. (HL)
- Describe sex determination.
- Explain and predict the inheritance of sex-linked traits.
- Identify benefits and limitations of Mendelian genetics.

LINKS TO:

Chapter 9	**Evolution and Natural Selection**
Chapter 16	**Cell Structure and Diversity**
Chapter 17	**Cell Continuity**
Chapter 21	**Plant Reproduction**
Chapter 24	**Information Flow in the Cell**
Chapter 26	**Genetic Technologies and Genomics**
Chapter 37	**Human Reproduction – From Fertilisation to Birth**

SUBJECT VOCABULARY

species, heredity, allele, locus, homozygous, heterozygous, gene expression, epigenetics, dominance, recessive, genotype, phenotype, incomplete dominance, linkage, sex linkage, law of segregation, independent assortment

Section 1: Chromosomes, Genes, Nuclear and Non-Nuclear Inheritance

At the end of this section, you should be able to:

- Describe the structure of a chromosome including protein and DNA, chromatin, coding and non-coding DNA.
- Explain the term 'histone'. HL
- Describe the role of a gene.
- Explain what is meant by nuclear inheritance.
- Compare nuclear inheritance of DNA and non-nuclear inheritance in mitochondrial and chloroplast DNA. HL

Unpacked from LOs 1.4.1, 1.4.2, 1.4.3

SPECIES AND HEREDITY

In Chapter 9, we learned that organisms that can naturally interbreed with each other to produce fertile offspring are considered **species**. Horses are one species of animal, and donkeys are another species. If a female horse were to breed with a male donkey, they would produce an animal called a mule. However, as a mule is infertile and unable to reproduce with another mule, it is **not** considered a new species of animal.

Figure 25.1: A mule is the infertile offspring of a horse and a donkey.

When animals or plants of the same species breed with each other, the offspring they produce will have some traits from their mother and some traits from their father. This is known as **heredity**. Heredity is the passing on of features from parents to their offspring using genes. Examples of heredity in humans include passing on of skin colour, earlobe type (attached or unattached) and hairline shape from parents to their children. Inherited genetic traits in other organisms such as plants include flower colour and seed type.

- A **species** is a group of organisms capable of interbreeding to produce fertile offspring.
- **Heredity** is the passing on of characteristics from one generation to the next by means of genes.

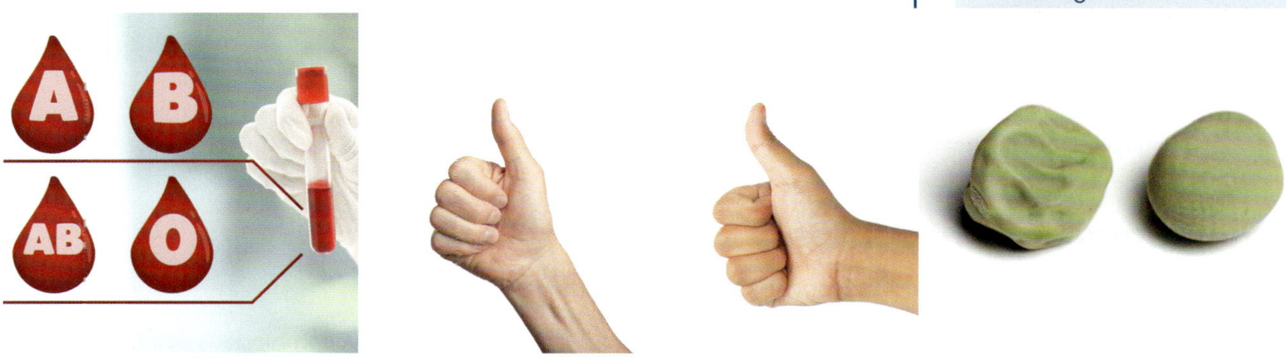

Figure 25.2: Inherited traits (L–R): blood type, straight or curved 'hitchhiker' thumb shape, seed texture.

CHROMOSOMES

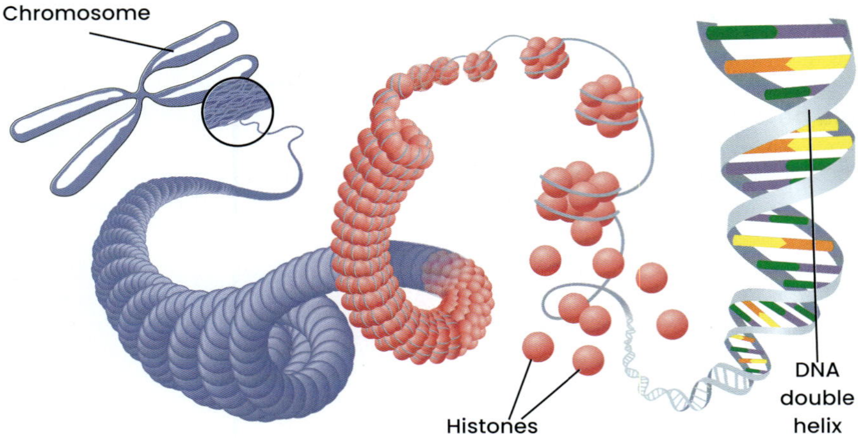

Figure 25.3: Chromosome structure.

> A **chromosome** is a coiled thread of DNA and protein that becomes visible in the nucleus during cell division.

We learned in Chapter 16 that **chromosomes** are found in the nucleus of eukaryotic cells. Prokaryotic cells also contain chromosomes, but they are not found inside a nucleus. Chromosomes play a very important role in heredity because they contain instructions needed to cause certain traits, such as eye colour, to appear in organisms. Chromosomes are made up of two substances: DNA (deoxyribonucleic acid) wrapped tightly around proteins.

HL The proteins that DNA wraps around are referred to as **histones**.

When a cell is not dividing, the chromosomes are spread out in the nucleus and referred to as **chromatin**. During cell division, the chromatin coils up into chromosomes. Meiosis results in offspring inheriting one set of chromosomes from each parent.

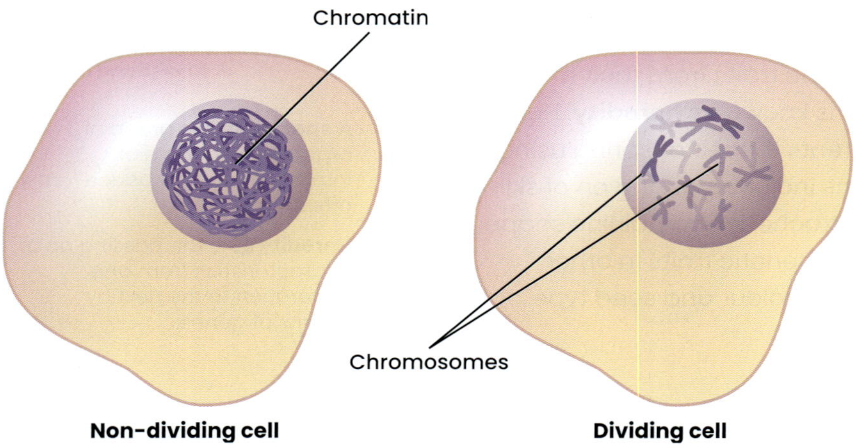

Figure 25.4: Chromatin vs chromosomes.

Homologous chromosomes

All human cells contain a total of 46 chromosomes (23 pairs), apart from the sex cells which have half (23 chromosomes in total). These chromosomal pairs, called **homologous pairs**, contain the same pattern of genes for the same traits. One of these chromosomes was inherited from the organism's mother and the other from its father.

> A **homologous pair** of chromosomes is a pair of chromosomes with the same pattern of genes.

GENES

In Chapter 24, we learned that 1–2% of DNA in a chromosome is made up of sections called **genes**, which contain the genetic code to make a particular protein. Human DNA contains approximately 20,000 genes. The remaining non-coding DNA controls when genes are expressed in a cell. Proteins coded by genes give rise to different traits, e.g. hair colour. Certain traits, such as height, are controlled by several different genes, whereas other traits, such as eye colour, are controlled by a single pair of genes.

A **gene** is a section of DNA that codes for a protein or trait.

The **locus** is the location of a gene on a chromosome.

The location of a gene on a chromosome is known as its **locus** (plural: loci).

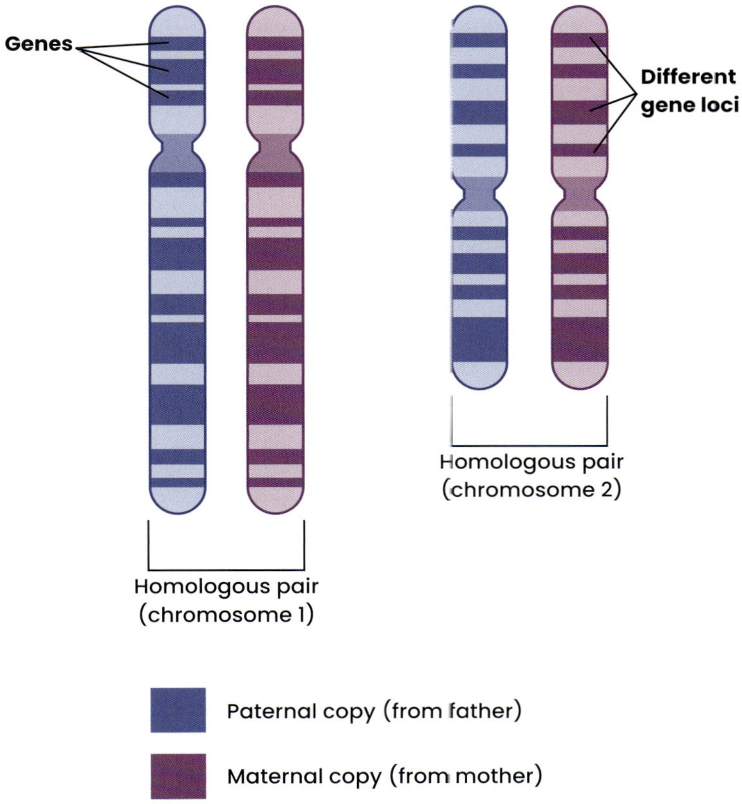

Figure 25.5: Homologous chromosomes.

Alleles

Each gene may exist in one of two different forms or **alleles** known as dominant or recessive. Dominant alleles are represented by a capital letter, e.g. **B**. Recessive alleles are represented by a small letter, e.g. **b**.

An **allele** is an alternative form of a gene, e.g. dominant or recessive.

For example, one chromosome in a pair might have a dominant gene for brown eyes (B) and the other homologous chromosome might have the recessive gene for blue eyes (b).

FOCUS ON ... HOMOZYGOUS AND HETEROZYGOUS ALLELES

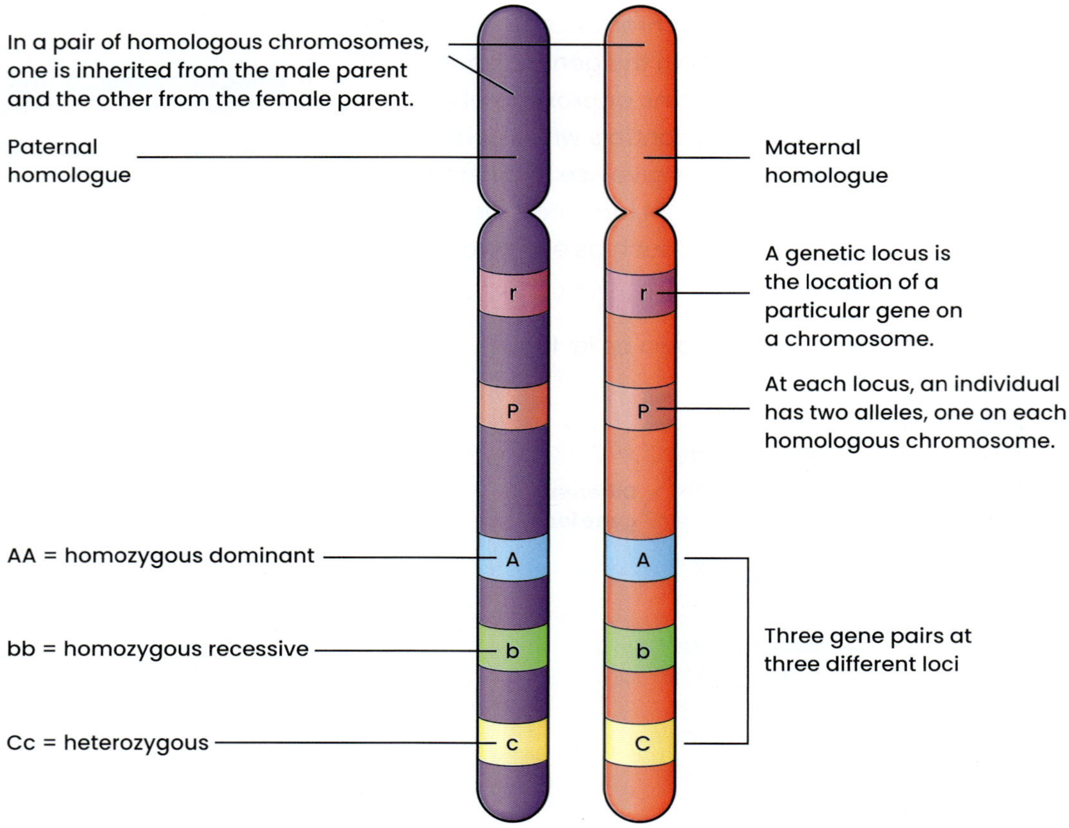

Figure 25.6: Allele forms.

In a pair of homologous chromosomes, if the two alleles for a particular gene are of the same form i.e. both dominant or both recessive, they are said to be **homozygous**. A pair of homozygous dominant genes is represented by two capital letters, e.g. **BB**. A pair of homozygous recessive genes is represented by two small letters, e.g. **bb**.

If the two alleles for a particular gene are different, i.e. one dominant gene and one recessive gene, they are said to be **heterozygous**. A pair of heterozygous genes are represented by one capital and one small letter, e.g. **Bb**.

Gene expression

We learned in Chapter 24 that protein synthesis involves two processes: transcription and translation. **Gene expression** refers to the process in which a particular gene is activated – or 'switched on' – to produce RNA and protein.

Every non-sex cell in an organism has the same set of chromosomes and collection of genes. However, cells in different locations within the body will express or activate different genes depending on their function, e.g. cells in the salivary glands in the mouth will express the gene to produce the enzyme salivary amylase. Cells can also determine how much of a protein is made. At any one time, only a fraction of the genes in a cell are expressed or 'switched on'; the remainder are 'switched off'. For example, a rise in blood glucose levels activates the genes in cells in the pancreas to secrete insulin. Once glucose levels return to normal, this gene is switched off.

- A **homozygous** pair of alleles consists of two genes of the same form, e.g. BB or bb.
- A **heterozygous** pair of alleles consists of two genes of different form, e.g. Bb.
- **Gene expression** is the process in which a gene is activated to make RNA and protein.

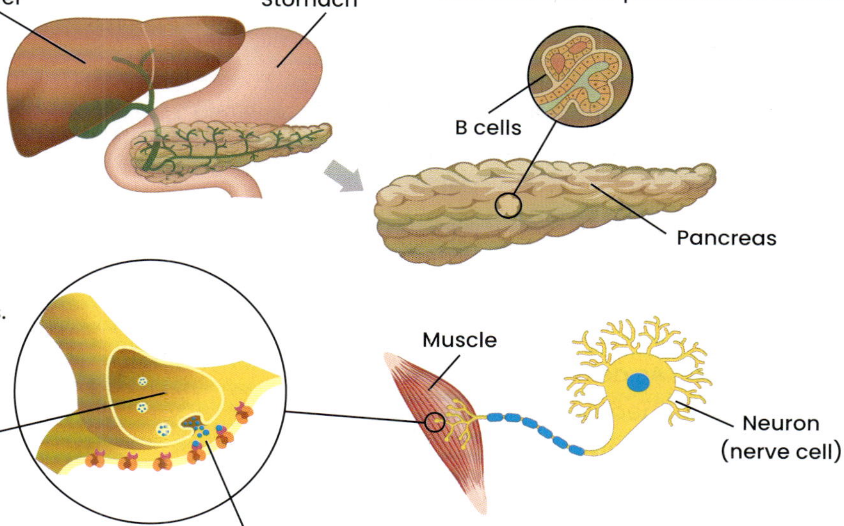

Figure 25.7: Gene expression.

Roles of a gene

- Genes contain the instructions (code) to make a particular protein.
- Many genes contain the instructions to make enzymes and hormones, which control the activities of the cell.
- Genes are the units of heredity and allow characteristics (traits) to be passed on from parents to offspring.

NUCLEAR INHERITANCE

In Chapter 17, we learned the importance of meiosis in gamete formation and creating variation in offspring. Meiosis and its transfer of genetic material to offspring involves the DNA found in the nucleus.

Nuclear inheritance refers to the transmission of DNA from parents to offspring.

NON-NUCLEAR INHERITANCE

HL

Two other cell organelles, the mitochondrion and the chloroplast, contain small amounts of their own DNA, distinct from that found in the nucleus. Each cell only has one nucleus with its set of DNA. As cells can have many mitochondria or chloroplasts, they contain multiple copies of mitochondrial or chloroplast DNA. Unlike nuclear DNA, which is mainly non-coding, the DNA found in mitochondria and chloroplasts has few non-coding regions.

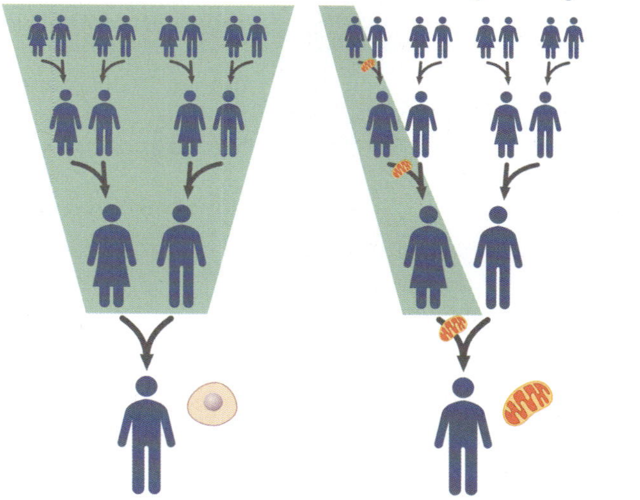

Figure 25.8: Nuclear inheritance vs mitochondrial inheritance.

Mitochondrial DNA

Mitochondrial DNA consists of a circular piece of DNA coding for about 37 genes, found in the mitochondria. Mutations within mitochondrial DNA can give rise to a group of disorders called mitochondrial diseases. During fertilisation of gametes, mitochondria from the male sperm do not enter the egg. This means that only mitochondria from the mother's egg are passed on to offspring. Consequently, most cases of mitochondrial disease are passed down from the mother to their child. As cells cannot function properly without healthy mitochondria, mitochondrial diseases can have a wide range of serious effects.

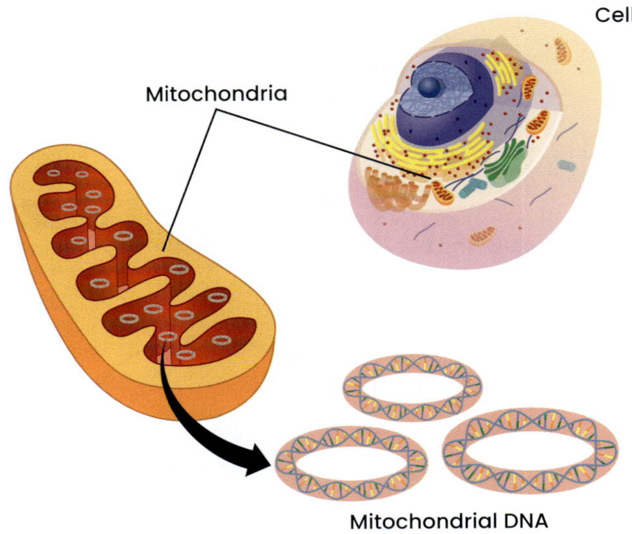

Figure 25.9: Mitochondrial DNA.

Chloroplast DNA

Another non-nuclear form of DNA is that found in the chloroplast in plants. This circular piece of DNA consists of about 120 genes that are involved in controlling photosynthesis. Like mitochondrial DNA, chloroplast DNA is believed to be inherited along the maternal line.

HL

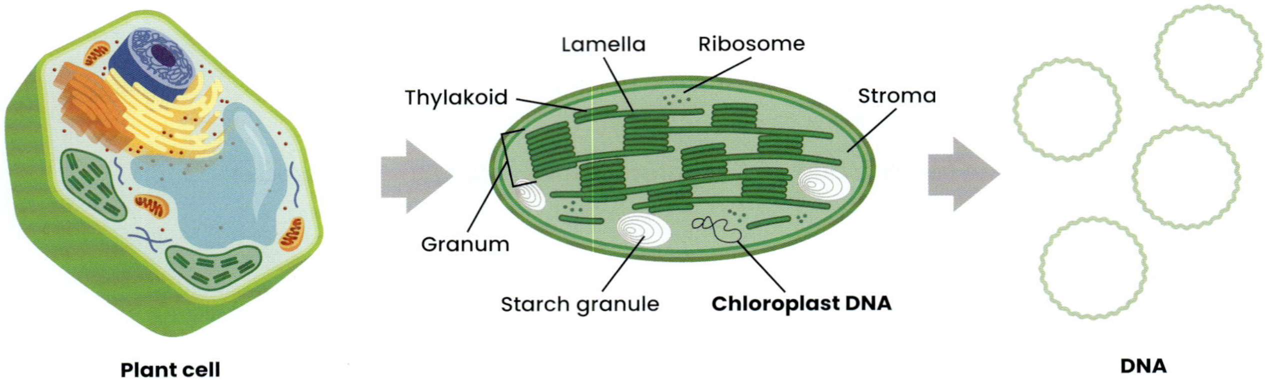

Figure 25.10: Chloroplast DNA.

	NUCLEAR DNA	NON-NUCLEAR DNA
LOCATION	Nucleus	Mitochondrion or chloroplast
INHERITED	From both mother and father	From mother
NUMBER OF GENES	Approx. 20,000 genes	37–120 genes
SHAPE	Two-stranded double helix	Circular
NUMBER OF COPIES IN A CELL	1 set of 46 chromosomes	Multiple copies per cell
CODING DNA	1–2%	93%

Figure 25.11: A summary of nuclear and non-nuclear DNA.

SCIENCE IN SOCIETY: HEALTH

Mitochondrial donation

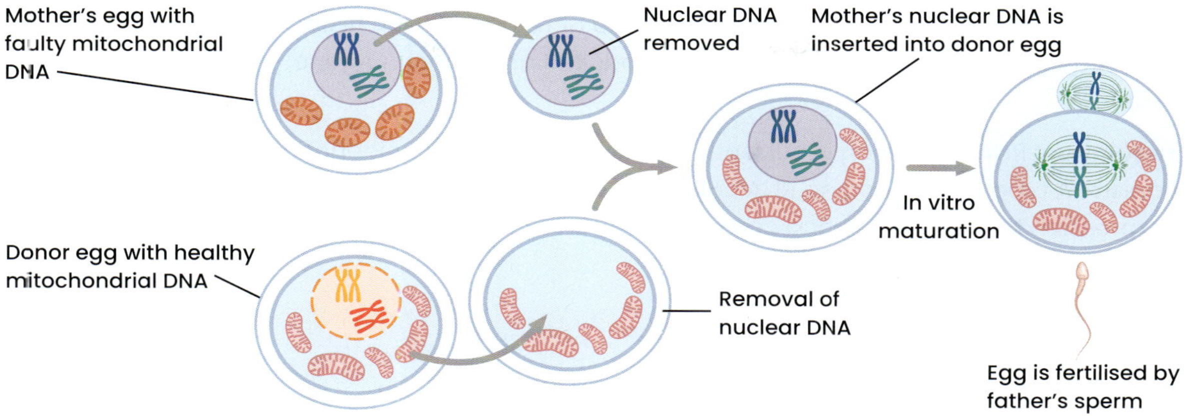

Figure 25.12: Mitochondrial IVF.

This novel (new) IVF procedure, sometimes called three-person in vitro fertilisation (IVF), involves removing the nuclear genetic material from an egg with faulty mitochondrial DNA and placing it into a healthy donor egg that has had its nuclear genetic material removed. In this way, the faulty mitochondrial DNA is replaced. This egg now has nuclear genetic material from the mother and mitochondrial DNA from another donor. This egg may be fertilised with the father's sperm.

Mitochondrial donation offers families with a history of mitochondrial disease a treatment to reduce the chance of transmitting the disease to their children.

PORTFOLIO LINK Go to your Learning Portfolio and complete Learning Check 1 to assess your knowledge and understanding of the topic so far.

Section 2: Genetic and Epigenetic Mechanisms

At the end of this section, you should be able to:

- Compare genetic and epigenetic mechanisms.
- Research one example of epigenetic inheritance in nature. (HL)

Unpacked from LOs U1.2, U3.2, 1.4.3

Nuclear DNA controls how cells develop through the production of proteins. However, an organism's behaviour and their environment can also affect which genes are expressed or 'switched on'. Behaviour such as diet, smoking or exercise and environmental factors such as stress and exposure to pollutants can affect gene expression. A person may have a set of genes that predisposes them to tallness, but their actual height will be influenced by factors such as diet.

The study of how an organism's behaviour and their environment affect their gene expression is known as **epigenetics**. Epigenetic changes result in changes in chromosomes but do not alter a person's DNA sequence; they are reversible and can vary as a person moves from birth to childhood to adulthood.

Epigenetics is the study of the effect of behaviour and the environment on gene expression in an organism.

The main mechanisms of epigenetics in cells are as follows:

- **DNA methylation:** A chemical methyl group is added to a specific site on DNA, generally a cytosine-guanine pair, blocking the expression of a gene.
- **Histone modification:** Chemical groups are added or removed from histone proteins, changing how tightly the histones are packed and affecting gene expression.
- **RNA regulation:** RNA regulation is currently the least understood epigenetic mechanism. A type of non-coding RNA called micro-RNA is thought to affect gene expression through regulating chromatin structure.

DISCUSSION

Discuss how epigenetic changes may be reversed.

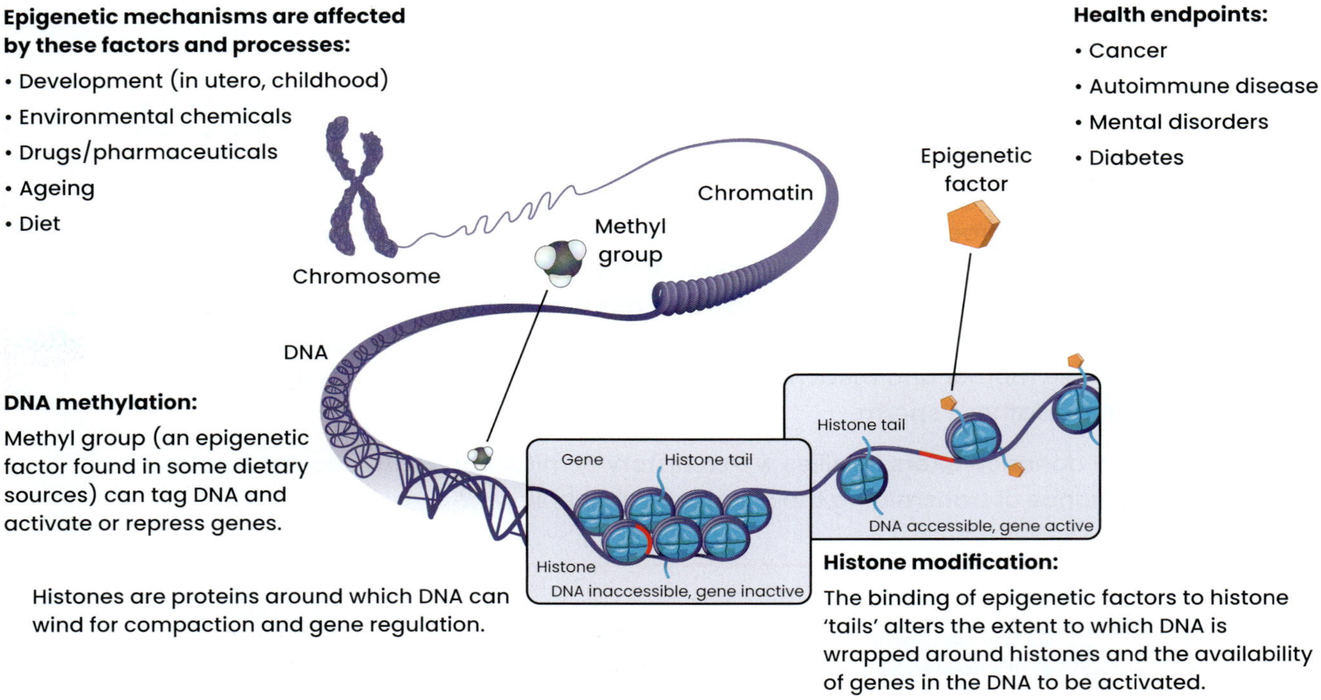

Figure 25.13: Epigenetic mechanisms: DNA methylation and histone modification.

Different cell types in an organism have different epigenetic markers. These epigenetic markers can change how a cell reads DNA instructions.

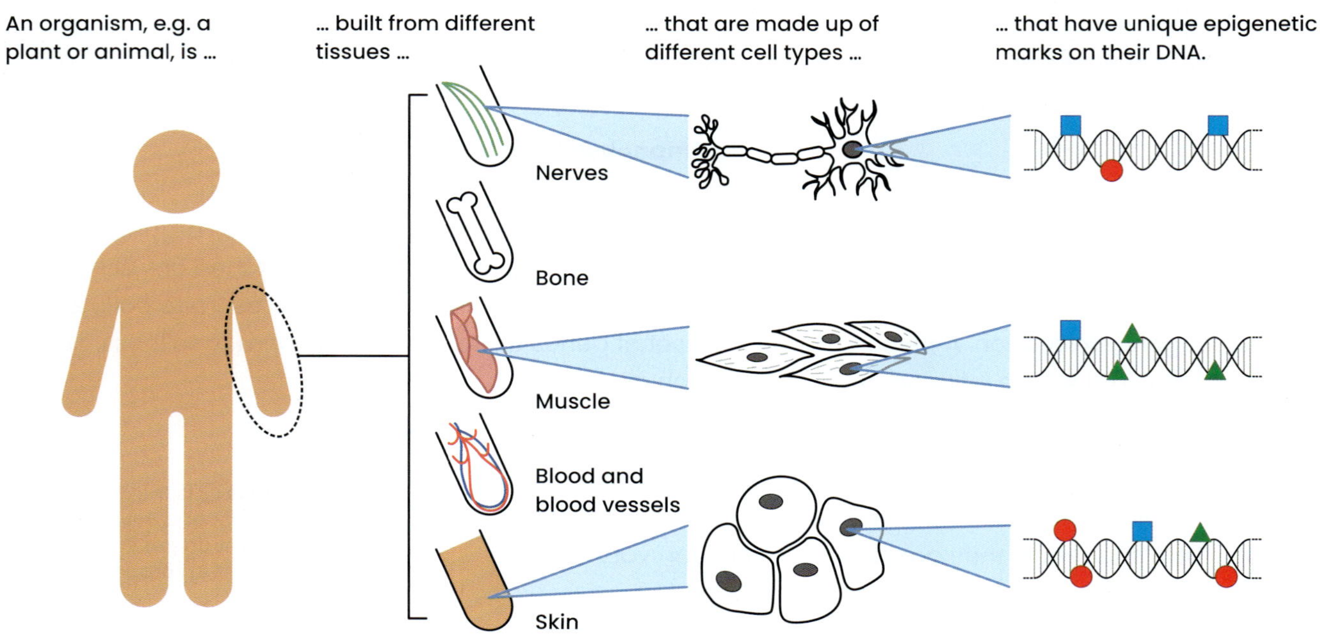

Figure 25.14: Epigenetic markers in different cell types.

Epigenetics can affect your health in different ways:

1. **Bacterial infections:** Bacteria can turn off genes in some immune cells, weakening the immune system and improving the bacteria's chance of survival.
2. **Nutrition during pregnancy:** A pregnant mother's environment and behaviour can affect their baby's epigenetics, e.g. children of women who were pregnant during a famine are more likely to develop diseases such as type 2 diabetes and heart disease.
3. **Cancer:** DNA methylation due to epigenetics can alter the expression of genes that cause cancers.

The table below compares genetic and epigenetic mechanisms:

Genetic mechanisms	Epigenetic mechanisms
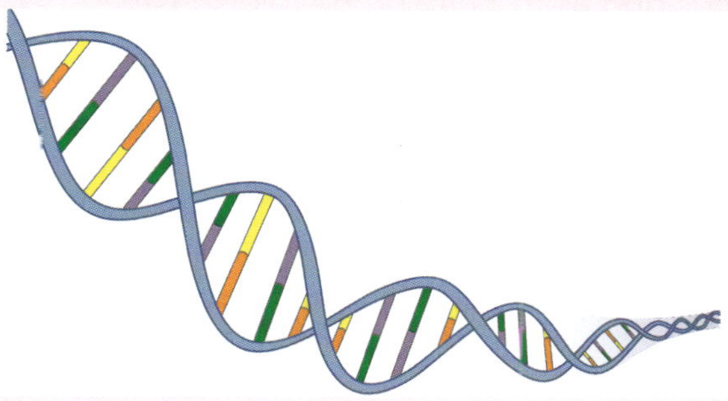	
Involves DNA sequence.	Involves changes to chromosomes, e.g. methyl tags to DNA, chromatin changes. The DNA sequence is not affected.
Genotype is resistant to environmental influences (except mutagens or DNA damage).	Epigenetic mechanisms are potentially susceptible to environmental influences, e.g. stress, diet, age.
DNA undergoes stable transmission and is highly conserved during mitosis and meiosis – any changes in DNA sequence are rarely reversed.	Epigenetic markers on DNA are normally removed during gamete formation. Epigenetic changes are reversible.
DNA is identical in all cell types.	Epigenetic markers are different in all cell types.

Figure 25.15: Genetic vs epigenetic mechanisms.

FOCUS ON … EPIGENETICS IN NATURE

Example 1: Smokers vs non-smokers

Smoking can trigger epigenetic changes. For example, smokers tend to have less DNA methylation in the 'AHRR' gene, which is related to tumours. Heavy and long-term smokers have greater reductions in their DNA methylation levels. Those who quit smoking can experience increased DNA methylation at this gene, reaching levels similar to non-smokers over a period of time. This can occur in under a year, depending on the individual's previous smoking history.

Example 2: Nurturing behaviour in rats

The level of nurturing rat pups receives from their mother triggers epigenetic differences that affect their stress responses later in life. Rat pups who were extensively licked and groomed by their mothers during their first week of life grew up to be happy and calm, whereas those who were ignored grew up to be anxious and were more prone to disease. Epigenetic science suggests that the effects of trauma such as neglect or abuse or other environmental factors could be passed from one generation to another. In the case of the rat pups, when the female pups grew and had their own offspring, the ones who were nurtured as pups themselves became high nurturing mothers. The ones who were neglected became low nurturing mothers.

RESEARCH ACTIVITY (HL)

LO 1.4.3

Research other examples of epigenetic inheritance in nature.

Go to YouTube.com and search for 'What is epigenetics?' (1:24) by The Jackson Laboratory.

PORTFOLIO LINK

Go to your Learning Portfolio and complete Learning Check 2 to assess your knowledge and understanding of the topic so far.

Section 3: First Generation Crosses and Mendel's Law of Segregation

At the end of this section, you should be able to:

- Carry out an unlinked monohybrid (single trait) cross involving homozygous and heterozygous parents to the first generation.
- Carry out a genetic cross involving incomplete dominance. (HL)
- Illustrate and state Mendel's law of segregation. (HL)
- Describe sex determination by X and Y chromosomes in humans.
- Identify benefits and limitations of Mendelian genetics to our understanding of heredity.

Unpacked from LOs 1.4.4, 1.4.5, 1.4.6, 1.4.9

GREGOR MENDEL

Gregor Mendel was born in 1822 in Austria. In university, he excelled in Physics and Mathematics, and afterwards became an Augustinian monk. While at the monastery, he carried out scientific experiments on pea plants. He studied the inheritance of seven characteristics of peas, such as stem height, flower colour and seed shape. Importantly, he studied traits with only two forms (e.g. plants were either tall or short, or the pods were either yellow or green). His experiments involved transferring pollen from one plant to another by hand. He repeated the experiment on hundreds of plants and counted the number of traits in the offspring, thus developing accurate ratios. Unfortunately, it wasn't until several decades after his death that his work was recognised. He is now known as the 'father of modern genetics'.

Figure 25.16: Gregor Mendel.

Sexual reproduction

During gamete formation, a selection of genes enters the sex cells. The male gamete inherits its genes from the father and the female gamete inherits its genes from the mother. When the male and female gametes fuse during fertilisation, they give rise to offspring with traits from each parent. We will learn about the formation of gametes – or sex cells – in Chapter 36.

HL

Mendel's law of segregation

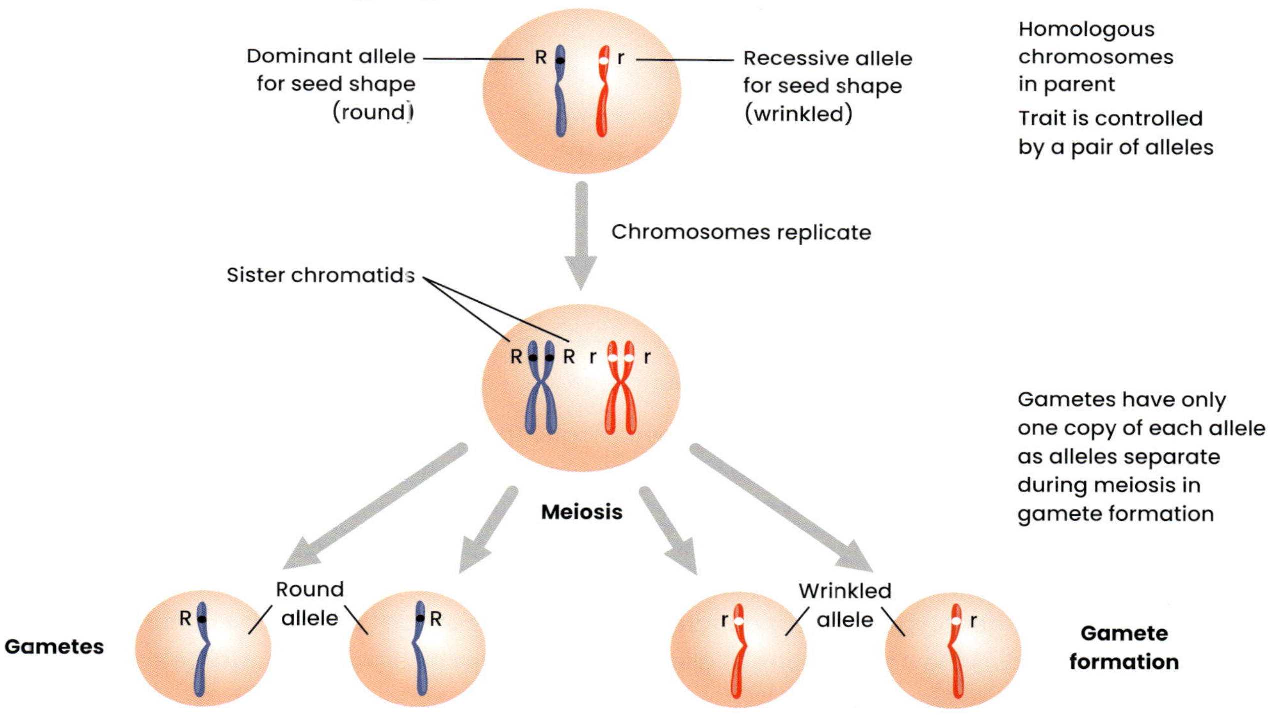

Figure 25.17: Mendel's first law, the law of segregation.

Mendel discovered that each trait, e.g. height, seed colour, is controlled by a pair of genes or alleles, one inherited from the mother and the other inherited from the father. During gamete formation, only one of these alleles enters each gamete. This is known as Mendel's first law – the **law of segregation**.

Mendel's law of segregation says that traits are governed by pairs of genes called alleles. These alleles separate during gamete formation, with each gamete receiving only one allele.

GENETIC CROSSES

We will now look at a series of genetic crosses. A genetic cross involves looking at the possible offspring when two organisms breed. Before carrying out a genetic cross, it is important to understand the following terms:

- **Phenotype** is the physical appearance of an organism.
- **Genotype** is the genetic make-up of an organism.
- A **recessive** allele is only expressed in the homozygous state.
- A **dominant** allele prevents a recessive allele from being expressed.

- **Alleles:** different forms of the same gene (e.g. B or b).
- **Dominant allele:** this allele prevents the recessive allele from being expressed (e.g. B).
- **Recessive allele:** this allele is prevented from being expressed by a dominant allele (e.g. b). When a dominant allele and a recessive allele appear together, it is the dominant allele that is expressed.
- **Homozygous:** both alleles are the same (e.g. BB or bb).
- **Heterozygous:** the alleles are different (e.g. Bb).
- **Genotype:** the genetic make-up of an organism (e.g. bb).
- **Phenotype:** the physical make-up or appearance of an organism (e.g. blue eyes).

FOCUS ON ... GENETIC CROSSES

We often use a table called a 'Punnett square' to illustrate genetic crosses. A Punnett square shows the genotype of the two parents, the alleles present in the gametes, and predicts the probability of offspring of different genotypes occurring. We will look at **monohybrid crosses**, where the parents differ in only one trait e.g. height. In the following examples, we will use a Punnett square to predict the inheritance to the first generation in crosses involving both homozygous and heterozygous parents.

- **Monohybrid cross** refers to genetic breeding between two parents who vary in only one trait.

In genetics, the offspring of a set of parents are referred to as the first filial generation, or F1. All the genetic crosses we examine in Section 2 will be to the first filial generation (F1).

Example 1: Heterozygous vs homozygous recessive

The genetic cross shown below is based on pea plants. The gene for tallness (T) is dominant over the gene for shortness (t). Each parent has two alleles for any trait, e.g. height. Of these two alleles, they can only donate one to each gamete. In this example, one parent plant has a heterozygous genotype (Tt) and the other has a homozygous recessive genotype for height (tt). The parent with the Tt genotype can donate a 'T' or a 't' allele to gametes. The parent with the tt genotype can only donate one type of allele – a 't' allele. The Punnett square below shows the combination of alleles that can form by crossing the gametes from these parents.

Example 1: Heterozygous vs homozygous recessive		Parent 1: Tt (tall) Gametes	
		T	t
Parent 2: tt (short) Gametes	t	Tt (tall)	tt (short)
	t	Tt (tall)	tt (short)
Offspring genotype		Tt, Tt, tt, tt	
Offspring phenotype		Tall, tall, short, short 50% tall : 50% short	

In example 1, the probability of the offspring being tall is 50% and the probability of the offspring being short is also 50%.

We will now look at two other monohybrid crosses using Punnett squares in the same way. Note that phenotype ratios reflect mathematical probabilities that may be different to actual offspring ratios.

Example 2: Homozygous dominant vs heterozygous

Example 2 illustrates the probability of the different potential offspring between a cross of homozygous dominant and heterozygous parents for a single trait (height). In this example, the probability that the offspring will be tall is 100%.

Example 2: Homozygous dominant vs heterozygous		Parent 1: TT (tall) Gametes	
		T	T
Parent 2: Tt (tall) Gametes	T	TT (tall)	TT (tall)
	t	Tt (tall)	Tt (tall)
Offspring genotype		TT, Tt, TT, Tt	
Offspring phenotype		Tall, tall, tall, tall 100% are tall	

Example 3: Heterozygous vs heterozygous

In this example, the probability that the offspring will be tall is 75% and the probability that they will be short is 25%.

As both the parents were tall, any short offspring are known as 'non-parental' types, as they have a different genotype and phenotype to their parents.

Example 3: Heterozygous vs heterozygous		Parent 1: Tt (tall) Gametes	
		T	t
Parent 2: Tt (tall) Gametes	T	TT (tall)	Tt (tall)
	t	Tt (tall)	tt (short)
Offspring genotype		TT, Tt, Tt, tt	
Offspring phenotype		Tall, tall, tall, short 75% tall : 25% short	

FOCUS ON ... PREDICTING UNKNOWN PARENTAL GENOTYPES

What if the genotypes of the parents are unknown? It is possible to predict their genotypes from the ratio of offspring produced, as shown in Figure 25.18.

If unknown is **homozygous** (BB)

	B	B
b	Bb	Bb
b	Bb	Bb

Phenotypic ratio: 100% black

If unknown is **heterozygous** (Bb)

	B	b
b	Bb	bb
b	Bb	bb

Phenotypic ratio: 50% black : 50% white

Figure 25.18: Predicting unknown parental genotype.

THINK, PAIR, SHARE

How might different genotypes give rise to the same phenotype?

SEX DETERMINATION

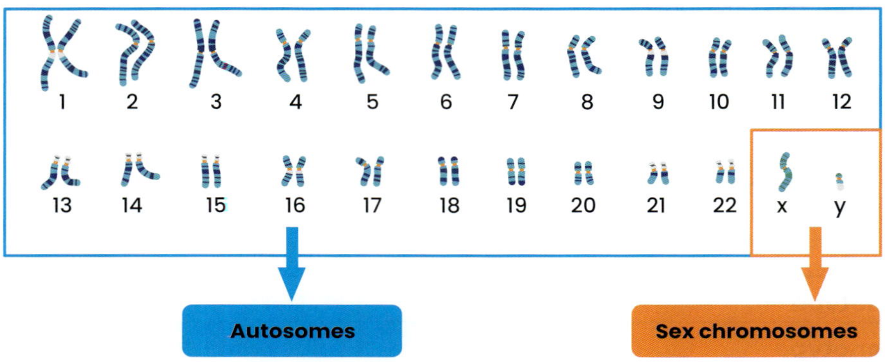

Figure 25.19: Male human karyotype showing X and Y sex chromosomes.

Of the 23 pairs of chromosomes that humans have, 22 are the same in males and females. These are called **autosomes**. The 23rd pair is different; they determine whether someone is male or female. These are called **sex chromosomes**. A female has two X chromosomes (XX) and a male has an X and a Y chromosome (XY). When a male and female organism reproduce, the female can only pass on an X to the gamete, whereas a male can pass on either an X or Y. As shown in the Punnett square here, there is a 50% chance the offspring will be male or female.

Sex determination		**Parent 1: Male (XY) Gametes**	
		X	**Y**
Parent 2: Female (XX)	**X**	XX (female)	XY (male)
Gametes	**X**	XX (female)	XY (male)
Offspring genotype		XX, XX, XY, XY	
Offspring phenotype		Female, female, male, male 50% female : 50% male	

INCOMPLETE DOMINANCE

In **incomplete dominance**, neither allele is dominant and both are expressed to produce offspring with an intermediate phenotype.

In most genetic crosses, there is a dominant allele which prevents the expression of a recessive allele. Occasionally, crosses occur in which neither allele is dominant or recessive with respect to the other and both alleles are expressed in the heterozygous genotype, producing an intermediate phenotype. For example, when a red rose is cross pollinated with a white rose, the rose produced is neither red nor white – it is pink. This is an example of incomplete dominance and is illustrated in the Punnett square below:

Incomplete dominance: **Red flower vs white flower**		**Parent 1: Red (RR)** **Gametes**	
		R	**R**
Parent 2: White (rr)	**r**	Rr	Rr
Gametes	**r**	Rr	Rr
Offspring genotype		Rr, Rr, Rr, Rr	
Offspring phenotype		Pink, pink, pink, pink 100% pink	
In this case, we still use two letter types to represent colour, e.g. R = red and r = white, but the use of a capital letter here does not signify dominance and lower case does not indicate recessiveness of that trait. You will always be told in an examination question if incomplete dominance exists between two traits.			

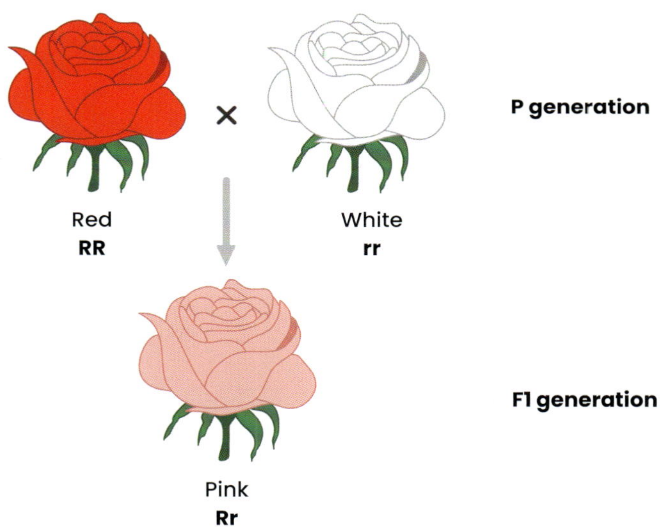

Figure 25.20: Incomplete dominance.

Go to YouTube.com and search for 'Video Number 1 – Genetic Crosses-Basic Introduction For Leaving Cert Biology Ireland' (7:10) by Biology Bugbears.

BENEFITS OF MENDELIAN GENETICS TO OUR UNDERSTANDING OF HEREDITY	LIMITATIONS OF MENDELIAN GENETICS
The predictive power of Mendelian genetics enhances our understanding of genetics.	Mendel's laws do not apply to complex traits that are controlled by multiple genes. In addition, Mendel's laws do not acknowledge environmental factors which may affect an organism's epigenetics.

PORTFOLIO LINK **Go to your Learning Portfolio and complete Learning Check 3 to assess your knowledge and understanding of the topic so far.**

Section 4: Second Generation Crosses and Linkage

At the end of this section, you should be able to:

- Explain and predict the inheritance of sex-linked traits from known examples. HL
- Illustrate and state Mendel's law of independent assortment. HL
- Explain how linkage affects Mendel's law of independent assortment. HL
- Carry out an unlinked dihybrid (two trait) cross involving homozygous and heterozygous parents to the second generation. HL

Unpacked from LOs 1.4.6, 1.4.7, 1.4.8, 1.4.10, 1.4.11

SEX LINKAGE

Sex linkage refers to genes that are carried on the X or Y (sex) chromosomes.

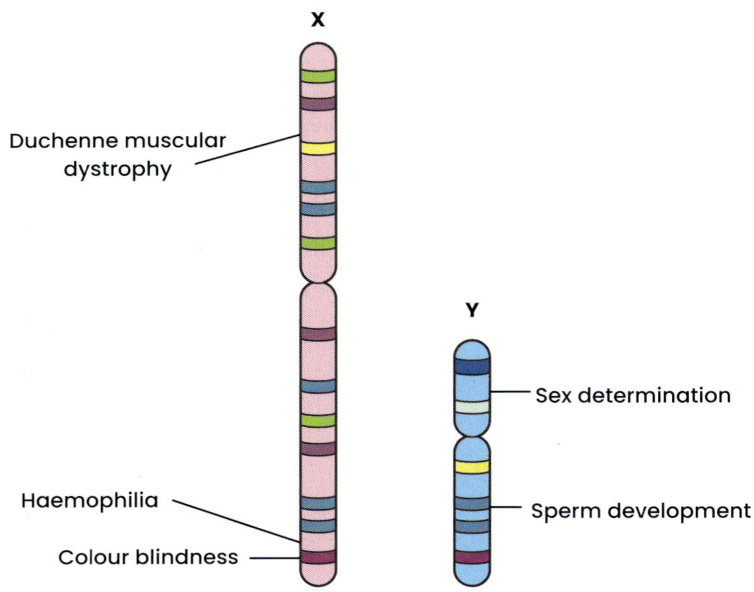

HL

Figure 25.21: Males have a Y chromosome that has genes for sperm production.

Sex linkage means that genes are carried on a sex chromosome. The Y chromosome is shorter than the X chromosome and doesn't contain some of the corresponding genes that are on the X chromosome. If a gene is found on the part of the X chromosome that has no corresponding part on the Y chromosome, the trait is often more commonly found in males than in females. This is because males only need to inherit one copy of the allele (from their mother), but females need to inherit the allele from both parents.

Sex-linked traits

Colour blindness and haemophilia are two examples of sex-linked traits caused by a recessive allele on the X chromosome. People who are colour blind have difficulty distinguishing different colours, e.g. red and green. People with haemophilia lack a blood clotting factor that stops bleeding and so they bleed for longer periods of time.

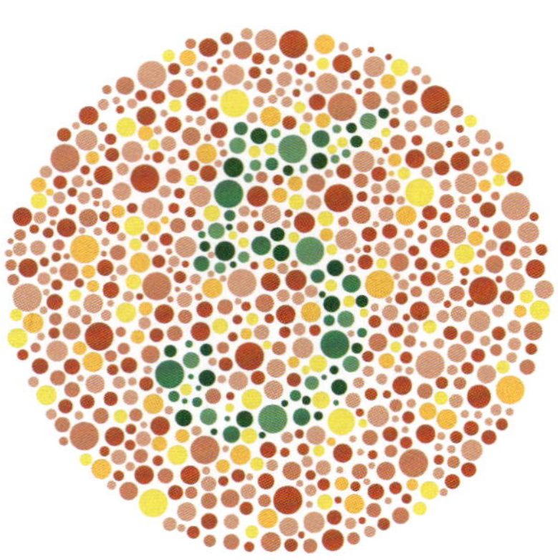

Figure 25.22: Sex-linked disorder: test for red-green colour blindness.

Inheritance of sex-linked traits

We will now look at the inheritance of a sex-linked trait for colour blindness. The gene for colour blindness is carried on the X chromosome. A normal X chromosome for vision is represented by X^N and an X chromosome with the gene for colour blindness is represented by X^n. As there is no corresponding allele on the Y chromosome for colour vision, males only need to inherit one X^n chromosome to have colour blindness. Females have two X chromosomes, both of which need to be X^n to have colour blindness. A female with a heterozygous genotype for colour blindness, X^NX^n, has normal vision but carries a gene for colour blindness which she could pass on to offspring.

Sex-linked trait **Example 1: Colour blindness**		**Parent 1: X^NY Male, normal for colour blindness** **Gametes**			
		X^N		**Y**	
Parent 2: X^NX^n **Female, carrier for colour blindness** **Gametes**	**X^N**	X^NX^N Female, normal		X^NY Male, normal	
	X^n	X^NX^n Female, carrier		X^nY Male, colour blind	
Offspring genotype		X^NX^N	X^NX^n	X^NY	X^nY
Offspring phenotype		Female, normal vision	Female, carrier for colour blindness	Male, normal vision	Male, colour blind
		75% normal vision : 25% colour blind			

HL

Pedigree charts

The genetic cross can also be represented using a pedigree chart.

The one below outlines the inheritance of haemophilia from the following parents: a mother who carries the gene for haemophilia (h) and a father who has the normal gene (H).

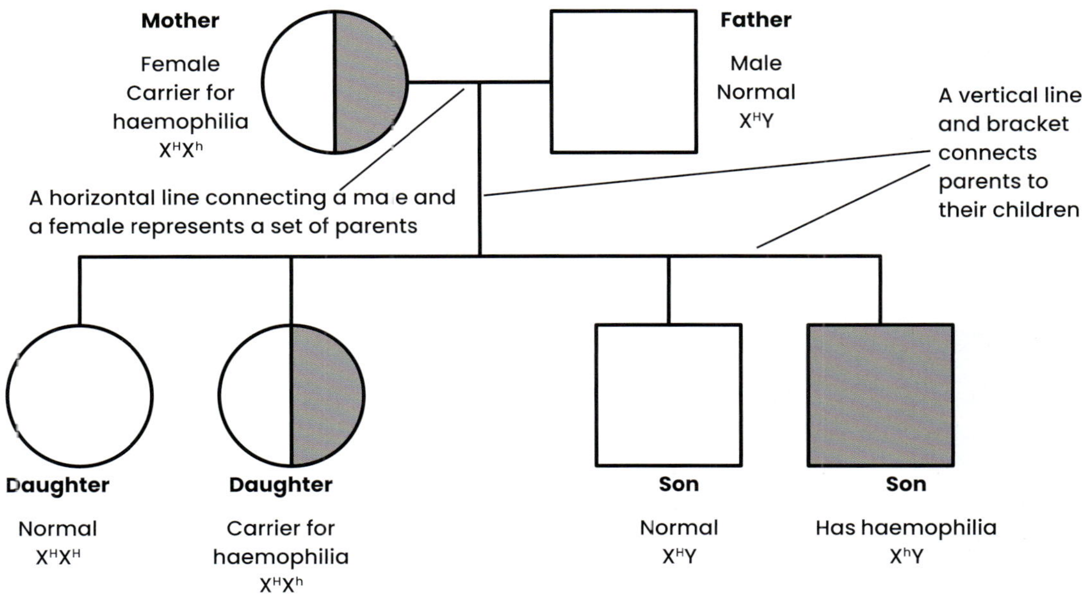

Figure 25.23: Pedigree chart for haemophilia.

A pedigree chart can be used to show a family's genetic history over several generations. Males are represented as squares and females as circles. A shaded symbol means that a person is affected by a condition e.g. haemophilia. An unshaded symbol means the person does not have the condition. People who are carriers for a condition are represented by a half-shaded symbol.

SCIENCE IN SOCIETY: HEALTH

X chromosome-linked gene targets viral infections

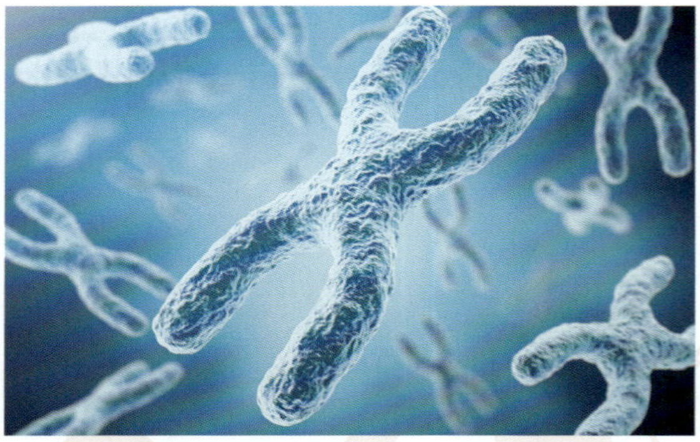
Figure 25.24: UTX is found on the X chromosome.

'Natural killer' (NK) cells in our immune system target viral cells. Although males have more of these NK cells, it is widely accepted that viral infections in males can be more severe than in females. New research by scientists may offer an explanation. Researchers found that NK cells in females have an extra copy of an X chromosome-linked gene called UTX. This UTX gene acts as an epigenetic regulator, enhancing NK cell antiviral function, while also repressing NK cell numbers.

(Source: UCLA Health)

DIHYBRID CROSSES AND INDEPENDENT ASSORTMENT

HL

All the crosses we have looked at so far were monohybrid crosses involving one trait. A **dihybrid cross** illustrates the inheritance of two different traits that are controlled by two different sets of alleles.

Mendel's **law of independent assortment** concerns dihybrid crosses involving two separate traits. For example, in peas, a round seed (R) is dominant to wrinkled seed (r) and yellow seed (Y) is dominant to green seed (y). A heterozygous round yellow seed has the genotype RrYy. If it were to produce gametes, either allele for seed shape (R or r) would have an equal chance of combining with either allele for seed colour (Y or y). The possible gametes would be RY or Ry or rY or ry, each with a 25% chance of being produced.

- A **dihybrid cross** illustrates the inheritance of two traits that are controlled by different sets of alleles.
- **Law of independent assortment**: Mendel's second law. It states that during gamete formation, either member of a pair of alleles can combine with either member of another pair of alleles.

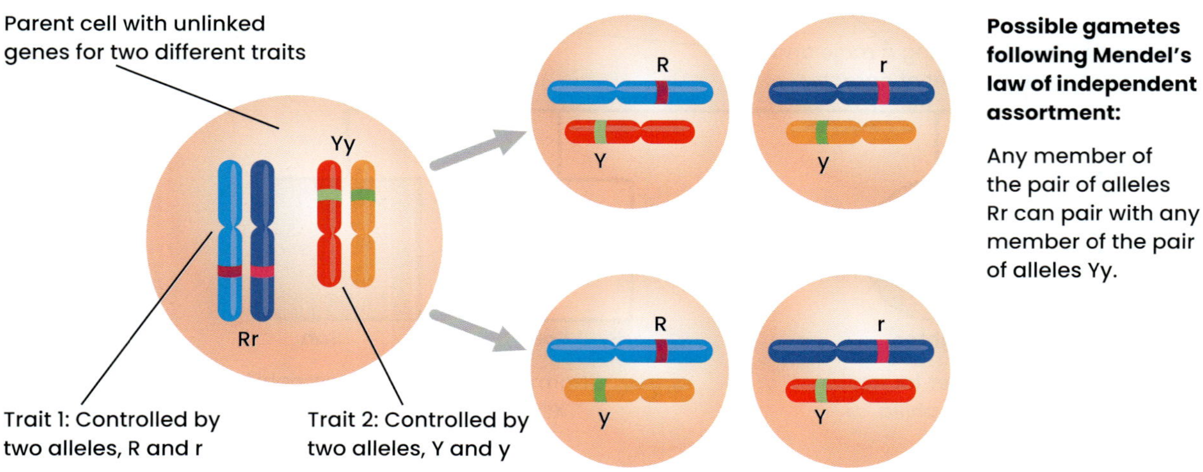

Figure 25.25: Mendel's law of independent assortment.

Dihybrid cross example 1

We can illustrate Mendel's law of independent assortment by examining the inheritance of two traits (seed colour and seed shape) when two pea plants with the following genotypes are crossed: parent 1 (homozygous recessive rryy wrinkled and green) and parent 2 (heterozygous RrYy round and yellow). This is an example of a dihybrid cross, as the inheritance of two traits is examined at the same time. As you can see from the Punnett square below, parent 1 can only donate one type of allele for each trait: wrinkled 'r' for seed shape and green 'y' for seed colour, to form gametes of type 'ry'.

Example 1 : Homozygous recessive vs heterozygous parents		**Parent 1** **Genotype:** rryy **Phenotype:** Wrinkled, green **Gametes**			
		ry			
Parent 2 **Genotype:** RrYy **Phenotype:** Round, yellow **Gametes**	**RY**	RrYy			
	Ry	Rryy			
	rY	rrYy			
	ry	rryy			
Offspring genotypes		RrYy	Rryy	rrYy	rryy
Offspring phenotypes		Round, yellow 25%	Round, green 25%	Wrinkled, yellow 25%	Wrinkled, green 25%

Dihybrid cross example 2

FIRST FILIAL GENERATION F1

We will now look at another dihybrid cross with the following parents: parent 1 (homozygous dominant RRYY round and yellow) and parent 2 (homozygous recessive rryy wrinkled and green).

Example 2: Homozygous dominant vs homozygous recessive parents **F1**		**Parent 1** **Genotype:** RRYY **Phenotype:** Round and yellow **Gametes**	
		RY	**RY**
Parent 2 **Genotype:** rryy **Phenotype:** Wrinkled and green **Gametes**	**ry**	RrYy	RrYy
	ry	RrYy	RrYy
Offspring genotypes		RrYy, RrYy, RrYy, RrYy	
Offspring phenotypes		100% round, yellow	

SECOND FILIAL GENERATION F2

If a pea seed with the genotype (RrYy) from the F1 generation of the dihybrid cross 1 or dihybrid cross 2 above is bred with another seed of the same genotype, we can use a Punnett square to predict the possible offspring genotypes and phenotypes of the second generation or F2 of peas.

Go to YouTube.com and search for 'Video Number 2 – Genetic Crosses Leaving Cert HL-Dihybrid Crosses-BASIC SUMMARY' (5:08) by Biology Bugbears.

Homozygous dominant vs homozygous recessive parents		**Parent 1** **Genotype:** RrYy **Phenotype:** Round, yellow **Gametes**			
		Possible gamete genotypes for parent 1			
F2		**RY**	**Ry**	**rY**	**ry**
Parent 2 **Genotype:** RrYy **Phenotype:** Round, yellow **Gametes** Possible gamete genotypes for parent 2	**RY**	RRYY Round, yellow	RRYy Round, yellow	RrYY Round, yellow	RrYy Round, yellow
	Ry	RRYy Round, yellow	RRyy Round, green	RrYy Round, yellow	Rryy Round, green
	rY	RrYY Round, yellow	RrYy Round, yellow	rrYY Wrinkled, yellow	rrYy Wrinkled, yellow
	ry	RrYy Round, yellow	Rryy Round, green	rrYy Wrinkled, yellow	rryy Wrinkled, green
Offspring phenotypes		Round, yellow	Round, green	Wrinkled, yellow	Wrinkled, green
Phenotype ratio		9/16	3/16	3/16	1/16

16 possible allele combinations from crossing parent 1 and parent 2 gametes

HL

LINKAGE

All the above genetic crosses refer to unlinked traits that are inherited independently and follow Mendel's law of independent assortment. **Linked genes** are located near each other on the same chromosome and tend to be passed on to gametes together. Genes on different chromosomes are never linked. Figure 25.26 shows a homologous pair of chromosomes. Genes A and B are on the same chromosome and are linked. This means that they will be inherited together in gametes as AB. Genes a and b on the other chromosome are linked and will be passed on to gametes as ab. As linked genes cannot undergo independent assortment, Ab or aB cannot form gametes, because the genes A and B are linked and cannot be separated. The ratios of gametes and offspring produced by linked crosses are different from those produced by non-linked crosses.

Linked genes are found on the same chromosome and tend to be inherited together. This is known as **linkage**.

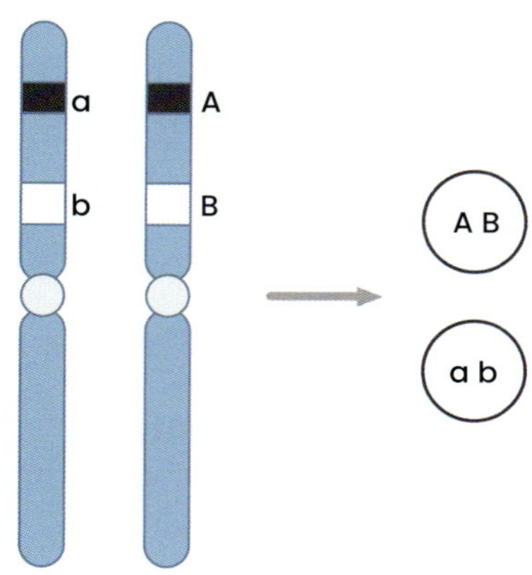

Figure 25.26: Formation of gametes by linked genes.

How linkage affects Mendel's law of independent assortment

The effects of linkage on Mendel's law of independent assortment can be seen in Figures 25.27 and 25.28 below. In Figure 25.27, the dominant gene for height (T) is linked with the dominant gene for purple flower (P). The recessive short gene (t) is linked with the recessive white flower gene (p). A cross of parents with the genotypes TtPp with linked genes results in offspring with the following phenotypes: tall and purple, short and white, in a ratio of 3:1. This is very different to the observed phenotypes and their ratio in the unlinked cross in Figure 25.28. When genes are linked, there is less variation in the offspring produced than when genes are not linked.

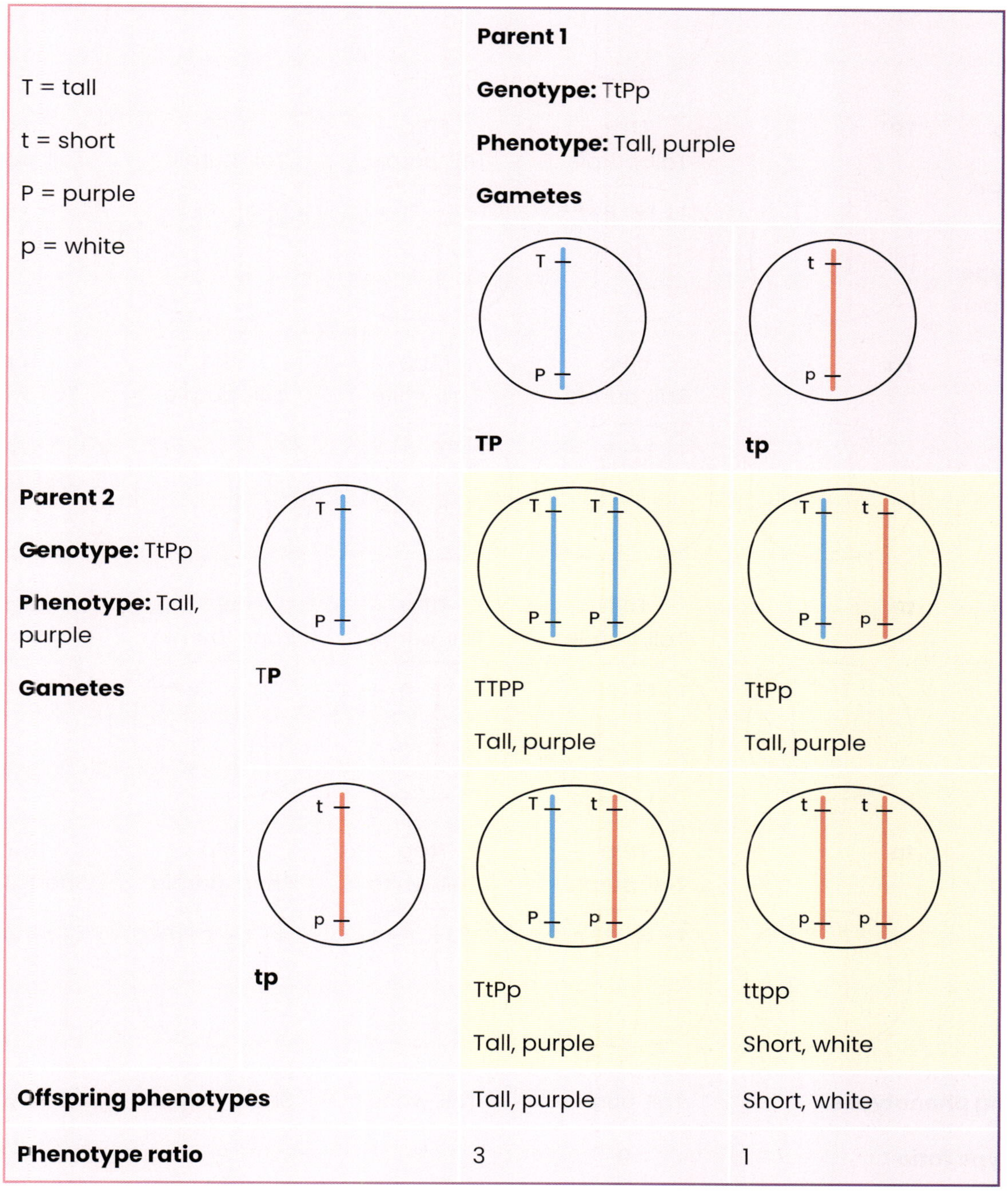

Figure 25.27: Linked cross.

T = tall t = short P = purple p = white		**Parent 1** **Genotype:** TtPp **Phenotype:** Tall, purple **Gametes**			
		TP	**Tp**	**tP**	**tp**
Parent 2 **Genotype:** TtPp **Phenotype:** Tall, purple **Gametes**	**TP**	TTPP Tall, purple	TTPp Tall, purple	TtPP Tall, purple	TtPp Tall, purple
	Tp	TTPp Tall, purple	TTpp Tall, white	TtPp Tall, purple	Ttpp Tall, white
	tP	TtPP Tall, purple	TtPp Tall, purple	ttPP Short, purple	ttPp Short, purple
	tp	TtPp Tall, purple	Ttpp Tall, white	ttPp Short, purple	ttpp Short, white
Offspring phenotypes		Tall, purple	Tall, white	Short, purple	Short, white
Phenotype ratio		9	3	3	1

Figure 25.28: Unlinked cross.

PORTFOLIO LINK Go to your Learning Portfolio and complete Learning Check 4 to assess your knowledge and understanding of the topic so far.

Chapter Summary

- A **species** is a group of organisms capable of interbreeding to produce fertile offspring.
- **Heredity** is the passing on of characteristics from one generation to the next by means of genes.
- **Chromosomes** are made up of two substances: DNA (deoxyribonucleic acid) and proteins called histones.
- A human has 46 (23 pairs) of chromosomes inside each of its cells (apart from sex cells, which have half).
- **Homologous pairs of chromosomes** are pairs of chromosomes which contain genes for the same traits.
- A **gene** is a section of DNA that codes for a protein which controls the development of a characteristic.
- **Roles of a gene:**
  - A gene contains the instructions (code) to make a particular protein.
  - Many genes encode enzymes.
  - Genes are the units of heredity and allow characteristics (traits) to be passed on from parents to offspring.

	NUCLEAR DNA	NON-NUCLEAR DNA
LOCATION	Nucleus	Mitochondrion or chloroplast
INHERITED	From both mother and father	From mother
NUMBER OF GENES	Approx. 20,000 genes	37-120 genes
SHAPE	Double-stranded spiral helix	Circular
NUMBER OF COPIES IN A CELL	1 set of 46 chromosomes	Multiple copies per cell
NON-CODING DNA	Mainly non-coding	Few non-coding sequences

- **Epigenetics** is the study of the effect of behaviour and the environment on gene expression in an organism.
- Some epigenetic effects are reversible.
- The main mechanisms of epigenetics in cells are DNA methylation and histone modification.
- **Gene expression** is the process by which a gene gets turned on in a cell to make RNA and proteins.
- **Alleles** are alternative forms of a gene e.g. dominant or recessive (e.g. B or b).
- A **dominant allele** prevents the recessive allele from being expressed (e.g. B).
- A **recessive allele** is only expressed in the homozygous state (e.g. b).
- **Genotype** describes the genetic make-up of an organism (e.g. bb).
- **Phenotype** describes the physical make-up or appearance of an organism (e.g. blue eyes).
- **Homozygous** refers to a pair of alleles for a gene that are of the same form (e.g. BB or bb).
- **Heterozygous** refers to a pair of alleles for a gene that are of different form (e.g. Bb).
- **Incomplete dominance** means that neither allele is dominant or recessive with respect to the other and both are equally expressed in the offspring to produce an intermediate phenotype.
- **Gregor Mendel** is known as the father of genetics.
- Mendel's law of segregation states that traits are governed by pairs of genes called alleles. These alleles separate during gamete formation, with each gamete receiving only one allele. HL
- Mendel's law of independent assortment states that during gamete formation, either member of a pair of alleles can combine with either member of another pair of alleles. HL
- **Linkage** means that genes are located on the same chromosome. HL
- **Sex chromosomes:** XX = female, XY = male.
- **Sex linkage** means that genes are carried on the X or Y (or sex) chromosome. HL
- **Example of sex-linked traits** include colour blindness and haemophilia.
- **Benefits of Mendelian genetics** include its predictive power that enhances our understanding of heredity.
- **Limitations of Mendelian genetics** include that Mendel's laws do not apply to traits which are controlled by multiple genes. In addition, Mendel's laws do not acknowledge environmental factors which may affect an organism's epigenetics.

END OF CHAPTER ASSESSMENT

REMEMBER

1. Define the following terms.
   (a) *Species*
   (b) *Heredity*
   (c) *Chromosome*
   (d) *Gene*
   (e) *Allele*
   (f) *Dominant*
   (g) *Recessive*
   (h) *Homozygous*
   (i) *Heterozygous*
   (j) *Genotype*
   (k) *Phenotype*
2. List **two** roles of a gene.
3. Define *epigenetics* and state the **two** main mechanisms of epigenetics.
4. HL Define the term *incomplete dominance* and give an example of a plant in which incomplete dominance occurs.
5. HL (a) State Mendel's law of segregation
   HL (b) State Mendel's law of independent assortment.

UNDERSTAND

6. If a couple were to have a child, what is the probability the child would be female? Explain your answer by displaying a genetic cross.
7. HL Are human males or females more likely to suffer from sex-linked diseases such as haemophilia? Refer to possible genotypes and phenotypes in your answer.
8. Distinguish between the following pairs of terms by writing a sentence about each term.
   (a) Homozygous and heterozygous
   (b) Genotype and phenotype
   HL (c) Linkage and sex linkage
   HL (d) Segregation and independent assortment
9. HL Compare nuclear and non-nuclear inheritance (mitochondrial/chloroplast DNA).

APPLY

10. Gregor Mendel studied the inheritance of various traits in pea plants. The results of some of his investigations are presented in the table.

PARENTAL CROSS	OFFSPRING GENOTYPE
Round × wrinkled seeds	All round seeds
Yellow × green seeds	All yellow seeds
Purple × white flowers	All purple flowers
Tall × dwarf plants	All tall plants

A dwarf pea plant with green seeds was crossed with a plant heterozygous for both height and seed colour.

(a) Illustrate, by means of a genetic cross, the possible genotypes and phenotypes of the progeny of this cross, if there is **no linkage** of genes.

(b) Explain how the results of the cross above would differ if the genes for height and seed colour were linked. HL

11. In a certain breed of cattle, there is incomplete dominance between the allele for red coat and the allele for white coat. Heterozygous individuals are roan in colour. In each of the following three crosses, calculate the percentage chance that a calf of the cross will have a red coat. HL
   (a) roan × white
   (b) roan × red
   (c) roan × roan HL
12. In the sweet pea plant the texture and colour of the testa (seed coat) are governed by two pairs of alleles, which are found on different chromosomes. The allele for smooth (S) is dominant to the allele for wrinkled (s) and the allele for yellow (Y) is dominant to the allele for green (y). HL
   (a) What term is used to describe alleles that lie on the same chromosome? HL
   (b) Why is it significant that the two pairs of alleles, mentioned above in relation to sweet pea plants, are located on different chromosome pairs? HL
   (c) Determine all the possible genotypes and phenotypes of the offspring of a cross between the following sweet pea seeds: wrinkled, heterozygous yellow × heterozygous smooth, green. HL

ANALYSE

13. In the Aberdeen Angus cattle breed, the allele for hornless condition is dominant to the one for the horned condition. The diagram shows part of the genotype of an Aberdeen Angus animal.

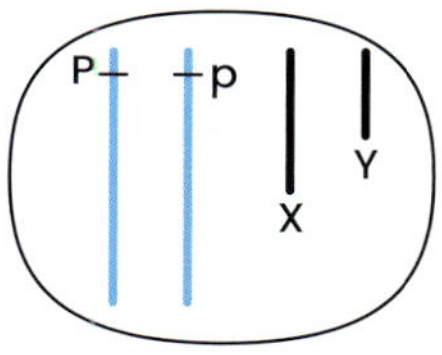

(a) What term is used to describe the Pp allele pair?

HL (b) Is this a sex-linked condition? Explain your answer.

(c) What is the phenotype and sex of the animal whose partial genotype is shown above?

14. The diagram shows part of the genotype of a mammal.

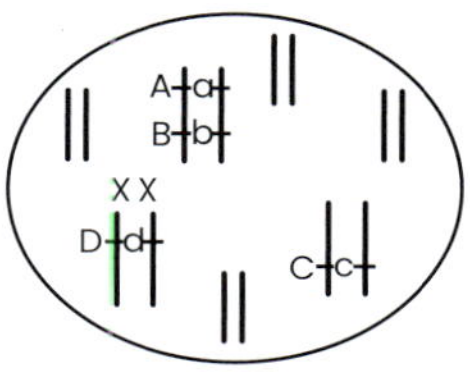

(a) What is the sex of this individual?

(b) Identify any pairs of linked genes present in the genotype.

(c) What term is used to describe the allele pair Cc?

(d) Draw a similar diagram for an animal of the opposite sex, in which A, B and C are homozygous.

HL **15.** Haemophilia is caused by a recessive allele found only on the X chromosome. The following pedigree chart shows the inheritance of this condition between parents and children in a particular family.

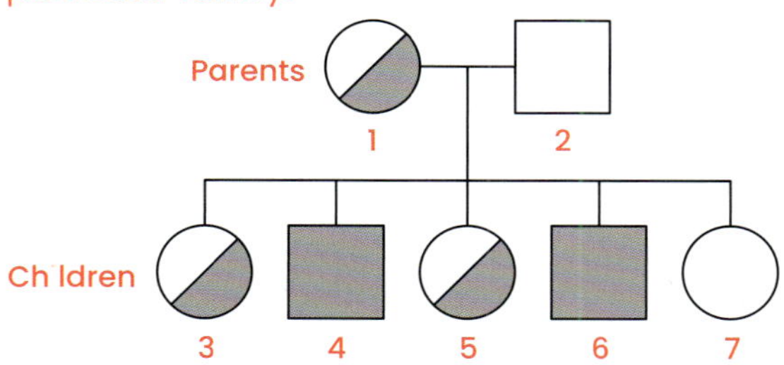

○ Non-sufferer female
◐ Heterozygous female
□ Non-sufferer male
■ Heterozygous male

HL (a) Using H to represent the dominant normal gene and h as the recessive gene for haemophilia, draw a labelled chromosome diagram for each parent, showing the position of each gene allele.

HL (b) Using a Punnett square, show how it is possible for this set of parents to have a male child who does not suffer from haemophilia.

HL (c) If person 7 were to have a child with a haemophiliac male partner, use a Punnett square to predict the potential genetic characteristics of any children they may have.

HL **16.** Red-green colour blindness in humans is caused by a recessive allele on the X chromosome. The following pedigree chart shows the inheritance of this condition between parents and their children in a particular family.

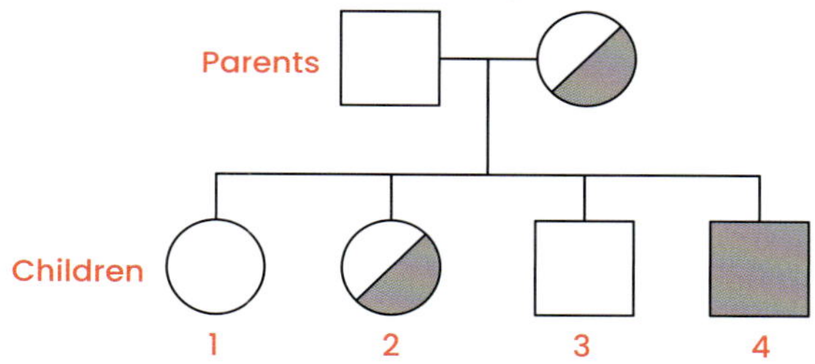

○ Female with red-green colour vision
◐ Heterozygous female
□ Male with red-green colour vision
■ Male with red-green colour blindness

HL (a) Using N to represent the dominant normal gene and n as the recessive gene for red-green colour blindness, state the genotype of (i) each parent (ii) all four children.

HL (b) Explain why males are more likely than females to be red-green colour blind.

HL **17.** In pea plants, yellow seeds are dominant over green and the tall allele is dominant over the dwarf allele. A dwarf pea plant with green seeds is crossed with a plant that is heterozygous for both seed colour and plant height.

HL (a) Draw a Punnett square to illustrate the genotypes phenotypes of the offspring if the genes for height and seed colour are not linked.

HL (b) How would you expect the offspring of this cross to vary if seed colour and plant height are linked?

EVALUATE

18. Evaluate the contribution of Mendel's genetic experiments on pea plants to our understanding of heredity.

CREATE

HL **19.** Design a poster to illustrate one of Mendel's laws using genetic crosses and Punnett squares.

26 Genetic Technologies and Genomics

LEARNING OUTCOMES: U2.4, U3.2, U4.1, 3.3.1, 3.3.2, 3.3.3, 3.3.4, 3.3.5, 3.3.6, 3.3.7

EYE-OPENER

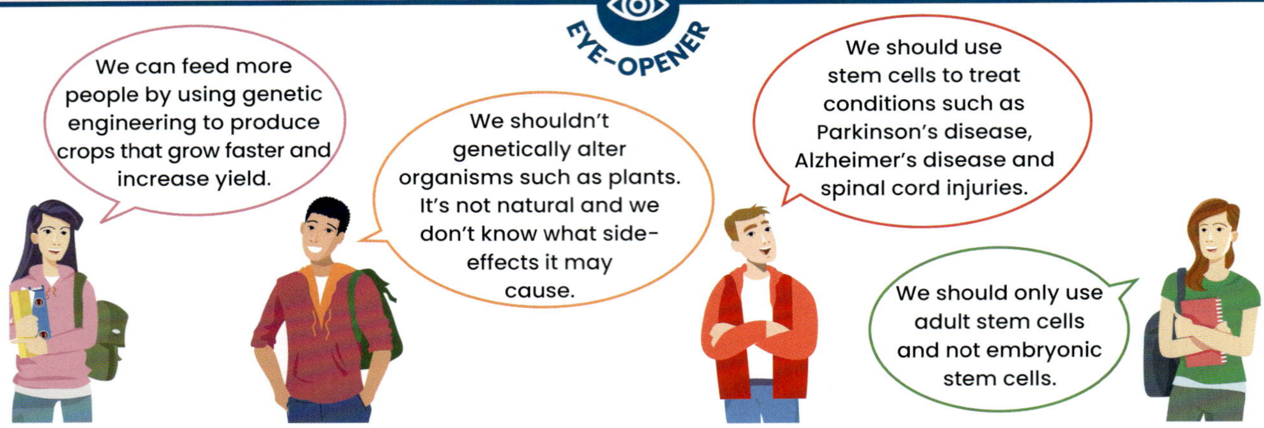

What do you think of each of the statements? Discuss in pairs.

At the end of this chapter, you should be able to:

- Outline the concept and process of genetic engineering.
- Outline the applications of genetic engineering.
- Outline what is meant by DNA profiling and its potential uses.
- Investigate patterns using a DNA profile.
- Model the steps involved in generating a DNA profile – DNA extraction, polymerase chain reaction (PCR) and gel electrophoresis. HL
- Outline the principle of DNA sequencing data (nucleotide bases). HL
- Use a genome database to search for alleles that are known to cause specific genetic diseases.
- Discuss the ethical issues arising from advancements in genetic technologies: genetic data access and use, genetic modification, DNA testing.

LINKS TO:

Chapter 2	**Scientific Knowledge and Research**
Chapter 7	**Bacteria and Fungi**
Chapter 16	**Cell Structure and Diversity**
Chapter 24	**Information Flow in the Cell**
Chapter 25	**Genetics and Inheritance**
Chapter 37	**Human Reproduction – From Fertilisation to Birth**

SUBJECT VOCABULARY

genetic engineering; vector; genetically modified organism; DNA profiling; DNA profile; polymerase chain reaction (PCR); gel electrophoresis; genome; bioinformatics; genome database; DNA sequencing

Section 1: What Is Genetic Engineering?

At the end of this section, you should be able to:

- Outline the concept and process of genetic engineering.
- Outline the applications of genetic engineering.

Unpacked from LOs 3.3.1, 3.3.7

In Chapter 24, we learned that DNA is a double-stranded molecule containing genes and non-coding regions. **Genetic engineering**, also called 'genetic modification', works on the principle of the artificial altering of an organism's DNA. This is not to be confused with random spontaneous changes that occur in DNA due to mutations, as discussed in Chapter 24. In genetic engineering, scientists transfer a gene for a desired characteristic from a donor organism into the DNA of a host organism. The organism now has an altered genotype and is known as a genetically modified organism (GMO).

Genetic engineering (genetic modification) involves the artificial altering or manipulation of an organism's DNA.

Genetic engineering can be carried out with plants, animals and micro-organisms. If the donor DNA comes from a different species to the host, the host organism is described as transgenic, e.g. the gene for making human insulin can be inserted into bacterial DNA. We will now take a closer look at the process of genetic engineering.

PROCESSES OF GENETIC ENGINEERING

1. **DNA isolation:** The section of DNA containing the gene of interest is identified e.g. the gene for human insulin.
2. **Cutting (restriction):** A restriction enzyme is used to cut the desired gene from human DNA. The restriction enzyme is also used to cut plasmids, circular pieces of DNA found in bacterial cells, at specific sites.
3. **Ligation:** Another enzyme called a DNA ligase is used to attach the ends of the human insulin gene to the cut ends of the bacterial plasmid.
4. **Transformation (introduction of base sequence changes):** The modified bacterial plasmids are inserted into bacterial cells. These genetically modified bacterial cells are now described as transformed. The base sequence in the bacterium's DNA has been changed as it now has a new gene, that to produce insulin. The plasmid acts as a **vector** – it is used to transfer DNA from one organism to another. In other cases, viral vectors are used.

A **vector** is a DNA molecule (often plasmid or virus) used in genetic engineering to carry a DNA segment or gene into a host cell.

5. **Expression:** When the genetically modified bacterial cells reproduce, millions of new bacterial cells capable of producing human insulin are formed. During protein synthesis in the bacteria, the human insulin gene is translated into insulin. Large amounts of human insulin are produced in this way. This can be removed and purified for use in the treatment of diabetics.

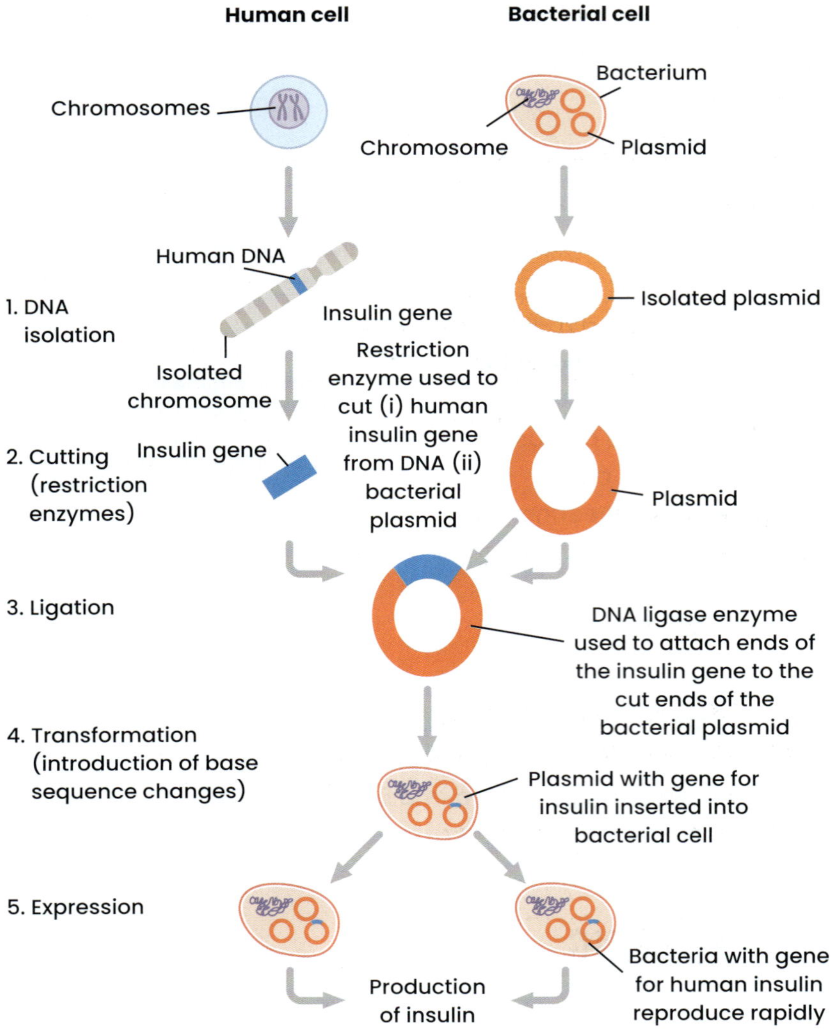

Figure 26.1: An example of genetic engineering.

Go to YouTube.com and search for 'Genetic engineering | Genetics | Biology' (4:58) by FuseSchool – Global Education.

GENETICALLY MODIFIED ORGANISMS

Examples of **genetically modified organisms (GMOs)** include:

- **Golden rice:** rice that is genetically modified with daffodil genes to produce more vitamin A.
- **Bt corn:** corn that is genetically modified with a bacterial insecticide gene. This allows the corn to produce toxins that protect it from insects.
- **Sheep** that have been genetically modified to produce human proteins in their milk that can be used to treat cystic fibrosis.
- **AquAdvantage salmon:** salmon that have been genetically modified to contain a growth promoter gene from another fish, so that the salmon grow to a larger size in a shorter time span.

Genetically modified organisms (GMOs) are organisms whose DNA has been altered using genetic engineering methods.

Figure 26.2: Genetically modified crop 'golden rice' on the right and normal rice on the left.

Ethical issues

There are ethical concerns regarding GMOs. We will examine the arguments for and against genetically modified crops in the following table.

	ARGUMENTS FOR	ARGUMENTS AGAINST
GM CROPS AND HUMAN HEALTH	GM crops with genes for vitamins or certain proteins inserted can increase the nutritional value of food e.g. GM oilseed crops with lower saturated-fat levels can benefit diets in developing countries.	There is concern that GM crops could trigger allergic responses in some people, although no evidence of this has been found to date.
GM CROPS AND FOOD SUPPLY	Over the next 30 years the world's population is expected to increase to over 9 billion, requiring food production to double. GM crops can have genes inserted that allow them to grow in harsh environments, grow larger more quickly, last longer without spoiling, be resistant to viruses and produce toxins to fight off pests.	The risk of GM herbicide-resistant crop seeds cross-pollinating wild plants could potentially lead to herbicide-resistant 'super weeds'. These 'super weeds' could outcompete other plants, reducing food supply and shelter for animals, leading to a loss of biodiversity.
GM CROPS AND THE ENVIRONMENT	GM crops can grow in a greater range of harsh environments, reducing the need to deforest land for farming. GM crops can have genes for resistance to pests, reducing the need for harmful pesticides.	GM crops can have genes to produce toxins against pests. These may also harm non-target species. Other insects that are resistant to the toxins could evolve and spread more quickly, upsetting the ecosystem.
GM CROPS AND ECONOMIC COSTS	The improved yields offered by GM crops reduces farming costs.	Biotech companies that produce GM crops own patents on the seeds and control their price. These companies can restrict the supply of seeds, forcing farmers to pay higher prices.

Figure 26.3: Arguments for and against the use of genetic modification in crops.

SCIENCE IN SOCIETY: TECHNOLOGY

CRISPR-Cas9

A new form of genetic engineering called CRISPR-Cas9 (pronounced 'crisper') was developed in the early 2000s. **CRISPR-Cas9** allows scientists to cut DNA at a specific sequence using an enzyme called Cas9. DNA can be added or removed at this point. This technique is adapted from a naturally occurring genome editing system that bacteria use to defend against attacking viruses. The applications of CRISPR being researched by scientists include gene therapy treatment of single-gene diseases such as cystic fibrosis and the genetic modification of crops and livestock. Scientists have discovered that CRISPR can cut out unintended sequences of DNA, raising ethical concerns about using the technique in editing human genes.

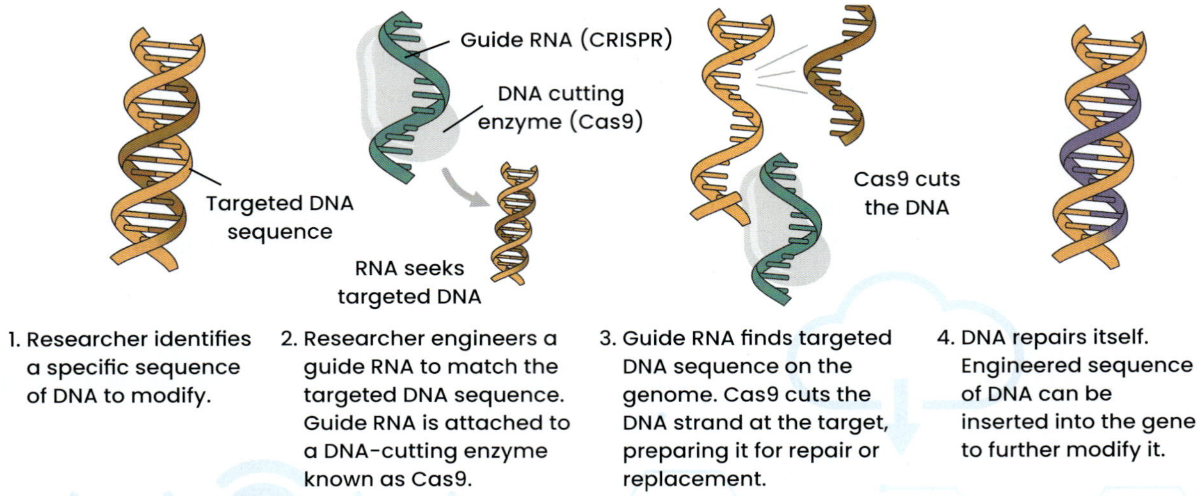

Figure 26.4: How CRISPR works.

PORTFOLIO LINK **Go to your Learning Portfolio and complete Learning Check 1 to assess your knowledge and understanding of the topic so far.**

Section 2: DNA Profiling

At the end of this section, you should be able to:

- Outline what is meant by DNA profiling and its potential uses.
- Investigate patterns using a DNA profile.
- Discuss the ethical issues arising from genetically modified organisms.
- Model the steps involved in generating a DNA profile – DNA extraction, polymerase chain reaction (PCR) and gel electrophoresis. HL

Unpacked from LOs 3.3.2, 3.3.3, 3.3.5, 3.3.7

Apart from genetically identical twins, each person's DNA is different. Non-coding regions of DNA that do not code for proteins are highly variable between people. These non-coding DNA regions are used in **DNA profiling**, a technique that reveals a unique pattern of bands from a person's DNA sample. This pattern of bands 'DNA profile' looks like a bar-code that may be used to identify the person.

DNA profiling is a technique that reveals a unique pattern of bands from an individual's DNA sample.

To reveal a person's DNA profile, a DNA sample is required – this may be taken from any cells of the body including blood, semen and hair follicles.

USES OF DNA PROFILING

1. **Paternity test:** to identify the biological parents of a child.

2. **Forensics:** to identify the perpetrator of a crime.

In Figure 26.5, we can see how the DNA profile from a crime scene can be compared against that of the three suspects. In this case, from the patterns in the DNA profiles, suspect 3 matches that obtained from the crime scene.

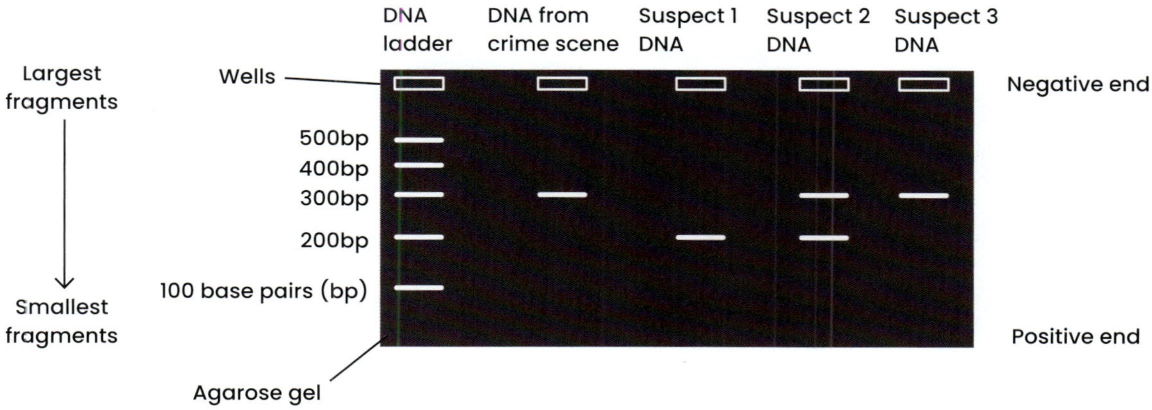

Figure 26.5: An example of a DNA profile in a forensic context.

Steps involved in generating a DNA profile

We will now look at the process involved in generating a DNA profile.

1. DNA extraction

DNA must be isolated from the cells in a sample, e.g. blood. Chemicals are used to break down the cell and nuclear membranes to release the DNA. The DNA needs to be separated from the rest of the cell.

2. Polymerase chain reaction (PCR)

If the amount of DNA obtained is very small, a technique called **polymerase chain reaction (PCR)** is used to produce millions to billions of copies of the DNA.

Polymerase chain reaction (PCR) is a laboratory technique for rapidly producing numerous copies of a segment of DNA.

HL

3. Cutting

Non-coding DNA is cut at specific base sequences using **restriction enzymes**. This produces fragments of DNA of different size. These fragments are unique to each individual.

4. Gel electrophoresis

The DNA fragments in the sample are separated on the basis of size using a technique called **gel electrophoresis**. The DNA sample is placed in a well at one end of an agarose gel in a tray. Another DNA sample with fragments of known lengths, called a DNA ladder, is placed in another well. When an electric current is passed across the gel, DNA fragments move towards the positive terminal as DNA is negatively charged. Longer fragments move more slowly than shorter fragments, separating them. This creates a unique pattern of bands on the gel.

Gel electrophoresis is a laboratory technique to separate DNA fragments on the basis of size using an electric current passed through a gel.

5. Pattern analysis

The DNA fragments are made visible using a fluorescent dye and UV light. The lengths of the sample's DNA fragments can be determined by comparison to those in the DNA ladder.

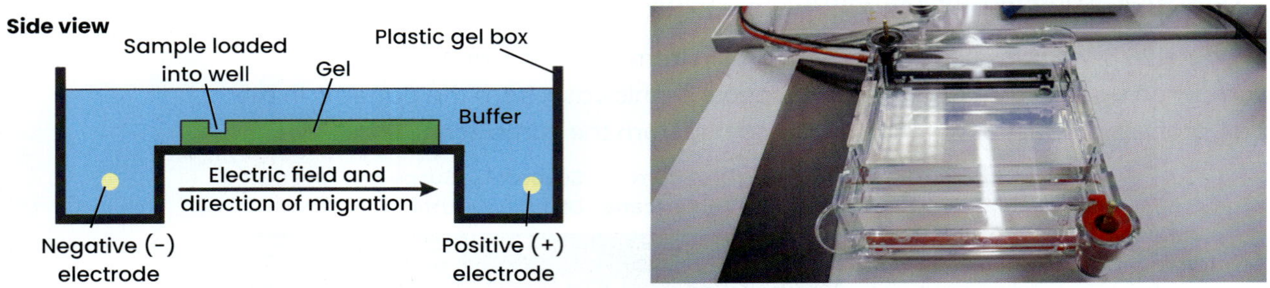

Figure 26.6: Gel electrophoresis.

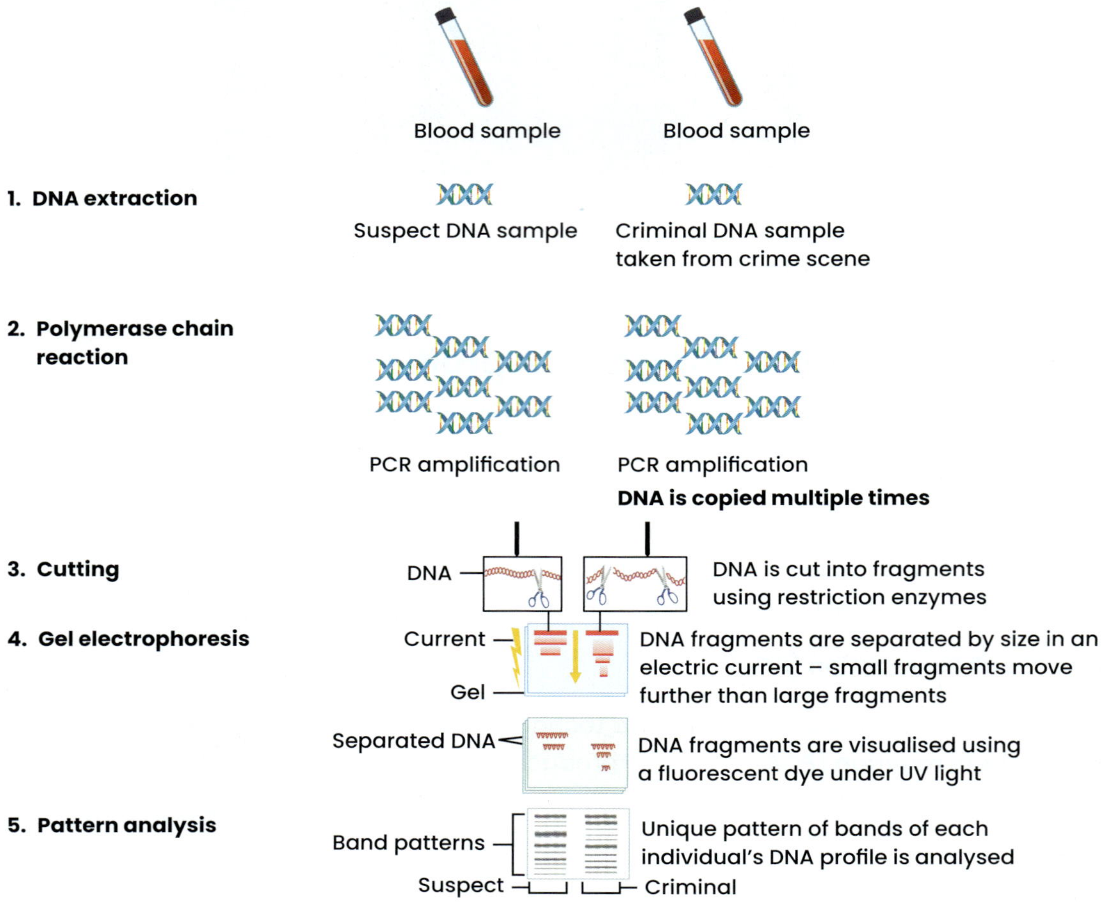

Figure 26.7: Process of DNA profiling.

Specified Investigation 12

Investigate patterns using a DNA profile using secondary data **(LO 3.3.6)**

MATERIALS

- DNA profiles (provided as secondary data)
- Ruler
- Highlighter
- Pen and paper for note-taking

METHOD

1. Crime scene DNA profiling

1. Obtain the DNA profiles for the crime scene sample, victim and potential suspects.
2. Place the ruler horizontally across all DNA profiles, covering each of the marker bands.
3. Slowly move the ruler down, revealing one row of bands at a time.
4. For each row, compare the crime scene DNA profile with each suspect's profile. If a band matches (in both position and thickness), highlight it on both profiles.
5. Continue this process for all rows.
6. Count the total number of matching bands for each suspect.
7. Calculate the percentage match for each suspect using the formula:

$$\frac{\text{number of matching bands}}{\text{total number of bands}} \times 100$$

Sample secondary data:

2. For paternity testing

1. Obtain the DNA profiles for the child, mother, and potential father or fathers.
2. Place the ruler horizontally across all DNA profiles, covering the marker bands.
3. Slowly move the ruler down, revealing one row of bands at a time.
4. For each row:
   (a) Identify the child's bands.
   (b) Determine which band came from the mother.
   (c) The remaining bands should match one of the potential father's markers.
   (d) Highlight matching bands (in both position and thickness) between the child and potential fathers.
5. Continue this process for all rows.
6. Count the total number of matching bands for each potential father.
7. Calculate the percentage match for each potential father using the formula:

$$\frac{\text{number of matching bands}}{\text{total number of bands}} \times 100$$

Sample secondary data:

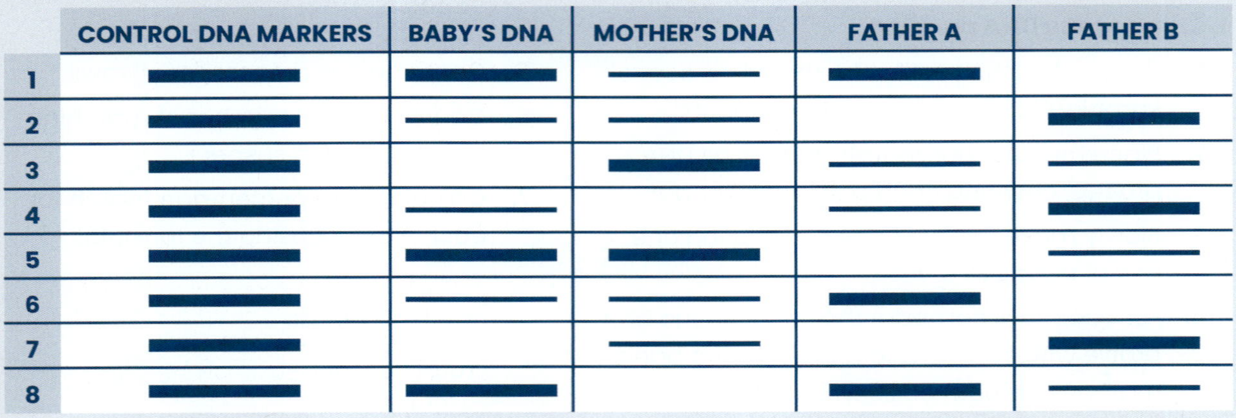

RESULTS

1. Crime scene DNA profiling

- Suspect 1 has two bands in common with the profile of the DNA found at the crime scene, giving them a 40% match ($2/5 \times 100 = 40\%$).
- Suspect 3 has one band in common with the crime scene DNA, giving them a 20% match ($1/5 \times 100 = 20\%$).
- The pattern of bands produced by the DNA found at the crime scene is a 100% match for Suspect 2's DNA profile ($5/5 \times 100 = 100\%$). This would indicate that Suspect 2 would be responsible for the crime, but remember – this is not definitive proof and should be considered alongside other evidence.

2. Paternity testing

- In the data provided, three bands are common in both the baby's and mother's DNA, leaving three further bands to analyse.
- Father A has three matches for those remaining bands giving them a 100% match ($3/3 \times 100 = 100\%$). This would indicate that Father A is the biological father of the baby.
- Father B has no matching bands with the baby's DNA, which would indicate he is not the father.
- In paternity testing, a match of 99% or higher typically indicates paternity, while a match below 90% usually rules out paternity. For paternity tests, the child should share approximately 50% of their DNA marker bands with each biological parent.

CONCLUSION

DNA profiling is a powerful and impactful technique which can be used in both criminology and paternity testing. Other uses of DNA profiling are found in agriculture, medicine, wildlife conservation and food security.

POSSIBLE EXTENSIONS

1. If your school has a gel electrophoresis kit, carry out a simulated DNA profile using DNA or dyes of different colours to represent fragments of different sizes. (Teachers, refer to your Teachers' Resource Book for instructions).
2. Use an online tool to simulate a DNA profile (e.g. Learn.Genetics Virtual Labs: https://learn.genetics.utah.edu/content/labs)
3. Carry out some research on how DNA evidence has led to the conviction or exoneration of suspected criminals.
4. Create your own DNA profiles for your classmates to compare, for a crime scene investigation, paternity test or any other reason.

INVESTIGATIVE LOG LINK

Go to your Investigative Log and complete your report on investigating patterns using a DNA profile using secondary data.

FOCUS ON ... PCR (HL)

Due to the minute size of a DNA sample available for DNA profiling, many copies of the DNA sample may need to be made. A technique called **PCR** is used to 'amplify' or produce many copies of the selected DNA in a laboratory. Each PCR cycle doubles the amount of DNA – a standard PCR sequence of 30 cycles creates over 1 billion copies, which can be used for analysis.

To carry out PCR on a DNA sample, the following are required:

- **DNA primers:** short strands of DNA that are complementary to the beginning and end of the DNA sequence to be copied – these identify the sequence to be copied.
- **Free nucleotides (dNTP deoxynucleotide triphosphate):** adenine (dATP), thymine (dTTP), guanine (dGTP) and cytosine (dCTP) to produce copies of the DNA.
- **Taq polymerase:** an enzyme isolated from a heat-loving bacterium (optimum 70–75°C). It can add nucleotides to the chain from the primers at the high temperatures used in PCR.
- **pH8 buffer:** stabilises pH at pH 8 to prevent Taq polymerase being denatured.
- **$MgCl_2$ Magnesium chloride:** activates DNA polymerase and stabilises DNA and primers.

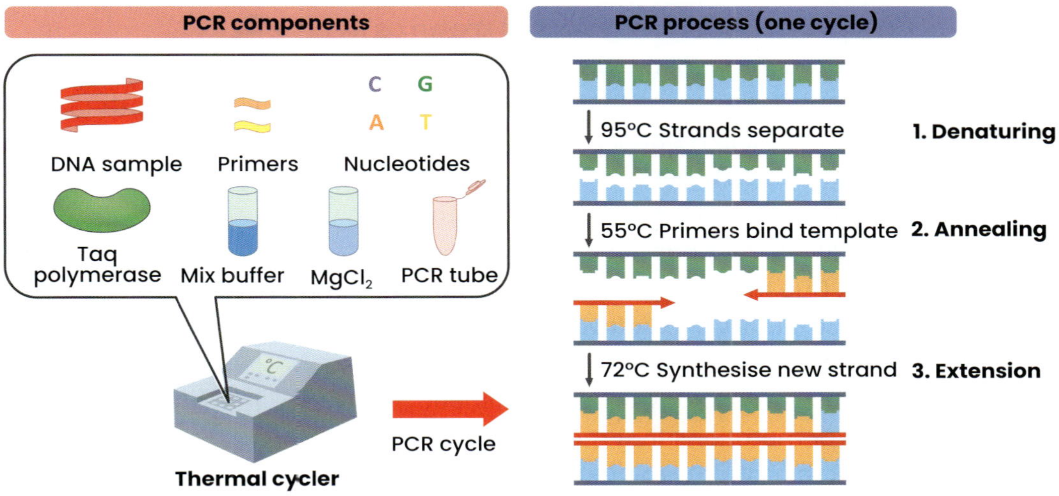

Figure 26.8: PCR copying a DNA sample.

The reaction occurs in three steps:

1. **Denaturation:** The sample is heated to 95°C to separate the two DNA strands.
2. **Annealing:** The DNA sample is cooled to 55°C to allow the primers to attach.
3. **Extension:** Sample is heated to the enzyme Taq polymerase's optimal temperature (70–75°C). Taq polymerase attaches the free nucleotides to form complementary copies of each of the exposed DNA strands. The two DNA molecules formed in a single PCR cycle can be used as templates in further cycles.

DNA samples amplified by PCR can be used:

- in forensics, to help solve crimes
- in paternity cases
- to detect mutations for inherited diseases.

RESEARCH ACTIVITY

Carry out research into another use of DNA profiling, apart from crime scene investigation and paternity testing, and present your findings to your class.

PORTFOLIO LINK

Go to your Learning Portfolio and complete Learning Check 2 to assess your knowledge and understanding of the topic.

Section 3: Genetic Technologies

At the end of this section, you should be able to:

- Understand the value of technology in analysing large quantities of genetic information to identify patterns and search for anomalies.
- Use a genome database to search for alleles that are known to cause (or be responsible for) specific genetic diseases.
- Discuss ethical issues arising from access to and use of genetic data in agriculture, health, industry, reproduction, forensics.
- Discuss ethical issues arising from DNA testing.
- Outline the principle of DNA sequencing data (nucleotide bases). HL

Unpacked from LOs 3.3.4, 3.3.5, 3.3.7

TECHNOLOGY IN ANALYSING GENETIC INFORMATION

The complete set of DNA in one cell of an organism is known as its **genome**. Establishing the exact order of the base pairs, e.g. (ATCGGCTA), in a DNA sample is known as **DNA sequencing**. Genomics is the study of an organism's entire genome including the interactions of genes with each other and their environment. It initially involves obtaining a sample of DNA from the organism, sequencing the DNA, identifying the genes in the sequence and determining their function. Due to the large size of DNA and protein biomolecules, the use of computers and software is required to analyse the enormous amounts of data involved – this is known as **bioinformatics**. Scientists use technology to search genetic information for anomalies or mutations that may give rise to genetic disorders or to look for patterns to predict a gene's function.

- The **genome** is the complete set of DNA (or RNA) in one cell of an organism.
- **Bioinformatics** is an interdisciplinary branch of science that uses computers and software to collect, store and analyse large amounts of biological data found in DNA genetic sequences or protein samples.
- A **genome database** is an online library of sequenced genomes of different species.

Genome databases

In 2003, scientists working in the Human Genome Project announced that they had determined the sequential order of the 3 billion base pairs in human DNA. Today, the genomes of thousands of other organisms have been sequenced including that of the chimpanzee, mosquito, rice, bacteria and yeast. When a genome is sequenced by scientists, they store the data collected in a public online **genome database**, such as:

- GenBank which is maintained by the National Center for Biotechnology Information (NCBI) in the USA
- European Bioinformatics Institute (EMBL-EBI)
- DNA Data Bank of Japan (DDBJ).

This genetic data can be used by other scientists in their research and is constantly being updated.

Case study: Using a genome database

We can use genome databases to search for alleles that are known to cause (or be responsible for) specific genetic diseases. As an example, we will use GenBank to search for the allele that causes the disease cystic fibrosis in humans.

Step 1: Open the website **www.ncbi.nlm.nih.gov**.

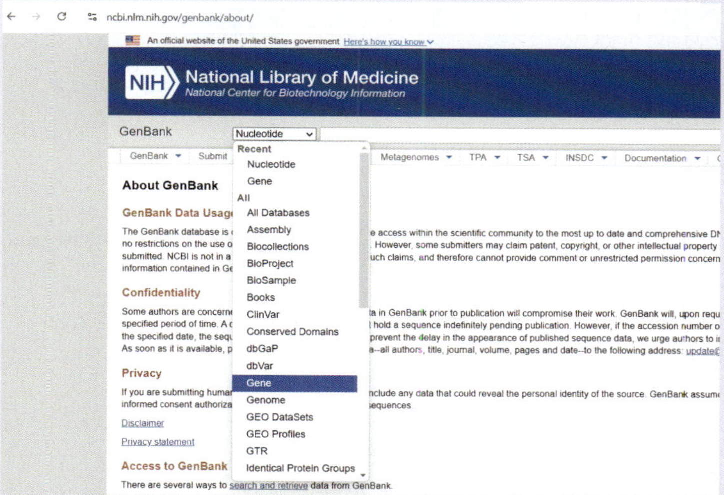

Step 2: Select **Gene** from the drop-down menu. Insert the name of the gene you wish to search for in the search bar, e.g. **CFTR homo sapiens** (cystic fibrosis transmembrane conductance regulator – the gene that causes cystic fibrosis in humans).

Step 3: Select the CFTR-CF transmembrane conductance regulator result to open a page with a summary of the CFTR gene.

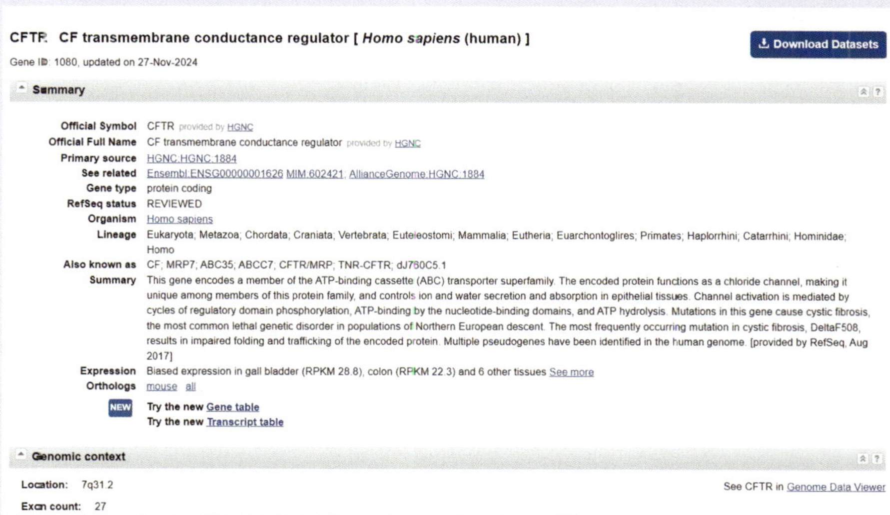

Step 4: Scroll down to the **Genomic context** section of the gene's page. This gives the chromosome location or 'locus' of the gene. The location of the CFTR gene is given as **7q31.2**, which means that it is found on:

- 7: Chromosome 7.
- q: Arm q. Each chromosome has a short arm (called p) and a long arm (called q) which extend from the chromosome's centromere. The gene CFTR is found on the long arm (q) of Chromosome 7.
- 31.2: Each chromosome is divided into regions and bands. The first number, in this case 3, indicates that the gene is found at region 3 of chromosome 7. The second number, in this case 1, indicates that the gene is located at band 1 of region 3. The number after the decimal point indicates that the gene is found at sub-band 2 within band 1.

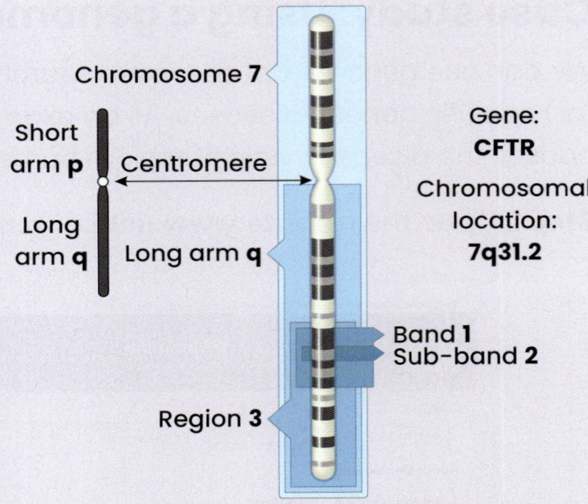

Figure 26.9: Naming of CFTR gene.

Step 5: Scroll further down to the **Genomic regions, transcripts, and products** section. From here, select the **Graphics** tab to view the CFTR gene, which is located at approximately 117 million base pairs on chromosome 7.

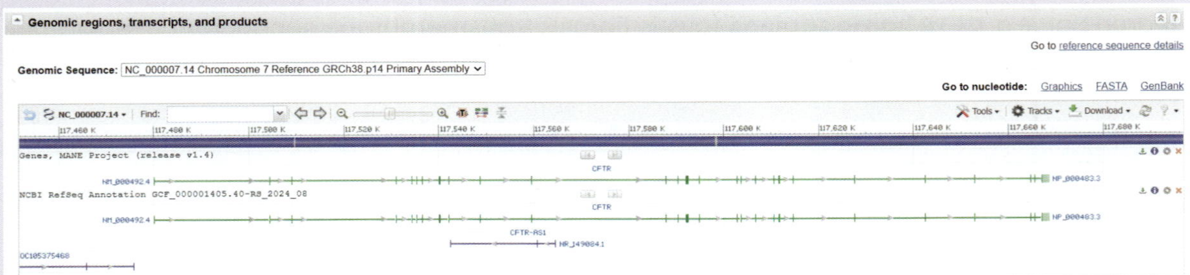

Step 6: Select the **FASTA** format to view the base sequence of the gene.

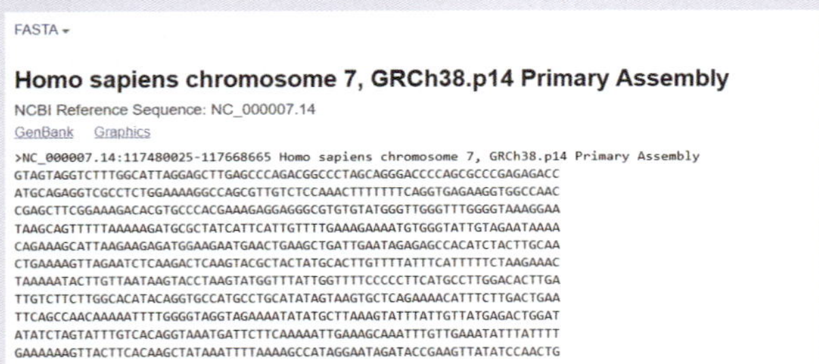

FASTA

Homo sapiens chromosome 7, GRCh38.p14 Primary Assembly

NCBI Reference Sequence: NC_000007.14

GenBank Graphics

```
>NC_000007.14:117480025-117668665 Homo sapiens chromosome 7, GRCh38.p14 Primary Assembly
GTAGTAGGTCTTTGGCATTAGGAGCTTGAGCCCAGACGGCCCTAGCAGGGACCCCAGCGCCCGAGAGACC
ATGCAGAGGTCGCCTCTGGAAAAGGCCAGCGTTGTCTCCAAACTTTTTTTCAGGTGAGAAGGTGGCCAAC
CGAGCTTCGGAAAGACACGTGCCCACGAAAGAGGAGGGCGTGTGTATGGGTTGGGTTTGGGGTAAAGGAA
TAAGCAGTTTTTAAAAAGATGCGCTATCATTCATTGTTTTGAAAGAAAATGTGGGTATTGTAGAATAAAA
CAGAAAGCATTAAGAAGAGATGGAAGAATGAACTGAAGCTGATTGAATAGAGAGCCACATCTACTTGCAA
CTGAAAAGTTAGAATCTCAAGACTCAAGTACGCTACTATGCACTTGTTTTATTTCATTTTTCTAAGAAAC
TAAAAATACTTGTTAATAAGTACCTAAGTATGGTTTATTGGTTTTCCCCCTTCATGCCTTGGACACTTGA
TTGTCTTCTTGGCACATACAGGTGCCATGCCTGCATATAGTAAGTGCTCAGAAAACATTTCTTGACTGAA
TTCAGCCAACAAAAATTTTGGGGTAGGTAGAAAATATATGCTTAAAGTATTTATTGTTATGAGACTGGAT
ATATCTAGTATTTGTCACAGGTAAATGATTCTTCAAAAATTGAAAGCAAATTTGTTGAAATATTTATTTT
GAAAAAAGTTACTTCACAAGCTATAAATTTTAAAAGCCATAGGAATAGATACCGAAGTTATATCCAACTG
```

Other genetic disorders that may be looked up in the same way include:

- **Duchenne muscular dystrophy (DMD):** a genetic disorder associated with the X chromosome. Search for DMD dystrophin [Homo sapiens (human)] in GenBank database.
- **Haemochromatosis:** a genetic disorder associated with chromosome 6 in humans. Search for HFE homeostatic iron regulator [Homo sapiens (human)] in GenBank database.

USE OF GENETIC DATA

Computer programs can be used to examine DNA-sequenced data for many potential uses.

Agriculture

- Access to plant and animal genetic data helps **breeding programmes** and to conserve **biological diversity**.
- Crop farmers can select seeds with genes that offer greater climate resilience, ensuring greater **food security**.
- Dairy farmers can select animals that emit lower levels of greenhouse gases for breeding, **reducing their carbon footprint**.
- Access to genetic data also allows confirmed parentage of animals, ensuring full **traceability within the food chain**.

Figure 26.10: DNA tests allow traceability within the food chain.

Health

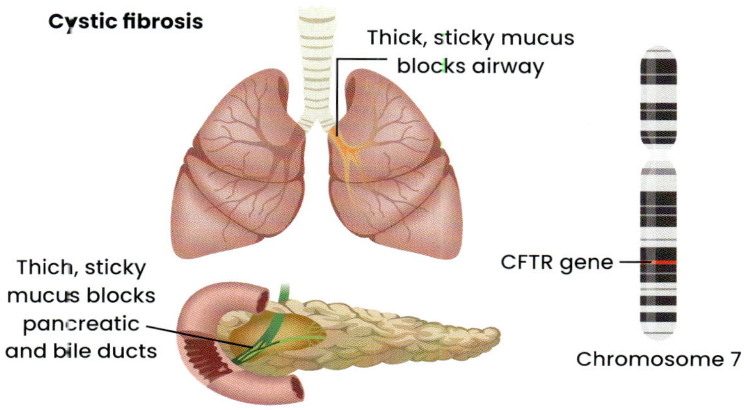

Figure 26.11: Gene tests for cystic fibrosis look for the presence of the CFTR gene.

Gene therapy has been approved in the EU for treating certain blood cancers, spinal muscular atrophy, haemophilia and sickle cell disease.

DID YOU KNOW?

- **DNA tests or gene tests** can be used to analyse a person's DNA for gene mutations that can cause or increase their risk of developing a genetic disorder, e.g. breast cancer, or to see whether they carry a specific mutated gene which they could pass on to their children, e.g. cystic fibrosis. Chromosome tests study chromosomes for changes in the order of genes or for an extra copy of a chromosome that could cause a genetic disorder, e.g. Down syndrome.
- A lot of research is being undertaken into **gene therapy** to treat genetic disorders – it is hoped that missing genes or disease-causing genes may be replaced with normal genes.

Industry

- Genetic data is used by pharmaceutical companies with a view to produce **personalised medicine**, tailored to an individual's genome, to treat diseases.
- Industrial biotechnology companies use genome databases to identify **genes coding for specific enzymes** that are more efficient for use in the **textile**, **detergent** and **food** industries.

Reproduction

- Genetic data is used in **paternity tests**.
- You will learn more about **antenatal screening, genetic screening and postnatal screening for genetic conditions** in Chapter 37.

THINK, PAIR, SHARE

Should a baby have its genome sequenced at birth?

Forensics

Genetic data is used in forensics to identify the **perpetrators of a crime**.

ETHICAL ISSUES RELATED TO DNA TESTING AND ACCESS TO GENETIC DATA

Go to YouTube.com and search for 'Genetic Screening' (7:01) by eusem1eu.

A person's DNA can be tested to see if they have genes that make them susceptible to a particular disease. However, many genetic disorders involve the contribution of several genes. Genetic predisposition alone does not mean that a person will develop a certain disease, e.g. coronary heart disease, as other environmental factors such as diet and exercise also play a role.

There are many ethical questions associated with DNA testing and access to genetic data.

1. **Access to genetic data:** Insurance companies or employers may wish to have access to a potential employee's genetic test. If a person has a gene allele that makes them susceptible to developing a disease, e.g. diabetes, could this person be denied insurance or employment?
2. **Embryo screening:** It is possible to screen or test embryos for many genetic diseases. A positive test result raises many ethical questions for potential parents. Should they continue with the pregnancy or have a termination?
3. **Medical treatment:** Suppose a patient undergoes a genetic test for an allele that causes a particular disease, e.g. muscular dystrophy. Their test results show the presence of an unrelated allele that makes the patient more susceptible to a different disease, e.g. breast cancer. Should their doctor inform them?
4. **Data ownership:** Who owns the data collected from a DNA test? Is it the scientist or company who collected the information, or the person with the genome? There are concerns about consent and control over one's genetic data. Could a person's genetic information be shared with a research or commercial facility without their consent, e.g. a pharmaceutical company researching a new product?
5. **Law enforcement:** DNA profiles can be used to identify perpetrators of a crime. Should law enforcement have access to genetic information of the general population? This raises questions about surveillance and privacy rights.

DNA SEQUENCING

HL

As mentioned earlier, establishing the exact order of the base pairs, e.g. ATCGGCTA, in a set of DNA is known as **DNA sequencing**. Scientists from across the world worked for 13 years to sequence the entire human genome at a cost of $2.7 billion (or €2.5 billion). Scientists working on the Human Genome Project used a method called Sanger sequencing, which has now been largely replaced by next-generation sequencing methods, which are faster, cheaper and more accurate. It now takes a couple of days and costs less than $1,000 to sequence a human genome.

DNA sequencing involves working out the exact order of base pairs (ATCG ... etc.) in a section of DNA. The principle of DNA sequencing is to generate a collection of DNA fragments, each of which is one nucleotide shorter than the next and separated on a gel electrophoresis.

FOCUS ON ... DNA SEQUENCING (HL)

The steps in DNA sequencing are as follows:

1. The enzyme DNA polymerase makes a new copy of the DNA sample by adding nucleotides (A, T, G, C) until it randomly adds a chain-terminating version of one of the nucleotides (dideoxy ddA, ddT, ddG or ddC), which is labelled with an identifiable coloured dye.
2. The above step is repeated, producing lots of fragments of DNA of different lengths, each ending with an identifiable chain-terminating form of a nucleotide (ddA, ddT, ddG or ddC).
3. These DNA fragments are separated based on size by gel electrophoresis and are lined up from longest to shortest. A collection of DNA fragments is generated, each one a nucleotide shorter than the next.
4. A laser identifies the terminating nucleotide of each chain from its coloured dye, revealing the sequence of nucleotides in the original DNA chain.

HL

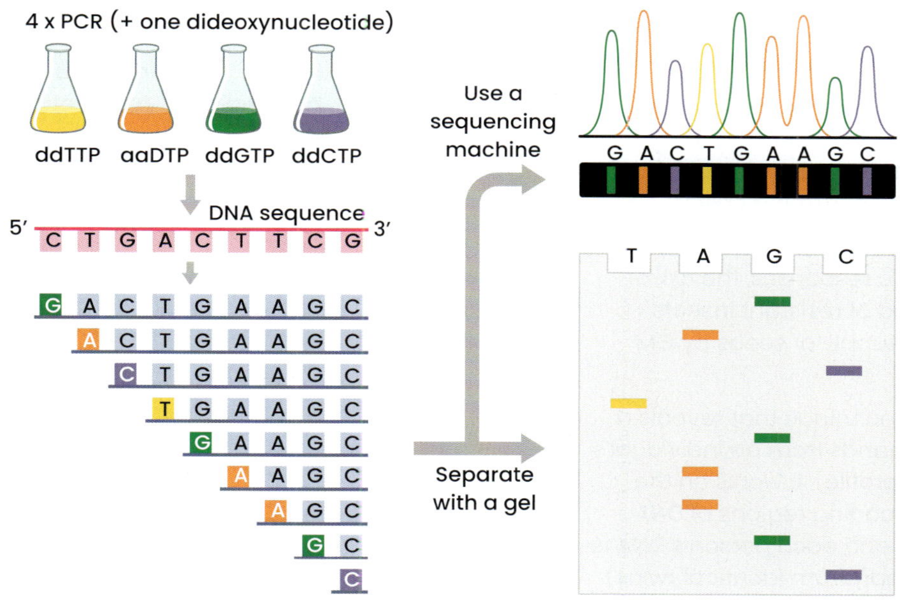

Figure 26.12: DNA sequencing.

Go to YouTube.com and search for 'What Is Sanger Sequencing?' (1:30) by Promega Corporation.

SCIENCE IN SOCIETY – TECHNOLOGY

Methods of DNA sequencing

An emerging method to sequence DNA involves threading single DNA strands through extremely tiny 'nanopores' in a membrane. As each of the DNA bases (A,T,G,C) passes through the pore, they have different effects on an electrical current flowing through the pore. This allows the DNA bases to be identified one at a time. This technique offers many potential advantages as it can be done cheaper and quicker than previous methods.

PORTFOLIO LINK Go to your Learning Portfolio and complete Learning Check 3 to assess your knowledge and understanding of the topic.

Chapter Summary

- **Genetic engineering** works on the principle of altering the DNA of an organism.
- Processes of genetic engineering:
  - **Isolation:** The desired gene is located, e.g. gene for human insulin.
  - **Cutting:** A restriction enzyme is used to cut the desired gene.
  - **Ligation:** DNA ligase enzyme is used to attach the ends of the insulin gene to the cut ends of the bacterial plasmid.
  - **Transformation:** Modified bacterial plasmids are inserted into bacterial cells.
  - **Expression:** The genetically modified bacterial cells reproduce and produce human insulin.
- Examples of genetically modified organisms (GMO) include golden rice and Bt corn.
- **Arguments for GM crops** include that they increase the nutritional value of food, increase crop yield, reduce the need for harmful pesticides and reduce farming costs.
- **Arguments against GM crops** include the possibility of allergic responses, the risk of superweeds, spread of resistant insects and the restricted supply of seeds by GM companies.
- **DNA profiling** is a technique that reveals a unique pattern of bands from an individual's DNA sample (DNA profile). It works on the principle that non-coding regions of DNA are highly variable and each person's DNA is unique to them (apart from identical twins).
- HL Processes in DNA profiling:
  - HL **DNA extraction:** DNA is isolated from the nuclei of cells from a sample such as blood.
  - **Polymerase chain reaction (PCR):** If the amount of DNA is very small, a technique called PCR can be used to copy the DNA up to a million times. HL
  - **Cutting:** Restriction enzymes cut non-coding DNA, producing different sized fragments, unique to the individual. HL
  - **Gel electrophoresis:** DNA fragments are separated on the basis of size using gel electrophoresis. HL
  - **Pattern analysis:** A fluorescent dye and UV light are used to make the DNA fragments visible. The lengths of the DNA fragments are worked out by comparison to those in the DNA ladder. HL
- **Applications of DNA profiling:** paternity test, forensics.
- **DNA sequencing** involves working out the exact order of base pairs (A, T, C, G, etc.) in a section of DNA. HL
- DNA sequencing involves generating a collection of DNA fragments, each of which is one nucleotide shorter than the next and separated on a gel electrophoresis. The last nucleotide in each strand may be identified to determine the sequence of the entire strand. HL
- **Applications of DNA sequencing** include agriculture, health, industry, reproduction and forensics. HL
- Data from a genome sequenced by scientists is stored in **public online genetic databases**, e.g. GenBank, EMBL-EBI or DDBJ.
- You can use a genome database to **search for alleles** that are known to cause (or be responsible for) specific genetic diseases, e.g. search GenBank database for the allele that causes cystic fibrosis in humans.

END OF CHAPTER ASSESSMENT

REMEMBER

1. Define the following terms.
   (a) *Genetic engineering* (b) *DNA profiling*
2. List the processes involved in the following.
   (a) Genetic engineering (b) DNA profiling
3. Give **two** examples of uses of DNA profiling.
4. What is a *plasmid*?
5. Outline the principle of *DNA sequencing*.
6. (a) What is a genome database?
   (b) Name a genome database that you used.
   (c) Name a disease-causing allele that you looked up using this genome database.
7. Describe how access to genetic data is useful in each of the following areas.
   (a) Agriculture (c) Industry (e) Forensics
   (b) Health (d) Reproduction

UNDERSTAND

8. Explain the following terms in relation to genetic engineering.
   (a) Isolation (c) Ligation (e) Expression
   (b) Restriction (d) Transformation
9. Give **two** examples of genetically modified crops.
10. (a) Why is non-coding DNA used in DNA profiling?
    (b) Why are restriction enzymes used in DNA profiling?
    (c) Why is gel electrophoresis used in DNA profiling?
    (d) In gel electrophoresis, why do the DNA fragments move towards the positive terminal when the electric current is turned on?
    (e) What is a DNA ladder and why is it used in gel electrophoresis?
11. Distinguish between the terms *DNA profile* and *DNA profiling*.

APPLY

12. DNA samples from a crime scene were analysed by a forensic scientist. DNA samples were taken from three people who were suspected of carrying out the crime. The different DNA profiles collected are shown below. Analyse the DNA profiles to determine which suspect is guilty of carrying out the crime.

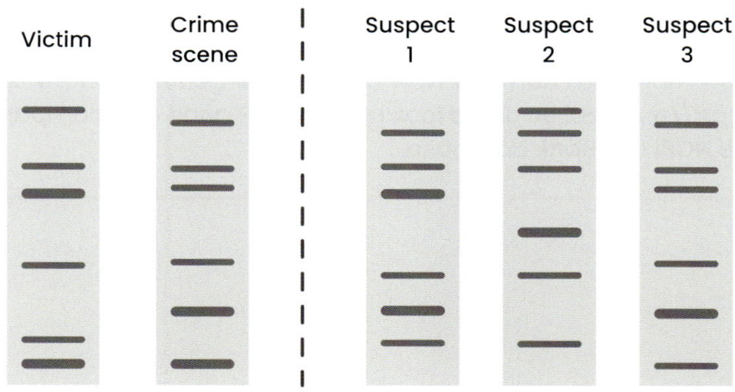

13. A student carried out gel electrophoresis on some DNA samples (A–D) and a DNA ladder. The DNA profile collected is shown below.

(a) Using the DNA ladder, determine the size of the DNA fragment(s) that formed from each of the DNA samples A–D.

(b) At which terminal of the gel electrophoresis chamber are the smallest DNA fragments found? Explain why this is the case.

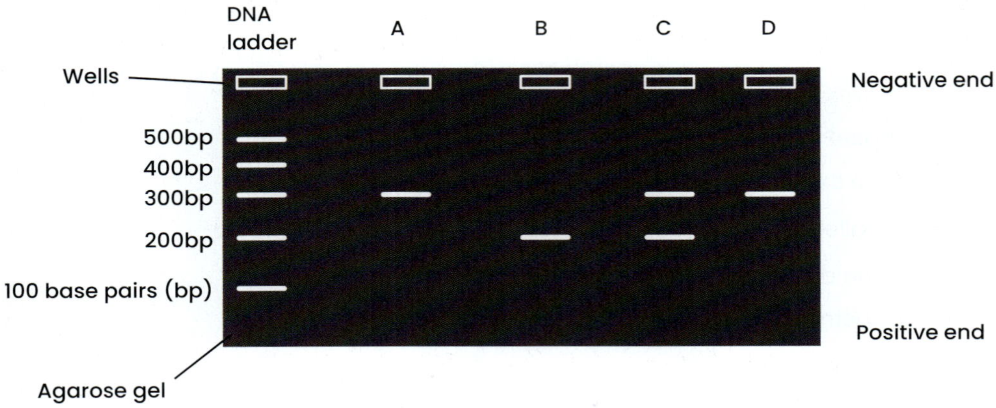

14. Based on your knowledge of using the NCBI GenBank database to look up the human cystic fibrosis gene, use the database to look up another gene responsible for causing disease, e.g. Duchenne muscular dystrophy, in humans. Summarise the results you obtain.

ANALYSE

HL 15. Compare and contrast DNA profiling and DNA sequencing.

16. Outline how each of the following could use genetic data.

(i) A farmer deciding on which barley variety to feed his sheep

(ii) A doctor treating a cancer patient

EVALUATE

17. Carry out a classroom debate to discuss the arguments for and against genetically modified crops. Divide into four teams to represent the following interest groups: consumers, farmers, environmentalists and GM crop manufacturers.
Present your arguments for/against GM crops on behalf of each of these groups.

CREATE

18. You have used the NCBI GenBank database to look up the CFTR gene – the gene that causes cystic fibrosis in humans. Now use another genome database (e.g. European EMBL-EBI database, https://www.ebi.ac.uk) to look up the CFTR gene and compare the results to those you obtained from the NCBI GenBank database.

UNIT 9

HUMANS – FUELLING AND FILTERING THE BODY

CHAPTERS

STRANDS

Unifying Strand: Nature of Science

Strand 2: Structures and Processes of Life

Strand 3: Interactions of Life

27 The Human Body

LEARNING OUTCOMES: 2.4, 2.5, 2.6

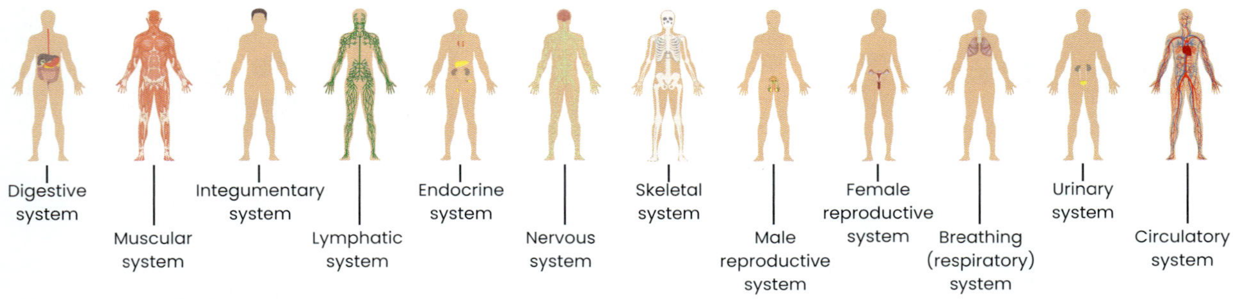

Figure 27.1: The systems of the human body.

The human body is a sophisticated living machine, comprising approximately 36 trillion cells forming at least 200 different types of tissue. These tissues form 78 organs, which in turn form 11 organ systems. The systems interact in many ways and no system is more important than the others.

In this chapter, we will provide a short introduction to each of the 11 organ systems. There is no assessment in this chapter, but you can use the Learning Portfolio to assess your understanding of the material covered.

THE BREATHING SYSTEM

The breathing system (sometimes called the respiratory system) enables **gas exchange** between the body and the air around us. The lungs are the primary organs involved in this process. They take in oxygen from inhaled air and exhale carbon dioxide, a waste product of **cellular respiration**. The diaphragm, a dome-shaped muscle located between the chest and abdomen, plays a vital role in the breathing process by contracting and relaxing to create pressure changes that draw air into and out of the lungs. We will learn more about the breathing system in Chapter 28.

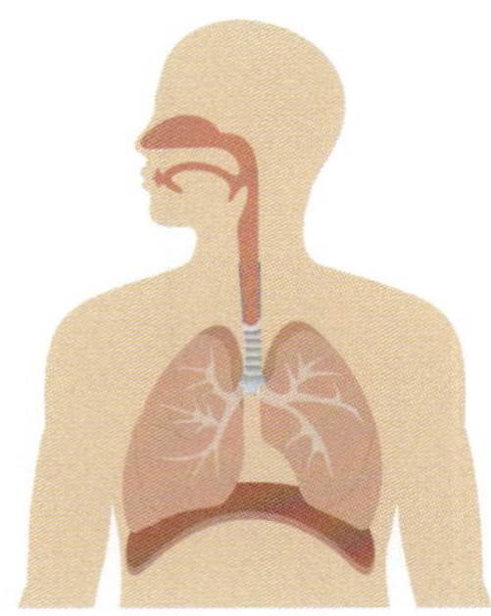

Figure 27.2: The breathing (respiratory) system enables gas exchange between the body and the air.

THE URINARY SYSTEM

The urinary system removes waste products and excess water from the body. The **kidneys**, two bean-shaped organs located on either side of the spine, filter waste products and water from the blood, producing urine. Urine flows from the kidneys through **ureters**, tubes that connect to the **bladder**, a muscular sac that stores urine until it is eliminated from the body through the **urethra**. We will learn more about the urinary system in Chapter 29.

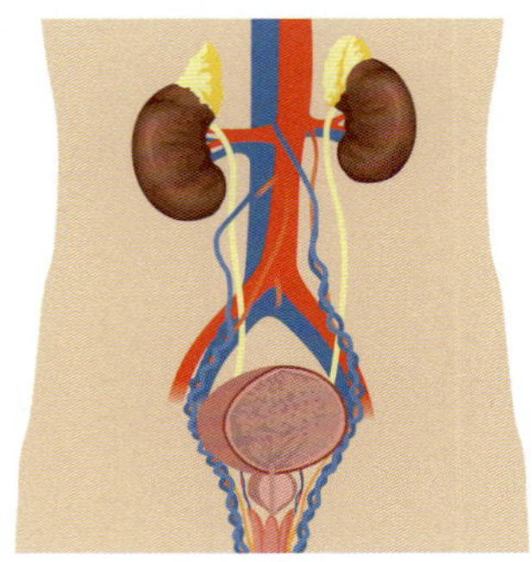

Figure 27.3: The urinary system removes waste products and excess water from the body.

THE DIGESTIVE SYSTEM

The digestive system breaks down food into its basic building blocks for **absorption** into the bloodstream. This system includes organs like the mouth, **stomach**, **intestines**, **liver** and **pancreas**, along with digestive **enzymes** and other secretions. The digestive process begins in the mouth, where food is **mechanically** broken down by **chewing** and mixed with saliva containing enzymes that start carbohydrate digestion. In the stomach, food is further **churned** and broken down by strong **acids** and enzymes. The small intestine is the primary site for nutrient **absorption**. The liver and pancreas play crucial roles in digestion by producing bile (aids in fat digestion) and digestive enzymes, respectively. Undigested materials are stored in the **rectum** before being eliminated through the **anus**. We will learning more about the digestive system in Chapter 30.

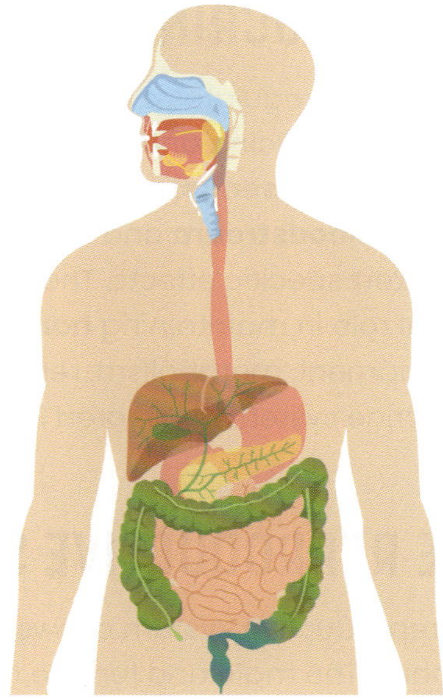

Figure 27.4: The digestive system breaks down food into its basic building blocks.

THE CIRCULATORY SYSTEM

The circulatory system is responsible for transporting **blood** throughout the body. Blood, a fluid tissue, carries **oxygen**, **nutrients**, **waste products**, **hormones** and **immune cells**. The heart acts as a muscular pump, propelling blood through a closed network of **arteries**, **veins** and **capillaries**. Arteries carry oxygen-rich blood away from the heart to tissues throughout the body. Veins return oxygen-depleted blood back to the heart. Capillaries, microscopic blood vessels, facilitate the exchange of gases, nutrients and waste products between the blood and tissues. Chapter 31 focuses on the circulatory system.

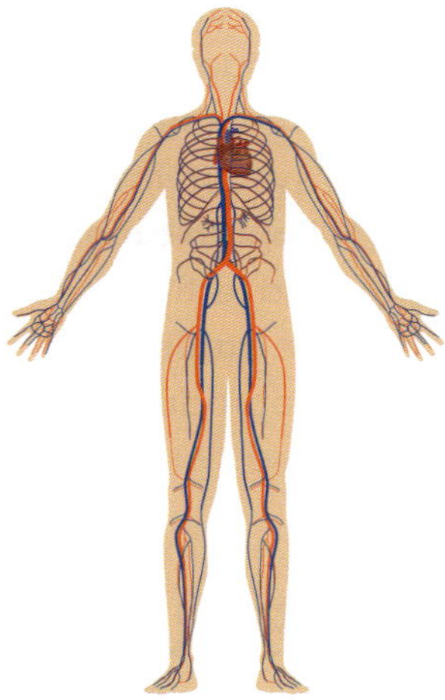

Figure 27.5: The circulatory system transports blood throughout the body.

THE NERVOUS SYSTEM

The nervous system acts as the body's control centre, gathering information from the environment and internal organs and choosing appropriate responses. It consists of the **brain**, **spinal cord** and a vast network of **nerves** that extends throughout the body. The brain is the most complex organ in the human body, responsible for consciousness, thought, memory and co-ordination. The spinal cord serves as a link, transmitting signals between the brain and the rest of the body. Nerves carry electrical impulses throughout the body, gathering information from our sense organs (eyes, ears, etc.), causing muscle movement and regulation of organ function. We will explore the nervous system in more detail in Chapter 33.

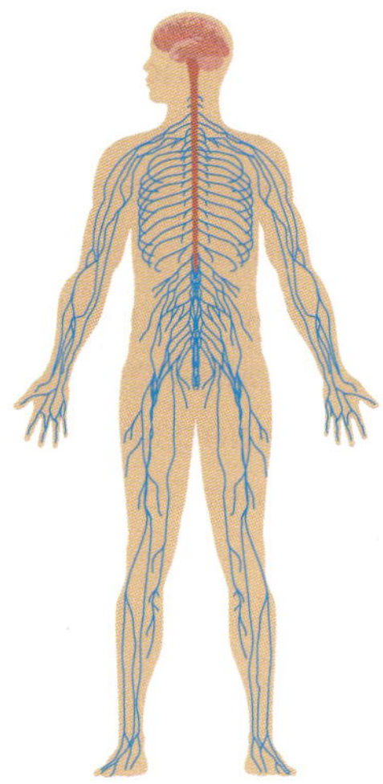

Figure 27.6: The nervous system acts as the body's control centre.

THE ENDOCRINE (HORMONAL) SYSTEM

This system regulates a wide range of processes in the body by producing and secreting **hormones**. Hormones are **chemical messengers** released by **glands**, small organs, into the **bloodstream** and travel to target organs, where they exert specific effects. The endocrine system plays a crucial role in maintaining **homeostasis**, regulating growth, development, metabolism, reproduction and mood. The endocrine system is explored in greater detail in Chapter 34.

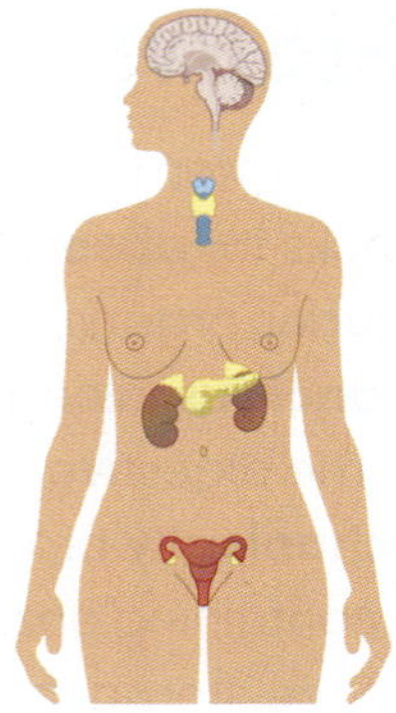

Figure 27.7: The female endocrine system. The endocrine system controls hormones.

THE REPRODUCTIVE SYSTEM

The reproductive system allows humans to create new humans. The male and female reproductive systems differ in structure and function but share the common goal of producing offspring. The male reproductive system consists of organs including the **testes** (which produce **sperm cells**), the **sperm ducts** (tubes that transport sperm), glands that produce seminal fluid and the **penis** (the organ for sexual intercourse and sperm delivery). The female reproductive system includes the **ovaries** (which produce **egg cells**), **fallopian tubes** (where fertilisation occurs), the **uterus** (where a fertilised egg implants and develops into a foetus) and the **vagina** (the canal for childbirth and sexual intercourse). Reproduction is studied in more detail in Chapters 36 and 37.

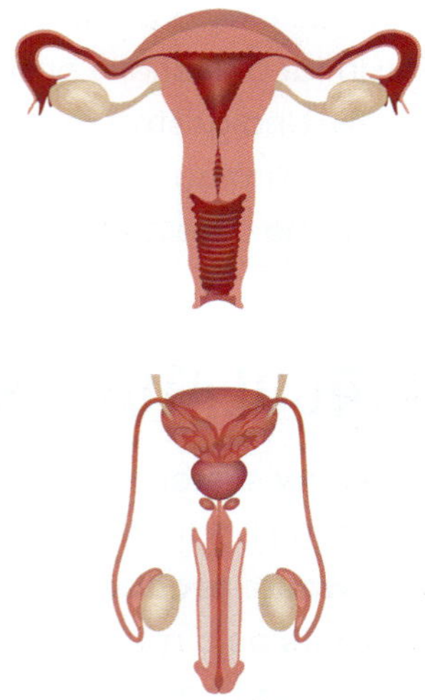

Figure 27.8: The female and male reproductive systems differ, but can work together to produce offspring.

THE SKELETAL SYSTEM

The skeletal system provides the body with structural **support** and a rigid framework for muscles to attach to, enabling **movement**. It also **protects** vital organs like the brain, heart and lungs from injury. The skeletal system is composed of **bones**, **cartilage** and **ligaments**. Bones are where blood cells are produced and they also store vital mineral nutrients like calcium and phosphorous. Cartilage acts as a cushion at joints, allowing for smooth movement. Ligaments are strong connective tissues that bind bones together, providing stability at joints. We will explore this further in Chapter 35.

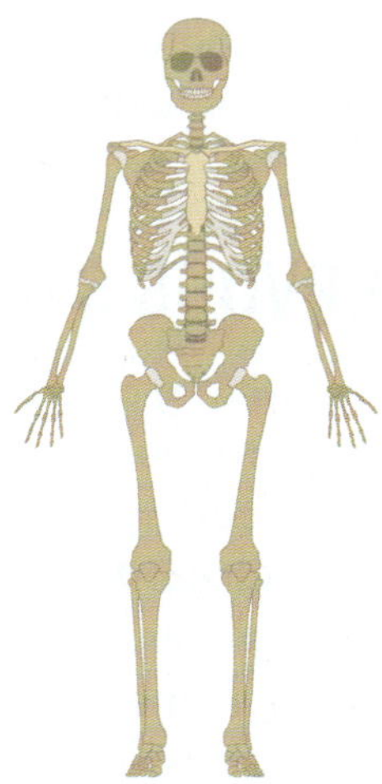

Figure 27.9: The skeletal system provides the body with structural support.

THE MUSCULAR SYSTEM

Muscles work with the skeletal system to enable **movement**, maintain **posture** and generate **heat**. The human body has three main types of muscle tissue: **skeletal muscle** (responsible for voluntary movements), **smooth muscle** (found in organs for involuntary control) and **cardiac muscle** (found only in the heart). Skeletal muscles are attached to bones and contract in response to signals from the nervous system. Smooth muscle lines the walls of organs like the stomach and intestines, propelling food along the digestive tract. Cardiac muscle is unique as it contracts rhythmically without conscious control, ensuring continuous blood flow throughout the body. The muscular system will also be studied further in Chapter 35.

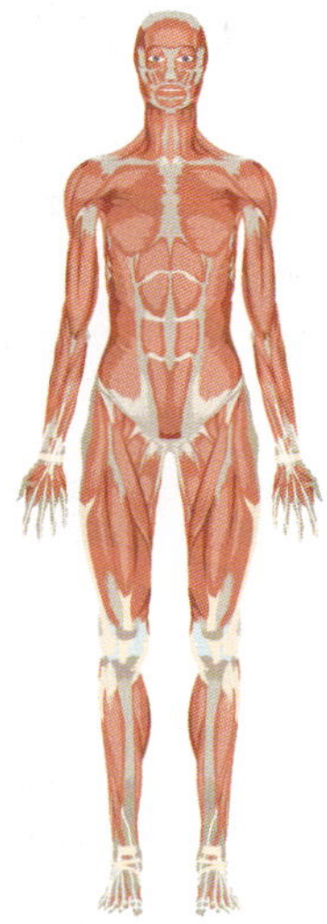

Figure 27.10: The muscular system enables movement and generates heat.

OTHER SYSTEMS

The following organ systems are not studied in detail in Senior Cycle Biology and have no dedicated chapters later, but they do interact with the other systems so it is important to be aware of them. They're also important to consider when evaluating your own health.

The integumentary system

Forming the body's external covering, this system consists of the **skin**, **hair** and **nails**. The skin serves as a **barrier** against the environment, **protecting** underlying tissues from physical and chemical injury, as well as **regulating body temperature** through sweating and blood flow adjustments. Hair and nails offer additional protection and enhance sensory perception.

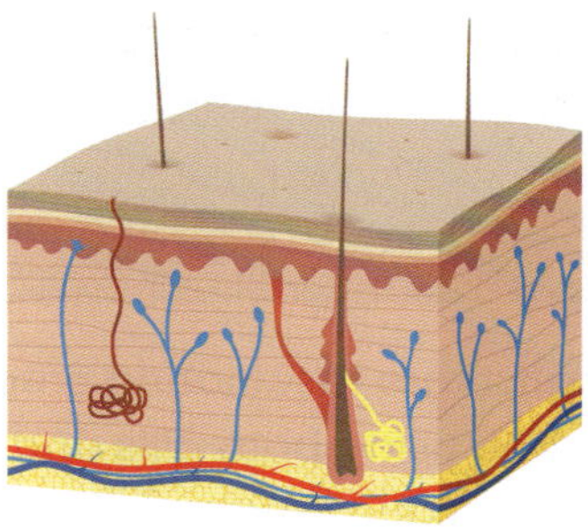

Figure 27.11: The integumentary system consists of the skin, hair and nails.

The lymphatic system

The lymphatic system, a connected system to the circulatory system, comprises a network of **lymphatic vessels** and tissues. It plays a vital role in nutrient transportation, fluid collection and immunity. It transports **lymph**, a fluid containing immune cells, throughout the body. **Lymph nodes**, bean-shaped structures scattered along the lymphatic vessels, act as filters, trapping pathogens (disease-causing organisms) and cell debris. White blood cells within these nodes attack and destroy these invaders, helping to defend the body against infection.

Note: The immune system, our means of protecting against infection and disease, isn't a stand-alone system and involves your circulatory, lymphatic, digestive and integumentary systems collaborating to fight disease. Chapter 32 takes a closer look at infectious disease and immune defence, exploring how these systems interact to protect your body from invaders.

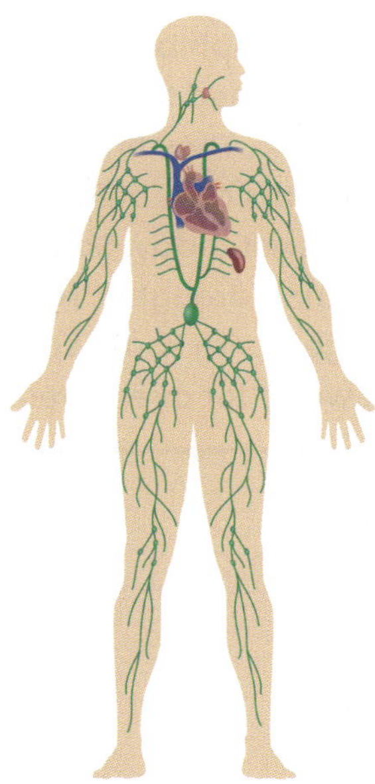

Figure 27.12: The lymphatic system plays a vital role in nutrient transportation, fluid collection and immunity.

PORTFOLIO LINK

Go to your Learning Portfolio and complete the Learning Check to assess your knowledge and understanding of the topic so far.

28 The Breathing System

LEARNING OUTCOMES: 2.6.7, 2.6.8

EYE-OPENER

Lungfish are fascinating creatures which breathe using both gills and primitive lungs. These unique creatures breathe air with a lung-like organ, a handy adaptation for when they find themselves in drier conditions. Lungfish are considered an important evolutionary link; their lobed fins resemble amphibian limbs, hinting at how fish transitioned to land, evolving into early amphibians. They first evolved over 400 million years ago, before life on land began, and still exist today. Some lungfish burrow in mud during droughts, relying on their lungs for oxygen, with some species able to survive out of water for over a year.

At the end of this chapter, you should be able to:

- Describe the anatomy of the breathing system.
- Outline how the body protects the lungs so they can carry out their function effectively.
- Describe the process of inhalation and exhalation, with reference to the ribs, intercostal muscles and diaphragm.
- Describe how the rate of breathing is controlled.
- Describe how gas exchange takes place in the lungs.

LINKS TO:

Chapter 18	**Movement Through Cell Membranes**
Chapter 23	**Respiration**
Chapter 31	**The Circulatory System**

SUBJECT VOCABULARY

upper respiratory tract, pharynx, larynx, epiglottis, lower respiratory tract, diaphragm, intercostal muscles, inhalation, exhalation, lung capacity, medulla oblongata, gas exchange, alveoli

Each of our cells requires oxygen for aerobic respiration – the efficient release of energy from our food. When carbohydrates are broken down, carbon dioxide is produced, which needs to be removed from the cell and later the body.

Our body utilises two systems to supply oxygen and remove carbon dioxide from our cells. Firstly, our breathing system consists of specialised organs designed to take oxygen from the air around us and move it into our blood; at the same time, carbon dioxide leaves the blood in our lungs and is removed from the body. Secondly, the circulatory system (which we will explore more in Chapter 31) transports the oxygen absorbed in our lungs to each one of our cells within the blood, picking up the waste carbon dioxide as it passes by.

In this chapter, we will explore the anatomy of the breathing system and how each of its organs plays a vital role in supplying the body's cells with oxygen. We will look at how oxygen and carbon dioxide are exchanged within the main organs of the breathing system – the lungs – and how the brain ensures it is all working efficiently.

Section 1: Anatomy of the Breathing System

At the end of this section, you should be able to:

- Describe the location of the lungs in the human body.
- List the major organs of the human respiratory system.
- Differentiate between the upper and lower respiratory tracts and identify key structures in each.
- Describe the pathway air takes as it travels into and out of the lungs.
- Explain the role of the diaphragm in breathing.

Unpacked from LO 2.6.7

You breathe, on average, 12 times per minute, although this can increase in frequency and intensity if you're exercising. Breathing is a process involving the various organs of the **respiratory tract**, with air flowing from the mouth and nose, into a long tube called the **trachea** and then into one of two **lungs** before flowing back out the way it came in. The lungs are enclosed within the **chest** (thoracic) **cavity** and protected by a set of 12 **ribs**.

The breathing system can be divided into two main regions: the **upper respiratory tract** and the **lower respiratory tract**.

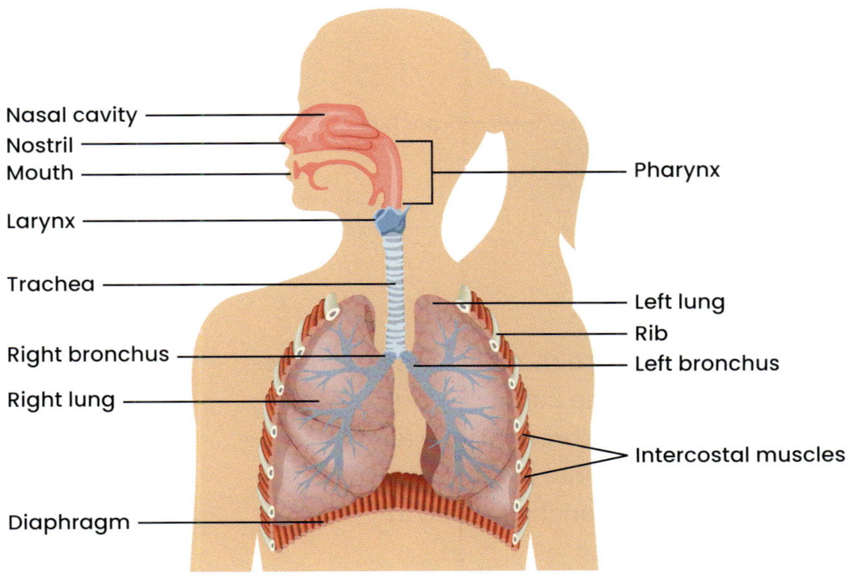

Figure 28.1: The breathing system.

UPPER RESPIRATORY TRACT

The **upper respiratory tract** acts as the entryway for air. As the air flows through the nose (nasal cavity), it is filtered, warmed and humidified before it reaches the lungs. This region comprises the nose, lined with a mucus membrane that traps dust and pathogens. Tiny hairs called **cilia** help propel these trapped particles upwards for elimination through sneezing or blowing your nose.

The **pharynx**, commonly referred to as the throat, is a muscular passageway that connects the nasal cavity and the mouth to the lower airways. The **larynx**, also known as the voice box, home to the vocal cords, is responsible for producing sound. A small flap of tissue called the **epiglottis** guards the entrance to the lower respiratory tract during swallowing, preventing food and drink from entering the **trachea**.

- The **upper respiratory tract** consists of the nose, pharynx and larynx.
- The **pharynx** is another name for the throat.
- The **larynx** is the voice box – a region where vocal cords are found allowing us to speak.
- The **epiglottis** is a flap of tissue that prevents food and drink entering the respiratory tract.
- The **lower respiratory tract** consists of the trachea, bronchi, bronchioles and lungs, including the alveoli.

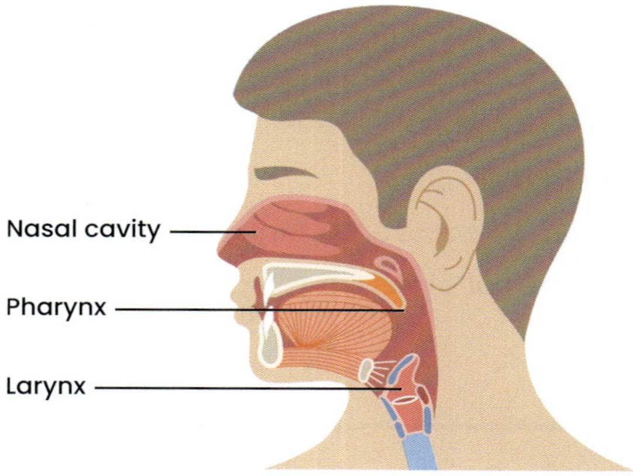

Figure 28.2: Upper respiratory tract.

LOWER RESPIRATORY TRACT

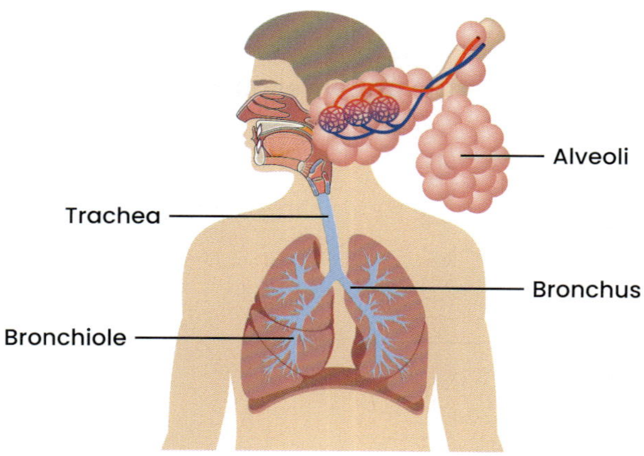

Figure 28.3: Lower respiratory tract.

The **trachea**, or windpipe, is a sturdy tube made of cartilage rings that provides a clear pathway for air. The cartilage rings keep the airway open. The trachea splits into two **bronchi**, one leading to each lung. These bronchi further divide into smaller and smaller tubes called **bronchioles**. Finally, the bronchioles end in microscopic air sacs called **alveoli**.

The alveoli are the site of gas exchange, with a large surface area and surrounded by tiny capillaries where oxygen diffuses into the bloodstream and carbon dioxide diffuses out. We will look at this in more detail later in the chapter.

The lower respiratory tract is located in the chest cavity and is surrounded by 12 pairs of protective ribs. Between the ribs are the **intercostal muscles**, which, along with the large muscular **diaphragm**, a dome-shaped muscle separating the chest and abdomen, control inhalation (breathing in) and exhalation (breathing out).

- The **diaphragm** is a dome-shaped muscle separating the chest from the abdomen.
- The **intercostal muscles** are found between the ribs.

FOCUS ON ... PROTECTING THE LUNGS

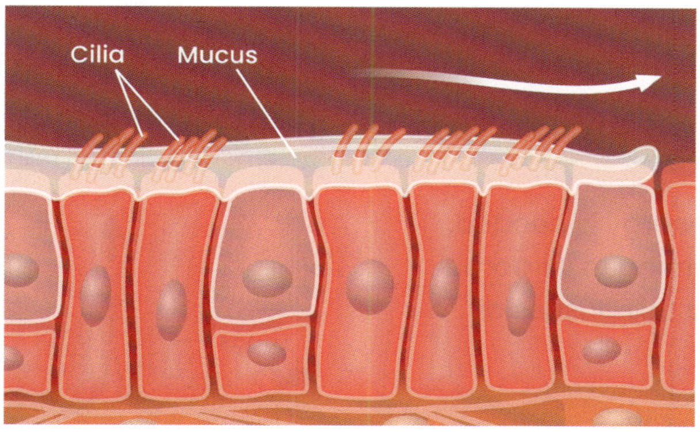

Figure 28.4: Cilia remove mucus from the lungs.

The ribcage acts like a bony shield, protecting the lungs from external impacts. Meanwhile, the **pleura**, a double membrane lining the lungs and chest cavity, produces lubricating fluid to reduce friction during breathing. Inside, the airways are lined with tiny hair-like projections called **cilia** that trap dust and particles. **Mucus** produced by specialised cells lining the respiratory tract captures these impurities, and the cilia sweep them upwards towards the throat to be swallowed or expelled through coughing. Wandering white blood cells, called macrophages, patrol the lungs capturing and eating bacteria, viruses or fungal spores.

Go to YouTube.com and search for 'Respiratory System Introduction – Part 1 (Nose to Bronchi) – 3D Anatomy Tutorial' (3:36) by AnatomyZone.

PORTFOLIO LINK **Go to your Learning Portfolio and complete Learning Check 1 to assess your knowledge and understanding of the topic so far.**

Section 2: Inhalation and Exhalation

At the end of this section, you should be able to:

- Outline the role of the diaphragm and intercostal muscles in inhalation and exhalation.
- Describe the changes in the chest volume during inhalation and exhalation.
- Describe how the rate of breathing is controlled.
- Outline what is meant by 'lung capacity'.
- Outline the role of carbon dioxide concentration as a controlling factor in the human breathing system. HL

Unpacked from LOs 2.6.7, 2.6.8

The diaphragm plays a central role in inhalation and exhalation, particularly during 'silent breathing' (when the body is resting or sleeping).

Inhalation

During inhalation, the diaphragm contracts (flattening out), increasing the chest cavity volume. This increase in volume draws air into the lungs. The lung tissue is highly elastic, meaning it can stretch as it fills with air.

Muscle contraction requires energy, so we say that inhalation is an **active process**.

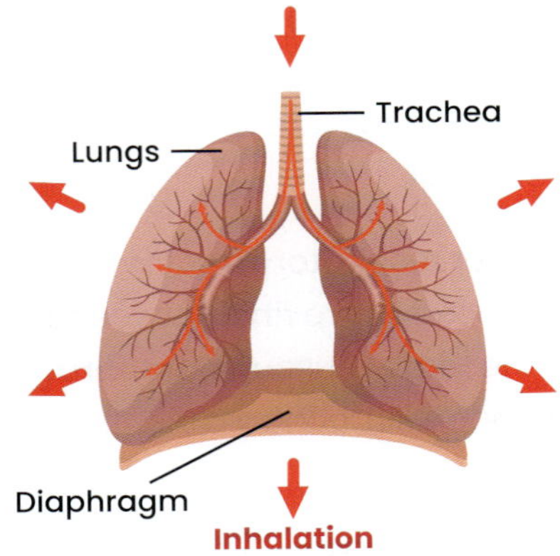

Figure 28.5: Pressure changes during inhalation.

Exhalation is the passive process of removing air from the lungs.

Exhalation

Exhalation is a passive process, meaning no energy is needed. During exhalation, the diaphragm relaxes, returning to its dome shape. The volume of the chest decreases and this pushes air back out through the airways.

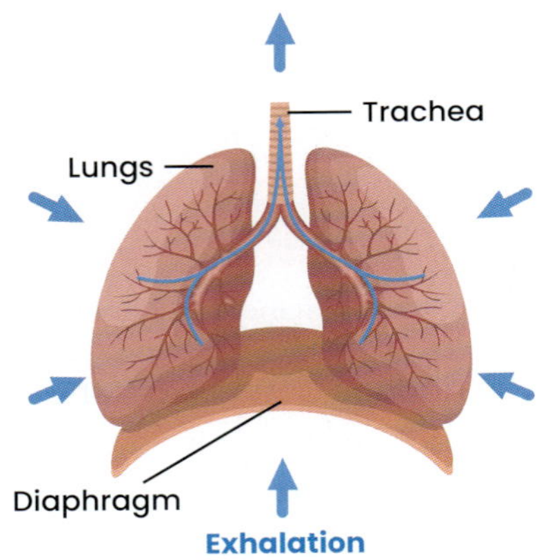

Figure 28.6: Pressure changes during exhalation.

Intercostal muscles

The muscles between the ribs can be recruited for deeper breaths or forced exhalation, such as during exercise or coughing. During 'silent breathing', the diaphragm is sufficient for making the necessary chest volume changes. However, during exercise, the intercostal muscles activate.

There are two sets of intercostal muscles – external and internal. During deep inhalation, the external intercostal muscles contract to lift the chest up and out, further expanding the chest cavity. During forced exhalation, the internal intercostal muscles contract, pulling the ribcage down and inward, further expelling air from the lungs.

Lungs are highly elastic and can expand to hold up to eight litres of air. The maximum amount of air the lungs can hold is called **lung capacity**.

Lung capacity is the maximum amount of air the lungs can hold.

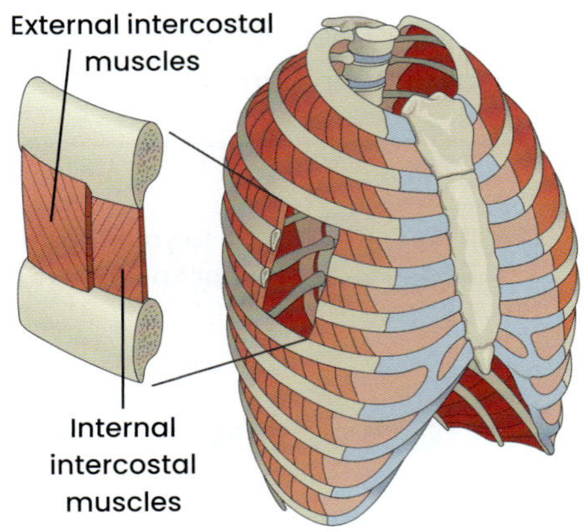

Figure 28.7: Internal and external intercostal muscles.

BREATHING RATE

While breathing is caused by contractions of the diaphragm and the intercostal muscles, ultimately it is the brain that controls how quickly and deeply you breathe. The **medulla oblongata**, found at the base of the brain, controls the rate of breathing. It forms part of our autonomic nervous system – where lots of our unconscious actions are controlled. The medulla oblongata sends regular impulses to the diaphragm, approximately 12 times per minute, instructing it to contract and relax. This is our resting breathing rate.

The **medulla oblongata** is a region of the brain that controls unconscious actions, e.g. breathing.

Carbon dioxide as a controlling factor

HL

During exercise, our cells, particularly our muscle cells, require more oxygen for their increased energy needs. As they burn more food, they increase the level of carbon dioxide (CO_2) in the blood. **This increase in carbon dioxide is detected by the medulla oblongata and it sends more frequent impulses to the diaphragm and intercostal muscles, increasing our breathing rate and volume.** We say that carbon dioxide is a controlling factor in our breathing rate.

CO_2 concentration is also the controlling factor for stomatal opening and closing, the tissues responsible for gas exchange in plants.

DID YOU KNOW?

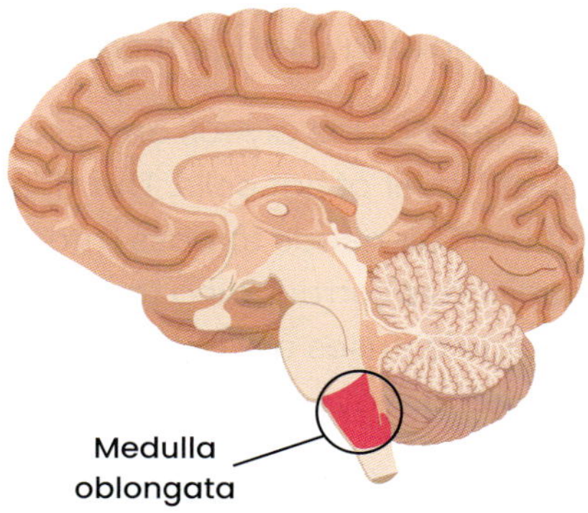

Figure 28.8: Location of medulla oblongata.

OTHER CONTROLLING FACTORS

Several other factors influence the rate and depth of breathing, including:

- **Physical activity:** Exercise increases the body's demand for oxygen, leading to faster and deeper breaths.
- **Metabolic rate:** Higher metabolic rate requires more oxygen, resulting in increased breathing.
- **Stress:** Anxiety or stress can cause rapid, shallow breathing.
- **Altitude:** At higher altitudes, the air is thinner, leading to faster breathing to compensate for reduced oxygen availability.

Breathing is not completely under unconscious control though; we also have some conscious control over our breathing when we need it. We can hold our breath, perhaps if we're swimming, or control the flow of air when talking or singing. Ultimately, however, the brain will take over if carbon dioxide levels increase too much.

PORTFOLIO LINK **Go to your Learning Portfolio and complete Learning Check 2 to assess your knowledge and understanding of the topic so far.**

Section 3: Gas Exchange

At the end of this section, you should be able to:

- Outline the structure of the alveoli.
- Detail the exchange of gases at the alveoli.
- Describe the role of diffusion in gas exchange.

Unpacked from LO 2.6.7

The core function of the breathing system is to facilitate **gas exchange**. Oxygen, from the atmosphere, is exchanged with carbon dioxide, produced by all cells during respiration. The blood, along with the heart and blood vessels, plays a central role in transporting the oxygen to the cells and the carbon dioxide to the lungs; in the blood, red blood cells carry oxygen while carbon dioxide is carried in the blood plasma. We will explore this role in more detail in Chapter 31.

Gas exchange takes place at the alveoli, where oxygen is exchanged for carbon dioxide.

ALVEOLI

Alveoli are tiny air sacs in the lungs where gas exchange take place. Each lung contains up to 400 million alveoli, providing the large surface area needed for efficient gas exchange.

Alveoli form in clusters at the end of each bronchiole, as illustrated in the diagram, and are surrounded by a network of thin blood vessels called **capillaries**. Capillaries are only one cell thick, meaning materials can diffuse easily in and out of the blood flowing through them.

The walls of the alveoli are also incredibly thin, composed of a single layer of flat cells. This thinness is crucial. Oxygen from inhaled air needs to pass through these cells and into the capillaries surrounding the alveoli. At the same time, carbon dioxide diffuses out of the capillary and into the alveolus. **Diffusion** plays a crucial role in gas exchange.

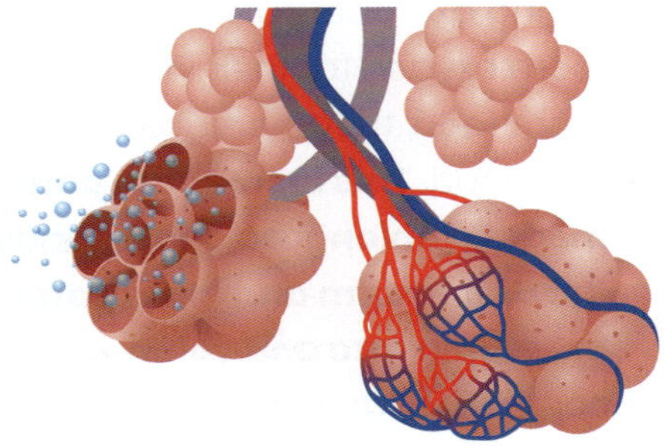

Figure 28.9: A cluster of alveoli.

Alveoli are tiny air sacs in the lungs; they are the site of gas exchange.

FOCUS ON ... DIFFUSION

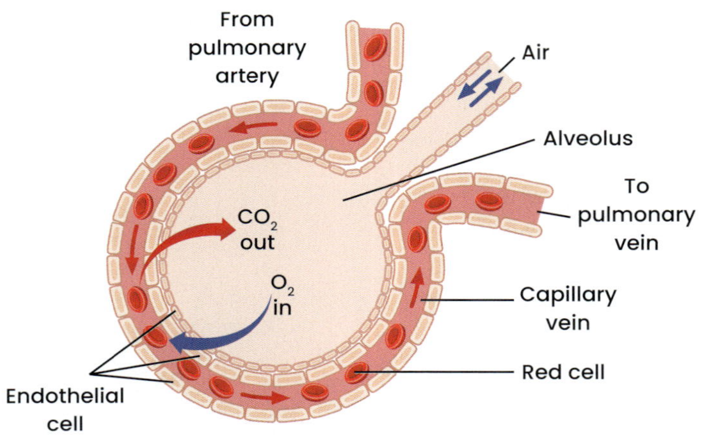

Figure 28.10: Diffusion of gases at the alveolus.

As we learned in Chapter 18, diffusion is the movement of molecules from an area of higher concentration to an area of lower concentration.

When we inhale each alveolus expands like a balloon. The concentration of oxygen is higher inside the alveolus than in the blood flowing through the capillaries. So, oxygen molecules readily diffuse across the thin cells of the alveolus and into the bloodstream.

Meanwhile, carbon dioxide travels in the opposite direction. Its concentration is higher in the blood than in the air within the alveolus. Therefore, carbon dioxide diffuses from the blood into the alveolar space, to be exhaled with the next breath.

Diffusion is a passive process; no energy is required for gas exchange.

RESEARCH ACTIVITY

Carry out some research on the effects of smoking and vaping on the respiratory system. How do they affect the various parts of the respiratory tract, including the alveoli?

Go to YouTube.com and search for 'How do lungs work? – Emma Bryce' (3:21) by TED-Ed.

SCIENCE IN SOCIETY – HEALTH

Asthma

Asthma is a chronic respiratory condition characterised by inflamed and overly sensitive airways. During an asthma attack, the muscles surrounding the bronchioles, the small airways leading to the alveoli, tighten. This narrowing, along with increased mucus production in the airways, significantly reduces airflow.

Due to the constriction of the airways, less air reaches the alveoli, the tiny air sacs responsible for gas exchange. This disrupts the exchange of oxygen and carbon dioxide, leading to shortness of breath, wheezing, coughing and chest tightness.

People with asthma often have airways that are more sensitive to triggers like allergens, dust or cold air.

Inhalers are used to both treat and prevent asthma attacks. During an asthma attack, quick-relief inhalers contain **bronchodilators**, medications that relax the muscles surrounding the bronchioles. This allows the airways to open up, easing airflow and relieving the immediate symptoms of an attack. Control inhalers contain medications that reduce inflammation in the airways. This helps prevent future asthma attacks and improve overall breathing control.

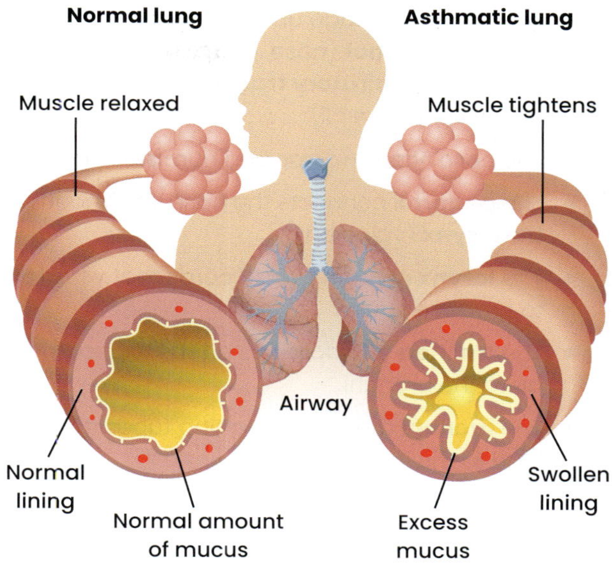

Figure 28.11: Comparing the airway in a healthy lung and an asthmatic lung.

Figure 28.12: Quick-relief inhaler.

PORTFOLIO LINK

Go to your Learning Portfolio and complete Learning Check 3 to assess your knowledge and understanding of the topic so far.

Go to YouTube.com and search for 'How does asthma work? – Christopher E. Gaw' (5:09) by TED-Ed.

Chapter Summary

- The function of the breathing system is to take in oxygen from the air and expel carbon dioxide.
- The system is made up of the **upper respiratory tract** (nose, pharynx, larynx) and the **lower respiratory tract** (trachea, bronchi, bronchioles, alveoli).
- The nose filters, warms and humidifies air.
- The **pharynx** connects the nose and mouth to the **larynx** (voice box).
- The **larynx** contains vocal cords that vibrate to produce sound.
- The **trachea** (windpipe) is a tube that carries air to the bronchi.
- The **bronchi** are two branching tubes that lead to the lungs.
- **Bronchioles** are thin tubes that further branch out from the bronchi.
- **Alveoli** are tiny air sacs where gas exchange takes place.
- The diaphragm is a muscle that separates the chest cavity from the abdominal cavity and plays an important role in breathing.
- During **inhalation**, the diaphragm contracts increasing the volume of the chest thus reducing the internal pressure; this draws air into the lungs.
- Inhalation is an **active process**.
- During exhalation, the diaphragm relaxes decreasing the volume of the chest thus increasing the internal pressure. This forces air out of the lungs.
- The intercostal muscles are used for deep breathing and forced exhalation.
- There are two types of intercostal muscles: internal and external.
- When inhaling deeply, the external intercostal muscles contract to move the chest up and out.
- When exhaling forcefully, the internal intercostal muscles contract to draw the chest down and in.
- The **medulla oblongata** is the region of the brain responsible for controlling the breathing rate.
- The **concentration of carbon dioxide** in the blood is a controlling factor in the rate of breathing.
- **Gas exchange** involves oxygen from the atmosphere exchanging for carbon dioxide in the blood.
- Alveoli have **extremely thin walls**, allowing **diffusion** take place.
- Diffusion is the movement of gases from an area of high concentration to low concentration.
- When we inhale, **oxygen diffuses from the alveoli into the blood**. At the same time, **carbon dioxide diffuses from the blood into the alveoli** and is then exhaled.
- **Asthma** is a chronic respiratory condition which narrows the bronchioles, reducing the airflow to the alveoli.

END OF CHAPTER ASSESSMENT

REMEMBER

1. Name the primary organs of the breathing system.
2. Which part of the breathing system is responsible for producing sounds?
3. How is the trachea supported to ensure it is kept open?
4. Name the tubes leading into each lung.
5. Name the tiny air sacs found in the lungs.
6. Name the large domed muscles that separate the chest cavity from the abdomen.
7. What is meant by *inhalation*?
8. Name the muscles found between the ribs.
9. Name the region in the brain responsible for controlling the breathing rate.
10. Name the gas, produced during respiration, that is a controlling factor in the breathing rate.

UNDERSTAND

11. Briefly describe the structure of the upper respiratory tract, outlining the role of each part.
12. Describe the role of the diaphragm in inhalation and exhalation.
13. *Intercostal muscles are involved in deep breathing and forced exhalation.* Comment on this statement.
14. How does the structure of the alveoli facilitate gas exchange?

APPLY

15. How does exercise affect the breathing rate? Make reference to respiration in your answer.
16. Explain how inhalers work to treat asthma.
17. Suggest a reason why some athletes train at higher altitude before competing.

ANALYSE

18. Study the data below and answer the questions that follow. Tidal volume refers to the amount of air inhaled during each breath while minute ventilation refers to the amount of air inhaled per minute.

STAGE OF EXERCISE	TIDAL VOLUME (ml)	RESPIRATORY RATE (breaths/ minute)	MINUTE VENTILATION (l/min)
Rest	500	12	6
Light exercise (Walking)	700	16	11.2
Moderate exercise (Brisk walking)	1000	20	20
Heavy exercise (Running)	1500	30	45

(a) Based on the data, describe the relationship between exercise intensity and tidal volume.

(b) Explain why tidal volume increases during exercise.

(c) How does the change in respiratory rate during exercise contribute to maintaining oxygen intake?

(d) Minute ventilation is calculated by multiplying tidal volume by respiratory rate. Considering both tidal volume and respiratory rate, explain why minute ventilation increases significantly during heavy exercise compared to rest.

(e) Draw a graph to represent the data in the chart.

EVALUATE

19. Evaluate the impact of air pollution on respiratory health.
20. Research vaping and its effects on the breathing system.

CREATE

21. Develop a presentation on different types of respiratory diseases.
22. Design a model to demonstrate the effects of smoking on the lungs.

29 The Urinary System

LEARNING OUTCOMES: 2.6.3

EYE-OPENER

The hummingbird attains water from the large amounts of nectar it consumes (nectar is a sugar-rich substance found in some flowers). When the hummingbird extracts the sugar from the nectar needed to fuel its metabolism, it is left with an abundance of water and risks illness if this is not removed. In lab tests, hummingbirds produced urine equivalent to 75–85% of their body weight per day, roughly the equivalent of a human excreting over 75 litres of water in urine daily.

At the end of this chapter, you should be able to:

- Describe the macrostructure of the human urinary system.
- Describe how the kidney filters the blood and, in doing so, forms urine.
- Describe the structure and function of the nephron and its blood supply. HL

LINKS TO:

Chapter 14	Biomolecules
Chapter 15	Metabolism and Enzymes
Chapter 18	Movement through Cell Membranes
Chapter 23	Respiration
Chapter 30	The Digestive System
Chapter 31	The Circulatory System
Chapter 34	The Endocrine System

SUBJECT VOCABULARY

excretion, homeostasis, urinary system, kidney, osmoregulation, deamination, nephron, renal artery, renal vein, glomerulus, sphincter muscles, urine, glomerular filtrate, tubular reabsorption, secretion

Section 1: Excretion and the Anatomy of the Urinary System

At the end of this section, you should be able to:

- Define what is meant by excretion.
- Describe the role of the urinary system in excretion.
- Identify the parts of the human urinary system, describing the role of each part.
- Discuss the role of the kidneys in filtering the blood.

Unpacked from LO 2.6.3

EXCRETION

As we learned in Chapter 1, all living things have a metabolism and produce waste. Animals produce more complex wastes than plants do (mainly due to the consumption of proteins) and often need a network of organs to deal with them. The removal of the waste products of metabolism is called **excretion**.

Excretion is the removal of the waste products of metabolism.

Humans produce different forms of waste. Cellular waste is produced within cells and is typically broken down proteins or organelles. It is first dealt with inside the cell by organelles called **lysosomes**, before being transported out of the cell. Later, our liver and kidneys remove these materials from the blood. We also produce metabolic waste from the chemical reactions which take place inside of our cells. All of our cells produce water and carbon dioxide when they break down their carbohydrates through respiration. They may also, at times, produce lactic acid if they break down the food without oxygen. These metabolic reactions also produce excess heat.

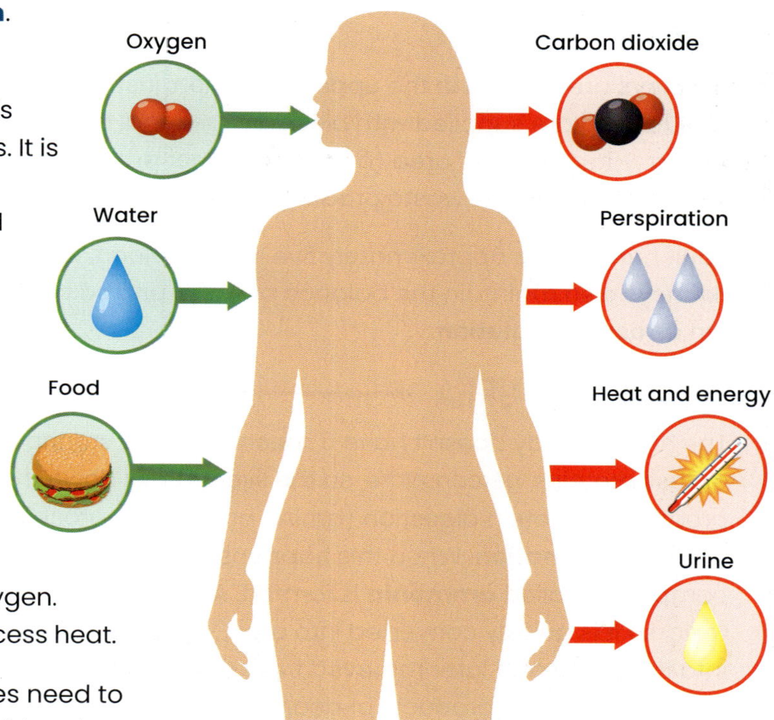

Figure 29.1: Metabolism produces waste products.

These various cellular and metabolic wastes need to be removed from the body or else they could make us ill. We have various organs responsible for this: the **lungs** remove carbon dioxide and water (see Chapter 28), our **skin** removes water, salt and excess heat, while our **kidneys** remove water, salt and urea. In this chapter, we are going to focus on the kidneys, which form part of our urinary system.

Excretion is a core feature of **homeostasis** – our body's ability to maintain its internal environment, ensuring metabolic reactions take place in their optimum condition. We will explore this further in Chapter 34.

Homeostasis is the maintenance of a steady state within the body.

THE URINARY SYSTEM

The human **urinary system** is composed of two bean-shaped organs called the **kidneys**, a storage area for urine called the **bladder**, a **blood supply** and a series of connecting **muscular tubes** (ureter and urethra).

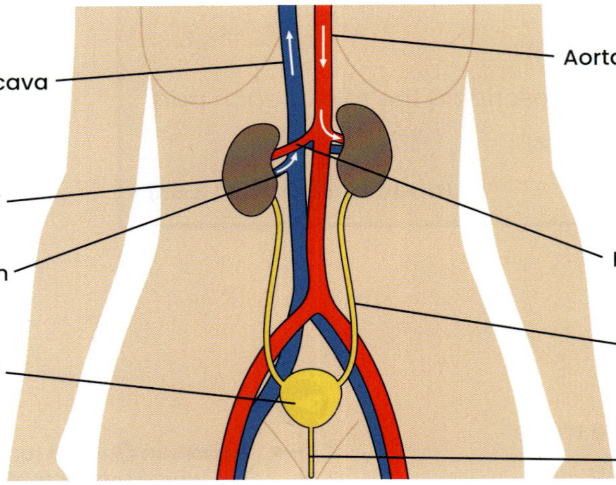

Figure 29.2: The urinary system.

Kidneys

The **kidneys** are located in the upper abdomen, just below your diaphragm. They are protected by a layer of fat and are supplied with blood by the renal artery. Kidneys have a major function in **excretion**, removing water, salt and urea (a waste product produced in the liver). The kidneys then act by filtering the blood, removing the waste products, before returning the blood to the vena cava via the renal vein.

An adult human has approximately five litres of blood, which is filtered by the kidneys 60 times a day. The kidneys also maintain the balance of water and salt in the blood and other body fluids – a process known as **osmoregulation**.

FOCUS ON ... UREA

The human body doesn't have a mechanism for storing any excess proteins we consume, so the liver breaks down the products of protein digestion (amino acids). This process is called **deamination**. When this happens, a nitrogen-based compound called **ammonia** is formed. Ammonia is highly toxic, so it is quickly converted into urea, which is released into the blood and later removed from the body by the kidneys. The other products of deamination can be converted into glucose in a process called **gluconeogenesis**. The glucose can then be used to fuel our cells.

- The **urinary system** is a collection of organs responsible for urine production.
- **Kidneys** are bean-shaped organs responsible for excretion and osmoregulation.
- **Deamination** is the breakdown of amino acids into sugars in the liver.

Inside the kidney, there are four distinct regions.

1. The **outer capsule** acts to protect the internal parts.
2. The **cortex** is where blood is filtered – the openings of millions of tiny tubes called **nephrons** are found here.
3. The **medulla** region contains approximately 12 cone-shaped structures called pyramids which form a network of collecting ducts which gather urine as it is formed.
4. The **renal pelvis** is a funnel-shaped cavity in the centre of the kidney which collects urine from the collecting ducts before it travels through the **ureter** and eventually to the **bladder**.

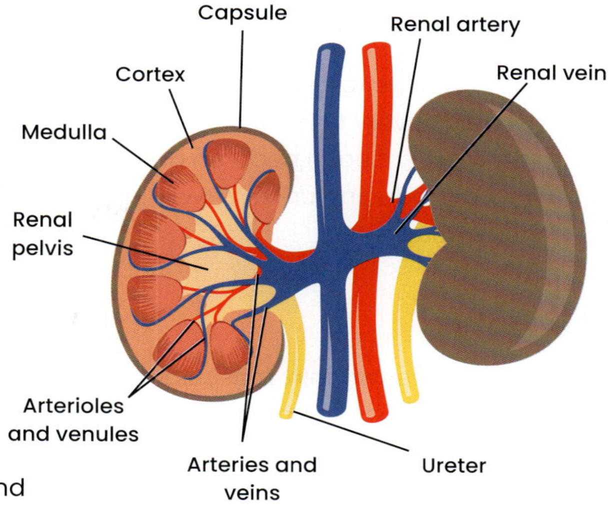

Figure 29.3: The internal structure of the kidney.

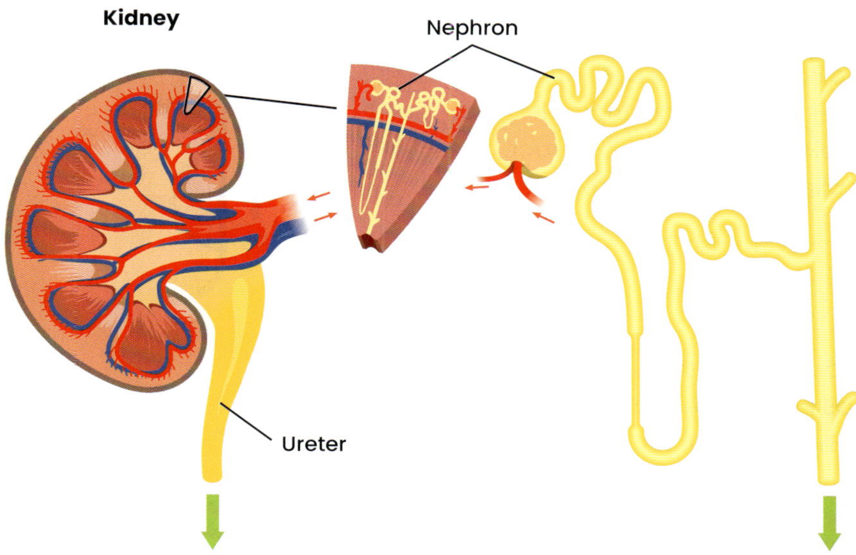

Figure 29.4: The kidney and the nephrons.

The kidneys have several functions other than excretion. They regulate the pH of the blood (how acidic it is) and control blood pressure by regulating the volume of the blood. The kidneys also control the production of red blood cells by producing the hormone erythropoietin.

DID YOU KNOW?

As shown in Figure 29.4, the kidney has approximately 1 million tiny functional units called **nephrons**. The nephrons filter the blood, removing wastes. They also ensure useful nutrients (e.g. sugars and amino acids) are maintained and control the amount of water in the blood. We will explore how the nephron works later in the chapter.

Renal blood supply

The **renal arteries** bring blood into each of the kidneys. As they enter, they branch off into smaller regional arteries. These snake through the kidney, branching off further into tiny arterioles that each supply a single nephron – the filtering units of the kidney.

After the blood has been filtered, the blood exits the kidney through a venule that feeds into a single large **renal vein**. This vein drains blood from the kidney and carries it back to the vena cava, the large vein that returns blood to the heart.

- **Nephrons** are tiny tubes found within the kidney – often referred to as the functional unit of the kidney.
- The **renal artery** supplies blood to the kidneys for filtration.
- The **renal vein** collects filtered blood from the kidney.

The glomerulus

HL

Each arteriole forms a structure known as a **glomerulus**, a network of very thin capillaries. The glomerulus is located at the entry point of each nephron and is the site of **filtration** – the first step in producing urine. We will learn more about this later on in the chapter.

The **glomerulus** contains a series of ultra-thin capillaries at the opening of the nephron.

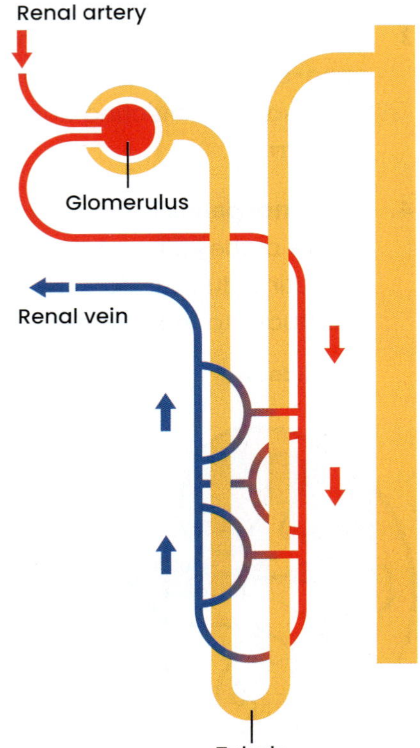

Figure 29.5: The glomerulus is an important feature of the renal blood supply.

Ureters

The **ureters** are two thin, muscular tubes that carry the filtered urine from the kidneys to the bladder. They use smooth muscle contractions, called **peristalsis**, to propel urine in a one-way direction, preventing backflow into the kidneys.

Bladder

The **bladder** is a muscular storage sac which acts as a reservoir for urine. As urine accumulates, the bladder walls stretch, triggering the urge to urinate. When you urinate, the bladder muscles contract, while a circular muscle at the base (called a **sphincter**) relaxes, allowing urine to flow out.

Urethra

The **urethra** carries urine from the bladder out of the body. In males, the urethra is longer and also serves as part of the male reproductive system. The urethra also has a sphincter muscle that controls urination.

SCIENCE IN SOCIETY – HEALTH

Kidney stones

A kidney stone is a tiny, rock-hard mineral deposit that can sometimes form within the kidney. Our blood carries waste products, and sometimes these can clump together, forming small stone-like masses. They can be as small as a grain of sand or, rarely, as big as a golf ball.

Most stones pass through your urinary system unnoticed, but if they get stuck, they can become very painful. To avoid kidney stones, hydration is key. Drinking plenty of water helps dilute the minerals in your urine and flush them out before they can form stones.

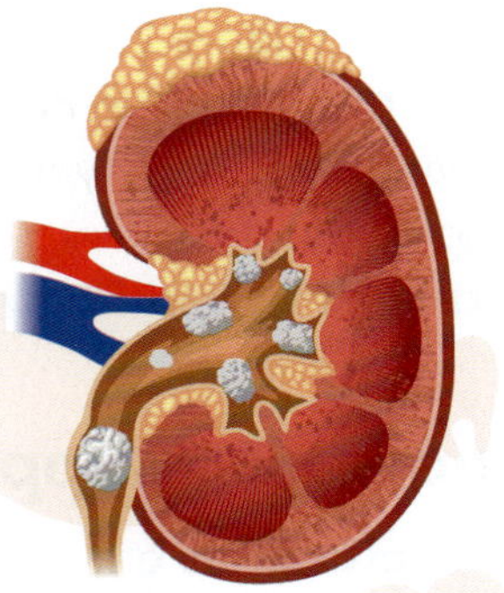

Figure 29.6: Kidney stones.

RESEARCH ACTIVITY

Research other kidney-related health issues. Why is good kidney health important?

Go to YouTube.com and search for 'What causes kidney stones? – Arash Shadman' (5:14) by TED-Ed.

PORTFOLIO LINK

Go to your Learning Portfolio and complete Learning Check 1 to assess your knowledge and understanding of the topic so far.

Section 2: Formation of Urine and the Nephron

At the end of this section, you should be able to:

- Describe how urine is formed in the kidney.
- Outline the structure of the nephron. HL
- Relate the parts of the nephron to the stages of urine production. HL
- Describe the role of ADH in osmoregulation and urine concentration. HL

Unpacked from LO 2.6.3

HOW URINE IS FORMED

Urine is the liquid waste produced by the kidneys. Urine is composed of mostly water, typically making up 95% of its volume. The remaining 5% consists of dissolved waste products and other substances including:

Urine is a liquid waste containing water, urea, uric acids and mineral salts.

- urea and uric acid (from the breakdown of proteins and nucleic acids)
- mineral salts like sodium, potassium and chloride
- other byproducts of metabolism and foreign substances (e.g. food additives or medicines).

Humans produce, on average, 1.5 litres of urine per day. Normal urine has a pale yellow colour, caused by pigments produced by the breakdown of haemoglobin in red blood cells. The concentration of urine can vary depending on your hydration status. When you're well hydrated, urine is diluted and pale yellow, but dehydration can lead to more concentrated urine with a deeper yellow colour.

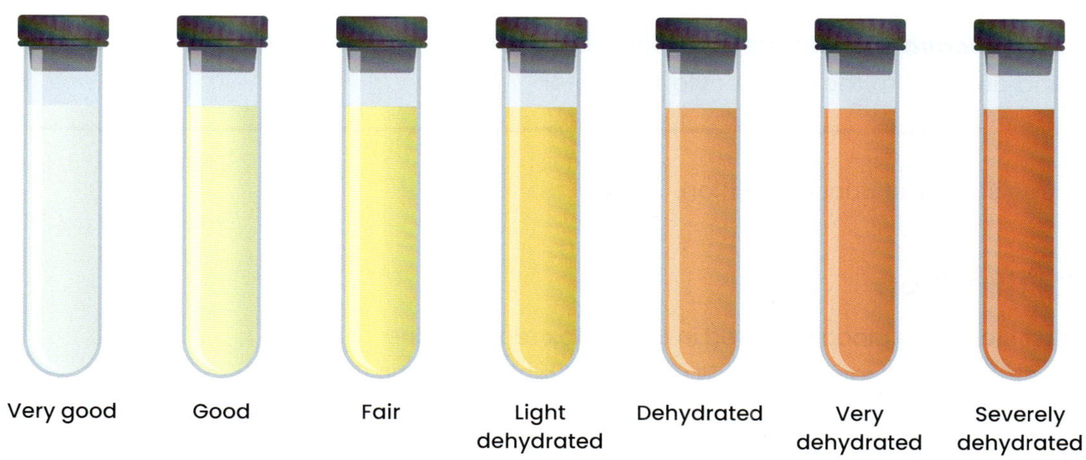

Figure 29.7: The colour of urine indicates your level of hydration.

Urine is formed as the blood is filtered by the kidney's functional units – the **nephrons**. For Higher Level, you will need to know the structure of the nephron before we get into the details of how urine is formed.

FOCUS ON ... THE NEPHRON (HL)

Nephrons are long thin tubes, approximately 3cm in length, and each kidney contains approximately 1 million of them. Each nephron has its own blood supply – a renal arteriole which branches from the renal artery. It initially flows into the Bowman's capsule, forming the **glomerulus**, then reforms and surrounds the nephron before connecting to a renal vein.

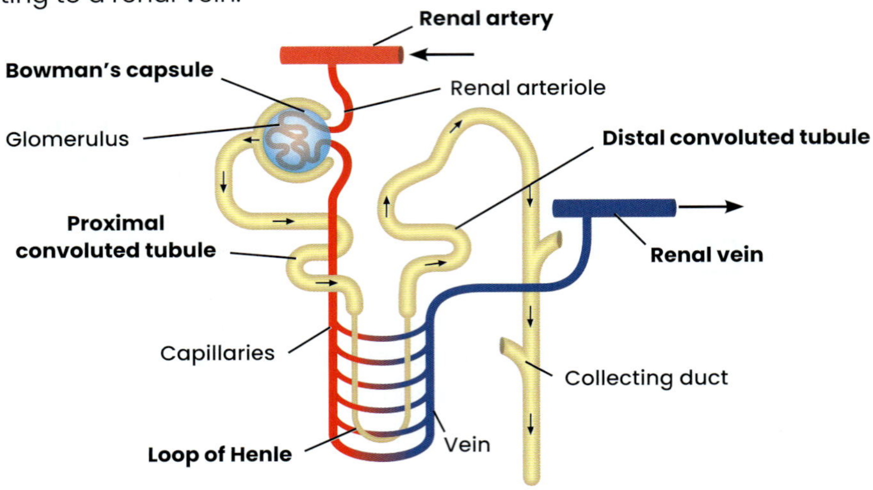

Figure 29.8: Basic structure of the nephron.

The nephron has four main sections:

1. **Bowman's capsule:** A wine glass-shaped opening which contains the glomerulus – a network of thin capillaries formed from a renal arteriole.
2. **Proximal convoluted tubule:** A small twisted tube, surrounded by blood capillaries.
3. **Loop of Henle:** A long straight section which extends down into the medulla region.
4. **Distal convoluted tubule and collecting duct:** These tubes collect urine as it forms and deliver it to the bladder via the ureter.

There are three stages of urine formation.

Stage 1: Filtration

During the first stage, the blood is filtered in the **cortex** region of the kidneys. This filtration separates the liquid part of the blood from the larger blood cells and plasma proteins. The liquid filtrate flows into the nephrons.

THE GLOMERULUS AND FILTRATION

HL

To be precise, the blood is filtered at the **glomerulus**, at the entrance to each of the 1 million nephrons. In these ultra-thin capillaries, the pressure of the blood increases (imagine narrowing the nozzle of a hose pipe and seeing the pressure increase). The high pressure forces the liquid part of the blood out of the capillaries where it flows into the nephron tubule. Larger items in the blood (the blood cells and plasma proteins) stay in the capillaries while the liquid part carries the dissolved nutrients (glucose and amino acids), salts and wastes (e.g. urea) into the tubule. We call this liquid the **glomerular filtrate**. At the end of filtration, the glomerular filtrate contains lots of blood wastes but also lots of useful material too.

Glomerular filtrate is the product of filtration – the first stage of urine formation.

Stage 2: Reabsorption

As the filtrate flows through the nephron tubules in the **cortex** and **medulla**, the kidney reclaims valuable resources like water, glucose and essential salts back into the bloodstream. This is called **reabsorption**.

Reabsorption happens very quickly, with over 80% of all useful materials (glucose amino acids, vitamins, etc.) reabsorbed back into the blood before it reaches the loop of Henle.

HL

THE NEPHRONS AND REABSORPTION

Reabsorption first takes place by diffusion, as materials move from high concentration to low concentration. The walls of the nephron also contain specialised transport proteins which use **active transport** to reabsorb these materials into the blood. Water is reabsorbed by **osmosis**. The nephron has a rich network of blood vessels surrounding it. This reabsorption is facilitated by thin capillary and nephron walls.

Stage 3: Secretion

While reabsorption is happening, the nephron walls can also remove any lingering waste products and excess water from the blood and move it into the filtrate. This stage is called **secretion**. It acts as a mechanism to remove any remaining wastes ensuring all are removed.

- **Reabsorption** involves moving valuable resources from the glomerular filtrate back into the blood.
- **Secretion** involves the removal of any remaining metabolic wastes from the blood into the nephron tubules.

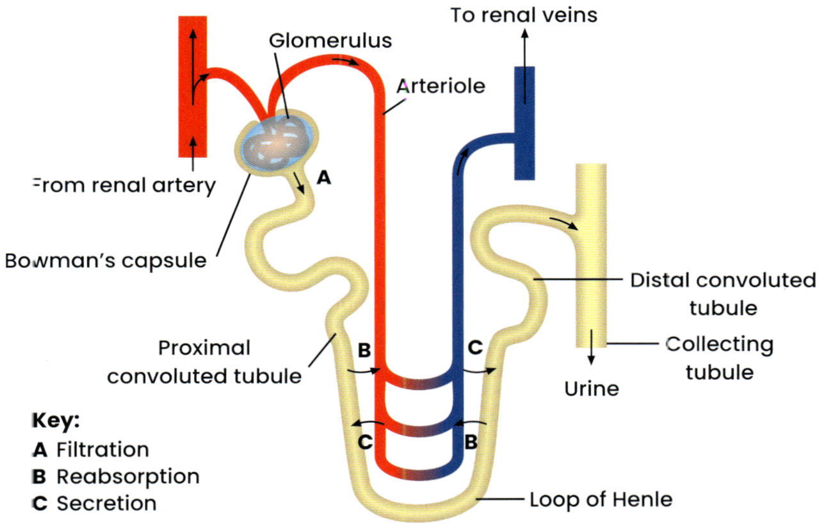

Figure 29.9: The stages of urine formation in the nephron.

URINE CONCENTRATION

By the time the filtrate reaches the distal tubule, all the useful materials should have been reabsorbed leaving just urine to flow into the collecting ducts. However, the concentration of the urine (whether it appears pale yellow or dark yellow) is controlled before the filtrate leaves the nephron. The loop of Henle is the region of the nephron responsible for reabsorption of water initially. Around 85% of water is reabsorbed in the loop of Henle.

If the body is dehydrated, it will instruct the **distal tubule** of the nephron to reabsorb the remaining water into the blood making the urine appear darker in colour. If the body is well hydrated, it allows all the water to remain in the nephron and flow into the bladder.

FOCUS ON ... ANTIDIURETIC HORMONE HL

Osmoregulation and the control of urine concentration is actually controlled by the brain. When the body is dehydrated or needs to conserve water, the **pituitary gland** releases a hormone called **antidiuretic hormone** or **ADH**.

HL This hormone makes the walls of the collecting ducts more porous, increasing water reabsorption into the bloodstream. This results in a more concentrated urine, with less water excreted.

Alcohol inhibits the production of ADH in the pituitary gland, so drinking alcohol actually causes dehydration.

Go to YouTube.com and search for 'How do your kidneys work? – Emma Bryce' (3:54) by TED-Ed.

PORTFOLIO LINK **Go to your Learning Portfolio and complete Learning Check 2 to assess your knowledge and understanding of the topic so far.**

Chapter Summary

- **Excretion** is the removal of waste products of metabolism from the body.
- **Osmoregulation** is the control of the water and salt balance in the blood and body fluids.
- The kidneys are organs of **excretion** and **osmoregulation**.
- Each kidney is supplied with blood by the **renal artery** which branches from the aorta.
- The blood in the **renal arteries** contains waste material including water, salt, urea and uric acid.
- The **renal vein** returns filtered blood to the vena cava.
- The **nephron** is the functional unit of the kidney.
- HL The nephron has four regions: **Bowman's capsule** (which contains the glomerulus), the proximal convoluted tubule, the loop of Henle and the distal convoluted tubule.
- **Urine** is composed of mostly water, dissolved salts and urea.
- Urine formation involves three stages: 1. filtration, 2. reabsorption, 3. secretion.
- Filtration takes places in the **cortex** region of the kidney.
- HL Specifically, filtration takes place at the **glomerulus**, encapsulated by **Bowman's capsule** of the nephron.
- **Filtration** removes waste material from the blood (e.g. water, salt, urea), but also many valuable substances (e.g. glucose, amino acids and vitamins), which flow into the nephron.
- **Glomerular filtrate** is the product of filtration, which flows into the nephron. HL
- Glomerular filtrate contains useful materials (e.g. sugars, amino acids, minerals) and wastes (water, salt and urea) but no blood cells or plasma proteins. HL
- **Reabsorption** involves bringing the useful material out of the nephron and back into the blood.
- Reabsorption happens mostly in the proximal tubule of the nephron by the processes of **osmosis**, **active transport** and **diffusion**. HL
- **Secretion** is the removal of any remaining metabolic wastes from the blood into the nephron tubules.
- The loop of Henle is tasked with conserving water, ensuring most water is reabsorbed. HL
- **Antidiuretic hormone (ADH)**, released from the pituitary gland when dehydrated, causes water to be reabsorbed from the nephron and collecting ducts into the blood. HL
- Urine leaves the collecting ducts, eventually passing into the **bladder** via a **ureter**.

END OF CHAPTER ASSESSMENT

REMEMBER

1. Define *excretion*.
2. Name **three** organs with some responsibility for excretion in the human body.
3. List the **four** organs that are part of the urinary system.
4. Briefly outline the function of each part named in your answer to question 3.
5. What is the name of the microscopic filtering unit in the kidney?
6. What **two** tubes carry urine from the kidneys to the bladder?
7. What muscular sac stores urine until it's released from the body?
8. What tube carries urine out of the body from the bladder?

UNDERSTAND

9. What is meant by *osmoregulation*?
10. Describe the internal layers of the kidney.
11. What is urine composed of?
12. Briefly describe the path that urine takes as it travels through the urinary system.
13. Distinguish between *excretion* and *secretion*.
14. Describe the **three** stages of urine production.
15. HL (a) Label the parts A, B, C shown in the diagram of the nephron.
    HL (b) Where in the diagram does reabsorption take place?
    HL (c) Name the blood vessels X and Y shown in the diagram.

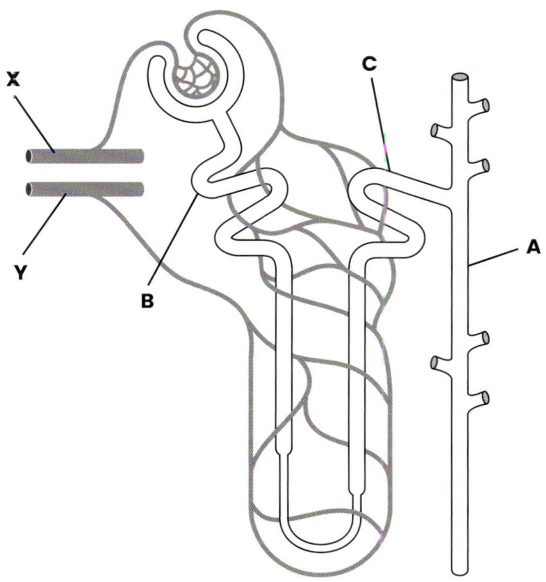

16. Why is maintaining the blood supply important for the kidneys and overall health?

APPLY

17. Why do you think excess amino acids are converted to sugar in the liver? What term describes this process?
18. Describe how ADH plays a role in osmoregulation. HL
19. Investigate the impact of diet and lifestyle choices on kidney health. Discuss how certain foods or habits might influence kidney function.
20. After drinking alcohol, people experience signs of dehydration. What symptoms would indicate dehydration? HL
21. Four students monitored their water intake, urine output and colour of their urine over the course of a day. Analyse the data below and answer the questions that follow.

STUDENT NAME	FLUID INTAKE (litres)	URINE OUTPUT (litres)	URINE COLOUR
Yashika	0.75	0.50	Dark yellow
Steve	2.50	1.50	Pale yellow
Seán	1.25	0.75	Yellow
Julie	1.50	1.25	Light yellow

(a) Is there a correlation between water intake and urine output in the data shown?

(b) Can you conclude that a high fluid intake directly causes a high urine output based on this data? Why or why not?

(c) Which of the above students is best hydrated? Why?

(d) Which of the above students is dehydrated? Explain your choice.

(e) Why is there a discrepancy between fluid intake and urine output?

(f) How might the investigation be improved?

HL **22.** During an experiment, the blood alcohol content of 10 people was measured along with their blood plasma water concentration. The results are shown below.

TEST SUBJECT	BLOOD ALCOHOL CONTENT (BAC) %	BLOOD PLASMA WATER CONCENTRATION %
1	0.00	91.0
2	0.05	90.5
3	0.10	90.0
4	0.00	91.5
5	0.15	89.5
6	0.20	89.0
7	0.00	92.0
8	0.25	88.5
9	0.30	88.0
10	0.25	89.0

Analyse the data, which shows how variations in blood alcohol levels can affect the concentration of water in blood plasma, and answer the questions that follow.

HL (a) Is there a direct correlation between blood alcohol content and water concentration in blood plasma, based on the data provided?

HL (b) From your knowledge of the urinary system, do you think changes in blood alcohol levels cause changes in water levels in blood plasma?

HL (c) If a person consumes a large amount of alcohol and doesn't drink enough water, how might this affect their blood plasma water concentration?

HL (d) What other factors might affect the blood plasma water concentration, apart from blood alcohol content?

ANALYSE

23. Siobhán monitored her water intake and urine output for a week. Her daily water intake averaged 2 litres, and her urine output averaged 1.5 litres. What does this data suggest about Siobhán's body's water balance? Explain your answer.

24. Create a graph to visually represent Siobhán's water intake and urine output data from Question 23. Label the axes and include a title.

25. You overexert yourself during a workout and experience dark-coloured urine afterward. What might explain this change in urine colour, and how can you prevent it from happening again?

EVALUATE

26. How do advancements in medical technology, like dialysis or kidney transplants, play a role in supporting people with kidney problems?

27. Explore the ethical considerations surrounding organ donation and transplantation for individuals with severe kidney dysfunction.

28. Urinary tract infections (UTIs) are common infections of the urinary system. Carry out some research on common UTIs in humans. What signs and symptoms might indicate a UTI?

CREATE

29. Create a visual flow chart that illustrates the stages of urine formation within a nephron. HL

30. Create a large model of the nephron, labelling each of the parts. HL

31. Imagine you're a doctor explaining the urinary system to a young patient. Develop a clear and age-appropriate explanation using simple language and analogies.

30 The Digestive System

LEARNING OUTCOMES: U1.2, U2.6, 2.6.4, 2.6.5, 2.6.6, 3.2.4

EYE-OPENER

Snakes have an unusual adaptation. They are able to swallow their prey whole by unhinging their jaws. This allows them to swallow food that is often larger than their own heads. The food is pushed down into their stomachs by their strong muscles before acids and enzymes slowly but surely digest it. It can sometimes take days to fully digest their prey, with only fur, feathers and bones remaining.

At the end of this chapter, you should be able to:

- Outline the main organs of the human digestive system.
- Define digestion and distinguish between mechanical and chemical digestion.
- Explain how carbohydrates, proteins and lipids are digested.
- Describe the absorption, transport and storage of the products of digestion.
- Discuss how dietary choices affect digestive health and efficiency.

LINKS TO:

Chapter 14	**Biomolecules**
Chapter 15	**Metabolism and Enzymes**
Chapter 18	**Movement through Cell Membranes**
Chapter 23	**Respiration**
Chapter 29	**The Urinary System**
Chapter 31	**The Circulatory System**
Chapter 34	**The Endocrine System**

SUBJECT VOCABULARY

digestion, physical digestion, chemical digestion, alimentary canal, amylase, pancreatic juice, fibre, peptides, pepsin, trypsin, lipase, bile, absorption, villi, lacteal, diffusion, active transport, hepatic portal vein

Section 1: Anatomy of the Digestive System

At the end of this section, you should be able to:

- Define what is meant by digestion.
- Distinguish between physical and chemical digestion.
- Name and identify the location of the principal organs and glands of the human digestive system.
- Briefly describe the role of the mouth, teeth, oesophagus, stomach, pancreas, gall bladder, liver, small intestine and large intestine.

Unpacked from LO 2.6.4

For the most part, our food is created from the cells of other living things, meaning it is made using biomolecules: proteins, lipids (fats), carbohydrates, vitamins, minerals and water. These biomolecules are then used to provide energy for our cells' activities (through cellular respiration), as well as providing the raw materials to create new cells for growth and repair. We discussed food and biomolecules in detail in Chapter 14, so it is worth revisiting that chapter before studying human nutrition.

THINK, PAIR, SHARE

With your partner, share what you already know about the digestive system. What do you want to learn more about? Feed back to your teacher.

WHAT IS DIGESTION?

The food we eat comprises carbohydrates, proteins, fats and other biomolecules. Some of these molecules are extremely large and are unable to travel efficiently through our blood. **Digestion** is the act of breaking these molecules down to the smallest possible units. Once broken down, these units are then **absorbed** into the blood so that they can be transported around the body, making their way to every one of our cells.

Digestion can be **physical** or **chemical. Physical digestion** breaks food into small pieces, increasing the surface area of the food. There is no change to the chemical nature of the food, i.e. the molecules aren't broken down. Physical digestion mainly takes place in the mouth and stomach. **Chemical digestion** involves enzymes, which break down the molecules of carbohydrates, fats and proteins into their smaller structural units (e.g. proteins are broken down into amino acids). Chemical digestion takes place in the mouth, stomach and small intestine.

The digestive system is a collection of organs and glands which work together to break down the food we eat, absorb the useful biomolecules and dispose of any material we can't break down. The digestive system works in partnership with the circulatory system (the heart, blood and blood vessels) and we will discuss this later in the chapter.

- **Digestion** is the breakdown of food (biomolecules) into smaller units.
- **Physical digestion** breaks food down mechanically into smaller pieces, increasing the surface area.
- **Chemical digestion** uses enzymes to break down biomolecules into their structural units.

Go to YouTube.com and search for 'How your digestive system works – Emma Bryce' (4:56) by TED-Ed.

The alimentary canal

The **alimentary canal** is the main part of the digestive system, consisting of a collection of connected organs along a long tube that extends from your mouth to your anus. It is approximately nine metres (30 feet) long in a fully grown adult. The alimentary canal consists of the following organs:

The **alimentary canal** is a series of connected organs along a long tube, extending from the mouth to the anus.

- mouth
- oesophagus
- stomach
- small intestine
- large intestine
- rectum
- anus

Found along the alimentary canal are a series of other organs and glands, assisting the digestion and absorption of food. These **accessory organs** include:

- salivary glands
- tongue
- liver
- gall bladder
- pancreas.

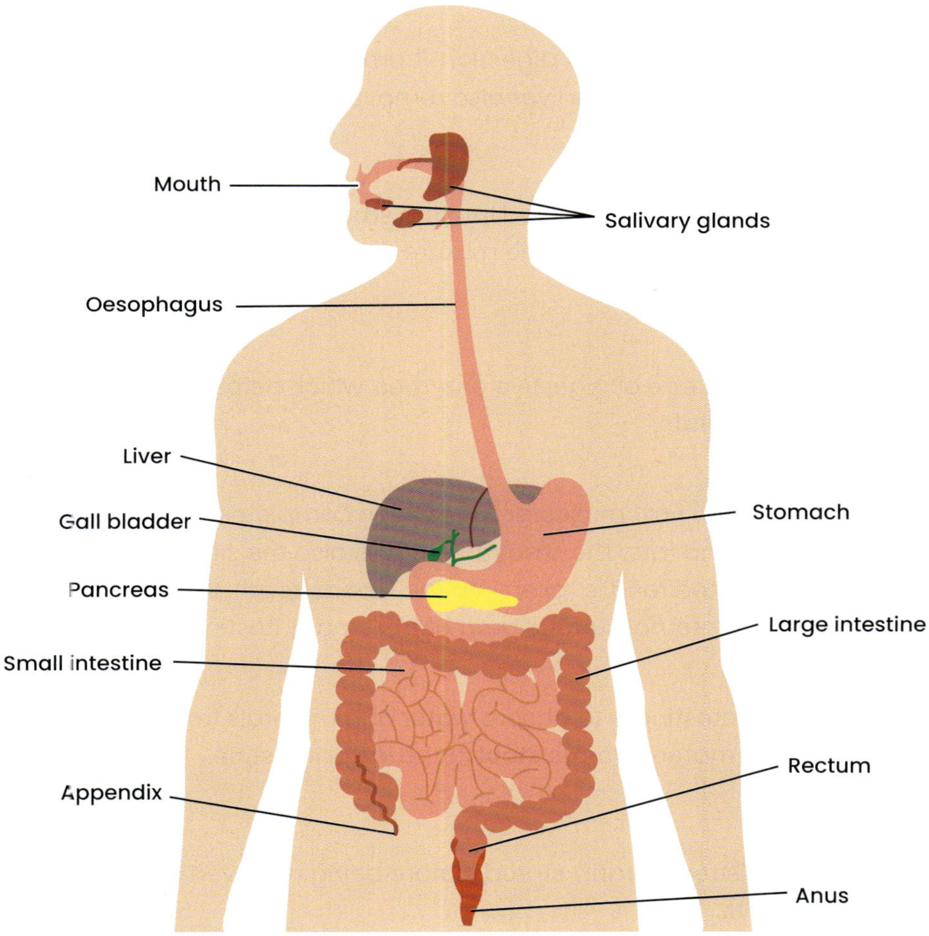

Figure 30.1: The digestive system.

DIGESTIVE SYSTEM: PARTS AND FUNCTIONS

PART	FUNCTION
Mouth	The mouth is the site of **ingestion**, where food is introduced to the body. **Teeth** break the food down into smaller pieces while it mixes with **saliva**.
Salivary glands	As the name suggests, saliva is produced in the salivary glands. Saliva is a slimy solution containing the enzyme **amylase**. Amylase begins the breakdown of carbohydrates.
Oesophagus	The oesophagus is a long muscular tube which pushes the food down into the stomach by a process called **peristalsis**. This process begins when food is swallowed.
Stomach	The stomach's **gastric juices** contain **hydrochloric acid** which loosens food and kills bacteria. The stomach juices also contain the enzyme **pepsin** which begins the process of breaking down proteins.
Liver	The liver is a large organ which supports digestion. It produces a chemical called **bile**, which helps digest lipids (fats). The liver also removes toxins from digested food.
Gall bladder	The gall bladder stores bile, which is released into the small intestine and helps digest fats. Bile also neutralises the acid food mixture that is released from the stomach.
Pancreas	The pancreas produces a mixture of digestive enzymes, which helps break down carbohydrates, proteins and fats.
Small intestine	The small intestine is a long (5.5m) muscular tube which completes the digestion of carbohydrates, proteins and fats. The small intestine is also the site of **absorption**, where the digested food is taken into the bloodstream. Small finger-like **villi** increase the internal surface area of the small intestine.
Large intestine	The main role of the large intestine is to absorb water and minerals from the ingested food. Undigested material forms **stool** in the large intestine.
Rectum	The rectum is where stool is temporarily stored, before being removed from the body.
Anus	The anus is the final opening of the digestive system, where stool is eliminated from the body. This is known as **egestion**.

SCIENCE IN SOCIETY – HEALTH

The appendix

The appendix is a small, finger-shaped pouch attached to the large intestine. Its exact function isn't fully understood, but theories suggest it may house beneficial gut bacteria or act as an immune system outpost. While not essential, its removal (appendectomy) is necessary if it becomes inflamed (appendicitis).

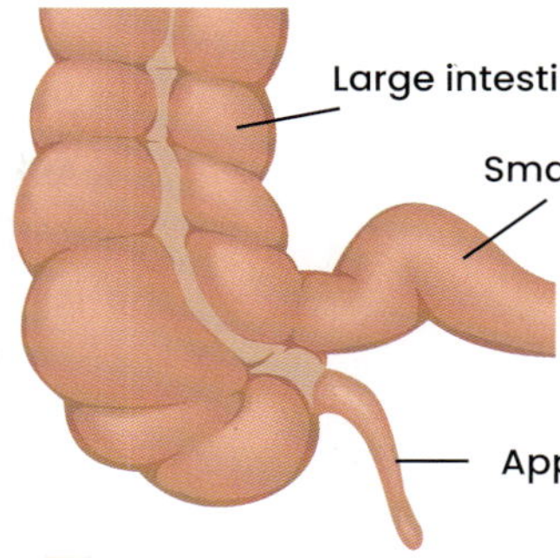

Figure 30.2: The appendix.

PORTFOLIO LINK Go to your Learning Portfolio and complete Learning Check 1 to assess your knowledge and understanding of the topic so far.

Section 2: Digestion of Carbohydrates, Proteins and Fats

At the end of this section, you should be able to:

- Describe digestion of carbohydrates, including the role of the mouth, pancreas and small intestine.
- Outline the role of teeth in the digestive process.
- Describe the digestion of proteins, including the role of the stomach, pancreas and small intestine.
- Describe the digestion of lipids (fats), including the role of the liver, gall bladder, pancreas and small intestine.

Unpacked from LO 2.6.4

CARBOHYDRATE DIGESTION

We learned in Chapter 14 that carbohydrates are composed of units called saccharides. Some carbohydrates contain just one unit (monosaccharides), some contain two (disaccharides) and others contain multiple units (polysaccharides). The ultimate goal of the digestion of carbohydrates is to break down the ingested carbohydrates into monosaccharides, like glucose. Only monosaccharides can be absorbed and transported in the blood to the cells that need them.

Stage 1: Mouth

Chewing food breaks it down into smaller pieces, increasing the surface area. This is physical digestion, the first step in the digestion of carbohydrates. The chemical digestion of carbohydrates also begins in the mouth, where the enzyme **amylase** (found in saliva) breaks down starch (polysaccharides) into maltose (a disaccharide).

Amylase is an enzyme that breaks down starch into maltose.

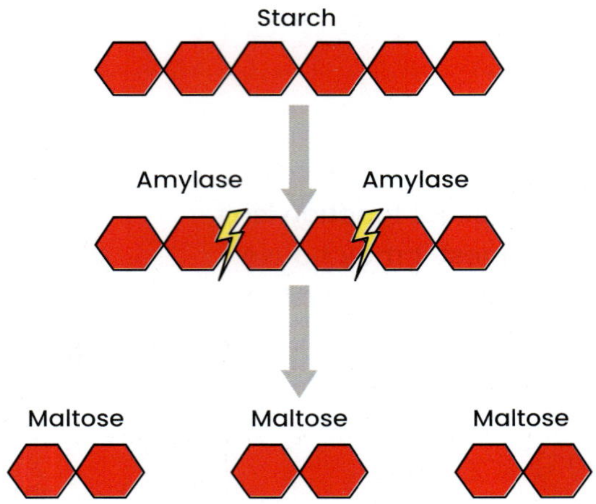

Figure 30.3: Digestion of starch by amylase.

Stage 2: Pancreas

The pancreas releases a mixture of enzymes called **pancreatic juice** into the small intestine. This mixture includes **amylase**, which breaks down any remaining starch in the food. Amylase works best in a neutral pH. Pancreatic juice contains alkaline bicarbonate ions that neutralise the acidic contents of the stomach, allowing amylase work at its optimum pH.

Stage 3: Small intestine

The wall of the small intestine contains a mixture of enzymes, too. These include **maltase**, which breaks down maltose, **lactase** which breaks down lactose and **sucrase** which breaks down sucrose. Maltose, lactose and sucrose are all disaccharides. The end products of the digestion of these disaccharides are simple sugars like glucose, which can be absorbed into the blood.

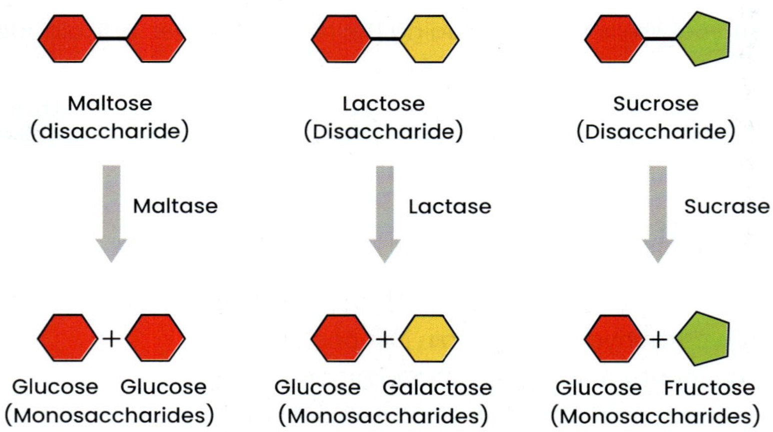

Figure 30.4: Digestion of disaccharides in the intestine.

Not all carbohydrates can be digested by the human digestive system. Very complex polysaccharides (e.g. cellulose) form strong internal bonds that can't be broken down by humans, so they pass through the digestive system unchanged. These indigestible carbohydrates are called **fibre**, and while they can't be digested and absorbed, they play an important role that we will look at later in the chapter.

Pancreatic juice is a mixture of enzymes and bicarbonate ions released from the pancreas into the small intestine.

Fibre is the collective term for indigestible carbohydrates.

FOCUS ON ... THE MOUTH AND TEETH

The mouth plays a major role in digestion. The palate and tongue contain taste buds, which give us a sense of taste. Being able to taste food is important: it allows us to detect if a food is sweet or salty and provides a sense of pleasure when eating. It can also detect foul tastes, signalling the food may be dangerous to consume. Taste also can stimulate saliva production.

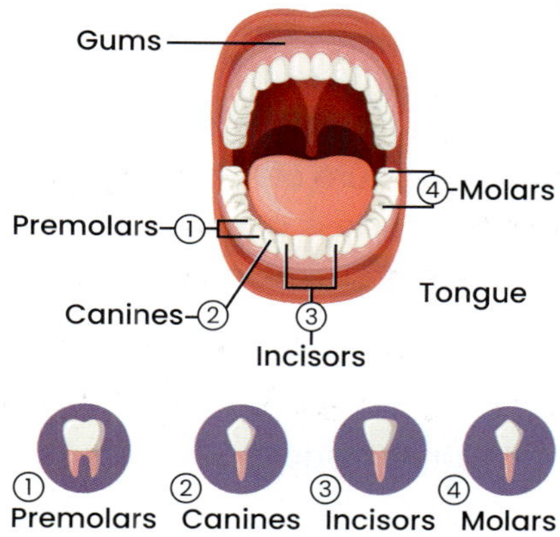

Figure 30.5: The mouth.

The mouth also contains teeth. Adult humans typically have 32 teeth, with four different kinds:

- **Incisors:** These sharp, front teeth are well-suited for biting and tearing off pieces of food, initiating the breakdown process.
- **Canines:** These pointed teeth are located next to the incisors and aid in tearing and shredding tougher foods.
- **Premolars:** Located behind the canines, premolars are responsible for grinding and crushing food particles into smaller pieces, increasing surface area for enzyme action.
- **Molars:** These flat-surfaced teeth located farthest back in the mouth are the strongest and perform the most powerful grinding and crushing of food particles.

The teeth play a vital role in the initial stage of digestion, known as physical digestion. By grinding and crushing the food, it increases the surface area making it easier to break down by enzymes later. Smaller food particles also make it easier to move food through the digestive system. When we swallow, the mouth forms the food into a smooth round mass called a bolus that can be easily transported through the oesophagus.

PROTEIN DIGESTION

As we discussed in Chapter 14, proteins are long and complex molecules, made from units called amino acids. A typical protein contains several hundred amino acids although some can contain thousands. The digestion of enzymes begins by breaking down proteins into shorter chains of amino acids called **peptides**. Later, the peptides are broken down into individual amino acids, which can then be absorbed by the blood.

Peptides are short chains of amino acids.

Stage 1: Mouth

There is no chemical digestion of proteins in the mouth but the chewing of food (physical digestion) is important for increasing the surface area for enzyme digestion later.

Stage 2: Stomach

The digestion of proteins begins in the stomach. First, the stomach churns your food for up to three hours, breaking it down into smaller pieces. This is physical digestion. The stomach also produces **gastric juices**, a mixture of hydrochloric acid, thick mucus and the enzyme **pepsin**. Pepsin is a **protease**, which breaks down proteins into peptides. This is chemical digestion.

Pepsin is an enzyme that breaks down proteins into peptides in the stomach.

Trypsin is an enzyme, found in pancreatic juice, that breaks down peptides into amino acids.

Pepsin works best in acid conditions (pH 2); this is where the hydrochloric acid comes in. It creates the optimum conditions for the pepsin to work. The acid also kills bacteria and loosens the food. The thick mucus protects the wall of the stomach from the acid.

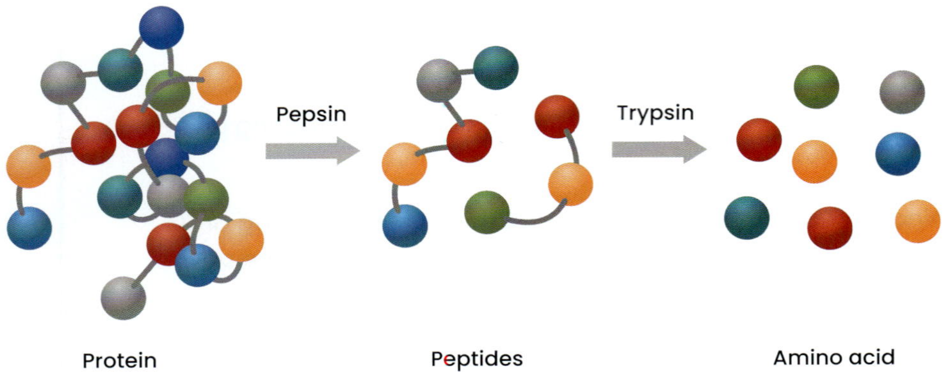

Figure 30.6: Pepsin breaks proteins down into peptides. Trypsin breaks down peptides into amino acids.

Stage 3: Pancreas and small intestine

Pancreatic juice contains more **proteases**, including the enzyme **trypsin**. Trypsin breaks down peptides into individual amino acids. Any remaining peptides are broken down by enzymes released by the small intestine wall.

LIPID (FAT) DIGESTION

We learned in Chapter 14 that lipids (fats) are composed of small units called fatty acids attached to a molecule of glycerol. The aim of lipid digestion is to break down the fats into individual fatty acid molecules and glycerol. The main challenge for lipid digestion is that lipids don't mix with water, which is the medium for enzymatic reactions.

Stage 1: Mouth

Again, chewing breaks the food down into smaller pieces increasing surface area for chemical digestion later. There is minimal digestion of lipids in the mouth although saliva contains a tiny amount of **lipase**, the enzyme which digests lipids.

Lipase is an enzyme that breaks down lipids (fats).

Stage 2: Gall bladder and bile

Bile is a yellowish-green fluid produced by your liver and stored in your gall bladder. When you're eating, the gall bladder releases bile into the small intestine, beginning the digestion of lipids. Bile contains **bile salts** but no digestive enzymes. The bile salts **emulsify** fats, turning them into smaller droplets. This increases the surface area for the enzymes to break down the lipid molecules later.

Bile emulsifies fats, turning them into smaller droplets

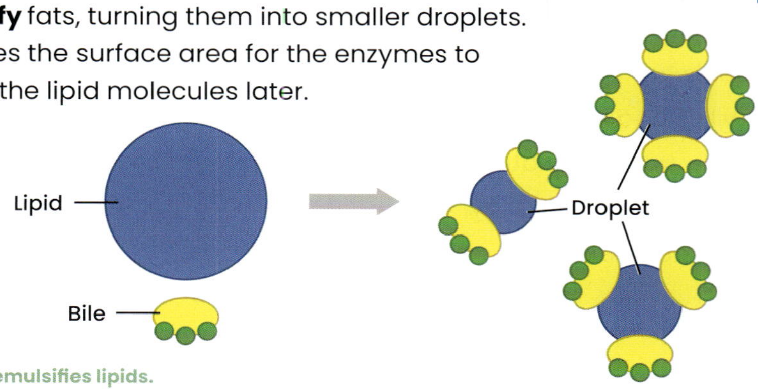

Figure 30.7: Bile emulsifies lipids.

Stage 3: Pancreas and small intestine

Pancreatic juice contains the enzyme **lipase**, which breaks down emulsified fats into their building blocks: fatty acids and glycerol. Lipase works best at neutral pH and, just like in protein digestion, the bicarbonate ions help neutralise the acidic stomach contents.

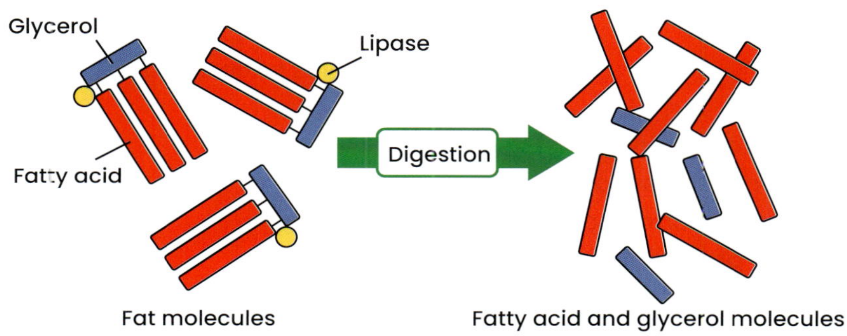

Figure 30.8: Lipase breaks down lipids.

FOCUS ON ... THE PANCREAS

The pancreas is a leaf-shaped organ that sits directly under the stomach. The pancreas has two main functions: digestion and sugar control.

We've already discussed how the pancreas creates pancreatic juice, which contains enzymes for digesting carbohydrates, proteins and lipids. These enzymes are released into the small intestine.

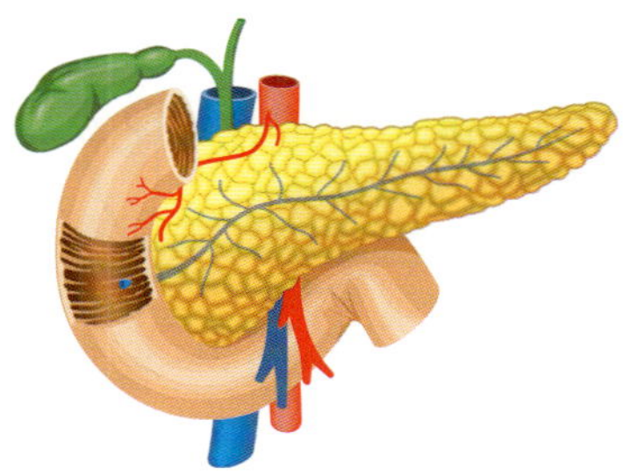

Figure 30.9: The pancreas.

The pancreas is also an **endocrine** gland, meaning it releases **hormones** into the blood. Specifically, it releases **insulin**, which helps to regulate sugar levels in the blood. We will learn more about insulin in Chapter 34.

PORTFOLIO LINK **Go to your Learning Portfolio and complete Learning Check 2 to assess your knowledge and understanding of the topic so far.**

Section 3: Absorption, Transport and Storage of Digested Food

At the end of this section, you should be able to:

- Define what is meant by 'absorption'.
- Discuss how the small intestine is adapted for efficient absorption of digested food.
- Describe the structure of the villus and how it facilitates absorption of digested food.
- Discuss the role of the circulatory and lymphatic systems in transporting digested food.
- Outline how digested foods are stored in the body.
- Discuss how dietary choices impact the digestive system.

Unpacked from LOs U1.2, U2.6, 2.6.5, 2.6.6, 3.2.4

ABSORPTION

The next step after food digestion is **absorption**. Absorption involves moving the products of digestion (monosaccharides, amino acids, fatty acids and water) out of the alimentary canal and into the blood. Absorption mostly takes place in the small intestine, although the absorption of water occurs in the large intestine.

- **Absorption** is the movement of digested nutrients from the small intestine into the blood.
- **Villi** are finger-like projections of the small intestine that increase surface area for absorption.

Small intestine

The small intestine is perfectly adapted for absorption of digested nutrients. Firstly, it is long (approximately 5.5m) and secondly, it has a very large surface area – over $30m^2$. The large surface area is mainly due to the presence of tiny finger-like projections called **villi** (singular: villus). This gives the inner lining of the small intestine a similar surface to a microfibre cloth.

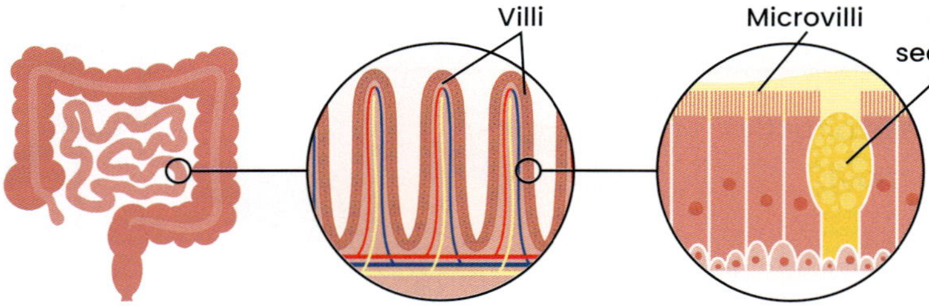

Figure 30.10: The structure of the small intestine.

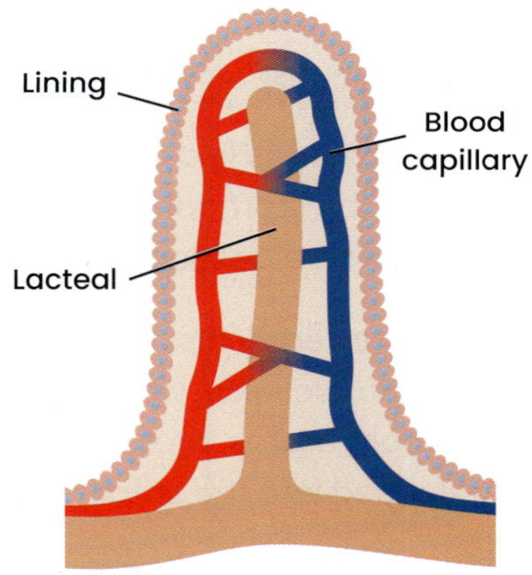

Figure 30.11: Internal structure of the villus.

Villi

The villi are the functional units of the small intestine. They are only around 1mm long and the small intestine contains billions of them. Each villus contains a blood capillary and a lacteal (a small extension of a lymphatic vessel). The villus wall is very thin, just one cell thick, meaning materials can easily move through the villus wall and into the capillary or **lacteal** within.

- A **lacteal** is an extension of a lymphatic vessel, where fats are absorbed in the villi.

The movement of digested nutrients from the intestine to the blood is carried out by **diffusion** and **active transport**. In diffusion, nutrients will move from an area of high concentration (the intestine) into the blood (low concentration). Diffusion does not require energy.

In active transport, energy (in the form of ATP) is used to move nutrients from the intestine into the blood using specific **transport proteins** on the membrane of the villus surface. This ensures all the digested nutrients are fully absorbed into the blood.

Go to YouTube.com and search for 'Animation 6.5 Absorption' (2:03) by Oxford Mastering Biology.

FOCUS ON ... ENTEROCYTES

The villus wall is composed mainly of cells called enterocytes, which are responsible for nutrient absorption. They are rich in mitochondria, which produce the ATP needed to absorb nutrients by active transport. Their outer membrane is heavily folded, forming tiny projections called microvilli, which greatly enhances the surface area of the villus further.

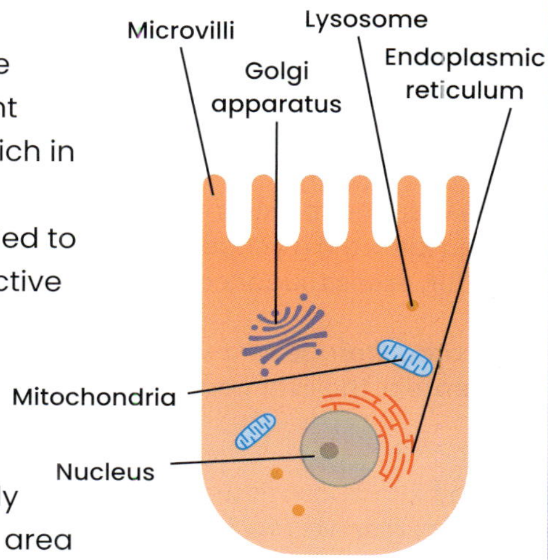

Figure 30.12: An enterocyte.

Digested carbohydrates (monosaccharides), proteins (amino acids), vitamins and minerals move from the intestine into the blood capillary within the villus and travel within the circulatory system. Digested lipids (fatty acids and glycerol) are passed to the lacteal and travel within the lymphatic system.

- **Diffusion** is the movement of materials from an area of high concentration to an area of low concentration.
- **Active transport** is the movement of nutrients across membranes using energy.
- The **hepatic portal vein** is a blood vessel bringing absorbed nutrients from the intestines to the liver.

TRANSPORTING NUTRIENTS

Once the digested nutrients are absorbed, they need to be transported around the body to the cells that need them or, if there is excess, to be stored in the body.

Digested carbohydrates and proteins travel within the blood, part of the circulatory system. All the capillaries converge forming a specialised blood vessel called the **hepatic portal vein**. The hepatic portal vein carries the digested food to the liver, where any toxins are removed before travelling to the heart. The heart pumps blood, with the nutrients, to every cell in the body.

Digested lipids are transported within the lymphatic system. The lymphatic system is a network of vessels that runs parallel to the circulatory system. (We don't study the lymphatic system in detail for Leaving Certificate Biology, but it is important to know it plays a role in digestion and immunity.)

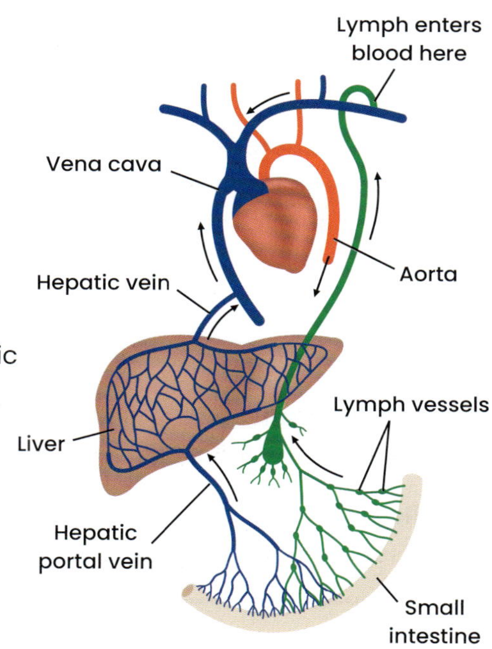

Figure 30.13: Transport of digested foods.

RESEARCH ACTIVITY

Crohn's disease is a digestive disease where the villi can be damaged or reduced in size. Carry out further research on Crohn's disease, its causes, symptoms and treatment.

Storage of digested nutrients

Many of the digested nutrients will be needed immediately by the cells. The products of carbohydrate digestion, simple sugars like glucose, will be used to produce ATP, the energy currency of the cell. A typical cell may require as much as 10 million ATP per second, so there is a high demand for sugar molecules. Digested proteins, amino acids, are needed to build new muscle, bone and other metabolic proteins like enzymes while digested fats, fatty acids and glycerol, can be used to form ATP or build new cell membranes.

However, the body will often have to store excess sugars, amino acids or fatty acids.

CARBOHYDRATES	Excess glucose is stored in the liver and muscles as **glycogen**, a readily available energy source. When energy is needed later, the glycogen can be broken back down into glucose for use. If glycogen stores are full, excess sugar can be converted to fatty acids.
PROTEINS	The body doesn't have dedicated storage for excess protein; however, if there's excess amino acids from digested protein, the liver can convert them to glucose for energy or store as glycogen.
LIPIDS	Excess fatty acids are stored within adipose tissue, specialised cells under the skin. This tissue acts as a long-term energy reserve, providing heat insulation and protection as well.

Figure 30.14: Storage of excess digested nutrients.

FOCUS ON ... THE LIVER

The liver is the largest and most metabolically active organ in the human body. It has over 500 metabolic functions, including:

Storage: The liver stores excess glucose (sugar) as glycogen. It also helps regulate blood sugar levels by releasing glucose when needed. It also stores fat soluble vitamins.

Detoxification: The liver acts as a detoxifying organ, removing toxins, waste products and harmful substances from the bloodstream before they can circulate throughout the body.

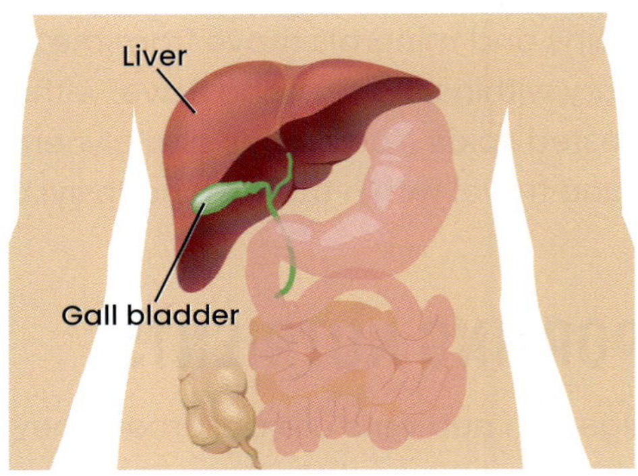

Figure 30.14: The human liver.

Deamination: The liver converts excess sugars into amino acids. This produces urea, which is later removed by the kidneys.

Protein synthesis: The liver is responsible for producing many essential proteins involved in various bodily functions, such as blood clotting, immune response and enzyme production.

Bile production: The liver produces bile, a digestive juice that aids in the breakdown and absorption of fats in the small intestine.

SCIENCE IN SOCIETY – HEALTH

Gut bacteria

The gut bacteria form part of our microbiome, the community of micro-organisms that live in and on us. The gut bacteria play an important role in the digestive system.

- The gut bacteria **produce enzymes** that help break down food. Some of these enzymes are only produced by bacteria, including some that can break down fibre.
- Gut bacteria can **produce vitamins**, including vitamins K and B.
- Gut bacteria keep your gut healthy and working efficiently by **competing with potentially harmful bacteria**.

Dietary choices and digestion

The foods we eat can affect how efficiently our digestive system works. In Chapter 14, we discussed the importance of a balanced diet, with reference to the food pyramid. The following factors also have implications for digestive health.

- **Variety of foods:** A varied diet which includes fruits, vegetables and wholegrains contributes to a healthy digestive system. These foods are generally high in **fibre** and are a source of vitamins and minerals.
- **Fibre:** Fibre is the collective term for indigestible carbohydrates, e.g. cellulose. It acts as a prebiotic, promoting the growth of beneficial gut bacteria, which contribute to a healthy digestive system. Fibre also adds bulk to stool, aiding smooth passage and preventing constipation. Males generally need more fibre than females.
- **Fats:** Healthy **unsaturated fats**, found in sources like olive oil and avocados, can aid digestion and nutrient absorption. However, excessive saturated and trans fats from processed foods can slow down digestion.
- **Sugar:** Excessive sugar intake can disrupt the delicate balance of good and bad bacteria in the gut. This can lead to digestive problems like bloating, gas and even diarrhoea.
- **Hydration:** Proper hydration is crucial for digestion. Water is the medium for enzyme reaction, helping to break down food. It also keeps stool soft for easier passage, and prevents constipation.
- **Age and gender:** As we get older, our digestive efficiency decreases. Older people produce less stomach acid and digestive enzymes and need more water, fibre and vitamins. Men and women have slightly different nutritional needs. Generally, males require more carbohydrates (for energy) and proteins than females while females require more mineral nutrients, especially iron (due to blood lost during menstruation). Pregnant women require additional nutrients, including folic acid (vitamin B_9) which is needed for embryonic spinal development.

Figure 30.16: Sources of dietary fibre, crucial for healthy digestion.

RESEARCH ACTIVITY

LO 2.6.6

Research the implications of dietary choices. Consider a dietary choice (e.g. veganism, low-carb diet) and research the potential benefits and disadvantages of that choice.

PORTFOLIO LINK

Go to your Learning Portfolio and complete Learning Check 3 to assess your knowledge and understanding of the topic so far.

Chapter Summary

- **Digestion** is the breakdown of food (biomolecules) into smaller units.
- Physical digestion is the breakdown of food **mechanically** (e.g. by teeth) to **increase surface area** of the food.
- Chemical digestion uses **enzymes** to break food molecules into smaller units.
- The alimentary canal is a series of connected organs along a long tube including the **mouth**, **oesophagus**, **stomach**, **small intestine**, **large intestine**, **rectum** and **anus**.
- Accessory organs which aid in the digestive process include: **salivary glands**, **tongue**, **liver**, **gall bladder** and **pancreas**.
- The mouth is the site of **ingestion**. **Teeth** break the food down into smaller pieces while it mixes with **saliva**.
- The oesophagus is a long muscular tube which pushes the food down into the stomach by a process called **peristalsis**.
- The stomach produces **gastric juices**, which contain **hydrochloric acid**, to loosen food and kill bacteria, and the enzyme **pepsin** which begins the process of breaking down proteins.
- The small intestine is the site of absorption, where digested food molecules enter the blood.
- The large intestine is the site of water absorption; undigested materials (**stool**) are stored in the **rectum** before being removed through the **anus**.
- Digestion of carbohydrates:
  - In the mouth, teeth increase the surface area; amylase, found in the saliva, breaks down starch into maltose.
  - The pancreas produces pancreatic juice which contains amylase to further digest starch.
  - The small intestine wall releases **maltase**, which breaks down maltose, **lactase** which breaks down lactose and **sucrase** which breaks down sucrose.
- Digestion of proteins:
  - In the stomach, pepsin breaks down proteins into peptides – shorter chains of amino acids.
  - Pancreatic juice contains **trypsin** which breaks down peptides into individual amino acids.
  - The wall of the small intestine also releases **protease** (enzymes that digest proteins).
- Digestion of lipids (fats):
  - Saliva contains small amounts of **lipase**, the enzyme that digests lipids.
  - The gall bladder releases **bile** (produced by the liver) which **emulsifies** fats, turning them into smaller droplets.
  - Pancreatic juice contains the enzyme **lipase** which breaks down emulsified fats into fatty acids and glycerol.
- The products of digestion:
  - Carbohydrates are broken down into **monosaccharides** (simple sugars).
  - Proteins are broken down into **amino acids**.
  - Lipids are broken down into **fatty acids** and **glycerol**.
- Absorption involves moving the products of digestion out of the alimentary canal and into the blood.
- Absorption takes place in the small intestine, in small finger-like structures called **villi** (villus, singular).
- A villus contains a **blood capillary** and a **lacteal**.
- Absorption of digested nutrients takes place by **diffusion** and **active transport**.
- Blood containing digested nutrients first travels to the liver via the **hepatic portal vein**.
- The liver removes toxins from the food, before they are sent all over the body.
- Excess carbohydrates are converted to **glycogen**, which is stored in the liver, or fats, which are stored under the skin.
- Excess proteins are converted to carbohydrates by the liver (**deamination**).
- Excess fats are stored in the **adipose tissue** under the skin.
- Egestion is the removal of indigestible waste.
- Gut bacteria produce vitamins, enzymes to aid digestion and compete with harmful bacteria.
- An efficient digestive system requires **high fibre**, **high unsaturated fats**, **low sugar** and sufficient **hydration**.

END OF CHAPTER ASSESSMENT

REMEMBER

1. List the major organs of the digestive system in the order in which food travels through them.
2. What is the role of teeth in the digestive system?
3. Briefly outline the role of saliva in the digestive system.
4. Name the enzyme in the stomach that breaks down proteins.
5. What is the function of the acid in the stomach?
6. Where does most of the chemical digestion of food take place?
7. What are the structures on the inner lining of the small intestine that increase surface area for absorption called?
8. What organ stores bile produced by the liver?
9. What is the function of the large intestine?
10. How does each of the following impact on a balanced diet?
    (a) Age (b) Sex (c) Activity level

UNDERSTAND

11. Explain the difference between physical and chemical digestion.
12. (a) Label the parts of the digestive system shown.
    (b) In which part shown does absorption take place?

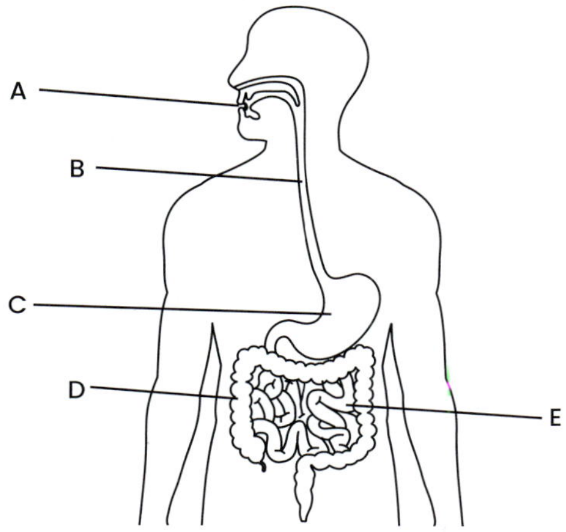

13. Describe the role of bile in fat digestion.
14. How does the structure of a villus aid in nutrient absorption?
15. In what form are carbohydrates stored in the body?
16. What is the main function of the liver in the digestive system?
17. Outline the role of fibre in the diet.
18. Explain the term *deamination*.
19. Write a note on each of the following types of enzymes under the following headings: (i) role and products formed, (ii) production site and (iii) optimum pH.
    (a) Amylase
    (b) Protease
    (c) Lipase

APPLY

20. Using your knowledge of the stomach, explain why indigestion tablets (e.g. Gaviscon) contain alkali salts.
21. You are planning a meal for someone recovering from surgery. What types of foods would be easiest for them to digest? Why?
22. Explain how gut bacteria can contribute to a healthy digestive system.

ANALYSE

23. Compare and contrast the digestion of carbohydrates, proteins and fats.
24. Why is it important for the digestive system to be able to break down complex molecules into simpler ones?
25. Analyse the role that enzymes play in the digestive process.

EVALUATE

26. Some people believe that consuming large amounts of probiotics (live bacteria) can improve digestion. Do you think this is a reliable approach to digestive health? Why or why not?
27. Research a specific digestive disorder (e.g. lactose intolerance). Explain the cause of the disorder and how it affects digestion.

CREATE

28. Create a model or diagram that illustrates the process of food moving through the digestive system.
29. Design a future technology that could help diagnose or treat digestive disorders. Explain how it would work and the benefits it could offer.

UNIT 10

HUMANS – BLOOD

CHAPTERS

STRANDS

Unifying Strand: Nature of Science

Strand 2: Structures and Processes of Life

31 The Circulatory System

LEARNING OUTCOMES: U2.3, U2.5, 2.6.9, 2.6.10, 2.6.11, 2.6.12, 2.6.13

EYE-OPENER

Glass frogs live in the rainforests of Central and South America and have a unique adaptation; their skin is almost completely transparent, allowing them blend seamlessly into their surroundings and avoid being eaten by predators. The skin on their bellies is so clear that you can see their hearts beating inside their bodies and their blood flowing through their arteries. Sadly, these amazing animals are under threat, as their forest habitats are being cut down.

SUBJECT VOCABULARY

blood, plasma, haemoglobin, closed circulatory system, double circulation, cardiac muscle, coronary arteries, SA node (pacemaker), pulse, blood pressure, artery, vein, arteriole, venule, capillary, lumen

At the end of this chapter, you should be able to:

- Explain the composition of blood; describe the functions of each component of blood.
- Describe the flow of blood through the body and the overall functions of the circulatory system.
- Describe the structure of the heart, using models and/or dissection, and relate the parts to their function.
- Describe the heartbeat; understand how and why heartbeat can fluctuate during activity.
- Distinguish between heart rate and blood pressure.
- Compare the structure of arteries, veins and capillaries and discuss their functions.
- Discuss the role of the circulatory system in the transport of digested nutrients, oxygen, hormones and the waste products of metabolism.

LINKS TO:

Chapter 14	**Biomolecules**
Chapter 18	**Movement through Cell Membranes**
Chapter 23	**Respiration**
Chapter 28	**The Breathing System**
Chapter 29	**The Urinary System**
Chapter 30	**The Digestive System**
Chapter 34	**The Endocrine System**

Section 1: Blood

At the end of this section, you should be able to:

- Describe the overall composition of blood.
- Outline the composition of plasma.
- Describe the functions of white blood cells, red blood cells and platelets.
- Understand why blood composition might fluctuate.
- Describe the flow of blood through the body, and the role of the heart, arteries, veins and capillaries.

Unpacked from LOs 2.6.10, 2.6.12

The human circulatory system is composed of three major components:

1. **Blood:** a deep red liquid suspension composed of water, proteins and three types of specialised cells: white blood cells, red blood cells and platelets.
2. **The heart:** a strong muscular organ which pumps the blood around the body.
3. **Blood vessels:** specialised tubes, including arteries, veins and capillaries, which transport essential materials to every cell in the body.

It is estimated that the combined length of the arteries, veins and capillaries in an average human is nearly 100,000km.

DID YOU KNOW?

In previous chapters, we have discussed the role of blood and the circulatory system as it interacts with other body systems, e.g. breathing, digestive and urinary systems. We have discussed how it plays a role in transporting materials like absorbed food molecules, metabolic wastes and gases like O_2 and CO_2 around the body. In this chapter, we will take a closer look at blood, the heart, blood vessels and those interactions with other body systems.

BLOOD

Blood is a vital human body fluid composed of specialised blood cells suspended in a liquid called **plasma**. Blood is categorised as a connective tissue with three main functions:

1. **Transportation** of nutrients, oxygen, hormones and metabolic wastes (e.g. urea, CO_2) through the body's vast network of blood vessels.
2. **Thermoregulation** – distributing heat, produced predominantly in the muscles, throughout the body.
3. **Immune defence** – white blood cells play a major role in fighting infection and disease.

An adult human will have approximately five litres of blood, although this will vary with age and sex.

Blood is a body fluid composed of specialised blood cells suspended in a liquid called **plasma**.

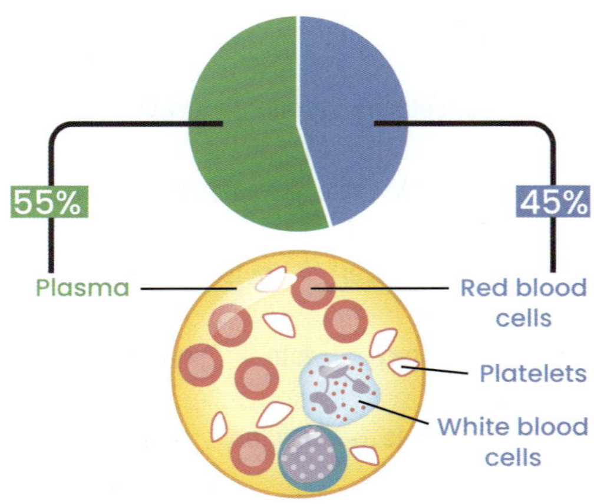

Figure 31.1: Blood is composed of a liquid called plasma and three types of blood cells.

Plasma

Plasma is the liquid portion of blood and accounts for approximately 55% of its volume. It is composed of the following parts:

Plasma is the liquid part of blood; it consists of water, plasma proteins and dissolved nutrients and wastes.

- **Water** (92%): An excellent solvent, water serves as a transport medium for various dissolved solutes. Water also has a high **specific heat capacity**, meaning it holds and distributes heat well.
- **Plasma proteins** (6%): These plasma proteins play major roles in maintaining blood volume and pressure, immune defence, transporting various molecules and in blood clotting.
- **Nutrients** (1%): Glucose (blood sugar), amino acids and lipids (fats) are transported to tissues for energy production and building blocks. Mineral nutrients like sodium, potassium and bicarbonate ions help regulate blood pH and pressure and other cellular functions.
- **Waste products** (1%): Carbon dioxide, a byproduct of cellular respiration, is transported to the lungs for elimination. Other waste products like urea are delivered to the kidneys for excretion.

Blood cells

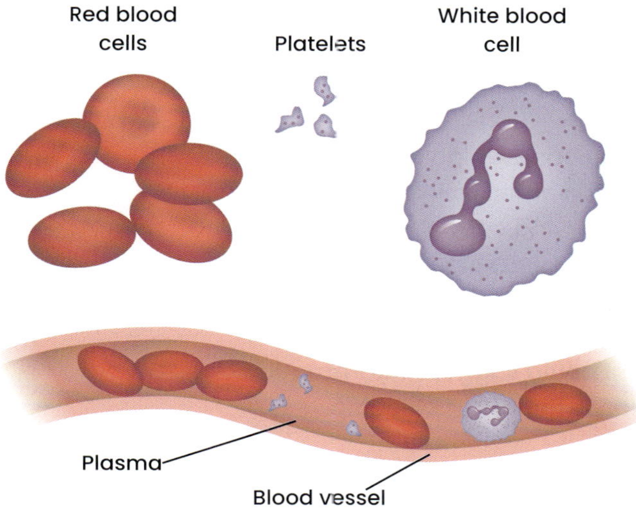

Figure 31.2: Blood is a suspension of blood cells in plasma, flowing through blood vessels.

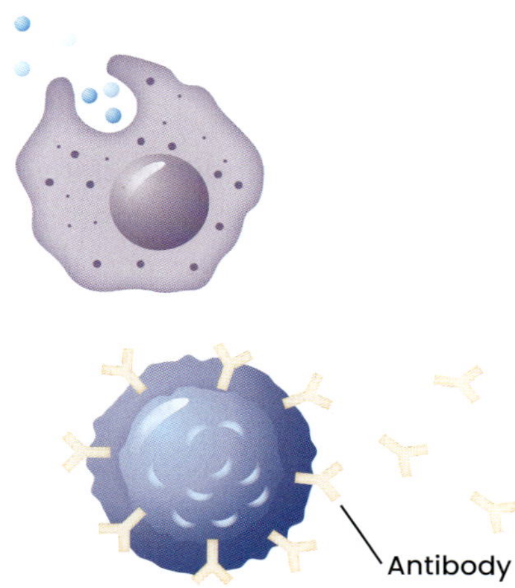

Figure 31.3: Monocytes (top) and lymphocytes (bottom), with antibodies.

Blood cells account for 45% of the volume of blood. They are suspended in the plasma, meaning they float in the liquid rather than dissolve. There are three main types of blood cell:

- **Red blood cells:** The most abundant cells in the blood, red blood cells are responsible for transporting oxygen around the body. They are **biconcave** in shape and contain the iron-rich protein **haemoglobin**, which binds and transports the oxygen. Mature red blood cells do not have a nucleus so they can carry as much haemoglobin (and therefore oxygen) as possible. The biconcave shape gives them the flexibility they need to push through the tiny capillaries delivering oxygen to the cells.
- **White blood cells:** These cells are responsible for fighting infection and disease. There are many different types of white blood cells, each with a different role in the body's immune system, although they are often first categorised as either **monocytes** or **lymphocytes**. Monocytes act by swallowing any foreign cells (a process called phagocytosis) while lymphocytes produce specialised proteins called **antibodies** which provide a more specific immune response. We will explore their role in greater detail in the next chapter.

Haemoglobin is an iron-rich protein found in red blood cells which binds and transports oxygen.

Antibodies are proteins produced by lymphocytes that mark pathogens for destruction.

- **Platelets:** Platelets are cell fragments rather than cells, and are responsible for blood clotting. Clotting is an important process which minimises blood loss when a blood vessel is damaged.

> The blood cells are produced in your bones, specifically in the red bone marrow found in flat bones, like the ribs or skull, or at the end of long bones, e.g. the femur.
>
> DID YOU KNOW?

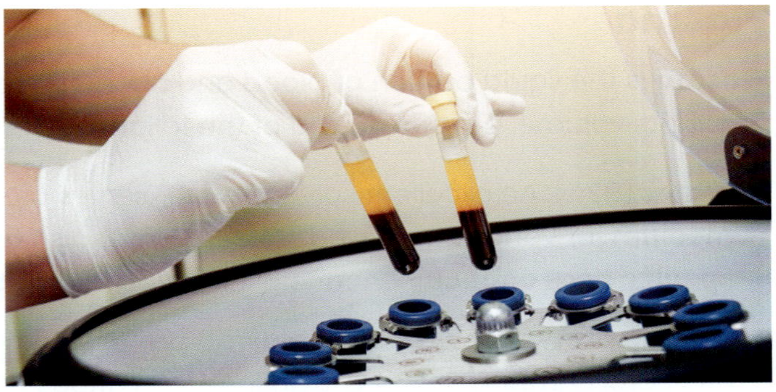

Figure 31.4: The components of blood can be separated by a centrifuge, which spins the blood at high speed, pushing the blood cells out of suspension. You can see the composition of blood more clearly once separated in this way.

SCIENCE IN SOCIETY – HEALTH

Blood tests and health

Blood composition is carefully balanced, but factors like dehydration, illness and even lifestyle choices can cause changes. Dehydration thickens the blood (less water, more cells), making it harder for the heart to pump effectively. Excessive fluids can dilute it, impacting its ability to carry oxygen.

Changes in blood cell numbers are also significant. A high red blood cell count might indicate dehydration or a blood disorder. Low red blood cells could signal anaemia, where oxygen-carrying capacity is reduced. Similarly, changes in white blood cell counts signal potential issues. High white blood cells often point to infection, while low levels might indicate an immune system issue or bone marrow disorder. Even platelet count changes matter. A high count could suggest inflammation or a blood clot risk, while a low count increases the risk of bleeding.

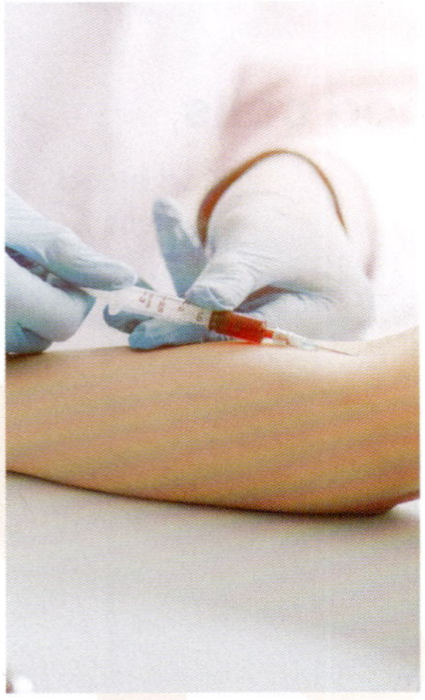

Figure 31.5: Blood tests are a common diagnostic tool.

Blood flow

The human circulatory system is a **closed circulatory system**, meaning blood stays within the blood vessels (for the most part) as it flows throughout the body. The heart is the pump ensuring the blood flows according to the nutrient and oxygen needs of the body. Some simple animals (e.g. insects, starfish) have an **open circulatory system** where the blood is pumped into the body cavity and surrounds the organs. Closed systems are much more efficient, ensuring each cell receives the nutrients and oxygen it needs while allowing for waste to be removed.

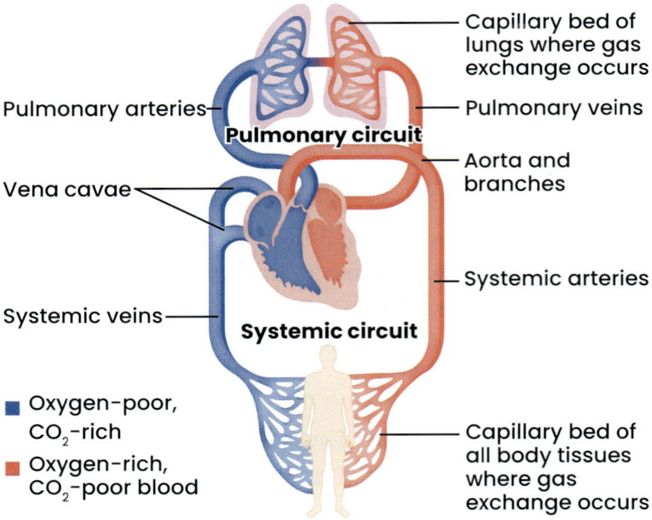

Figure 31.6: Blood flow in humans.

A **closed circulatory system** is a system in which blood stays within the blood vessels.

In humans, blood flows in two connected circuits:

1. **Systemic circuit:** Oxygenated (oxygen-rich) blood is pumped to the other organ systems, delivering nutrients and oxygen and collecting metabolic waste. The circuit begins in the heart, flows through the **aorta** (the main artery in the body) and branches to a series of smaller arteries. These arteries then branch into capillaries in each organ, which later form veins. The blood from the veins all flow into the **vena cava** (the main vein in the body) which delivers the now deoxygenated (oxygen-poor) blood back to the heart.

2. **Pulmonary circuit:** This deoxygenated blood is pumped to the lungs (breathing system) where **gas exchange** takes place (carbon dioxide diffuses out of the blood while oxygen diffuses in – see Chapter 28 for more detail). The oxygen-rich blood is returned to the heart where it then enters the systemic circuit. Again, this circuit begins and ends in the heart.

This **double circulation** system ensures oxygen-rich blood is always available for the body's organs. It is important to note that blood will flow through both systems, completing a figure of 8 as it moves through the body. The heart provides the connection between the two circuits, pumping the blood through both circuits. It is sometimes therefore called a double pump. Our body's five litres of blood circulates through the body roughly once per minute, although blood flows faster in arteries than it does in veins.

As discussed in Chapter 30, blood can also flow through a portal pathway or vein. In our digestive system, blood carrying absorbed food molecules flows from the small intestine to the liver via the **hepatic portal vein**, rather than flowing through the heart.

We will discuss the structure of arteries, veins and capillaries in more detail in Section 3.

FOCUS ON ... BLOOD TYPES

There are four main blood types (A, B, AB and O), although each can be also categorised as positive or negative. Blood types relate to small molecules called antigens found on the surface of red blood cells. There are three such antigens found in human blood. Blood type A has the A antigen, B has the B antigen, AB has both while O has neither. A third molecule, called the Rhesus factor, creates the positive and negative; if it is present then the person is said to have positive blood type.

Knowing your blood type is important for blood transfusions, as receiving the wrong blood type via transfusion can be fatal. People with blood type O negative are often called the *universal donor* as they can give blood to a person with any blood type.

Visit the Irish Blood Transfusion Service website (www.giveblood.ie) to find out more about donating blood.

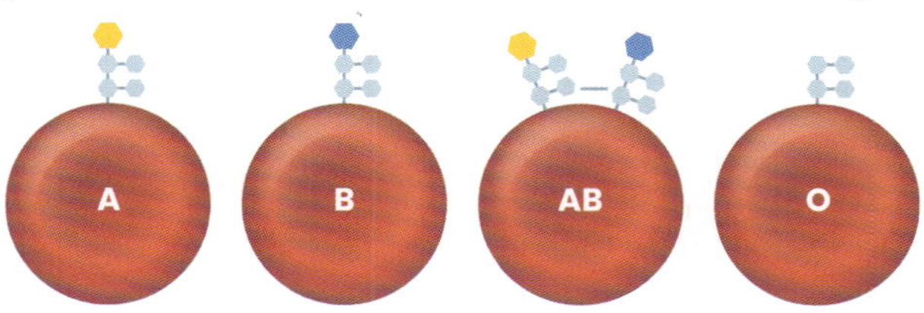

Figure 31.7: The four main blood types.

Double circulation refers to the presence of the pulmonary and systemic circuits in the body.

Go to YouTube.com and search for 'Why do blood types matter? – Natalie S. Hodge' (4:41) by TED-Ed.

PORTFOLIO LINK **Go to your Learning Portfolio and complete Learning Check 1 to assess your knowledge and understanding of the topic so far.**

Section 2: The Heart

At the end of this section, you should be able to:

- Describe the internal and external structure of the heart.
- Draw a diagram of the internal structure of the heart.
- Briefly outline the function of valves in the heart.
- Briefly outline the role of the pacemaker in controlling the heartbeat.
- Distinguish between pulse and blood pressure; explore the causes of increased pulse and blood pressure.

Unpacked from LOs 2.6.9, 2.6.11

The heart is a muscular organ about the size of a fist, located behind the breastbone and slightly left of centre. It's the main pump in the circulatory system, responsible for keeping blood flowing through the pulmonary and systemic circuits.

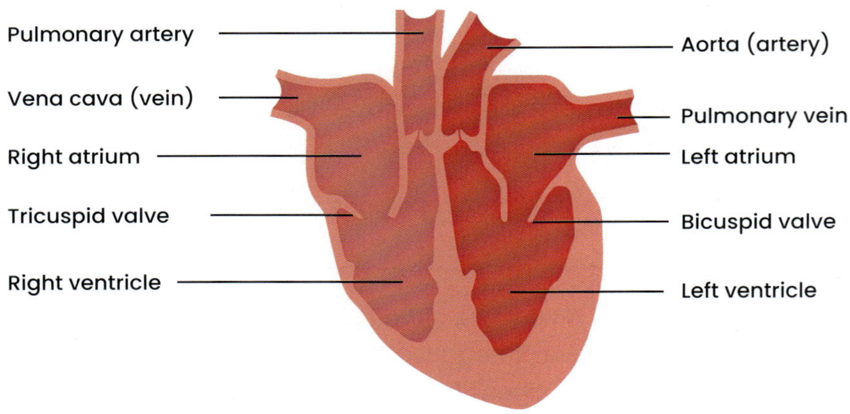

Figure 31.8: The anatomy of the heart. Note the left side appears on the right – imagine you're a surgeon looking at a person's heart from above, with their left on your right.

STRUCTURE OF THE HEART

The heart has four chambers: two upper atria (singular: atrium) that receive blood, and two lower ventricles that pump blood. The right atrium receives deoxygenated blood returning from the body via the **vena cava**, while the left atrium receives oxygenated blood from the lungs via the **pulmonary vein**.

The right ventricle pumps deoxygenated blood to the lungs through the **pulmonary artery** for oxygen pickup, and the left ventricle, the strongest chamber, pumps oxygenated blood throughout the body through the **aorta**. The left and right side of the heart are separated by a muscular wall called the **septum**.

Valves separate the atria and ventricles, ensuring blood flows in one direction. The **tricuspid valve** is found on the right side while the **bicuspid valve** is found on the left side. Smaller valves, called **semi-lunar** valves, are found at the base of the pulmonary artery and aorta, preventing blood flowing back into the heart. Figure 31.9 shows the flow of blood through the heart.

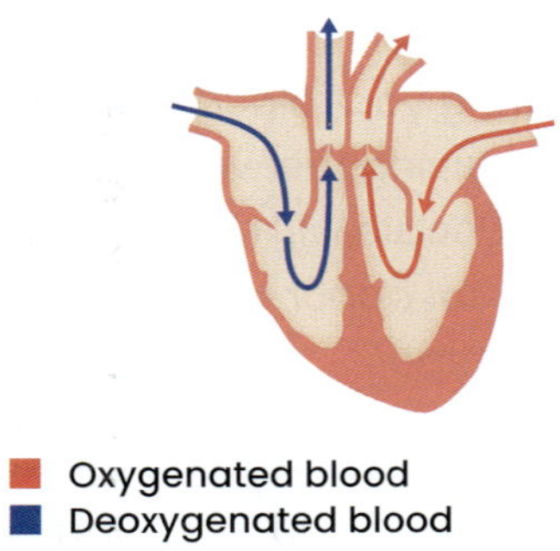

Figure 31.9: Blood flow through the heart.

Go to YouTube.com and search for 'Circulatory System and Pathway of Blood Through the Heart' (8:13) by Amoeba Sisters.

The heart is made from specialised **cardiac muscle** cells, which don't tire or grow like the muscles attached to our skeleton. They're under unconscious control, meaning we can't decide consciously when to contract the heart.

The heart's muscle cells do need lots of nutrients and oxygen to ensure they have sufficient energy; therefore, the heart has its own blood supply. **Coronary arteries**, which branch off from the aorta, supply the heart muscle cells with the blood they need.

The heart beats at a typical rate of 72 beats per minute, controlled by a natural pacemaker within the organ. This steady rhythm ensures a constant supply of oxygen and nutrients to all tissues of the body.

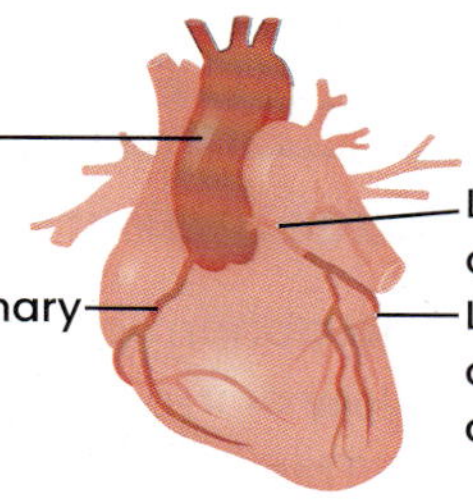

Figure 31.10: The coronary arteries provide the muscle cells of the heart with oxygen and nutrients.

- **Cardiac muscle** is the specialised muscle cells of the heart; they don't tire or grow and are under unconscious control.
- **Coronary arteries** supply the cardiac muscle cells with blood (oxygen and nutrients).

THE HEARTBEAT

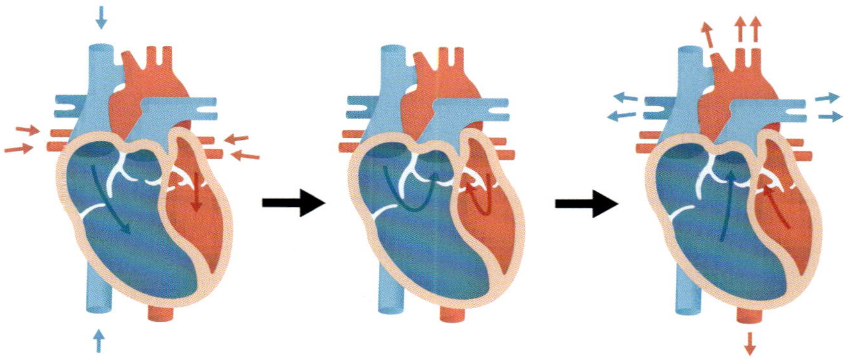

Figure 31.11: The heartbeat has a number of stages.

RESEARCH ACTIVITY

Cardiovascular disease occurs when the coronary arteries become blocked or narrowed. Carry out some research on cardiovascular disease. Report back on the causes, symptoms and treatments for cardiovascular disease.

The heartbeat is controlled by a natural pacemaker called the **sinoatrial (SA) node** which can generate its own electricity to stimulate the muscle cells of the heart to contract. A second pacemaker, the **atrioventricular (AV) node** is also involved.

- The **sinoatrial (SA) node** is the heart's natural pacemaker, controlling the heartbeat.

The heartbeat has a number of repeating stages, which collectively are called the cardiac cycle. We do not need to know the detail on this cycle for Leaving Cert Biology, but it is useful to know that there is a contraction stage (called the systole) and a resting phase (called the diastole). The sound of the heartbeat ('lub dub') is caused by the opening and closing of the valves within the heart.

The rate of the heartbeat changes depending on factors like temperature, physical exercise, stress or sleep. When we exercise, the muscle cells require more sugars and oxygen so the heartbeat increases to meet the demand. When we sleep, our heart rate generally falls.

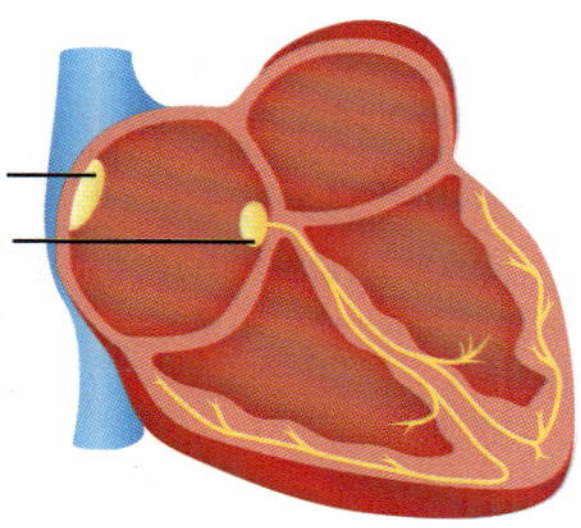

Figure 31.12: The SA and AV nodes of the heart – the heart's natural pacemaker.

SCIENCE IN SOCIETY: TECHNOLOGY AND HEALTH

Artificial pacemakers

For people with abnormal heart rhythms, particularly those that are too slow or irregular, an artificial pacemaker can be a lifesaver. This medical device, implanted near the heart, has electrodes that constantly monitor the heartbeat. If the heart beats too slowly, the pacemaker sends electrical pulses to stimulate the heart muscle back to a healthy rhythm.

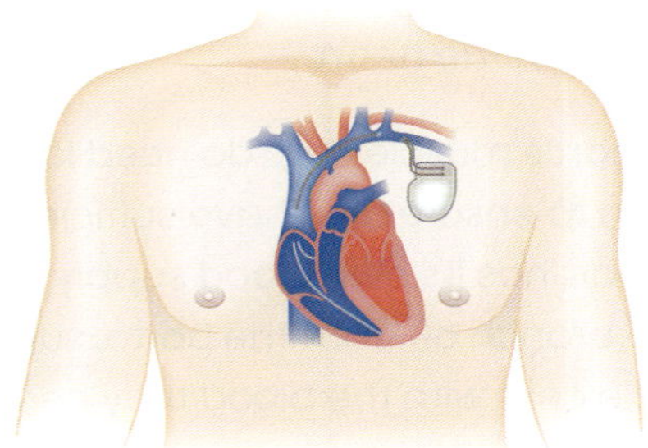

Figure 31.13: Artificial pacemaker.

Artificial valves

Sometimes, the natural valves of the heart can become damaged and the flow of blood through the heart can slow down or even reverse. In these cases, artificial valves can be used to replace the faulty valves. These can be made from carbon or titanium or even living tissue. Traditionally, valves from pigs were used to replace faulty valves in humans but now new valve tissue can be cultured in labs. Carry out a little further research on the pros and cons of each type of material.

Heart rate monitors

Many people routinely wear smart watches with a built-in heart monitor. These watches typically have a green light source, which is absorbed easily by haemoglobin. As blood flows through the arteries of the wrist, the level of light absorbed changes and the pulse rate is recorded.

Pulse and blood pressure

The **pulse**, often felt at the wrist or the neck, is the number of times your heart beats per minute. It reflects how fast your heart is pumping blood at a given moment. Only arteries have a pulse. When your heart ventricles contract (ventricular systole), they push blood through the arteries, stretching their elastic walls. When the heart stops pumping (diastole), the walls of the arteries collapse. The repeating expansion and contraction of the walls of the arteries creates a wave that can be felt as a pulse.

Blood pressure refers to the force of blood pushing against your artery walls when the heart beats. It's measured in two numbers: systolic (pressure when the heart contracts) and diastolic (pressure when the heart relaxes between beats).

Both the pulse and blood pressure are useful metrics for human health and higher or lower than normal figures could indicate heart or cardiovascular problems.

RESEARCH ACTIVITY

Research other technologies that contribute to heart and circulatory system health.

- **Pulse** is the repeating expansion and contraction of the walls of an artery.
- **Blood pressure** is the force of blood pushing against the artery walls.

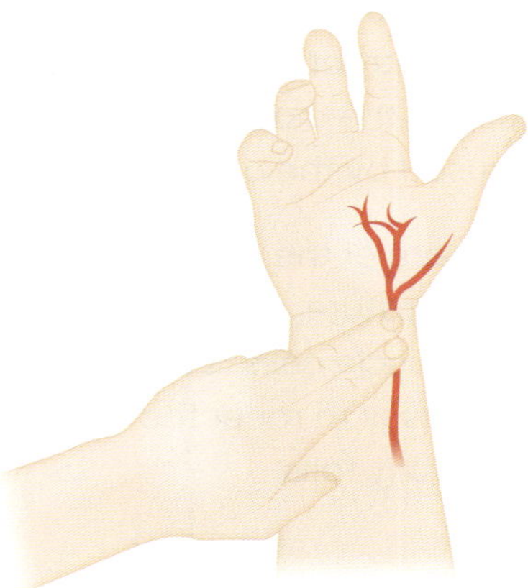

Figure 31.14: How to find your pulse.

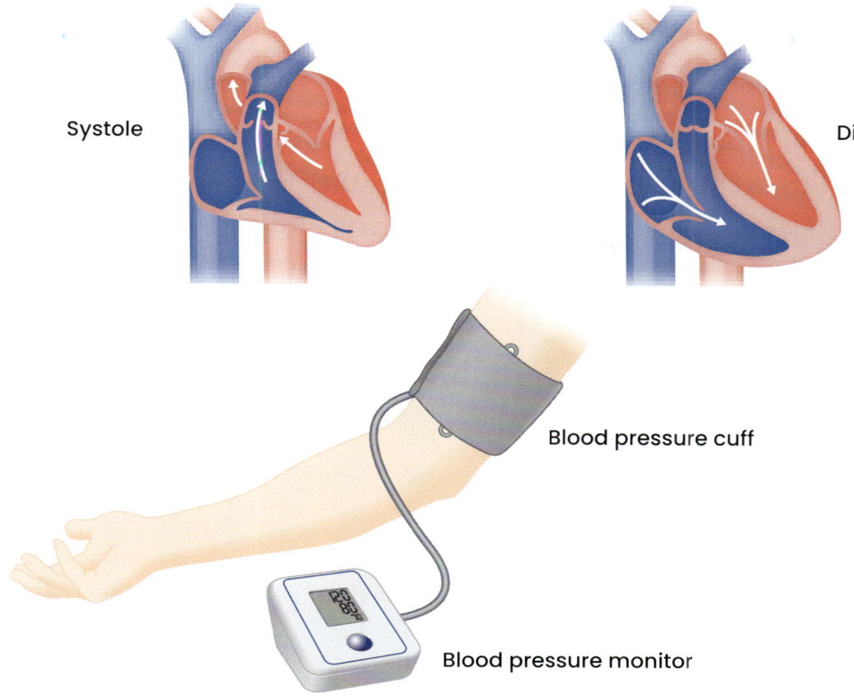

Figure 31.15: Normal blood pressure is 120 over 80, measured using a blood pressure monitor.

FOCUS ON ... FACTORS AFFECTING PULSE AND BLOOD PRESSURE

The pulse and blood pressure can change, depending on a number of factors. These include:

Figure 31.16: You can monitor your blood pressure, among other things, using a smart watch.

- **Physical activity:** Exercise naturally increases your heart rate (pulse) to deliver more oxygen to working muscles. Blood pressure might also rise slightly during exertion, but should return to normal afterwards.
- **Age:** As we age, our arteries can become less flexible and stiffen, leading to slightly higher blood pressure readings.
- **Emotions:** Stress, anxiety or excitement can cause temporary spikes in both pulse and blood pressure.
- **Body position:** Generally, lying down leads to a lower pulse and blood pressure compared to standing. This is because gravity affects blood flow.
- **Medications:** Certain medications, like stimulants or beta-blockers, can influence heart rate and blood pressure.
- **Hydration:** Dehydration can increase heart rate and blood pressure as the heart works harder to circulate thicker blood. Drinking plenty of fluids helps maintain healthy levels.

PORTFOLIO LINK

Go to your Learning Portfolio and complete Learning Check 2 to assess your knowledge and understanding of the topic so far.

Section 3: Arteries, Veins and Capillaries

At the end of this section, you should be able to:

- Compare the structure of arteries, veins and capillaries.
- Discuss how the structure of the blood vessels relates to their function.
- Describe how capillaries allow for the efficient exchange of materials.
- Discuss how the circulatory system works with other body systems to transport materials around the body.

Unpacked from LOs 2.6.10, 2.6.13

BLOOD VESSELS

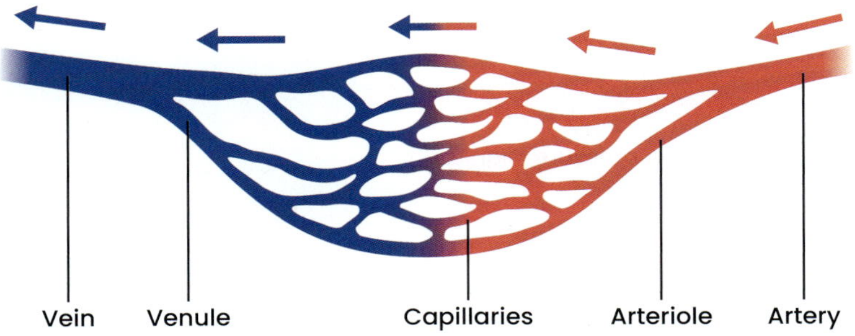

Figure 31.17: The internal structure of arteries, veins and capillaries.

Go to YouTube.com and search for 'GCSE Biology – Blood Vessels' (4:01) by Cognito.

There are three types of blood vessel found in the human body: **arteries**, **veins** and **capillaries**. Arteries carry blood away from the heart to all the major organs, branching off into small arteries called **arterioles**. Veins bring blood back to the heart; they form from smaller veins called **venules**. Capillaries connect arterioles and venules; they are microscopic tubes thin enough to allow materials move between the body tissues and the capillaries themselves.

Structure of arteries

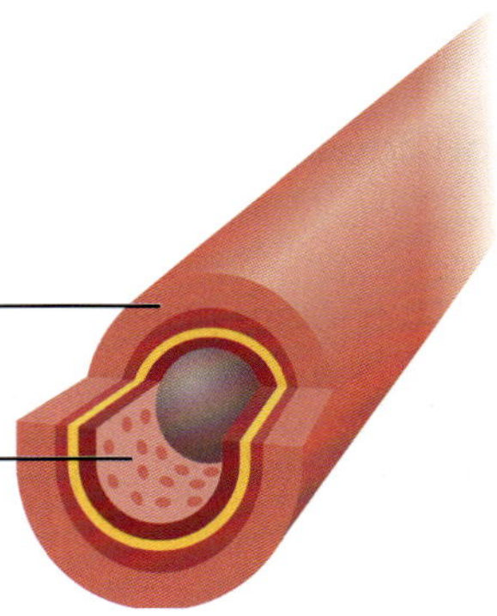

Figure 31.18: Internal structure of an artery.

- **Arteries** are blood vessels that carry blood away from the heart.
- **Arterioles** are small arteries.
- **Veins** are blood vessels that carry blood to the heart.
- **Venules** are small veins.
- **Capillaries** are thin-walled blood vessels that allow for exchange of materials between the blood and body cells.
- The **lumen** is the internal space in an artery or vein.

Arteries have a thick elastic wall resulting in a small internal space (called the **lumen**). As we discussed already, arteries have a pulse. Most arteries carry oxygenated blood (the pulmonary artery is the exception) and all travel away from the heart. For protection, arteries are found deeper under the skin.

Structure of veins

Veins have thinner walls, which results in a larger lumen. Veins lack a pulse, so blood moves more slowly through the veins. When we move, our muscles squeeze the veins resulting in the blood inside moving. Veins contain valves which prevent the blood from flowing backwards. Veins are found closer to the surface of the skin, allowing heat escape when needed.

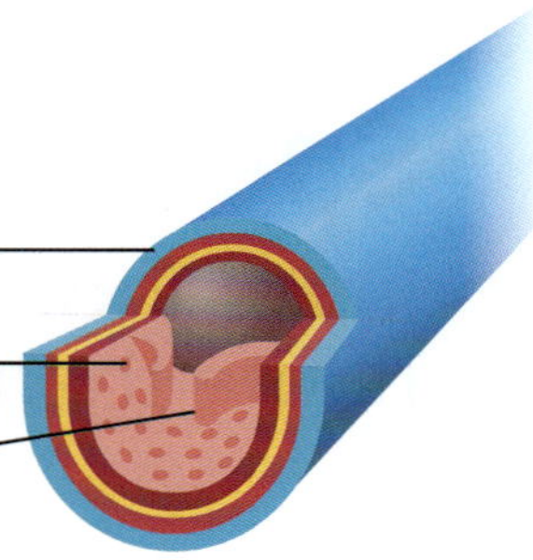

Figure 31.19: Internal structure of a vein.

Structure of capillaries

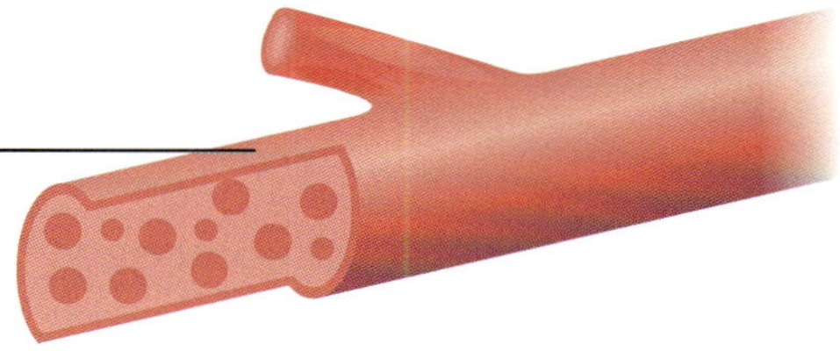

Figure 31.20: Internal structure of a capillary.

The walls of capillaries are only one cell thick. This allows for easy movement of material from the blood to the cells and vice versa. Movement mainly occurs via diffusion with materials (e.g. oxygen or nutrients) moving from an area of high concentration (the blood) to an area of low concentration (the cells).

The network of capillaries is vast (thousands of kilometres long). This creates an enormous **surface area** which, along with the thin walls, creates the perfect conditions for diffusion to occur.

FOCUS ON ... INTERACTIONS WITH OTHER BODY SYSTEMS

The circulatory system works in close collaboration with other body systems to transport materials throughout the body. Each of the following connections are explored in greater detail in the relevant chapters.

Breathing system: Deoxygenated blood from the circulatory system travels to the lungs. In the lungs, carbon dioxide (waste) diffuses out of the blood into the air sacs (alveoli) for exhalation. Simultaneously, oxygen from the inhaled air diffuses into the blood, carried by red blood cells. This oxygenated blood then travels back to the heart for distribution. See Chapter 28, Section 3 for more detail.

Urinary system: Waste products generated by cellular activity (like urea) enter the bloodstream. The circulatory system carries these waste products to the kidneys for filtration. The filtered waste products are then eliminated from the body through urine formation. See Chapter 29, Section 2 for more detail.

Digestive system: The digestive system breaks down food into nutrients. These absorbed nutrients enter the bloodstream through capillaries in the small intestine. The circulatory system then transports these nutrients to all body cells for energy and growth. See Chapter 30, Section 3 for more detail.

PORTFOLIO LINK **Go to your Learning Portfolio and complete Learning Check 3 to assess your knowledge and understanding of the topic so far.**

Specified Investigation 13

Investigate the structures of the heart and relate them to their functions. (LO 2.6.9)

Scan the QR code to watch a video demonstration of this investigation.

In this experiment, we are going to carry out a dissection of a mammalian heart.

Equipment: Fresh or preserved sheep heart, dissecting tray or chopping board, disposable gloves, scalpel, pointing needle and/or forceps, scissors, paper towels.

Safety precautions:

- Always wear disposable gloves to protect yourself from potential pathogens.
- Use scalpels and other sharp instruments with caution.
- Dispose of sharps containers appropriately.
- Wash your hands thoroughly after the dissection.

METHOD

1. External observation:

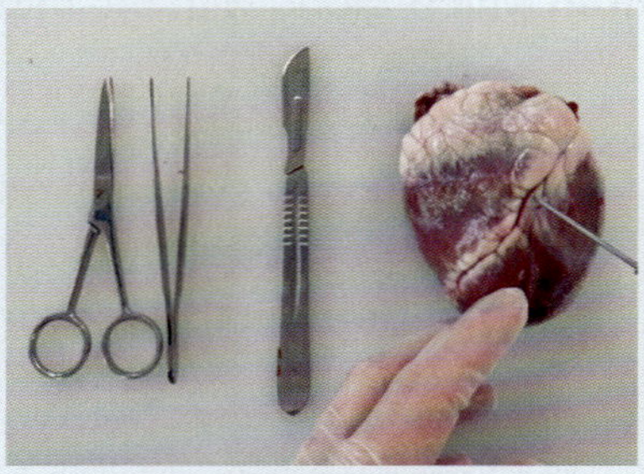

- Place the sheep's heart on the dissecting tray.
- Identify the apex (pointed end) and base (broad end).
- Locate the large blood vessels entering and leaving the heart: vena cava (deoxygenated blood), aorta (oxygenated blood), pulmonary artery (deoxygenated blood to lungs) and pulmonary vein (oxygenated blood from lungs).
- Locate the coronary arteries.

2. Initial incisions:

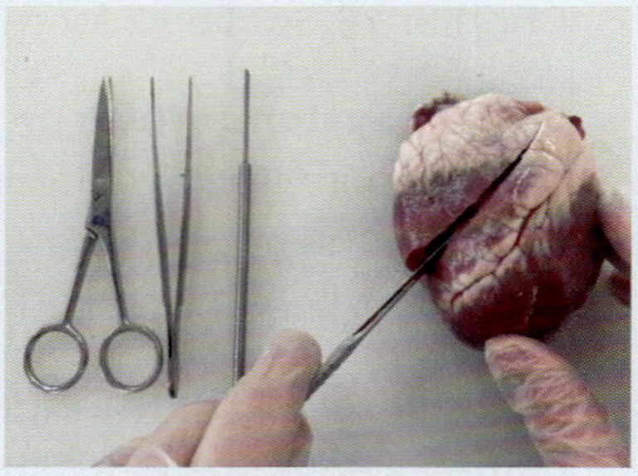

- Using a scalpel, make a shallow cut 1cm left of the septum, into the right atrium and ventricle.
- Be careful not to cut too deeply and damage internal structures.
- Fold open the chambers.
- Repeat on the other side of the heart.

3. Second incision:

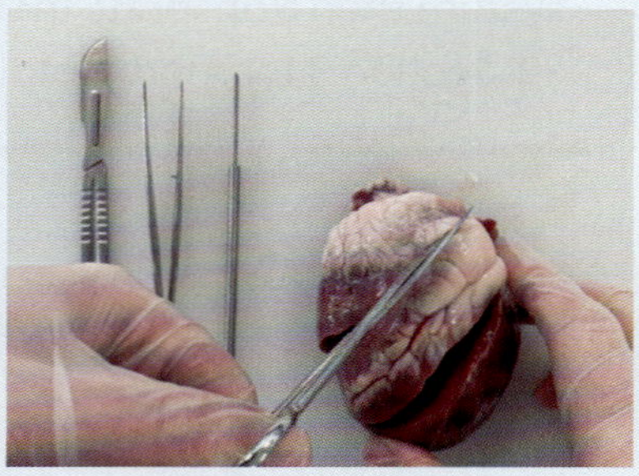

- Using scissors, cut up into the pulmonary artery (on the right side of the heart) and aorta (on the left).
- Carefully spread the cut open with your fingers or blunt probe, exposing the four chambers of the heart: two upper atria (receiving chambers) and two lower ventricles (pumping chambers).

4. Examining the atria:

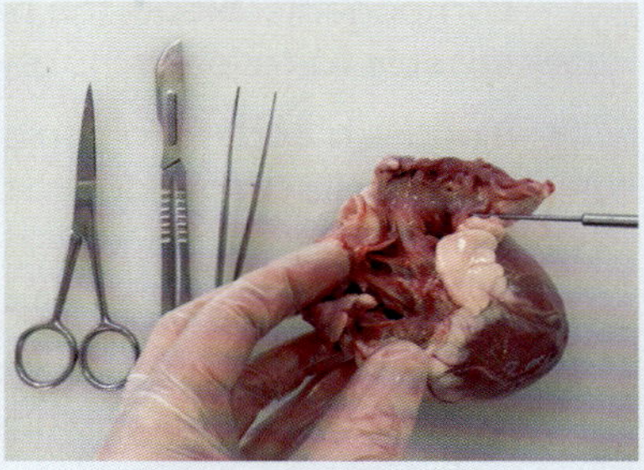

- Identify the right atrium and left atrium.
- Look for the openings of the vena cava in the right atrium and pulmonary veins in the left atrium.

5. Exploring the ventricles:

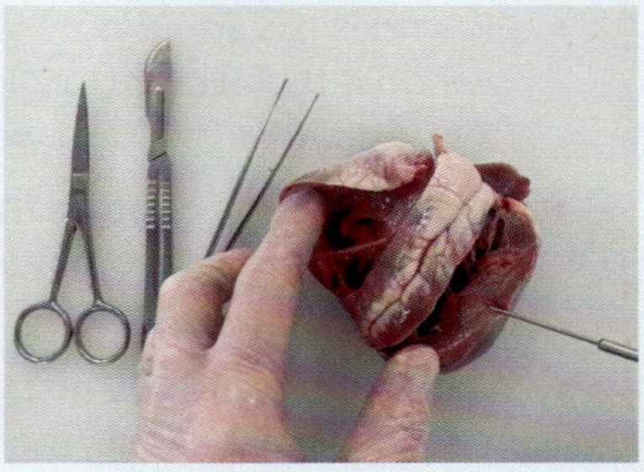

- Open the ventricles fully, exposing their muscular walls and internal surfaces.
- Notice the difference in wall thickness between the right ventricle (thinner) and the left ventricle (thicker). The left side needs to pump all around the body while the right side just pumps to the nearby lungs.

6. Valves:

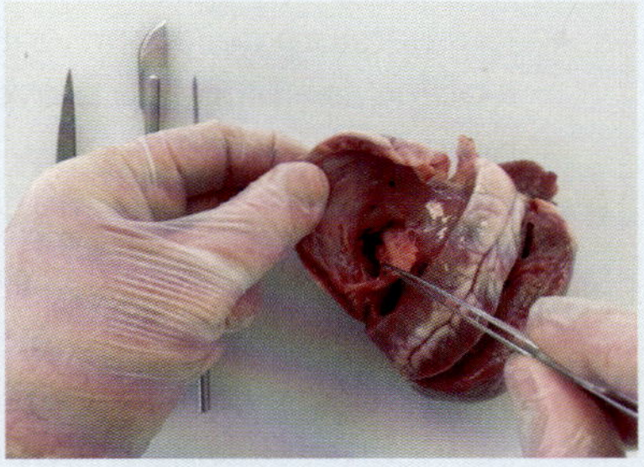

- Locate the tricuspid valve (on the right) and bicuspid valve (on the left) separating the atria and ventricles. These valves prevent backflow of blood.
- Look for the semi-lunar valves guarding the openings of the aorta and pulmonary artery from the ventricles, respectively.

7. Further exploration (optional):

- You can use a probe to gently explore the coronary arteries supplying blood to the heart muscle itself.
- Identify any fatty deposits that might be present in the coronary arteries.

8. Cleaning and disposal:

- Rinse the dissected heart under running water.
- Dispose of the specimen according to your teacher's instructions.
- Wash your hands thoroughly and clean all instruments used in the dissection.

Tips:

- Take your time and observe the structures carefully before making any cuts.
- Use your fingers or blunt probes to gently separate tissues instead of forceful cutting.
- Create a series of labels and label each part of the heart.

INVESTIGATIVE LOG LINK

Go to your Investigative Log to record your lab work and answer questions on the experiment in this chapter.

Chapter Summary

- The human circulatory system has three main components: **blood**, the **heart** and **blood vessels** (arteries, veins and capillaries).
- Blood is a body fluid composed of specialised blood cells suspended in a liquid called plasma.
- Blood has three main functions: transportation of materials (e.g. oxygen and nutrients), thermoregulation and immune defence.
- Plasma is the liquid part of blood and is composed of water (92%), plasma proteins (6%), nutrients (1%) and wastes (1%).
- Red blood cells contain the protein **haemoglobin** which binds with and transports oxygen.
- White blood cells are responsible for fighting infection and disease; there are two main types: **monocytes** and **lymphocytes**.
- Platelets are cell fragments responsible for clotting the blood, to prevent blood loss from injury.
- Humans have a **closed circulatory system**, meaning blood remains in the arteries, veins and capillaries as it flows around the body.
- In humans, blood flows through two connected circuits: the **pulmonary circuit** and the **systemic circuit**.
- Blood also can flow through a portal pathway, where blood flows from one organ to another; the **hepatic portal vein** is an example of a portal pathway where blood flows from the small intestine to the liver.
- There are four main blood types in humans: A, B, AB and O. Blood groups can be Rhesus positive or Rhesus negative.
- The heart is a double pump, pushing blood through both the pulmonary and systemic circuit.
- The heart has four chambers: left atrium, left ventricle, right atrium, right ventricle.
- The heart is made from **cardiac muscle** cells which don't tire or grow and are under unconscious control.
- **Coronary arteries** supply the heart cells with the oxygen and nutrients they require.
- The **heartbeat** is controlled by the **SA node**, the heart's natural pacemaker.
- The **pulse** is the repeating expansion and contraction of the walls of the arteries.
- **Blood pressure** is the force of the blood pushing against the artery walls.
- Blood flow begins in the heart then flows into an **artery**, then into an **arteriole**, **capillaries**, **venules**, **veins** then back to the heart.
- Capillaries provide the thin walls and large surface area for materials to diffuse easily between the blood and the body cells.
- The circulatory system interacts with other body systems:
  - The breathing system allows for exchange of gases between the blood and inhaled air.
  - The digestive system sees absorbed foods transfer to the blood for transportation.
  - The urinary system filters blood at the kidneys, removing metabolic waste.
  - The endocrine system releases hormones into the blood for transportation.

END OF CHAPTER ASSESSMENT

REMEMBER

1. What are the **three** main components of the human circulatory system?
2. What are the different types of blood cells and their main functions?
3. What is plasma?
4. Define each of the following terms.
   (a) Artery
   (b) Vein
   (c) Capillary
5. Name the iron-rich protein found in red blood cells.
6. Define *closed circulatory system*.
7. Name the **four** main blood types.
8. Name the **four** chambers of the heart.
9. What is the function of valves in the heart?
10. What are coronary arteries?

UNDERSTAND

11. What are the functions of blood?
12. Describe the composition of plasma.
13. Distinguish between monocytes and lymphocytes.
14. Distinguish between the pulmonary circuit and the systemic circuit.
15. What is the role of the heart in the circulatory system?
16. Explain the pathway of blood flow through the heart.
17. How do valves in the heart ensure proper blood flow direction?
18. Describe how exercise affects your heart rate and blood pressure.
19. Distinguish between pulse and blood pressure.
20. Using simple diagrams, outline the difference between the structure of arteries, veins and capillaries.
21. Distinguish between arterioles and venules.

APPLY

22. Label the parts (A to K) in the diagram of the heart below.

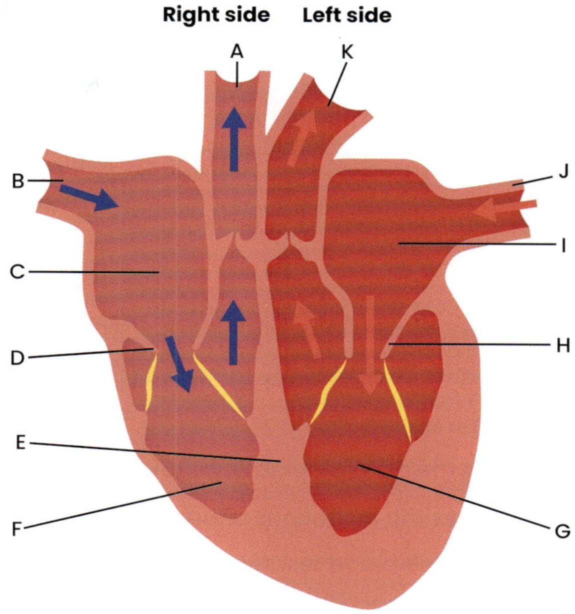

23. Why is the left ventricle of the heart thicker than the right ventricle?
24. Describe the role of blood and blood vessels in body temperature control (thermoregulation).
25. Design a simple diagram illustrating the different blood types (A, B, AB and O) and their compatibility for blood transfusions.
26. Suggest a reason why high pulse and blood pressure can lead to health problems.
27. Evaluate the pros and cons of different lifestyle choices (e.g. diet, exercise) on cardiovascular health.
28. Research a specific cardiovascular disease (e.g. coronary artery disease, stroke) and explain its causes, symptoms and treatment options.
29. What effects might a blockage of the coronary arteries have?

ANALYSE

30. Study the data chart below and answer the questions that follow.
Three students ran on a treadmill for 15 minutes, with their heart rates recorded every five minutes.

	RESTING HEART RATE (BEFORE EXERCISE)	HEART RATE AFTER 5 MINUTES	HEART RATE AFTER 10 MINUTES	HEART RATE AFTER 15 MINUTES	HEART RATE 5 MINUTES AFTER EXERCISE
John	80	120	145	168	120
Mary	72	110	130	150	100
David	65	100	120	135	70

(a) Based on the data, what is the effect of exercise on the heart rate during the exercise period?

(b) Suggest a reason why the heart rate is different between each of the individuals at the 15-minute mark.

(c) Comment on the heart rate five minutes after exercise (the recovery period).

(d) Does the data provide any indication of the fitness levels of the three students?

(e) Discuss the limitations of using heart rate as the sole measure of exercise intensity or fitness level.

31. Below are the blood test results for three individuals who visited a doctor's surgery one day. Mo's results were deemed normal by the doctor.

	RED BLOOD CELLS (trillion cells/l)	WHITE BLOOD CELLS (billion cells/l)	HAEMOGLOBIN (HGB) (g/dl)	PLATELETS (million cells/µl)
Mo	4.8	7.5	14.2	250
Cara	4.2	12.0	12.5	280
Siobhan	3.8	4.0	10.0	200

(a) Which one of the individuals tested might have an infection? Use the data to support your answer.

(b) Siobhan reported symptoms including fatigue, tiredness and pale skin. Suggest a reason for her symptoms.

(c) Suggest a reason why Mo's haemoglobin figures are higher than Cara's and Siobhan's.

(d) Use the data above to comment on the relative proportions of red blood cells, white blood cells and platelets in a blood sample.

32. Compare and contrast the structures and functions of arteries, veins and capillaries.

EVALUATE

33. Analyse the ethical considerations of using new technologies (e.g. wearable health monitors) to track and manage cardiovascular health.

CREATE

34. Design a poster to show the interactions between the circulatory system and other body systems.

32 Infectious Disease and the Human Immune Response

LEARNING OUTCOMES: 2.4.12, 2.4.14, 2.4.15, 2.4.16

EYE-OPENER

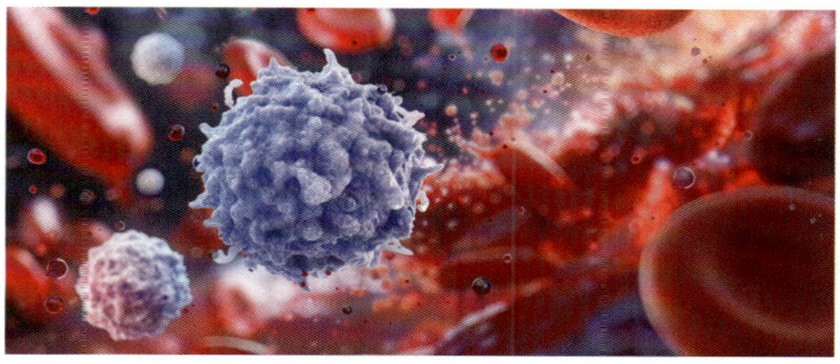

Did you know that 70–80% of immune system cells are found in the gut? These immune system cells are affected by the bacteria in our gut. Our diet affects these gut bacteria and, in turn, our immune cells. Fibre, as well as fermented foods like yoghurt and kimchi, helps maintain a healthy environment for our gut bacteria and our immune systems.

At the end of this chapter, you should be able to:

- Describe how the innate immune system works.
- Describe how the acquired immune system works.
- Explain the strategies used to prevent microbial diseases.
- Explain the strategies used to treat microbial diseases.
- Describe the role of different white blood cells in the immune response – monocytes, natural killer cells, B lymphocytes and T lymphocytes (helper, killer, suppressor and memory). HL
- Explain how infectious diseases emerge in animals and plants.
- Assess the importance of knowledge of emerging diseases in society.

SUBJECT VOCABULARY

immune system, pathogen, prion, innate immune system, acquired immune system, antigen, antibody, immunity, monocyte, natural killer cells, B lymphocyte, T lymphocyte, helper T cell, memory T cell, killer T cell, suppressor T cell, active immunity, vaccine, vaccination, passive immunity, emerging disease, autoimmune disease, pathogen persistence, basic reproduction number R_0 value (R-zero)

LINKS TO:

Chapter 7	**Bacteria and Fungi**
Chapter 8	**Viruses**
Chapter 9	**Evolution and Natural Selection**
Chapter 31	**The Circulatory System**
Chapter 37	**Human Reproduction – From Fertilisation to Birth**

Section 1: Innate and Acquired Immunity

At the end of this section, you should be able to:

- List different types of pathogens.
- Describe how the innate immune system works.
- Describe how the acquired immune system works.
- Describe the antigen-antibody response.
- Describe the role of different white blood cells in the immune response – monocytes, natural killer cells, B lymphocytes and T lymphocytes (helper, killer, suppressor and memory). HL

Unpacked from LOs 2.4.12, 2.4.14

Our **immune system** constantly protects us against disease-causing **pathogens** such as **prions**, bacteria, viruses, fungi, protists and parasitic animals. An important feature of the immune system is that it can distinguish between tissue that belongs to the organism ('self' cells) and foreign ('non-self') cells.

Organs of the immune system include the thymus, bone marrow, spleen, tonsils and lymph nodes. The immune system is divided into two parts: the **innate immune system** and the **acquired immune system**. We will examine each of these in turn.

- The **immune system** consists of a network of cells, organs and the chemicals they produce that protect the body against infection.
- A **pathogen** is an organism or agent that causes disease in its host.
- A **prion** is a type of protein that can cause disease in animals and humans by triggering normal proteins in the brain to fold abnormally.

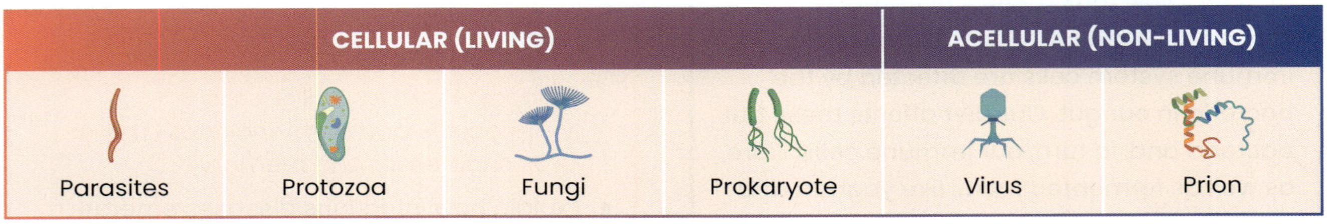

Figure 32.1: Disease-causing pathogens.

INNATE IMMUNE SYSTEM

Humans are born with an **innate immune system**. This innate immune system provides non-specific immunity – it does not distinguish between different types of pathogens and responds to all pathogens in the same way.

- The **innate immune system**, present at birth, provides short-term, non-specific defence against pathogens.

The innate immune system consists of multiple elements.

The innate immune system's first line of defence consists of barriers to prevent the entry of pathogens:

- The **skin** acts as a physical barrier to prevent infection.
- **Lysozyme enzyme** in saliva and tears kills certain bacteria by breaking down their cell wall.
- **Mucous membrane linings** of the digestive, respiratory and urinogenital systems secrete mucus that traps small foreign particles and pathogens preventing their entry. Cilia – hair-like projections in the lining of the respiratory tract – help to move any trapped particles back up the throat so that they can be expelled.

- **Gastric juice** in the stomach contains hydrochloric acid, which can kill many bacteria.

If a pathogen gets past these barriers, the innate immune system's second line of defence will act to protect the body against infection:

- **Fever**, an increase in body temperature, helps suppress microbial growth.
- **Antimicrobial proteins** destroy foreign cells.
- **Interferon proteins** inhibit the replication of many viruses.

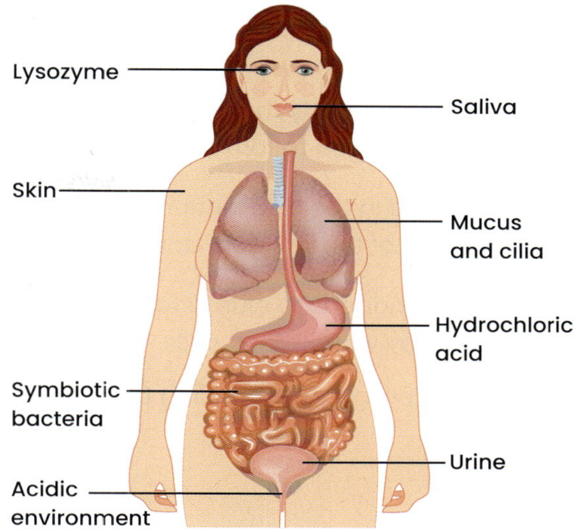

Figure 32.2: The innate immune system.

ACQUIRED IMMUNE SYSTEM

The innate immune system responds rapidly to infection, but it does not provide long-term immunity to specific pathogens. Innate immune responses give the body time to develop **acquired immune responses** which are specific to pathogens. Unlike the innate immune system, **acquired immunity** – also called adaptive immunity – is not present at birth and must be developed. White blood cells called lymphocytes are responsible for acquired immunity. Only vertebrates, organisms with a backbone, have an acquired immune system.

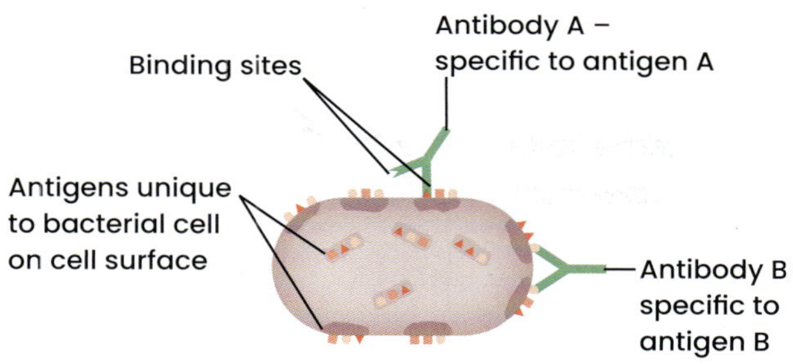

Figure 32.3: Each pathogen has its own unique antigens on its surface.

ANTIGENS

As mentioned earlier, the immune system distinguishes between tissue that belongs to the organism ('self' cells) and pathogenic 'non-self' bodies. This is because cells have molecules on their surface called **antigens,** which identify them as 'self' or 'non-self'. Antigens may be protein, carbohydrates, lipids or nucleic acids. Antigens on 'non-self' bodies such as bacteria, fungi and viruses trigger the formation of a specific protein molecule called an **antibody** by the acquired immune system.

- The **acquired immune system** develops over a person's lifetime. It provides long-term specific immunity in response to antigens.
- An **antigen** is a molecule on the surface of foreign bodies that triggers the formation of antibodies.
- An **antibody** is a protein molecule produced on detection of a particular antigen.

Antigen-antibody complex

When the acquired immune system detects a non-self antigen, it produces an antibody that is specific to the antigen. The antibody binds to the antigen to form an **antigen-antibody complex**.

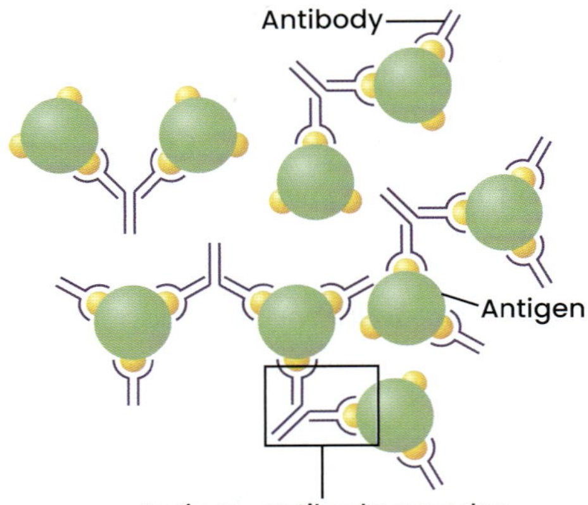

Figure 32.4: An antigen-antibody complex.

Antibodies aid in the destruction of pathogens in different ways. On formation of the antigen-antibody complex, antibodies:

- trigger pathogens to clump together, highlighting them for ingestion by white blood cells.
- perforate pathogens, causing them to burst.

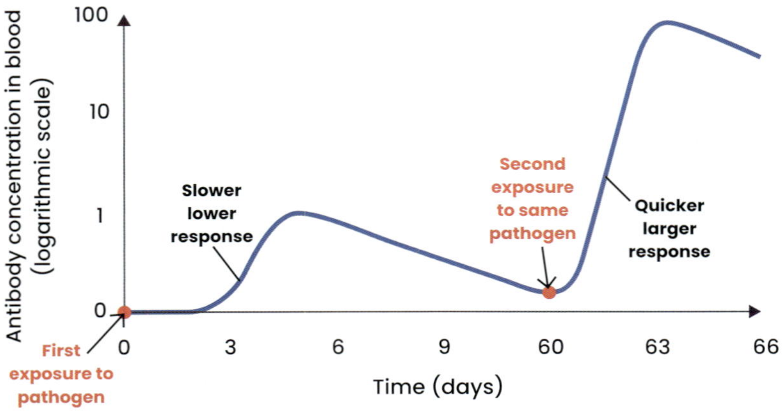

Figure 32.5: Memory of immune response.

Scientists believe the human immune system can produce a quintillion, or one million trillion, unique antibodies.

DID YOU KNOW?

This initial response takes time. Once the immune system produces an antibody for a specific antigen, it remembers this response. If a second infection with the same antigen occurs in the future, a larger number of antibodies will be produced more quickly. This results in **immunity** against the pathogen.

Immunity is protection against an infectious pathogen resulting from production of antibodies.

WHITE BLOOD CELLS

White blood cells of the innate immune system

MONOCYTES

HL

When pathogens get past the barriers of the innate immune system, white blood cells called **monocytes** work to remove them. Monocytes are part of the innate immune system because they are non-specific, responding to all pathogens in the same way. Monocytes are the largest type of white blood cell. They are produced in the bone marrow and circulate in the blood and lymph. Monocytes ingest pathogens by phagocytosis. Some monocytes mature to form cells called macrophages that settle in the tissue of the spleen and lymph nodes for months or possibly years.

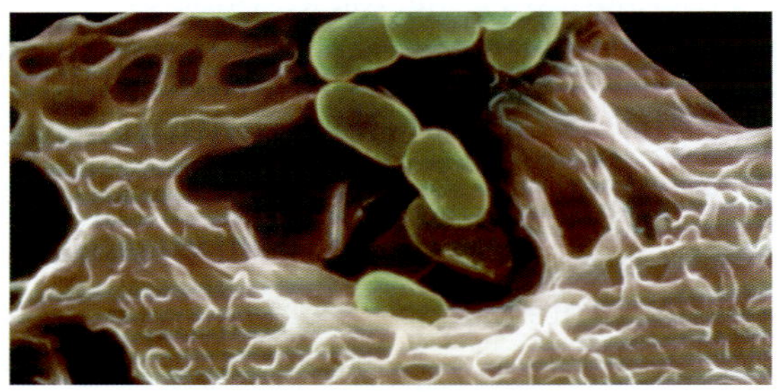

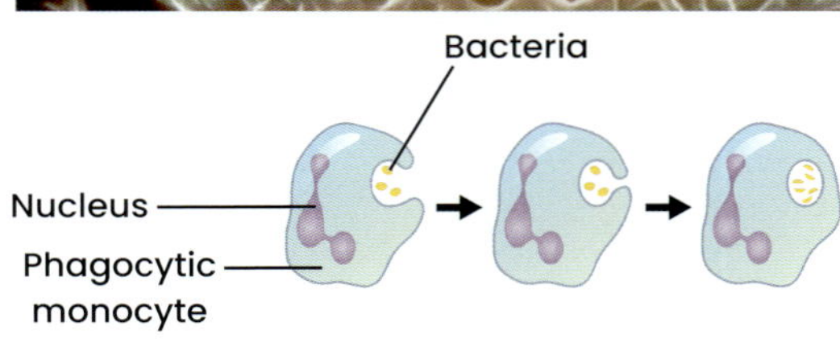

Figure 32.6: Macrophage engulfing bacteria in a process called phagocytosis.

Monocytes are non-specific phagocytic white blood cells that form part of the innate immune system.

NATURAL KILLER (NK) CELLS

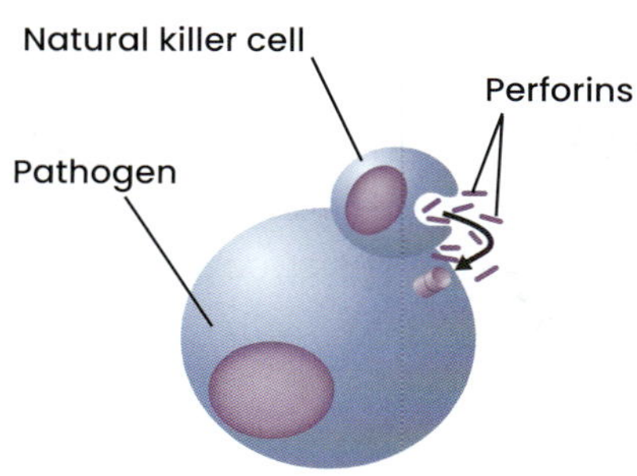

Figure 32.7: NK cell attacking a cancer cell.

Natural killer (NK) cells are white blood cells that target cancer and virus-infected cells without being activated by antigens.

Another white blood cell of the innate immune system is the **natural killer (NK) cell**. These cells do not need to be activated by antigens and provide a rapid, non-specific response to pathogens. NK cells mainly target cancer cells and virus-infected cells. The NK cells kill pathogens in a variety of ways including releasing chemicals (perforins) that create holes in the pathogen's cell membrane.

White blood cells of the acquired immune system

LYMPHOCYTES

Lymphocytes are a type of white blood cell formed in the bone marrow. Some lymphocytes travel in the blood to the **thymus** where they differentiate to form T lymphocytes, or **T cells**. (Think of **T** in **t**hymus and **T** cell.)

HL

Other lymphocytes remain in the **bone marrow** and differentiate to form B lymphocytes, or **B cells**. (Think of B in bone marrow and **B cell**.)

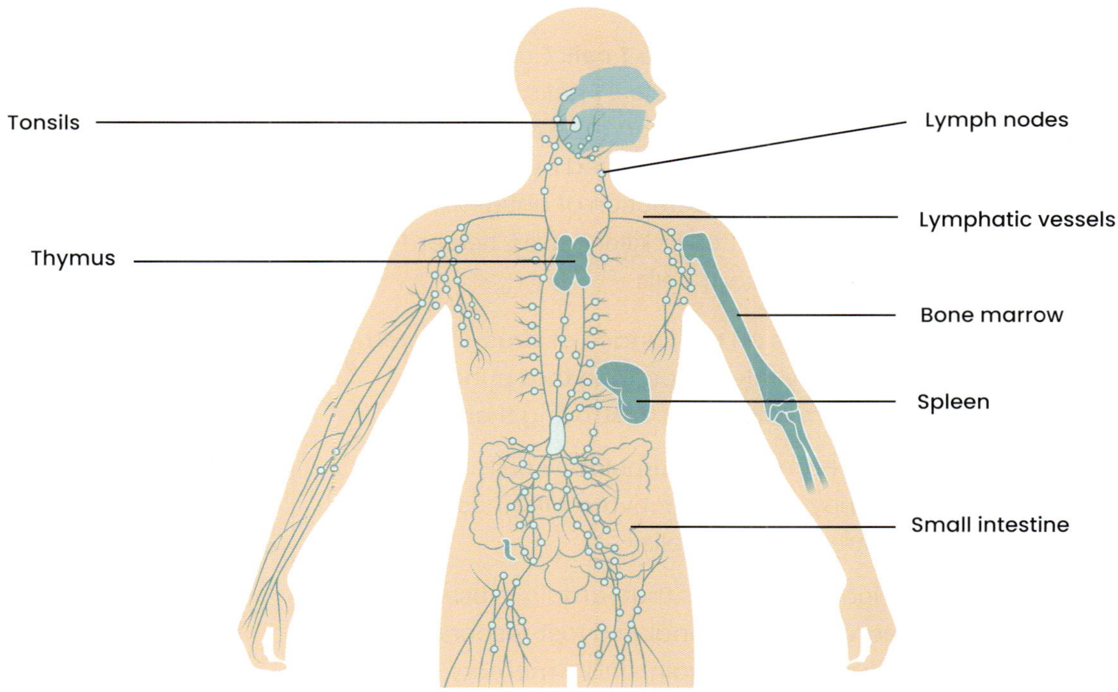

Figure 32.8: Organs of the immune system.

Lymphocytes circulate in the bloodstream, but also concentrate in the tissues of the lymphatic system, such as the spleen, tonsils and lymph nodes.

B LYMPHOCYTES

The human body forms millions of different B **lymphocytes** or B cells, each capable of responding to a different antigen. On detection of a foreign antigen, B cells divide rapidly by mitosis to form antibody-secreting plasma cells and memory B cells. Each antibody has a complementary shape to its particular antigen, meaning that each antibody can only bind to one specific antigen.

- **B lymphocytes** are part of the acquired immune system and form a complementary antibody on detection of a specific antigen.

MEMORY B CELLS

When B cells detect an antigen, they divide in large numbers to produce antibodies to remove the antigen. This initial response takes time. If a second infection with the same antigen occurs in the future, memory cells will already be present in large amounts and will react rapidly, causing antibodies to be produced more quickly and in greater numbers. Memory B cells offer long-term immunity against the antigen.

T LYMPHOCYTES

HL

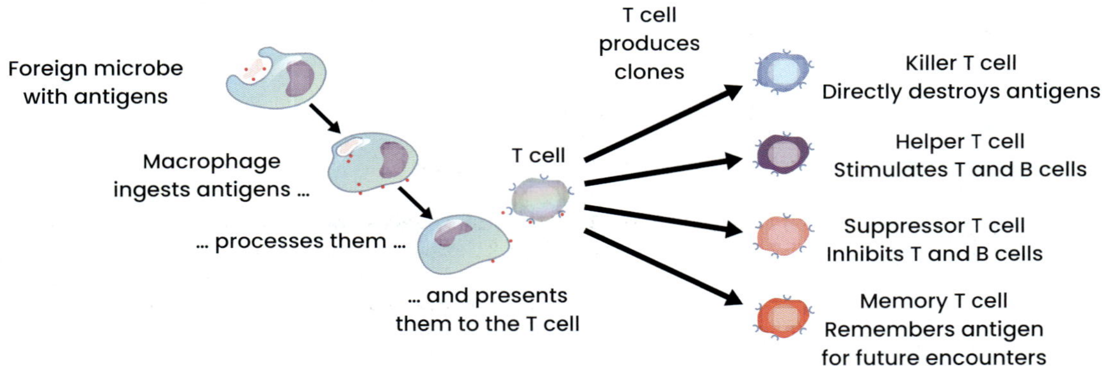

Figure 32.9: Activation of T cells.

The other main form of lymphocyte is the **T cell**. T cells mature in the thymus and do not produce antibodies. T cells are part of the acquired immune system, as they recognise and respond to 'non-self' antigens. Unlike antibodies produced by B cells, T cells do not directly bind to antigens. On detection of antigens, T cells divide to form four main types of T cells: killer T cells, helper T cells, memory T cells or suppressor T cells.

- **Killer T cells** detect antigens on the surface of virus-infected cells and cancer cells. They release perforin that punches holes in the target cells' membranes, destroying them.
- **Helper T cells** release chemicals to stimulate B cells to produce antibodies. They also help activate killer T cells and macrophages to destroy infected cells.
- **Memory T cells** are long-living cells that remain in tissue after an infection has cleared up. They enable a fast response if the same pathogen is encountered again. Unlike memory B cells, Memory T cells are not involved in antibody production.

- **T lymphocytes** or T cells are white blood cells that mature in the thymus and respond to antigens without producing antibodies.
- **Killer T cells** kill cancer cells or virus-infected cells by producing perforin which creates pores in their cell membranes.
- **Helper T cells** activate macrophages, B cells and killer T cells. Helper T cells stimulate B cells to form antibody producing plasma cells.
- **Memory T cells** provide long-term immunity against previously encountered pathogens.

- **Suppressor T cells** regulate the immune response to stop once an infection has cleared up. They inhibit the action of other B lymphocytes and T lymphocytes to prevent the immune system becoming overactive.

Suppressor T cells inhibit the action of other lymphocytes once an infection has cleared up.

PORTFOLIO LINK Go to your Learning Portfolio and complete Learning Check 1 to assess your knowledge and understanding of the topic.

Go to YouTube.com and search for 'Immune System: Innate and Adaptive Immunity Explained' (7:00) by Science ABC.

Section 2: Active and Passive Immunity

At the end of this section, you should be able to:

- Describe active immunity as acquired by infection or vaccination.
- Describe passive immunity acquired by donor transfer of antibodies and breastfeeding.
- Explain the strategies used to prevent and treat microbial diseases.

Unpacked from LO 2.4.12

ACTIVE IMMUNITY

When the body is exposed to a pathogen's antigens, the acquired immune system responds by producing antibodies that are specific to the antigens. For example, antibodies produced on detection of the measles virus will only defend against the measles virus and not the mumps virus. The type of immunity that develops from the production of antibodies is known as **active immunity**. Memory cells are produced during active immunity offering protection from pathogenic infection that can last a lifetime.

Active immunity can occur in one of two ways: naturally after an infection or via **vaccination**.

Active immunity develops from the production of antibodies by the immune system on detection of an antigen.

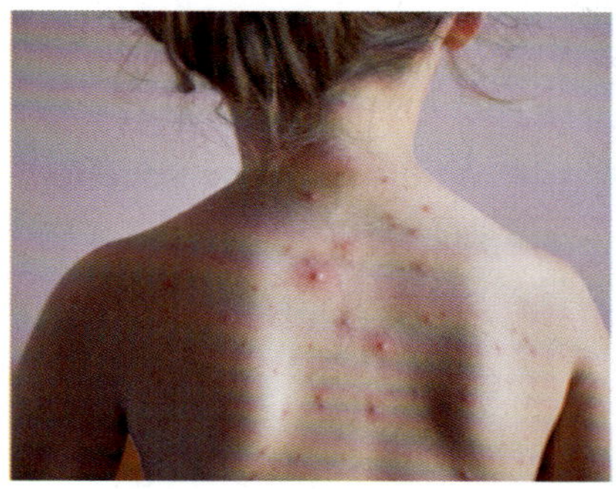

Figure 32.10: The varicella-zoster virus causes the chickenpox infection.

Natural immunity after infection

Natural immunity occurs when a person produces antibodies following infection by a pathogen. If a person becomes infected with chickenpox, B cells produce antibodies and memory B cells specific to the chickenpox virus. If the person is exposed to the chickenpox virus again, their immune system will produce antibodies to defend them, and they don't develop the disease again.

Vaccination

A **vaccine** contains a weakened or inactivated form of a pathogenic virus or bacteria, that stimulate the production of antibodies.

A **vaccine** is a biological substance that stimulates the production of antibodies against a particular bacteria or virus.

Vaccination is the medical process of introducing a vaccine to the body, usually by injection.

There are three main ways to produce a vaccine. These differ in whether they use:

- a whole virus or bacterium
- parts of the virus or bacterium that trigger an immune response, or
- the genetic material (DNA or RNA) that contain instructions for making particular proteins.

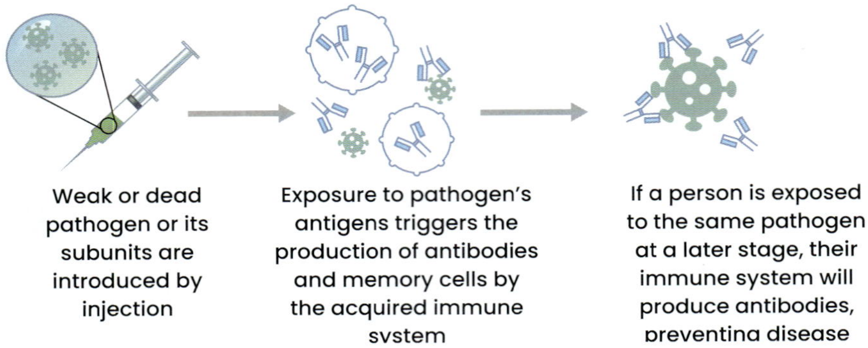

Figure 32.11: How a vaccine works.

A vaccine exposes a person's immune system to pathogenic antigens, stimulating the production of antibodies and memory cells against the pathogen. If the person is exposed to the actual pathogen at a later stage, their memory cells will trigger the formation of a much larger number of antibodies. This secondary response is much quicker and does not allow time for disease symptoms to develop – the person is now immune to the disease. Vaccines provide long-term immunity against disease. Immunity developed by vaccination is the same as that acquired by infection.

Vaccines offer active immunity as the person produces the antibodies themselves. Most vaccines such as the MMR (measles, mumps, rubella) vaccine, the Covid-19 or flu vaccines act against viral infections but some, such as the tuberculosis vaccine, act against bacterial infections.

SCIENCE IN SOCIETY: HEALTH AND TECHNOLOGY

Types of vaccine

- **Whole microbe vaccines** contain a weakened form of a disease-causing virus; often the protein capsid coat with the nucleic acid material removed.
- **Subunit vaccines** contain specific parts of a virus or bacterium that cause an immune system response, e.g. proteins. Most childhood vaccines are of this type, offering protection against diseases such as whooping cough and meningitis.
- **Nucleic acid vaccines** use a particular section of the microbe's genetic material (DNA or RNA). For example, an mRNA vaccine delivers instructions to cells on how to manufacture the pathogen protein that we want our immune system to produce antibodies against.

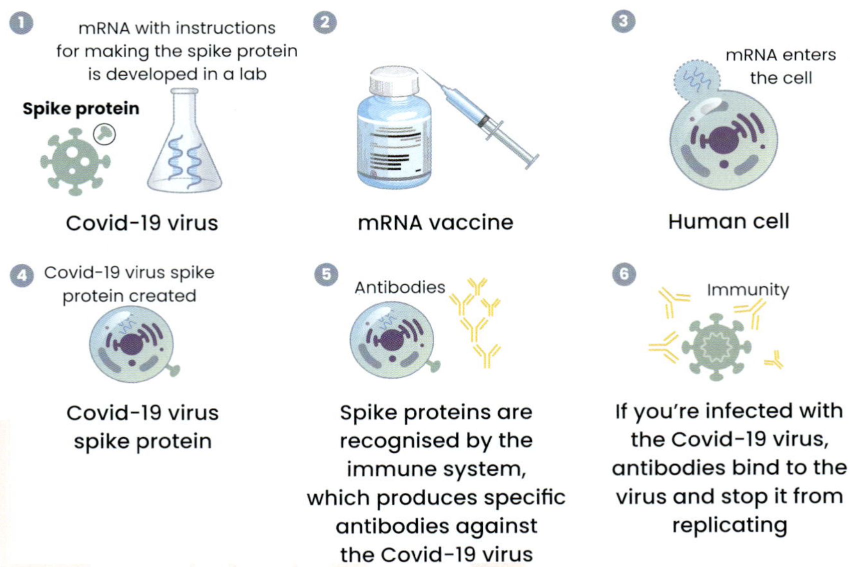

Figure 32.12: Action of an mRNA Covid-19 vaccine.

FOCUS ON ... VACCINES

English doctor Edward Jenner (1749–1823) is credited with developing the first vaccine. Jenner was aware of folklore that milkmaids who caught cowpox from their cows did not contract smallpox, one of the most fatal diseases at the time. In 1796, Jenner tested this theory by transferring pus from a cowpox pustule to a scratch made on the arm of an eight-year-old boy, James Phipps. Phipps became mildly ill with cowpox but recovered a few days later. Jenner subsequently exposed Phipps to smallpox, but the boy never developed the disease – he now had immunity against it.

Jenner published the results of his study and coined the term 'vaccine'. Today, Jenner's experimental method would be considered unethical.

Figure 32.13: Edward Jenner, father of immunology.

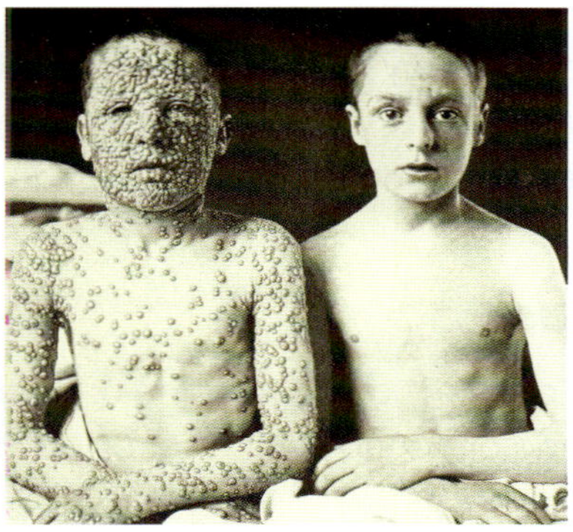

Figure 32.14: A photograph taken by a doctor in Leicester in 1901. Both boys were exposed to smallpox, but only the boy on the right was vaccinated.

DISCUSSION

What do you think of Jenner's method? Did the results justify the testing on an eight-year-old child?

The term 'vaccine' comes from the Latin word ***vacca***, meaning 'cow'.

DID YOU KNOW?

Go to YouTube.com and search for 'How mRNA Vaccines Work | Penn Medicine Explains' (2:23) by Penn Medicine.

PASSIVE IMMUNITY

While active immunity occurs when the body produces antibodies itself and memory cells develop, **passive immunity** occurs when antibodies from an immune donor are given to a non-immune person. In this case, memory cells do not develop as the person's immune system did not produce the antibodies itself.

Passive immunity occurs when antibodies are introduced to the body and memory cells do not develop.

Examples of the donor transfer of antibodies include:

- **Antibody booster injections:** Antibodies can be directly administered to a patient when they need immediate protection against a potentially dangerous infection for example in the treatment of rabies or tetanus. This is an artificial form of passive immunity.
- **Breastfeeding:** An example of natural passive immunity is the transfer of maternal antibodies to a baby during breastfeeding. While the protection offered by passive immunity is immediate, it is short term and reinfection by the same pathogen can occur.

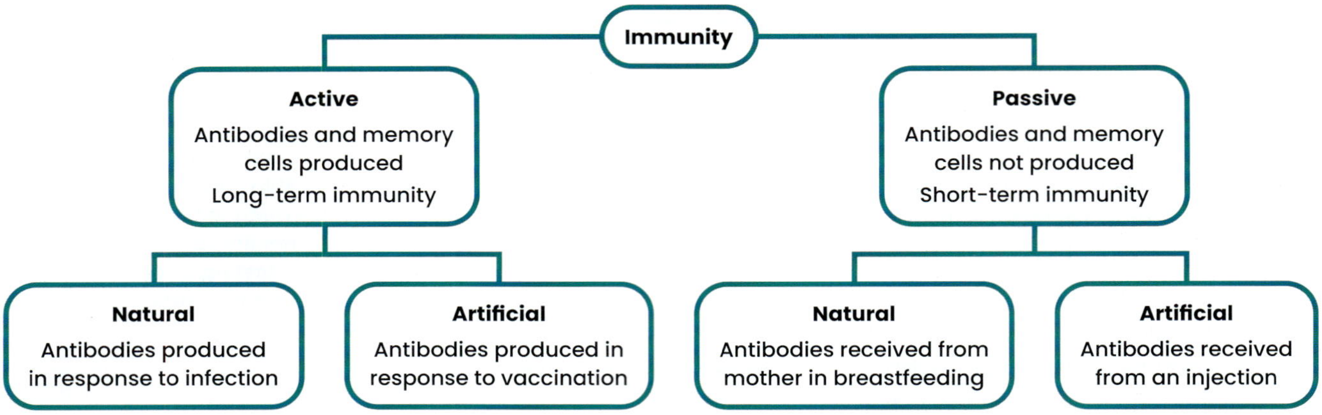

Figure 32.15: Active vs passive immunity.

MICROBIAL DISEASE

Microbial infections can be spread in several ways, e.g. by inhaling airborne droplets, eating or drinking contaminated food or water and unprotected sexual activity. To prevent microbial infections, it is necessary to block the microbes from entering the body.

Preventing microbial disease

- **Good personal hygiene** helps prevent an infection developing or spreading to others. It is important to wash hands thoroughly before preparing or eating food, after using the bathroom, coughing or sneezing.
- **Food safety** involves preparing and storing food safely to prevent infection by food-borne pathogens.
- **Vaccination** helps prevent the spread of many microbial diseases.
- **Practising safe sex** means that condoms should be worn to prevent the transfer of sexually transmitted diseases.
- **Antiseptics** are chemical treatments that can be used on wounds to kill microbes and prevent infection.
- **Disinfectants** such as bleach can be used on surfaces to kill microbes and prevent the spread of infection.

DISCUSSION

Discuss the pros and cons of the use of antibiotics.

Treating microbial disease

- **Antibiotics:** In Chapter 7, we learned that antibiotics are chemicals produced by micro-organisms such as bacteria and fungi that are toxic to bacteria. Antibiotics have no effect on viral infections. It is very important that a person completes a course of antibiotics to prevent the spread of antibiotic resistance.
- **Antiviral drugs:** These may be used to treat viral diseases such as influenza, herpes, hepatitis and HIV. Antiviral drugs inhibit the ability of the virus to reproduce.
- **Phage therapy:** Bacteriophages are viruses that kill specific bacteria. As more bacteria become resistant to antibiotics, it is hoped that 'phage therapy' may offer an alternative treatment.

Go to YouTube.com and search for 'CDC NERD Academy Student Quick Learn: How does disease spread?' (12:42) by Centers for Disease Control and Prevention.

PORTFOLIO LINK **Go to your Learning Portfolio and complete Learning Check 2 to assess your knowledge and understanding of the topic.**

Section 3: Emerging Diseases

At the end of this section, you should be able to:

- Explore factors that contribute to the emergence of infectious diseases in plants and animal.
- Discuss the importance of a knowledge of emerging diseases in society.

Unpacked from LOs 2.4.15, 2.4.16

SPREAD OF INFECTIOUS DISEASE

Emerging diseases

Emerging, or novel, diseases are infections that have recently appeared within a population or that may have existed previously but are now spreading rapidly.

About 40 infectious diseases have been discovered since the 1970s.

Emerging disease refers to an infection that has not occurred in a population before or has previously occurred but only affected a small number of people in isolated places.

NOVEL DISEASES IN ANIMALS

Novel diseases that affect humans or other animals include SARS, Ebola, swine flu and the Zika virus. Many emerging diseases occur when infections in animals caused by bacteria, fungi, viruses or parasites pass on to humans. This is known as **zoonosis**. As the human population increases and spreads to new geographical locations, the chance of humans coming into close contact with animal species carrying infectious disease may increase. The 2014–2016 Ebola virus outbreak in West Africa is believed to have been transmitted to humans through close contact with bodily fluids of infected bats or monkeys.

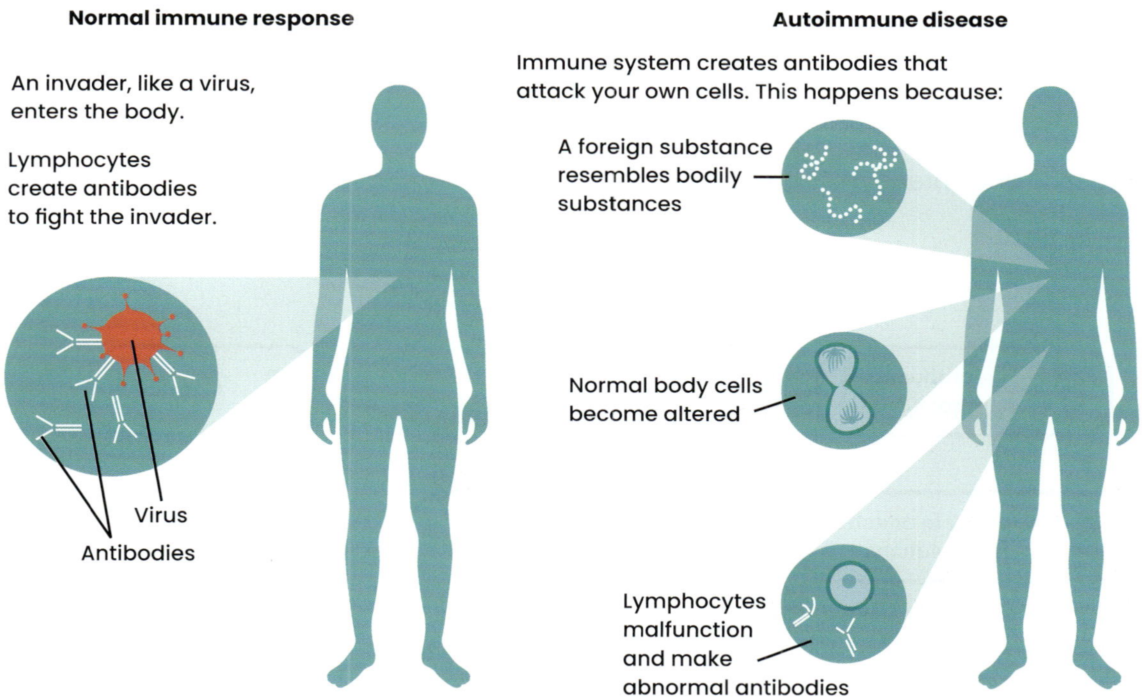

Figure 32.16: Autoimmune disease.

AUTOIMMUNE DISEASES

Autoimmune diseases occur when a person's immune system is overactive, causing it to mistakenly attack and damage their body's own healthy cells. Examples of autoimmune diseases include rheumatoid arthritis, psoriasis and Crohn's disease. A person's genes along with factors such as infections and other environmental conditions can affect the development of autoimmune diseases. Novel diseases can sometimes lead to autoimmune disorders as seen in some Covid-19 patients.

Autoimmune disease occurs when a person's immune system mistakenly attacks and damages their body's own healthy cells.

NOVEL DISEASES IN PLANTS

Figure 32.17: Tomato brown rugose fruit virus.

RESEARCH ACTIVITY

Carry out some research on another autoimmune disease. Summarise its symptoms and treatments.

Tomato brown rugose fruit virus is a highly contagious plant virus that affects tomatoes and peppers, and was first discovered in 2014. Novel or new plant viruses can cause significant economic losses due to damage to crops and reduced yields. Furthermore, they can infect native plant species, affecting biodiversity. Early detection and management are essential to controlling the spread of novel plant viruses.

EMERGING DISEASES AND SOCIETY

Understanding the factors that lead to increased numbers of emerging diseases enables strategies to control their increase and spread.

The emergence and spread of diseases are dependent on interrelated factors:

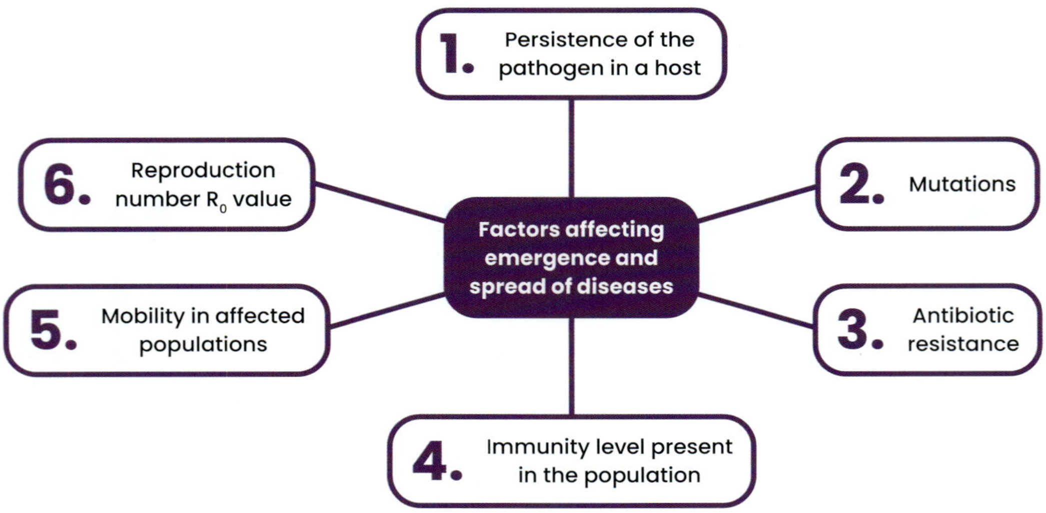

1. Persistence of the pathogen in a host

Certain pathogens can evade the host's immune system and remain for long periods of time without causing symptoms. Persistent viruses continue to replicate slowly, increasing the risk of spreading disease to others.

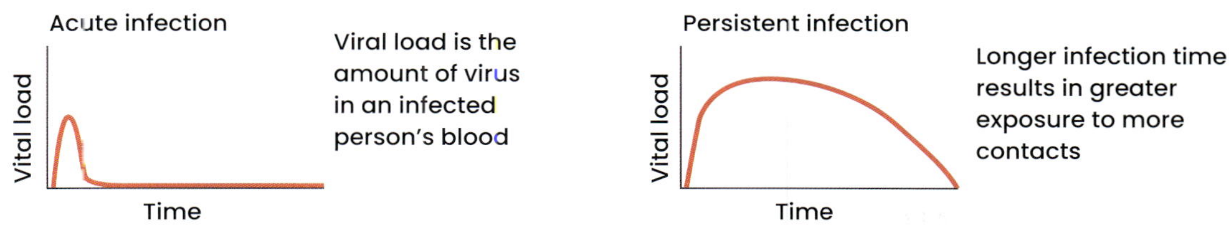

Figure 32.18: Knowledge of viral persistence informs infection control.

2. Mutations

Pathogens, particularly viruses, can mutate, changing their surface proteins. Mutations can enhance a pathogen's ability to spread, evade immune responses or resist treatments. This is the reason that a different influenza vaccine is needed each year.

Pathogen persistence refers to how long a pathogen can survive in a host while remaining infectious.

3. Antibiotic resistance

Bacteria are evolving over time and developing ways to withstand antibiotics. MRSA (methicillin-resistant *Staphylococcus aureus*) is a bacteria that has become resistant to several antibiotics. Overuse or misuse of antibiotics can result in antibiotic resistance developing, making it harder to control infections.

4. Immunity level present in the population

The number of people immune to a disease either because of previous infection or vaccination affects the ability of a disease to spread. If large numbers of the community are vaccinated, they are less likely to be a source of infection to non-vaccinated people. This is known as **herd immunity**.

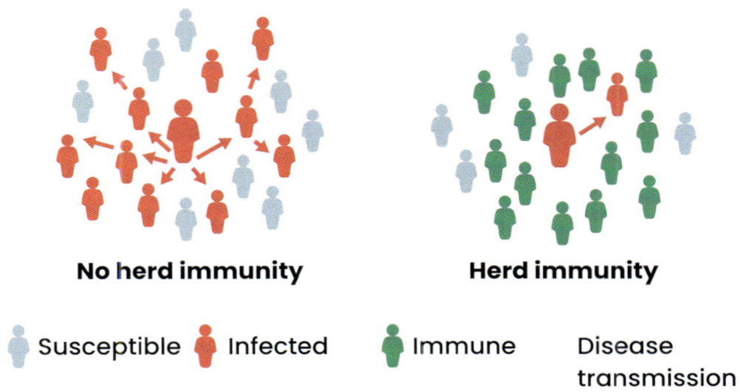

Figure 32.19: Herd immunity helps reduce the spread of disease.

5. Mobility in affected populations

Transmission and spread of infectious disease is facilitated by local, regional and global movement of organisms. Air travel for the transport of humans, food products and animals has meant that infectious disease agents can spread rapidly from one part of the world to another part where populations may have little or no immunity.

The Covid-19 pandemic illustrated how global travel can enable the rapid spread of a novel virus. Measures to reduce the spread of disease due to travel include the use of quarantine, travel restrictions, health screenings at borders and vaccination requirements. Biosecurity measures at each stage in food production and trade help prevent and reduce the transmission of infectious diseases in agriculture. Infected livestock may be culled or contaminated crops may be destroyed to prevent the spread of disease. Food traceability helps to prevent the spread of disease through food products. Restrictions on the import and export of wildlife help to prevent the spread of emerging diseases from animals to humans.

6. Reproduction number (R_0 value)

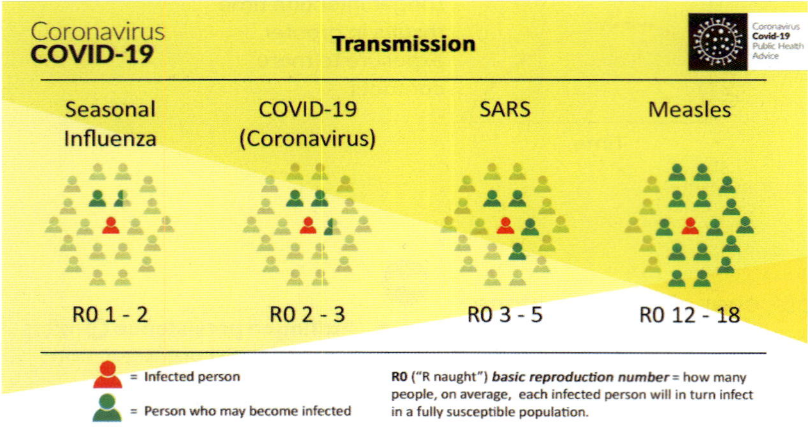

Basic reproduction number R_0 value (R-zero) is the average number of infections directly caused by an infected individual during their infectious period.

Figure 32.20: Basic reproduction number R_0 value (R-nought or R-zero) during the Covid-19 pandemic.

The **basic reproduction number R_0 value (R-zero)** refers to the average number of infections directly caused by an infected individual during their infectious period. If one person develops the infection and passes it on to three others, the R_0 is 3. R_0 is used to determine the ability of a disease to spread within a given population. The higher the value of R_0, the faster an epidemic will progress. Measles is highly contagious and has an R_0 value of 12–18.

Controlling an outbreak of an infectious disease often depends on reducing the R_0 through vaccinations, quarantine and social distancing.

SCIENCE IN SOCIETY: HEALTH

Disease surveillance

The Health Protection Surveillance Centre (HPSC) is Ireland's service for the surveillance of contagious diseases. All medical practitioners in Ireland are required to report an occurrence of certain diseases. This information is used to enable early identification of outbreaks, monitor changing levels of the disease and help prevent spread of infection. The HPSC also works with health service providers around the world to provide information to help control and prevent infectious diseases.

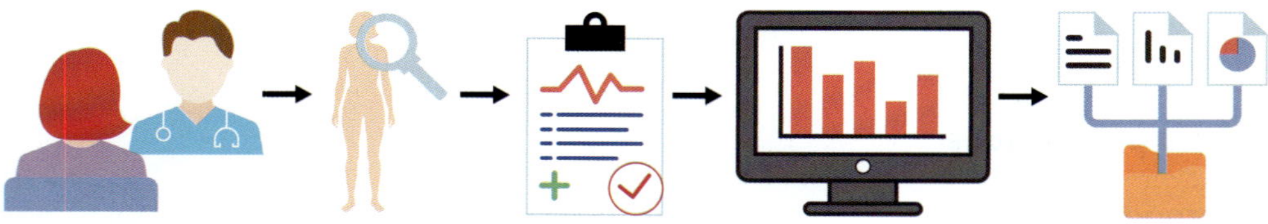

Figure 32.21: Stages of disease surveillance.

PORTFOLIO LINK

Go to your Learning Portfolio and complete Learning Check 3 to assess your knowledge and understanding of the topic.

Go to YouTube.com and search for 'Emerging Infectious Diseases – 3 Minutes Microlearning' (2:27) by BioTech Whisperer.

Chapter Summary

- The immune system consists of a network of cells, organs and the chemicals they produce that protect the body against infection.
- A **pathogen** is an organism or agent that causes disease in its host e.g. prions, bacteria, viruses, fungi, protists and parasitic animals.
- A **prion** is a type of protein that can cause disease by triggering normal proteins in the brain to fold abnormally.
- The **innate immune system**, present at birth, provides short-term, non-specific defence against pathogens.
- The **acquired immune system** which develops over life, provides long-term, specific immunity in response to antigens.
- An **antigen** is a molecule on the surface of foreign bodies that triggers the formation of antibodies by B cells.
- An **antibody** is a protein molecule produced by B cells on detection of a particular antigen.
- **Immunity** is protection against an infectious pathogen resulting from production of antibodies.
- The antibody binds to the antigen to form an **antigen-antibody complex** causing them to clump together and marking them for destruction.

HL

TYPES OF WHITE BLOOD CELLS IN IMMUNE RESPONSE	
INNATE IMMUNE SYSTEM	**ACQUIRED IMMUNE SYSTEM**
Provides short-term immunity and no memory cells are formed.	Provides long-term immunity and memory cells are formed.
Monocytes are a non-specific phagocytic white blood cell that form part of the innate immune system. Natural killer cells are white blood cells that target cancer and virus-infected cells without being activated by antigens.	**B lymphocytes** form a complementary antibody on detection of a specific antigen. **T lymphocytes** are white blood cells that respond to antigens without producing antibodies. T lymphocytes form four main types of T cells: • Killer T cells kill cancer cells or virus-infected cells. • Helper T cells stimulate B cells to form antibodies. • Memory T cells provide long-term immunity. • Suppressor T cells regulate the immune response.

- **Active immunity** develops from the production of antibodies on detection of an antigen.
- A **vaccine** is a biological substance that stimulates the production of antibodies.
- **Passive immunity** occurs when antibodies are introduced to the body and memory cells do not develop.
- **Microbial disease may be prevented** by good personal hygiene, food safety, vaccination, practising safe sex and use of antiseptics or disinfectants.
- **Microbial disease** may be treated using antibiotics, antiviral drugs and phage therapy.
- **Factors that affect emerging diseases** include persistence of the pathogen in a host, mutations, antibiotic or chemical resistance, immunity level present in the population, mobility in affected populations and reproduction number (R_0 value).
- **Emerging disease** refers to an infection that has not occurred in a population before.
- **Pathogen persistence** refers to how long a pathogen can survive in a host.
- **Basic reproduction number R_0 value (R-zero)** is the average number of infections directly caused by an infected individual throughout his infectious period.

END OF CHAPTER ASSESSMENT

REMEMBER

1. List **six** different types of pathogens.
2. What is a prion?
3. Name and define the **two** parts of the immune system.
4. Identify **three** physical barriers in the immune system.
5. Define each of the following terms.
   (a) *Antigen*
   (b) *Antibody*
   (c) *Antigen–antibody complex*
6. How do antibodies act to remove antigen-presenting cells?
7. List strategies used to prevent microbial disease.
8. List strategies used to treat microbial disease.
9. Identify **three** factors that contribute to emerging diseases.
10. What is an autoimmune disease?
11. Explain each of the following terms.
    (a) Pathogen persistence
    (b) Reproductive number R_0
12. HL What are monocytes?
13. HL What are natural killer (NK) cells?
14. HL List the four types of T lymphocytes and give a function of each in immunity.

UNDERSTAND

15. Distinguish between the innate and acquired immune systems.
16. Distinguish between active and passive immunity.
17. Explain the principle upon which vaccines work.

APPLY

18. (a) Copy and complete the graph below to illustrate the primary and secondary antibody response to a particular antigen infection.

Primary and secondary immune response to a particular antigen

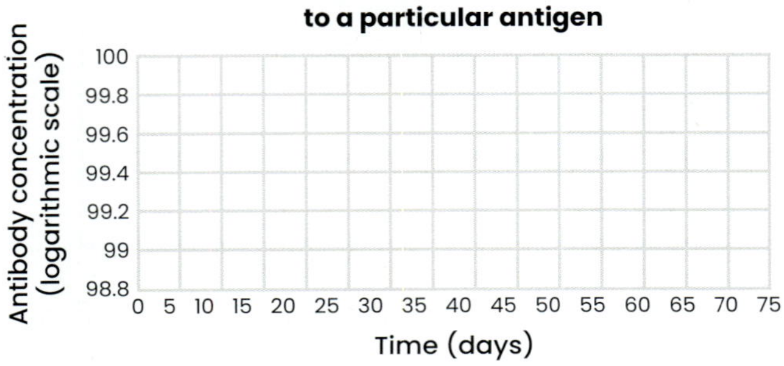

(b) How would the above graph compare to that for the immune response to a vaccination and subsequent infection by the same antigen?

19. (a) Use the infographic below to illustrate the concept of herd immunity.
    (b) Who in society is most at risk of infection if a disease outbreak occurs? Explain your answer.

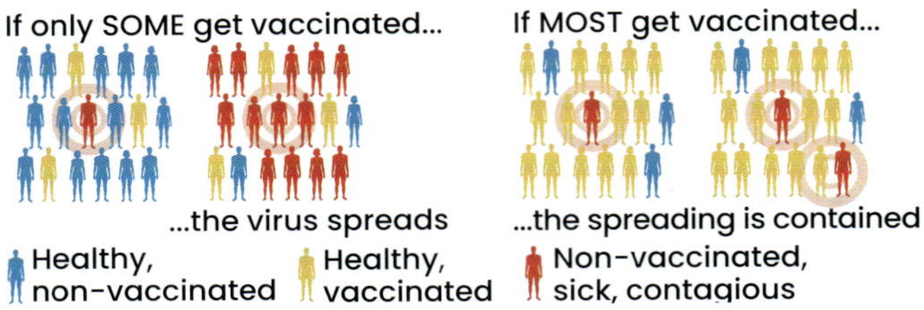

ANALYSE

20. Compare B lymphocytes and T lymphocytes under the following headings:
    (a) Site of formation
    (b) Site of maturation
    (c) Production of antibodies
    (d) Role in immunity
21. What could happen if a person's immune system was lacking suppressor T cells?

EVALUATE

22. How may knowledge of each of the following factors be used to prevent the spread of infectious disease?
    (a) Pathogen persistence
    (b) Mutations
    (c) Antibiotic resistance
    (d) Immunity level present in the population
    (e) Mobility in affected populations
    (f) Reproduction number (R_0 value)
23. Why would Edward Jenner's experiments into vaccinations not be allowed today?

CREATE

24. Create a poster for a doctor's surgery to explain why it is necessary to get the flu vaccine every year.

UNIT 11

HUMANS – SENSING AND RESPONDING

CHAPTERS

STRANDS

Strand 2: Structures and Processes of Life

33 The Nervous System

LEARNING OUTCOMES: 2.4.4, 2.4.6, 2.4.7, 2.4.8, 2.4.9

EYE-OPENER

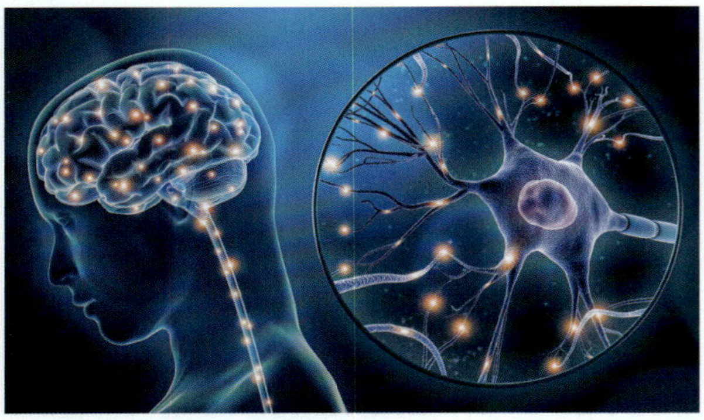

In the human brain, 86 billion nerve cells form 100 trillion connections to each other. Scientists from Harvard Medical School are working on a new field of neuroscience called 'connectomics', which aims to comprehensively map connections between neurons in the brain.

At the end of this chapter, you should be able to:

- Relate the structure of the parts of the central nervous system to their functions.
- Relate the structure of the parts of the peripheral nervous system to their functions.
- Relate the structure of a motor and sensory neuron to their functions.
- Describe the role, structure and mechanism of a reflex action.
- Explain the role of neurotransmitters at a synapse.
- Describe the conduction of nerve impulses across the synaptic cleft. HL
- Consider the impact of disruptions to impulse travel. HL
- Discuss the roles of the neurotransmitters dopamine and endorphins in humans.

LINKS TO:

Chapter 16	**Cell Structure and Diversity**
Chapter 28	**The Breathing System**
Chapter 31	**The Circulatory System**
Chapter 34	**The Endocrine System**
Chapter 35	**The Musculoskeletal System**

SUBJECT VOCABULARY

stimulus, receptor, effector, central nervous system, peripheral nervous system, neuron, sensory neuron, motor neuron, interneuron, meninges, grey matter, white matter, cerebrospinal fluid, synapse, neurotransmitter, neurotransmission, impulse, reflex action, reflex arc, synaptic cleft, threshold stimulus, refractory period, endorphin, dopamine, psychoactive drugs

Section 1: Central and Peripheral Nervous System

At the end of this section, you should be able to:

- Name the parts of the central nervous system (CNS) and briefly outline their functions.
- Describe the structure of the peripheral nervous system (PNS) and outline the function of each part.
- Briefly describe the function of each part of a neuron.
- Compare and contrast the structure of motor neurons, sensory neurons and interneurons.

Unpacked from LOs 2.4.4, 2.4.6

The ability of an organism to respond to changes in its environment is vital for its survival. Our sense organs (eye, ear, nose, tongue and skin) constantly take in information about our external environment. Information about our internal environment (e.g. body temperature, blood pH, water content and carbon dioxide concentration) is also picked up by internal receptors. A change in our environment is called a **stimulus**. For example, if a person driving a car meets a red traffic light, **receptor** cells in their eyes detect this stimulus and send a message or impulse to the brain. The brain processes this information, and a response is carried from the brain to **effector** muscles in the foot to use the brakes to stop the car.

- A **stimulus** is a change in the environment that brings about a response.
- A **receptor** is a group of cells that detect a change in the environment and generate an impulse.
- An **effector** is a muscle or gland that carries out a response to a stimulus.

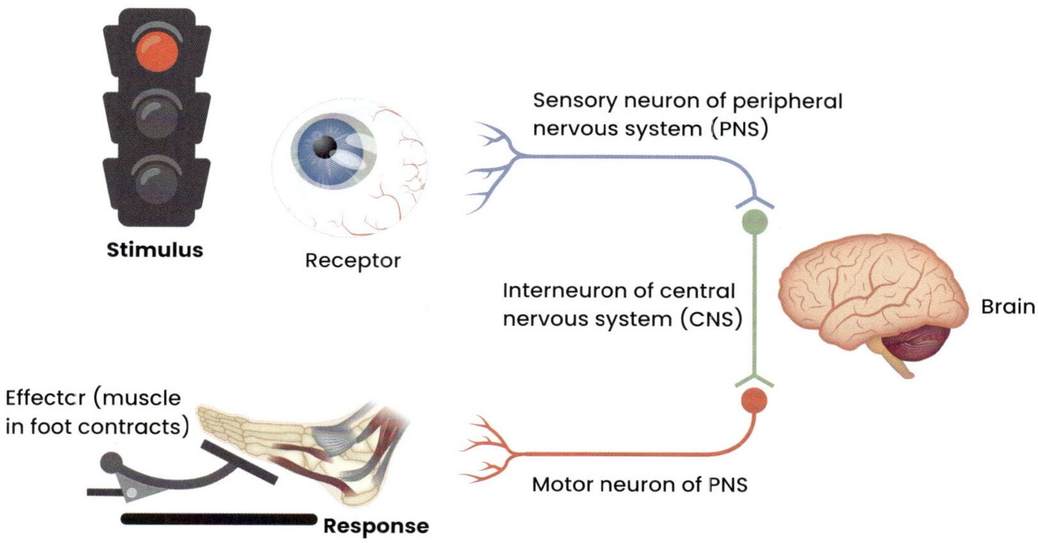

Figure 33.1: Nerve response.

Other examples of nerve responses include:

- Salivary gland in the mouth releasing saliva in response to a tasty smell.
- Release of the hormone insulin by the pancreas in response to rising blood sugar levels.
- Muscle in the arm contracting to catch a falling ball.

In humans, the nervous system allows fast-acting, short-lived responses to stimuli. Along with the endocrine system (hormonal system), the nervous system is responsible for maintaining constant internal conditions in the body (homeostasis). We will learn more about the hormonal system in the next chapter.

The brain does not have any pain receptors!

DID YOU KNOW?

The nervous system has two main parts:

- **Central nervous system (CNS)** consisting of the brain and spinal cord.
- **Peripheral nervous system (PNS)** consisting of the nerves that carry impulses to and from the CNS.

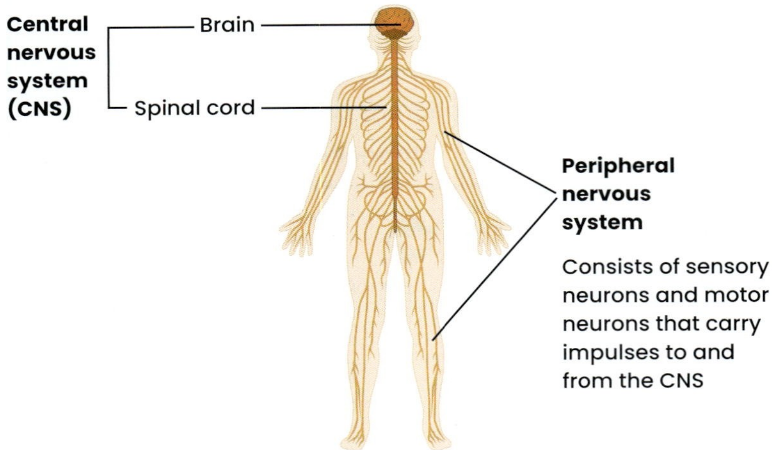

Figure 33.2: The human nervous system.

The **central nervous system (CNS)** consists of the brain and spinal cord.

The **peripheral nervous system (PNS)** consists of the nerves that carry impulses to and from the CNS (brain and spinal cord).

NERVE STRUCTURE

A nerve consists of a 'bundle' of nerve cells called **neurons**. Neurons grouped together in a nerve may be compared to individual electrical wires grouped together in electrical cabling. Nerve cells do not undergo mitosis so if a nerve cell is damaged, it cannot be replaced.

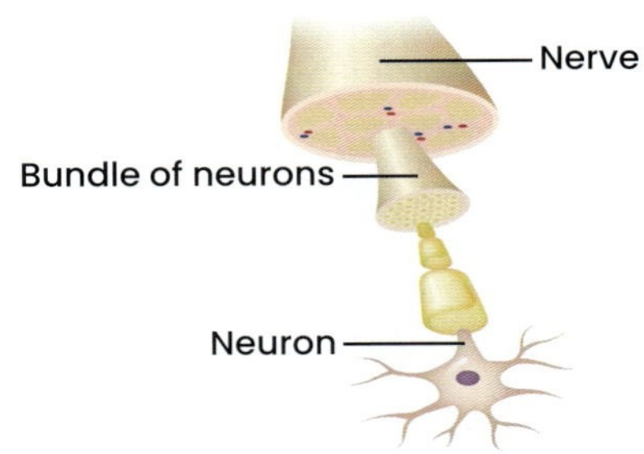

Figure 33.3: A nerve consists of a bundle of neurons.

Neuron structure

Each neuron (nerve cell) has three main parts: dendrites, a cell body and an axon.

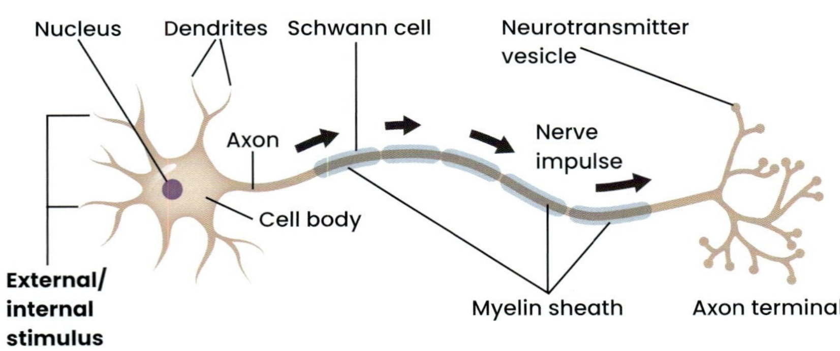

Figure 33.4: Neuron structure.

A **neuron** is an individual nerve cell.

We will now look at each part of the neuron's structure.

- **Dendrites:** Dendrites are short fibres that convert chemical messages received from other neurons to electrical impulses and direct them towards the cell body. Dendrites are branched to increase the surface area across which they can receive impulses.

Go to YouTube.com and search for '2-Minute Neuroscience: The Neuron' (1:47) by Neuroscientifically Challenged.

- **Cell body:** The cell body contains the nucleus and other organelles. It produces neurotransmitter chemicals which carry the impulse from one neuron to the next.
- **Axon:** The axon consists of a long fibre that conducts electrical impulses away from the cell body. The axon is sometimes referred to as a nerve fibre.

Other important parts of the neuron include:

- **Myelin sheath:** The myelin sheath is a white fatty layer made up of Schwann cells that surrounds the axon. It protects and electrically insulates the axon, speeding up the conduction of electrical impulses.
- **Schwann cell:** Schwann cells are cells that wrap around the axon in single file to form the myelin sheath.
- **Neurotransmitter vesicle:** Neurotransmitter vesicles are membrane-bound sacs found at the axon terminal that store chemicals called neurotransmitters. These neurotransmitters are released into the gap between neurons, the **synapse**, to carry the impulse from one neuron to the next neuron.

Types of neuron

There are three main types of neurons: sensory neurons, interneurons and motor neurons.

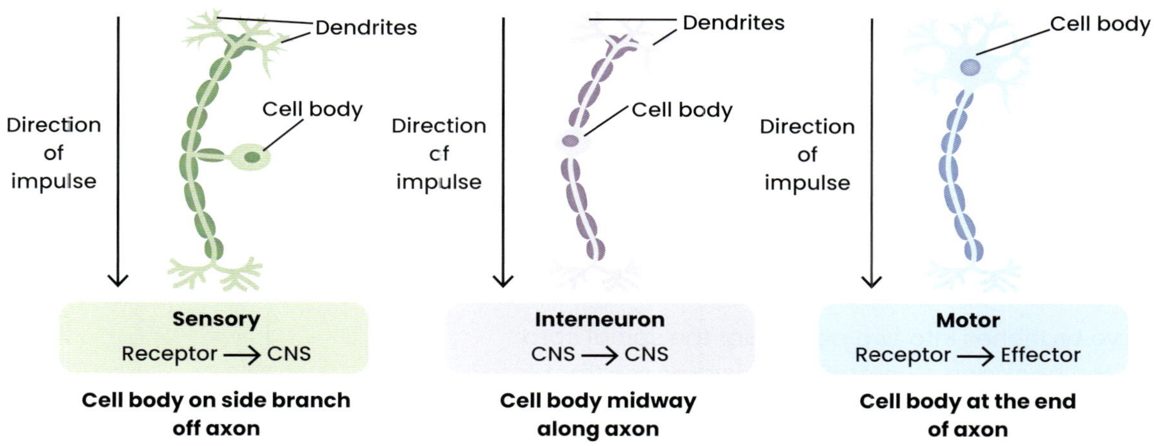

Figure 33.5: Structure of sensory neurons, interneurons and motor neurons.

Sensory neurons carry impulses to the central nervous system (CNS), whereas **motor neurons** carry impulses away from the CNS to an effector muscle or a gland. An example of an effector gland is a salivary gland in the mouth which releases saliva on receiving an impulse from a motor neuron. **Interneurons** carry impulses between sensory and motor neurons within the CNS.

The three main types of neurons differ according to the position of their cell body; the cell body in sensory neurons is located on a side branch off the axon, in interneurons, the cell body is located midway along the axon and in motor neurons, the cell body is found at the end of the axon.

- A **sensory neuron** is a nerve cell that carries an impulse to the CNS.
- An **interneuron** is a nerve cell that carries impulses between sensory and motor neurons within the CNS.
- A **motor neuron** is a nerve cell that carries an impulse from the CNS to an effector muscle or gland.

PERIPHERAL NERVOUS SYSTEM (PNS)

The PNS consists of sensory nerves and motor nerves. Later in the chapter, we will learn in detail how impulses are carried by these nerves to and from the CNS.

CENTRAL NERVOUS SYSTEM (CNS)

The CNS, consisting of the brain and spinal cord, processes information received from the PNS and co-ordinates responses.

The brain is protected by the skull and the spinal cord is protected by the vertebral column of the skeleton. The brain and spinal cord are surrounded by a layer of three membranes called the **meninges** and **cerebrospinal fluid**, which helps protect the CNS against shock. The CNS has an outermost layer of **grey matter** and an inner region of **white matter**. White matter contains the axons where nerve impulses are carried. Grey matter contains the cell bodies where information is processed.

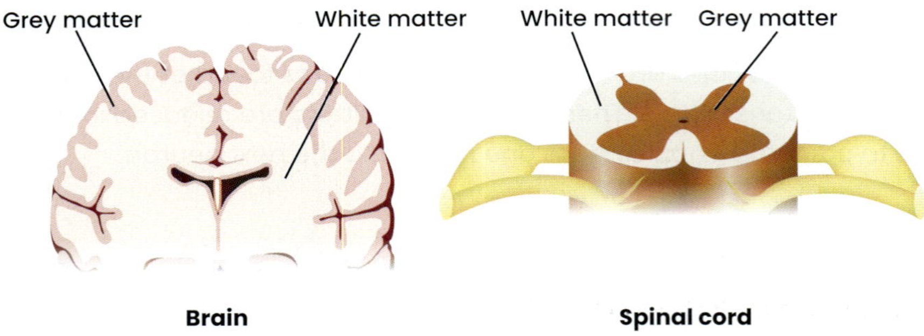

Figure 33.6: Grey matter and white matter in the brain and spinal cord.

- **Meninges** consist of a protective layer of three membranes surrounding the brain and spinal cord.
- **Cerebrospinal fluid** is a clear liquid that surrounds and protects the brain and spinal cord and provides them with nutrients.
- **Grey matter** is made up of cell bodies in the CNS.
- **White matter** is made up of axons in the CNS.

Spinal cord structure

Thirty-one pairs of nerves carry impulses in and out of the spinal cord. Each pair of nerves connects the spinal cord to a specific region of the body. If a region of the spinal cord is damaged, e.g. in a car accident, it may not be able to transfer impulses up towards the brain.

Each spinal nerve branches into two parts near the spinal cord: a dorsal root at the back and a ventral root towards the front.

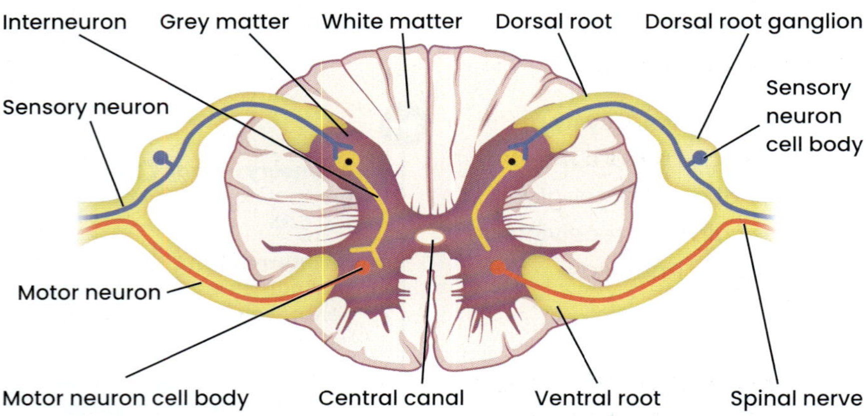

Figure 33.7: Transverse section of spinal cord.

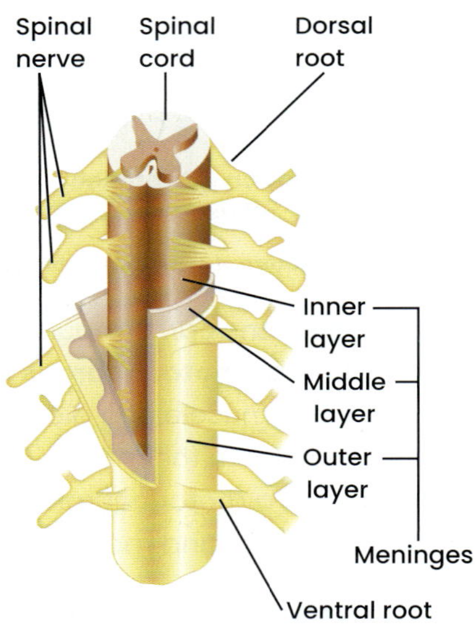

Figure 33.8: Longitudinal section of spinal cord with spinal nerves.

The dorsal root carries sensory neurons to the spinal cord and the ventral root carries motor neurons from the spinal cord. The cell bodies of the sensory neurons lie outside the spinal cord in an area called the dorsal root ganglion. Cell bodies of the motor neurons lie within the spinal cord. Sensory and motor neurons are connected by interneurons within the spinal cord. Unlike in the brain, white matter in the spinal cord lies to the outside and the centre consists of a butterfly-shaped region of grey matter. At the centre of the spinal cord is a central canal filled with cerebrospinal fluid.

PORTFOLIO LINK **Go to your Learning Portfolio and complete Learning Check 1 to assess your knowledge and understanding of the topic.**

Section 2: Neurotransmission

At the end of this section, you should be able to:

- Explain the role of neurotransmitters at a synapse.
- Describe the role, structure and mechanism of a reflex action.

Unpacked from LO 2.4.7

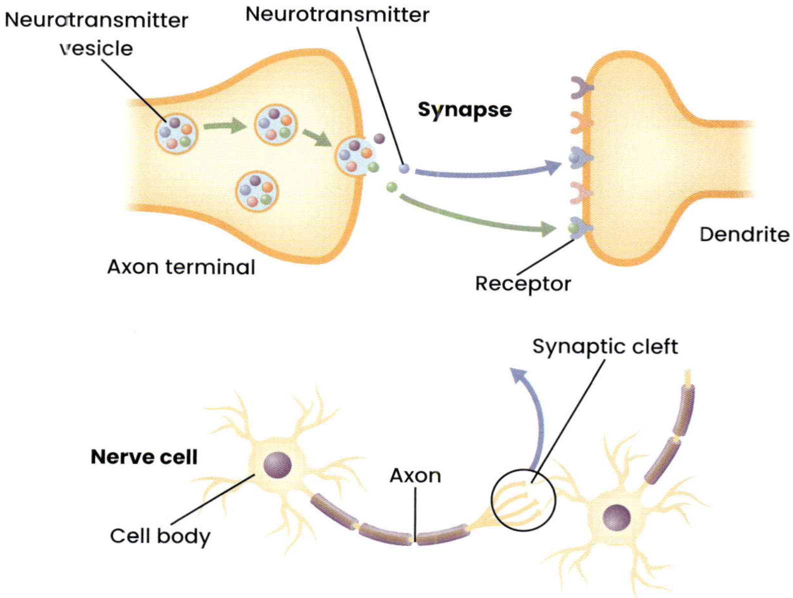

Figure 33.9: Neurotransmission is the transfer of impulses between neurons.

The movement of nerve impulses along and between sensory, motor and interneurons or to a muscle or gland is known as **neurotransmission**.

- Neurons do not physically touch each other. A gap called the **synaptic cleft** is found between two neurons. The region where two neurons are close to each other is called a **synapse**.
- Once a stimulus is received, it triggers an inrush of charged particles called ions along the axon.
- This change in charge moves along the axon until it reaches the neurotransmitter vesicles at the axon terminal.
- When an electrical impulse reaches the neurotransmitter vesicles, it triggers **neurotransmitter** chemicals to be released into the synaptic cleft.
- The neurotransmitter chemicals diffuse across the synaptic cleft and lock into receptors in the membrane of the dendrite of the next neuron.
- This stimulates an inrush of ions in the second neuron and the **impulse** is carried along its axon.

THINK, PAIR, SHARE

Discuss how the structures of nerve cells are adapted to carry nerve impulses.

- **Neurotransmission** is the movement of a nerve impulse from one nerve to another or to a muscle or gland.
- The **synaptic cleft** is the gap between two neurons.
- The **synapse** is the region where two neurons are close to each other.
- A **neurotransmitter** is a chemical messenger that carries an impulse across a synapse between two neurons, e.g. acetylcholine.
- An **impulse** is an electrical signal triggered by a stimulus that travels the length of a nerve fibre and ends in the release of a neurotransmitter.

ROLE OF REFLEX ACTION

A special example of neurotransmission is that of **reflex action**. A reflex action is a rapid, involuntary and automatic movement response to a stimulus. In certain situations, e.g. if you touch a very hot object, your sensory nerves do not carry the impulse to the brain. To protect you from harm, the impulse is carried to the spinal cord and a response is directed along a motor neuron to a muscle to drop the hot object. This quicker response is necessary to our survival and is called a 'reflex action'.

> A **reflex action** is a rapid, involuntary and automatic movement response to a stimulus.

Sometimes, the brain can override a reflex action, e.g. to hold onto a hot plate to prevent dropping your dinner on the floor, but this response must be learned by the brain. Most of the reflex actions that occur in the body are controlled by the spinal cord.

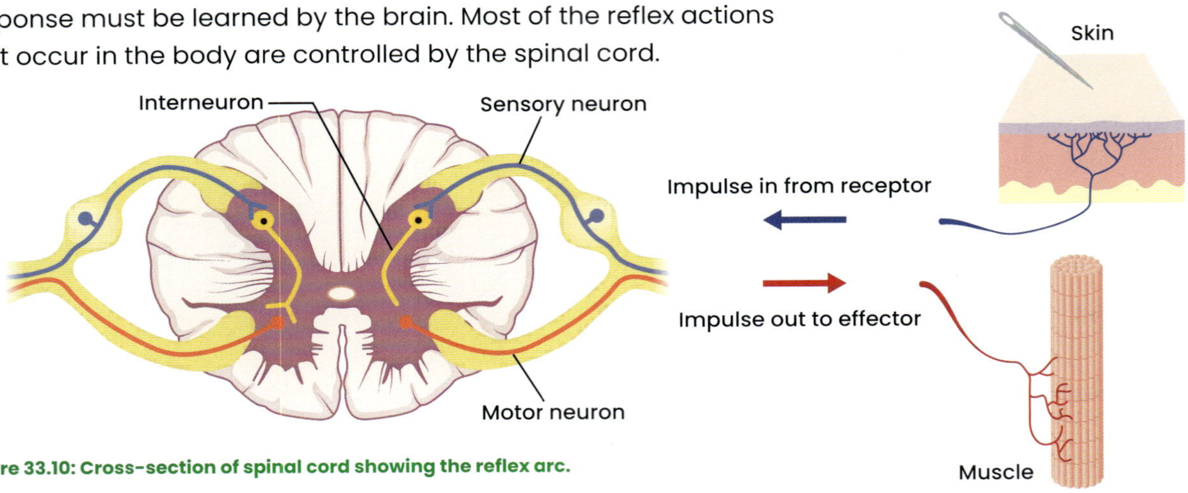

Figure 33.10: Cross-section of spinal cord showing the reflex arc.

MECHANISM OF REFLEX ACTION

The **reflex arc** or pathway followed in a reflex action is illustrated in Figure 33.10.

> The **reflex arc** is the nerve pathway followed in a reflex action.

1. A sensory receptor detects a stimulus, e.g. pain receptor in the skin.
2. A sensory neuron carries an electrical impulse to the spinal cord via the dorsal root.
3. The sensory neuron synapses with an interneuron within the CNS.
4. The interneuron synapses with a motor neuron.
5. The motor neuron carries an electrical impulse out of the spinal cord via the ventral root.
6. The motor neuron carries the impulse to an effector muscle causing a response, e.g. muscle contracts causing you to move away from the sharp object.

PORTFOLIO LINK **Go to your Learning Portfolio and complete Learning Check 2 to assess your knowledge and understanding of the topic.**

Section 3: Neurotransmitters

At the end of this section, you should be able to:

- Discuss the roles of the neurotransmitters dopamine and endorphins in humans.
- Describe the conduction of nerve impulses across the synaptic cleft. HL
- Consider the impact of disruptions to impulse travel. HL

Unpacked from LOs 2.4.8, 2.4.9

Neurotransmitters like dopamine and endorphins affect mood and behaviour.

ENDORPHINS

Endorphins are neurotransmitters released by the pituitary gland and hypothalamus in the brain, during pleasurable activities such as eating a meal, laughing or having sex. Known as the body's natural painkiller, endorphins help relieve pain. When you feel pain, impulses travel from pain receptors to the brain, causing the release of endorphins to block nerves receiving pain signals. This allows the body to continue functioning in challenging or stressful situations. Endorphins also help to boost mood.

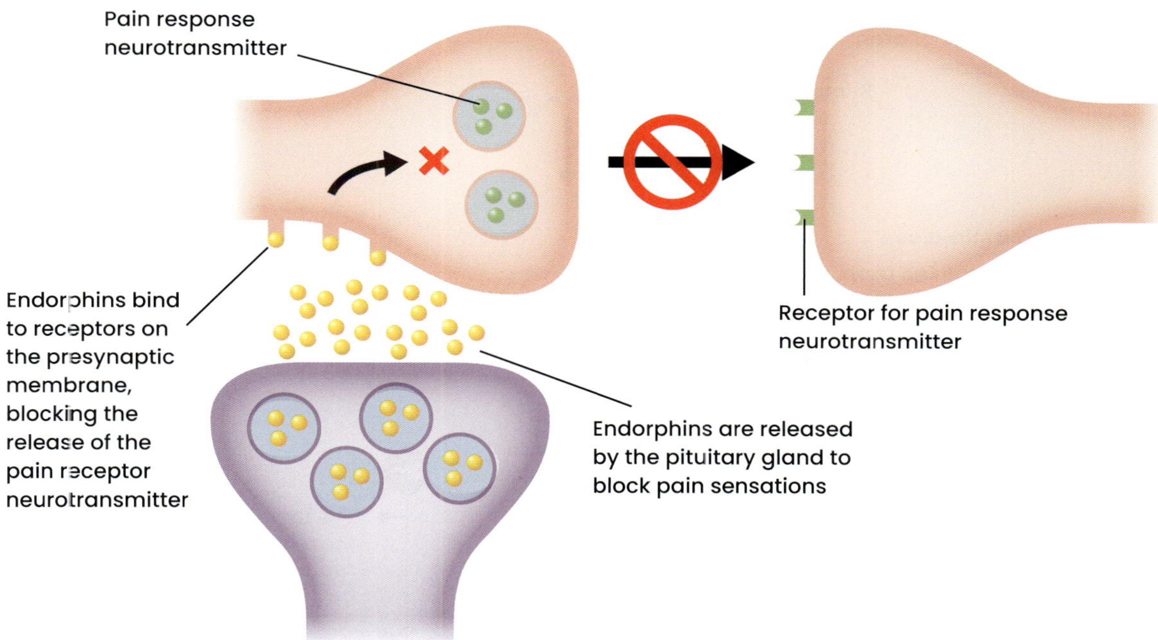

Figure 33.11: Action of endorphins.

DOPAMINE

Dopamine is a 'feel-good' neurotransmitter that is released following pleasurable activities such as eating. Our bodies can crave this temporary feeling or 'dopamine rush'. When the brain doesn't produce enough dopamine, Parkinson's disease, which causes involuntary shaking of the limbs, can result. The main form of treatment is the drug L-dopa, which is converted to dopamine in the body.

EFFECT OF LIFESTYLE CHOICES ON DOPAMINE AND ENDORPHIN LEVELS

- **Non-medical drug use:** Drugs such as cocaine cause the release of abnormally large amounts of dopamine. Over time, drug abuse results in a decrease in the number of dopamine receptors as the brain tries to normalise the level of stimulation. As a result, the person is less able to experience pleasure naturally and becomes addicted to the drug.
- **Exercise:** Exercise stimulates the production of endorphins, the body's natural painkillers and mood elevators. Dopamine and endorphins work together – for example, when a runner experiences muscle pain, it may be relieved by the release of endorphins. This is followed by a surge in dopamine which can lead to a 'runner's high', which encourages them to run again.
- **Dietary choices:** It is thought that eating protein-rich foods such as chicken and eggs can boost dopamine production. Foods rich in antioxidants, such as berries and green leafy vegetables, are also thought to increase dopamine levels. While eating triggers pleasure and dopamine release, studies have shown that diets high in saturated fat reduce dopamine release, resulting in overeating to achieve the same dopamine reward.

FOCUS ON ... CONDUCTION OF NERVE IMPULSES (HL)

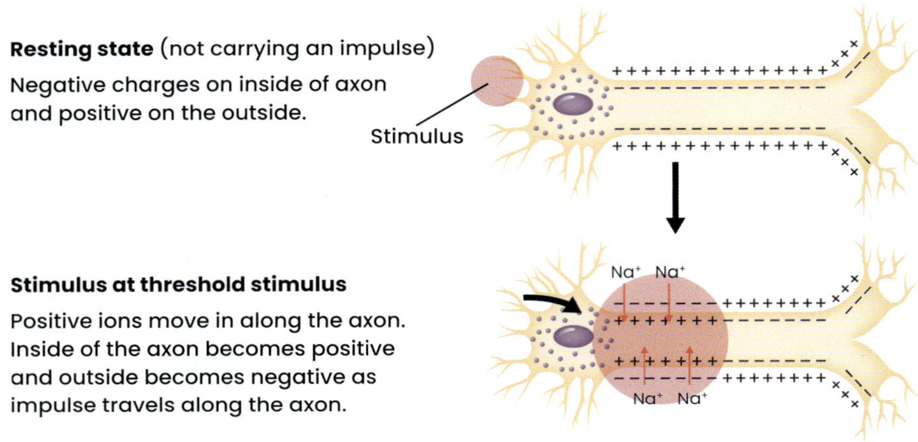

Figure 33.12: Neurotransmission: movement of nerve impulse along an axon.

(HL)

- When a neuron is not carrying an impulse, there is a difference in electrical charge across the neuron's membrane; the inside is negative, and the outside is positive.
- For an impulse to be carried, a stimulus of a certain strength is required – the **threshold stimulus.** A stimulus below the threshold will not cause an impulse to be carried. This is known as the **'all or nothing' rule**.

> The **threshold stimulus** is the minimum stimulus required for an impulse to be carried.
>
> The **'all or nothing' rule** means that an impulse is only carried if the threshold stimulus is reached. The same impulse is carried regardless of the strength of the stimulus.

- Once the threshold stimulus is reached, movement of ions, e.g. sodium ions (Na^+), causes a change in the charge across the dendrite with the inside becoming positive and outside becoming negative. This change in charge moves along the axon until it reaches the neurotransmitter vesicles. When the ion movement reaches the neurotransmitter vesicles, it activates the neurotransmitter vesicles to release neurotransmitter chemicals into the synaptic cleft.
- The neurotransmitter chemicals diffuse across the synaptic cleft. This takes a short period of time – a millisecond or less.

- The neurotransmitters lock into receptors in the dendrite of the next neuron (post-synaptic neuron). This stimulates an inrush of ions in the postsynaptic neuron and the impulse is carried along its axon.
- Once the impulse has been transmitted to the postsynaptic neuron, enzymes in the synaptic cleft break down the neurotransmitter chemicals, stopping the response.
- The presynaptic neuron reabsorbs the products of this neurotransmitter breakdown to form new neurotransmitters.

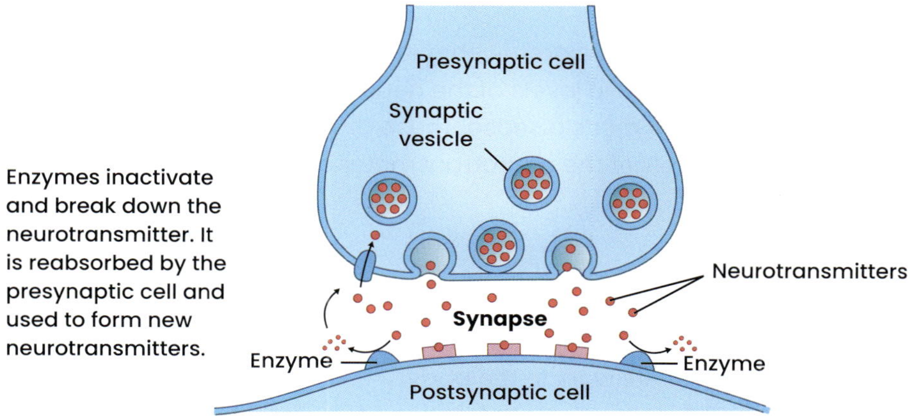

Figure 33.13: Breakdown and inactivation of neurotransmitters by enzymes.

- The neuron returns to its original resting charges. The minimum length of time to reverse the changes of the last impulse is known as the **refractory period**. During this time, the neuron will not respond to a new stimulus.

The **refractory period** is the minimum length of time required between two impulses being carried along a neuron.

HL

Nerve impulses are electro-chemical in nature. The movement of an impulse along an axon is electrical in nature, whereas the movement of the impulse from one neuron to the next is by means of chemical neurotransmitters.

Go to YouTube.com and search for '2-Minute Neuroscience: Synaptic Transmission' (1:51) by Neuroscientifically Challenged.

DISRUPTIONS TO NERVE IMPULSE TRANSMISSION

When neurotransmitters don't function properly, this can lead to acute disorders such as a seizure or chronic disorders such as Alzheimer's disease. We saw earlier that following an impulse transmission, enzymes in the synaptic cleft break down the neurotransmitters. Disorders that disrupt the transmission of nerve impulses may be treated using drugs that either:

1. inhibit enzymes that break down neurotransmitters, or
2. inhibit reuptake of neurotransmitters at the synapse.

This increases the levels of the neurotransmitter and the duration of the response, creating an enhanced effect.

Acute disorders

Acute disorders are severe and occur suddenly.

EXAMPLE: SEIZURES

Some seizures are caused by temporary conditions that irritate the brain, such as an infection, head injury or fever. Antiseizure drugs increase the levels of a neurotransmitter such as GABA by inhibiting enzymes that degrade the neurotransmitter.

Chronic disorders

Chronic disorders develop and worsen over time.

EXAMPLE: ALZHEIMER'S DISEASE

Alzheimer's disease is a chronic disease associated with memory loss, confusion and cognitive decline. The disease is linked to a decrease in levels of the neurotransmitter acetylcholine. An enzyme inhibitor is used in treating Alzheimer's disease. These drugs block enzymes that break down acetylcholine, enhancing the levels of the neurotransmitter and increasing communication between nerve cells.

Psychoactive drugs

Psychoactive drugs are chemicals that affect how the brain works and cause changes in mood, thoughts, feelings or behaviour.

HL

We saw earlier how non-medical drugs can lead to addiction. Drugs interfere with how neurotransmitters work in the body. Some drugs, such as marijuana and heroin, mimic the action of natural neurotransmitters by attaching to neurotransmitter receptors. They cause an increase in the levels of the feel-good neurotransmitter dopamine. Other drugs such as cocaine cause the release of abnormally large amounts of natural neurotransmitters or prevent their normal reuptake.

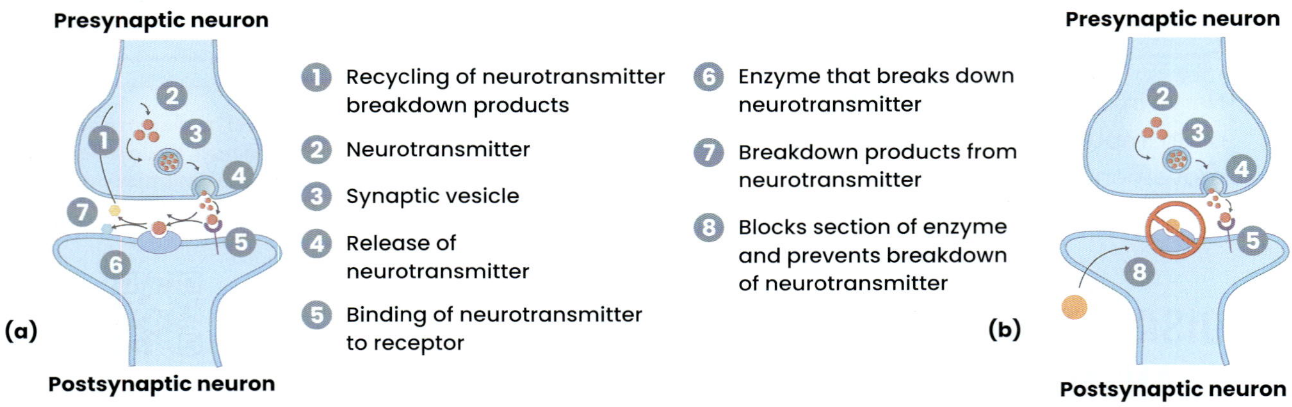

Figure 33.14: (a) Normal neurotransmission; (b) Treatment of acute and chronic disorders (similar process for psychoactive drugs and painkillers).

Painkillers

Painkillers such as morphine bind to receptors in the presynaptic nerve terminal to prevent neurotransmitters from being released and carrying the pain signal. Morphine also causes a release of the feel-good neurotransmitter dopamine in the brain.

SCIENCE IN SOCIETY: HEALTH AND TECHNOLOGY

Deep brain stimulation treatment for Parkinson's disease

Parkinson's disease is caused by a loss of brain nerve cells that produce the neurotransmitter dopamine, resulting in a loss of control of movement. Deep brain stimulation (DBS) is a surgical treatment in which electrodes are implanted in the dopamine producing regions of the brain. An impulse generator is implanted under the skin in the upper chest. This is connected to the electrodes in the patient's brain and used to control the level of current delivered to the electrodes. The patient has a controller allowing them to turn the device on or off. DBS addresses the movement symptoms of Parkinson's disease but does not cure or stop the disease progressing.

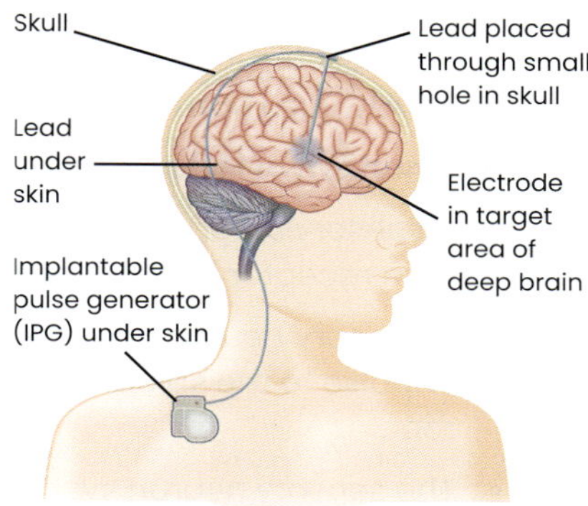

Figure 33.15: Deep brain stimulation.

PORTFOLIO LINK Go to your Learning Portfolio and complete Learning Check 3 to assess your knowledge and understanding of the topic.

Chapter Summary

- The **central nervous system (CNS)** consists of the brain and spinal cord.
- **White matter** is made up of axons in the CNS. **Grey matter** is made up of cell bodies in the CNS.
- **Cerebrospinal fluid** is a clear liquid that surrounds and protects the brain and spinal cord.
- **The peripheral nervous system (PNS)** consists of nerves that carry impulses to and from the CNS.
- **Sensory neuron** is a nerve cell that carries an impulse to the CNS.
- **Motor neuron** is a nerve cell that carries an impulse from the CNS to an effector.
- **Dendrites** convert chemical messages to electrical impulses and direct them towards the cell body.
- The **axon** is a long fibre that conducts electrical impulses away from the cell body.
- **Cell body** contains the nucleus and other organelles. It produces neurotransmitter chemicals.
- **Schwann cells** wrap around the axon in single file to form the myelin sheath.
- **Myelin sheath** protects and electrically insulates the axon, speeding up conduction of electrical impulses.
- **Neurotransmitter vesicles** store neurotransmitter chemicals.
- The **synapse** is the region where two neurons are in close proximity. The **synaptic cleft** is the gap between two neurons.
- **Neurotransmission** is the movement of a nerve impulse from one nerve to another or to a muscle or gland.
- A **neurotransmitter** is a chemical messenger that carries an impulse across a synapse, e.g. acetylcholine.
- **Mechanism of impulse transfer:**
  - Once the threshold stimulus is reached, movement of ions triggers an impulse to be carried.
  - Change in charge moves along the axon to the neurotransmitter vesicles.
  - Impulse causes neurotransmitter chemicals to be released into the synaptic cleft.
  - Neurotransmitter chemicals diffuse across the synaptic cleft and lock into receptors of the next neuron.
  - This stimulates an inrush of ions in the second neuron and the impulse is carried along its axon.
- Enzymes in the synaptic cleft digest and recycle the neurotransmitter chemicals, stopping the response. HL

Chapter Summary

- A **reflex action** is a rapid, involuntary and automatic movement response to a stimulus. Impulse is not carried to the brain but only to the spinal cord – this is necessary to protect the body from harm.
- **Mechanism of reflex action:**
  - A sensory receptor detects a stimulus e.g. pain receptor in the skin.
  - A sensory neuron carries an electrical impulse to the spinal cord via the dorsal root.
  - The sensory neuron synapses with an interneuron within the CNS.
  - The interneuron synapses with a motor neuron.
  - The motor neuron carries an electrical impulse out of the spinal cord via the ventral root.
  - The motor neuron carries the impulse to an effector muscle causing a response, e.g. dropping a hot plate.
- **Endorphins** are neurotransmitters released by the brain that help relieve pain.
- **Dopamine** is a 'feel-good' neurotransmitter released following pleasurable activities.
- **Effect of lifestyle choices on dopamine and endorphin levels:**
  - Drugs such as cocaine cause the release of abnormally large amounts of dopamine. Over time the person is less able to experience pleasure naturally and becomes addicted to the drug.
  - Exercise stimulates the production of endorphins, the body's natural painkillers and mood elevators.
  - Foods rich in protein or antioxidants increase dopamine levels. Diets high in saturated fat reduce dopamine release, resulting in overeating to achieve the same dopamine reward.
- **Disruptions to nerve impulse transmission** can lead to acute disorders such as a seizure or chronic disorders such as Alzheimer's disease. Such disorders may be treated using drugs that either inhibit enzymes that degrade neurotransmitters or inhibit reuptake of neurotransmitters at the synapse. HL
- Psychoactive drugs interfere with how neurotransmitters work. Some drugs attach to neurotransmitter receptors, causing an increase in dopamine levels. HL
- Painkillers such as morphine bind to receptors in the presynaptic nerve terminal to prevent neurotransmitters from being released and carrying the pain signal. HL

END OF CHAPTER ASSESSMENT

REMEMBER

1. Name the **two** parts of the nervous system and give the function of each part.
2. (a) Identify the type of nerve cell A, B and C below, give a reason for your choice in each case.
   (b) State the function of each type of nerve.

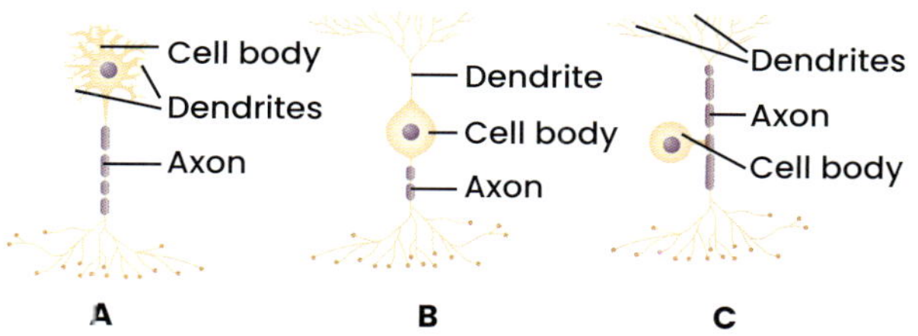

3. (a) Copy the diagram and add the labels.
   (b) Label the direction of the nerve impulse.

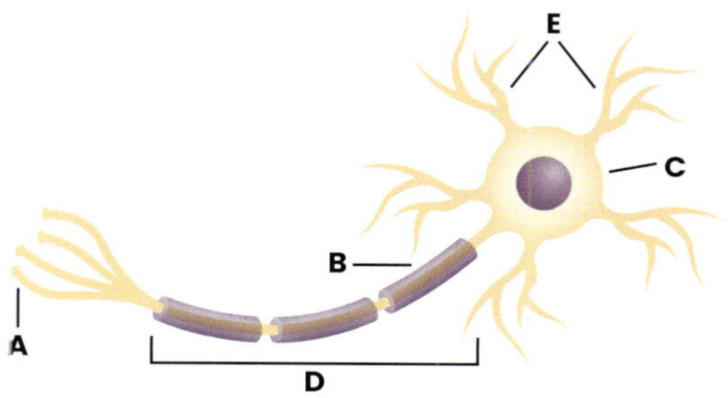

UNDERSTAND

4. Describe how the nervous system responds to a stimulus, such as a bell ringing at the end of a lesson in school.
5. Summarise the mechanism of impulse transfer.
6. Explain how the CNS is protected against damage.
7. Outline the pathway followed by an impulse during a reflex action.
8. Explain why messages carried by the nervous system are described as *electro-chemical*.
9. (a) What are endorphins?
   (b) What is dopamine?
   (c) How do endorphins and dopamine affect mood and behaviour?
10. Describe the impact of each of the following on dopamine and endorphin levels.
    (a) Non-medical drug use
    (b) Exercise
    (c) Dietary choices
11. HL (a) Name an acute disorder resulting from disruption to the conduction of nerve impulses.
    HL (b) Describe the ways in which drugs used to treat this disorder work.
12. HL (a) Name a chronic disorder resulting from disruption to the conduction of nerve impulses.
    HL (b) Describe the ways in which drugs used to treat this disorder work.
13. HL Describe the relationship between the conduction of nerve impulses and (a) psychoactive drugs (b) painkillers.

APPLY

14. The threshold stimulus for a particular muscle to contract in a certain person is 10mA. Explain what would happen if each of the following stimuli were applied.
    (a) 8mA (b) 10mA (c) 15mA
15. HL What would happen if neurotransmitters were not digested or recycled following the passage of an impulse from one neuron to the next?
16. What could happen if a person's spinal cord was damaged at the lower back region?

ANALYSE

17. Distinguish between each of the following.
    (a) Synapse and synaptic cleft
    (b) Neurotransmitter and neurotransmitter vesicles
    (c) White matter and grey matter
    (d) Nerve and neuron
    (e) Receptor and effector
18. How is a neuron adapted to receiving, carrying and transferring nerve impulses?

EVALUATE

19. Why are reflex actions important to the human body?
20. What is the advantage of having a myelin sheath around an axon?
21. Why are endorphins and dopamine important to the body?

CREATE

22. HL Research how drugs can interfere with the transmission of nerve impulses. Present your findings to your class.
23. Research a disorder of the nervous system and prepare a report for your class.

34 The Endocrine System

LEARNING OUTCOMES: 2.4.5, 2.4.10, 2.4.11

At the end of this chapter, you should be able to:

- Illustrate the location of each of the main endocrine glands.
- Describe a function for the hormones produced by the main endocrine glands.
- Compare nervous and hormonal co-ordination.
- Illustrate the use and impact of hormonal manipulation on organisms in agriculture, sport and health.
- Explore the interaction of the hormonal and nervous systems to maintain homeostasis.
- Explain how the hormonal system uses negative feedback to maintain homeostasis.

LINKS TO:

Chapter 14	Biomolecules
Chapter 21	Plant Reproduction
Chapter 27	The Human Body
Chapter 29	The Urinary System
Chapter 30	The Digestive System
Chapter 31	The Circulatory System
Chapter 33	The Nervous System
Chapter 36	Human Reproduction – Anatomy
Chapter 37	Human Reproduction – From Fertilisation to Birth

SUBJECT VOCABULARY

endocrine system | hormone | gland | endocrine gland | exocrine gland | hormonal manipulation | homeostasis | negative feedback

EYE-OPENER

The human body has an internal clock controlled by an area of the brain called the SCN (suprachiasmatic nucleus), located in the hypothalamus. The SCN is sensitive to light levels. In the morning, the optic nerve in your eyes senses the morning light. This message is relayed to the SCN which triggers the release of the hormone cortisol from the adrenal glands to help you wake up. At night, when it becomes dark, the SCN sends messages to the pineal gland in the brain. The pineal gland triggers the release of the hormone melatonin, making you feel sleepy.

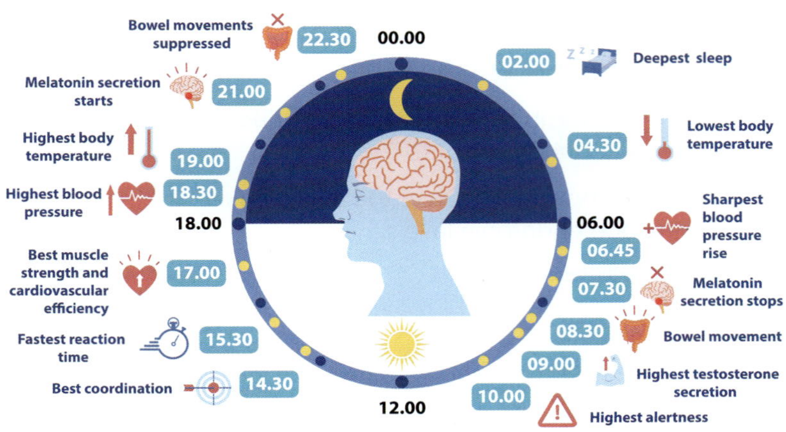

Section 1: Endocrine Glands and Hormones

At the end of this section, you should be able to:

- Compare nervous and hormonal co-ordination.
- Illustrate the location of each of the main endocrine glands.
- Describe a function for the hormones produced by the main endocrine glands.

Unpacked from LOs 2.4.5, 2.4.10

THE ENDOCRINE SYSTEM

The **endocrine system** along with the nervous system are the two main communication systems in the body. The endocrine system consists of a group of organs called **glands** around the body that produce and secrete chemicals called hormones. **Hormones** are involved in maintaining homeostasis and regulating many functions in the body including metabolism, growth, sexual development and reproduction. In Chapter 14, we learned that hormones may be either protein or steroid in nature. Steroid hormones are composed of lipids.

Hormones are secreted directly into the blood by ductless **endocrine glands**. For example, the islets of Langerhans in the pancreas are an endocrine gland which secretes the hormone insulin into the blood. Figure 34.1 compares ductless endocrine glands with **exocrine glands** that have a duct and carry their secretion directly to their destination. Examples of exocrine glands include tear ducts, sweat glands, salivary glands and the pancreas.

The pancreas is classified as both an endocrine gland and an exocrine gland. In addition to producing the hormone insulin, it secretes digestive enzymes directly into the duodenum via the exocrine pancreatic duct. We learned about the role of pancreatic digestive enzymes in Chapter 30.

- The **endocrine system** is a group of glands that produce and secrete hormones.
- A **gland** is an organ that produces a chemical with a specific function in the body and releases it into the bloodstream or a duct.
- A **hormone** is a chemical messenger produced by an endocrine gland that travels in the bloodstream to act on a specific tissue.
- An **endocrine gland** is a ductless gland that secretes a hormone directly into the bloodstream.
- An **exocrine gland** is a gland with a duct that carries its secretion to its destination.

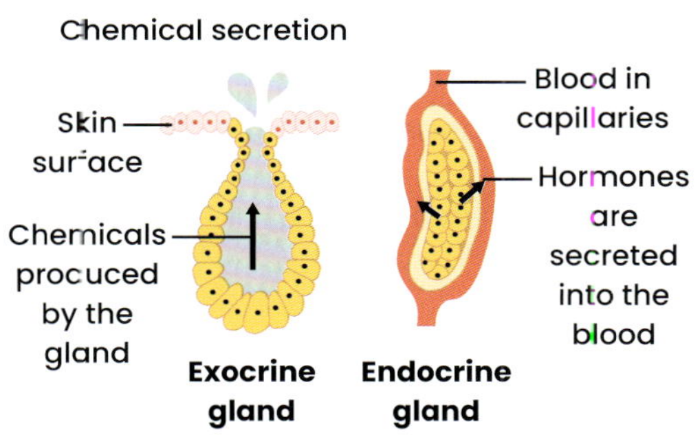

Figure 34.1: Exocrine glands and endocrine glands.

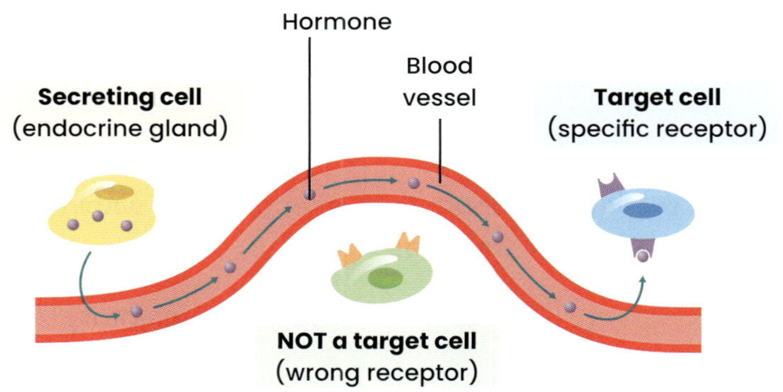

Figure 34.2: Hormones travelling in bloodstream to receptor cells.

Hormones travel in the bloodstream to act on specific target cells that have a particular receptor. When the hormone reaches its target effector cell, it binds to its receptor and activates the cell to produce a particular response or effect. Each hormone can only act on cells that have the correct receptor. For example, during childbirth, the hormone oxytocin travels from the pituitary gland in the mother's brain to the uterus, where it binds to oxytocin receptors, causing the uterus to contract.

COMPARING NERVOUS AND HORMONAL CONTROL

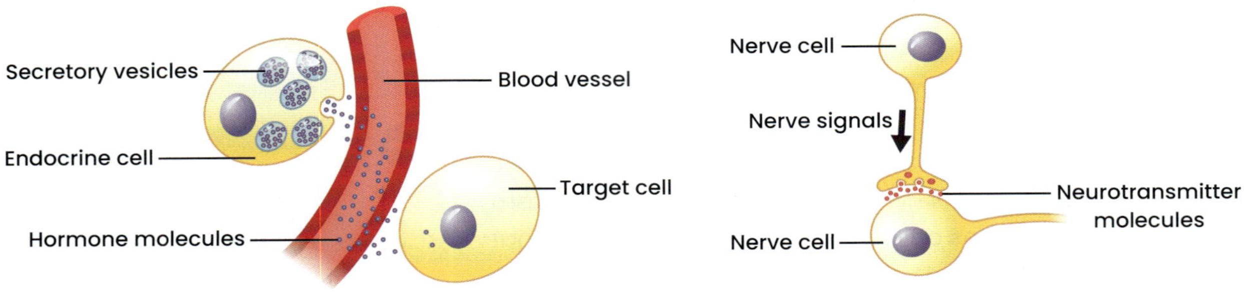

Figure 34.3: Hormonal vs nervous control.

The pituitary gland in the brain links the nervous and endocrine systems. While the endocrine system along with the nervous system are the two main communication systems in the body, differences exist between them. These are outlined in Figure 34.4.

NERVOUS AND HORMONAL CO-ORDINATION		
Similarities		
Both systems allow the body to communicate with its internal and external environments and co-ordinate the response. Both systems have a chemical component to their response message		
Differences		
	Nervous system	**Endocrine system**
Type of message	Nerve impulses are electro-chemical messages	Hormones are chemical messengers
How messages are transmitted	Carried in nerves	Carried in blood
Speed of response	Faster response	Slower response
Length of response	Short-term effect: usually within seconds	Longer-term effect
Action	Can be voluntary or involuntary	Involuntary
Response target	Response is localised to target cells to which nerves are connected	Response often more widespread, may be more than one target organ
Example of response	Movement of arms or legs, reflex actions, e.g. blinking	Growth of the skeleton

Figure 34.4: Comparison of nervous and hormonal co-ordination.

MAIN ENDOCRINE GLANDS AND THEIR HORMONES

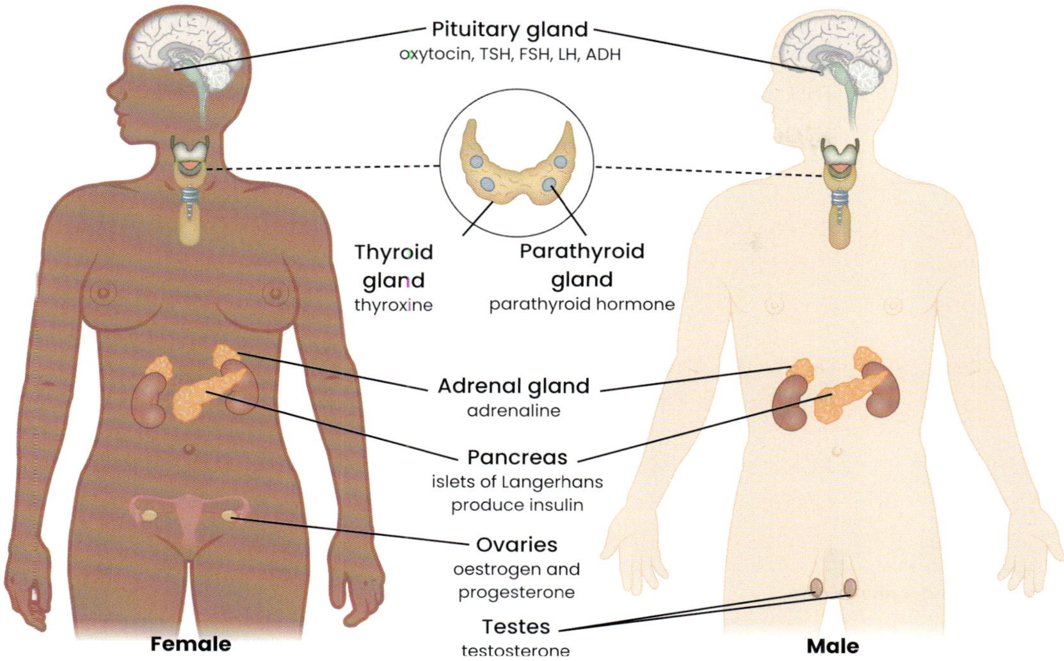

Figure 34.5: Main endocrine glands and the hormones they produce.

Figure 34.5 illustrates the location of the main endocrine glands in the human body. The location and role of the main endocrine glands along with the hormones produced by each gland are outlined in Figure 34.6.

> Males produce oestrogen, too.
> **DID YOU KNOW?**

	HORMONE(S) PRODUCED	FUNCTION OF HORMONE
ADRENAL GLANDS	Adrenaline	Produced in response to fear or stress 'fight or flight' situations, where a rapid response is needed. Adrenaline causes the heart rate and breathing rate to increase.
PANCREAS (ISLETS OF LANGERHANS)	Insulin	Insulin regulates glucose levels in the blood. When glucose levels are too high, insulin targets the liver and other cells to convert glucose to glycogen for storage.
THYROID	Thyroxine	Thyroxine controls metabolic rate and is involved in controlling growth and development.
PARATHYROID	Parathyroid hormone	Parathyroid hormone regulates calcium levels in the blood by stimulating the release of calcium from bones back into the blood.
OVARIES	Oestrogen Progesterone	Oestrogen controls female secondary sexual characteristics. Oestrogen targets many cells including those in the breasts, bones and the uterus. Oestrogen and progesterone both target the uterus, causing the uterine lining, the endometrium, to mature.
TESTES	Testosterone	Testosterone controls the development of male secondary sexual characteristics and the production of sperm.

Figure 34.6: Overview of the main endocrine glands.

Continued overleaf

PITUITARY GLAND	• Oxytocin	Oxytocin triggers the walls of the uterus to contract during labour and the milk ducts in the breasts to eject milk during breast-feeding.
	• TSH (thyroid stimulating hormone)	TSH triggers the production of thyroxine by the thyroid gland.
	• FSH (follicle stimulating hormone)	FSH targets the ovary, where it triggers the development of the Graafian follicle. HL
	• LH (luteinising hormone)	LH targets the ovary, where it causes ovulation – the release of an egg from a Graafian follicle. HL
	• ADH (anti-diuretic hormone)	ADH controls water concentration in the blood. It targets the collecting ducts in the kidney, causing them to become more porous and reabsorb more water. HL

Figure 34.6: Overview of the main endocrine glands.

PORTFOLIO LINK Go to your Learning Portfolio and complete Learning Check 1 to assess your knowledge and understanding of the topic.

Section 2: Hormone Manipulation and Homeostasis

At the end of this section, you should be able to:

- Illustrate the use and impact of hormonal manipulation on organisms in agriculture, sport and health.
- Explore the interaction of the hormonal and nervous systems to maintain homeostasis.
- Explain how the hormonal system uses negative feedback to maintain homeostasis.

Unpacked from LOs 2.4.5, 2.4.11

HORMONAL MANIPULATION

Scientists have developed methods of artificially altering hormone levels in an organism. We will look at how increasing or decreasing hormone levels in an organism (plant or animal) can affect the organism by examining examples from three different areas: agriculture, sport and health.

Hormone manipulation refers to the artificial altering of hormone levels in an organism.

Agriculture

PLANT GROWTH REGULATORS

In Chapter 21, we saw that plant seeds naturally produce growth regulators. These growth regulators or plant hormones may be used artificially in agriculture to promote growth and increase crop yields. Gibberellins are a class of plant hormone used to promote the formation of seedless fruits, e.g. seedless grapes. Gibberellins are also used in the flower industry to promote flowering, resulting in larger, more attractive flowers. Ethylene, another plant hormone, is used to control the ripening of bananas in the food industry. Bananas are harvested when they are still green and may be sprayed with ethylene before delivery to shops, to promote ripening. The plant hormone auxin is used in rooting powder to promote root growth from cuttings. Other plant hormones are used as selective weedkillers.

Figure 34.7: Seedless grapes.

Health

HORMONE REPLACEMENT THERAPY (HRT)

When a woman reaches menopause, her menstrual cycle stops and there is a large drop in the production of oestrogen and progesterone. The drop in levels of these hormones causes symptoms such as hot flushes, vaginal dryness and mood swings. Hormone replacement therapy (HRT) is used in medicine to treat the symptoms of menopause by replacing the hormones oestrogen and progesterone. As oestrogen and progesterone keep bones healthy, HRT can also help reduce the risk of osteoporosis and heart-related diseases.

Figure 34.8: HRT patch.

CONTRACEPTIVE PILL

In Chapter 36, we will learn more about how the female hormones oestrogen and progesterone are used in the contraceptive pill to prevent pregnancy. The hormones in the contraceptive pill prevent ovulation (the release of an egg from the ovaries). This means that there is no egg for sperm to fertilise and pregnancy cannot occur.

STEROID INHALERS FOR ASTHMA

Some inhalers used to reduce the symptoms of asthma contain steroids. These steroids, an artificial version of hormones normally produced by the adrenal glands, reduce inflammation in the airways.

Sport

ANABOLIC STEROIDS

Anabolic steroids such as the hormone testosterone build up muscle and strength. Some athletes may illegally use testosterone to increase muscle mass and reduce the time needed to recover after a training session, allowing them to train harder for longer periods of time. When used in higher doses than required for medical reasons, anabolic steroids can increase the risk of stunted growth in teenagers, heart disease and liver damage. Males taking anabolic steroids may experience shrunken testicles, breast growth and infertility due to a decrease in sperm production. Females taking anabolic steroids may experience an increase in body or facial hair, deepening of the voice and infertility due to abnormal menstrual cycles.

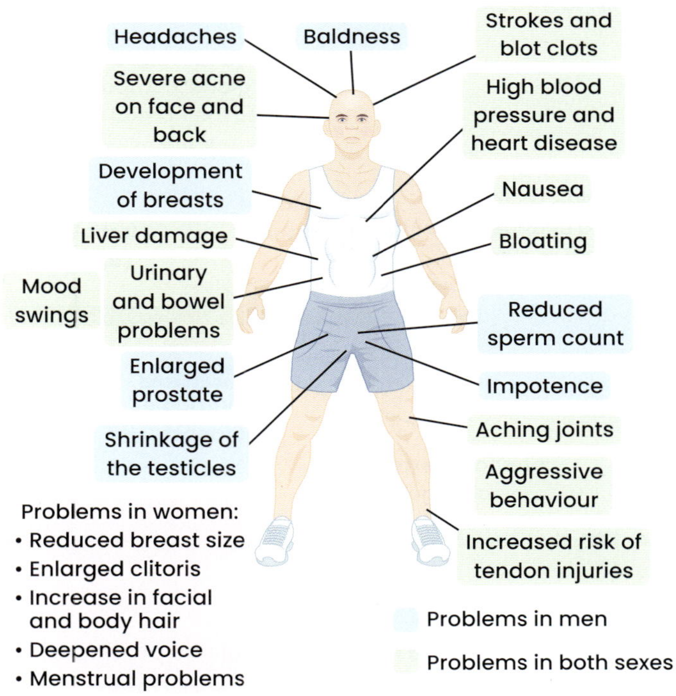

Figure 34.9: Effects of the illegal use of anabolic steroids on the human body.

HOMEOSTASIS

Maintaining a constant internal environment in the body is essential to the proper functioning of cells. This is known as **homeostasis**. The endocrine and nervous system are both responsible for maintaining homeostasis. The brain continually monitors changes internal and external to the body including temperature, blood pressure, hunger and thirst amongst other variables. Sensory nerves carry information on these variables to the brain. If any of these factors rise too high or fall too low from their normal value, the brain responds by stimulating nerves to carry messages to the relevant organ to correct it. This returns the body to its normal steady state. The brain also controls the secretion of hormones by the pituitary gland, helping to maintain homeostasis. In this way, the pituitary gland acts as a link between the nervous and endocrine systems. Certain hormones can also stimulate or inhibit the flow of nerve impulses.

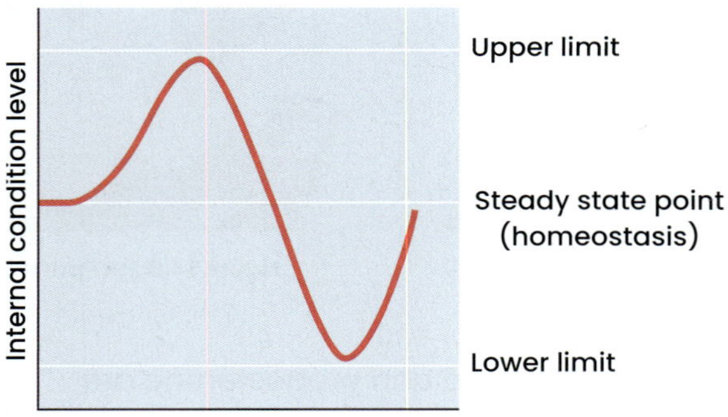

Homeostasis is the maintenance of a steady state within the body.

Figure 34.10: The body's internal environment must be kept at a steady set state to function properly.

Negative feedback

One way in which the body maintains homeostasis, or a state of balance, is by using negative feedback loops. **Negative feedback** is a biological response in which the effects of a reaction slow or stop that reaction. In Chapter 35, we will study the control of hormone levels in the menstrual cycle via negative feedback. We will now look at the control of the levels of two hormones insulin and thyroxine by negative feedback.

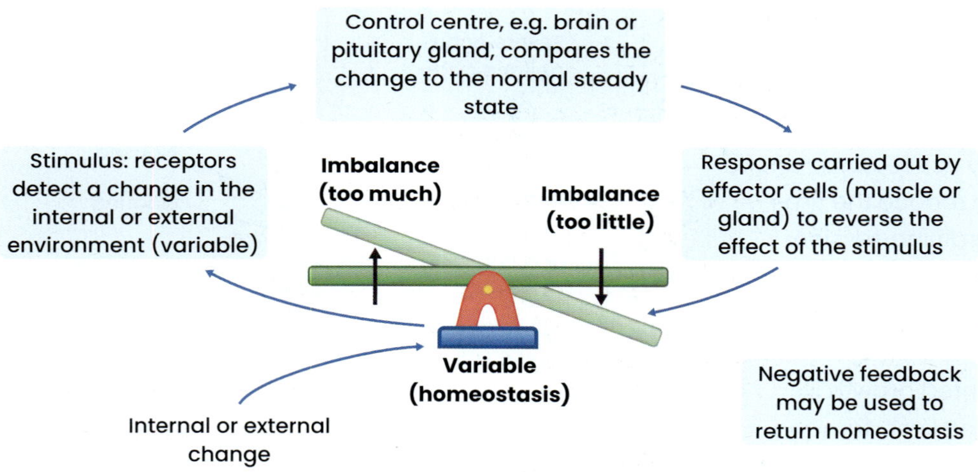

Negative feedback is a biological response in which the effects of a reaction slow or stop that reaction. Negative feedback loops help to maintain homeostasis in cells.

Figure 34.11: The body's internal and external environment are continually monitored to maintain homeostasis.

EXAMPLE 1: INSULIN

As outlined earlier, the hormone insulin is produced by the islets of Langerhans in the pancreas. Insulin helps control glucose levels in the blood. During digestion of food, glucose is absorbed through the walls of the small intestine into the blood. This causes blood sugar levels to rise – some of this sugar is used in respiration. Blood sugar levels are maintained at normal levels by the hormone insulin through a negative feedback mechanism.

High blood sugar levels

- If the sugar (glucose) concentration in the blood gets too high, insulin is released by the islets of Langerhans in the pancreas.
- Insulin sends a message to the liver, muscles and other cells to take up and store the excess sugar. Some of the excess sugar is stored as body fat.
- This causes the blood sugar levels to drop and return to normal levels. A return to homeostasis stops the islets of Langerhans releasing insulin.

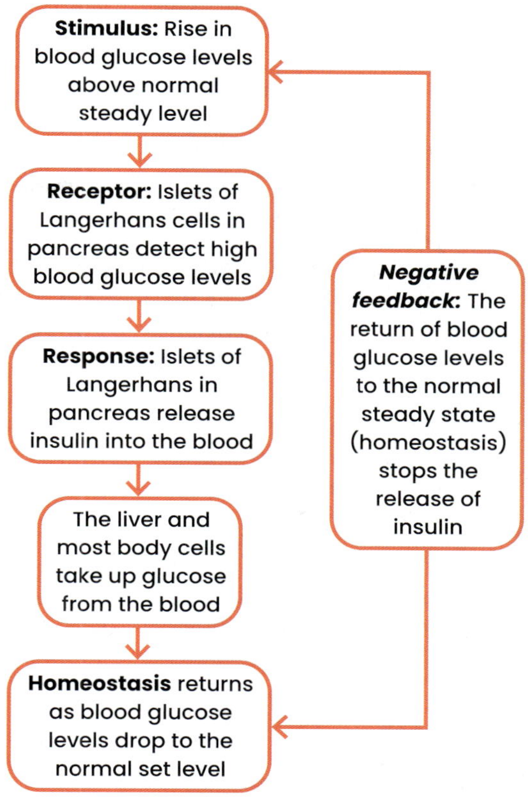

Figure 34.12: The negative feedback loop.

Insulin deficiency: diabetes

Diabetes is a condition that occurs when the pancreas does not produce insulin, or cells do not respond to insulin. This results in too much glucose in the blood.

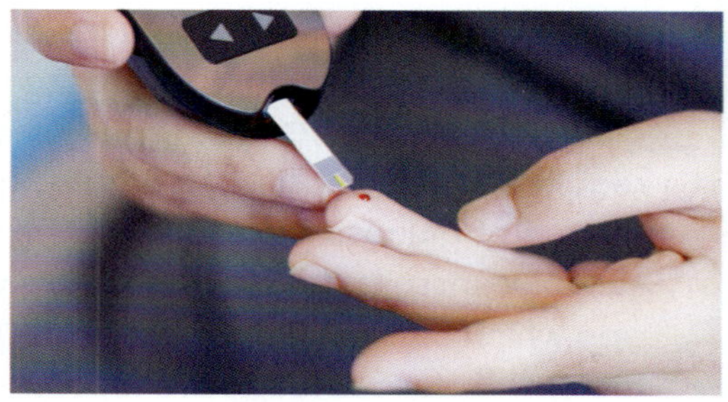

Figure 34.13: A diabetic checking their blood glucose levels.

EXAMPLE 2: THYROXINE

Thyroxine is a hormone produced by the thyroid gland in the neck that controls metabolism and has a role in growth and development. A negative feedback loop also controls the levels of thyroxine in the blood.

- Low levels of thyroxine in the blood trigger the release of TSH (thyroid stimulating hormone) from the pituitary gland.
- This in turn triggers the release of thyroxine from the thyroid gland, causing thyroxine levels in the blood to return to normal.
- When thyroxine levels in the blood are normal, it inhibits the release of TSH from the pituitary gland. In this way, thyroxine levels are prevented from rising above normal and homeostasis is maintained.

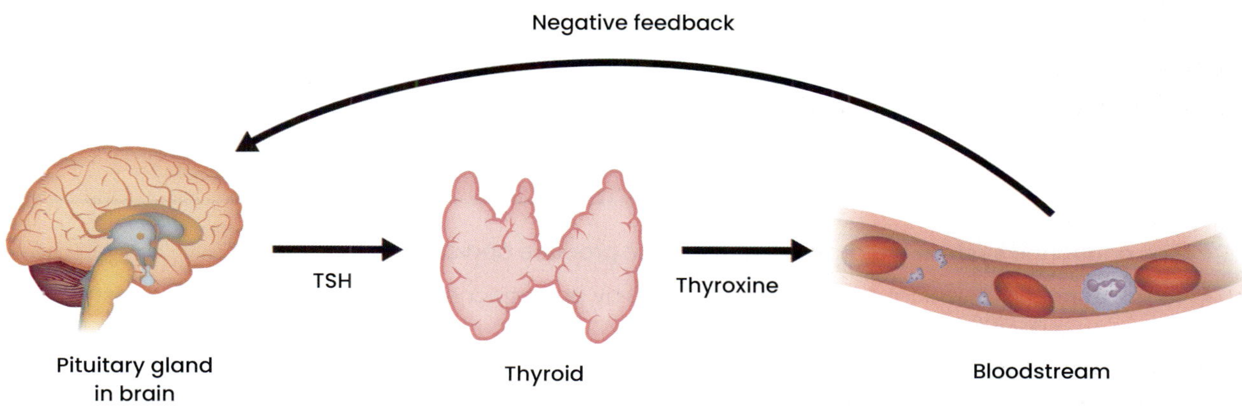

Figure 34.14: Negative feedback mechanism – thyroxine.

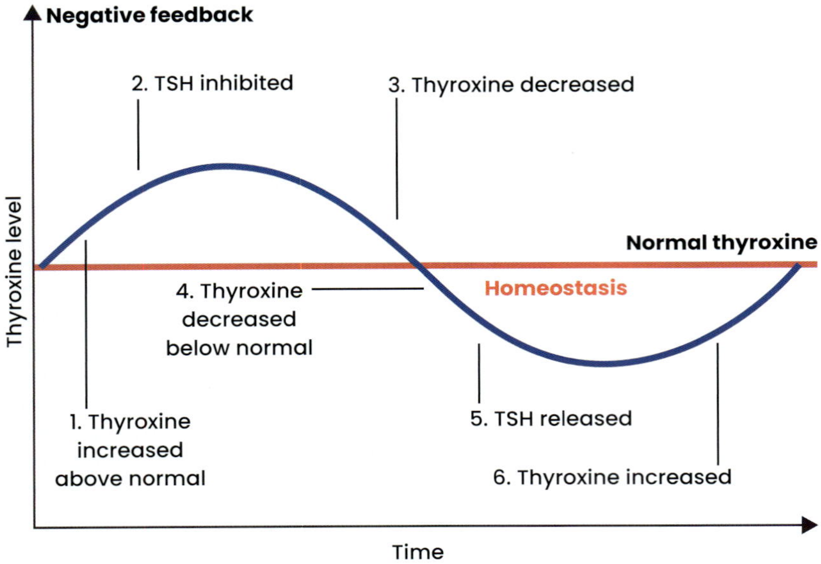

Figure 34.15: A negative feedback loop maintains homeostasis, keeping thyroxine levels in the blood constant.

THINK, PAIR, SHARE

How do hormones help to maintain homeostasis in the body?

Go to YouTube.com and search for 'Negative Feedback & Hormonal Response-Leaving Cert Biology 2024' (3:28) by Biology Bugbears.

SCIENCE IN SOCIETY: HEALTH AND TECHNOLOGY

Technology in managing diabetes

Diabetics may need to test their blood sugar levels several times a day. Traditionally, this has been done using a drop of blood from a finger prick to check glucose levels at that moment with a blood glucose meter. Some people living with diabetes may choose to use a continuous glucose monitor (CGM) to reduce the number of finger prick tests required. CGMs consist of a small disposable sensor under the skin which measures glucose levels around the body's cells every five minutes. The sensor transmits the glucose results wirelessly to a smart device with an app which can be read by the user. The sensor can send alarms alerting the user if glucose levels are too high or too low. CGM data can also be shared with family members or with medical professionals.

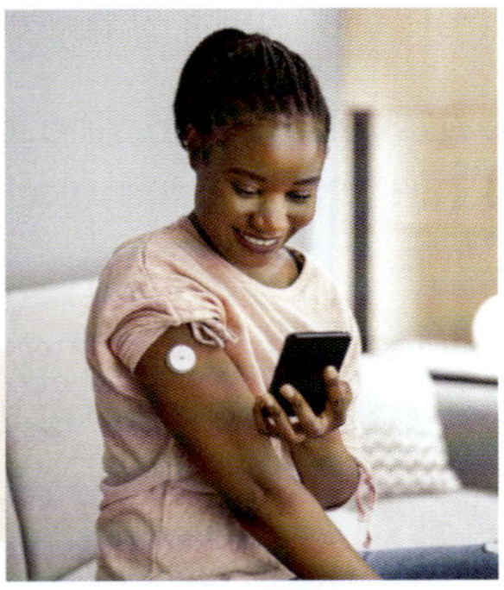

Figure 34.16: Continuous glucose monitor.

PORTFOLIO LINK **Go to your Learning Portfolio and complete Learning Check 2 to assess your knowledge and understanding of the topic.**

Chapter Summary

- The **endocrine system** is a group of glands that produce and secrete hormones.
- A **hormone** is a chemical messenger produced by an endocrine gland that travels in bloodstream to act on a specific tissue.
- An **endocrine gland** is a ductless gland that secretes a hormone directly into the bloodstream.

ENDOCRINE GLANDS AND THEIR HORMONE(S)	
Adrenal glands (on top of the kidneys)	
Hormone	*Function*
Adrenaline	Produced in response to 'fight or flight' situations.

Pancreas (behind the stomach in the upper left abdomen)	
Insulin	Insulin regulates glucose levels in the blood.
Thyroid (front of the neck)	
Thyroxine	Thyroxine controls metabolic rate.
Parathyroid (behind the thyroid gland)	
Parathyroid hormone	Parathyroid hormone regulates calcium levels in the blood.
Pituitary gland (base of the brain)	
Oxytocin	Oxytocin triggers the walls of the uterus to contract during labour. Triggers milk ducts to eject milk during breast-feeding.
TSH	Thyroid stimulating hormone triggers the thyroid gland to produce thyroxine.
FSH	Follicle stimulating hormone triggers development of the Graafian follicle. HL
LH	Luteinising hormone causes ovulation. LH stimulates the formation of the corpus luteum from the remains of the Graafian follicle. HL
ADH	Anti-diuretic hormone ADH controls water concentration in the blood. HL
Ovaries (lower abdomen)	
Oestrogen Progesterone	Oestrogen controls female secondary sexual characteristics. Oestrogen and progesterone involved in controlling the menstrual cycle.
Testes (in the scrotum)	
Testosterone	Controls male secondary sexual characteristics and sperm production.

	NERVOUS SYSTEM	ENDOCRINE SYSTEM
Type of message	Electro-chemical messages	Chemical messengers
Message transmission	Carried in nerves	Carried in blood
Speed of response	Rapid response	Slow response
Length of response	Short-term effect	Longer-term effect
Action	Can be voluntary or involuntary	Involuntary

- **Hormone manipulation** is the artificial altering of hormone levels in an organism:
  - **Agriculture:** plant growth regulators are used in producing seedless grapes, ripening bananas, rooting powders and selective weedkillers.
  - **Health:** hormone replacement therapy for menopausal women.
  - **Sport:** illegal use of the hormone testosterone.
- **Interaction of the hormonal and nervous systems to maintain homeostasis:** The brain controls the secretion of hormones by the pituitary gland, helping to maintain homeostasis.
- **Negative feedback** is a biological response in which the effects of a reaction slow or stop that reaction. Negative feedback loops help to maintain homeostasis in cells.

END OF CHAPTER ASSESSMENT

REMEMBER

1. (a) Identify each of the endocrine glands in the diagram below.

(b) Name a hormone produced by each of the endocrine glands (A–F).

(c) Give a function for each of the hormones you have identified above.

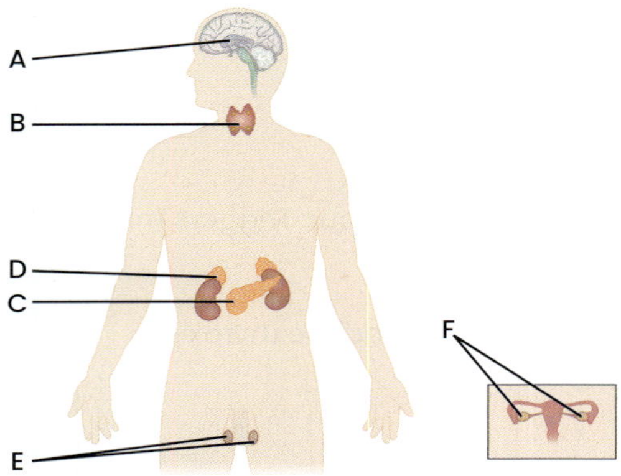

2. Define the term *hormone*.

3. What is the chemical nature of hormones?

UNDERSTAND

4. *The pancreas may be described as a dual-function gland, as it functions as both an endocrine and exocrine gland.*
Explain your understanding of this statement.

5. Distinguish between endocrine and exocrine glands. Give an example of each.

6. What is meant by hormone manipulation? Describe the effect of hormone manipulation on organisms in each of the following settings.

(a) Agriculture

(b) Health

(c) Sport

APPLY

7. Normal blood glucose levels before a meal should be within the range 4–7mmol/l.

(a) Explain why blood glucose levels two hours after a meal range from 5–9mmol/l.

(b) How would you expect the blood glucose levels of a diabetic to compare to normal levels both before and after a meal?

8. How does the human body respond to high glucose levels?

ANALYSE

9. Compare and contrast hormonal and nervous control.

10. How do negative feedback loops help maintain homeostasis in the human body? Refer to hormonal examples in your answer.

11. Why do you think the pituitary gland is sometimes referred to as the 'master gland'?

12. Study the blood sugar level chart below and answer the following questions.

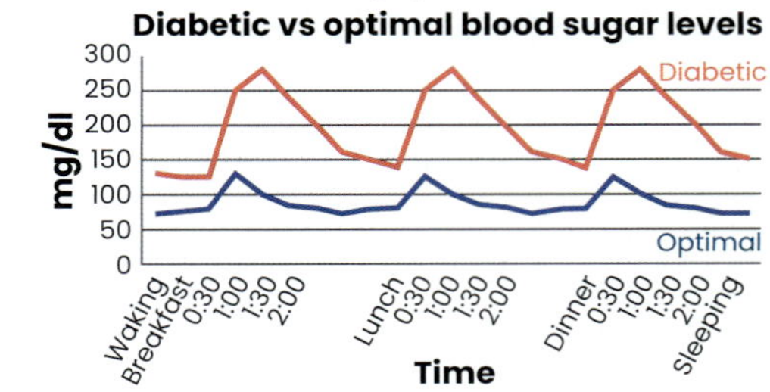

(a) What is the optimal range of blood sugar levels throughout a day?

(b) How do the blood sugar levels of a diabetic compare to the optimal levels?

(c) What happens blood sugar levels after a meal?

(d) How does the above chart illustrate homeostasis within the body?

(e) Copy the above chart and draw a graph to illustrate normal blood insulin levels throughout the day.

EVALUATE

13. Evaluate the importance of maintaining blood sugar levels within a normal range.

14. Which of the graphs below best represents normal blood sugar level?

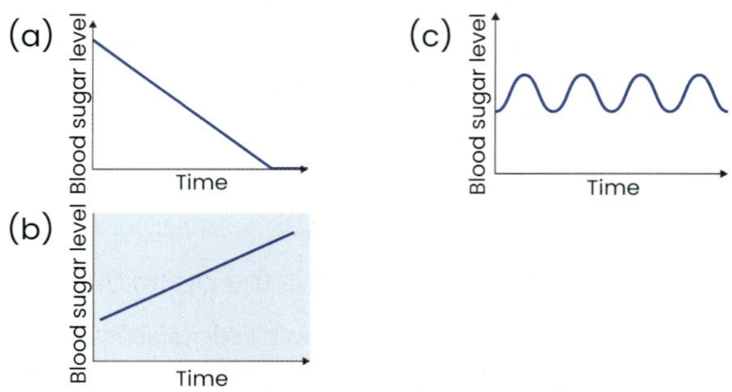

CREATE

15. Create a poster to illustrate the interaction of the hormonal and nervous systems to maintain homeostasis.

35 The Musculoskeletal System

LEARNING OUTCOMES: 2.4.2, 2.4.3

EYE-OPENER

Crustaceans such as lobsters and crabs lack the internal bony skeleton that humans and other mammals have. Instead, they have hard outer skeletons called exoskeletons. These serve as their protective outer covering and support structure. They also allow them blend into their surroundings. Crustacean exoskeletons are reinforced with calcium carbonate, providing strong protection against predators and underwater pressures. However, exoskeletons make growing a little more difficult. They must shed their old exoskeleton to reveal a larger one underneath. This process is crucial for their growth but leaves them temporarily vulnerable to predators as the new skeleton is initially softer. Insects and spiders also have exoskeletons (crustaceans, insects and spiders are all classified as arthropods) but their chemical composition differs.

At the end of this chapter, you should be able to:

- Describe the anatomy of the musculoskeletal system.
- Describe the main features of the skeleton, bones and joints.
- Outline the different kinds of muscle found in humans and their roles.
- Describe the different kinds of joints.
- Describe how muscles act in pairs to support the movement of joints.

LINKS TO:

Chapter 14 Biomolecules
Chapter 27 The Human Body
Chapter 28 The Breathing System
Chapter 30 The Digestive System
Chapter 31 The Circulatory System
Chapter 33 The Nervous System

SUBJECT VOCABULARY

bone, osteocytes, compact bone, spongy bone, muscle, myocytes, skeletal muscle, cardiac muscle, smooth muscle, ligaments, tendons, cartilage, axial skeleton, appendicular skeleton, joint, synovial fluid, antagonistic muscle pair

Section 1: The Musculoskeletal System

At the end of this section, you should be able to:

- Describe the structure and functions of the musculoskeletal system.
- Outline the functions of the musculoskeletal system.
- Briefly outline the structure of bone.
- Name the three types of muscle and briefly outline their functions.

Unpacked from LO 2.4.2

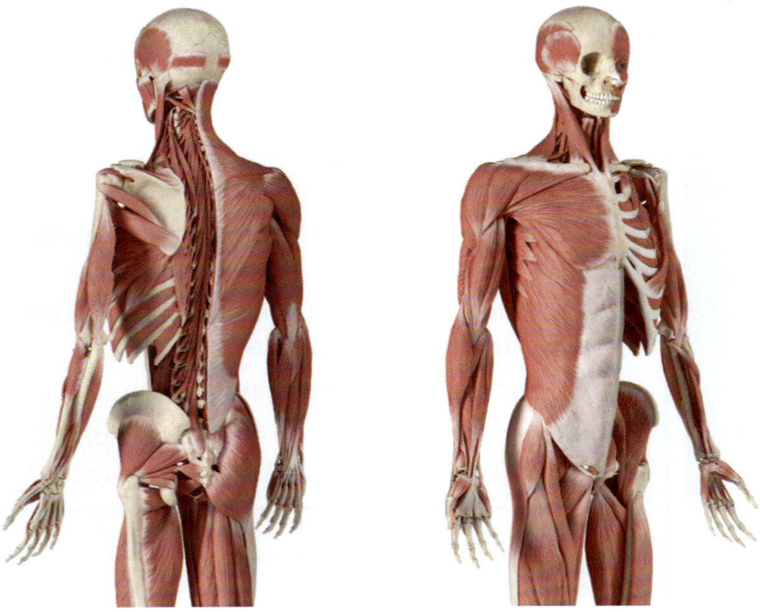

Figure 35.1: The musculoskeletal system is composed of our 206 bones and over 600 muscles.

The musculoskeletal system comprises our bones and the muscles that attach to them. The adult human has 206 bones, over 600 muscles and other connective tissues called ligaments, tendons and cartilage. All of these work together, and in collaboration with other body systems, to fulfil a number of key functions.

FUNCTIONS OF THE MUSCULOSKELETAL SYSTEM

1. **Movement:** Strong muscles and rigid bones allow the body to move. Movement is controlled by the nervous system (brain and spinal cord) and is a form of response.
2. **Support:** The skeleton supports the internal organs and gives the body shape.
3. **Protection:** The brain, lungs and heart are all protected by the skeleton.
4. **Heat generation:** The contraction of muscles creates heat, as their cells break down glucose to produce energy.
5. **Blood cell production:** Red and white blood cells, as well as platelets, are produced within the bones.
6. **Storage:** Bone stores minerals like calcium and phosphorous, and high energy fat is stored in the bone marrow. Muscle is used to store carbohydrates in the form of glycogen.

WHAT IS BONE?

Bone is a specialised connective tissue. It is a living tissue with various types of bone cells embedded within a framework called bone matrix. **Osteocytes** are the most common bone cell, making up 95% of bone. The matrix is composed of collagen fibres (a flexible protein) with calcium phosphate salts providing strength.

There are two main categories of bone tissue: **compact bone** and **spongy bone**. These combine, in different ways, to form the various bones of the human skeleton. Compact bone has a very compressed matrix, making it very hard and durable. It contains blood vessels and nerves. Spongy bone has a more open matrix and contains **red bone marrow**, which contains undifferentiated stem cells to produce red and white blood cells. It also contains yellow bone marrow, which holds fat.

A typical bone in our skeleton will contain both compact and spongy bone. Compact bone is generally found in the long section of our bones while spongy bone is found at the ends.

Figure 35.3: A typical bone in our skeleton contains both compact and spongy bone tissue.

- **Bone** is a specialised connective tissue composed of bone cells embedded in a matrix of collagen and calcium salts.
- **Osteocytes** are bone cells.
- **Compact bone** contains a compressed matrix, making it very hard and durable.
- **Spongy bone** has an open matrix and contains blood cell producing red bone marrow.

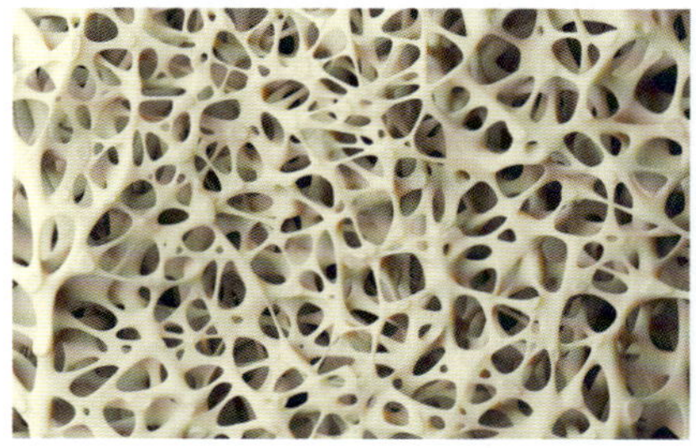

Figure 35.2: The bone matrix.

Go to YouTube.com and search for 'What Are Your Bones Made Of? | Bang Goes The Theory | Earth Science' (2:43) by BBC Earth Science.

WHAT IS MUSCLE?

Muscle is another specialised tissue found in the bodies of almost all animals. It is made of cells called **myocytes**, which contain proteins called **myosin** and **actin**. These proteins give muscle cells their unique trait; they can contract, becoming shorter and tighter when stimulated (the stimulation usually comes from a nerve). Muscle cells come together to form fibres, which then build the 600 separate muscles of your body.

There are three types of muscle tissue found in humans:

1. **Skeletal muscle** – the muscles that attach to our skeleton to form the musculoskeletal system.
2. **Cardiac muscle** – the muscle found only in the heart. It is under unconscious control and doesn't grow or tire like skeletal muscle.
3. **Smooth muscle** – found in the digestive tract and in arteries and veins.

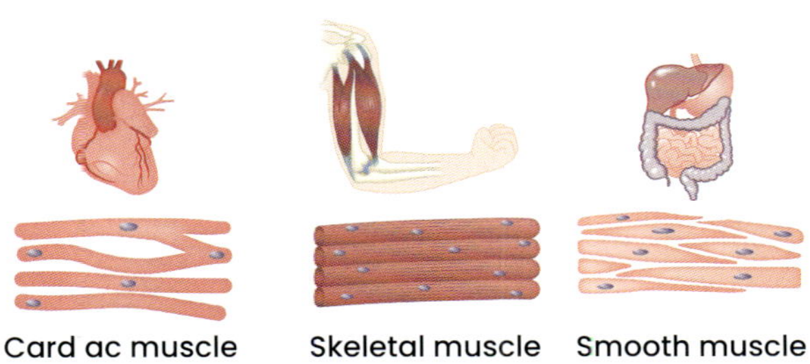

Figure 35.5: Types of muscle.

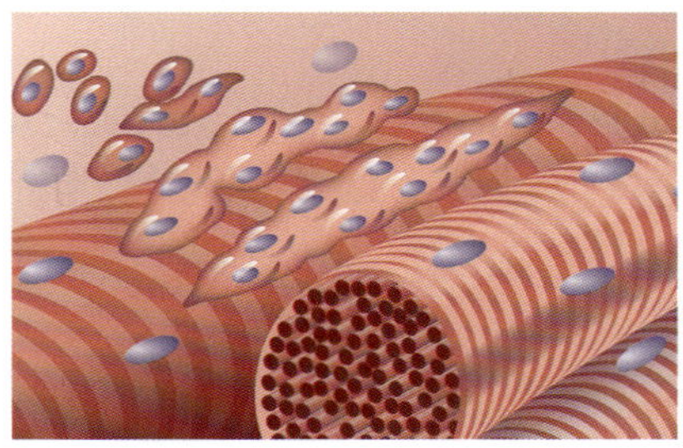

Figure 35.4: Muscle cells bundle together to form fibres, which then form the various muscles of the human body.

- **Muscle** is a specialised tissue made from cells that can contract when stimulated.
- **Myocytes** are muscle cells.
- **Skeletal muscle** is a type of muscle that attaches to our skeleton to form the musculoskeletal system.
- **Cardiac muscle** is muscle found only in the heart.
- **Smooth muscle** is found in our digestive tract, arteries and veins.

FOCUS ON ... LIGAMENTS, TENDONS AND CARTILAGE

The musculoskeletal system doesn't just consist of bones and muscle. Three other important tissues connect everything together. These are **ligaments**, **tendons** and **cartilage** – referred to as connective tissues.

Ligaments connect bone to bone. They are made from specialised cells that form elastic proteins, making them flexible and stretchy.

Tendons connect muscle to bone. Again, they are made from specialised cells but the protein they produce is not flexible or elastic. This makes tendons exceptionally strong.

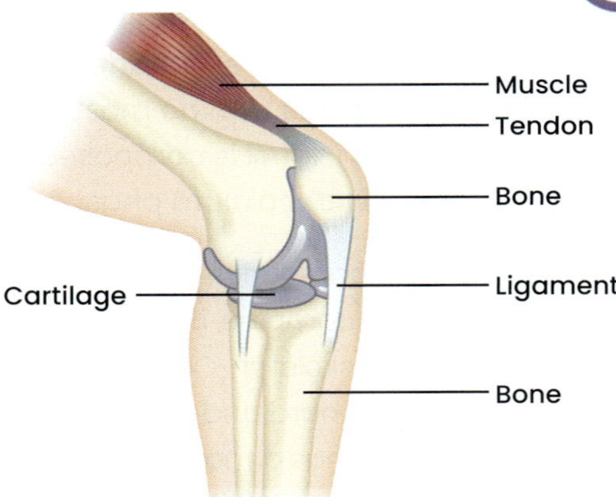

Figure 35.6: Ligaments, tendons and cartilage help form joints with muscles and bones.

Cartilage is another flexible tissue that **surrounds bones** and acts as a cushion or shock absorber and reduces friction.

In the diagram shown, we can see how ligaments, tendons and cartilage, along with muscle and bones, form joints (in this case, the knee). A joint exists where two or more bones meet. We will discuss joints in more detail in Section 3.

Go to YouTube.com and search for 'How your muscular system works – Emma Bryce' (4:44) by TED-Ed.

- **Ligaments** connect bone to bone.
- **Tendons** connect muscle to bone.
- **Cartilage** surrounds and cushions bone.

PORTFOLIO LINK **Go to your Learning Portfolio and complete Learning Check 1 to assess your knowledge and understanding of the topic so far.**

Section 2: The Structure of the Skeleton

At the end of this section, you should be able to:

- Describe the structure of the axial skeleton.
- Discuss the role of the vertebral discs.
- Relate the structure of the skeleton to its functions.
- Describe the structure of the appendicular skeleton and the attached limbs.

Unpacked from LO 2.4.2

THE SKELETON

The human skeleton consists of 206 bones. It is divided into two divisions: the **axial skeleton** and the **appendicular skeleton**.

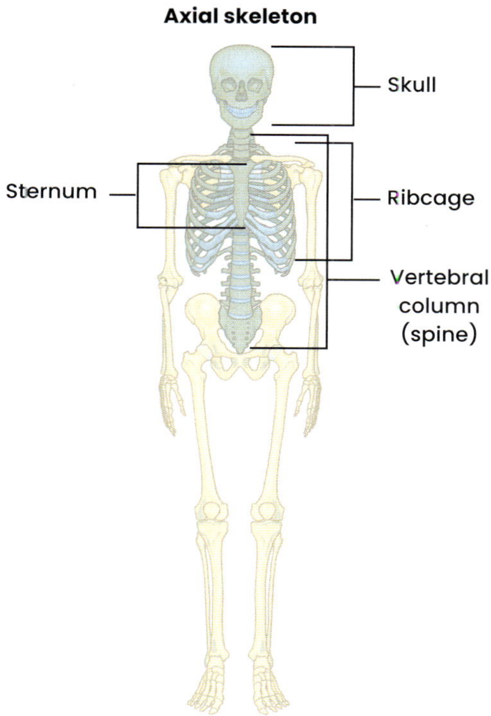

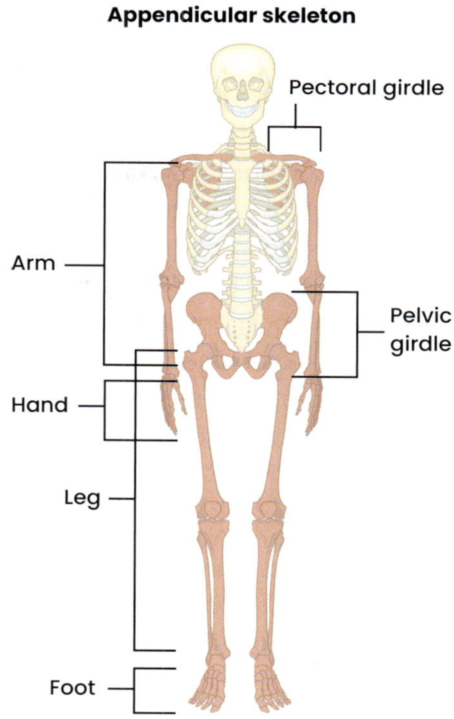

Figure 35.7: The axial and appendicular skeleton.

THE AXIAL SKELETON

The **axial skeleton** forms the central axis of the body and consists of four main structures:

1. **Skull:** Made of 22 fused bones, the skull protects the brain and sense organs within (eyes, ears, olfactory region of the nose and the tongue).
2. **Vertebral column (spine):** A stack of 33 **vertebrae** protects the spinal cord and provides attachment points for the muscles of the neck and back.
3. **Ribcage:** 12 pairs of ribs protect the lungs and heart inside the chest. The muscles between the ribs, called intercostal muscles, play a role in the breathing mechanism.
4. **Sternum:** This strong bone at the centre of the chest strengthens the ribcage.

In addition to their protective roles, the bones of the axial skeleton are rich in red bone marrow, where blood cells are produced.

The **axial skeleton** consists of the skull, vertebral column, sternum and ribcage.

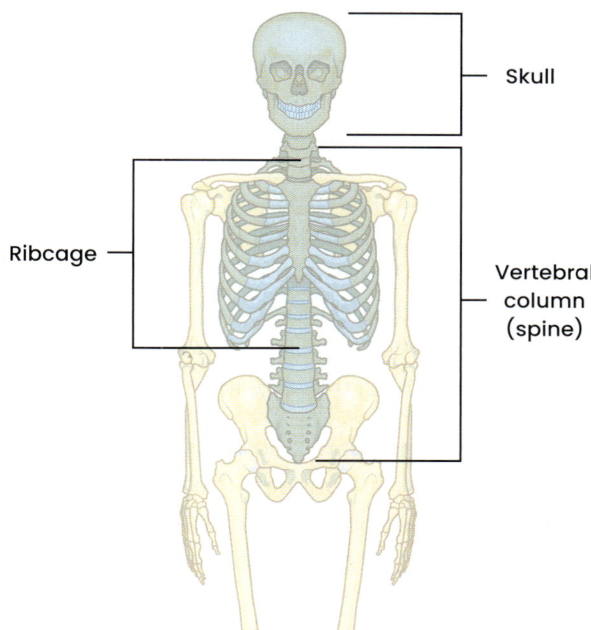

Figure 35.8: The axial skeleton consists of the skull, vertebral column, sternum and ribcage.

FOCUS ON ... THE VERTEBRAL COLUMN

The vertebral column or spine is a flexible structure that consists of 33 vertebrae. The spine has five distinct regions: 7 cervical vertebrae in the neck, 12 thoracic vertebrae in the chest, 5 lumbar vertebrae in the lower back, 5 fused sacral vertebrae in the sacrum and 4 fused vertebrae in the coccyx (or tailbone). Each of the vertebrae in the neck, chest and lumbar regions is separated by a disc of cartilage, which acts as a shock absorber. By creating a small space between the vertebrae in these regions, the spine has room to move. The discs also connect each of the vertebrae.

As we age, the discs can lose water and become less flexible. This causes the discs to flatten and bulge, which can limit movement and cause pain.

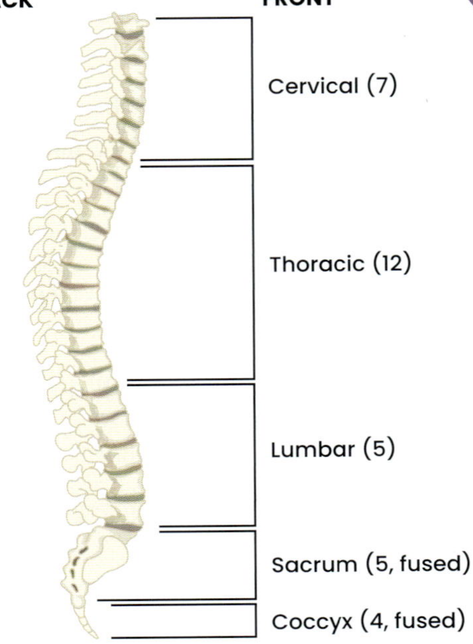

Figure 35.9: The spine has five distinct regions.

The three smallest bones in the body are found in the ear. The hammer, anvil and stirrup, as they are known, play a role in hearing.

DID YOU KNOW?

DISCUSSION

Carry out some research on the factors that affect disc degeneration and report back to your teacher.

THE APPENDICULAR SKELETON

The **appendicular skeleton** has two component parts, which provide attachment points for our four limbs (arms and legs) to the axial skeleton. These are:

1. **The pectoral girdle:** Also called the shoulder girdle, it consists of two bones: the clavicle (collar bone) and scapula (shoulder blade). The pectoral girdle connects the upper limbs (arms) to the axial skeleton and allows for a wide range of movement in the shoulders and arms.
2. **The pelvic girdle:** A fusion of multiple bones, the pelvis or hip connects the lower limbs to the axial skeleton. In humans, the pelvis is firmly attached to the sacrum, a lower region of the spine. The pelvis contains red bone marrow and is a production site for blood cells also.

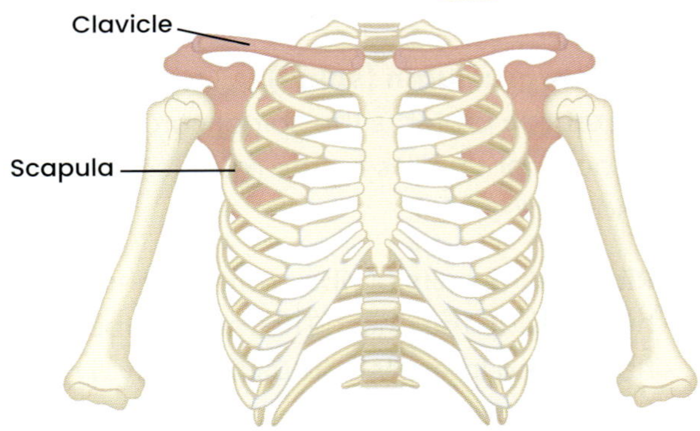

Figure 35.10: The pectoral girdle consists of the clavicle and scapula.

The **appendicular skeleton** consists of the pectoral and pelvic girdles.

Both the pectoral and pelvic girdles help maintain posture and balance.

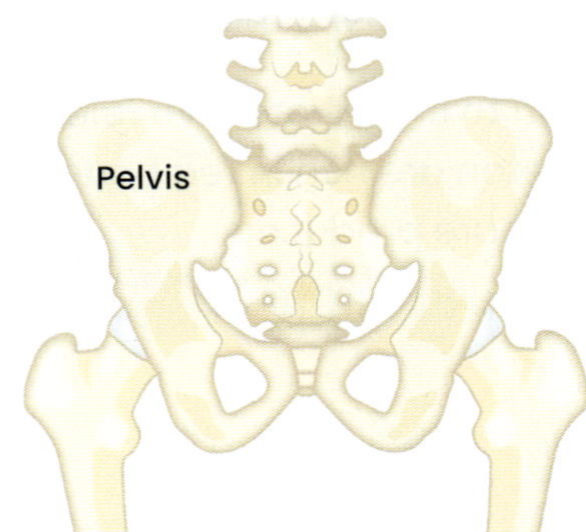

Figure 35.11: The pelvic girdle connects the lower limbs to the axial skeleton (vertebral column).

LIMBS – ARMS AND LEGS

The upper limbs (arms) and lower limbs (legs) are connected to the axial skeleton by the pectoral and pelvic girdle respectively.

The arms consist of three main bones: the **humerus**, **radius** and **ulna**, as shown in Figure 35.12.

The humerus forms a joint with the scapula. The end of the humerus is shaped like a ball and the scapula has a bowl shaped depression, which the humerus fits neatly into. As mentioned earlier, this joint has a wide range of movement and is called a **ball and socket joint**, due to the shape of the connecting points (we will learn more about this later in the chapter). Joints also form at the elbow and at the wrist.

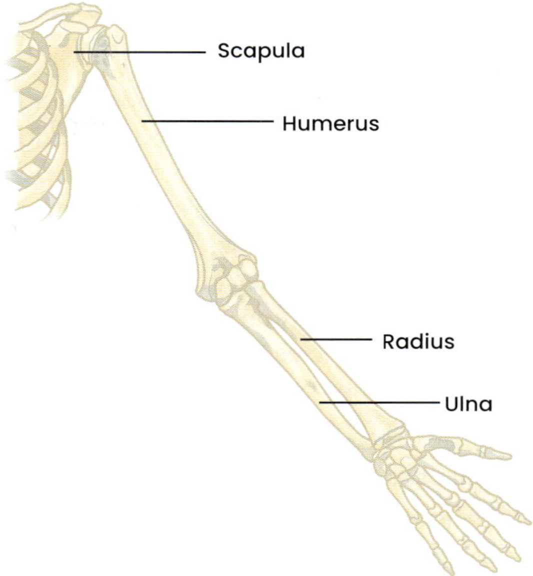

Figure 35.12: The arm consists of the humerus, radius and ulna.

The hands are found at the end of each arm and each contains 27 bones. Together, the arms and hands allow for lifting, gripping and performing many other careful movements like writing or typing.

The legs also consist of three main bones: the **femur**, **tibia** and **fibula** along with a smaller bone, the **patella** (or kneecap), which protects the knee joint.

The end of the femur is ball shaped and fits into another bowl shaped depression in the pelvis, again forming a ball and socket joint. There are more joints at the knee and ankle.

Each foot consists of 26 bones. Together, the legs and feet allow the body stand, walk, run and jump. They also play a role in distributing our mass and maintaining balance.

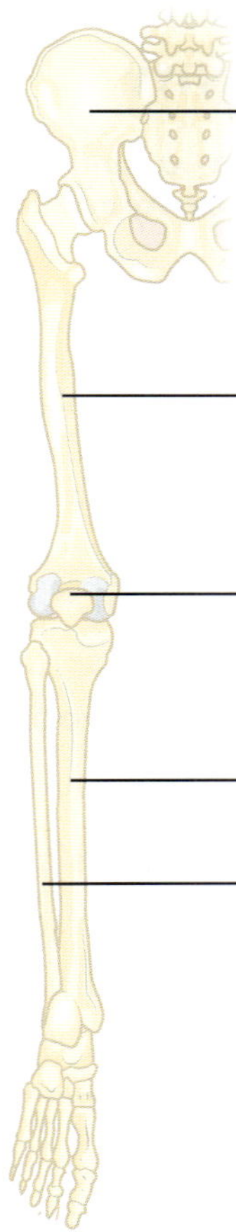

Figure 35.13: The leg consists of the femur, tibia, fibula and patella (kneecap).

The bones of the skeleton are constantly being broken down and reformed. The complete process of skeleton renewal takes around 10 years, although some bones renew more quickly than others. As we age, bone renewal slows down and bone can become more brittle.

DID YOU KNOW?

SCIENCE IN SOCIETY: HEALTH AND TECHNOLOGY

Imaging technologies

X-rays and magnetic resonance imaging (MRI) are two important medical imaging technologies used to detect broken bones or fractures. X-rays use ionising radiation to create images of dense structures like bones. They are quick, widely available and cost-effective. X-rays can easily detect clear breaks or inconsistencies in the bone structure. However, they may miss subtle fractures or those in complex anatomical areas.

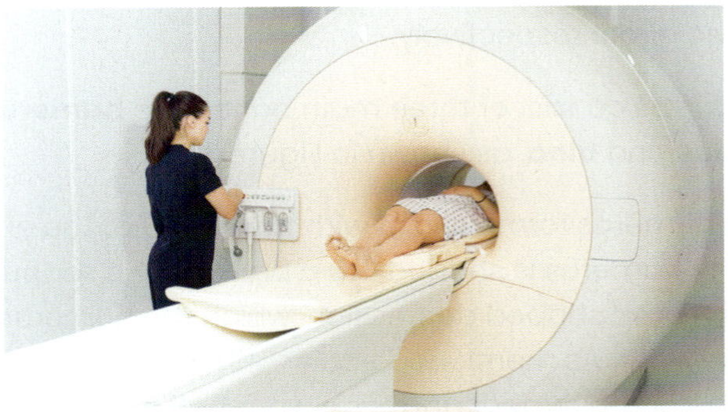

Figure 35.14: An MRI machine.

MRI uses powerful magnets and radio waves to create detailed images of both bones and soft tissues (muscle, ligaments, tendons and cartilage). It is particularly useful for detecting stress fractures, hairline fractures and injuries to surrounding tissues and for imaging the spine to check for flat or bulging discs. MRI devices are more expensive and time-consuming than X-rays but don't expose the patient to harmful radiation.

RESEARCH ACTIVITY

Carry out a search of X-ray and MRI scans of various body parts to see how they differ in clarity and detail.

PORTFOLIO LINK

Go to your Learning Portfolio and complete Learning Check 2 to assess your knowledge and understanding of the topic so far.

Section 3: Joints and Antagonistic Muscles

At the end of this section, you should be able to:

- Define the term 'joint'.
- Briefly describe how joints are categorised.
- Give examples of synovial joints.
- Describe how antagonistic muscle pairs work to allow for movement.

Unpacked from LO 2.4.3

JOINTS

A **joint** is a part of the skeleton where two or more bones meet, usually to facilitate movement. Almost all bones are connected to at least one other bone in the human skeleton.

The hyoid bone, located in the neck, is the only bone in the body not connected to any other bone.

DID YOU KNOW?

A **joint** exists where two or more bones meet.

Types of joint

Joints can be classified by how much movement they permit. There are three main categories: **immoveable**, **semi-moveable** and **freely moveable (synovial)**.

IMMOVEABLE	SEMI-MOVEABLE	FREELY MOVEABLE (SYNOVIAL)
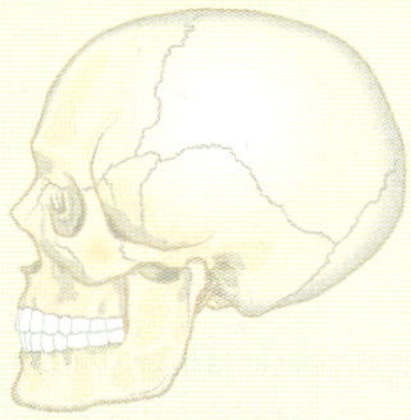	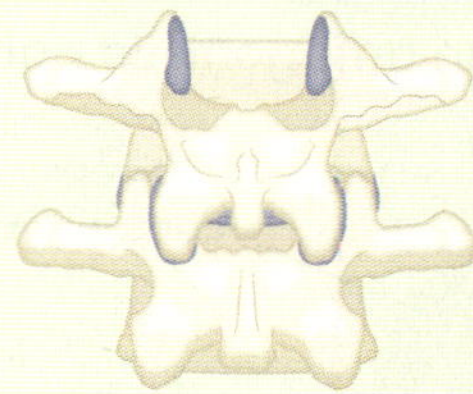	
Immoveable or fused joints permit minimal or no movement. Examples include the skull and pelvis.	Semi-moveable joints allow some limited movement. Cartilage is usually the material binding the bones in semi-moveable joints. An example is the vertebrae in the spine.	Freely moveable joints permit a large range of movement. The bones in these joints are surrounded by a slippery fluid called synovial fluid, which reduces friction in the joint. Examples include the shoulder, knee and hip joints.

SYNOVIAL JOINTS

There are a number of different kinds of synovial joint and we've mentioned some already. A ball and socket joint is found where the femur meets the hip and where the humerus meets the scapula. Hinge joints are found in the elbow and knee.

A typical synovial joint is shown in Figure 35.16. The ligaments hold the bones together and form a sealed **joint cavity** which is filled with slippery **synovial fluid**. This fluid lubricates the joint and allows the bones to move freely.

Cartilage is attached to the ends of each bone, providing cushioning and preventing the bones from rubbing off each other. Tendons connect muscles to the bones, allowing for movement.

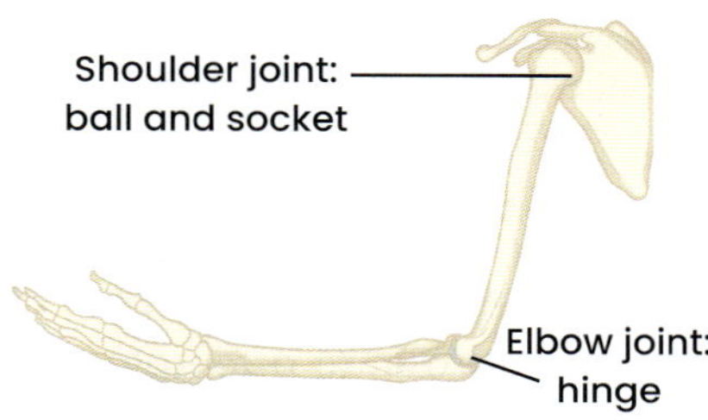

Figure 35.15: Synovial joints in the arm.

Synovial fluid is a slippery fluid found in freely moveable joints, which lubricates the joint and aids movement.

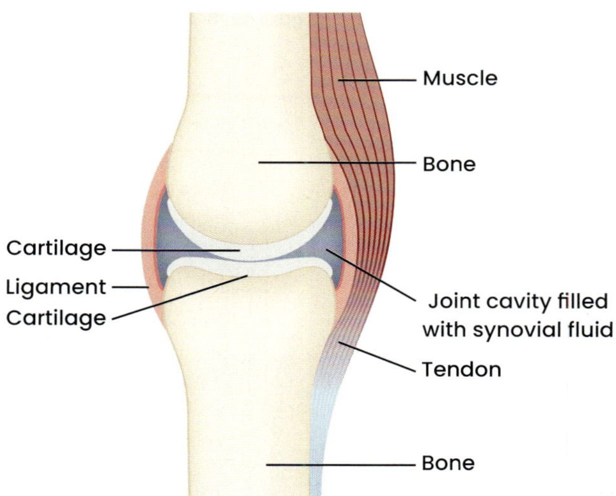

Figure 35.16: A synovial joint.

FOCUS ON ... ANTAGONISTIC MUSCLE PAIRS

Muscles are found at every semi-moveable or freely moveable joint, and the joint can't move without them. In most cases, a pair of muscles – called **antagonistic muscles** – work in opposite directions to facilitate the movement. One muscle works while the other rests. The biceps and triceps of the upper arm are a good example of an antagonistic pair.

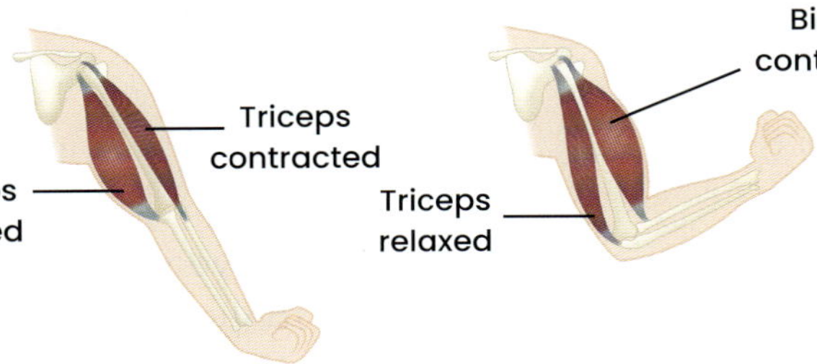

Figure 35.17: The biceps and triceps are an antagonistic muscle pair.

When bending the arm, the biceps contracts while the triceps relaxes (they act in opposite directions). When straightening the arm, the biceps relaxes while the triceps contracts. This system allows for more careful control and smoother movements of the joint.

Antagonistic muscle pairs are muscles which work with opposite actions to allow movement of the skeleton.

Go to YouTube.com and search for 'Muscles – How Biceps and Triceps Work Together' (2:00) by Children's Classroom TV.

SCIENCE IN SOCIETY: HEALTH AND TECHNOLOGY

Artificial joints

Sometimes a joint or part of a joint needs to be replaced, usually due to age-related wear and tear or injury. Modern joint replacements began in the 1960s, but advances in material science and 3D printing are revolutionising this important medical practice.

In Ireland, approximately 6,000 hip replacement surgeries take place every year. These involve creating a new ball and socket joint using advanced materials, e.g. titanium, carbon fibre, plastics or foams. Knee replacements are also common.

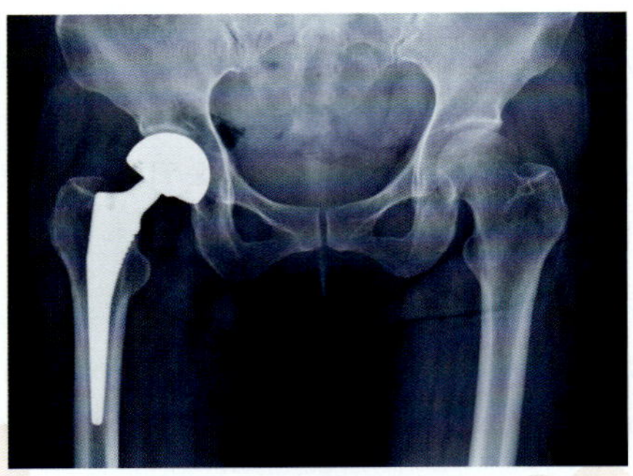

Figure 35.18: An x-ray image of an artificial hip joint.

PORTFOLIO LINK Go to your Learning Portfolio and complete Learning Check 3 to assess your knowledge and understanding of the topic so far.

Go to YouTube.com and search for 'Total Hip Replacement (3D animation)' (2:01) by ViewMedica.

- The musculoskeletal system consists of our **bones** and the **muscles** that attach to them.
- An adult human has 206 bones and over 600 muscles.
- **Ligaments**, **tendons** and **cartilage** are connective tissues that also form part of the musculoskeletal system.
- The functions of the musculoskeletal system include movement, support, protection, heat generation, blood cell production, storage (of minerals, fat and glycogen).
- Bone is a specialised connective tissue composed of bone cells (mostly osteocytes) embedded in a matrix of collagen and calcium salts.
- There are two types of bone:
  - **Compact bone** – strong bone with a compressed matrix, making it hard and durable.
  - **Spongy bone** – softer bone with an open matrix full of red bone marrow, which contains stem cells to produce blood cells.
- Muscle is a specialised tissue made from cells (called myocytes) that can contract when stimulated.
- There are three types of muscle tissue:
  - **Skeletal muscle** – attached to the skeleton
  - **Cardiac muscle** – only in the heart
  - **Smooth muscle** – in the alimentary canal and in arteries and veins.
- **Ligaments** connect bone to bone.
- **Tendons** connect muscle to bone.
- Cartilage is a connective tissue that acts as a protective cushion for bone, preventing wear and tear.
- **The axial skeleton** – consists of the **skull**, **vertebral column**, **sternum** and **ribcage**.
- **The appendicular skeleton** – consists of the **pectoral** and **pelvic girdles**.
- The skull protects the brain and sense organs.
- The vertebral column (spine) consists of 33 vertebrae (cervical, thoracic, lumbar, sacrum, coccyx); it protects the spinal cord.
- **Vertebral discs** connect each of the vertebrae and act as shock absorbers.
- The ribcage consists of 12 pairs of ribs; it protects the lungs and heart and facilitates breathing.
- The sternum is a strong bone which strengthens the ribcage.
- The pectoral girdle consists of the **clavicle** (collar bone) and **scapula** (shoulder blade); it connects the arms to the axial skeleton.
- The pelvic girdle is a fusion of multiple bones; it connects the legs to the axial skeleton.
- The arm is connected to the scapula via a ball and socket joint; it consists of the humerus, radius, ulna and bones of the hand.
- The leg is connected to the pelvis via a ball and socket joint; it consists of the femur, tibia, fibula and bones of the feet.
- A **joint** exists where two or more bones meet.
- There are three types of joint:
  - Immoveable joints – bones are fused together and unable to move e.g. skull, pelvis.
  - Semi-moveable – bones are bound together by cartilage and can move slightly e.g. the vertebrae in the spine.
  - Freely moveable (**synovial**) – large range of movement; joint surrounded by slippery synovial fluid.
- There are different types of freely moveable joints:
  - Ball and socket joints, e.g. femur to hip, humerus to scapula
  - Hinge joints, e.g. knee and elbow.
- Muscles are found at semi-moveable and freely moveable joints; they usually act in pairs called **antagonistic muscle pairs** that work in opposite directions.
- The biceps and triceps are examples of an antagonistic pair.
- When bending the arm, the biceps contracts and the triceps relaxes.
- When straightening the arm, the biceps relaxes and the triceps contracts.

END OF CHAPTER ASSESSMENT

REMEMBER

1. Name the main tissues found in the musculoskeletal system.
2. Name **three** organs protected by the skeleton.
3. Name **three** items stored by the musculoskeletal system.
4. What is *bone*?
5. Name the protein found mainly in the skeleton.
6. What are *myocytes*?
7. How many pairs of ribs are there?
8. Name the bones in the pectoral girdle.
9. Name the **three** categories of joint found in the body.
10. Give an example of an antagonistic muscle pair.

UNDERSTAND

11. Describe **three** functions of the musculoskeletal system.
12. Distinguish between ligaments and tendons.
13. What is the role of cartilage?
14. Briefly describe the structure of bone matrix.
15. Distinguish between spongy and compact bone.
16. Name the **three** types of muscle. Where in the body would you find each of them?
17. Describe the role of the vertebral column.
18. What is an antagonistic muscle pair?

APPLY

19. Briefly describe the role of two proteins in the musculoskeletal system.
20. Compare the axial and appendicular skeleton.
21. Suggest how the musculoskeletal system and the circulatory system interact as they carry out their functions.
22. The biceps and triceps are examples of antagonistic pairs. Suggest another location in the body for an antagonistic pair (the names of the muscles are not required).
23. Apart from the skull, give another location of an immoveable or fused joint in the skeleton.
24. One function of the musculoskeletal system is heat generation. Which process produces this heat?
25. Suggest what might happen if the cartilage in a joint began to wear away.

ANALYSE

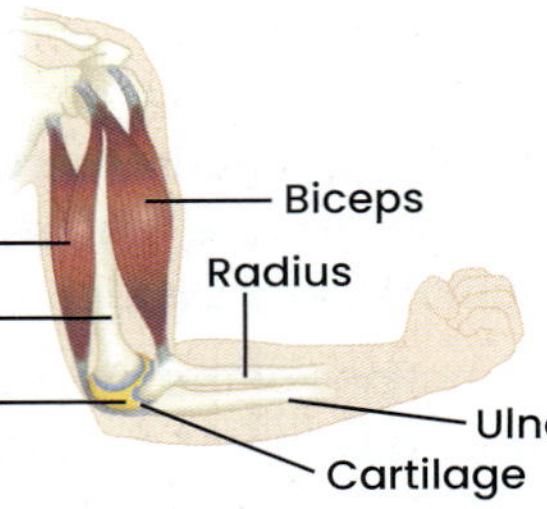

26. The diagram above shows a joint and the muscles that support it.
    (a) Identify the joint. Give reasons for your answer.
    (b) What is the purpose of the cartilage in the joint?
    (c) Which tissue connects bone to bone in this joint?
    (d) What is the purpose of the synovial fluid in the joint?
    (e) Which muscle is contracting in this image? Give a reason for your answer.
27. Suggest a reason why tendon tissue isn't as elastic as ligament tissue.
28. Suggest a reason why longer bones contain more compact bone that shorter bones.

EVALUATE

29. In humans, some of the vertebrae of the lower spine are fused together. Suggest a reason for this.
30. Unlike most other animals, humans walk on two legs. Consider the biological advantages and disadvantages of walking on two legs.
31. Consider the importance of consuming protein in the diet to the health of the musculoskeletal system.
32. While in space or other zero-gravity situations, an astronaut's bone density decreases. Suggest a reason for this and the effects this may have on the astronaut when they return to Earth.

CREATE

33. Draw a diagram of a synovial joint showing the location of ligaments, tendons and cartilage.
34. Build a model of a joint between two bones.
35. Design an experiment to test the strength of chicken bones.

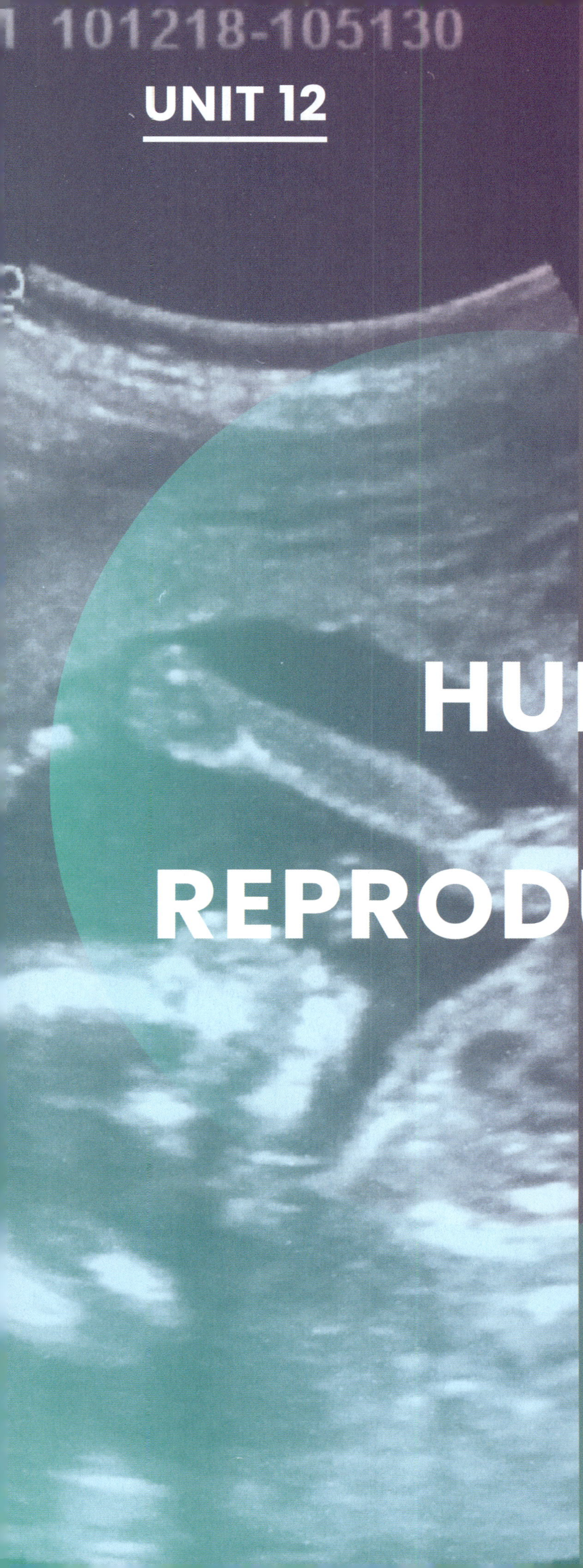

UNIT 12

HUMANS – HUMAN REPRODUCTION

CHAPTERS

STRANDS

Strand 2: Structures and Processes of Life

36 Human Reproduction – Anatomy

LEARNING OUTCOMES: 2.5.1, 2.5.2, 2.5.4, 2.5.6

EYE-OPENER

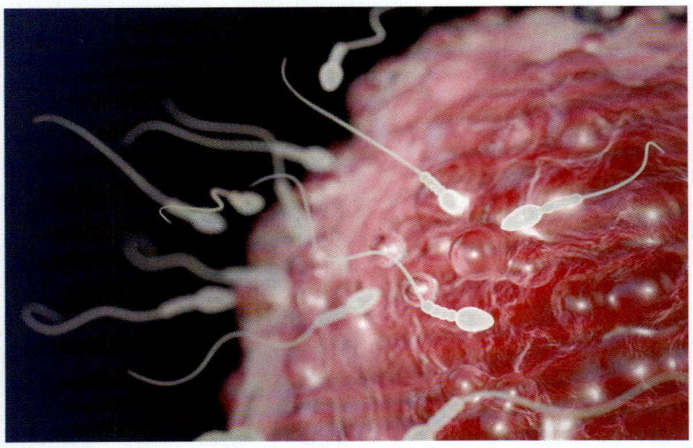

The human egg releases chemicals called chemo-attractants to attract sperm. Research carried out in 2020 at Stockholm University and the University of Manchester found that when a female's egg was exposed to sperm from different male donors, the egg was better at attracting sperm from one male compared to another. An egg from a different woman was better at attracting a different male's sperm, meaning that the attraction between the sperm and the egg depended on the particular people involved in the test. They also found that the egg did not always attract the male partner's sperm. It is thought that the egg may preferentially attract sperm that are more genetically compatible. It is hoped that this research will help scientists to help couples with 'unexplained' causes of infertility.

At the end of this chapter, you should be able to:

- Relate the structure of the male reproductive system to its function.
- Relate the structure of the female reproductive system to its function.
- Understand the role of hormones in sperm and egg production, puberty, the menstrual cycle and the menopause.
- Outline four phases of the menstrual cycle: menstruation, ovulation, follicular phase, luteal phase.
- Outline the changes in hormone levels during each stage of the menstrual cycle. (HL)
- Identify the fertile period in the menstrual cycle.
- Understand that the male's fertile period is continuous.

LINKS TO:

Chapter 9	**Evolution and Natural Selection**
Chapter 17	**Cell Continuity**
Chapter 34	**The Endocrine System**
Chapter 37	**Human Reproduction – From Fertilisation to Birth**

SUBJECT VOCABULARY

secondary sexual characteristics | puberty | menstrual cycle | menopause | menstruation | ovulation | follicular phase | luteal phase | fertile period

Section 1: Male and Female Reproductive Systems

At the end of this section, you should be able to:

- Relate the structure of the male reproductive system to its function.
- Relate the structure of the female reproductive system to its function.
- Understand that the male's fertile period is continuous and the female's is cyclical.
- Model the role of FSH and LH in the male reproductive system. HL

Unpacked from LO 2.5.1, 2.5.4, 2.5.6

A characteristic of all living things is the ability to reproduce. Sexual reproduction involves two parents: a male and a female. In Chapter 9, we learned that when a human male sperm cell and a female egg cell combine, the zygote that forms is one of over 64,000,000,000,000 possible combinations that could form from the parents' genes! Meiosis and sexual reproduction are responsible for this potential variation.

MALE REPRODUCTIVE SYSTEM

Figure 36.1 illustrates the structure of the male reproductive system. We will look at the function of each of the parts below.

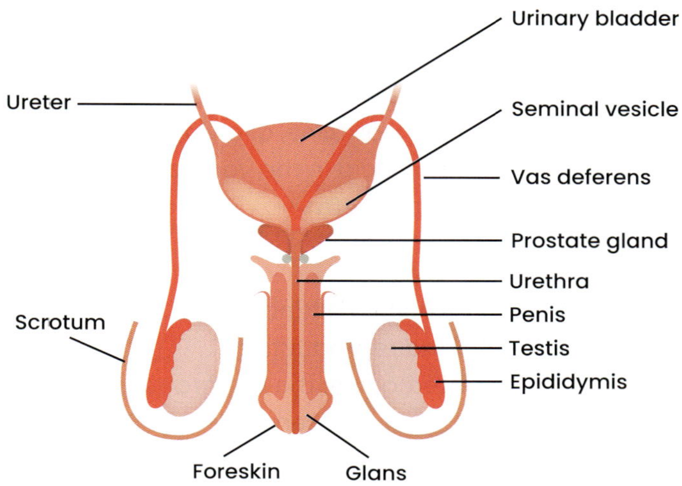

Figure 36.1: Male reproductive system (front view).

Penis

The penis transfers semen from the male to the female's vagina during intercourse.

Glans

The glans is the rounded tip or head of the penis. The glans contains the opening of the urethra which allows semen to be released.

Foreskin

The loose skin that covers the glans is called the foreskin. The foreskin protects the glans.

Scrotum

The scrotum is a sac that holds the testes outside the male body at 35°C. This temperature, which is 2°C lower than normal body temperature, is needed for proper sperm production.

Testes

Males have two testes which produce sperm, the male gamete (sex cell), and the hormone testosterone. Testosterone controls the development of male secondary sexual characteristics. The testes contain highly coiled tubes called 'seminiferous tubules' where diploid (2n) cells divide by meiosis to form haploid (n) sperm cells (Figure 36.2). This process begins when a boy reaches puberty and continues throughout his life, meaning that male fertility is continuous from puberty.

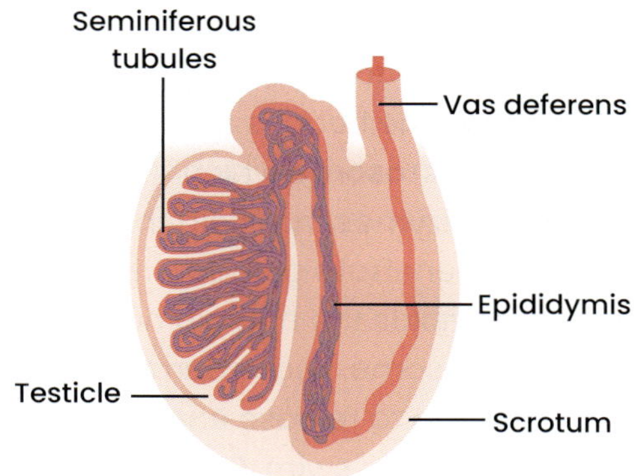

Figure 36.2: Testis.

Epididymis

The immature sperm cells formed in the testes are not capable of moving on their own and are described as non-motile. These immature sperm are incapable of fertilising an egg. These sperm cells are transferred to the epididymis, a highly coiled tube outside the testes, where they mature and become motile.

Vas deferens (sperm duct)

Males have two sperm ducts, one connecting each testis to the urethra. The sperm duct is a muscular tube which moves sperm by peristalsis during ejaculation from the testes to the urethra.

Glands associated with the male reproductive system:

The seminal vesicles and the prostate gland are exocrine (ducted) glands linked to the male reproductive system. These glands secrete seminal fluid which enables sperm to swim. This fluid along with sperm is known as semen.

- **Seminal vesicle:** The two seminal vesicles secrete about 60% of the seminal fluid into the urethra. Seminal fluid has a high fructose sugar content which provides energy for the sperm cells.
- **Prostate gland:** The prostate gland secretes seminal fluid into the urethra. This seminal fluid is alkaline, allowing it to neutralise any acid in the vagina which could kill the sperm.

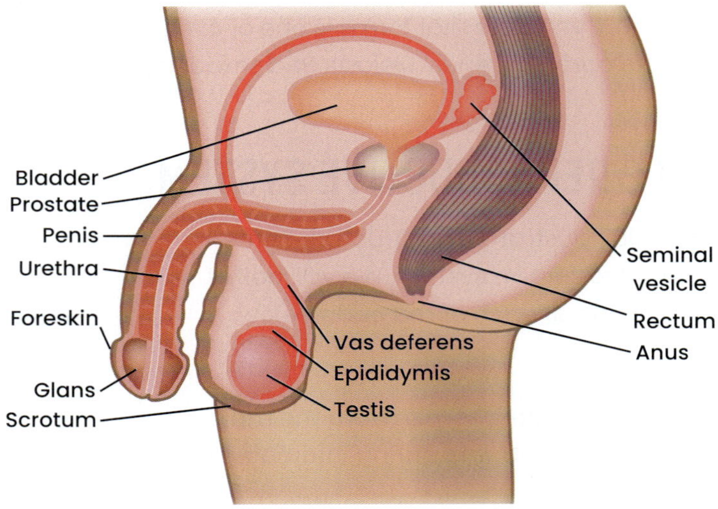

Figure 36.3: Male reproductive system (side view).

Urethra

In males, the urethra has a function in both the urinary and reproductive system. During intercourse, the urethra carries semen (sperm and seminal fluid) out of the penis. A sphincter muscle prevents urine from the bladder entering the urethra during ejaculation. Urine is acidic and would kill the sperm.

THINK, PAIR, SHARE

How are sperm adapted to their function?

MALE SEX CELL

The male sex cell, the sperm, is the smallest human cell. It has three main parts: head, middle piece and a tail. Inside the head is a structure called the **acrosome** which contains enzymes that help the sperm to penetrate an egg. The head also holds the haploid nucleus, containing 23 chromosomes. Sperm cells contain either an X or Y sex chromosome. Each sperm is genetically different. The midpiece of the sperm cell contains many mitochondria to release energy during aerobic respiration. This energy is required to move the sperm's tail, allowing it to swim. Fertile men have 20–150 million sperm per millilitre (ml) of semen. Sperm can survive up to five days in the female reproductive system.

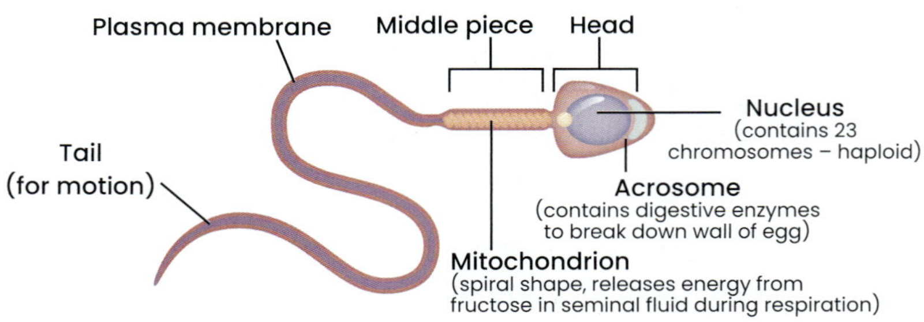

Figure 36.4: Male sex cell, sperm.

Hormonal control of sperm production

The hormone testosterone controls the completion of sperm production and stimulates the seminal vesicles and prostate gland to secrete seminal fluid.

Hormonal control of sperm production – A closer look

HL

Two other hormones – **follicle stimulating hormone** (FSH) and **luteinising hormone** (LH), released by the pituitary gland – are involved in sperm production. FSH initiates sperm production in the testes. LH triggers the production of testosterone. The production of testosterone is controlled by a negative feedback system.

Go to YouTube.com and search for 'Male Reproductive System Overview' (1:39) by Merck Manuals.

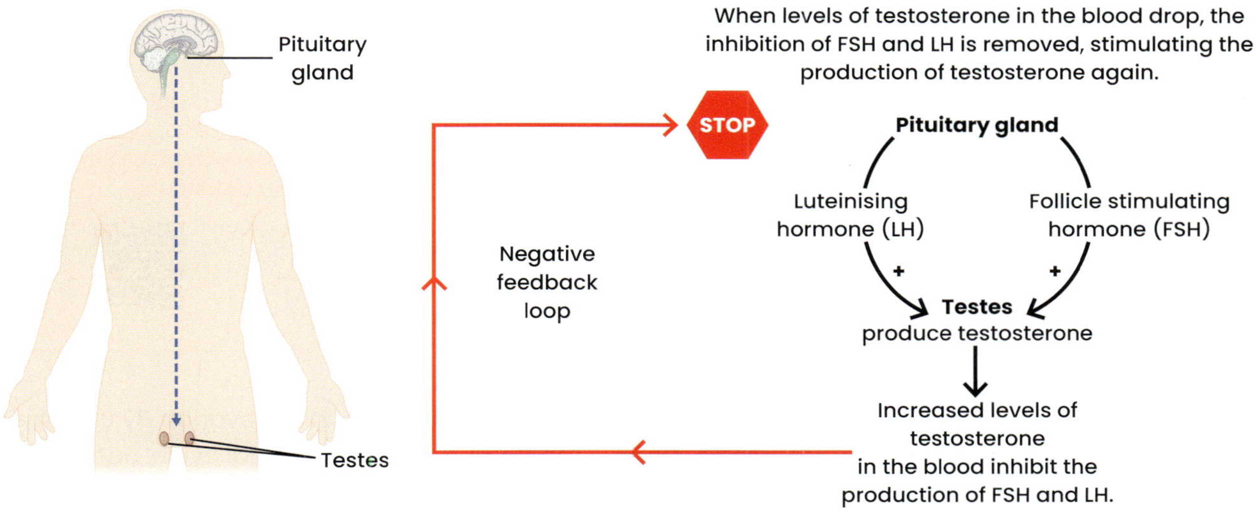

Figure 36.5: Negative feedback control of testosterone levels.

FEMALE REPRODUCTIVE SYSTEM

The female reproductive system is mainly internal. It consists of two ovaries, Fallopian tubes, the uterus and vagina. Externally, surrounding the vaginal and urethral openings, are folds of tissue called the vulva.

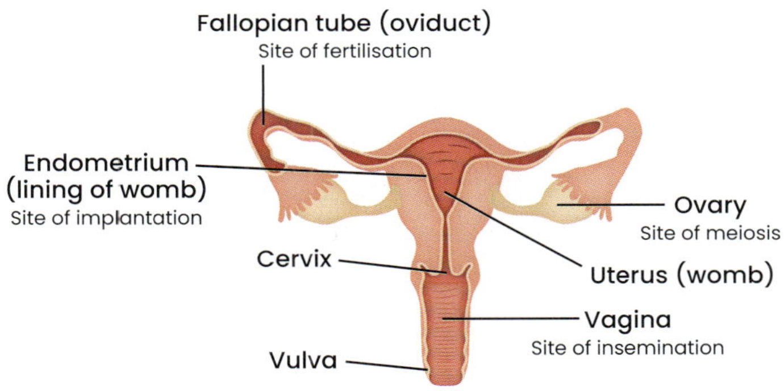

Figure 36.6: Female reproductive system (front view).

Vulva

The vulva is the collective term for the external parts of the female reproductive system. They consist of the external lip-like folds of skin called **labia** that protect the openings to the vagina and the urethra.

Labia

The outer and inner folds of the vulva are called the labia. These folds of skin protect the opening of the vagina and the urethra.

Clitoris

Below the inner labia folds is the clitoris. The clitoris consists of sensitive tissue that swells with blood when stimulated.

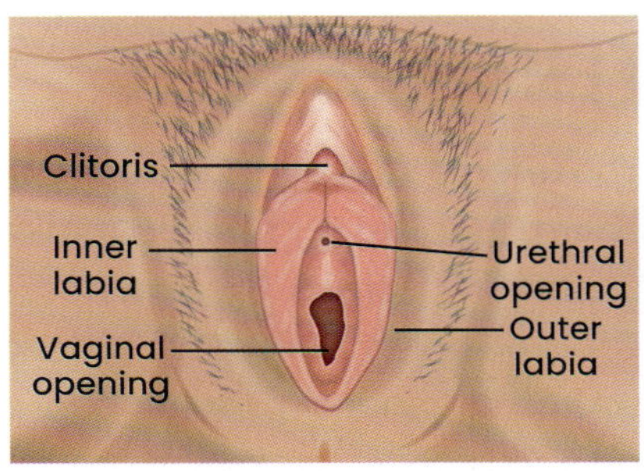

Figure 36.7: The vulva of the female reproductive system.

Vagina

The vagina is a muscular tube that holds the penis during intercourse. It is also the birth canal along which the baby is delivered during labour. The endometrium, lining of the womb, is shed from the body via the vagina during menstruation. The vagina has an acidic pH, which helps protect it from bacterial infection and forms part of the innate immune system.

Cervix

The cervix is the neck of the uterus, which opens into the vagina.

Uterus (womb)

The uterus is where the foetus develops during pregnancy.

Endometrium

The inner lining of the uterus is rich in blood vessels and is called the endometrium. If fertilisation does not occur, the endometrium is shed during menstruation.

Fallopian tube (oviduct)

Long, narrow Fallopian tubes connect the ovaries to the uterus (womb). The Fallopian tube collects the egg during ovulation. Tiny hair-like cilia along the lining of the Fallopian tubes move the egg down to the uterus. If sperm are present, the egg may become fertilised. Fertilisation usually occurs in the Fallopian tube.

Go to YouTube.com and search for 'The Female Reproductive System' (1:51) by Merck Manuals.

Ovary

A female baby is born with about 1–2 million immature diploid (2n) eggs in her ovaries. By the time she reaches puberty, only about 300,000 remain. Once a month during her menstrual cycle, one or two eggs will undergo meiosis and mature. The ovary is an endocrine gland, producing the female sex hormones: oestrogen and progesterone.

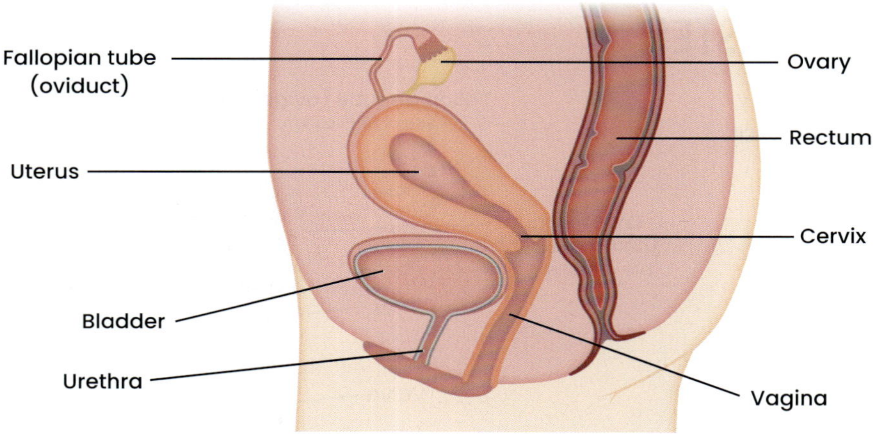

Figure 36.8: Female reproductive system (side view).

THINK, PAIR, SHARE

How is the female reproductive system adapted to its functions?

FEMALE SEX CELL

The female sex cell is called the **egg** or ovum. The egg is the largest human cell and is 30 times wider than a sperm. Haploid egg cells, containing 23 chromosomes, are produced by meiosis. Egg cells contain only X sex chromosomes. The egg possesses mitochondria with their own DNA that is passed on to offspring. Children only inherit mitochondria and hence mitochondrial DNA from their mother. This DNA can be the source of genetic mitochondrial disorders. The egg can survive for up to 24 hours following ovulation. When an egg is penetrated by a sperm, a chemical change is triggered that prevents entry of any other sperm.

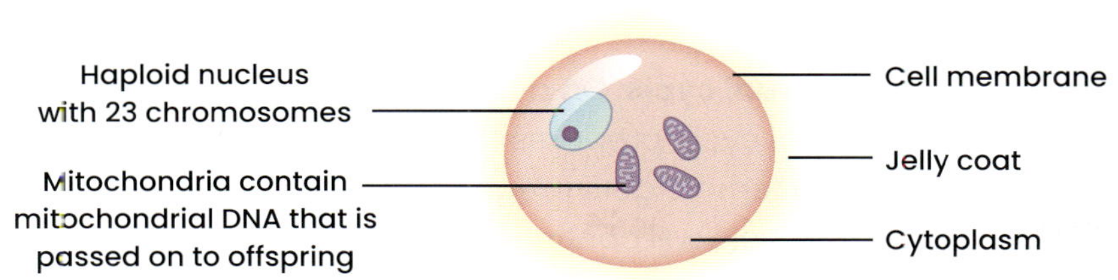

Figure 36.9: Human female egg.

PORTFOLIO LINK Go to your Learning Portfolio and complete Learning Check 1 to assess your knowledge and understanding of the topic.

Section 2: Reproductive Hormones and the Menstrual Cycle

At the end of this section, you should be able to:

- Outline four phases of the menstrual cycle: menstruation, ovulation, follicular phase, luteal phase.
- Outline the changes in hormone levels during each stage of the menstrual cycle. HL
- Identify the fertile period in the menstrual cycle.

Unpacked from LOs 2.5.2, 2.5.4, 2.5.6

ROLE OF HORMONES IN THE PRODUCTION OF SECONDARY SEXUAL CHARACTERISTICS

When a person reaches puberty, they start to mature sexually and develop **secondary sexual characteristics**. Puberty generally starts in females around 11 years of age and in males around 12 years of age. In males, secondary sexual characteristics include the growth of facial and pubic hair, deepening of the voice and broadening of the shoulders. The development of male sexual characteristics is controlled by the hormone **testosterone**. In females, secondary sexual characteristics include widening of the hips, growth of the breasts and growth of pubic hair. The development of these female characteristics is controlled by the hormone **oestrogen**.

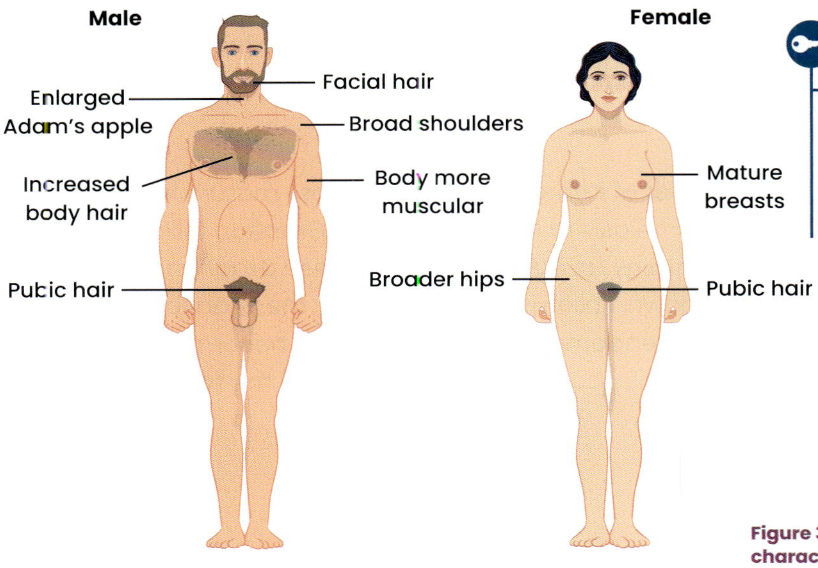

Secondary sexual characteristics are physical features that appear during puberty in response to sex hormones and distinguish between males and females. They are not directly involved in reproduction.

Figure 36.10: Secondary sexual characteristics in males and females.

MENSTRUAL CYCLE

When a female reaches **puberty**, she begins her **menstrual cycle**. On average, the menstrual cycle lasts 28 days but can be shorter or longer. The menstrual cycle prepares the reproductive system to receive a fertilised egg in anticipation of pregnancy. If a female gets pregnant, her menstrual cycle stops for the duration of the pregnancy. The menstrual cycle will continue until she reaches the **menopause**, around 50 years of age.

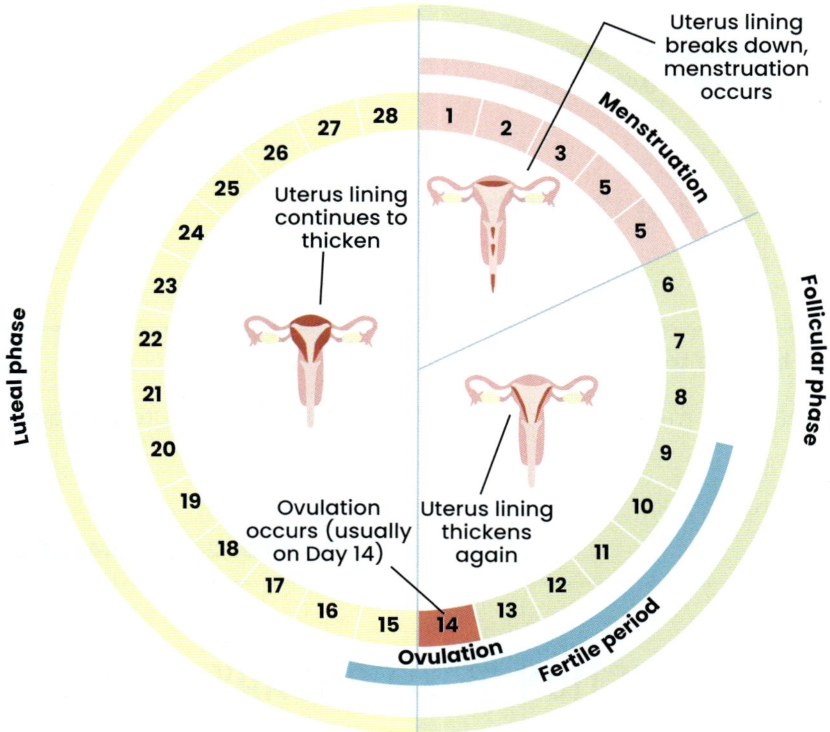

Figure 36.11: The menstrual cycle.

- **Puberty** is the period in life when a child undergoes physical and hormonal changes to become sexually mature and capable of sexual reproduction. Puberty is controlled by sex hormones.
- The **menstrual cycle** is a series of changes that happen in the ovaries and lining of the uterus in a fertile, non-pregnant female, every 28 days.
- **Menopause** is the time in a woman's life when her ovaries stop producing hormones and the menstrual cycle stops.
- **Menstruation (Days 1–5)** is the shedding of the endometrium (lining of the womb).
- **Follicular phase (Days 1–14)** of the menstrual cycle involves the maturation of immature eggs in the ovary in preparation for ovulation.

There are four phases in the menstrual cycle: menstruation, follicular phase, ovulation and the luteal phase. We will look at each of these in turn.

The four phases of the menstrual cycle

MENSTRUATION (DAYS 1–5)

Day 1 of the menstrual cycle begins on the first day of **menstruation,** when the endometrium, lining of the uterus, is shed and released through the vagina. This is also called **having a period**. Menstruation lasts five days on average and is triggered by low levels of the hormones oestrogen and progesterone.

FOLLICULAR PHASE (DAYS 1–14)

At the start of each menstrual cycle, a small number of immature eggs (follicles) mature in the ovaries. The immature eggs are diploid ($2n$) and undergo meiosis to form haploid (n) Graafian follicles. The Graafian follicle is a fluid filled structure which contains an egg. Only one of these Graafian follicles continues to grow and mature, the others break down. Following menstruation, the Graafian follicle releases the hormone oestrogen, which causes the endometrium to thicken and become enriched with blood and nutrients in preparation for pregnancy. This occurs during Days 6–14 of the **follicular phase**.

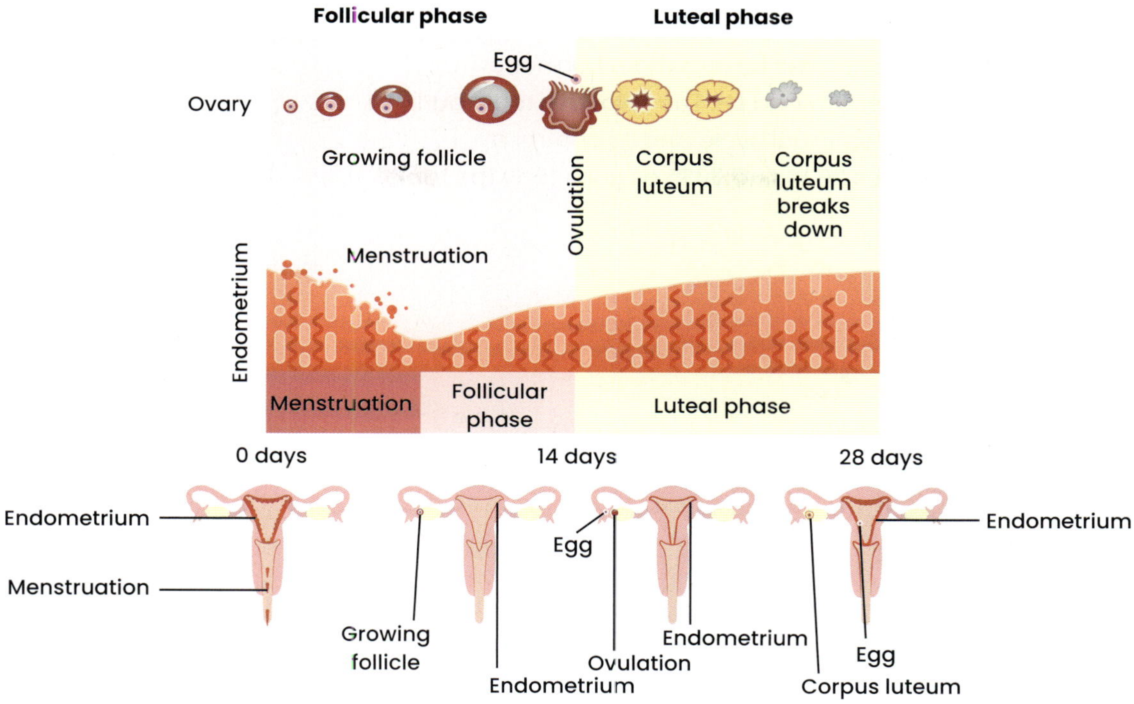

Figure 36.12: The phases of the menstrual cycle.

OVULATION (DAY 14)

The mature Graafian follicle releases its egg from the ovary during **ovulation**. Ovulation occurs on Day 14 of an average menstrual cycle.

LUTEAL PHASE (DAYS 15–28)

The **luteal phase** is the time after ovulation until the start of the next menstrual cycle or pregnancy occurs. The remains of the Graafian follicle form a structure called the **corpus luteum**. The corpus luteum releases the hormones oestrogen and progesterone that maintain and thicken the endometrium further in preparation for pregnancy. If fertilisation does not occur, the corpus luteum starts to break down and oestrogen and progesterone are no longer produced.

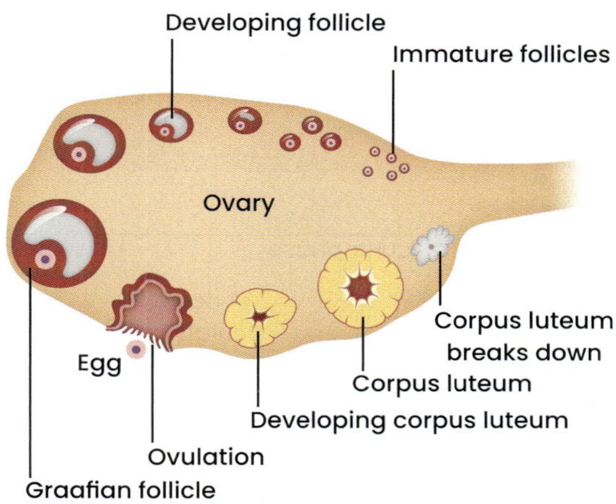

Figure 36.13: The life cycle within the ovary.

Fertile period

Unlike males, whose fertility is continuous following puberty, fertility in females is **cyclical**. The **fertile period** is the time in the menstrual cycle when a female is most likely to get pregnant if she has sexual intercourse. Sperm can survive in the female reproductive system for up to five days. The egg survives for 24 hours after ovulation. If ovulation occurs on Day 14 of the menstrual cycle, fertilisation is most likely if intercourse occurs up to five days before ovulation and up to one day after ovulation, i.e. Days 9-15 of a 28-day cycle.

- **Ovulation (Day 14)** is the release of an egg from the ovary.
- **Luteal phase (Days 15–28)** is when the endometrium thickens further in preparation for pregnancy.
- **Fertile period** is the time in menstrual cycle when a female is most likely to get pregnant if she has sexual intercourse.

Hormonal control of the menstrual cycle

The menstrual cycle is controlled by four main hormones: **oestrogen**, **progesterone**, **FSH** and **LH**. The following table outlines the main stages in the menstrual cycle and its control by hormones. Refer to Figures 36.14 and 36.15 as you study the table.

Go to YouTube.com and search for 'The Menstrual Cycle | 3D Animation (2/2)' (2:25) by Dr. Paulien Moyaert.

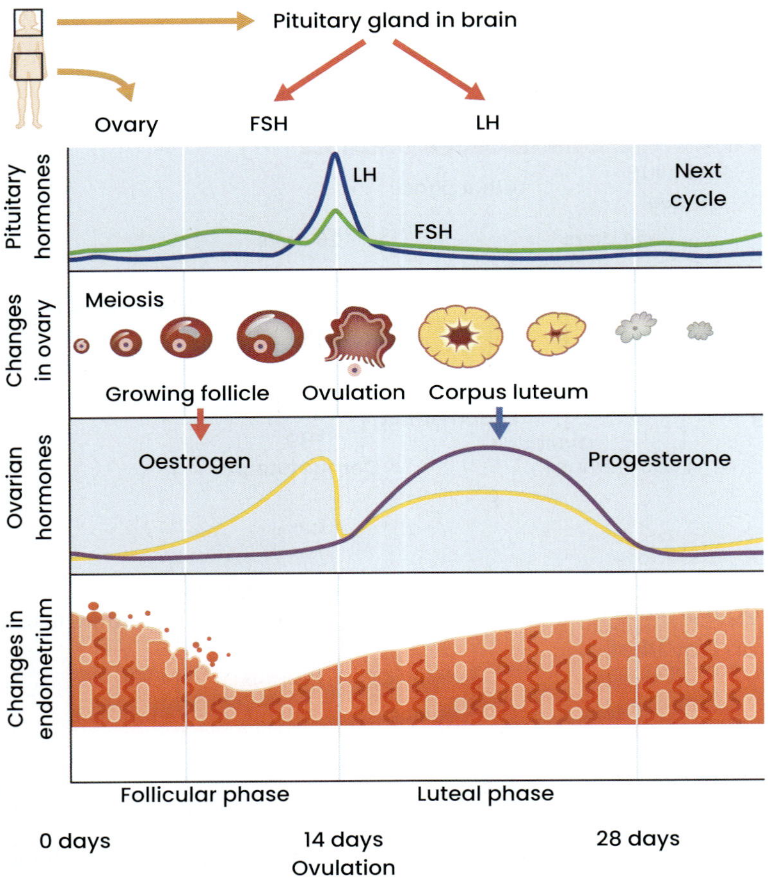

Figure 36.14: Changes during the menstrual cycle.

FOCUS ON ... HORMONAL CONTROL OF THE MENSTRUAL CYCLE (HL)

	Days	HORMONES AND THE MENSTRUAL CYCLE
FOLLICULAR PHASE	1–5	• Low levels of the hormones oestrogen and progesterone trigger **menstruation**. • The pituitary gland in the brain secretes follicle stimulating hormone (FSH) and luteinising hormone (LH) which travel in the blood to the ovaries. The levels of FSH increase more quickly than those of LH. • FSH stimulates the growth of several Graafian follicles (immature eggs) in the ovary. A haploid egg develops inside the Graafian follicle by meiosis.
	6–13	• The Graafian follicle releases the hormone oestrogen which causes the endometrium to thicken. • As the levels of oestrogen build up, it inhibits FSH being produced. This is an example of negative feedback. • The reduced levels of FSH cause the smaller Graafian follicles to die off.

	Days	HORMONES AND THE MENSTRUAL CYCLE
FOLLICULAR PHASE	6–13	• The oestrogen released by the Graafian follicle causes a surge in the levels of LH released from the pituitary gland. This is an example of positive feedback.
	14	• **Ovulation**: A surge in LH causes a mature egg to be released from the Graafian follicle into the Fallopian tube. The egg can survive for up to 24 hours after ovulation. • If the female has sexual intercourse during the fertile period, sperm may be present to fertilise the egg. If the female becomes pregnant, the menstrual cycle does not continue.
LUTEAL PHASE	15–26	• LH causes the remains of the Graafian follicle to be converted into the corpus luteum. • The corpus luteum releases oestrogen and progesterone. Progesterone causes the lining of the womb to thicken further and maintains it, in preparation to receive a fertilised egg. • If the egg is not fertilised or the fertilised egg does not implant, progesterone and oestrogen secreted by the corpus luteum inhibit the release of FSH and LH from the pituitary gland (negative feedback). • As the levels of LH decline, the corpus luteum starts to break down and stops releasing oestrogen and progesterone.
	27–28	• Progesterone and oestrogen levels drop rapidly as the corpus luteum has broken down. • As the levels of progesterone and oestrogen drop, FSH and LH production is no longer inhibited.
The cycle starts again as the drop in oestrogen and progesterone levels triggers menstruation and FSH and LH are released again.		

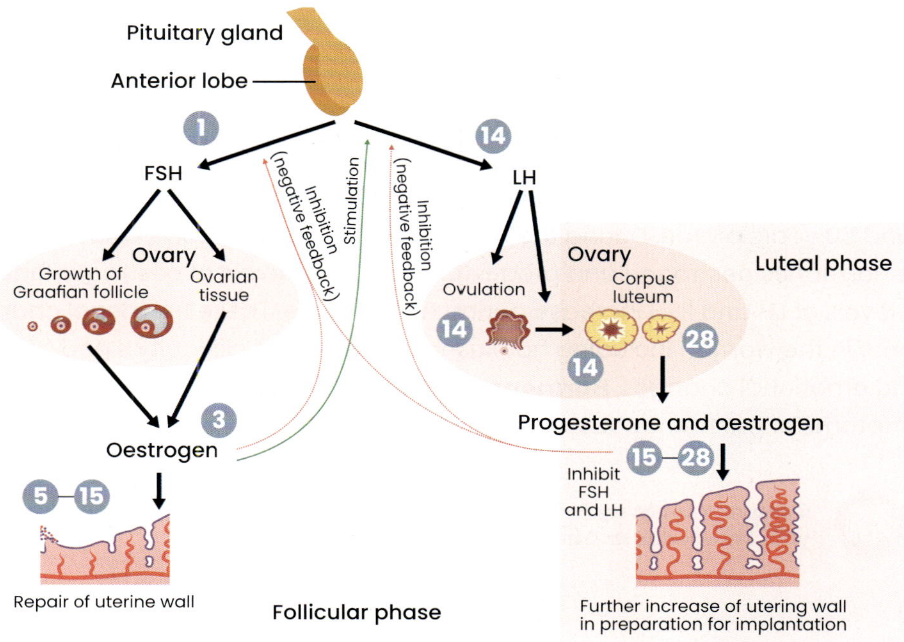

Figure 36.15: Hormonal control of the menstrual cycle.

HL

FOCUS ON ... HORMONAL CONTROL OF THE MENSTRUAL CYCLE (HL)

HORMONE	GLAND	ROLE
Follicle stimulating hormone (FSH)	Pituitary gland	In the follicular phase, FSH stimulates the growth of several Graafian follicles (immature eggs) in the ovary.
Oestrogen	Ovaries	In the follicular phase, oestrogen is responsible for thickening and maintaining the endometrium, inhibiting the release of FSH from the pituitary gland (to allow only one egg to fully mature) and stimulating the release of LH from the pituitary gland.
Luteinising hormone (LH)	Pituitary gland	LH triggers ovulation, the release of an egg from the ovary. In the luteal phase, LH causes the remains of the Graafian follicle to become the corpus luteum.
Progesterone	Ovaries	In the luteal phase of the menstrual cycle, progesterone causes the endometrium to thicken further and maintains it. Progesterone inhibits the release of FSH and LH.

SCIENCE IN SOCIETY: HEALTH

Cervical cancer

Cervical cancer is cancer of the cells lining the cervix. Cervical cancer is the second most common female cancer in Europe. Most cervical cancers are caused by the human papillomavirus (HPV). HPV is spread through skin-to-skin contact and by sexual intercourse. The risk of cervical cancer can be reduced by getting the HPV vaccine, having regular cervical screening tests from the age of 25 and not smoking.

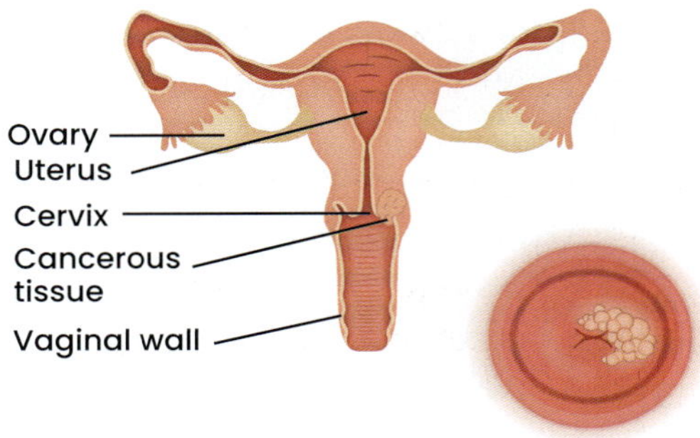

Figure 36.16: Cervical cancer is the second most common female cancer in Europe.

CHANGING HORMONE LEVELS DURING MENOPAUSE

The **menopause** is when the menstrual cycle in a female stops for longer than 12 months. This usually happens around 50 years of age. During the menopause, ovulation stops, and the ovaries stop producing the hormones oestrogen and progesterone. As the levels of oestrogen and progesterone decrease, the levels of LH and FSH increase during menopause. These hormonal changes can cause many symptoms in the woman including hot flushes, headaches, difficulty sleeping, tenderness around the breasts and emotional changes. **Hormone replacement therapy (HRT)** is a treatment used to relieve the symptoms of the menopause.

PORTFOLIO LINK **Go to your Learning Portfolio and complete Learning Check 2 to assess your knowledge and understanding of the topic.**

Chapter Summary

- The **scrotum** is a sac holding testes outside the body at 35°C for sperm production.
- The **testes** produce sperm and testosterone.
- The **glans** contains the opening of the urethra which allows semen to be released.
- The **foreskin** protects the glans.
- **Sperm** mature and are stored in the **epididymis**.
- The **vas deferens** transports sperm from the testes to the urethra.
- The **urethra** transports semen (sperm and seminal fluid) during intercourse.
- The **penis** transfers semen to the female's vagina during intercourse.
- The **seminal vesicle** and **prostate gland** produce seminal fluid, allowing sperm to swim. **Seminal fluid** contains fructose sugar, providing the sperm with energy to sperm.
- An **egg** is released from the **ovary** during ovulation. The ovaries produce the hormones oestrogen and progesterone.
- The **Fallopian tube** collects the egg from the ovary after ovulation. It is the site of fertilisation.
- The **uterus** is where the foetus develops during pregnancy.
- The **endometrium** is the inner lining of the uterus that is rich in blood vessels.
- The **cervix** is the neck of the uterus that opens into the vagina.
- The **vagina** is a muscular tube that holds the penis during intercourse. It is the birth canal for the delivery of a baby.
- The **labia** are folds of skin that protect the opening of the vagina and the urethra.
- The **clitoris** consists of sensitive tissue that swells with blood when stimulated.
- The **vulva** are external folds of tissue that protect the openings to the vagina and urethra.
- **Puberty** is the period in life when a child undergoes physical and hormonal changes to become sexually mature.
- The **menstrual cycle** is a series of changes that happen in the ovaries and endometrium in a fertile, non-pregnant female, every 28 days.
- **Menstruation (Days 1–5)** is the shedding of the endometrium (lining of the womb).
- **Follicular phase (Days 1–14)** of the menstrual cycle involves the maturation of immature eggs in the ovary in preparation for ovulation.
- **Ovulation (Day 14)** is the release of an egg from the ovary.
- **Luteal phase (Days 15–28)** is when the endometrium thickens further in preparation for pregnancy.
- **Fertile period** is the time in menstrual cycle when a female is most likely to get pregnant if she has intercourse.
- Hormonal control of the menstrual cycle: HL
  - **Days 1–5:** Low levels of oestrogen and progesterone trigger shedding of endometrium (menstruation). A new Graafian follicle develops in the ovary. Egg undergoes meiosis. Pituitary gland secretes FSH and LH. FSH stimulates growth of Graafian follicles. HL
  - **Days 6–13:** Graafian follicle releases oestrogen, causing endometrium to thicken. Only one Graafian follicle matures fully. Build-up of oestrogen inhibits release of FSH and causes a surge in LH. HL
  - **Day 14:** Surge in LH triggers ovulation. Intercourse during fertile period can lead to pregnancy (egg can survive 24 hours). HL
  - **Days 15–26:** LH causes corpus luteum to form from remains of Graafian follicle. Corpus luteum releases oestrogen and progesterone. Progesterone maintains and thickens endometrium. Progesterone and oestrogen inhibit secretion of FSH and LH. Levels of LH decline, causing corpus luteum to break down. If implantation does not occur, corpus luteum breaks down and stops releasing oestrogen and progesterone. HL
  - **Days 27–28:** Progesterone and oestrogen levels drop rapidly as the corpus luteum has broken down. Inhibition of secretion of FSH and LH is removed. HL
- **Menopause** is when a female's menstrual cycle stops, the ovaries stop producing oestrogen and progesterone and the levels of LH and FSH increase.

END OF CHAPTER ASSESSMENT

REMEMBER

1. Name and give a function for each of the labelled parts A–H of the male reproductive system.

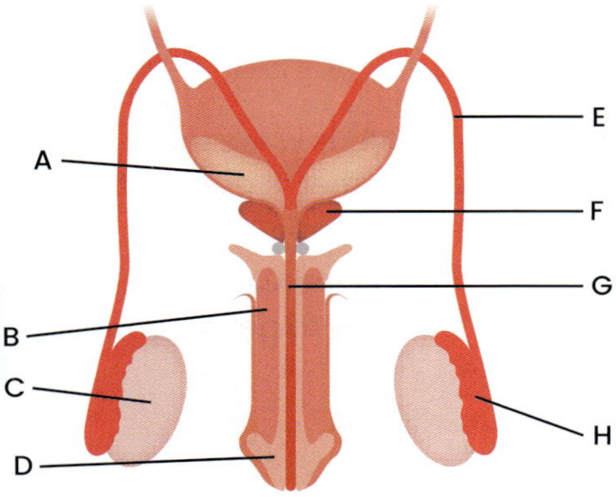

2. Name and give a function for each of the labelled parts A–F of the female reproductive system.

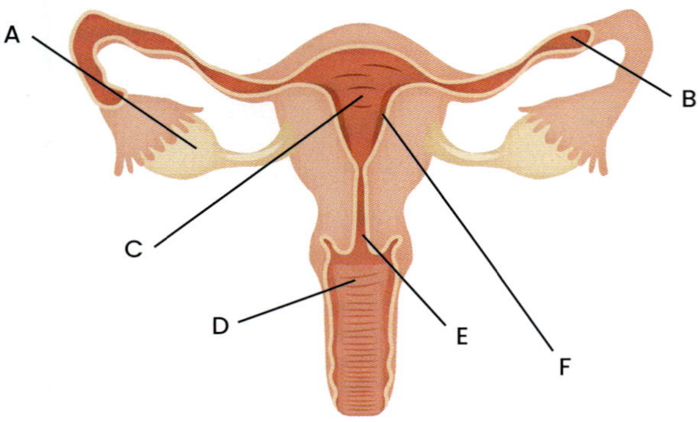

3. Where does meiosis occur in the (a) male (b) female reproductive systems?

4. State the location of (a) fertilisation (b) ovulation in the female reproductive system.

5. Draw a labelled diagram of the (a) male (b) female gamete.

6. Identify the location in the male reproductive system where the male gamete (a) is formed (b) matures (c) is stored.

7. (a) Define the term *puberty*.

(b) Identify the hormones that control puberty in (i) males (ii) females.

8. (a) What is the average length of the menstrual cycle?

(b) State the days in the menstrual cycle when the following events occur.

(i) Corpus luteum forms

(ii) Graafian follicle develops

(iii) Lining of the womb rebuilds

(iv) Menstruation

(v) Ovulation

9. Name the hormones that control the menstrual cycle in females.

10. Define each of the following terms.

(a) Menstrual cycle

(b) Menstruation

(c) Ovulation

(d) Fertile period

UNDERSTAND

11. Why are the testes held outside the body in the scrotum?

12. Outline each of the following stages of the menstrual cycle.

(a) Menstruation

(b) Follicular phase

(c) Ovulation

(d) Luteal phase

13. Copy and complete the following table to outline the role of the hormones oestrogen, progesterone, FSH and LH in controlling the menstrual cycle. HL

HORMONE	SECRETED BY	FUNCTION
Oestrogen		
Progesterone		
FSH		
LH		

14. Explain how negative feedback by the hormones oestrogen and progesterone helps control the menstrual cycle. HL

APPLY

15. (a) Copy and complete the following graph to illustrate the changes that occur to (i) the lining of the uterus (endometrium) during the menstrual cycle (ii) levels of the hormones oestrogen and progesterone.

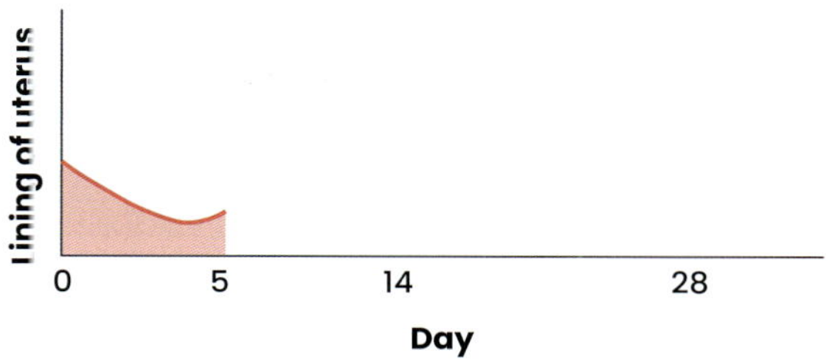

(b) On your graph, identify when the following events occur:

(i) Menstruation (ii) Ovulation

HL 16. Sketch a graph to illustrate the levels of oestrogen, progesterone, FSH and LH at each stage in the menstrual cycle.

ANALYSE

17. Relate each of the following parts of a sperm cell to its function.

(a) Acrosome
(b) Mitochondria
(c) Tail
(d) Haploid nucleus

HL 18. The graph below shows the levels of two female reproductive hormones in the blood at different stages of the menstrual cycle. Study the graph and identify the female reproductive hormones A and B.

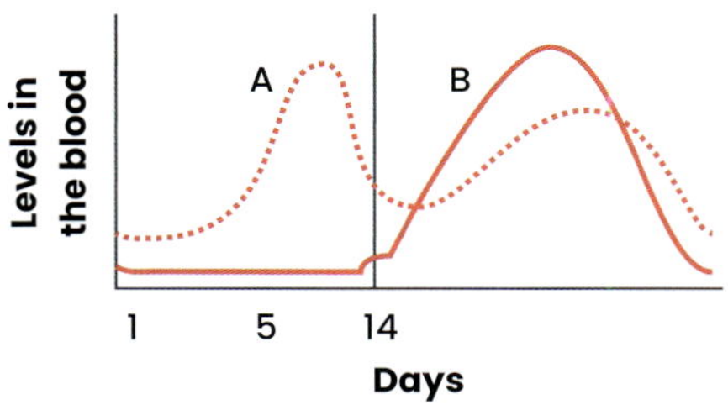

EVALUATE

19. The vagina has an acidic pH and semen is alkaline.

(a) What is the benefit to the female of an acidic pH in the vagina?

(b) Why do you think semen is alkaline?

CREATE

20. Prostate cancer is the most common cancer affecting men in Ireland. Research the disease and present your findings to your class.

37 Human Reproduction – From Fertilisation to Birth

LEARNING OUTCOMES: 2.5.3, 2.5.4, 2.5.5, 2.5.6

EYE-OPENER

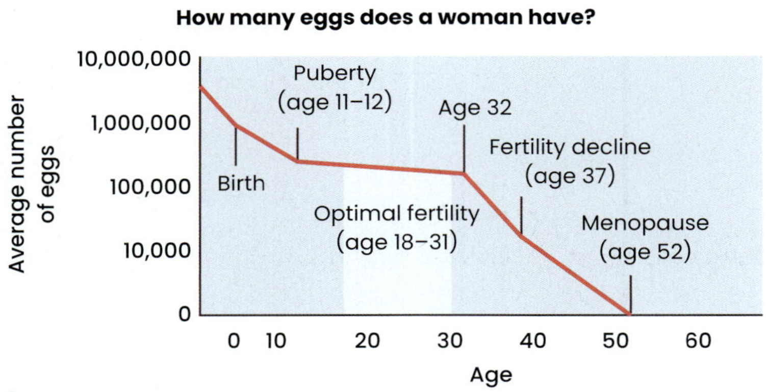

This graph illustrates the numbers of eggs in an average female's ovaries from birth to menopause. Some women are now choosing to freeze their eggs for future use.

At the end of this chapter, you should be able to:

- Model the role of hormones in implantation, growth of the foetus and birth (oestrogen, progesterone, oxytocin).
- Model the role of FSH and LH in implantation, growth of the foetus and birth. HL
- Describe pregnancy from the development of the fertilised embryo to birth.
- Describe germ layers in foetus development. HL
- Model the role of hormones in the production of colostrum and milk (oxytocin, prolactin).
- Appreciate the use of technology in prenatal care – antenatal screening, genetic screening.
- Appreciate the use of technology in postnatal care – postnatal screening including PKU.
- Describe different forms of contraception.
- Describe infertilty treatments.

LINKS TO:

Chapter 9	**Evolution and Natural Selection**
Chapter 17	**Cell Continuity**
Chapter 34	**The Endocrine System**
Chapter 36	**Human Reproduction – Anatomy**

SUBJECT VOCABULARY

insemination, fertilisation, morula, blastocyst, implantation, pregnancy, amnion, placenta, germ layers, labour, prenatal care, postnatal care, infertility, in vitro fertilisation, artificial insemination, intracytoplasmic sperm injection, contraception

Section 1: Copulation, Fertilisation and Implantation

At the end of this section, you should be able to:

- Describe copulation.
- Describe fertilisation and implantation.
- Describe the development of the morula and blastocyst. HL

Unpacked from LOs 2.5.3, 2.5.4

COPULATION

Copulation is the biological term for sexual intercourse. When a male is sexually aroused, blood flow to the penis increases causing it to become rigid and erect. This allows it to be inserted into the female's vagina during sexual intercourse. Sexual arousal in the female causes the vagina to become moist and wider, allowing it to accommodate the penis. At sexual climax or orgasm in the male, semen is ejaculated into the vagina, just below the cervix. This is known as **insemination**. Muscles in the vagina contract to tighten around the penis. The alkaline semen helps protect sperm from the acidic environment in the vagina. Contractions of the uterus help to move the sperm into the Fallopian tubes.

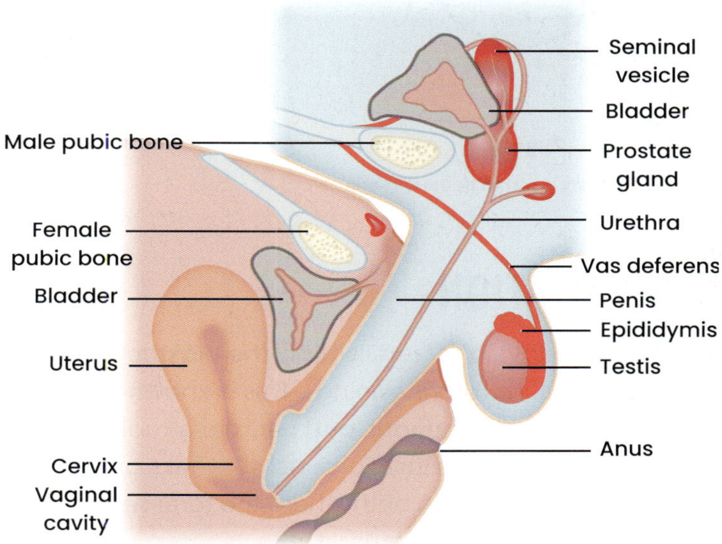

Figure 37.1: Sexual intercourse.

FERTILISATION

Following ovulation, cilia along the walls of the Fallopian tubes help to move the released egg along. If sperm are present in the Fallopian tubes, they may fuse with the egg. This fusion of the haploid sperm and haploid egg is known as **fertilisation**. The resultant diploid cell is called a zygote. Half the zygote's DNA came from the mother's egg and the other half came from the father's sperm.

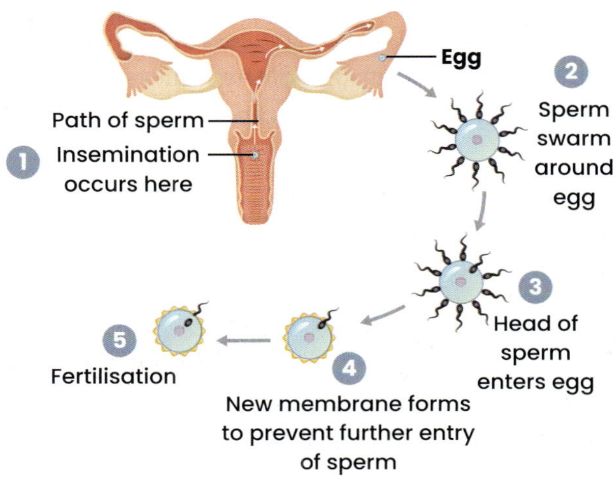

Figure 37.2: Fertilisation.

EARLY EMBRYO DEVELOPMENT

Following fertilisation, the diploid zygote undergoes mitosis several times. After four days, it has formed a ball of 32 undifferentiated cells, known as a **morula**. This undergoes mitosis again to form a hollow, fluid-filled ball of about 64 cells called a **blastocyst**, around Day 5–6 after fertilisation.

- **Insemination** is the release of semen into the vagina.
- **Fertilisation** is the fusion of a haploid egg cell and a haploid sperm cell to form a diploid zygote.
- The **morula** is a mass of 32 undifferentiated cells in early embryo development. HL
- The **blastocyst** is a fluid-filled ball of about 64 cells in early embryo development. HL

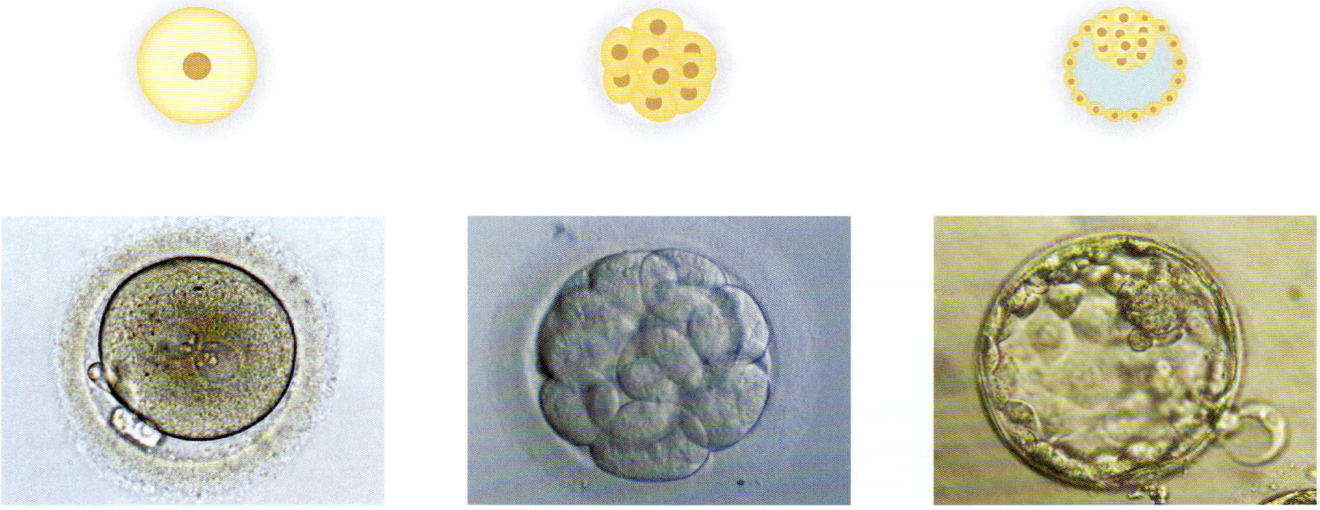

Figure 37.3: Early embryo development.

IMPLANTATION

About seven days after fertilisation, the blastocyst may embed in the endometrium (uterine lining). This is known as **implantation**. The female is now pregnant, and her menstrual cycle stops. **Pregnancy** in humans lasts around 40 weeks and describes the time from fertilisation to birth where the fertilised egg develops into an embryo and then a foetus.

- **Implantation** is the embedding of a fertilised egg in the lining of the womb.
- **Pregnancy** describes the period of development of the fertilised zygote to birth.

Role of hormones in implantation

Progesterone and oestrogen secreted by the corpus luteum in the ovary assist implantation of the fertilised egg by causing the endometrium to thicken and become enriched with blood vessels.

Go to YouTube.com to search for 'Fertilization' (5:42) by Nucleus Medical Media.

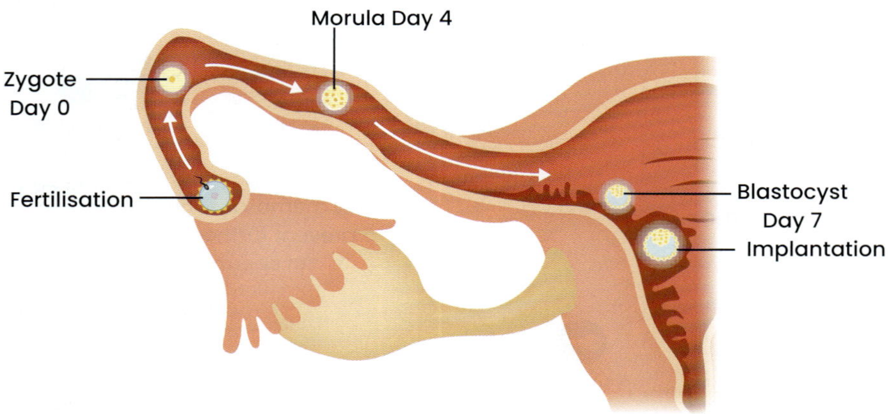

Figure 37.4: Fertilisation to implantation.

PORTFOLIO LINK Go to your Learning Portfolio and complete Learning Check 1 to assess your knowledge and understanding of the topic.

Section 2: Embryo Development, Labour and Birth

At the end of this section, you should be able to:

- Describe pregnancy from the development of the fertilised embryo to birth.
- Describe germ layers in foetus development. HL
- Relate the structure of the placenta to its functions.
- Describe the role of hormones in the production of colostrum and milk (oxytocin, prolactin).

Unpacked from LOs 2.5.3, 2.5.4

EMBRYONIC DEVELOPMENT (UP TO WEEK 12)

Blastocyst structure

HL

The inner mass of cells in the blastocyst becomes the embryo. The outermost cells in the blastocyst, the 'trophoblast' has two layers: an outermost layer called the 'chorion' that will form the placenta and an innermost layer called the 'amnion'.

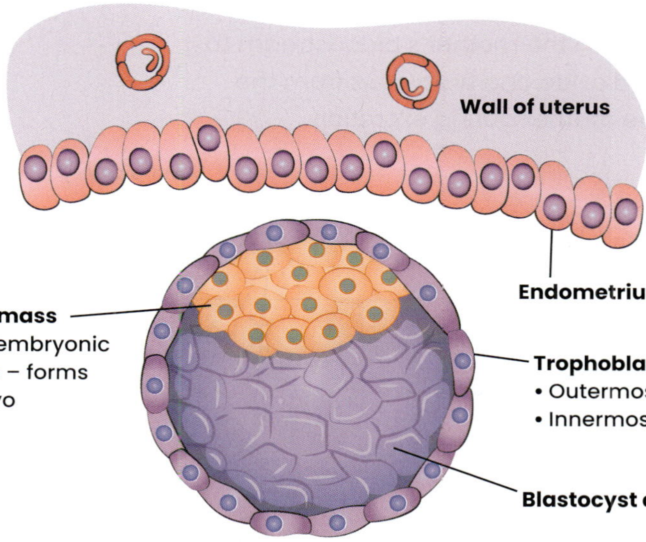

Figure 37.5: Blastocyst structure.

AMNIOTIC SAC AND AMNIOTIC FLUID

Within days of a woman conceiving, the amniotic sac forms from the **amnion**. The amniotic sac is a membrane that surrounds and protects the developing foetus in the womb. It secretes amniotic fluid, a straw-coloured fluid, in which the developing foetus floats. Amniotic fluid acts as a shock absorber, cushioning the foetus from outside impact.

Amnion is a membrane which forms the amniotic sac and amniotic fluid that surrounds and protects the developing embryo.

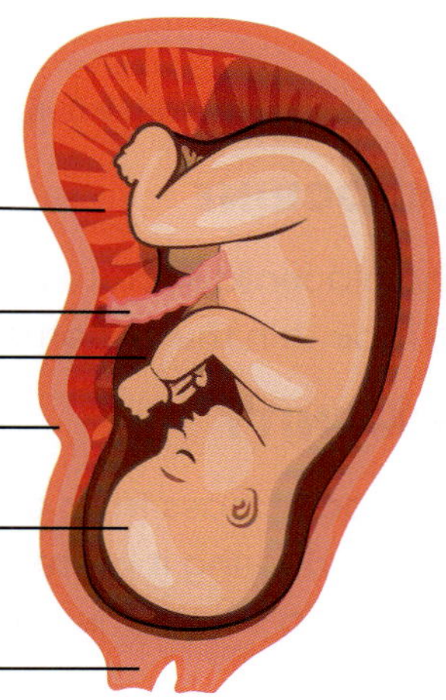

Figure 37.6: Foetus.

PLACENTA FORMATION

The **placenta** is a disc shaped structure formed from tissue from the embryo and tissue from the endometrium. It is connected to the foetus by the umbilical cord. The placenta is fully formed after 12 weeks but continues to grow throughout pregnancy. It secretes the hormones oestrogen and progesterone, helping to maintain the endometrium throughout pregnancy.

The **placenta** is a disc shaped structure formed from tissue from the embryo and the endometrium.

The placenta allows the exchange of materials between the mother's bloodstream and that of the foetus, without the bloodstreams mixing. The placenta is adapted to its function as it has a large surface area and rich blood supply. Harmful substances taken in by the mother can also pass through the placenta to the developing foetus, e.g. drugs, alcohol and chemicals in cigarettes. When the baby is born, the placenta is expelled as 'afterbirth'.

Functions of the placenta

EXCHANGE OF MATERIALS BETWEEN MOTHER'S AND FOETUS'S BLOODSTREAM

Nutrients, water and oxygen pass from the mother's bloodstream to the foetus's bloodstream. Carbon dioxide and urea pass from the foetus's bloodstream to that of the mother during excretion.

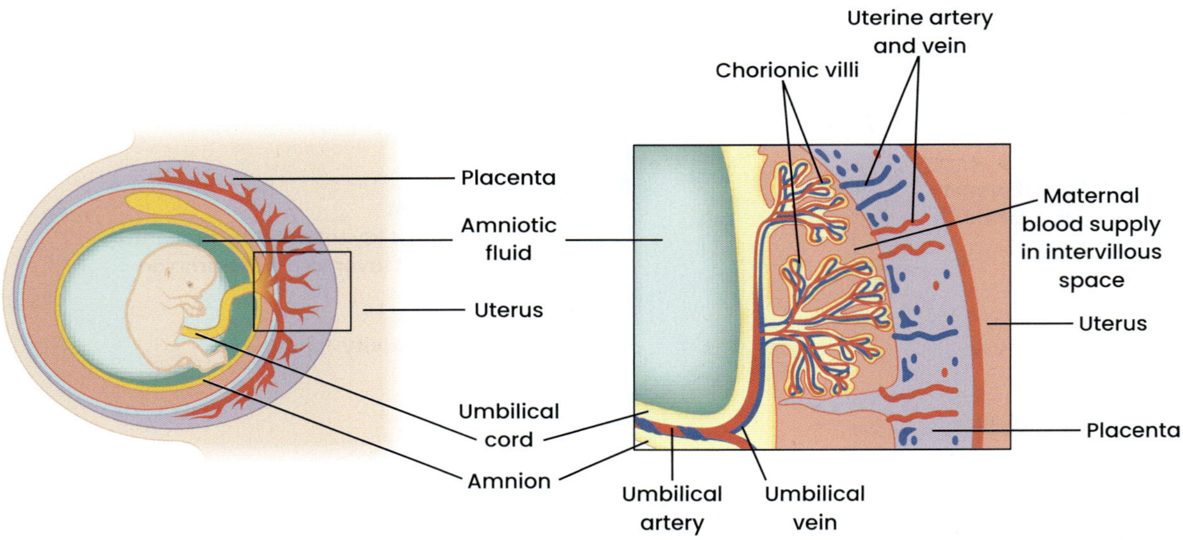

Figure 37.7: Placenta structure.

ENDOCRINE

The placenta secretes the hormones oestrogen and progesterone to maintain the endometrium during pregnancy.

IMMUNITY

Antibodies pass from the mother's bloodstream to that of the foetus, offering it protection against diseases. This passive immunity lasts for a short time after the birth of the baby.

GERM LAYERS

Once the blastocyst implants in the endometrium, its inner mass of cells begins to form three embryonic tissue layers called **germ layers**: the top ectoderm layer, middle mesoderm layer and inner endoderm layer. The germ layers give rise to specific tissues and organs in the embryo.

Germ layers are layers in the embryo that give rise to specific tissues and organs.

HL

- **Ectoderm:** gives rise to the skin, central nervous system and eyes.
- **Mesoderm:** gives rise to the heart, circulatory system, skeleton, kidneys and reproductive system.
- **Endoderm:** gives rise to the embryo's liver, pancreas and digestive system.

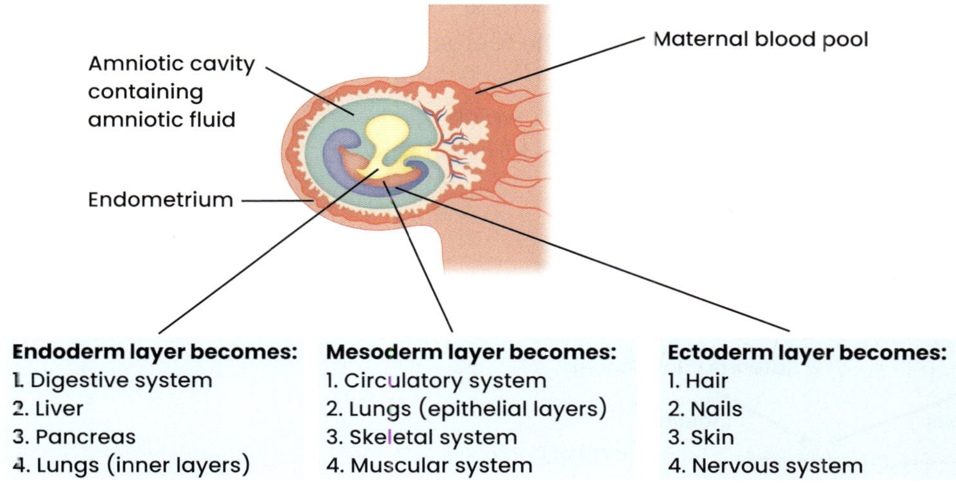

Figure 37.8: Embryonic germ layers.

EMBRYONIC DEVELOPMENT (UP TO WEEK 12)

The embryonic stage continues to Week 12 of pregnancy. During the embryonic stage:

- Most of the baby's organs begin to form and the embryo increases in length.
- The heart and main blood vessels form, and the heart starts beating by the sixth week.
- Sex organs begin to differentiate.
- The brain and spinal cord begin to form.
- Arms and legs also form, and the embryo is about two inches long by Week 12.

FOETAL DEVELOPMENT (WEEKS 12–40)

From Week 12 onwards, the developing embryo is now recognisable as human and is referred to as a foetus. During the foetal period, the organs continue to grow and develop, and the foetus puts on weight. The foetus's sex can be identified by Weeks 12–14. The brain and spinal cord continue to develop. Bones continue to develop and become harder. Hair and fingernails grow. The foetus begins to move, hear and swallow.

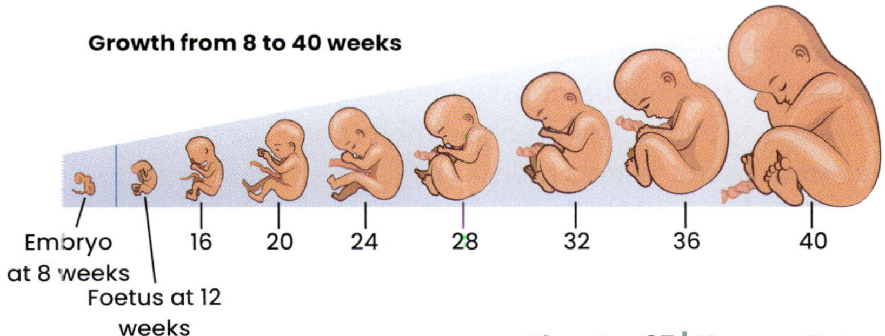

Figure 37.9: Embryo-foetal development.

Role of hormones in foetal development

The corpus luteum starts to break down when the placenta is fully formed and secreting enough progesterone and oestrogen to maintain the endometrium. This allows the foetus to develop and grow. Progesterone and oestrogen levels increase throughout pregnancy. The high levels of progesterone during pregnancy prevent ovulation.

LABOUR AND BIRTH

A full term human pregnancy lasts approximately 40 weeks; the baby is now ready to be delivered. During pregnancy, progesterone secreted by the placenta inhibits the production of the hormone oxytocin by the pituitary gland. At the end of pregnancy, the foetus produces oestrogen, which stops the placenta producing progesterone. This in turn promotes the release of oxytocin, which causes the uterus to contract and eventually brings on the birth of the baby.

Labour is the process by which the baby and the placenta are delivered from the uterus to the outside world.

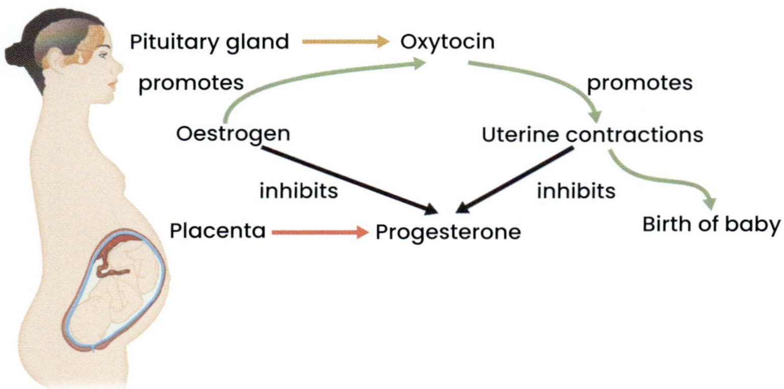

Figure 37.10: Hormonal control of birth.

Childbirth or **labour** can be divided into three stages. We will look at each of these in turn.

Stages in labour

STAGE 1: DILATION

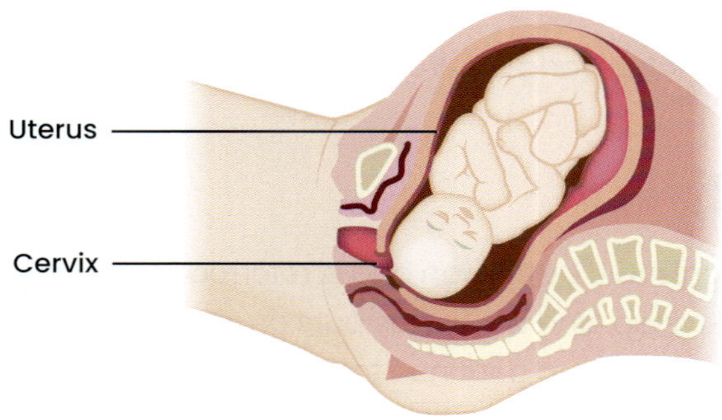

Figure 37.11: Stage 1 of labour – the cervix relaxes, causing it to dilate and thin out.

Levels of oestrogen and progesterone fall, causing labour to begin. Another hormone oxytocin, secreted by the pituitary gland, travels in the bloodstream to the uterus, causing muscles in the uterine walls to contract. Contractions cause the cervix to widen or 'dilate'. The amniotic membrane bursts releasing amniotic fluid through the vagina – this is known as the 'waters breaking'.

STAGE 2: DELIVERY

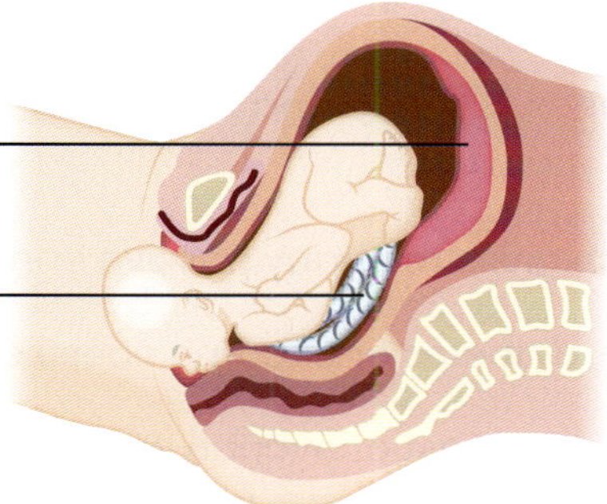

Figure 37.12: Stage 2 of labour – uterine contractions increase in strength and the infant is delivered.

Oxytocin causes the cartilage between the mother's two hipbones to relax and widen. When the cervix is fully dilated to 10cm, the baby will pass head-first from the uterus into the vagina. The baby passes through the vagina, and out to the outside world. The umbilical cord is clamped and cut.

STAGE 3: PLACENTAL STAGE

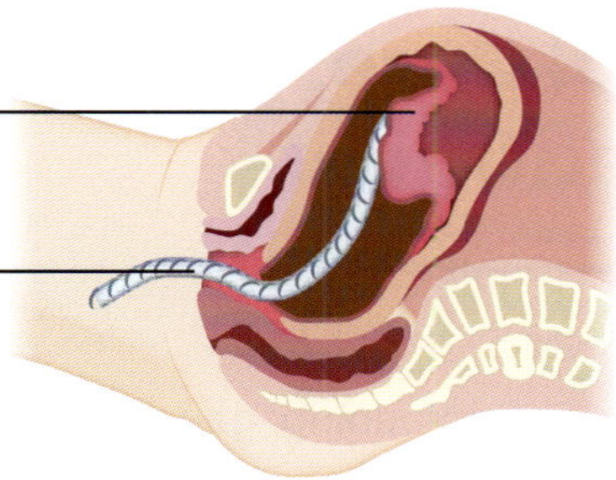

Figure 37.13: Stage 3 of labour – the placenta is expelled.

Go to YouTube.com and search for 'Patient Education Animation: Labor and Vaginal Birth' (2:53) by Nucleus Medical Media.

Stage 3 of labour happens after the baby is born. Further contractions of the uterus cause the placenta or 'afterbirth' to come out the vagina.

MILK PRODUCTION

The mother's levels of oestrogen and progesterone fall following the birth of the baby. This triggers the secretion of another hormone 'prolactin' from the pituitary gland in the brain. Prolactin travels in the bloodstream to the mammary glands in the breasts, causing milk to be produced. This process usually takes two to three days. In the first few days after birth, a form of milk called **colostrum** is produced. Colostrum is a yellow, highly nutritious liquid that is rich in antibodies. Every time the infant suckles, the pituitary gland secretes prolactin and oxytocin. Oxytocin causes muscles in the mammary glands to contract and eject milk out of the milk ducts.

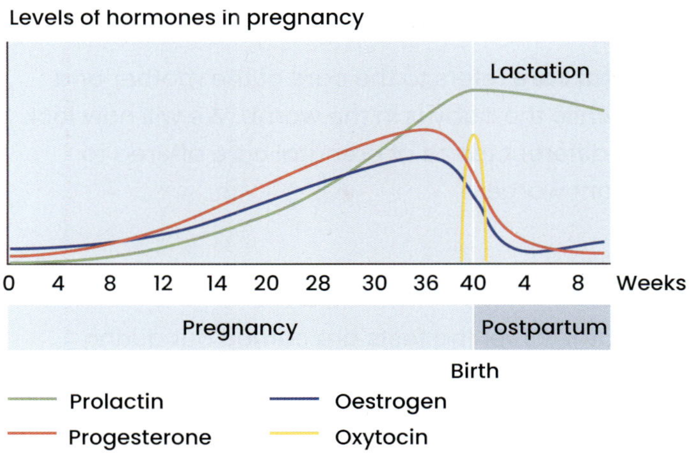

Figure 37.14: Hormone levels before and after birth.

There are a number of biological benefits of breastfeeding:

- Breast milk contains all the nutrients that the baby needs, is sterile and at body temperature.
- Breast milk contains antibodies from the mother which gives the baby passive immunity.
- Breast fed babies are less likely to have heart disease, diabetes or become obese as adults.
- Breastfeeding reduces the risk of the mother developing breast or ovarian cancer or osteoporosis.
- Breastfeeding creates a bond between the mother and baby.

DID YOU KNOW?

PORTFOLIO LINK **Go to your Learning Portfolio and complete Learning Check 2 to assess your knowledge and understanding of the topic.**

Section 3: Prenatal and Postnatal Care, Contraception and Infertility Treatment

At the end of this section, you should be able to:

- Appreciate the use of technology in prenatal care – antenatal screening, genetic screening.
- Appreciate the use of technology in postnatal care – postnatal screening including PKU.
- Describe different strategies of controlling fertility including freezing eggs and different forms of contraception (natural, mechanical, surgical and chemical methods and their use including prevention of STIs).
- Describe infertility treatments.

Unpacked from LOs 2.5.5, 2.5.6

PRENATAL CARE

Prenatal care refers to the care of the mother and baby while the baby is in the womb. We will now look at the different forms of prenatal care offered to pregnant women.

Prenatal care refers to the care of the mother and baby while the baby is in the womb.

Antenatal screening

Antenatal screening tests are carried out during pregnancy to determine if the mother or the baby are at risk of any health issues. Antenatal screening includes blood tests and ultrasound scans.

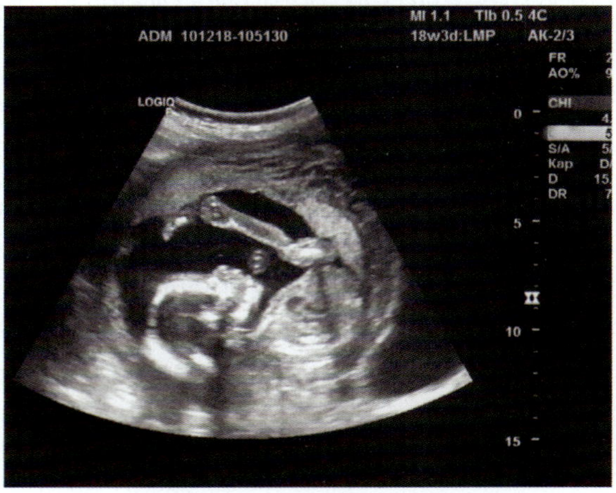

Figure 37.15: Prenatal ultrasound screening.

- **Blood tests:** A pregnant mother's blood is tested for conditions such as anaemia and gestational diabetes, a form of diabetes that can occur during pregnancy. Blood tests for the Rhesus factor are also carried out to determine if the mother's blood is Rhesus positive or negative. If the mother is Rhesus negative, she will be given anti-D injections to prevent the production of anti-D antibodies by her immune system, which could affect any future pregnancies.
- **Ultrasound scans:** Pregnant women are usually given two ultrasound scans. The first is a 'dating scan' between 8 and 14 weeks to determine the baby's due date. The second, a 'foetal anatomy scan', is carried out between 18 and 22 weeks. This scan detects any serious physical abnormalities such as spina bifida or heart problems.

Genetic screening

Prenatal genetic screening tests can be carried out to check for chromosomal and genetic conditions such as Down syndrome, Edwards' syndrome and Patau's syndrome in unborn babies. There are two main types of tests.

- **Amniocentesis:** A small sample of the amniotic fluid surrounding the baby is used to test for genetic conditions. This test is usually done after 15 weeks and is very accurate.
- **Chorionic villus sampling (CVS):** A small sample of cells from the mother's placenta is tested for chromosomal and genetic conditions. This test is usually carried out between 11 and 12 weeks.

There is a risk of miscarriage associated with both tests, with CVS having an increased risk compared to amniocentesis. These tests are generally offered to mothers who may have a higher risk of having a baby with a chromosomal or genetic condition.

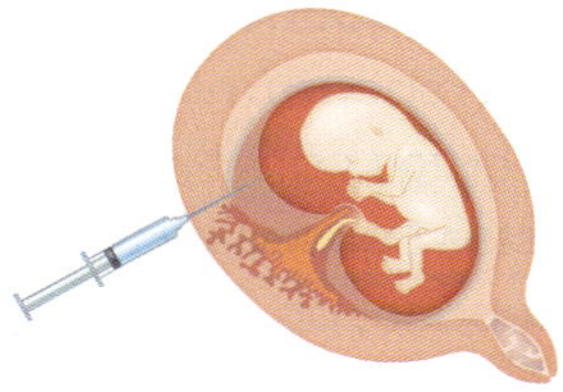

Amniocentesis
Needle used to withdraw foetal cells from amniotic fluid

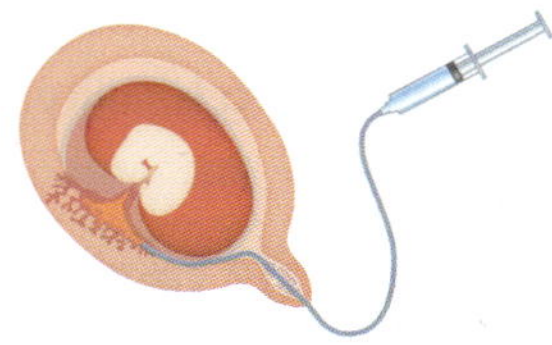

Chorionic villus sampling
Suction tube used to remove foetal cells from chorion (where placenta will develop)

Figure 37.16: Prenatal genetic screening.

POSTNATAL CARE

Postnatal screening

Newborn babies in Ireland undergo two main postnatal tests: hearing screening and the 'heel prick test'.

Postnatal care refers to the care of the mother and baby after the birth of the baby, generally for the first 6–8 weeks.

The heel prick test is usually carried out when the baby is three to five days old. A small sample of blood is taken from the baby's heel and checked for nine rare but serious conditions including cystic fibrosis and phenylketonuria (PKU). PKU is an inherited disease where the body is unable to break down the amino acid phenylalanine. This causes a build-up of phenylalanine in the blood and brain, leading to brain damage. Treatment for PKU involves a diet avoiding high protein foods such as meat, eggs and dairy.

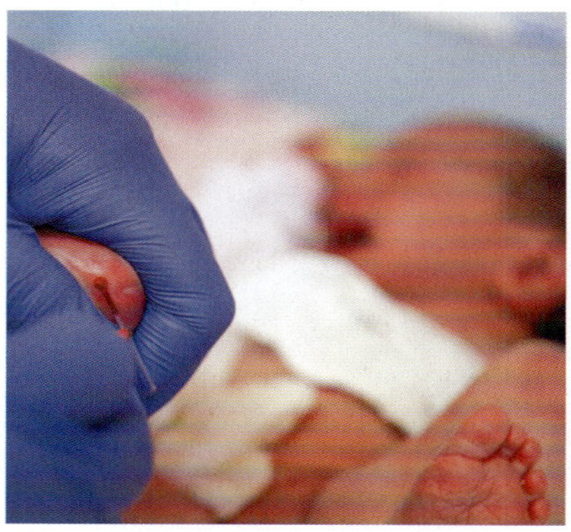

Figure 37.17: Heel prick test.

Go to YouTube.com and search for 'Amniocentesis and CVS Animation' (1:16) by Homework Clinic.

INFERTILITY

A couple are described as **infertile** if they do not conceive a child within a year after having regular unprotected sexual intercourse.

Infertility is an inability to conceive a child.

One in six couples worldwide suffer from infertility.

DID YOU KNOW?

Causes of male infertility

- **Low sperm count:** A man is described as having a low sperm count if his semen contains less than 15 million sperm/ml. Excessive drinking of alcohol or smoking, obesity, use of anabolic steroids or taking drugs can lower testosterone levels, decreasing sperm production. Elevated temperature of the testes caused by tight-fitting clothing can also reduce sperm count. Treatment for male infertility is dependent on identifying the underlying cause. If a lifestyle factor e.g. smoking or excessive alcohol consumption is the cause, stopping smoking or reducing alcohol intake may return fertility levels to normal.
- **Low testosterone levels:** Low levels of the hormone testosterone can decrease sperm production, lowering sperm count. Hormone replacement therapy may be recommended if testosterone levels are affecting fertility.

Causes of female infertility

- **Polycystic ovary syndrome (PCOS):** PCOS is a condition where the ovaries become enlarged, and many fluid-filled sacs (follicles) develop around the eggs, preventing ovulation.
- **Endometriosis:** This occurs when tissue from the endometrium grows outside the uterus, e.g. around the ovaries and Fallopian tubes. This displaced endometrial tissue continues to follow the menstrual cycle – breaking down during menstruation and thickening again. The broken-down tissue builds up, causing pain. The Fallopian tubes can become blocked preventing fertilisation. The endometrium can also be affected, preventing implantation. Surgery may be used to remove the displaced tissue.
- **Fibroids** are non-cancerous growths that occur in the uterus or Fallopian tubes. They may block a Fallopian tube, preventing fertilisation or implantation. Surgery may be used to remove fibroids.
- **Infections:** Sexually transmitted infections such as chlamydia and gonorrhoea can damage or block Fallopian tubes, preventing fertilisation. Antibiotics may be used to treat these infections.

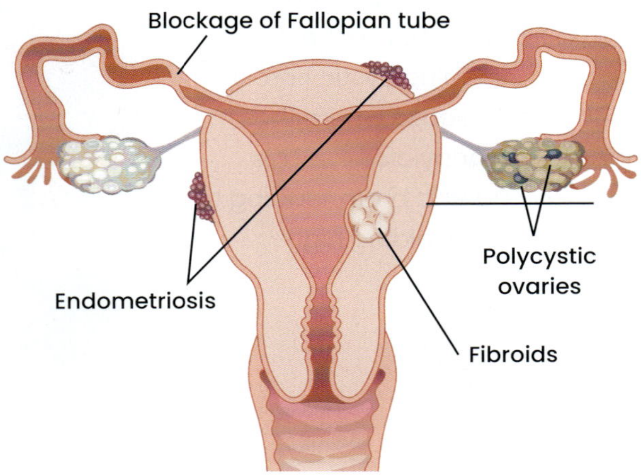

Figure 37.18: Possible causes of female infertility.

Treatments for infertility

- **Stimulating ovulation:** In cases where infertility is caused by an ovulation disorder, e.g. PCOS, fertility drugs that work like follicle stimulating hormone (FSH) or luteinising hormone (LH) are used to regulate ovulation.
- **Artificial insemination:** In artificial insemination, a doctor inserts sperm directly into a woman's cervix, Fallopian tubes or uterus.

Artificial insemination is a medical procedure where sperm is inserted directly into a woman's cervix, Fallopian tubes or uterus.

IVF is a technique where eggs are fertilised by sperm in a glass dish.

FOCUS ON ... IN VITRO FERTILISATION

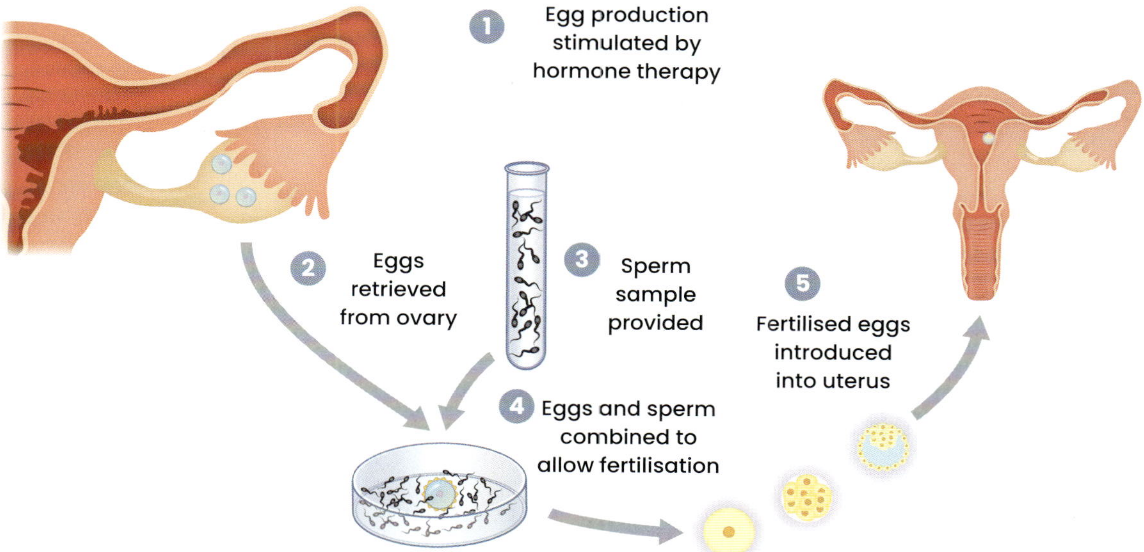

Figure 37.19: Stages of in vitro fertilisation.

In vitro fertilisation (**IVF**) is another treatment for infertility. During IVF, eggs are removed from a woman's ovaries and fertilised by sperm in a glass dish. The fertilised egg or 'embryo' is placed in the woman's uterus to grow.

IVF involves the following stages:

1. **Boost egg supply (hyper-ovulation):** The woman is given follicle stimulating hormone (FSH) to cause her to produce multiple eggs.
2. **Egg collection:** After about 12 days, the eggs are collected using a needle guided by ultrasound.
3. **Fertilising the eggs:** The eggs are mixed with sperm in a glass dish in a laboratory. Any fertilised eggs (embryos) are left to grow for up to six days.
4. **Embryo transfer:** The fertilised eggs are collected at the blastocyst stage and one or more are transferred to the woman's uterus. If the IVF cycle is successful, the fertilised egg will implant in the endometrium and develop as during a normal pregnancy. Unused embryos may be frozen for use at a later stage.

Go to YouTube.com and search for 'How IVF works | 3D Animation' (1:15) by Dr. Paulien Moyaert.

- **Intracytoplasmic sperm injection:** Intracytoplasmic sperm injection (ICSI) is like IVF in that eggs and sperm are collected from each partner. The procedures differ in how fertilisation occur. In ICSI, a single sperm is injected directly into an egg using a needle. This assists fertilisation by helping sperm cells to pass through the membrane of the egg. ICSI is useful in cases where the sperm may have difficulty reaching or penetrating the egg on its own, or when IVF has not succeeded.

Intracytoplasmic sperm injection (ICSI) is a treatment for infertility where a sperm is injected directly into an egg using a needle.

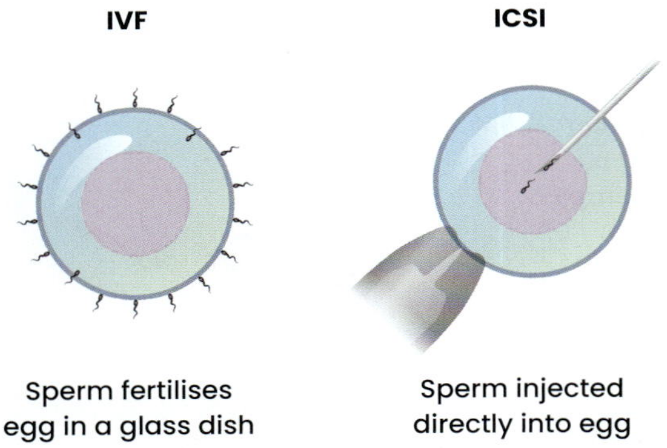

Figure 37.20: Fertilisation in IVF compared to ICSI.

CONTROLLING FERTILITY

Freezing eggs

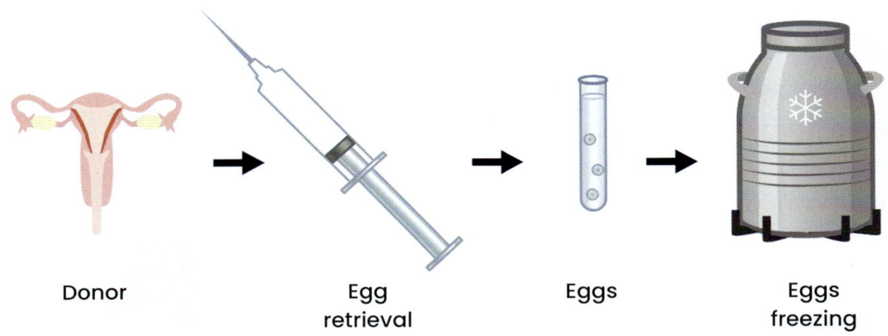

Figure 37.21: Controlling fertility by freezing eggs.

The number and quality of eggs a female has decreases rapidly from around her mid 30s. Eventually, around 50 years of age, ovulation stops when menopause is reached. Women may wish to enhance their future chances of conceiving by freezing their eggs. Ideally, it is recommended that a woman freeze her eggs before 35–36 years of age to have a greater chance of conceiving with those eggs in the future. The steps in retrieving eggs are the same as in IVF, except that the eggs are stored in liquid nitrogen tanks in a lab.

Birth control

Birth control or **contraception** involves methods to prevent pregnancy in a female. Contraceptive methods work by preventing ovulation, fertilisation or implantation. There are four main categories of contraceptive methods: natural, mechanical, surgical or chemical.

Contraception refers to methods of preventing pregnancy in a female.

Methods of contraception

NATURAL

- **Abstinence:** Abstinence means that a person doesn't have intercourse. This is 100% effective at preventing both STIs and pregnancy.
- **Rhythm method:** The rhythm method means that a couple avoids intercourse during the female's fertile period of her menstrual cycle. The rhythm method is around 75% effective at preventing pregnancy.

Go to YouTube.com and search for 'GCSE Biology – Contraception #61' (7:07) by Cognito.

MECHANICAL

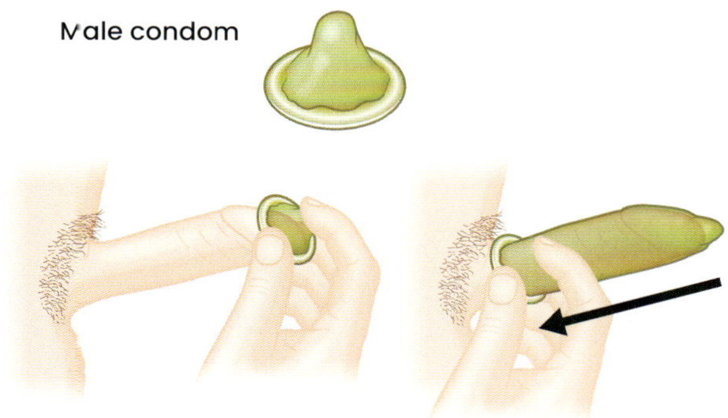

Figure 37.22: How to use a male condom.

- **Male condom:** A condom is a rubber sheath that is rolled over the erect penis before sexual intercourse. The condom stops semen entering the vagina, preventing fertilisation. Condoms are 98% effective when used correctly.

Apart from abstinence, the only form of contraception that protects men and women from sexually transmitted infections (STIs) is the male condom.

DID YOU KNOW?

Go to YouTube.com and search for 'Sexual Wellbeing : Condom Explainer Video' (0:58) by HSE Ireland.

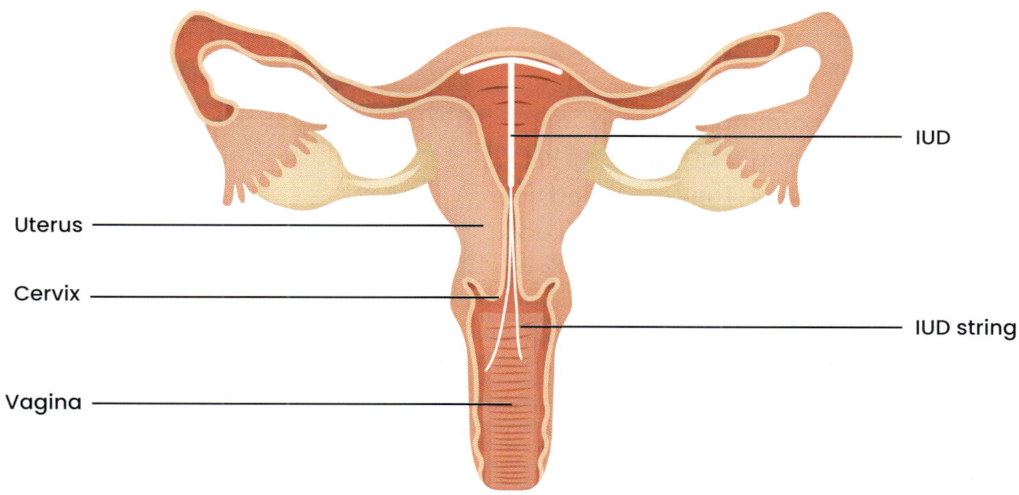

Figure 37.23: IUD.

Intra-uterine device (IUD): An IUD or 'coil' is a small T-shaped device that is inserted into the uterus, where it can be left for five or more years. Some IUDs release hormones. An IUD stops the sperm reaching the egg and prevents fertilisation. The IUD is more than 99% effective against pregnancy.

CHEMICAL

Contraceptive pill: The contraceptive pill contains female hormones (oestrogen and/or progesterone). The pill prevents ovulation. When taken correctly, the pill offers 99% protection against pregnancy.

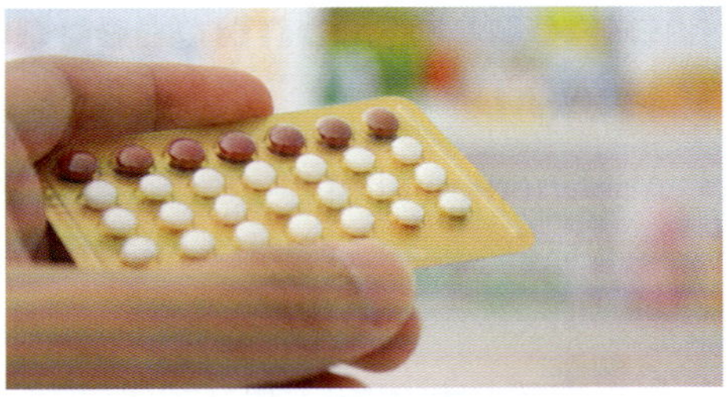

Figure 37.24: Contraceptive pill.

SURGICAL

Vasectomy: Vasectomy is a surgical method of birth control, in which a man's sperm ducts (vas deferens) are cut so that no sperm is released during intercourse. Generally, a permanent form of contraception, the procedure can sometimes be reversed. It is almost 100% effective.

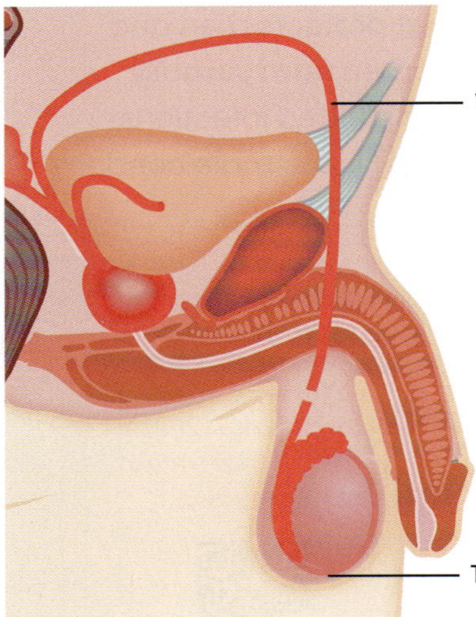

Figure 37.25: Vasectomy.

SCIENCE IN SOCIETY: SUSTAINABILITY

World population

In 2019, 11,000 scientists around the world signed a declaration stating that the world is facing a climate emergency. They identified six critical steps that governments, businesses and all of humanity can take to lessen the effects of climate change. One of these steps was to stabilise or gradually reduce world population by making contraception and sexual health education available to all.

PORTFOLIO LINK Go to your Learning Portfolio and complete Learning Check 3 to assess your knowledge and understanding of the topic.

- In **sexual intercourse**, the penis becomes erect and rigid so that it can be inserted into the female's vagina. Semen is ejaculated into the vagina, below the cervix.
- **Fertilisation** is the fusion of the sperm and egg to form a diploid zygote.
- **Pregnancy** describes the period of development of the fertilised zygote to birth.
- HL The diploid zygote undergoes mitosis to form a mass of 32 cells, the **morula**, and then a 64-cell fluid-filled ball called a **blastocyst**.
- HL **Germ layers** are layers in the embryo that give rise to specific organs.
- HL The **ectoderm** forms the skin, central nervous system and eyes.
- HL The **mesoderm** forms the heart, circulatory system, skeleton, kidneys and reproductive system.
- HL The **endoderm** gives rise to the embryo's liver, pancreas and digestive system.
- **Implantation** is the embedding of a fertilised egg in the endometrium.
- The **placenta** is formed from tissue from the embryo and the endometrium.
- Placenta functions:
  - Transport of nutrients from mother to foetus.
  - Transport of waste from foetus to mother.
  - It secretes the hormones oestrogen and progesterone.
  - Antibodies pass from the mother's bloodstream to that of the foetus.
- The **amniotic sac** is a membrane that surrounds and protects the developing foetus. It secretes amniotic fluid.
- **Stage 1 of labour:** Levels of oestrogen and progesterone fall. Oxytocin causes muscles in the uterus to contract and the cervix to dilate. Amniotic membrane bursts releasing amniotic fluid.
- **Stage 2 of labour:** Oxytocin causes the pelvic cartilage to relax and widen. When the cervix is fully dilated, the baby passes head-first from the uterus into the vagina and to the outside world. The umbilical cord is clamped and cut.
- **Stage 3 of labour:** Further uterine contractions cause the placenta to pass out of the vagina.
- **Milk production:** Prolactin hormone causes milk to be produced. In the first 2–3 days after birth, highly nutritious colostrum is produced.
- **Infertility** is an inability to conceive a child.
- **Causes of infertility in males:** low sperm count due to smoking, excessive alcohol consumption, obesity or taking drugs, low testosterone levels.
- **Causes of female infertility:** polycystic ovary syndrome (PCOS), endometriosis, fibroids, blocked Fallopian tube due to infections.
- **Infertility treatments:**
  - **Stimulating ovulation:** Fertility drugs that work like follicle stimulating hormone (FSH) or luteinising hormone (LH) can be used to regulate ovulation.
  - **Artificial insemination** is a medical procedure where sperm is inserted directly into a woman's cervix, Fallopian tubes or uterus.
  - **In vitro fertilisation:** In IVF, eggs are removed from a woman's ovaries and fertilised by sperm in a glass dish or test tube. The fertilised egg is placed in the woman's uterus to grow. Any unused embryos may be frozen.
  - **Intracytoplasmic sperm injection (ICSI)** is an infertility treatment where a sperm is injected directly into an egg using a needle.
- **Contraception** refers to methods of preventing pregnancy in a female.
  - The **contraceptive pill** is a chemical method of contraception that prevents ovulation.
  - A **condom** is a is a mechanical method of contraception that prevents fertilisation and STIs.
  - A **vasectomy** is a surgical method of contraception that prevents fertilisation.
  - **Abstinence** is a natural method of contraception that prevents fertilisation.
- **Prenatal care** refers to the care of the mother and baby while the baby is in the womb.
  - **Antenatal screening** tests include blood tests and ultrasound scans during pregnancy.
  - Prenatal **genetic screening tests** include amniocentesis on the amniotic fluid and CVS on placenta cells.
- **Postnatal care** refers to the care of the mother and baby for the first 6–8 weeks after the birth of the baby.
- The heel prick test checks a newborn baby for phenylketonuria (PKU).

END OF CHAPTER ASSESSMENT

REMEMBER

1. State the location of (a) fertilisation (b) insemination (c) implantation in the female reproductive system.
2. (a) Identify the tissues from which the placenta forms.
   (b) State the functions of the placenta.
3. Identify structures A–E in the following diagram and give a function of each.

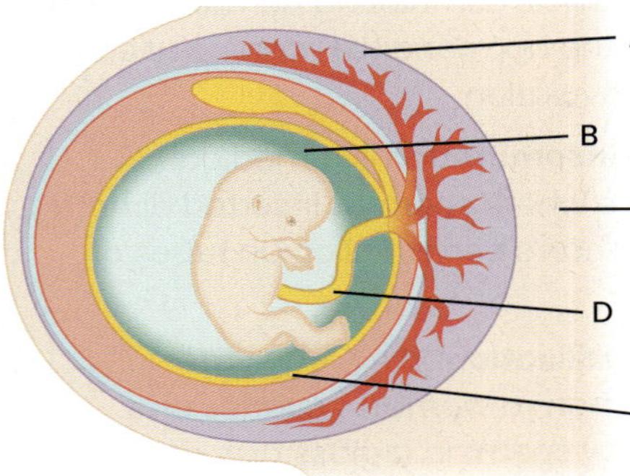

HL 4. Define each of the following terms.
   (a) *Germ layer* (c) *Mesoderm*
   (b) *Endoderm* (d) *Ectoderm*

HL 5. Define each of the following terms.
   (a) *Morula* (b) *Blastocyst*

6. Define each of the following terms.
   (a) *Fertilisation* (d) *Infertility*
   (b) *Pregnancy* (e) *Contraception*
   (c) *Labour*

UNDERSTAND

7. Describe **four** methods of birth control under the following headings:
   (a) How it works (b) Effectiveness
8. Discuss a potential cause and a potential treatment for (a) male infertility and (b) female infertility.
9. Outline the stages in in vitro fertilisation (IVF).

HL 10. Describe the development of a fertilised egg to a blastocyst, to an embryo, to a foetus.

11. Outline the role of hormones in each of the following processes:
   (a) Implantation (c) Birth
   (b) Growth of the foetus (d) Milk production
12. Describe the three stages of labour.
13. What is colostrum and what is its importance?

APPLY

14. Why would a person use contraception?
15. Why is it recommended that a female choosing to freeze her eggs do so before the age of 35–36 years?
16. Which fertility treatments may be used to assist with difficulties with (a) ovulation (b) fertilisation?

ANALYSE

17. Distinguish between the following terms.
   (a) Fertilisation and infertility
   (b) Implantation and infertility
   (c) Embryo and foetus
18. Distinguish between the following procedures.
   (a) Amniocentesis and chorionic villus sampling (CVS)
   (b) Artificial insemination and IVF
   (c) IVF and ICS
19. Relate the effect of the hormone oxytocin on muscles to its role in (a) labour (b) breastfeeding.
20. Fertility rate is the average number of children had per woman of childbearing age during their lifetime. In 1965, the fertility rate in Ireland was 4.03. By 2015, this had dropped to 1.86 (Source: CSO). Why do you think the fertility rate in Ireland dropped between 1965 and 2015?
21. A fertility rate of 2.1 children per woman ensures a generally stable population in a country (Source: WHO). The current fertility rate in Ireland is lower than this level. What impact do you think this will have on Ireland's population in the future?

EVALUATE

22. Discuss the (i) advantages and (ii) disadvantages of using each of the following forms of contraception.
   (a) Male condom
   (b) Contraceptive pill
   (c) Intra-uterine device (IUD)
   (d) Vasectomy

CREATE

23. Create a poster to be placed in a doctor's surgery for expectant mothers to encourage them to breast-feed their child once it is born.

Biology in Practice Investigation

INTRODUCTION

The Biology in Practice Investigation is an important element of the Leaving Certificate Biology course. It is your chance to use your new knowledge, research and laboratory skills to conduct your own scientific research investigation on an aspect of biology. You will apply a lot of what you learned in Chapters 2 and 3 when carrying out the investigation, but the knowledge you've gained across the whole course will help you in the process. It's also a chance for you to expand your knowledge of the living world beyond the scope of the textbook. Your report on your Biology in Practice Investigation will be worth 40% of your final grade, so it is important to understand the stages involved in the process and the features of an excellent research project.

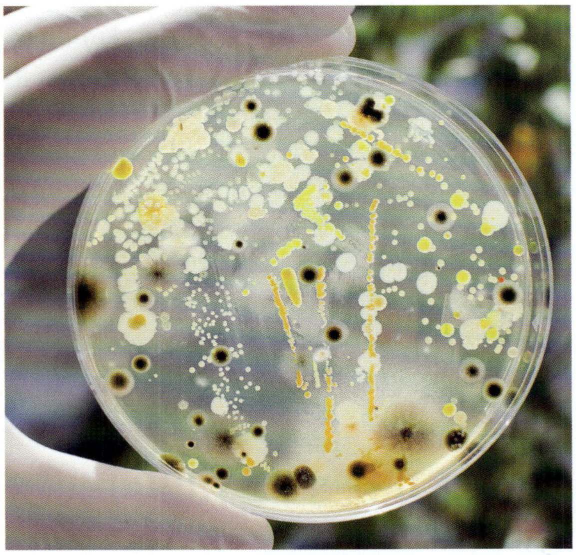

Figure 1: Growing bacteria on agar plates.

It is estimated that the Biology in Practice Investigation will take approximately 20 hours to complete, spread out over the course of your studies. The time spent in lessons, where you might acquire new biological knowledge or practical skills, will help you prepare for completing the research investigation.

TIMELINE

In the second term of 5th Year (January to April), the State Examinations Commission will publish an **investigation brief**, which will aim to set context for the investigation. The investigation brief will also outline the deadline for the completion of the project, which will likely be in the second term of 6th Year.

You'll have about 20 hours of class time to work on your Biology in Practice Investigation and, in the section to follow, there are suggested time allocations for each stage. Once you've completed all aspects of your investigation, and written your report, you will submit your work to the State Examinations Commission. Your report will be corrected alongside your exam paper in the summer.

INVESTIGATIVE LOG

The **investigative log** is your way of documenting your journey through the Biology in Practice Investigation. It is a space for you to record your thoughts, ideas and decision-making processes throughout each stage of your investigation. You'll use it to jot down your initial responses to the brief, document your research findings, sketch out experimental designs and note any challenges or changes you encounter along the way. The log isn't just for your benefit – it also allows your teacher to monitor your progress and provide timely guidance. Moreover, it serves as evidence of the authenticity of your work, demonstrating that the investigation is truly your own. While the log itself isn't submitted for assessment, the information you gather in it will be invaluable when you're writing your final report. Remember to update your log regularly, date your entries and be honest in your reflections – it's a record of your learning process, not just your final outcomes. You can use the Investigative Log workbook that accompanies this textbook, a separate notebook or scrapbook, or your investigative log can take a digital format (e.g. a Google Doc, shared with your teacher).

Figure 2: Your investigative log will be used to record your journey through your Biology in Practice Investigation.

THE SIX STAGES OF YOUR INVESTIGATION

1. Initial response to the investigation brief

The investigation brief will be published in the second term of 5th Year. Firstly, you should read the brief carefully and jot down your first thoughts and questions in your investigative log. These thoughts and questions can form the basis of an initial discussion with your classmates and teacher.

Some questions to consider:

- What do I already know about the topic or issue in the investigation brief?
- Do I understand the topic or issue, or do I need to learn more about it?
- What research and experimental activities in Biology connect to the topic or issue in the investigation brief?
- What aspect of the topic am I interested in finding out more about?
- What can I use to find out more about the topic?
- Have I completed an experiment related to the theme before? Could I extend or adapt a completed experiment, or develop an original approach to an experiment?

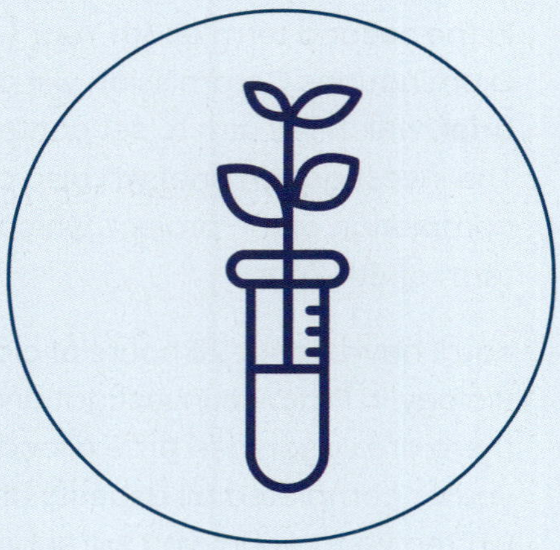

Stage 1 of the process is expected to take 1–2 hours.

2. Background research

During this stage, you will conduct your own research to gain a deeper understanding of the theme of the investigation brief. Use your investigative log to keep track of your research, carefully noting down what you've learned and the source of your findings. You can reference books, scientific journals, newspaper or magazine articles, websites, podcasts and videos, but you should ensure the research you gather is **relevant**, **unbiased** and from reputable sources. All sources need to be referenced carefully (more on this later). This stage is expected to take around 2–3 hours to complete and at the end of the research phase you should be able to formulate your research question. You can refer to Chapter 2, Section 2 for guidance on evaluating sources.

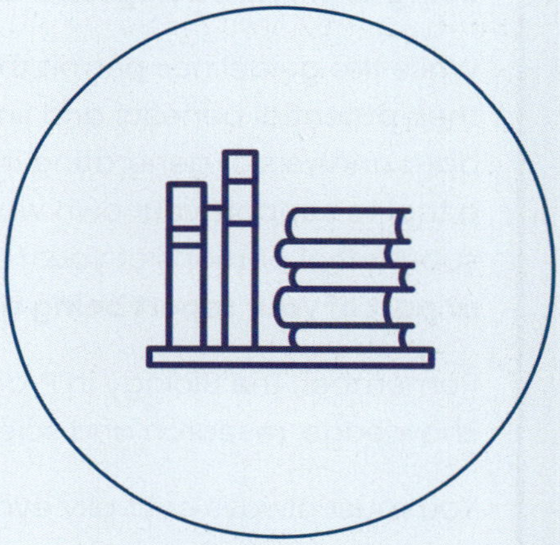

Sources of information		
Books		In-depth, detailed coverage of a topic and background information.
Articles	Scholarly journals	Up-to-date and highly specific for scholars and researchers.
	Trade publications	Targeted towards professionals in a discipline or industry.
	Magazines	Broad summaries of issues for a general audience.
Newspapers		Up-to-date national and regional information for a general audience.
Internet		Wide variety of information. Evaluate websites carefully.

Figure 3: Sources of information for background research.

3. Designing and planning the experiment

Once you have a research question, you can start to design your experiment. The first step is deciding on what you want to test (i.e. the hypothesis). It is important now to consider sample size, variables (independent and dependent) and **controls**. You should note in your investigative log how you will carry out the investigation (step by step), list all the equipment you will need, discuss safety considerations and the time required. You should also consider how to make your experiment more **reliable**, **valid**, **accurate**, **precise** and **fair**. You should consult with your teacher to see if your experiment can be conducted safely in the school or if certain specialist equipment is required. Sections 1 and 2 in Chapter 3 have more information on all aspects of experimental design. This is an important stage as you want to ensure everything is planned carefully before beginning the experiment to test your hypothesis. This stage should take 2–3 hours to complete.

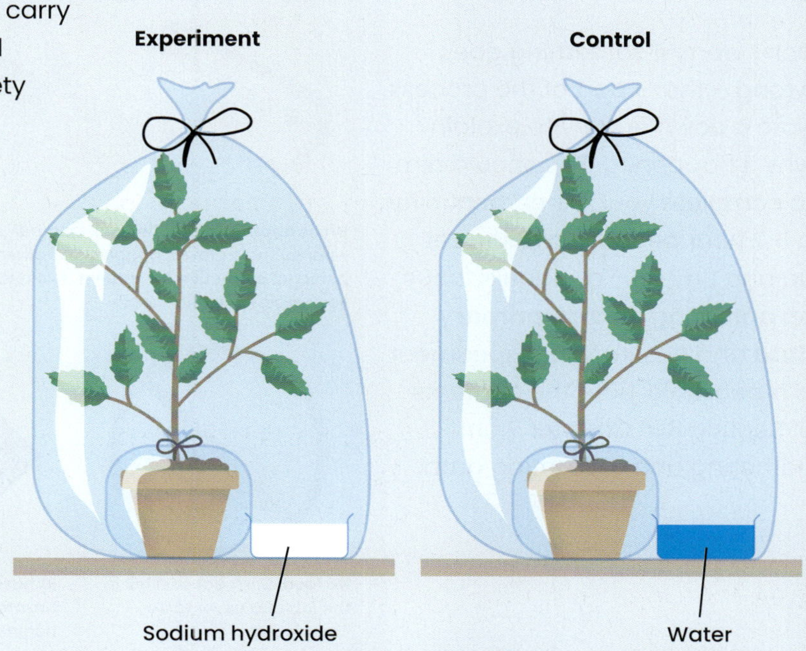

Figure 4: It is important to design a control for your experiment, for comparison.

Using artificial intelligence tools

While the guidelines permit the use of AI tools for research purposes, you should be aware of both their potential benefits and limitations. AI tools can be helpful for tasks such as literature searches, data analysis or generating images. However, it's crucial to understand that AI should be used as a tool to support your own work, not to replace it. Misuse of AI would include having it generate substantial portions of your report or design your experiment. **The misuse of AI could result in all or part of your report being deemed invalid.**

Remember, the Biology in Practice Investigation is about using and developing your own knowledge, research and scientific skills, rather than having a tool do that for you.

You must always critically evaluate any AI-generated content and ensure it aligns with your own understanding and research. The importance of referencing AI use correctly cannot be overstated. Students should clearly indicate which parts of their work involved AI assistance and explain how the AI was used. This transparency is essential for maintaining academic integrity and allows teachers and markers the opportunity to evaluate your own contributions accurately. Proper referencing of AI use also demonstrates your understanding of the role of technology in scientific research and your ability to use it responsibly.

4. Conducting the experiment

Your experiment must be carried out in the presence of your teacher, either in a school lab or as a field-based experiment. Each student must complete their own experiment but you can ask your classmates to help you assemble your apparatus and collect your data at this stage. You should check with your teacher to ensure your apparatus is safe before beginning your investigation. In your investigative log, you should note all relevant observations and data collected, both **qualitative** and **quantitative**. You should take photos or videos as you assemble your apparatus, conduct your experiment and gather your data; these will enhance your report later and validate your work.

Don't worry if something goes wrong – that's part of the process. Note it down and try to explain why it happened. You should aim to complete your experiment over a 1–2 hour period, including set up and tidy up. You should focus on gathering reliable **primary data** and ensure your experiment is repeatable. Note the contents of Section 3 in Chapter 3 on gathering and evaluating data.

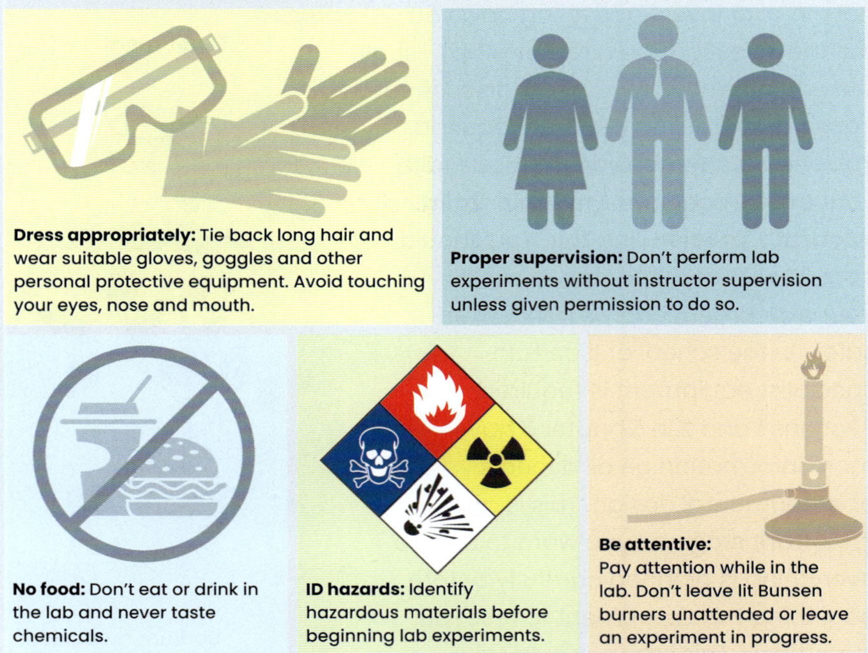

Figure 5: Laboratory safety is an important consideration when conducting your experiment.

5. Data analysis and conclusions

After your experiment is complete you should organise your data into tables or graphs, recording them in your investigative log. Share the data with your teacher. Look for patterns or trends in your results and try to explain what your results mean. You should critically evaluate the quality of your data. Is it accurate, precise and repeatable? Are there **anomalies** in the data? If so, try to explain why they arose. You can use tools like Microsoft Excel or Google Sheets to analyse your data and draw graphs. Finally, decide if your hypothesis was supported or not. Although, it's important to note that isn't the goal of the experiment. This stage should take you around 1–2 hours to complete. You can find out more about data analysis in Section 3 of Chapter 3.

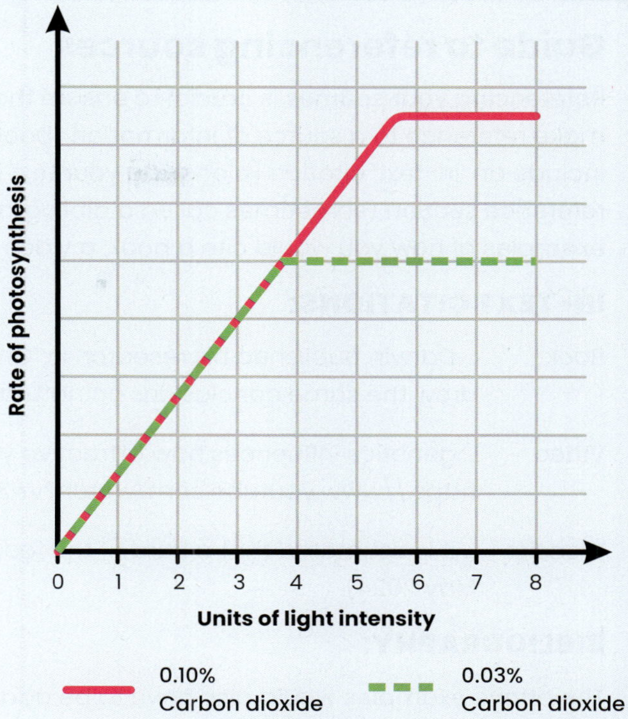

Figure 6: Charts and graphs help you visualise your data more clearly to find patterns or trends.

6. Finalising the report

This will be the longest stage in the process, and is expected to take approximately 4 hours. Using a template provided by the State Examinations Commission (SEC), you will write up all parts of your investigation. The report will provide space for an introduction, background research, experimental procedure, data collected and analysis of the results. The SEC will provide word counts for each section and you must adhere to these. You will also be permitted to include a limited number of photos, sketches, tables and graphs. You should proofread your document carefully; your teacher is not allowed to correct this report or make suggested changes before you submit it. Finalising your report will be much easier if you have a complete investigative log that has been regularly updated.

It is extremely important that all of the work you submit for assessment is your own. The submission of work not entirely completed by you could lead to you losing some or all of your marks for your Biology in Practice Investigation. Remember, it is worth 40% of your total grade.

The report will also have space for referencing your sources (this will not have a word count limit so include every source). Use the referencing guide to make sure you've referenced all your sources correctly.

Guide to referencing sources

Referencing your sources is crucial to ensure the validity of your research investigation. If you make reference to a source of information (book, magazine, website, podcast, etc.) you should include an 'in-text' citation (alongside your text in your report) and a reference within your reference section (sometimes called a bibliography) at the end of your report. Below are some examples of how you would cite a book, a video and a website in your text.

IN-TEXT CITATIONS:

Book ... Darwin published his research in 'On the Origins of Species' after Alfred Russel Wallace drew the same conclusions on natural selection (p. 9, Darwin, 1859).

Video ... genetics influences how attractive you are to mosquitos (3:40 to 4:30, https://www.youtube.com/watch?v=38gVZgE39K8).

Website ... it is estimated that a third of the food people eat is a result of pollination (RTÉ News, 20 May 2024).

BIBLIOGRAPHY:

The above examples would also have to be added to your references at the end of your report as follows:

Darwin, C. (1859) *On the Origins of Species*.

Veritasium (2018, 7 Feb.) *Why mosquitos bite some people more than others*. Video, available at https://www.youtube.com/watch?v=38gVZgE39K8.

RTÉ News (2024, 20 May) *Call to monitor honey bee colonies on World Bee Day.* Available at https://www.rte.ie/news/ireland/2024/0520/1450226-world-bee-day. Accessed 12 September 2024.

You must acknowledge the use of material generated by AI tools or applications. The reference should include the AI tool's name, the date the content was created and a brief explanation of how it was used. If the AI tool provides a shareable URL for the chat session, include it in the research sources. If no URL is available, include the tool's name and the prompt used. Be aware that no marks will be awarded for material provided that has come from an AI tool, even if it is properly referenced.

IN-TEXT CITATION:

... honey is often considered a natural antibiotic due to its unique properties (chatgpt.com/share/6761d4d0-7be4-800f-a7f7-3f1fc5341e21) [or without a link] (ChatGPT, 11 November 2024).

BIBLIOGRAPHY:

OpenAI (2024) ChatGPT response to a question about honey as a natural antibiotic. Available at: chatgpt.com/share/6761d4d0-7be4-800f-a7f7-3f1fc5341e21. Accessed 11 November 2024.

Or, without a link:

OpenAI (2024) ChatGPT response to a question about honey as a natural antibiotic [Large Language Model]. Accessed 11 November 2024.

DESCRIPTORS OF QUALITY

The following table describes what are known as the **descriptors of quality** – a way for you to self-assess your report to see if it warrants a top grade or a lower grade. It can be useful to reference this table throughout your research investigation to ensure you remain focused on achieving a top grade.

	Students demonstrating a high level of achievement ...	Students demonstrating a moderate level of achievement ...	Students demonstrating a low level of achievement ...
Knowledge and understanding	engage thoroughly with the concepts being investigated; explain clearly and accurately, using appropriate means, the biological phenomena involved; where applicable, pose a testable hypothesis that is underpinned by biological theory and clearly describe the purpose of the investigation.	have a good engagement with the concepts being investigated; describe the biological phenomena involved; where applicable, pose a testable hypothesis and outline the purpose of the investigation.	have a limited engagement with the concepts being investigated; outline the biological phenomena involved; are provided with a testable hypothesis to investigate.
Demonstrate investigative skills (design and method)	use clear, thorough and appropriate investigative design and methods to collect primary data, leading to high-quality data, presentation and analysis.	use clear and appropriate investigative design and methods to collect primary data, leading to good quality data, presentation and analysis.	use unclear investigative design and methods to collect primary data, leading to limited data, presentation and analysis.
Demonstrate investigative skills (analysis and conclusions)	draw valid conclusions justified by the data and relating to any hypotheses made; thoroughly evaluate the investigation, acknowledging sources of error in the investigative design.	draw conclusions that relate to any hypotheses made, identify potential sources of error in the investigative design.	draw limited conclusions. Potential sources of error in the investigative design are not identified.
Relate their investigative work to the work of scientists in society	offer a considered reflection that locates the outcomes of the investigation within broader issues relating to the work of a scientist or science and society.	reflect on how the outcomes of the investigation relate to the work of a scientist or science and society.	make limited links between the outcomes of the investigation and the work of a scientist or science and society.

Figure 7: The descriptors of quality for the Biology in Practice Investigations.

TIPS FOR SUCCESS

Here are some tips and advice for completing a high-quality scientific research investigation.

1. **Choose a clear and focused research question.**
   - Start with a specific, clear and answerable research question or hypothesis.
   - Avoid overly broad or vague topics; focus on a specific aspect that can be investigated within a short time frame.
   - Consider adapting or expanding an investigation you've done already in class.
2. **Conduct a thorough literature review.**
   - Firstly, write what you already know about the topic and identify gaps.
   - Use reliable sources like peer-reviewed journals, scientific databases (e.g. PubMed, Google Scholar) and reputable websites.
   - Cite all relevant research to provide a solid foundation for your study.
3. **Develop a strong research plan.**
   - Write clear goals for your investigation, with a focus on appropriate research methods (e.g. experimental, observational, qualitative or quantitative).
   - Plan how data will be collected, measured and analysed.
4. **Ensure your methods are robust.**
   - Use reliable and trusted tools or techniques.
   - Maintain consistency in how experiments or data collection are performed.
   - Control variables carefully to reduce bias.
   - If possible, include a control for your investigation.
5. **Be systematic and organised.**
   - Keep detailed and organised notes, including observations, procedures and results.
   - Document everything in your investigative log or in a lab notebook.
   - Use charts, tables or spreadsheets to organise raw data for easy analysis.
6. **Collect high-quality data.**
   - Consider ways to improve accuracy and precision.
   - Repeat measurements to reduce errors.
   - Avoid cherry-picking data; record and report **all** findings, even unexpected results.
7. **Analyse data effectively.**
   - Use appropriate statistical methods or software (e.g. SPSS or Excel) to analyse data.
   - Ensure your sample size is large enough for meaningful conclusions.
   - Present results clearly using graphs, tables or figures.
8. **Interpret result objectively.**
   - Discuss findings based on evidence, not personal biases or expectations.
   - Compare your results with existing research to identify agreements or discrepancies.
   - Be transparent about limitations or uncertainties in your study.
9. **Maintain scientific integrity.**
   - Avoid plagiarism. Always credit other researchers' work.
   - Report all data truthfully, including errors or unexpected findings.
   - Follow ethical guidelines, such as minimising harm to the environment.
10. **Proofread your report for clarity, grammar and logical flow.**
11. **Seek feedback and revise.**
    - Share your work with your teacher and classmates for constructive feedback.
    - Scientific research often involves trial and error. Be patient and open to unexpected results.

Glossary

1/t is a measure of rate of enzyme activity, where t = time taken.

Abiotic factors are the non-living components of an ecosystem.

Absorption is the movement of digested nutrients from the small intestine into the blood.

Accuracy describes how close a measured value is to the actual value.

The **acquired immune system** develops over a person's lifetime. It provides long-term specific immunity in response to antigens.

Activation energy is the minimum amount of energy needed for a reaction to occur.

Active immunity develops from the production of antibodies by the immune system on detection of an antigen.

The **active site** is the region in an enzyme to which its particular substrate attaches and from which the product(s) are released.

Active transport is the movement of a substance from an area of low concentration to an area of high concentration across a semi-permeable membrane, using energy.

An **adaptation** is a feature of an organism that allows it to thrive and survive in its ecosystem.

Adenosine triphosphate (ATP) is a molecule used to store and transfer energy in cells.

Adhesion is the ability of water to stick to surfaces.

Aerobic respiration is the enzyme-controlled release of energy from food using oxygen.

Afforestation is the planting of trees.

The **alimentary canal** is a series of connected organs along a long tube, extending from the mouth to the anus.

Alkaloids are chemicals produced by plants to protect them from predators and to control their growth.

The **'all or nothing' rule** means that an impulse is only carried if the threshold stimulus is reached. The same impulse is carried regardless of the strength of the stimulus.

An **allele** is an alternative form of a gene, e.g. dominant or recessive.

Alveoli are tiny air sacs in the lungs; they are the site of gas exchange.

An **amino acid** is the basic structure in a protein.

Amnion is a membrane which forms the amniotic sac and amniotic fluid that surrounds and protects the developing embryo.

Amylase is an enzyme that breaks down starch into maltose.

In an **anabolic reaction**, small molecules are converted into larger molecules, using energy.

Anaerobic respiration (fermentation) is the controlled release of energy from food in the absence of oxygen.

Angiosperms are green plants that produce flowers to aid their reproduction.

Animals are multicellular eukaryotic and heterotrophic organisms. Animals are multicellular, with cells without walls.

An **anomaly** is an irregular piece of data that does not fit the pattern of a particular data group.

Antagonistic muscle pairs are muscles which work with opposite actions to allow movement of the skeleton.

Antibiotic resistance occurs when bacteria are no longer affected by antibiotics which could once be used to kill them.

Antibiotics are chemicals produced by micro-organisms such as bacteria and fungi that are toxic to bacteria.

Antibodies are proteins produced by lymphocytes that mark pathogens for destruction.

An **anticodon** is a group of three nucleotide bases on one end of a tRNA molecule that is complementary to a codon in mRNA.

An **antigen** is a molecule on the surface of foreign bodies that triggers the formation of antibodies.

The **appendicular skeleton** consists of the pectoral and pelvic girdles.

Archaea are an 'ancient' type of bacteria that are more closely related to eukaryotes than other 'true' bacteria.

Arteries are blood vessels that carry blood away from the heart.

Arterioles are small arteries.

Artificial insemination is a medical procedure where sperm is inserted directly into a woman's cervix, Fallopian tubes or uterus.

In **artificial selection**, humans intervene in animal or plant breeding by selecting parents with desirable traits to produce offspring with these traits.

If something is **aseptic**, it is free from pathogenic disease-causing micro-organisms.

Asexual reproduction involves one parent and results in genetically identifcal offspring.

Assimilation is the use of nutrients in a living cell.

Autoimmune disease occurs when a person's immune system mistakenly attacks and damages their body's own healthy cells.

Autotrophic organisms have the ability to produce their own food, e.g. photosynthesis.

The **axial skeleton** consists of the skull, vertebral column, sternum and ribcage.

B lymphocytes are part of the acquired immune system and form a complementary antibody on detection of a specific antigen.

A **bacteriophage** is a virus that replicates inside a bacterial cell.

A **balanced diet** contains the correct amount of each of the different nutrient groups to keep a person healthy.

Basic reproduction number R_0 value (R-zero) is the average number of infections directly caused by an infected individual during their infectious period.

A **belt transect** gathers quadrat data in a straight-line sequence.

Bias refers to any deviation from the truth in data collection, analysis, interpretation and publication that leads to false conclusions.

Bile emulsifies fats, turning them into smaller droplets.

Binary fission is the division of bacterial cells in a form of asexual reproduction.

Binomial nomenclature is a system of using two names (a genus and a species) to name and classify an organism.

Biodiversity is the variety of life on Earth or in an ecosystem.

Bioinformatics is an interdisciplinary branch of science that uses computers and software to collect, store and analyse large amounts of biological data found in DNA genetic sequences or protein samples.

Biology is the study of life and living things.

Biomass is the total weight of the community in an ecosystem.

Biomolecules are the chemical building blocks of cells.

A **bioreactor** is a sterile stainless-steel vessel in which biochemical reactions are carried out using enzymes.

The **biosphere** is the part of the Earth where life exists.

The **blastocyst** is a fluid-filled ball of about 64 cells in early embryo development.

The **blastula** is a hollow ball of cells found in the embryo of all animal life.

Blood is a body fluid composed of specialised blood cells suspended in a liquid called plasma.

Blood pressure is the force of blood pushing against the artery walls.

Bone is a specialised connective tissue composed of bone cells embedded in a matrix of collagen and calcium salts.

Botany is the study of plants.

Bryophytes are mosses: non-vascular green plants that reproduce using spores.

A **bud** is a site of new stem, leaf or flower growth.

Cancer is a group of disorders characterised by the loss of control of mitosis.

Capillaries are thin-walled blood vessels that allow for exchange of materials between the blood and body cells.

Capillary action is the movement of water upwards in thin tubes against the force of gravity.

A **capsid** is a protein coat that surrounds the nucleic acid in a virus.

A **carbohydrate** is a biomolecule formed from the elements carbon, hydrogen and oxygen in the ratio $C_x(H_2O)_y$.

A **carbon sink** is anything that absorbs more carbon than it releases. Forests are natural carbon sinks.

A **carcinogen** is a chemical known to cause cancer.

Cardiac muscle is the specialised muscles cells of the heart; they don't tire or grow and are under unconscious control.

The **carpel** is the female part of the flower. It consists of the stigma, style and ovary.

The **carrying capacity** is the maximum population of a species an ecosystem can support.

Cartilage surrounds and cushions bone.

In a **catabolic reaction**, large molecules are broken down into smaller ones, releasing energy.

Catalysis is the acceleration of chemical reactions by catalysts, including enzymes.

A **catalyst** is a chemical which generally speeds up the rate of a chemical reaction but is unchanged in the reaction.

Causation indicates that one event is the result of the occurrence of another event.

The **cell** is the basic unit of life.

The **cell cycle** refers to the changes that occur over the life of a cell. It involves growth, DNA synthesis and division.

The **cell wall** provides strength and support to the cell.

The **central nervous system (CNS)** consists of the brain and spinal cord.

Cerebrospinal fluid is a clear liquid that surrounds and protects the brain and spinal cord and provides them with nutrients.

Chemical digestion uses enzymes to break down biomolecules into their structural units.

Chlorophyll is the green pigment found in the chloroplast in plant cells that absorbs light to provide energy for photosynthesis.

Chloroplasts are the organelles responsible for photosynthesis; they contain the pigment chlorophyll.

Chromatin consists of DNA wrapped around proteins called histones.

A **chromosome** is a thread-like structure of DNA and protein that becomes visible in the nucleus during cell division.

A **chromosome mutation** is a change in the number or structure of chromosomes.

The **citric acid cycle** is a series of reactions in the matrix of the mitochondrion, breaking down Acetyl-CoA to release carbon dioxide, ATP and NADH.

A **clade** is a grouping on a phylogenetic tree that share a common ancestor.

A **closed circulatory system** is a system in which blood stays within the blood vessels.

A **codon** is a sequence of three consecutive bases of DNA or RNA that codes for a specific amino acid.

Co-evolution is evolution of two or more species, creating an interdependent relationship.

Combustion is the burning of fossil fuels, releasing heat and carbon dioxide into the atmosphere.

A **community** is a group of different species living in a particular ecosystem.

Compact bone contains a compressed matrix, making it very hard and durable.

Competition is the struggle between organisms for essential resources.

Concentration gradient describes the difference in the amount of a substance per unit volume between two different areas.

A **conclusion** summarises how the results of an investigation support or contradict your original hypothesis.

Conservation is the management of Earth's resources, ecosystems and biodiversity.

Consumers (heterotrophs) are organisms that cannot make their own food.

Contest competition is a confrontational struggle between two organisms; the winner takes all.

Continuity of life explains that all life can only derive from other living things.

Contraception refers to methods of preventing pregnancy in a female.

A **control group** is a group not exposed to the variable under study and used as a comparison to the experimental group.

A **controlled experiment** (or fair test) is one where only one variable is changed at a time and all other conditions are kept constant.

Coronary arteries supply the cardiac muscle cells with blood (oxygen and nutrients).

Correlation is a numerical measure of the degree of a relationship between two variables.

The **cortex** is the tissue found between the dermal and vascular tissue in a plant's stem or roots.

Cristae are folds on the inner membrane of the mitochondrion.

Cytoplasm describes the contents of the cells, except the nucleus.

The **cytosol** is the liquid part of the cytoplasm.

Data refers to observations or measurements recorded in an experimental test.

Daughter cells are the cells produced from the result of mitosis or meiosis.

Deamination is the breakdown of amino acids into sugars in the liver.

Decomposition is the breakdown of the remains of dead animals and plants.

A **denatured** enzyme is an enzyme that has permanently lost its shape and can no longer carry out its function.

Denitrification is the conversion of nitrates to nitrogen gas by denitrifying bacteria in the soil.

A **dependent variable** is a variable that is observed or measured during the experiment.

Dermal tissue is formed from protective skin-like cells in plants.

The **diaphragm** is a dome-shaped muscle separating the chest from the abdomen.

Dichotomous keys use a series of questions to help identify unknown species.

Differentiation refers to the variety of cell types in an organism.

Diffusion is the passive movement of a substance (solute) from an area of high concentration to an area of low concentration.

Digestion is the breakdown of food (biomolecules) into smaller units.

A **dihybrid cross** illustrates the inheritance of two traits that are controlled by different sets of alleles.

A **diploid** cell contains two sets of chromosomes, e.g. a cheek cell.

DNA – deoxyribonucleic acid – is a two-stranded double helix molecule containing genetic information for the development and functioning of an organism.

DNA polymerase is an enzyme that joins DNA nucleotides together to form a strand of DNA.

DNA profiling is a technique that reveals a unique pattern of bands from an individual's DNA sample. This is called their DNA profile. DNA profiling works on the principle that the non-coding regions of DNA are unique to a person.

DNA replication is the process by which DNA in each chromosome makes an exact copy of itself.

DNA sequencing involves working out the exact order of base pairs (ATCG etc.) in a section of DNA. The principle of DNA sequencing is to generate a collection of DNA fragments, each of which is one nucleotide shorter than the next and separated on a gel electrophoresis.

A **dominant** allele prevents a recessive allele from being expressed.

In **double blind testing**, neither the researcher nor the test subject knows who is receiving the test drug or the placebo in a clinical trial.

Double circulation refers to the presence of the pulmonary and systemic circuits in the body.

Ecology is the study of living things and their interactions with their environment.

An **ecosystem** is the unit of ecology – a group of organisms interacting with each other and their environment.

An **effector** is a muscle or gland that carries out a response to a stimulus.

The **electron transport chain** is a series of protein complexes and mobile electron carriers embedded in the mitochondrial membrane (respiration) and thylakoid membrane (photosynthesis).

Emerging disease refers to an infection that has not occurred in a population before or has previously occurred but only affected a small number of people in isolated places.

An **endocrine gland** is a ductless gland that secrete a hormone directly into the bloodstream.

The **endocrine system** is a group of glands that produce and secrete hormones.

Endoplasmic reticulum is an organelle responsible for protein and lipid synthesis.

The **endosperm** is a specialised tissue which acts as a food store in seeds.

An **enzyme** is a selective catalyst controlling biochemical reactions.

Enzyme specificity describes how each enzyme will only act on one particular substrate.

The **enzyme-substrate complex** is the structure formed when an enzyme binds to its particular substrate.

Epigenetics is the study of the effect of behaviour and the environment on gene expression in an organism.

The **epiglottis** is a flap of tissue that prevents food and drink entering the respiratory tract.

Error refers to the difference between a measurement and its true value.

Ethics is the study of standards of right and wrong – dealing with moral conduct and judgement.

A **eukaryote** is an organism (or cell) with a membrane-bound nucleus and organelles.

Eukaryotic cells contain a nucleus and membrane-bound organelles.

Eutrophication is the addition of nutrients to a waterway, which often results in an algal bloom.

Evolution refers to inheritable changes in the characteristics of a species over generations due to natural selection and mutations, which may result in the formation of a new species.

Excretion is the removal of the waste products of metabolism.

Exhalation is the passive process of removing air from the lungs.

An **exocrine gland** is a gland with a duct that carries its secretion to its destination.

An **exoskeleton** is a strong supportive layer outside of the body of an animal.

An **experimental group** is a group exposed to the variable under study.

Extremophiles are micro-organisms that thrive in harsh environments, e.g. extreme heat or pH.

A **fat** is a lipid that is solid at room temperature.

The **fertile period** is the time in the menstrual cycle when a female is most likely to get pregnant if she has sexual intercourse.

Fertilisation is the fusion of two haploid gamete cells forming a diploid zygote.

Fibre is the collective term for indigestible carbohydrates.

Flaccid describes cells that contain little water and are not firm.

Follicular phase (days 1–14 of the menstrual cycle) of the menstrual cycle involves the maturation of immature eggs in the ovary in preparation for ovulation.

A **food chain** illustrates what eats what in an ecosystem.

Food security means everyone has access to safe and nutritious food.

A **food web** is a set of interconnected food chains.

'Form fits function' exists when a structure of a body part is well designed for the job it carries out.

A **fruit** is a structure that develops from an ovary and contains a seed or seeds.

Fully permeable means all substances pass through it.

Gas exchange takes place at the alveoli, where oxygen is exchanged for carbon dioxide.

Gel electrophoresis is a laboratory technique to separate DNA fragments on the basis of size using an electric current passed through a gel.

A **gene** is a section of DNA that codes for a protein or trait.

Gene expression is the process in which a gene is activated to make RNA and protein.

Generalisation refers to an overall conclusion about a population based on a sample.

The **genetic code** is the sequence of triplets of nucleotides (codons) in DNA or RNA that determines the amino acid chain in a protein.

Genetic diversity is a measure of the variation of the alleles in a population in an ecosystem.

Genetic engineering (genetic modification) involves the artificial altering or manipulation of an organism's DNA.

Genetically modified organisms (GMOs) are organisms whose DNA has been altered using genetic engineering methods.

A **genome** is the complete set of genes present in an organism's cells. Humans have over 20,000 genes.

A **genome database** is an online library of sequenced genomes of different species.

Genotype is the genetic make-up of an organism.

Germination occurs when the embryo begins to regrow after a period of dormancy.

Germ layers are layers in the embryo that give rise to specific tissues and organs.

A **gland** is an organ that produces a chemical with a specific function in the body and releases it into the bloodstream or a duct.

Glomerular filtrate is the product of filtration – the first stage of urine formation.

The **glomerulus** contains a series of ultra-thin capillaries at the opening of the nephron.

Glycolysis is the breakdown of six-carbon glucose to two molecules of three-carbon pyruvate.

Golgi apparatus is responsible for packaging protein and lipids manufactured in ER.

The **greenhouse effect** is the natural process that warms the Earth's surface when gases, like CO_2, reabsorb heat back into the atmosphere.

A **greenhouse gas** is a gas, such as carbon dioxide or methane, that traps heat in the Earth's atmosphere, contributing to the greenhouse effect and influencing global climate patterns.

Grey matter is made up of cell bodies in the CNS.

Ground tissue are groups of cells in plants that are not categorised as dermal or vascular – they have many functions.

Guard cells are crescent-shaped cells that actively control the size of stomata.

Gymnosperms are conifers: vascular green plants that produce seeds in cone-like structures.

A **habitat** is a place where an organism lives.

Habitat diversity is a measure of the number of different habitats in an ecosystem.

Haemoglobin is an iron-rich protein found in red blood cells which binds and transports oxygen.

A **haploid** cell contains just one set of chromosomes, e.g. a sperm cell.

Helper T cells activate macrophages, B cells and killer T cells. Helper T cells stimulate B cells to form antibody producing plasma cells.

The **hepatic portal vein** is a blood vessel bringing absorbed nutrients from the intestines to the liver.

Heredity is the process of passing information, using DNA, from one generation to the next.

Heterotrophic organisms must consume food to obtain energy and the raw materials for growth.

A **heterozygous** pair of alleles consists of two genes of different form, e.g. Bb.

Homeostasis is the maintenance of a constant internal environment within an organism.

A **homologous pair** of chromosomes is a pair of chromosomes with the same pattern of genes.

A **homozygous pair** of alleles consists of two genes of the same form, e.g. BB or bb.

A **hormone** is a chemical messenger produced by an endocrine gland that travels in the bloodstream to act on a specific tissue.

Hormone manipulation refers to the artificial altering of hormone levels in an organism.

Hydrogen bonds are weak bonds between molecules.

An external solution is **hypertonic** when its water concentration is lower than that inside a cell.

A **hypothesis** is an educated guess based on observation.

An external solution is **hypotonic** when its water concentration is higher than that inside a cell.

An **immobilised enzyme** is an enzyme that is held in place on an inert surface or trapped in a gel.

The **immune system** consists of a network of cells, organs and the chemicals they produce that protect the body against infection.

Immunity is protection against an infectious pathogen resulting from production of antibodies.

Implantation is the embedding of a fertilised egg in the lining of the womb.

An **impulse** is an electrical signal triggered by a stimulus that travels the length of a nerve fibre and ends in the release of a neurotransmitter.

In **incomplete dominance**, neither allele is dominant and both are expressed to produce offspring with an intermediate phenotype.

An **independent variable** is the variable that is changed by the scientist.

Infertility is an inability to conceive a child.

The **innate immune system**, present at birth, provides short-term, non-specific defence against pathogens.

Insemination is the release of semen into the vagina.

Integrity is maintaining honesty at all stages of scientific practice.

The **intercostal muscles** are found between the ribs.

Interdependence describes how all species rely on each other for their survival.

An **interneuron** is a nerve cell that carries an impulse between sensory and motor neurons within the CNS.

Interphase is the period of a cell's life when it is not dividing.

Interspecific competition occurs between different species.

Intracytoplasmic sperm injection (ICSI) is a treatment for infertility where a sperm is injected directly into an egg using a needle.

Intraspecific competition occurs between members of the same species.

An **invasive species** is a non-native plant or animal that outcompetes native species.

An external solution is **isotonic** when its water concentration is equal to that inside a cell.

IVF (in vitro fertilisation) is a technique where eggs are fertilised by sperm in a glass dish.

A **joint** exists where two or more bones meet.

Kick sampling is a technique used in stream studies, where the stream bed is kicked to disturb then collect invertebrates.

Kidneys are bean-shaped organs responsible for excretion and osmoregulation.

Killer T cells kill cancer cells or virus-infected cells by producing perforin which creates pores in their cell membranes.

Labour is the process by which the baby and the placenta are delivered from the uterus to the outside world.

The **lacteal** is an extension of a lymphatic vessel, where fats are absorbed in the villi.

The **larynx** is the voice box – a region where vocal cords are found allowing us to speak.

Lenticels are pores on the surface of stems which allow for gas exchange.

Life is a condition whereby organic matter has the capacity to use energy for its development.

Ligaments connect bone to bone.

The **light-dependent stage**, the first stage of photosynthesis, involves the absorption of light energy by chlorophyll and the production of NADPH, ATP and oxygen.

The **light-independent stage**, the second stage of photosynthesis, involves the conversion of carbon dioxide to carbohydrate using protons, electrons and ATP from the light-dependent stage.

Line transects use a long rope or tape to collect data in a straight-line sequence.

Linked genes are found on the same chromosome and tend to be inherited together. This is known as linkage.

Lipase is an enzyme that breaks down lipids (fats).

The **locus** is the location of a gene on a chromosome.

The **lower respiratory tract** consists of the trachea, bronchi, bronchioles and lungs, including the alveoli.

LUCA – the last universal common ancestor – is the organism/cell that all life derived from.

The **lumen** is the internal space in an artery or vein.

Lung capacity is the maximum amount of air the lungs can hold.

The **luteal phase** (days 15–28 of the menstrual cycle) is when the endometrium thickens further in preparation for pregnancy.

Mammals are animals that can produce milk, have hair and regulate their body temperatures.

The **matrix** is the fluid-filled inner space in the mitochondrion.

The **medulla oblongata** is a region of the brain that controls unconscious actions, e.g. breathing.

Meiosis is the type of nuclear division associated with the production of sex cells or gametes.

Memory T cells provide long-term immunity against previously encountered pathogens.

Mendel's first law, the law of segregation, says that traits are governed by pairs of genes called alleles. These alleles separate during gamete formation, with each gamete receiving only one allele.

Mendel's second law is the **law of independent assortment**. It states that, during gamete formation, either member of a pair of alleles can combine with either member of another pair of alleles.

Meninges consist of a protective layer of three membranes surrounding the brain and spinal cord.

Menopause is the time in a woman's life when her ovaries stop producing hormones and the menstrual cycle stops.

The **menstrual cycle** is a series of changes that happens in the ovaries and lining of the uterus in a fertile, non-pregnant female, every 28 days.

Menstruation (days 1–5 of the menstrual cycle) is the shedding of the endometrium (lining of the womb).

Messenger RNA (mRNA) is a single-stranded molecule complementary to DNA that carries the genetic code from the nucleus to the ribosome.

Metabolism is the collective term for all the chemical reactions that occur within an organism.

Metastasis is the spreading of cancer from one part of the body to another.

The **microbiome** is the community of micro-organisms that live in a particular environment.

The **middle lamella** is the space between the cell walls of two cells.

A **mineral** is a chemical element needed by organisms to carry out functions necessary for life.

Mitochondria are organelles responsible for respiration, the release of energy from food.

Mitosis is the copying and division of the nucleus of a cell.

Monocytes are non-specific phagocytic white blood cells that form part of the innate immune system.

Monohybrid cross refers to genetic breeding between two parents who vary in only one trait.

The **morula** is a mass of 32 undifferentiated cells in early embryo development.

The **most recent common ancestor (MRCA)** of a set of organisms is the most recent individual from which all organisms in that set are descended from.

A **motor neuron** is a nerve cell that carries an impulse from the CNS to an effector muscle or gland.

Multicellular refers to organisms which contain many cells, usually of different types.

Murein is a complex of protein and sugars which forms the cell walls of eubacteria.

Muscle is a specialised tissue made from cells that can contract when stimulated.

A **mutation** is a spontaneous, random change in the nucleotide sequence of an organism's DNA.

Myocytes are muscle cells.

Natural killer (NK) cells are white blood cells that target cancer and virus-infected cells without being activated by antigens.

Natural selection is the process by which species members with genetic characteristics better adapted to their environment tend to survive and have more offspring to whom they pass on their genetic traits: 'survival of the fittest'.

Negative feedback is a biological response in which the effects of a reaction slow or stop that reaction. Negative feedback loops help to maintain homeostasis in cells.

Nephrons are tiny tubes found within the kidney – often referred to as the functional unit of the kidney.

A **neuron** is an individual nerve cell.

Neurotransmission is the movement of a nerve impulse from one nerve to another or to a muscle or gland.

A **neurotransmitter** is a chemical messenger that carries an impulse across a synapse between two neurons, e.g. acetylcholine.

A **niche** is the role of an organism in its ecosystem.

Nicotinamide adenine dinucleotide (NAD^+) is a low energy molecule that acts as an electron and proton (H^+ hydrogen ion) carrier in metabolic reactions to form a high-energy molecule NADH.

Nicotinamide adenine dinucleotide phosphate ($NADP^+$) is a low-energy molecule that acts as an electron and proton (H^+) carrier in cells, forming high-energy NADPH.

Nitrification is the conversion of ammonia and nitrite compounds to nitrates by nitrifying bacteria in the soil.

Nitrogen fixation is the process by which atmospheric nitrogen is converted into nitrates or other compounds.

Non-coding DNA does not provide instructions for making proteins but rather contains regulatory sequences that determine when and where genes are turned on and off in a cell.

Nuclear inheritance refers to the transmission of DNA from parents to offspring.

A **nucleoid** is a structure within prokaryotic cells where the DNA is found.

A **nucleotide** is a repeat unit in a nucleic acid (DNA or RNA) consisting of a phosphate molecule, a sugar and a nitrogenous base.

The **nucleus** is the control centre of the cell. It contains DNA found in structures called chromosomes.

Observation is the collection of information or data from a primary source, e.g. an experiment.

An **oil** is a lipid that is liquid at room temperature.

The **optimum pH** is the pH at which an enzyme works best.

The **optimum temperature** is the temperature at which the enzyme works best.

An **organ** contains a group of different tissues working together to perform a specific function.

An **organ system** consists of multiple organs co-operating to carry out a specific function.

An **organelle** is a structure within eukaryotic cells with a particular function, e.g. mitochondria.

Organic matter refers to the remains of living things or their by-products and wastes.

Osmoregulation is the control of water and salt levels in the blood and body fluids.

Osmosis is the movement of water from an area of high concentration to an area of low concentration through a semi-permeable membrane.

Osmotic pressure (or root pressure) is the upward movement of water caused by the movement of water (by osmosis) into xylem vessels.

Osteocytes are bone cells.

Ovulation (day 14 of the menstrual cycle) is the release of an egg from the ovary.

Pancreatic juice is a mixture of enzymes and bicarbonate ions released from the pancreas into the small intestine.

A **parasite** is an organism that lives off another organism causing it harm.

Passive immunity occurs when antibodies are introduced to the body and memory cells do not develop.

Passive transport does not require energy for the movement of a substance.

A **pathogen** is an organism or agent that causes disease in its host.

Pathogen persistence refers to how long a pathogen can survive in a host while remaining infectious.

Peer review refers to the evaluation of a scientist's work by others who are experts in the same field.

Pepsin is a protease that breaks down proteins into peptides in the stomach.

Peptides are short chains of amino acids.

Percentage cover is a measure of how much of the ground is covered by a species.

Percentage frequency is a measure of how common a species is in an ecosystem.

The **peripheral nervous system (PNS)** consists of the nerves that carry impulses to and from the CNS (brain and spinal cord).

The **pharynx** is another name for the throat.

Phenetics (or trait-based) classification organises organisms into groups according to similarities in their structure, i.e. number of legs or petals.

Phenotype is the physical appearance of an organism.

Phloem (pronounced *flow-em*) is a plant tissue that transports dissolved food in a plant.

A **phospholipid** is a lipid unit consisting of glycerol, two fatty acids and one phosphate group.

Photosynthesis is the process by which green plants make their own food by converting carbon dioxide and water to glucose and oxygen, using energy from sunlight.

Phylogeny (phylogenetic classification) classifies organisms based on their evolutionary development (genetics).

Physical digestion breaks food down mechanically into smaller pieces, increasing the surface area.

Pictorial keys use images to identify unknown species.

A **placebo** is a fake treatment that is used with a control group in a medical trial.

The **placenta** is a disc shaped structure formed from tissue from the embryo and the endometrium.

Plant gas exchange is the diffusion of carbon dioxide and oxygen through stomata or lenticels.

A **plant growth regulator** is a chemical messenger produced in small amounts which affects a plant's growth development.

Plants are multicellular, eukaryotic and autotrophic organisms. Their cells contain cell walls and chloroplasts.

Plasma is the liquid part of blood; it consists of water, plasma proteins and dissolved nutrients and wastes.

A **plasmid** is a small ring of DNA found in prokaryotic cells.

Plasmolysis is the shrinking of the contents of a plant cell away from the cell wall, due to loss of water.

A **point mutation** is a mutation that occurs due to incorrect base pairing during replication resulting in a change in nucleotide sequence.

Pollination is the transfer of pollen from the anther to the stigma.

Pollution is a harmful addition to an ecosystem by human activity.

Polymerase chain reaction (PCR) is a laboratory technique for rapidly producing numerous copies of a segment of DNA.

A **polypeptide** is a long chain of amino acids linked by peptide bonds.

A **population** is the number of each type of species in an ecosystem.

Postnatal care refers to the care of the mother and baby after the birth of the baby, generally for the first 6–8 weeks.

Precision describes how close two or more measurements are to each other.

Predation is the killing of one organism by another for food.

A **prediction** is the expected result of a test that is derived from a hypothesis.

Pregnancy describes the period of development of the fertilised zygote to birth.

Prenatal care refers to the care of the mother and baby while the baby is in the womb.

Primary data is data collected by the researcher themselves.

A **prion** is a type of protein that can cause disease in animals and humans by triggering normal proteins in the brain to fold abnormally.

Producers (autotrophs) are organisms that can make their own food.

The **product** is the substance formed when an enzyme acts upon its substrate.

A **prokaryote** is an organism (or cell) without a membrane-bound nucleus or organelles.

Prokaryotic cells do not contain a nucleus or membrane-bound organelles.

Protein synthesis is the two-step process of converting the triplet nitrogenous base code from DNA to mRNA to amino acids to form a protein chain.

Psychoactive drugs are chemicals that affect how the brain works and cause changes in mood, thoughts, feelings or behaviour.

Pteridophytes (pronounced *ter-id-o-fights*) are ferns: vascular green plants that reproduce using spores.

Puberty is the period in life when a child undergoes physical and hormonal changes to become sexually mature and capable of sexual reproduction. Puberty is controlled by sex hormones.

Pulse is the repeating expansion and contraction of the walls of the artery.

A **pyramid of biomass** illustrates the collective mass of the organisms on each trophic level.

A **pyramid of numbers** illustrates the number of organisms on each trophic level.

Qualitative data is non-numerical data.

Qualitative studies do not collect numerical data during investigations.

Quantitative data is data that can be expressed in numerical form.

Quantitative studies collect numerical data during investigations.

Random sampling means that every subset of a population has an equal chance of being included in the study.

Random selection means that all individuals have an equal chance of being selected to take part in a study.

Range = highest data value – lowest data value.

Reabsorption involves moving valuable resources from the glomerular filtrate back into the blood.

Reactants are substances required and used up during a chemical reaction.

A **receptor** is a group of cells that detect a change in the environment and generate an impulse.

A **recessive** allele is only expressed in the homozygous state.

A **reflex action** is a rapid, involuntary and automatic movement response to a stimulus.

The **reflex arc** is the nerve pathway followed in a reflex action.

The **refractory period** is the minimum length of time required between two impulses being carried along a neuron.

Relevant research contributes something valuable to scientific knowledge.

Reliability is the ability of an instrument or method to measure stable and consistent results.

The **renal artery** supplies blood to the kidneys for filtration.

The **renal vein** collects filtered blood from the kidney.

Repeatability describes getting the same result when a scientist makes a particular measurement under the same conditions.

A **replicate** is a repeat of an experiment.

A **representative** sample is a subset of the population that reflects the characteristics of the entire population.

Reproducibility describes whether an investigation can be carried out by a different research team using the same method outlined in a scientific paper and obtain the same results.

Reproduction is the formation of individuals of the same species.

Respiration is the enzyme-controlled release of energy from food.

Response is the ability of an organism to detect changes in its surroundings and react to increase their chance of survival.

Rewilding allows unused farmland to return to its natural state.

Ribonucleic acid (RNA) is a single-stranded polynucleotide molecule that is complementary to DNA.

Ribosomal RNA (rRNA) forms part of the ribosome. It forms weak bonds with mRNA during translation.

Ribosomes are the organelles responsible for protein synthesis.

RNA polymerase is an enzyme that joins RNA nucleotides together to form a strand of mRNA.

Root hairs are narrow extensions of dermal cells which absorb water and minerals.

The **SA node** is the heart's natural pacemaker, controlling the heartbeat.

Safety is keeping researchers and participants free from harm in an investigation.

A **sample group** is a subset with characteristics reflective of its population.

Sample size refers to the number of individuals or items tested in an experiment.

Saprophytes are organisms that release enzymes onto organic matter for decomposition.

The **scientific community** is the global interacting network of scientists working on research.

Scientific practice is the set of steps followed by scientists when carrying out investigative research.

Scramble competition is a non-confrontational struggle between two organisms; each get an equal share.

Secondary data is data collected by another researcher.

Secondary sexual characteristics are physical features that appear during puberty in response to sex hormones and distinguish between males and females. They are not directly involved in reproduction.

Secretion involves the removal of any remaining metabolic wastes from the blood into the nephron tubules.

A **seed** contains a young plant embryo, a food store and a protective outer coat called the testa.

Seed dispersal is the movement of seeds away from the parent plant.

Seed dormancy is a period when a seed does not grow, even though conditions might be favourable.

Selectively permeable describes how a cell membrane actively controls the entry and exit of materials in a cell.

A **semi-permeable membrane** is a barrier layer that allows certain substances to pass through but not others.

A **sensory neuron** is a nerve cell that carries an impulse to the CNS.

Sex linkage refers to genes that are carried on the X or Y (sex) chromosomes.

Sexual reproduction is the formation of new individuals by the fusion of haploid sex cells called gametes.

SI units are standard scientific measurement units based on the metric system.

Sinks are the cells in a plant where excess foods are moved and stored.

Skeletal muscle is a type of muscle that attaches to our skeleton to form the musculoskeletal system.

Smooth muscle is found in our digestive tract, arteries and veins.

Soil microbiome describes the community of bacteria and fungi in the soil.

Sources are the cells in a plant where excess foods are produced.

Specialised cells have altered structures to match their specific function.

A **species** is defined as a population of organisms that can interbreed successfully and produce fertile offspring.

Species abundance measures the number of a particular species in an ecosystem.

Species distribution records how spread out a plant or animal is, and can be recorded by measuring species abundance data from multiple points in an ecosystem.

Species diversity is the number of and relative abundance of different species in an ecosystem.

Species population density is a measure of the number of a species per square metre.

Spongy bone has an open matrix and contains blood cell producing red bone marrow.

The **stamen** is the male part of the flower. It consists of the anther and the filament.

Statistical uncertainty is a measure of the range of random variation in the results obtained when repeated measurements of the same event are recorded.

Stem cells are unspecialised cells in the body.

If something is **sterile**, it is free from all micro-organisms.

A **stimulus** is an internal or external detectable change which prompts a response in a living thing.

Stomata are tiny leaf pores that allow for gas exchange and transpiration.

The **substrate** is the particular molecule an enzyme acts on.

Suppressor T cells inhibit the action of other lymphocytes once an infection has cleared up.

Sustainability is the ability of biological systems to remain healthy, diverse and productive over time.

Symbiosis is a close relationship between two organisms where at least one benefits.

The **synapse** is the region where two neurons are close to each other.

The **synaptic cleft** is the gap between two neurons.

Synovial fluid is a slippery fluid found in freely moveable joints, which lubricates the joint and aids movement.

A **system** is the organisation of interdependent or interrelated objects working together as a whole, e.g. a biological system or the classification system.

Systematic uncertainty arises from defects in the measurement apparatus, equipment or method used by the experimenter.

T lymphocytes or T cells are white blood cells that mature in the thymus and respond to antigens without producing antibodies.

Taxonomy is the science of classification.

Tendons connect muscle to bone.

Tetrapods are four-limbed animals.

A scientific **theory** is a well-established explanation for a natural phenomenon that is supported by scientific evidence over a long period of time.

The **threshold stimulus** is the minimum stimulus required for an impulse to be carried.

A **tissue** is a group of similar specialised cells, carrying out a specific function.

Topography describes the lie of the land: slope and aspect.

Transcription is the production of mRNA using the sequence of DNA bases as a template.

Transfer RNA (tRNA) carries amino acids to the ribosomes to be assembled into protein.

Translation involves linking amino acids together to make a protein using the mRNA code.

Translocation is the process by which food molecules are transported from one part of a plant to another.

Transpiration is the loss of water from the leaves, mainly through tiny pores called stomata.

Transpiration pull is a force, caused by water's great cohesion, that draws water up through a plant.

The **transpiration stream** is the continuous flow of water from the roots to the leaves.

Transport proteins move materials across biological membranes and are involved in active transport.

A **triglyceride** is the basic unit in a lipid consisting of glycerol linked to three fatty acids.

A **trophic level** is a step (feeding level) on a food chain.

A **tropism** is a directional growth response in a plant to a stimulus.

Trypsin is a protease, found in pancreatic juice, that breaks down peptides into amino acids.

A **tuber** is a modified organ for food storage and asexual reproduction.

Turgid describes cells that contain lots of water and are firm.

Turgor is the internal pressure exerted by the large vacuole on the cell walls.

Uncertainty refers to the range of possible values within which the true value of the measurement lies.

Unicellular describes an organism consisting of just one cell.

The **upper respiratory tract** consists of the nose, pharynx and larynx.

The **urinary system** is a collection of organs responsible for urine production.

Urine is a liquid waste containing water, urea, uric acids and mineral salts.

Vaccination is the medical process of introducing a vaccine to the body, usually by injection.

A **vaccine** is a biological substance that stimulates the production of antibodies against a particular bacteria or virus.

Validity indicates how well a test measures what it intends to measure.

A **variable** is a factor or condition that can be changed or controlled in an investigation.

Variation refers to differences between members of the same species.

Vascular bundles contain xylem and phloem in the stem and leaves.

Vascular tissue refers to cells that transport materials, e.g. xylem transports water in plants.

A **vector** is a DNA molecule (often plasmid or virus) used in genetic engineering to carry a DNA segment or gene into a host cell.

Vegetative reproduction is asexual reproduction in plants which results in the formation of clones.

Veins are blood vessels that carry blood to the heart.

Venules are small veins.

Vertebrates are animals with a bony spinal column, or vertebrae.

Villi are finger-like projections of the small intestine that increase surface area for absorption.

A **virus** is a microscopic obligate parasite that can only replicate in a living host cell.

A **vitamin** is an organic molecule needed in small amounts by organisms to function properly.

White matter is made up of axons in the CNS.

Xylem (pronounced *zai-lum)* is a plant vascular tissue that transports water from root to leaf.

Zoology is the study of animals.

Index

M

N

O

R

S